DIE EIGENART DER KINDLICHEN HIRNTÄTIGKEIT

CEREBRAL FUNCTION IN INFANCY AND CHILDHOOD

Other volumes in

The International Behavioral Sciences Series:

SLEEP THERAPY IN THE NEUROSES
by B. V. Andreev

OLIGOPHRENIA – Mental Deficiency in Children
by M. S. Pevzner

Research on the ETIOLOGY OF SCHIZOPHRENIA
by Prof. G. Yu. Malis

ANIMAL BEHAVIOR – Its Normal and Abnormal Development
by Prof. L. V. Krushinskii

BIOPOTENTIALS OF CEREBRAL HEMISPHERES IN BRAIN TUMORS
by N. P. Bekhtereva

The
International Behavioral Sciences
Series

editor Joseph Wortis, M.D.

CEREBRAL FUNCTION IN INFANCY AND CHILDHOOD

by

Albrecht Peiper, *M.D.*

Professor Emeritus
Formerly Chief, Department of Pediatrics
Leipzig University

Translation of the third revised German edition
by
Benedict Nagler, *M.D.* and Hilde Nagler, *Ph.D.*

Springer Science+Business Media, LLC

1963

The original German text was
published by VEB Georg Thieme
in Leipzig, in 1961

ISBN 978-1-4899-5141-0 ISBN 978-1-4899-5139-7 (eBook)
DOI 10.1007/978-1-4899-5139-7

Library of Congress Catalog Card Number: 62-12856

Originally published by Consultants Bureau Enterprises, Inc. in 1963.
Softcover reprint of the hardcover 1st edition 1963

TRANSLATORS' NOTE

In the English edition of Professor ALBRECHT PEIPER's book, *Die Eigenart der kindlichen Hirntätigkeit*, the translators have tried to retain the essential characteristics of his style and his vivid expressions. Particularly impressive are his strength of conviction and the intriguing combination of the scientist's objective and most critical attitude and the true physician's warmth, missionary spirit (for example, in stressing the importance of the mother for her child, this "highest unity"), and concern for his patients. The task of conveying these aspects in another language has not always been easy; the translators hope that this flavor of the original book has been retained without doing too much harm to the correct usage of the English language.

To Dr. HANNAH S. DAVIS, Chief Psychologist, and Mrs. JANET F. NEAL, Executive Secretary at the Lynchburg Training School and Hospital, the translators express deep appreciation for their assistance in preparing and editing the manuscript, their very valuable advice, and many excellent suggestions. We also join with the publisher in acknowledging the invaluable contributions of Dr. JOSEPH WORTIS to the final editing of the translation.

Colony, Virginia, June, 1962

Hilde Nagler
Benedict Nagler

PREFACE TO THE FIRST EDITION

Ever since pediatrics first emerged as a specialty, it has particularly concerned itself with the role of diet in infancy. It has given comparatively minimal attention to the characteristics of cerebral function in infancy, although this function certainly presents as many distinctive differences between infancy and maturity as does the digestive function. This difference in cerebral function does not merely involve the presence of the Babinski sign in early infancy, as some statements in the literature would lead us to believe; on the contrary, cerebral function in the newborn, in whom an extensive and important portion of the brain — the cerebral hemispheres — is not yet functioning, cannot be judged by adult neurological standards. Other neurological standards must be applied to the newborn, standards, moreover, that are continuously revised as the infant grows older. Many disease states are related to the immaturity of the brain and therefore will be misunderstood if the infant is judged according to adult standards, as is so often done.

With few exceptions, even the neurologists have not paid heed to the emergence of brain function, since they usually deal with adults, seldom with children, only under exceptional circumstances with infants, and never with premature infants. In contrast, the anatomic development of the brain has certainly been thoroughly investigated. However, though more difficult than the dissection of dead matter, observation of the infant's developing brain function is more rewarding; and such information may even explain important neurological phenomena in the adult. The deeper centers that express their activity only in early infancy are merely suppressed during later stages; they are not fully extinguished and may later at any time and under specific conditions again assert themselves.

Child psychology, too, has been completely unproductive in this area of knowledge. No psychological theory can make provable assertions on the newborn or on still earlier phases of development because the neurological characteristics, when compared with the adult, are the most conspicuous in these early stages. It would be a rewarding task for a child psychologist to demonstrate how, with increasing brain development, phenomena gradually appear that can be attributed with more certainty to processes of consciousness going beyond the purely neurological events.

Although these sciences have thus far contributed little to our knowledge of the development of brain function, many physiological experiments and findings concerning animals can be applied to the human infant, for the brain function of early human infancy corresponds far more to that of the animal than to later human stages. Thus, I endeavored to apply the research of PAVLOV, R. MAGNUS, RADEMAKER, VON HOLST, VON UEXKÜLL, etc. to the infant and to develop special experimental methods appropriate for this stage of life in order to gain knowledge of the laws governing the emergence of human brain function.

I am grateful to HANS VOLKELT for the information he has supplied.

The quality of reproduction of the illustrations is limited by the difficulties of our times.

Leipzig, July 28, 1949 *Albrecht Peiper*

PREFACE TO THE THIRD EDITION

In the present edition I have made use of my own investigations and have tried to take into consideration the research of recent years.

Electroencephalography has not been discussed because I have not acquired sufficient experience in this field.

Leipzig, April 17, 1961 *Albrecht Peiper*

TABLE OF CONTENTS

1. THE EMERGENCE OF CEREBRAL FUNCTION AS DEPICTED IN THE SCIENCES

Man knows his own brain function from inner experience and self-observation, the brain function of others only from observation. As infants and animals are not able to observe themselves and report their observations, insight into infant and animal psychology can be gained only by observing their behavior. This methodology has been accepted in the field of animal psychology and has been found very productive, but it still remains very foreign to the field of infant psychology. Infant psychology should therefore model itself on animal psychology in this respect. "The soul of infants differs, so to speak, in nothing from the soul of animals" (ARISTOTLE, *Natural History of Animals*, VIII, 1).

ANIMAL PSYCHOLOGY

> "The investigation of instinctive action is not a field for wide scientific speculation but a sphere where, at least at present, only individual experimental research is justified."
>
> K. LORENZ 1937

Primitive man and the child are apt to ascribe to animals, plants, and even inanimate objects sensations and thoughts similar to those of their own experience. But science meets impregnable obstacles when it tries to gain a clear understanding of the consciousness of animals.

Because the facial expressions of anthropoid apes are so similar to those of man and because these animals are able to solve by cogitation certain easy mental tasks, W. KÖHLER (1915) would like to ascribe to them mental functions that, even though less developed, are in principle similar to those of man. But what becomes of this theory as one descends progressively from the higher to the lower animals?

According to HEMPELMANN (1926), the following characteristics are cited as proof of the existence of a psyche or consciousness in animals: presence of a nervous system (FOREL), protoplasmic brain activities (YERKES, DRIESCH), memory (RICHET, EDINGER), learning ability (BETHE, BRUN), associative capacity (LOEB, 1894, 1899;

BOHN), plasticity of the instincts (DEMOLL), discriminative ability (BINET), and understanding similar relations (SZYMANSKI). However, none of these characteristics is uniform. JENNINGS (1910), HEMPELMANN (1926), and ALVERDES (1932) therefore concluded — and rightly so — that there is not and cannot be any outwardly recognizable indication of consciousness in animals.

Thus, no agreement has been reached on the answer to the question: When in the course of the development of the species (phylogenesis) did consciousness appear? On the basis of the information given by VON BIERENS DE HAAN (1935), LUKAS draws the boundary line above Protozoa and fungi; HEMPELMANN does not consider Protozoa to be animal organisms, but he does not wish to delineate the boundary. LOEB draws the line below Arthropoda and Cephalopoda, DAHL between Echinodermata and worms, and EDINGER between fish and Amphibia. According to EBBEKE (1958), there is no exact demarcation line in the course of phylogenesis:

> "There are all stages of transition. From the smallest beginnings, consciousness develops and, on the other hand, descends downward, losing itself in the abyss of the unconscious. There are stages at which the unconscious blossoms into the conscious, stirs itself, and strains upward; and there are other stages in which the conscious lapses into the unconscious."

The viewpoints, in very abbreviated form, of certain philosophers on the content of consciousness in animals are as follows:

According to ARISTOTLE (384—322 B.C.), animals are able to perceive but only rational beings can think (*On the Soul*).

DESCARTES (1596—1650) regarded animals as machines:

> "If machines could exist that conform in shape and in all their limbs to an ape or any other animal, they could not be differentiated in any way from these animals. But if there could be any machines similar to our bodies and which could imitate all our actions, we would still see signs that they are not human. First of all, they would have no language..."

LEIBNIZ (1704) differentiates between human and animal intellect by the fact that animal intellect is based solely on experience and follows only example while man is capable of scientific thought based on logical arguments. The conclusions of animals are only shadows of reasoned deliberation.

In his *Critique of Judgment* (1790) KANT writes:

> "By analogy we can conclude that animals act also on the basis of conceptions and are not machines as DESCARTES stipulates."

According to SCHOPENHAUER (1819), animals have intelligence but no reason; therefore, they have perceptive but no abstract knowledge. They apprehend and comprehend immediate connections, but, properly speaking, they do not think, for they lack concepts, i.e., abstract ideas.

HEGEL (*System of Philosophy*, 1817) considers the belief that man differs from the lower animals because he is able to think an old prejudice.

According to WUNDT (1919) and RENSCH (1954), even the lowest animals (Protozoa, Coelenterata) display characteristics which would indicate that they have conceptions and the faculty of volition. One could not make the most elementary deduction on animal psychology without proceeding from the facts of human consciousness. ERNST HAECKEL assumed an uninterrupted sequence of mental capabilities from the one-cell animal to man.

This concept seemed to be confirmed in 1904 by the Berlin wonder horse. "Der kluge Hans" ("Clever John") could read and calculate, and recognize sounds, playing cards, etc. As its means of expression the horse tapped with the right forefoot. For four years his owner, a retired teacher, had instructed him, and for this purpose had transformed into numbers many concepts, such as the alphabet, the notes of the scale, etc. Experienced teachers attributed to "Clever John" the intelligence of a thirteen- to fourteen-year-old child and well-known scholars admired his skills until PFUNGST (1907) demonstrated that the behavior of "Clever John" conformed to involuntary, almost imperceptible movements of his teacher.

In the meantime, cybernetics invented electronic computers, the faculties of which had so far been credited only to the human brain. These machines have memory and are able to solve very complicated problems in a minimum amount of time; to translate languages; to play chess; to learn to regulate, adapt, and coordinate; and to solve problems which are goal-directed. Thus, they possess faculties that formerly were thought to be the hallmarks of human consciousness.

A new trend in research that developed since the turn of the last century is the study of the behavior of animals left to themselves under as natural conditions as possible, observing their behavioral response to stimuli, while avoiding special experiments contrived to explore the content of their consciousness.

This type of investigation was introduced by THORNDIKE (1898), in the United States. He shut hungry cats or dogs in a locked box, placed food within their reach, and measured the time that had elapsed until they freed themselves. In order to do this the animals had to perform a certain action whereby the door would be opened, such as moving a bolt or pulling down a loop. With frequent repetition of the same experiment, the animal escaped from the box with increasing rapidity. This ability to learn through experiences of success and failure was considered the result of associating given stimuli with useful movements.

BEER, BETHE, and VON UEXKÜLL (1899) introduced terms that describe the behavior of animals without using the language of psychology. "The growing science of comparative physiology proved to be a deadly enemy to comparative psychology... In the face of objective research, the sensations, memory, and thoughts of animals disappeared like fluttering nebulous forms... We do not maintain that animals have no psyche, but we do maintain that empirical knowledge of this question is impossible" (VON UEXKÜLL, 1902).

JENNINGS (1906, U.S.A.), who described the behavior of lower organisms in a classic manner, stressed, in the preface to his book, that from his observations and experiments he did not derive any insight into the consciousness of animals.

Under the influence of THORNDIKE and JENNINGS, behaviorism (the theory of behavior) developed in the United States. "Behaviorism considers consciousness a concept that is neither explainable nor useful. 'Consciousness' is only a different word for the former expression 'psyche'. The behaviorist has eliminated from his vocabulary all such terms as sensation, perception, conception, desire, and purpose and even thinking and feeling." (WATSON, 1930.)

At the turn of the century PAVLOV intensively studied the digestion of dogs, and in the course of this investigation he encountered the phenomenon of the so-called "psychic" stimulation of the digestive glands. In attempting to understand the nature of the discovered phenomena, he applied psychologic methods of interpretation, i.e., tried to deduce what the experimental animal might think and feel during the experiments; in this his co-worker disagreed and they reached an impasse. Both retained their individual viewpoints on this question, and neither could convince the other by means of specific experiments. This caused PAVLOV to renounce in principle the psychologic approach and instead to investigate such phenomena on a purely objective basis. It was by this means that the theory of conditioned reflexes evolved. This theory was utilized in investigations into the nature of infancy by, among others, PAVLOV's co-worker KRASNOGORSKI (1909) (see p. 512).

With a procedure consisting essentially of the production of conditioned reflexes, O. KOEHLER (1953, 1954, 1955) demonstrated "nonverbal thinking" in young children and higher animals (see p. 529). In children it constitutes the first step to language; in animals it does not develop any further.

The concept of animal instinct has always been very difficult to comprehend. For we are sometimes confronted with very complicated instances of behavior that are characteristic for the species and that, because of their great usefulness, seem planned, but we are not able — at least thus far — to attribute to animals any insight into their actions. Some instinctive actions are performed only once so that practice is not a possible factor, and many such actions provide for the future, which must be an unknown concept to the individual animal. Again and again the complexity and usefulness of their behavior are surprising, for example, the social organization of ants, termites, and bees; care for their offspring; etc.

According to CUVIER (1817), one can gain a clear concept of the instinct if one accepts the premise that the consciousness of animals contains constant images that force the animals to act. These images are a kind of dream or vision that always pursues the animal. They can be characterized as somnambulistic. Instinct can cause certain actions that are characteristic for the species and necessary for its preservation but that are alien to its present needs. These actions do not result from intelligence, for they often demand far more foresight and knowledge than the species possesses; often they cannot have occurred by imitation, because the animals carrying out these actions have never seen others perform them.

The new science of behavior (ethology), which started with O. HEINROTH, KONRAD LORENZ, OTTO KOEHLER, and TINBERGEN, provided a new basis for investigating instinct. LORENZ (1935, 1937, 1942/43) believed that the purely intelligence-oriented methods of animal psychology cannot yield knowledge about the learning ability and intelligence of animals without, at least, a general understanding of the instincts and action patterns of a species under natural conditions. When observing animal behavior we are never able to determine the degree to which the behavior ensues from inherited instinctual drives or from learned responses and intellect. Without exact knowledge of an animal's instinctual constitution we never know how difficult a given task will be for it to perform.

In most higher animals, instinct, conditioned reflexes, and insight combine to create a uniform course of action. In animal behavior goal-directed activity plays a role and emerges in association with instinctive behavior, directing the animal toward a certain goal or away from it. It cannot be regarded as due to an innate instinct because it is not consistent. Purposeful movement is both the simplest and the most widespread form of noninstinctive behavior encountered in the animal organism. It is the genetic root of every conative type of behavior (striving for release). It differs from instinctive behavior and should be considered as separate from it, but this is difficult since both forms often occur together. A characteristic of instinctive actions is that they occur even in a *Leerlauf* situation ("vacuum activity"), i.e., induced without external provocation or by a deceptive device.

Expressive movements and sounds are inborn characteristics of higher animals. Similarly inborn is the faculty to emit these signals and to comprehend them correctly, i.e., to answer them in a purposeful way that will preserve the species. Vital stimuli and innate reactive patterns correspond to each other as does the key to the keyhole. Since the response in turn serves as a signal for the partner, instinctive actions such as mating or care of offspring will be achieved in complicated combinations.

In his observations of animals, LORENZ (1935) found that instinctive behavior must be distinguished from learned and rational behavior. These two forms of behavior do not represent either ontogenetically or phylogenetically successive stages; on the contrary, they represent two divergent developmental directions. Wherever one of the two behavior patterns is especially highly developed, it markedly suppresses the other. When instinctive behavior is suppressed by a higher development of perception and insight, LORENZ believes there is no coalescence of both behavioral forms. The rigid instinctive links of a chain of actions do not become more flexible with higher development nor can they be influenced by experience; these "links" drop out one by one and are replaced by acquired or rational forms of behavior.

THEORY OF EVOLUTION

> "Light will fall on the origin of man and on his history."
>
> C. DARWIN 1859

Based in part on the work of previous investigators, CHARLES DARWIN developed in 1859 the theory of evolution. Since then, innumerable contributions in the various branches of natural science have served to confirm its correctness. That this theory also applies to man is demonstrated by the account of fossils, which is continuously being augmented by new evidence.

To be sure, of all human endeavors natural science has been declared as especially incapable of furnishing decisive explanations on the essence and genesis of man. Thus, F. RÜSCHE (1937), professor at the Philosophical Theological Academy in Paderborn, writes:

"Even if the ascent of man from the animal background of nature is proved, this in no way resolves the question of the origin and descent of man as man; in its true essence it is hardly touched upon. For man is made man by his soul. The question of its origin goes beyond the competence of natural science and is primarily and essentially a task for philosophy."

Similarly, ERICH WASMANN (1907) maintains:

"The soul of man cannot have originated through evolution."

The metaphysical concept of the soul proved unproductive in natural science. This was shown, for example, when the physiologist PFLÜGER (1853) wanted to prove that the site of the frog's soul is in its central nervous system (see p. 264). Since then, physiologists have no longer been looking for the "soul" but, rather, devote their attention entirely to the action of the brain and spinal cord. They thereby stay within the realm of natural science and thus avoid sterile speculations.

Our only question, therefore, is whether brain activity could have ascended phylogenetically from the animal to the human stage. We are not in a position to observe this phenomenon anymore, but we do see a similar occurrence today. However, it is so common that everybody accepts it as self-evident without giving it any further thought: The human newborn does not possess functioning cerebral hemispheres, and therefore his brain activity is lower than that of many animals. Should the development through which each child passes before our very eyes have been impossible in the evolution of the species?

PSYCHOLOGY OF INFANCY

> "Modern scientific investigation attaches no importance to the most ingenious inventions of the mind; it considers as its task the acquisition of knowledge which can be attained only through indefatigable work and effort."
>
> J. VON LIEBIG 1844

HERODOTUS reports that the Egyptian king Psammetichos wanted to know which was the most ancient people. For this purpose he gave two newborn infants to a shepherd for upbringing. Nobody was allowed to speak a word in their presence; they were to be kept completely alone in an empty room, to be fed from goats let into the room at appropriate times, and to be provided with all essentials. By these means the king wanted to find out which word the children would utter first.

When the shepherd entered their room two years after the beginning of the experiment, the children, extending their hands, greeted him pleadingly with the word "bekos." As this occurred repeatedly, the shepherd reported it to his master. Then Psammetichos also heard the word being uttered by the children, and he searched for the language to which it belonged. It was discovered that the Phrygians called bread "bekos" and therefore should be regarded as a more ancient people than the Egyptians.*

This old legend shows how far back man meditated on the development of brain functions in the infant and tried to distinguish between that which is innate and that which has been acquired through experience.

The psychology of infancy in our day reflects the fact that psychology based on the natural sciences developed in a continuous struggle against psychology based on the humanities. There are thus now two different currents in child psychology: The one oriented toward the natural sciences tries to discover and establish laws that derive from the study of the child itself. The other does so deductively, by basing its considerations on a philosophical or psychological system without first observing the child's behavior.

Thus PAVLOV (1932) writes with justification:

> "The psychologist, who has distinguished himself from the philosopher only a short time ago, has not yet given up his preference for the philosophical method, i.e., deduction by purely logical means without checking every step of this theoretical system for its conformity with the facts."

The same applies to the psychology of infancy. As the philosopher LANGE (1866) said:

> "It is amazing how tediously our good philosophers keep arguing about the origin of consciousness without ever feeling any need to step into the nursery just once to see for themselves what is actually happening there that might be related to this problem."

From the beginning the mother supposes that her child possesses her own sensations, feelings, and conceptions, at least in an attenuated form. She talks to her newborn as a child does to her doll and would be astonished and incredulous if told that her child understands hardly more of her words than would a doll. But who wants to argue with her? In a short while the child will acquire the understanding that he initially lacks. Science originally went the same way as the mother does and today still quite often insists on this approach.

Again and again the fate of the newborn has been lamented. EMPEDOCLES (490—430 B.C.) recollected: "I cried and lamented at birth when I saw the unaccustomed place." And EURIPIDES (480—406 B.C.), in a fragment of a tragedy, says no less than:

> "The newborn should be greeted by a hollow sound
> Of mourning for the pain which soon will come.
> Only the funeral rites when death descends
> Will bring the festive songs in unison."

ARISTOTLE has Silen express the same thought in *Eudemos:*

> "It is not all best for man to be born and to participate in the Nature of Highest; the best for all, man and woman, is not to be born, and the next best that man can attain is to die as soon as possible after birth."

---

* For a medieval version of this legend, see p. 627.

In the *Eudemian Ethics* it is said:

"If one could choose, one would prefer not to have been born. What a life one leads in childhood! No reasonable person would desire to return to it."

LUCRETIUS (98—55 B.C.) has similar feelings on the matter:

"Look at the boy, similar to the sailor whom angry waves
Have thrown against the land; naked he lies on the ground
And in need of every help, after nature has forced him
Through painful labor from his mother's womb into the light.
Whimpering and moaning he lies there now, for it is an augury
To every one that in life so many evils await."

PLINY THE ELDER (23—79 A.D.) continued this train of thought:

"Nature places only man naked on the naked earth where he greets the day of his birth with crying, whereas no other animal has to cry; and this happens right at the beginning of life... Man knows nothing without instruction. He cannot do anything but cry."

PLUTARCH (50—120 A.D.), too, bemoans the infant's fate:

"Nothing is so incomplete, nothing so needy, nothing so naked and dirty as man, when he comes from the mother's womb. Bespattered with blood, full of filth, more like a murdered than a born creature, he enters the light, so that only those in whom love is imprinted by nature can touch him, kiss him and take him in their arms."

In FRIDANK's *Bescheidenheit* (Modesty) (written between 1225—1240) it is stated:

"Zer werlde komen wir âne wât
In swacher wât ouch sie uns lât."
(We come into the world without clothing,
and with little clothing it lets us go.)

SHAKESPEARE expresses the same thought (King Lear; IV, 6):

"When we are born, we cry that we are come
To this great stage of fools."

BUFFON (1707—1788) sees in the immediate crying of the newborn nature's warning that he is born only for pain, slated to participate in all the toils and troubles of the human race. Similarly, KANT (1798) says: "The newborn child cries because he regards his inability to use his limbs as restraint, and thus he announces his right to freedom immediately." According to HEGEL (1817), the child shows by his crying "that he is fully aware that he has a right to demand the satisfaction of his needs from the world around him — this shows that the independence of the outside world from man is a fiction." For MICHELET (1801—1893), a pupil of HEGEL, this cry is "the terror of the mind for being subjected to nature," with which the newborn has to conduct a continuous struggle. According to BELL (1774—1842), the expression of the newborn betrays the desires of the body in early childhood and the suffering of the mind in later years.

KANT's erroneous notion that the newborn cannot bestir itself was probably derived from the philosophical system that had its origin in a remark of PLINY THE ELDER:

"Furthermore, right after his birth man is placed in chains and fetters, which does not happen to any other creature. And thus, the being which later is supposed to reign over all the others, lies there crying, his hands and feet bound from the beginning."

PLINY evidently refers here to the swaddling of infants, once a common practice, as we know, for example, from a textbook for midwives by SORANOS OF EPHESOS (about 100 B.C.).

Although now the custom, or rather the bad habit, of swaddling has been discontinued, the immobility of the newborn has been asserted in philosophy until today. As late as 1958 GEHLEN in his philosophical anthropology characterizes the human newborn as being incapable of motion; every mother knows the opposite.

To THOMAS AQUINAS (1225—1274) "it is evidently natural that newborns do not possess sufficient strength to move their limbs — because of the excessive humidity of the brain."

According to FREUD, birth itself causes psychological damage. For the first time in his life the child, that until now had been protected in his mother's womb, is presumably seized by fear. Later mental development is said to be so unfavorably influenced by this event that at every experience of anxiety the same physical phenomena experienced at birth will recur. Thus, every state of anxiety repeats the birth act psychologically and physically, the respiratory and circulatory changes that accompany anxiety especially corresponding to those at birth. The multiform anxiety of the neurotic individual is therefore ultimately rooted in the anxiety of birth; on the other hand, every pleasure drive is a striving to restore the original intrauterine pleasure. Many creations of mythology and religion and of art and philosophy can be traced to this original anxiety.

A report of HUG-HELLMUTH (1922) illustrates how adult concepts are projected into the world of the newborn: A woman surprised by the onset of labor had to give birth alone. The baby, born on the floor, clung frantically to the hem of his mother's gown. This supposedly demonstrates that the newborn's usual reaction of fear to unaccustomed light and temperature sensations was in this instance compounded by anxiety arising from the new awareness of hardness and space. By this mode of reasoning, a reaction that possibly may be the result of anxiety states in adults is assumed to be caused by the same type of anxiety in the newborn. Actually, the infant's reaction represented the palmar grasp reflex (see p. 158), which corresponds to the weaker plantar grasp reflex. How can one prove that the newborn feels anxiety when this reflex is elicited?

According to HUG-HELLMUTH, a dim inferiority feeling in combination with an intensive overrating of oneself plays an important role in the infant. HUG-HELLMUTH even recognizes — merely from the facial expression — that occasion when the infant first experiences a presentiment of independent creativity: the observation of the products of his own defecation.

DOROTHY GARLEY (1924), without further ado, affirms that the newborn is able to remember his birth. For example, if someone is more unpleasantly affected by a slap on the back than expected, it is postulated that a birth memory is responsible; for wasn't he slapped the same way to start his breathing? Although no one has ever tried to specifically prove this assertion, it was demanded that Cesarean section be generally used in order to spare the infant the psychological trauma of natural birth.

Psychoanalysis projects the experience and views of neurotic adults into the soul of the newborn and in this way distorts the actual experience of the newborn.

Individualanalyse proceeded similarly. In the *Handbook of Individual Psychology* G. KRAUS (1926) writes:

"The first experience of the infant is that in order to satisfy his needs he requires the help of another, that of an adult person. The first clash with the outside world gives him the recognition of his own helplessness and of the helpfulness of the nearest human being, usually the mother."

She has to write this way to conform to the dogma.

Similarly, for E. STERN (1950) the infant is the picture of total helplessness; he could not subsist for a single day without the care which adults bestow upon him. Somehow he must become aware of this fact, and this "with absolute certainty" is the starting point of his inferiority complex.

The infant psychology formulated by BERNFELD (1925) is rooted in the tenets of psychoanalysis and advances nothing original; it does not deserve to be considered a part of the natural sciences. In animal psychology it has its counterpart possibly in the work of HECKER (1939), *The Animal Soul Based on the Basic Scientific Philosophy of J. Rehmke.*

It becomes evident how much the concepts of the newborn state are based on theory and not at all on the behavior of the infant itself, particularly in the schools of thought of psychoanalysis and individual psychology. The anthroposophists differ even among themselves.

According to the anthroposophist ZUR LINDEN (1958), the mind and soul of the sleeping newborn are not yet functioning, the child being in a state similar to that in his mother's womb — without consciousness. The incorporation of the mind or spirit into the infant's body, fashioned and derived from his parents, depends on nutrition:

"From the prenatal nutrients and, later, from the mother's milk the child derives, together with protein, living formative forces from the mother. The fat of the mother's milk serves as a means for the mental forces of the newborn to enter the child's organism and thus to become acclimated to it. The special task of the milk sugar consists in arousing the ego." The ear, being an exposed organ contributes most to the physical and mental assessments.

In contradistinction, MÜLLER-ECKHARDT (1956), who is closely allied with anthroposophy, maintains:

"The child comes to the world asleep, but we know that this sleep is not unconscious, but that the child is able to participate in all the events of his environment clearly and as if fully awake... In reality, he (the newborn) lives with a very strong consciousness, which, of course, cannot be compared to that of the adult, in a world of light and dreams which is full of mystic and magic secrets..."

The anthroposophist KARL KÖNIG (1957) urges that sensory physiology adopt a completely new orientation toward the mental development of the child. A typical product of this new science is quoted here in KÖNIG's own words:

"Downward divides the trunk of the trachea into the legs of the great bronchi; these separate later and become innumerable small lung vesicles; these touch like very tender tactile organs the unfolding surface of the surrounding blood mirror. The flow of the blood will be touched by the air organism of the speech apparatus in the pulmonary space...

Upward, however, there spread, like two arms, the two Eustachian tubes and reach with their ends, like hands, the vicinity of the ears. There they seize the ear and hold fast onto it, so that hearing and speech, talk and countertalk can be synchronized. In the middle ear the air organism palpates like a hand the membrane of the tympanum... Finally, in the inner ear there lies the mysterious cell in which the names of things and beings are hidden. There, the eternal ideas from which all being and becoming originate touch the earth space of the human organism and pronounce their true names. There, reaches out the aerial hand of the speech organism; there, it touches the word imagined for every thing and being."

Gradually a special science developed that studied the development of the child without preconceived notions and independent of an already established philosophical-psychological doctrine by utilizing the methodology of the natural sciences. Since it was formerly impossible to raise infants in institutions, the first observations are those of fathers who observed their own children and kept diaries on their observations. DIETRICH TIEDEMANN in 1787 was the first to publish such a paper on the mental faculties of children. F. SCHWARZ (1804), SIGISMUND (1856), DARWIN (1877), and STRÜMPELL (1880) used the same approach. In these papers the *observation* of infants was emphasized.

Other publications summarizing the available knowledge on infant psychology obtained in a similar manner are those of the following authors: POSEWITZ (1795), following TIEDEMANN's work; FRIEDLANDER (1815), the director of the obstetric hospital in Leipzig JÖRG (1826); the physiologist BURDACH (1838); the physician STIEBEL (1856, 1865); the Viennese professor of gynecology and pediatrics LÖBISCH (1851); and—in a prize-winning paper — the physician HEYFELDER (1857).

JÖRG already knew, as did the French physiologist MAGENDIE (1826), of the taste reactions of the newborn, the discovery of which has erroneously been attributed to KUSSMAUL (1859):

"He takes sweet fluids with the expressions and sounds of well-being; everything else with the expression of disgust. Particularly, salty, spicy, and bitter things absolutely displease the baby."

Reactions to sound stimuli "become effective in the first days after birth." The response of the child consists in fright and startle reactions. Thus, JÖRG proved himself to be an experienced observer. Although he specifically cautions against using the adult as a yardstick for the child, he himself commits this common error when he expresses his convictions "that no child can feel himself 'complete' as already the concept 'child' contradicts physical and mental completeness." Incidentally, many subsequent errors in child psychology could have been avoided had this absolutely unphilosophical approach not been forgotten.

As did JÖRG, RUSSEL (1948) regards the human infant completely from the standpoint of the adult. According to him, the first three months are a rather sad experience for the infant except for the moments when he is being fed. If he is feeling well, he is sleeping; if he awakes he is usually irritated by something. The happiness of a human being depends on his mental faculties, but they can barely express themselves in an infant who is not even three months old. This is due to the lack of experience and of muscle control. In general, the first three months are filled with boredom for the infant.

KUSSMAUL's investigations into the sensations of the newborn, published in three editions (1859—1895), aroused general attention and had a lasting effect. Examining thoroughly the different senses, KUSSMAUL discovered — or we may have to say today rediscovered — the taste reactions. These findings have not been forgotten. In 1873 GENZMER wrote a doctoral thesis confirming KUSSMAUL's findings. While JÖRG still regarded the eye as the most important sensory organ of the newborn, GENZMER arrived at the remarkable conclusion that no sense is at first as well developed as the tactile sense and that tactile stimuli are best perceived in the face, where they play a role in the intake of food. KUSSMAUL and GENZMER succeeded in causing the newborn to turn his head sideways by stroking his cheeks; the baby would then seize the finger with his lips and suck on it. KUSSMAUL left the question of whether or not the newborn can already hear unanswered, but GENZMER claimed that sound sensitivity appears on the first, or at the latest, on the second day of life.

PREYER (1882) went much farther than these investigations in his fundamental book, *The Soul of the Child.* By profession a physiologist, he set himself the task of observing his son during the child's first three years of life and recording the results in a diary in order to compile information that would shed light on the origin of the phenomena of life. The abundance of material and the meticulousness of the observations and descriptions in PREYER's book do not have their equal. It would be worthwhile to consult his observations for every single question. He supplemented this work by the still important *Special Physiology of the Embryo* (1885). In PREYER's time there did not yet exist a scientific child psychology. He therefore interpreted his findings according to a kind of popular psychology that today is no longer acceptable.

PREYER had a very stimulating effect. Many similar diaries appeared in rapid succession, even more in English than in German. The findings were summarized by HEUBNER (1919).

CANESTRINI (1913) was the first to record continuously the movements of the infant on a recording device such as is generally used in physiology. He reported on his findings in his book *Sensory Function of the Newborn.* He recorded respiratory rates and brain pulsations, the latter by leads from the anterior fontanel, and observed the oscillations that occurred under the influence of sensory stimuli. He thereby obtained permanent proof of the excitability of the newborn; until that time only observations had been described.

PREYER and CANESTRINI hardly said anything about the content of the consciousness of the children they examined so thoroughly. Many child psychologists do make very detailed statements about this content of consciousness, but they fail to give their understanding of such terms as "soul," "mind," and "psyche."

As early as his second day of life PREYER's child without hesitation took cow's milk diluted with water, but refused the same food stubbornly on the fourth day. SIKORSKI (1908) draws the following conclusion from this:

> "The child already recognizes the qualities of the milk; he remembers how it tasted yesterday and the day before yesterday and exhibits a genuine will that watches over higher goals and higher interests of self-preservation. All this proves to be true expressions of the psyche: memory of the past, ability to recognize, and a higher will."

According to KOFFKA (1925), we must not deny to the newborn every form of consciousness. He is not a purely paleencephalic being to whom a restricted type of consciousness should be attributed. On the contrary, the anencephalus described by EDINGER and FISCHER (1913) is evidence that the normal newborn already differs from those without cerebral hemispheres. Since the hypothesis that cerebral hemispheric function is lacking in the newborn cannot be proved, the existence of every form of consciousness in the newborn cannot be refuted. As an example, the expressive movements present at a very early age contradict such an allegation. KOFFKA "construes" the following experiences of the newborn:

"Against a relatively neutral background the baby sees a luminous spot. He feels on his hand, which usually lies passively, an interference in one area. Briefly and generally: From the background of the vague and indefinite there emerges a circumspect and specific phenomenon, a quality."

STERN (1923) maintains that the newborn already possesses two instincts – the sucking instinct and, in a rudimentary form, the orienting instinct.* STERN attributes to the newborn the first traces of consciousness. He bases his opinion on the histological studies of FLECHSIG (1876) (see p. 539), according to which a number of nerve pathways are already in the process of formation at birth. However, it cannot be histologically observed whether or not these nerve pathways are able to function. STERN believes that vague states of consciousness whose sensory and emotional components are inseparably fused exist in the newborn. That the newborn in the first day of life experiences sensations of comfort and discomfort, as indicated by posture and facial expression, is generally recognized.

K. BÜHLER (1922) differentiates three stages in the mental development of the child: instinct, training, and intellect. Instinctive actions represent the inherited behavioral characteristics of the species and arise spontaneously. In the newborn they appear in the form of activities related to the intake and transmission of food, such as sucking, biting, licking, swallowing, vomiting, and excreting stools and urine, and to normal respiratory functioning, such as sneezing, yawning, and coughing. Grasping and walking are learned later, but so quickly that there must exist a certain preformed coordination that can be regarded as a kind of instinct although not as a pure one.

K. BÜHLER finds no proof whatsoever that consciousness depends on the functioning of the cerebral hemispheres. Certainly there are no compelling arguments against the assumption of a deep-seated consciousness in the newborn; just as certainly there are no decisive arguments in favor of such a hypothesis.

C. BÜHLER (1928) even wrote a *Genesis of Consciousness*, according to which consciousness is not yet present in the first month of life. The reactions during this period are so rigid in their immediacy that, compared with the second and third months, no aftereffect or memorizing can be discovered. Actually, however, one can prove that an aftereffect already exists in the unborn child (see p. 90).

---

* A literal translation of STERN's term "Hinwendungsinstinkt" would be "turning-toward instinct"; the meaning is probably similar to the term "rooting reflex" as used in the English literature – The translators.

STIRNIMANN (1940) prescribes for psychology the task of investigating the principle underlying the unity inherent in the individual's development in relation to the external world and his own body. In the newborn there already exists a structured "'condition' that is at least similar to consciousness and from which the later consciousness evolves, gradually increasing its content." The soul in its complexity, even the intellect, are already in existence in an "anlage." Otherwise, the newborn would not be able to utilize his experiences.

According to REMPLEIN (1952), the centers and connections of the cerebral hemispheres are already partially matured at birth, and therefore exert a certain influence on the behavior of the child. Correspondingly, the newborn already possesses consciousness and from the first day reflects pleasure and displeasure states in his facial expression. These expressions reveal that the psyche attempts to express itself from the beginning, and this leads to the presumption that certain emotions and a practical tendency to withdraw from what is disagreeable and harmful exist in the newborn. By his crying the baby distinctly announces that he wants to be released from a disagreeable situation.

AENGENENDT (1954) believes that consciousness develops in the course of the first year, i.e., "The newborn already possesses an emotionality, on the basis of which he produces his signals." This emotional responsiveness, this awareness of existing remains a vague feeling until the beginning of the ninth month, when numerous sentiments begin to color various drives. The awareness of existing has reached a higher level. The "sensation of oneself" has become an "awareness of oneself." The child experiences emotion. While the awareness of existing evolves into an awareness of matter, the consciousness adapts itself to still higher levels of personality.

As this review indicates, there is great diversity in the conceptions of the nature of consciousness in the newborn. In contrast to the German child psychologists, two Americans who edited handbooks on child psychology, MURCHISON (1933) and CARMICHAEL (1946, 1954), have occupied themselves extensively with the physiology of the maturing brain. In both of his books CARMICHAEL deals with the prenatal stages of development. K. C. PRATT writes about the newborn. CARMICHAEL (1954, p. 157) finds no generally accepted scientific definition of consciousness. For studying the developing consciousness he does not object to using the same approach ontogenetically and phylogenetically. However, it should be realized that we are dealing with complex deductions and not with facts obtained through observation.

The usual personal relations between adults are only possible by assuming from the behavior of others that the same phenomena of consciousness exist in others as known to oneself by one's own experience. We may assume that they exist in the child at the earliest after the first year of life. Only by this age does the child attain the faculties that differentiate man from the preceding higher animals, i.e., the erect unsupported gait which frees the hands for the use of tools and the beginnings of language and reason. Only at this stage of development, not clearly defined as to time, may we ascribe to the child phenomena of consciousness similar to our own and therefore understandable to us.

The largest and most important portions of the human brain, the cerebral hemispheres, are not yet functioning in the young infant. The early developmental stages differ neurologically from the brain of the adult in direct relation to the difference in age. The behavior of the newborn is still on a level lower than that of many animals. The contradictions that arose in the attempt to determine at what point in evolution the beginning of consciousness arose and how it made itself outwardly recognizable were bound to be repeated when investigating the same problem in the infant.

Self-observation – the only method possible to perceive consciousness – cannot be utilized in the animal nor, since memory does not reach back far enough, in the infant. Therefore, we do not know one single certain fact relating to the consciousness of the newborn. Even future investigations cannot change this aspect of the problem. However, even if the degree to which the animal can progress in levels of consciousness remains unknown, there are no doubts about the child's becoming a conscious being.

Strictly speaking, it is not possible to prove that the newborn experiences the same sensation from a pain stimulus as an adult. Nevertheless, we have continued to use this generally accepted way of referring to such an experience. If we discontinue the use of the term "pain," as WATSON has suggested, the description of this phenomenon would become quite complicated. ZIEHEN (1921) considers it absolutely justifiable to refer to seeing and hearing in animals.

> "One has to be aware that this seeing cannot with certainty be interpreted as a psychological process in the sense of a visual sensation... Without a compelling reason we thus depart too far from everyday language and in many cases arrive at a clumsy and unintelligible way of expression."

BETHE (1952) feels that language becomes inadequate, boring, and circumstantial if we omit from scientific terminology all subjective expressions that are considered to be transferable from human thinking and feeling to the animal: We would have to reject words such as purpose, adaptation, light, color, and pain and many others. However, if we use these and similar words, it has to be left to the individual to consider these terms as short-cuts. The investigators of behavior use psychological terms without fear of being accused of anthropomorphism. Their success justifies this procedure (JUNG 1953).

What applies to the animal surely applies to the infant, for we know that sooner or later he will acquire consciousness.

SENSES AND REASON

"Only the eye creates the world."
C. MORGENSTERN 1928

Even philosophers in ancient times realized that sensory functions are prerequisites of the thinking process and that thinking depends on sensory perception. In *Theaetetus* PLATO (427–347 B.C.) imagines memory as a kind of wax tablet located in the soul. We imprint our visual, auditory, or pure thought memories on this tablet. As long as the imprints remain, we remember them; if they fade, we forget them.

ARISTOTLE uses in his own treatise on the soul the image of a wax tablet but in another context: The mind has the ability to comprehend the totality of the thinkable before it thinks. We have to picture this as a writing tablet on which in reality nothing has yet been written. The same applies to the mind. Like anything else which can be thought, the mind itself can be the object of thinking. To be sure, HEGEL disagrees that a *tabula rasa* should represent what ARISTOTLE had in mind, for to HEGEL reason does not possess the passivity of a writing tablet but is a reality itself.

Philosophy has held on to the picture of a *tabula rasa*. According to the stoic theory, the original soul amasses images and simultaneously symbols of the script by perceiving external objects and inward conditions. It cannot be established who was the first to make the statement: *Nihil est in intellectu, quod non prius fuerit in sensu.* ("There is nothing in the intellect that has not first been in the senses.") We find the essence of this statement in the writings of the materialist philosophers, namely, DEMOCRITUS (460—360 B.C.), EPICURUS (341—270 B.C.), and LUCRETIUS (98—55 B.C.), and even in the doctrine of the followers of materialism in the last century.

According to these philosophers, our knowledge is only a sensory perception or a mechanical rearrangement of parts of our sensory perception. EPICURUS teaches according to a Latin dictum that: *Quidquid animo cernimus, id oritur a sensibus.* ("Whatever we perceive with the soul enters through the senses.") According to THOMAS AQUINAS: *Est autem naturale homini, ut scientiam per sensus acquirat.* ("It is natural for man to acquire knowledge through the senses.")

For the sensationalist JOHN LOCKE (1623—1704) sensations are the primary but by no means the sole source of our concepts; next in importance is "reflection," which in modern German literature is usually rendered as "self-observation" or "inner perception." In addition, LOCKE regarded reason as a specific creative faculty that depends on rational and not sensory function.

The quoted sentence is in itself the nucleus of sensationalism, which is represented, among others, by CONDILLAC (1715—1780) and BONNET (1720—1793). They regard all psychological occurrences as transformed sensations. Supposedly nothing is added to sensation, neither to its content nor to its function, when it becomes knowledge. The conceptual thinking which follows is nothing more than a more refined and complicated sensation and its reproduction. CONDILLAC aspires in his *Traité des sensations* to understand clearly every psychological function on the basis of its origin, conceiving these functions as transformations of sensory perceptions.

LEIBNIZ discussed LOCKE's concept in his *New Treatise on Understanding*:

"To be sure, experience is necessary so that the soul will be predestined to this or that thought and will be on the alert for the ideas within us... The acknowledged philosophical axiom that nothing is present in the soul that is not derived from the senses will be held up to me. But the soul itself and its condition should be excepted from this axiom: *Nihil est in intellectu, quod non prius fuerit in sensu, excipe: nisi ipse intellectus.*"

LEIBNIZ says specifically that he believes that this conforms to LOCKE's opinion.

KANT (1781) shared the opinion that there is no mental activity without the senses:

"Observation and concepts form the elements of all our knowledge... Without sensory function no object would be apparent to us and without intellect none would be imagined.

Thoughts without content are empty. Views without concepts are blind. Intellect is not able to view and the senses cannot think. Only from their combination can knowledge result."

Seeing facilitates learning. Thus we find again in the Latin preface of the most famous book on visual instructions, *Orbis sensualium pictus* by AMOS COMENIUS (1562—1670), the sentence: *In intellectu autem nihil est, nisi prius fuerit in sensu.* COMENIUS was not a pure sensationalist but believed in an innate intellect in man that digests the impressions of the senses.

The senses form the foundation for every higher brain activity. They alone mediate between knowledge and the world around us. LICHTENBERG (1742—1799) expressed this relation as follows: "The eye creates light and ear, sounds. Without us they are nothing; we lend them these organs."

The senses overwhelmingly influence the development of brain functioning, especially in the first years of life. They are the prerequisites for all newly emerging faculties. CARUS (1831) called them "awakeners" of the life of the soul.

According to BURDACH (1838), the world offers the newborn material for the formation of his body and phenomena that when mediated by the senses contribute to his psychologic development. "Through the doors of the senses the world enters into the soul." (LOEBISCH, 1851.)

However, according to the psychologist K. BÜHLER (1922—1949), instinct is "the lowest stage and at the same time the fertile soil from which all higher functions grow." With this statement BÜHLER places himself in opposition to the cited philosophers who since ancient times have unanimously based mental development on the functions of the senses. Most frequently and independently of these philosophers, this opinion has also been expressed by the natural scientists.

In his textbook on pediatrics JÖRG (1826) already pointed to the connection between the gathering of sensory impressions and the development of thinking. According to the physiologist SECHENOV (1866), the psychological functions of man are sustained by the effect of external stimuli on his senses and by sensations emanating from his own body. Psychological functions would be impossible without the presence of all the senses.

The physiologist PREYER (1882) begins his book on the soul of the infant with the words: "The basic condition for every mental development is the functioning of the senses. Without these, no psychogenetic occurrence can be imagined." In another place: "The only material available to the intellect is received by the senses. This material enters the mind only through sensations." PREYER restores the image of the wax tablet:

"The tablet is not blank but already before birth has been inscribed with many illegible signs, the traces of the inscription of innumerable sensory impressions of long past generations. Whoever observes the child attentively can read this at first incomprehensible writing. Of course, no function of the soul can exist without the function of the senses, but neither can it exist without the inherited anlage."

According to the physiologist VERWORN (1920), who specifically refers to LOCKE, all mental functions issue from a single source, the sensations. Our entire imagination consists of memories of sensations. Only from the memorized impression of sensation are we able to construct our thoughts and arrive at the most complicated conclusions from logical reasoning. The entire rich activity of our consciousness would not exist at all had not our brain, fertilized by the stimuli of the world around us, first given birth to the many-colored multitude of sensations. KRETSCHMER's (1930) medical psychology reflects the same viewpoint. He asserts that a man who is blind and deaf and without the sense of touch would never be capable of developing what could be called "mental life," or an ego, and that, therefore, the entire psyche is nothing but a function of the senses.

KLAATSCH (1922) gave the following phylogenetic explanation: The nervous system originates from the ectoderm. Our brain was originally an external sensory organ. Then it was enveloped within the organism and became "overgrown" but maintained connections with the surface through several brain portions in "advanced" positions, its scouts reporting on outside events, while the "central" portion assumed the role of transforming the stimuli and responding to them by appropriate actions.

Sensory stimuli are carefully registered even by the newborn. He adapts himself to these stimuli by reflex actions and increased sensitivity in distinguishing between welcome and unwelcome stimuli. The facial expressions accompanying emotions develop from the movements of the muscles surrounding the eyes, mouth, and nose that regulate the perceptive ability of the sensory organs (see p. 108).

Originally, sensory stimuli elicited only simple reflexes, but in the course of phylogenetic evolution the reactions became increasingly varied. The cerebral mechanism that effects a stimulus response may considerably outlive the stimulus. For example, the newborn cries because of an unpleasant stimulus and continues to cry for quite some time after the stimulus has ceased to operate.

The cerebral mechanisms that respond to sensory stimulation become increasingly complicated during the first years of life, improving the organism's ability to evaluate sensory perceptions. This applies equally to the conditioned reflexes and to activities based on understanding, which always initiate originally from direct sensory impressions.

SUMMARY

Originally, animal psychology tried, in vain, to find out at what stage of phylogenetic development consciousness began and what constitutes animal consciousness. The contradictory explanations that have been offered can be neither validated nor refuted and therefore cannot be considered of scientific merit. There is no method by which consciousness in animals can be recognized, and there will be none in the future. Therefore, modern animal psychology occupies itself exclusively with animal behavior.

Infant psychology today still uses the wrong approach, as animal psychology did in the past. Actually, we have to assume that a consciousness similar to our own and therefore understandable to us does not exist in the child before the end of the first year of life. None of the widely diverging views on consciousness during the first year of life can be proved, and therefore they are all unscientific. Only the behavior of the infant can be investigated.

When adapted, with certain changes, to the infant, the methods of neurophysiology used in animal experiments yield very good information. Only after the first year of life may these methods be supplemented by research on consciousness.

Since ancient times philosophers have arrived at the knowledge, later confirmed by natural scientists, that the higher mental functions of man depend on his senses.

REFERENCES

AENGENENDT, J.: Psychologie des ersten Lebensjahres (Psychology of the First Year of Life). p. 193. Bonn, 1954.

ALVERDES, F.: Die Tierpsychologie in ihren Beziehungen zur Psychologie des Menschen (Animal Psychology in Its Relation to the Psychology of Man). Forschungen zur Völkerpsychologie u. Soziologie (Research on the Psychology and Sociology of Nations). Vol. 12, p. 17. Leipzig, 1932.

AMOS COMENIUS: Orbis sensualium pictus. 1658. Präfatio.

ARISTOTLE: De anima. Über die Seele (On the Soul). German translation by A. Busse. 2nd ed., 430 a., p. 78. Leipzig, 1922.

– Hauptwerke (Principal Writings). German translation by W. Nestle. pp. 5 and 65. Stuttgart, 1953. Über die Seele (On the Soul). p. 176.

BEER, T., A. BETHE, and J. VON UEXKÜLL: Zbl. Physiol. *13*, 137 (1899).

BELL, C.: Anatomy and Philosophy of Expression. 6th ed. London, 1872.

BERNFELD, W.: Säuglingspsychologie (Psychology of the Infant). Vienna, 1925.

BETHE, A.: Allgemeine Physiologie (General Physiology). p. 138. Berlin-Göttingen-Heidelberg, 1952.

BIERENS DE HAAN, H. A.: Die tierpsychologische Forschung (Research in Animal Psychology). Bios. Vol. 2. Leipzig, 1935.

BUFFON, G. L. L.: Sämtliche Werke (Collected Works). German edition, Vol. 4, p. 243. Cologne, 1840.

BÜHLER, C.: Kindheit und Jugend. Genese des Bewußtseins (Childhood and Youth. Genesis of Consciousness). Leipzig, 1928.

BÜHLER, K.: Die geistige Entwicklung des Kindes (The Mental Development of the Child). 3rd ed., p. 2. Jena, 1922.

– Abriß der geistigen Entwicklung des Kindes (Compendium on the Mental Development of the Child). 7th ed., p. 10. Heidelberg, 1949.

BURDACH, K. F.: Physiologie als Erfahrungswissenschaft (Physiology as an Empirical Science). 2nd ed. 1838. 3, 196.

CANESTRINI, S.: Über das Sinnesleben des Neugeborenen (Sensory Function of the Newborn) Berlin, 1913.

CARMICHAEL, L.: Manual of Child Psychology. New York, 1946, 2nd ed., 1954.

CARUS, C. G.: Vorlesungen über Psychologie (Lectures on Psychology). Reprint. Leipzig, 1931.

CONDILLAC: Nach Überwegs Grundriß der Geschichte der Philosophie (In Überweg: Compendium on the History of Philosophy). 11th ed., Vol. 3, p. 260. (1914).

– s. Abhandlung über die Empfindungen. (Treatise on the Sensations). Philosophische Bibliothek (Philosophical Library), Vol. 31. Berlin, 1870.

CUVIER, G.: Le règne animal distribué après son organisation (The Animal Kingdom Distributed According to Its Organization). I, 52, 53. Paris, 1817.

DARWIN, C.: Die Entstehung der Arten (Origin of the Species). German translation by Stuttgart o. J. p. 296. (1859).

DESCARTES, R.: Nach I. E. Erdmanns Geschichte der neueren Philosophie (In I. E. Erdmann: History of Modern Philosophy). I./II. Vol. 1, p. 207. Stuttgart, 1932.

EBBECKE, U.: Physiologie des Bewußtseins in entwicklungsgeschichtlicher Betrachtung (Physiology of Consciousness in the Light of Phylogenesis). p. 17. Stuttgart, 1958.
EDINGER, L., and B. FISCHER: Neur. Zbl. *1913*, p. 876.
EMPEDOCLES, bei H. DIELS: Die Fragmente der Vorsokratiker (In Diels: The Fragments of the Pre-Socratic School). 5th ed., Vol. 2, p. 49. Berlin, 1935.
EPICURUS: nach Eislers Wörterbuch der Philosophie (In Eisler: Dictionary of Philosophy). Vol. 3, p. 61. (1930).
FREUD, S.: Drei Abhandlungen zur Sexualtheorie (Three Contributions to the Theory of Sex). 5th ed. Leipzig and Vienna, 1922.
FRIDANK: Bescheidenheit (Modesty). W. Bezzenberger, Halle, 1872.
FRIEDLANDER: De l'éducation physique de l'homme (The Physical Education of Man). Paris, 1815.
GARLEY, D.: Internat. Zschr. Psychoanal. *10*, 134 (1924).
GEHLEN, A.: Der Mensch (Man). 6th ed., p. 141. Bonn, 1958.
GENZMER, A.: Untersuchungen über das Seelenleben des neugeborenen Menschen (Investigation on the Mental Processes of the Newborn). Diss. Halle, 1873.
GOMPERTZ, T.: Griechische Denker (The Greek Thinkers). 3d ed., 3, 126. Berlin-Leipzig, 1931.
HECKE, B.: Die Tierseele auf der Grundlage der grundwissenschaftlichen Philosophie von J. Rehmke (The Animal Soul Based on the Basic Scientific Philosophy of J. Rehmke). Greifswald, 1939.
HEGEL, G. W. F.: Sämtliche Werke. System der Philosophie (Collected Works: System of Philosophy). Vol. 8, p. 24. Stuttgart, 1927–29.
– Sämtliche Werke. Geschichte der Philosophie (Collected Works: The History of Philosophy). Vol. 18, p. 386. Stuttgart, 1928.
– Sämtliche Werke: Philosophie des Geistes (Collected Works: Philosophy of the Mind). Vol. 7, Sec. 3. Berlin, 1845.
HEMPELMANN, F.: Tierpsychologie (Animal Psychology). p. 613. Leipzig, 1926.
HERODOTUS: Historien (Histcry). German translation by Aug. Horneffer. Book 2, p. 131. Leipzig, 1910.
HEUBNER, O.: Erg. inn. Med. *16*, 1 (1919).
HEYFELDER, O.: Die Kindheit des Menschen (The Childhood of Man). 2nd ed. Erlangen, 1858.
JENNINGS, H. S.: Das Verhalten der niederen Organismen (The Behavior of Lower Organisms). German edition. Leipzig and Berlin, 1910.
JÖRG, J. C. G.: Handbuch der Kinderkrankheiten (Textbook of Pediatrics). Leipzig, 1826.
JUNG, R.: Handbuch d. Inneren Medizin (Handbook of Internal Medicine). 4th ed. V. 1, p. 153. Berlin-Göttingen-Heidelberg, 1953.
KANT, I.: Anthropologie (Anthropology) (1798). 2nd ed., p. 230. Königsberg, 1800.
– Kritik der reinen Vernunft (Critique of Pure Reason). Reclam. pp. 77 and 264.
– Ges. Werke. Kritik der reinen Vernunft (Collected Writings: Critique of Pure Reason) (1781). Berlin, 1911. 3, 74. Transcendentale Elementarlehre II. T. Transcendentale Logik. Einleit.; Werke Berlin 1914. Kritik der Urteilskraft (Fundamental Principles of Elementary Teaching, part II. Fundamental Principles of Logic. Introduction; Berlin Press, 1914. Critique of Judgement) (1790), 5, 546.
KAUS, G.: Die seelische Entwicklung des Kindes, in: Wexbergs Handbuch der Individualpsychologie (The Psychologic Development of the Child, in Wexberg: Handbook of Individual Psychology). Munich, 1926.
KLAATSCH, H.: Der Werdegang der Menschheit (The Development of Man). 2nd ed., p. 50. Stuttgart, 1922.
KOFFKA, K.: Die Grundlagen der psychischen Entwicklung (The Foundations of Psychologic Development). 2nd ed. Osterwieck, 1925.
KÖHLER, W.: Aus der Antropoidenstation auf Teneriffa (The Anthropoid Station in Tenerife). II.–IV. Abh. preuß. Akad. Wiss. Berl., Physik.-math. Kl. Sec. (1915–18).

KÖNIG, K.: Die ersten drei Jahre des Kindes (The First Three Years of Childhood). pp. 30, 79. Stuttgart, 1957.

KRASNOGORSKI, N.: Jb. Kinderhk. *69*, 1 (1909); *78*, 373 (1913); *144*, 255 (1935).

KRETSCHMER, E.: Medizinische Psychologie (Medical Psychology). 4th ed. Leipzig, 1930.

KUSSMAUL, A.: Über das Seelenleben des neugeborenen Menschen (The Mental Processes of the Newborn). Leipzig, 1859.

LANGE, F. A.: Geschichte des Materialismus (History of Materialism). p. 473. Iserlohn, 1866.

LEIBNIZ, G. W.: Die Hauptwerke. Kröners Taschenausgabe (Principal Writings. Kröner's Pocket Edition). Vol. 112. Leipzig, 1933. Neue Studien (New Studies). p. 123. (1704).

– Neue Abhandlungen über den menschlichen Verstand (New Treatises on Human Reason). Book. II, Chapter 1.

LIEBIG, J. v.: Chemische Briefe (Letters on Chemistry) (1844). p. 12. Hamburg and Berlin, 1913.

LINDEN, W. ZUR: Geburt und Kindheit (Birth and Childhood). 2nd ed., pp. 91, 130, 139. Frankfort on the Main, 1958.

LOCKE, J.: Versuch über den menschlichen Verstand (An Essay Concerning Human Understanding) (1689). Translation 2nd ed. Leipzig, 1894.

LÖBISCH: Entwicklungsgeschichte der Seele des Kindes (Ontogenesis of the Soul of the Child). p. 12. Vienna, 1851.

LOEB, J.: Pflügers Arch. *56*, 247 (1894); Einleitung i. d. vgl. Gehirnphysiologie u. vgl. Psychologie (Introduction to Comparative Brain Physiology and Comparative Psychology). p. 142. Leipzig, 1899.

LORENZ, K.: J. Ornith. *83*, 137 (1935); Naturwiss. *25*, 289 (1937); Z. Tierpsychol. *5*, 235 (1942/43).

LUCRETIUS: Von der Natur der Dinge (On the Nature of Things). German translation. Vol. 2, Book 3, Verse 223. Stuttgart, 1868.

MAGENDIE, F.: Lehrbuch der Physiologie (Textbook of Physiology). German translation. Vol. 1, p. 109. Tübingen, 1826.

MICHELET, G. L.: Anthropologie und Psychologie (Anthropology and Psychology). p. 151. Berlin, 1840.

MORGENSTERN, C.: Die Schallmühle (The Sound Mill). p. 22. Munich, 1928.

MÜLLER-ECKHARDT, H.: Das unverstandene Kind (The Misunderstood Child). 4th ed., pp. 19 and 251. Stuttgart, 1956.

MURCHISON, C. A.: A Handbook of Child Psychology. 2nd ed. Worcester, Mass., 1933.

PAVLOV, I. P.: Psychol. Review *39*, 91 (1932).

– Vorlesungen über die Arbeit der Großhirnhemisphären (Lectures on the Function of the Cerebral Hemispheres). Leningrad, 1932.

PFLÜGER, E.: Die sensorischen Functionen des Rückenmarks der Wirbeltiere (The Sensory Function of the Spinal Cord of the Vertebrates). Berlin, 1853.

PFUNGST, O.: Das Pferd des Herrn von Osten (The Horse of Sir von Osten). Leipzig, 1907.

PLATO: Ausgewählte Werke (Selected Writings). German translation by Schleiermacher. Vol. 2, Theaitetos, p. 385. Munich, 1918.

PLINY THE ELDER: Naturgeschichte (Natural History). Book 7. Einleitung: nach Dannemann, P. und seine Naturgeschichte (Introduction: after Dannemann, P. and his Natural History. Jena, 1921). p. 71.

PLUTARCH: Vermischte Schriften. Über die Liebe der Eltern zu ihren Kindern (Selected Writings: On the Love of Parents for Their Children). Vol. 3, p. 100. Munich and Leipzig, 1913.

POSEWITZ: J. f. Med., Chir. u. Geb.-Hilfe. 1. Heft. Herborn, 1799.

PREYER, W.: Die Seele des Kindes (The Mind of the Child) (1882). 4th ed., pp. VIII, 230. Leipzig, 1895; Spezielle Physiologie des Embryo (Special Physiology of the Embryo). Leipzig, 1885.

REMPLEIN, H.: Die seelische Entwicklung in der Kindheit und Reifezeit (The Psychologic Development in Childhood and Adolescence). 3rd ed. Munich-Basel, 1952.

RENSCH, B.: Neuere Probleme der Abstammungslehre (Newer Problems of the Theory of Descent). 2nd ed. Stuttgart, 1954.
RÜSCHE, F.: Geist und Blut (Mind and Blood). p. 55. Paderborn, 1937.
RUSSEL, B.: Erziehung (Education). p. 58 of German translation. Düsseldorf and Frankfort, 1948.
SANCTIS, S. DE: Die Mimik des Denkens (The Expression of Thought). German translation. Halle, 1906.
SCHOPENHAUER, A.: Sämtliche Werke – Die Welt als Wille und Vorstellung (Collected Works: The World as Will and Idea). Vol. 2, Ch. 5, p. 62. Munich, 1911.
SCHWARZ, F. H. C.: Erziehungslehre (Theory of Education). Vol. 2, p. 336. Leipzig, 1804.
SECHENOV, I.: Selected Works. Moscow and Leningrad, 1935.
SIGISMUND, B.: Kind und Welt (Child and World) (1856). Altenburg, 1897.
SIKORSKI, I. A.: Die seelische Entwicklung des Kindes (The Psychological Development of the Child). 2nd ed. Leipzig, 1908.
STERN, E.: Jugendpsychologie (Psychology of Youth). 4th ed. 1950.
STERN, W.: Psychologie der frühen Kindheit (Psychology of Childhood). 3rd ed. Leipzig, 1923.
STIEBEL, S. T.: Gehirndiätetik der Säuglinge (Brain Dietetics of Infants). J. Kinderkrkh. *26*, 249. Berlin, 1856.
– Des alten Dr. Stiebel Jubeldissertation (Old Dr. Stiebel's Festival Thesis). Frankfort on the Main, 1865.
STIRNIMANN, F.: Psychologie des neugeborenen Kindes (Psychology of the Newborn Child). pp. 10, 98, 103. Zürich and Leipzig, 1940.
STRÜMPELL, L.: Physiologische Pädagogik (Educational Physiology). Leipzig, 1880.
THOMAS AQUINAS: Summa theologica. Vol. 7, p. 156 (99, 1); p. 169 (101, 1) of the German edition. Munich-Heidelberg, 1941.
THORNDIKE, E. L.: Animal Intelligence. New York, 1898.
TIEDEMANN, D.: Beobachtungen über die Seelenfähigkeit von Kindern (Observations on the Psychologic Ability of Children) (1787). C. Ufer, Altenburg, 1897.
UEXKÜLL, J. v.: Ergeb. Physiol. *1*, II, 212 (1902); Theoretical Biology, p. 123. Berlin, 1920.
VERWORN, M.: Die Entwicklung des menschlichen Geistes (The Development of the Human Mind). 4th ed. Jena, 1920.
WASMANN, E.: Der Kampf um die Entwicklungsprobleme in Berlin (The Berlin Debate on Problems of Development). p. 35. Freiburg i. Br., 1907.
WATSON, J. B.: Psychology from the Standpoint of a Behaviorist. 2nd. ed. Philadelphia and London, 1924; Der Behaviorismus (Behaviorism). Berlin and Leipzig, 1930.
WUNDT, W.: Vorlesungen über die Menschen- und Tierpsychologie (Lectures on Human and Animal Psychology). 6th ed., p. 389. Leipzig, 1919.
ZIEHEN, T.: Die Beziehungen der Lebenserscheinungen zum Bewußtsein (The Relations of Life Phenomena to Consciousness). Abh. z. theoretischen Biologie. H. 13, p. 64. Berlin, 1921.

2. THE DEVELOPMENT OF SENSORY FUNCTION

GENERAL REACTIONS

*(1, 6, 7, 8, 12, 13, 14, 17, 51)**

A stimulus affecting an infant can produce two different types of response: sensory reactions and general reactions. Sensory reactions take place in the stimulated sensory organs and, in general, serve to increase or decrease the readiness to receive the stimulus. For this purpose, the sensory organs, particularly the eyes and the mouth, are surrounded by two layers of muscles, radiating and circular. Contraction of the radiating muscles enlarges the place of entrance for the stimulus. Contraction of the circular muscles narrows it. For example, bright light produces a closing of the lids and narrowing of the pupil, while the eye opens widely and the pupil becomes enlarged in dim light. Other such responses to stimuli serve the same purpose, namely, to minimize or to keep away unwelcome stimuli and to strengthen the effect of those that are welcome. For example, sweet-tasting substances produce licking and sucking movements in the newborn; bad-tasting substances, on the contrary, produce facial grimacing, choking, and flow of saliva. Similarly, lid blink, pupillary reactions, and the eye-neck reflex prevent the blinding of the eyes; and KRETSCHMER's reflex (arrest of breathing because of a bad odor) prevents further inhalation of irritating gases. These sensory reactions are described in detail later in this volume.

There are reactions of a general nature that can be elicited from every sensory organ and whose symptomatology does not proceed from a specific type of irritating stimulus even though they may be influenced by the strength of the stimulus. There is no evidence that these reactions may change the receptive capacity for the stimulus, and it is even doubtful that they are of any use at all. In this group belong changes in cerebral pulsation and the respiratory rhythm, galvanic skin reflexes, fright reactions, and motor restlessness.

CANESTRINI (1919) examined the irritability of the newborn by recording their respirations and cerebral pulsations from the anterior fontanel. Since these tracings change greatly under the influence of various stimuli, he concluded — and rightly so — that irritation had occurred. Based on my own experience, I can agree only with

* The numbers refer to the list of my own publications at the end of the book (see p. 654).

reservations with his interpretation of these findings. Under the influence of effective stimuli, the infant easily shows fright reactions or at least general motor restlessness. Because of this, the tracings of the respirations and cerebral pulsations are of necessity so greatly altered that they do not represent the true picture. For example, a very minute motion of the head is sufficient to displace or shift the very sensitive recording device, which is attached to the skull, and, thus, a change in the tracing of the cerebral pulsation may occur due to an external factor. To avoid this error, the child must be made to lie completely motionless before and after the stimulus is applied, and to preserve this state of immobility the baby must be kept under continuous and strict observation. The best way to accomplish this is by means of a special instrument for the purpose. As I have shown, we can comply satisfactorily with these requirements in the sleeping child. However, the children examined by me were not infants; thus, so far there have been no objective investigations of cerebral pulsation in infancy.

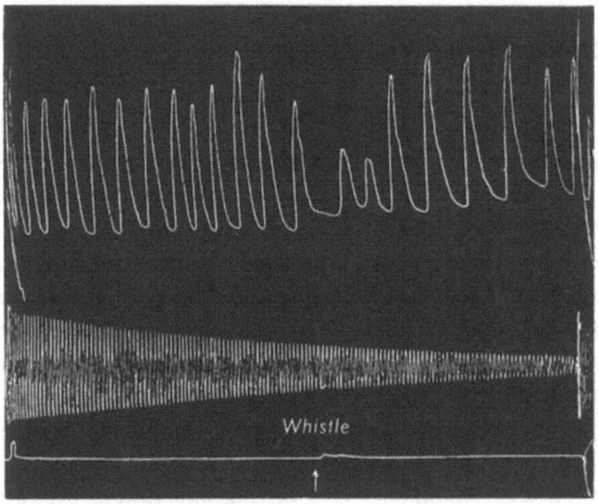

Fig. 1. Changes in respiration of a premature infant during sleep caused by sound stimulation. Time 1/5 second.

WATSON (1924) was satisfied to record the respiratory and hand movements of the newborn under the influence of external stimuli. It seems to me that no less can be deduced from his tracings than from those of cerebral pulsation by CANESTRINI, which, after all, for the most part represent only head movements. To measure the infant's reaction time I placed him on an inflated cuff and registered the pressure changes caused by his movements. If special precautionary measures are not taken, external stimuli will always cause pronounced general movements.

It is easier to record respiration than the cerebral pulsation. To demonstrate changes in the respiratory rhythm it is necessary that the child have a normal rhythm prior to the experiment, and this occurs only in sleep. If a stimulus only suffices to produce a respiratory reaction without interfering with sleep, we frequently encounter transient acceleration, slowing, or shallowness of respiration (Fig. 1); occasionally deeper expirations are noted. In most cases, however, one deep inspiration followed by a period of respiratory arrest or shallow breathing occurs. If the child awakes after the stimulus has been applied or cries or moves, the breathing becomes completely irregular, as can be seen in many of CANESTRINI's tracings.

MAYR (1862) gave a good description of this pattern. If the sensory organs of the newborn or older infant are affected by an unexpected or sudden stimulus, the child, stricken by fear, stops breathing; then, some very fast inhalations take place, usually followed by a deepening of the respiration.

Exactly the same responsiveness of the respiratory center to external stimuli can be found in a well-developed form in many animals, as has often been described. STRAUB (1913) found that the breathing of rabbits is influenced by noise. In SIEFERT's experiments on lizards (1896), breathing increased after minor skin stimulation, such as blowing or light touch, and was sometimes depressed after stronger stimuli were applied. In the frog the respiratory rhythm changes after a forced change of position (BROWN, 1916) and sound stimuli (YERKES, 1905). BAGLIONI (1907, 1911) noted respiratory changes in fish when he touched the oral mucosa. The respiratory tracings reported by him have a surprising similarity to those obtained from the infant. Evidently very basic reflexes are involved.

In the newborn the pupil dilates in response to an external stimulus of any kind, though not with the same regularity as later in life. This reactive pattern is treated in greater detail on page 57.

The galvanic skin reflex (VERAGUTH's psychogalvanic reflex), which according to GILDEMEISTER (1915, 1922) is automatically associated with pupillary dilatation, is illustrated by the following experiment: If two points on a person are connected to an electric circuit and a galvanometer, a galvanometric deflection will occur a second after the person is subjected to a sensory stimulus of any kind. The stimulus causes sudden and transitory stimulation of the sympathicus. This reflex has already been demonstrated in lower animals, such as the frog and the toad.

According to my investigations, the galvanic skin reflex is generally absent in the infant. The reflex begins to appear at the end of the first year of life. Probably the central part of the reflex arc or its centrifugal pathway in the sympathicus is not yet functioning or completely functioning. The infant's parasympathetic nervous system is in general more sensitive than that of the sympathicus (DRESEL, 1921). In the course of later development the sympathetic system gradually becomes dominant.

I gather from a report by M. C. JONES that H. E. JONES * was able to demonstrate the galvanic skin reflex even in infants, although it was weaker and less easily elicited. This difference in findings is probably due to a more sensitive experimental instrument. I used the experimental design of MÜLLER-VERAGUTH.

In older (after infancy) sleeping children the galvanic skin reflex depends on the depth of sleep. Since the stimulus is the same, we must conclude that the reflex is present in light sleep and absent in deep sleep. Stimuli that decrease the depth of sleep elicit the reflex. In the ordinary somnambulistic hypnosis the reflex is always present; it does not even disappear when one tries, through appropriate suggestion, to inactivate the sensory sphere to be subjected to the stimulus.

A general reaction that I term the "fright reaction" (startle) is very clearly present in the infant. I do not wish to imply that the infant consciously experiences the same feeling as does the adult when he is frightened. Rather, this is a reaction that occurs under the same conditions in the infant as in the frightened adult.

---

* I had no access to this paper.

The fright reaction of the infant was mentioned frequently even in the earlier literature of this century, e.g., MÜLLER (1911), CANESTRINI (1913), and CEMACH (1920). The films of STRAUSS (1929) and especially of LANDIS and HUNT (1936, 1938, 1939) demonstrate – contrary to my previously held opinion – a difference between the fright reaction (startle) and the *Umklammerung* reflex of MORO (see p. 151). According to LANDIS and HUNT, the startle reaction is originally a flexor reaction, the MORO clasping reflex an extensor reaction. In the startle reflex the arms are extended outward and upward but not to the same degree as in the clinging reflex. The fingers are flexed, and the head and trunk are bent forward. In the MORO reflex the arms are extended sideways at a right angle to the body. The fingers and frequently also the trunk or head are likewise extended. Both reactions cause flexion of the legs. As slow-motion films have demonstrated, the same stimulus often causes both responses in succession, so that the fast startle reaction can still be distinguished from the subsequent slower MORO reflex. The startle reaction and clasping or clinging reflex can be elicited in the infant by a sudden loud sound stimulus. According to CLARKE (1939), the MORO reflex is the first to appear in the course of the individual's development. Later, some features of this reaction disappear; extension becomes flexion, as in the startle reaction. According to HUNT (1939), both reflexes can be elicited in the newborn.

WIESER, DOMANOWSKY, and HEINEN (1957) and WIESER (1958) also investigated the behavior caused by fright during the first two years of life; they did this by using slow-motion film (400–600 frames per second). In response to the same stimulus two patterns of movement, one following the other, appear, i.e., the initial flexor phase and then the secondary spreading phase. They differ from each other in regard to the provoking stimulus, the duration of the latency period and reaction time, and the muscle area involved in the spread of the stimulation.

1. The initial flexor phase is elicited by strong, suddenly-appearing irritating stimuli: A cold breeze or a pinprick is just as effective as jarring of the base on which the infant is lying. The latency period is only 0.02 to 0.07 second. The spine and the neck are flexed; the arms are raised, turned inward, and adducted; the forearms are flexed and pronated; the fingers are equally and slightly bent; the legs are flexed at the joints; and the feet and toes are dorsally flexed. At the same time, every other movement is suppressed. This shows that flexion dominates all other reactions or reflexes in the struggle for the final common path.

In the course of the individual's development the reaction becomes less general. At the age of six months the leg movements are less pronounced while the face and shoulders become more involved: The blinking reflex always occurs, the mouth closes, the facial expression becomes frozen, and the slightly bowed head is drawn into the shoulders. Angular flexion of the forearms may complete the picture. Sound stimuli gain increasing importance while those coming from the body surface become less effective. Finally, this general nonspecific reflex that forms the basis of the startle reaction becomes part of the adult movement path.

2. In the secondary spreading phase the spinal column assumes an extended position and the neck is overextended backward while the limbs are spread apart. The arms are raised and are spread apart, the elbow joints are in a slight flexor

position, and the fingers are spread and extended. The legs are abducted from the hips and are slightly flexed at the knees, the feet are dorsally flexed and supinated, and the toes are spread and dorsally flexed. Often the head is turned sideways, and this is followed by a similar turn of the trunk, or at least of the upper part of the body, so that the trunk assumes the form of a spiral. In addition, frequently the arm and leg on the occipital side are flexed and the other arm and leg are extended. In this way a complicated chain reflex results from several postural and righting reflexes,* i.e., a labyrinthine righting reflex of the body, an asymmetrical tonic neck reflex of the limbs, and "a readiness of the limbs to jump." Thus, the secondary spreading phase corresponds in almost all details to the reflex turning of apes, cats, guinea pigs, and other mammals during a free fall.

During the first year of life this general reflex disappears because higher patterns of the nervous system replace the various righting and posture reflexes.

WOLFF (1959) uses the term "startle" for all incidents whereby suddenly and out of complete or nearly complete inactivity the infant executes a rapid abduction and extension or flexion of both arms, usually associated with spreading of the fingers and arching of the back. During normal sleep and when breathing, the newborn frequently exhibited these extensor and flexor startle reactions without any external cause, i.e., without preceding spontaneous movements or accompaniment by any other movements of significance. During irregular sleep and irregular respiration, i.e., changing from rapid shallow to slow deep breathing, these startle reactions appeared less frequently in the absence of an external cause. Instead, various other types of movements occurred. Startle reactions occurring without external causation were less frequent in states of drowsiness (irregular breathing, the eyes sometimes open and sometimes closed), and motor restlessness was more pronounced in these states than in normal sleep and occurred more often than in irregular sleep.

STRÖDER and SANDEN (1955) regard the toe fanning reflex elicited by the introduction of a nasal catheter as one of the fright reactions of infancy. About half of the premature and normal newborns exhibited this phenomenon until about the ninth to eleventh week of life. Insertion of a rubber tube through the mouth did not produce the same effect. According to these authors, the fright reaction first spreads to the toes.

Even in the first days of life the MORO reflex can be clearly demonstrated. It usually disappears after the fourth month. The reflex seems to be associated with the immaturity of the brain (HUNT and LANDIS, 1938); perhaps this explains its persistence in the feeble-minded (MORO). The startle reaction, however, remains after maturation of the brain and can still be observed in the adult. A. STRAUSS (1929) filmed the startle reaction in the adult elicited by the sound of a shot and compared it with the corresponding reaction of the infant. The adult's startle reaction consists of: flexion of the legs and trunk, forward flexion of the head, raising and forward movement of the shoulders, abduction, forward raising and inward turning of the arms, pronation of the forearms, closing of the hands, lid blinking, facial grimacing, and contraction of the abdominal muscles. Not all the features of this

---

* The reflexes mentioned here are described on page 147.

motor reaction could be observed in each instance, but the lid blink was invariably present. Just as in the adult, the infant's fright reaction is most pronounced if the stimulus occurs while he is at complete rest, for example, during sleep. If the same stimulus is repeated several times at short intervals, the fright reactions will soon subside (refractory phase) (see p. 86). Also, it is impossible to frighten the infant if he is greatly distracted by other stimuli. For instance, in noisy surroundings it is more difficult to frighten a child by sound stimuli than it would be after a prolonged period of silence. This is no different in the adult. The fright reaction is generally associated with a change in the respiratory rhythm, as has been described on page 24, and therefore its demonstration can be aided by recording respiration. By this means it is easy to recognize the refractory phase.

Although the MORO reflex can be elicited successively as often as desired by shaking of the head and without a concomitant refractory phase, the same is not possible for the fright reaction. According to YLPPÖ (1919), it is possible to elicit in a small immature infant a jumping-jack-like jerking of the arms and legs by tapping his breastbone with a small hammer. YLPPÖ regarded this finding as an indication of cerebral hemorrhage due to birth trauma. However, ISBERT and PEIPER (1927 [*17*]) were always able to elicit identical movements in mature newborns during sleep; frequently the head also jerked backward. The fact that this phenomenon can be elicited successively as often as desired shows that it is a form of the fright reaction.

According to KELLAWAY and FOX (1952), the electroencephalographic waves in sleeping children in response to single stimuli have certain qualities in common with the regular pattern: Maximum voltage is noted in both temporal regions. The wave form is similar and both patterns are seen most distinctly in the early sleep states. In general, the wave occurring as a reaction is higher and has a sharper peak than the spontaneous wave. The pattern can be elicited only at certain intervals because every reaction is followed by a rather long refractory phase, as in the fright reaction.

SENSORY FUNCTIONS

Skin Sensations

The anatomic basis for the different senses is known only incompletely. According to JALOWY's findings (1939), all forms of nerve ending on the finger and toe tips develop simultaneously during the fetal period. The intraepithelial fibers and the first anlage of MERKEL's, MEISSNER's, and VATER-PACCI's corpuscles appear at about the end of the fourth fetal month. The number of intraepithelial fibers grows continuously until the seventh fetal month and then diminishes, one portion degenerating and the remaining fibers spreading over a larger surface. At about the end of the seventh fetal month MERKEL's tactile bodies are completely developed. At about the end of the eighth fetal month the structure of VATER-PACCI corpuscles corresponds to that of the structure of the newborn. HOGG (1941) found in the twelfth week formations in the region of the mouth similar in appearance to MERKEL's corpuscles (see p. 34).

The pressure points are restricted to those places on the skin that correspond to the surface projections of the slanted hair follicles. The receptors are the nets of the nerve tissue around the sheath of the hair roots. In the areas not covered by hair the pressure stimulus is received by MEISSNER's tactile bodies (VON FREY, 1926). BECKER (1922) states that these reach the final stage of development in the sixth month of life; according to JALOWY this takes place even later, in the twelfth or thirteenth month. From the time of birth to the final stage of development the nerve endings continue to branch out and form additional connections. The VATER-PACCI corpuscles, the GOLGI-MAZZONI corpuscles, and the free nerve endings are regarded as the receptors of muscle sensation (VON FREY). According to BECKER, VATER-PACCI corpuscles are smaller in the newborn than in the adult and contain less fluid.

Free intraepithelial nerve endings are believed to be the receptors of sensations of pain. They were shown by BECKER to be comparatively numerous in the skin of the newborn although they had not yet reached their peak of development. BECKER stresses the wealth of neural end formations and of free nerve fibers in the lips, especially in the deeper layers of the formations known as "sucking pads," and, as indicated by him, these anatomic findings give no support to the formerly widely held opinion that prematurely born infants are not sensitive to pain. The tickling sensation is caused by repeated irritation of the pressure sense (THÖLE, 1912), and itching can be traced to stimulation of the pain nerves in the epidermis (THÖLE; VON FREY, 1926).

According to GOLDSCHEIDER (1926), nothing certain is known about the anatomic structure of the temperature points. HENSEL (1955) feels that all statements on formations that could be regarded as thermoreceptors lack any verification.

Pain

(1, 9, 13, 31, 59)

The infant's sensitivity to pain can be examined electrically, or mechanically by pinpricks.

GENZMER (1873) stated that the pain sensation occurring after a mechanical stimulus is very weakly developed in the newborn. In examining 60 infants he found that during the first few days of life they were almost insensitive to pinpricks. He stuck immature infants with thin pins in the most sensitive areas, such as the nose, upper lip, and hand, and so severely that a little drop of blood came from the wound opening, yet the infants showed no signs of discomfort. In his experiments normal newborns responded with localized movements to pinpricks of medium strength one to two days after birth, rarely earlier. The reaction time was as much as two seconds. During the first week of life the sensitivity increased considerably.

KRONER (1882) confirmed GENZMER's findings, although he observed in mature or nearly mature newborns pronounced pain reactions, such as facial grimaces and loud crying, after strong mechanical or electrical stimulation. According to CANESTRINI (1913), the newborn's pain reaction to pinpricks is minimal and the reaction time is long. He stated this in spite of the fact that in eight out of ten trials distinct responses had occurred. REMPLEIN (1952) similarly states that the pain threshold is much higher in the newborn than in later life. Even beyond infancy the child is said to experience less pain from blows than is generally assumed by adults. TUMLIRZ (1955) concurs that pain sensitivity is least in the first year of life.

Actually, there is no normal or immature newborn who could not be easily aroused by pinprick. I could elicit by pinprick twitching in the facialis area of a human fetus 23 cm long. The experimental findings of SHERMANN and FLORY (1936) indicate that, as a rule, the face is the most sensitive part of the body. As always, the degree of irritability depends on the infant's mood and general state of health. Fatigue, distraction, hunger and satiation, crying, and sickness may considerably influence the reaction. Therefore, a definite figure for the depth of a pinprick necessary to produce a reaction cannot be given. For example, the sole of an infant has to be pricked to a depth of 2 to 4 mm for a reaction to ensue. According to SHERMANN and FLORY, sensitivity to a painful stimulus increases with age and for a given age the number of reactions is directly proportional to the strength of the stimulus.

The reaction time cited by GENZMER is too long. According to my measurements, the healthy infant when awake will usually react with movements after 0.12 to 0.70 second, sometimes a bit later. The reaction in the adult is about the same. It is true that the infant starts to cry only after 2 to 5 seconds. The reaction time is somewhat longer in immature and sleeping infants. When vaccinating infants, PETROSELLI (1929) found the reaction time to be very long, up to 7 seconds. The expression of pain observed by him occurred considerably later than the flinching reaction.

Some mentally defective infants are remarkably insensitive to pinprick, a finding which can be of diagnostic importance, but others are especially sensitive and respond to every pain stimulus with fright reactions.

The responses of the newborn and the young infant to painful stimuli consist of protective skin reflexes (PEIPER, 1936 [*59*]), i.e., general or focal reactions that are precisely related to the site of stimulation, but it is often not possible to decide with certainty the extent to which the tactile sense has been simultaneously stimulated. There are various ways for an irritating object to produce an effect: Either the object's position in space is left unchanged so that it cannot be displaced by the movements of the infant but its position is changed in relation to the skin, or it is fastened to the infant so that it is displaced in space with his motions while the same area of the skin is continuously being irritated.

The first approach is used when the pin is kept motionless so that its point presses slightly against the skin of the infant. Motor restlessness, which usually starts in the irritated body part but which may involve the whole body, follows. This motor restlessness ceases only when the irritated skin area is inadvertently removed from the pin point. If an area in the face is being irritated, the head is turned sideways and often moves several times to and fro. Should these movements accidentally bring the skin of the face again in touch with the pin point, a rare occurrence, the motor restlessness continues. If the excitement increases, the child may thrash about, moving his body, arms, and legs in a manner characteristic of the infant. Weak stimuli of the sort mentioned affect the centers of expressive movements to a lesser degree. The withdrawal reflex (flexion of the extended leg at the knee and hip joints caused by pricking the sole) is only an especially conspicuous portion of this protective reflex mechanism, which can be elicited at any time in every healthy infant from any part of the body — face, trunk, arms, or legs.

To effect continuous irritation of a specific skin area, I used clothespins, which are kept in place by the spring. In the adult they sometimes cause a disagreeable feeling of skin tension, sometimes a mild sensation of pain.

Exactly as observed with the previously described stimuli, the simplest response of the infant consists of more pronounced movements of the irritated limb. I term this response "localized movement reaction." When the face is irritated, the head moves to and fro. If the clothespin is fixed to the toes, the foot and soon afterward the whole leg alternately flexes and extends, a movement which soon spreads to the other leg. But even with increasing excitement, the directly irritated body part will be moved more violently than the others. Thus, the extension and flexion of the nonaffected leg is distinctly less pronounced. Similar movements appear in the arms. If the clothespin is fixed to the trunk, there are, of course, less distinct specific movements.

Not infrequently an additional response is noted, i.e., the irritated area is touched by another part of the body, such as the hand or the foot. This reaction can be most easily obtained if the clothespin is fixed on the medial surface of the leg. Even in the usual kicking movements occurring without local stimulation, the inner edge of the foot of one leg often glides along the inner side of the other leg. With appropriate stimulation the clothespin may be stripped off by this movement; but if it is fixed to the outer aspect of the leg, the irritated area will only occasionally be reached by the foot of the opposite side. If the clothespin is fixed to the abdomen, the anterior surface of the thigh during these leg movements may frequently touch it.

Corresponding movements of the arms and hands are also present in the newborn, but they are less distinct. If the clothespin is fastened to the face, it is often reached by the hand. But sometimes this does not occur even though the hand is very near the irritated area at the beginning of the experiment. Of course, if the child is sucking his fingers, one cannot expect such a reaction.

Irritated skin areas of the upper part of the body are occasionally reached by the hands. But rarely in the first months of life does the hand reach the opposite arm. If a clothespin is fixed to the outer aspect of the thigh, a half-turn of the infant takes place, so that the irritated area sometimes comes in contact with the mattress. A last and frequently occurring reaction of the infant is crying.

Watson (1924) described similar reactions. The newborn was placed so that he lay on his back with his legs extended; Watson pinched the child's knee lightly and then observed that after a short while his own hand was hit by the infant's opposite foot. The purposefulness of this movement was evident and occurred as regularly as in a reflex frog. When Watson irritated the nose of the newborn with pinpricks, his hand was hit by the fingers of the newborn after 2 to 4 seconds. The same reflexes can also be elicited by heat stimuli. In his experiments on heat sensitivity of the newborn, Crudden (see p. 42) observed that a capsule through which warm water was circulating and which was fastened to one of the infant's legs was "scratched" with the other leg. Corresponding reactions may appear with touch (EBBECKE, p. 38). Similar reactions of the newborn were reported by ANDRÉ-THOMAS (1952) and referred to as "réflexe d'allongement" and "réflexes adaptés de dérobement."

These reactions are found in healthy infants during the first months of life. They are most pronounced when the infant lies on his back, and they become indistinct when the infant is in the prone position. The prerequisite for the occurrence of these reflexes is a "readiness to react," which is not always present to the same extent. This readiness to react is markedly decreased in sleep, in poor states of health, or when the nervous energy is diverted by other activities, such as taking the bottle, finger sucking, staring (at one object), lifting the head while in the abdominal position, crying, etc. The defensive capacity can also be altered when the same stimulus elicits responses in various reflex groups.

There may be a change in the sequence of the reflexes that aid the infant in his attempt to free himself from the irritating stimulus. Often the local movement reaction is the first to appear. Then the irritated skin area is touched with the arm or leg, and finally increased excitement and loud crying occur. The infant does not always follow this sequence. For example, a child who is already somewhat grouchy will cry immediately when irritated; another one will start with kicking, etc.

Protective skin reflexes (which include scratching movements) may in the same manner be elicited by painful hot (see p. 40), tactile (see p. 34), itching, and tickling stimuli. These reflex actions are completely capable of removing an ordinary stimulus. Through the movements of the irritated limbs, the specific skin area is withdrawn from the source of irritation or perhaps a stinging mosquito is frightened away. In many instances touching of the irritated skin area has the same effect, although such movements will be unsuccessful in dealing with such an unusually persistent irritation as a clothespin. Finally, crying protects the infant by alerting the mother.

The reported observations absolutely contradict the assertion of BÜHLER and SPIELMANN (1928) that only sucking, crying, and purposeless movements are available in dangerous situations to the "completely helpless" human newborn.

The protective skin reflexes remind us in principle of the so-called method of trial and error that JENNINGS (1910) demonstrated first in the unicellular animals and later in higher animals, although the basic nervous mechanisms may be quite different. This method of trial and error consists of somehow finding a means of escape from an irritating or movement-preventing situation by tentative actions. The exploratory movements of an ameba crawling away from a narrow gap is an illustration of such activity. The objectionable irritation causes a reaction that creates new and at times more favorable conditions. When through this maneuvering a situation that is free of damaging influences is finally found, the behavior no longer changes and the favorable situation is maintained. The movements cause various situations from which one is "selected." Justifiably, JENNINGS has warned against drawing any conclusions as to conscious patterns that could occur during these testing movements of the lower animals. In young infants this question must also remain unanswered.

Protective skin reflexes do not depend on functioning cerebral hemispheres. The wiping movements in the spinal preparation of the frog, by which one leg removes a drop of acid from the other or from the body or repulses a stabbing instrument with the foot (GOLTZ, 1869), correspond to the described protective skin reflexes of the young infant. As his cerebral hemispheres are not yet functioning, the centers of the protective skin reflexes can lie only in the brain stem.

The exploratory movements in the young infant represent unconditioned reactions, and on this basis the scratch reflex develops as a conditioned reflex. The successful reaction, which at first occurs together with unsuccessful ones, finally supplants the other movements by conditionally being successfully combined with the stimulus.

The infant and small child show surprisingly little pain sensitivity to electric current, as can be seen in Table I. This phenomenon, also noted in treatment with

*Table I: Relation between Chronologic Age and Sensitivity to Primary Induction Currents**

Age of Child	Smallest strength of primary current eliciting recognizable sensations (ma)	Age of Child	Smallest strength of primary current eliciting recognizable sensations (ma)
1 day	400	9 months	250
6 days	300	9 months	300
11 days	400	2 years	200
17 days	300	3 years	200
23 days	250	4¾ years	150
24 days	250	6 years	50
3 months	250	6 years	50

faradic currents, has so far not been explained. CZERNY (1892) used the awakening threshold of the electric current as a standard for measuring the depth of sleep in children (see p. 499). WOLOWIK (1927) similarly tested the influence of food intake on the sensitivity of the infant; he measured the threshold of the electric current before and during feedings. It was found that the infant became significantly less sensitive during breast or bottle feeding so that considerably stronger currents were needed to elicit a response to the stimulus (see p. 464).

The adult reserves for himself the right to be protected against pain during surgical interventions. Because of their sensitivity to pain, newborns and infants have the same right but, contrary to the adult, they are defenseless when their rights are violated. Surgeons who declare infants to be insensitive to pain and – as I have witnessed – perform operations for pylorospasm without anesthesia, do so contrary to everyday experience and scientific knowledge.

* Based on CZERNY, 1892.

Touch and Pressure
(1, 9)

The question whether tactile sensations are only weak pressure sensations or whether these two types of sensation comprise separate categories is answered differently by various physiologists. While VON FREY (1926) considers both as belonging to the same category, STRÜMPELL (1904) and others consider them separate in principle. STIRNIMANN (1940), in his *Physiology of the Newborn*, differentiated between the two senses because in the newborn touch stimuli cause reactions different from those induced by pressure stimuli. However, a principal difference could hardly be deduced from this observation, as intense and weak stimuli generally differ in their effects.

The infant is very sensitive to touch. From all skin areas it is possible to produce general and frequent focal reactions, i.e., reactions originating from the point of irritation. The localized reactions were described in detail as far back as the last century by KUSSMAUL (1859), GENZMER (1873), and PREYER (1895).

As in adults, the individual parts of the newborn's body differ in their sensitivity to touch: The hands, soles of the feet, and face react quickly; however, it is difficult to stimulate the skin of the forearms and lower legs; even more insensitive is the skin of the shoulders, chest, abdomen, back, and thighs (PREYER and GENZMER).

HOOKER (1954) examined with a fine hair the tactile excitability of human fetuses surviving for from 5 to 15 minutes in a salt solution. It was possible to elicit the first reaction at the fetal age of seven and a half weeks. The circumscribed area of excitability was in the region of the mouth and by the tenth week expanded to the area of the second and third branch of the trigeminal nerve. This excitability appeared much later in the area supplied by the ophthalmic branch. Palms and soles became excitable only by ten to ten and a half weeks of fetal life. The earliest stimulus responses consisted of movements of the neck muscles and effected a turning away from the stimulus; these responses spread cervicocaudally toward the long trunk muscles and the muscles of the limbs. The general reaction to touch became less pronounced at the fetal age of thirteen to fourteen weeks compared with local responses. Local reactions of the oral zone are described on page 404.

According to HOGG (1941), no nerve fibers have reached the epithelial cells at the time when the first reactions from the oral zone can be elicited. These cells have no contact with any nerve fibers until the twelfth fetal week so that only the undifferentiated growing tips of nerve fibers can be considered as receptors of the trigeminal stimulation. Formations resembling MERKEL's corpuscles appear in the reflexogenic region of the mouth during the twelfth week.

The turning of the hungry infant toward the stimulus of touching the oral region (rooting reflex, p. 404) and the turning away of the satiated infant from the same stimulus serve the food intake function. Touching and tapping the lips without simultaneously stimulating the taste sense can provoke sucking movements. The reflexes associated with food intake are described in more detail elsewhere.

PREYER vividly described the effect of touch stimuli on the tongue. Tickling the top of the tip of the tongue with a smooth glass stick causes sucking movements: The edges of the tongue curve upward toward both sides of the stick and surround it, and the lips protrude snoutlike. At the same time the facial expression characteristic of that produced by a "sweet" sensation appears. Touching the middle region of the top of the tongue causes tight shutting of the eyes and raising of the edges of the nostrils and of the corners of the mouth but no sucking. When tickling the root of the tongue and the palate PREYER noted: retching, opening of the mouth, protruding of the tongue, raising of the larynx, increased salivation, and the facial expression produced by a "bitter" sensation and corresponding to the adult expression of nausea.

On tapping of the young infant's musculus orbicularis oris we at times see a snoutlike pushing forward of the lips. This reflex was first described by ESCHERICH (1888) as a "mouth phenomenon" and later was given various other designations (see p. 407).

The eyes of the newborn are already very well protected; touching the eyelids, conjunctiva, cornea, or eyelashes or blowing at them cause immediate shutting of the eyes. The reflexogenic zone is even more extended; simply touching the tip of the nose (PREYER) or the forehead (MORO, 1906) causes the infant to immediately close his eyelids. In this way from the very beginning the eyes are protected from foreign matter. Since the eyes and mouth are the first parts of the face to be reached by the hands of the infant and since the blinking reflex triggered by an approaching object is still absent at this time of life, this protective device is all the more important.

If one irritates the nasal mucosa with a small brush, the infant becomes restless, blinks, turns his head, or even moves his hands toward the nose. WATSON (1924) even intensified the stimulus by holding and closing the infant's nose. Thereupon the infants became very restless and began to vigorously thrust their arms in the air. In newborns the reaction sometimes lasted only for 3 seconds after the finger of the experimenter was reached.

PRATT, NELSON, and SUN (1930) searched for defense movements in the newborn by pressing a finger on his nose. The infants responded with backward bending of the head and back, general motor restlessness, and aimless movements of the body. The hand of the infant reached his own nose or the finger of the experimenter in only 1% of the trials.

On being tickled in the external ear canal, some newborns become restless and raise their hands to the face but not to the ears. PREYER observed a peculiar expression of tension when he closed the ear canal of a crying infant.

The spine reflex and the tonic palmar and plantar grasp reflexes belong to the group of tonic skin reflexes, which are discussed in greater detail elsewhere (see p. 158). The BABINSKI sign and the cremasteric and abdominal reflexes are also described later (see p. 285).

ANDRÉ-THOMAS, CHESNI and AUTGAERDEN (1954) were able to elicit reactions on the fingers of the newborn: If the inner edge of the flexed little finger is stroked (repeated stimulation is often necessary), the finger becomes more or less extended. On continuous stimulation, the extension gradually spreads to the ring, middle, and index fingers; only the thumb hesitates to move. If the stimulus subsides, the fingers are flexed again. If the stimulus effect spreads to the arm, supination of the forearm takes place and the arm is abducted. Stimulation of the back of the hand causes slight inward rotation; pinpricks readily lead to lively reactions.

NASSAU and BORTNIK (1938) described a "maturation reflex" of the hand of the newborn: As a result of general stimulation, the thumb will be strongly flexed in the metacarpophalangeal and interphalangeal joints so that it will almost touch the thenar region. The same movement occurs in the index finger, but, in contrast, the other three fingers are spread and extended. In some infants the middle finger may also participate in the flexion. This movement occurs in both hands symmetrically. The wrist joint remains in extension, and the forearm is slightly flexed at the elbow joint. The reflex is not associated with the Moro Reflex and is independent of the palmar grasp reflex as the palms are not stimulated. Each stimulus of somewhat increased intensity affecting the infant, such as waking him up, taking him out of the crib, undressing him, etc., triggers this movement. This reflex was found to be present in 99 out of 130 newborns and was particularly distinct in 10 prematurely born infants. Only a few infants still exhibited this reflex at the age of four weeks. ROSENBAUM (1958) did not quite succeed in demonstrating it.

STIRNIMANN (1940) saw grasp attempts in the newborn occurring when the child's closed fist was touched from the outside. If he placed his finger across the middle phalanges of the infant's fingers that were in a closed-fist position, the fist opened and the fingers were extended, usually with the exception of the thumb. Then the infant's hand enclosed the finger of the experimenter, and as soon as the palm touched it the hand grasped the finger vigorously. When STIRNIMANN used a little stick instead of his finger, this reaction was less pronounced and occurred far less frequently. The manner in which true grasping of a visualized object develops in the course of the first year of life is described elsewhere (see p. 255).

STIRNIMANN (1937) also described an "extrication" reaction in the newborn, incited by inhibition of free movements. When infants were fitted with mittens to protect them from occasional injuries, 23 of 150 newborns tried to free themselves from the mittens. By means of rubbing motions they gradually loosened the ribbons that were tied around the wrists and finally extricated first one and then the other hand from the mittens. The movements were not constant but were interrupted at intervals of various duration.

STIRNIMANN (1947) characterized another defense movement of the newborn as a "heart protection phenomenon." He noted that peculiar movements appeared when he listened to infants' hearts with the stethoscope. Placing the stethoscope over the cardiac region elicited directed defensive movements in 106 of a group of 200 newborns, and nondirected movements in 30. The goal of the movements. was the stethoscope, which was thereby to be pushed away. The children used either both hands simultaneously or, much more often, the left hand for these attempts. STIRNIMANN explained this heart protection phenomenon on the basis of phylogenetic memory — an "unconscious memory of the hard man-to-man battle for life." The objection should be raised that we do not know of any similar pattern for the protection of the heart in man or animal and that such a movement could not really protect the heart. I prefer to classify the so-called heart protection phenomenon as one of the skin protection reflexes (see p. 30).

In considering these reports the fact that almost always only newborns have been tested should be borne in mind. Older infants become increasingly skillful in their attempts to remove the irritated skin area from the stimulus. The reaction apparently does not depend on the stimulus type, and the identical skin protection reflexes are always elicited, be it through painful, tickling, tactile, or pressure stimuli.

The scratch movement is not an inborn reflex, but is acquired. This has been demonstrated by the detailed investigation of SZYMANSKI (1922). Newborns respond to an irritating stimulus only with general motor restlessness, weeping, and crying; they do not scratch themselves. The scratch reflex appears later and at different times for various skin areas. It appears in the second month when the eyelids are touched. As the infant becomes older, stimulation of other bodily parts becomes successively effective: the eyelids and ear lobes (four to six months), the edges of the nostrils (seven to twelve months), the other parts of the face, the trunk, and the extremities. The stimulation threshold decreases with increasing age. Stimulation of the upper arms has the weakest effect, evidently because the scratching has to be done with the opposite extremity. The ability to scratch with fingernails and finger tips is fully developed only at the age of eighteen months. In the feeble-minded, SZYMANSKI found all stages in the development of the scratch reflex from its absolute absence to a normal pattern, depending on the degree of the mental retardation.

PREYER (1895) described the development of the scratch reflex during the first six months in two children suffering from eczema. He found that a successful accidental contact between head and hand, i.e., relief of itching, caused the preference of this type of hand movement over all other movements. According to my observations, scratching develops sooner in infants suffering from an itching skin disease than in normal infants. In some infants the itching becomes so unbearable that, unless they are prevented from doing so, they scratch themselves repeatedly until bleeding occurs. I noted even in a ten-day-old infant with this type of skin irritation the distinct ability to scratch himself. Another infant, age six months, suffering from an itching eczema of his face was tied down so that he was unable to reach his face with his hands; he succeeded in scratching his face with his feet.

HARTMANN-KARPLUS (1931) tested the effect of itching powder on infants. Here, too, the scratch reflex was clearly related to the infant's age. The reflex appeared in its fully developed form after the sixth month of life. Younger children became very restless. The varying excitability of different body areas has not been examined in detail; however, the face proved to be more irritable than the abdominal skin. The scratch reflex appeared prior to the infant's ability to correctly localize the area of itching; the healthy infant does not succeed in correct localization before the eleventh to twelfth month.

The ability to scratch oneself is maintained during sleep if the stimuli, for example, pinpricks, are sufficiently strong. The reactions in the older child are not distinguishable from those that have been so far described in infants. The face is the most sensitive area. In general, the hands scratch the homolateral side of the body. The median area will be fairly equally scratched from both sides. Even if the contralateral hand lies near the irritated body part, the sleeping child ordinarily uses the homolateral hand. Frequently the scratch movement does not reach its goal; for example, the hand may rub the face but will not make exact contact with the irritated spot.

Sometimes the movement is not completed: All stages may take place, from a mere jerk of the hand to partial elevation to real scratching. Afterward the scratching hand sometimes assumes its initial position and sometimes remains near the irritated body area.

General restlessness occurs with stronger pain stimuli. The movement is not restricted to the irritated body areas but will spread to other regions. When the face is irritated, the legs are flexed or the arms extended or the whole body is rolled onto one side. These reactions indicate that the depth of sleep has been diminished and that awaking will soon take place.

A single mechanical stimulation of a pressure point conveys a touch sensation, usually mixed with a slight sensation of tickling. Repeated stimulation of the same spot does not lead to an intensification of the tickling sensation. This means that tickling is a passing sensation; it lasts longer and becomes more distinct only if the stimulus glides across the skin (VON FREY, 1926).

NASSAU (1938) tested the tickle reaction of the infant with light stroking motions of a brush or the finger tips on the neck, axilla, or sole, that is, in the skin regions that are the most susceptible in the adult. The tickle reaction was absent in all infants in the first two months of life and appeared in increasing numbers of infants in the succeeding months. In the ninth month the tickle reaction was present in all tested infants. The course of development of this reaction was as follows: It appeared first in the axillae, then on the soles, and finally on the neck. However, deviations from this sequence were not uncommon. The reaction began with expressions of uneasiness and defense; in a few weeks a pleasure-accentuated defensive expression or expressions of comfort appeared until finally the infant smiled or laughed and kicked his legs happily, thus showing that this type of contact was welcomed by him. The tickle reaction appeared only in the healthy child. It was absent or delayed in instances in which the general condition was severely disturbed, particularly in chronic nutritional disturbances or in severe rickets.

Although he has distinct touch sensations, the tickle reaction is absent in the young infant. According to NASSAU, the touch sensation becomes a tickle sensation only when a particular conscious psychologic process has developed, i.e., a certain distinction between the ego and the world around him has been made. The conscious relation to the outside world is prerequisite for the tickle sensation. We cannot laugh and we perceive tactile stimuli as a tickling sensation to a lesser degree or not at all if we try to tickle ourselves.

EBBECKE (1951) tickled newborn children with a bristle, moving it over the skin; the bristle was attached to the prong of a vibrating tuning fork. During this procedure the newborn showed a distinct wipe reflex when touched in the region of the eyes, the nostrils, the entrance to the ear canals, and particularly the lips. Facial grimaces and purposeful hand movements occurred as in the adult. Entrance into the portals of the sensory organs is prevented in this manner. The reactive wipe motion is a broad soft rubbing; it will stop a tickling sensation that otherwise would last for a long time, and it drives away stinging and blood-sucking insects. In addition, the nose, eyes, and ears are protected externally by special hairs; touching of these hairs is felt as an unpleasant tickling sensation and causes lively defense motions.

Infants vary greatly in their reactions to the contact stimulus of flies crawling on the skin. If flies are not warded off, e.g., by a net covering, they tend to gather in moist and warm spots, for example, around the eyes. Most children in the first, second, and possibly even third year of life are unable to defend themselves, but there are exceptions. For example, there may be a child who can defend himself in the tenth month and this so skillfully that no fly remains on his face, whereas perhaps the face of a two-year-old child lying in the next bed just seems to attract the flies. General reflex excitability declines in dying children and with this the defensive power against the flies. Therefore, it is always a bad sign if flies start to accumulate on the face of a sick child. Incidentally, today there are methods for effectively keeping flies out of a pediatric ward.

The pilomotor reflex, popularly known as "goose flesh," is probably caused by the combination of cooling or warming with tactile stimuli. HARTMANN-KARPLUS (1931) investigated this reflex in infants of various ages in two series of 15 each. It was much easier to irritate the naked infants who were in a cool room (17° to 22°C) than those who were in a warmer room (20° to 23°C). These stimuli consisted of stroking with a pin, spraying with ethyl chloride, and rubbing with a 1% Adrenalin solution. As in the adult, the reflex would either remain limited to the irritated skin area or would be extended. Occasionally half of the body was involved. The skin of the chest was found to be the most sensitive area, particularly the region around the nipples. In the newborn the reactions were less distinct and very fleeting, as was also noted by CAREDDU (1935). The reaction to Adrenalin was more frequently demonstrated.

Certain skin sensations, namely, those responding to pressure, temperature, and pain, can be localized. The simplest way to test this ability, which is attributed to a special "spatial sense," in the older child and in the adult is as follows: The dull points of a compass are placed at various distances on the skin region to be examined, and the smallest distance between which two points can be perceived separately is established.*

CZERMAK (1885) stated that in children the threshold of the spatial sense of the skin is lower and the "delicacy" of the spatial sense, or spatial sensitivity, is higher than in the adult. He explains this phenomenon by the fact that skin growth is not accompanied by an increase in nerve fibers. Similar investigations were made by VIERORDT, CAMERER (1889), and FRIEDMANN (1927). As can be seen in Table II, the skin of an adult's finger has a finer sense of touch than that of a child's, maybe owing to greater practice. Palms and forearms do not show significant differences. However, the child is definitely more sensitive in the elbow, thigh, and acromion. Therefore, the difference in sensitivity between the finger tip, which is the most sensitive area, and the acromion region, which is the least sensitive, is less pronounced in the child than in the adult.

---

* This method is known in the English neurological literature as "two-point discrimination."

Table II: Smallest Distances of Two-Point Discrimination in the Various Skin Areas (mm) *

Body Part (Flexor Side)	Girls			Adult Women	Boys †		Adult Men §
	6 Years Old	8 Years Old	Median Value of the Two		Min.	Max.	
Tip of middle finger	3.82	2.83	3.08	2.47	1.65	1.65	2.26
Middle of first phalanx of middle finger	7.83	8.07	7.95	7.0	3.30	4.12	6.78
Center of palm . .	9.67	9.19	9.43	11.7	—	—	—
Wrist	19.09	22.81	20.95	21.4	—	—	—
Forearm (middle) .	33.05	29.59	31.32	29.65	28.5	37.3	40.6
Elbow joint . . .	32.55	30.32	31.43	41.2	—	—	—
Upper arm (middle)	32.99	28.54	30.76	48.5	24.2	48.4	36.1–67.7
Acromion	41.93	33.05	37.47	60.1	—	—	—

Temperature Sensation

(1, 6, 15a)

Warming and cooling have a stimulating effect. Depending on the situation, i.e., whether a warm or cold stimulus affects the whole body or only a part of the body, there is a general or a local response to the stimulus.

As soon as the newborn leaves the mother's warm womb and enters the cold outside world, he starts to cry, tremble, and turn pale; but he will calm down in a warm bath and his skin will regain its color. KUSSMAUL a century ago was able to conclude from this observation that the newborn possesses a well-developed sensitivity to hot and cold temperatures. It may be that cold stimuli excite the respiratory center. When dealing with a newborn who breathes poorly, the obstetrician uses alternating hot and cold baths in order to stimulate the respiration. PREYER's infant (1895) began to cry each time the water in his bath was cooled beyond a specific temperature, namely, 32.5 °C. BLANTON (1917) observed yet another reaction in the first quarter hour after delivery: When cold, the newborn started to tremble and to "chatter" with his still toothless lower jaw. As soon as he was warm again, this stimulus response disappeared.

Furthermore, infants are already able to regulate their body temperature according to their needs. KAWURA (1944) exposed infants who were more or less naked to low temperatures and measured the skin and body temperatures in order to determine the effect of cooling. If completely naked infants were placed in a room at 20 °C, the rectal temperature dropped only 0.2 ° to 0.3 °C and the axillary temperature 0.2 ° to 0.5 °C. This occurred after 45 minutes of cooling. The skin temperature of the chest and hand fell after only 15 minutes of cooling, but returned to approximately initial values during the exposure. At the beginning of the cooling process

* According to VIERORDT. † According to CZERMAK. § According to WEBER.

the peripheral blood flow is markedly decreased, while the body temperature (taken rectally) does not show any changes for the first 30 minutes of the cooling period. If the exposure to cold was continued beyond this time, the rectal temperature fell while the skin temperature rose, an effect observed in a healthy infant. The infant is able to regulate his body temperature almost as well as the adult so that one could be deceived into thinking that an ill child is well.

Local stimuli have been applied in various ways. GENZMER (1873) could elicit vigorous movements by touching the newborn with a cold metal stick. If he stimulated the sole of the foot in this manner, the foot was withdrawn. The head was turned if the cheek was touched. Blowing at a moistened part of the body caused the children to move — the children behaving, as KRONER (1882) correctly stated, as if these regions had not been previously moistened. CANESTRINI (1913) observed the patterns of respiration and cerebral pulsation after having touched the infants with metal sticks and after applying an ethyl chloride spray. Of course, the tactile sense was also thus stimulated, so that the observed responses cannot be specifically attributed to temperature sensation.

Incontestable results can be obtained only if nothing but the temperature sense is being stimulated. In the newborn I found the following method useful: A flat metal capsule, equipped with afferent and efferent tubing for the flow of water of various temperatures, was placed on the chest of the infant. The infant responded with general restlessness. To prove that the flow of water in itself did not cause this effect, water of body temperature was first passed through the tubing and capsule and this was followed by a sudden flow of cold water. The infant responded with vivid movements to the last stimulus, which could have excited only cold sensitivity; he started to cry if this experiment was prolonged. STIRNIMANN (1939) raised the objection that, even with this experimental design, the tactile sense was not completely eliminated since tactile stimuli that at first can be tolerated without any discomfort later have some effect on the infant. However, this applies only to heavy pressure, which can be avoided by the use of a lightweight instrument. Besides, at the start of each experiment we make sure that the infant is not annoyed by the capsule, that he is resting quietly, and that the irritation is caused only by the cold stimulus.

My co-worker SCHENDEL (1944 [*15a*]) examined 100 infants (up to the age of eight months) by means of this procedure in order to determine the threshold for hot and cold water stimulation. All children responded at a temperature of 43° to 30 °C, that is, at a temperature difference of 6 or 7 degrees from the average body temperature. When he decreased the stimulus threshold, the reactions became more and more infrequent; at a difference of 2 °C, only 4% of the infants responded. When he increased the threshold, he found that half of the infants responded at a decrease of temperature of 3.7° and at an increase of 2.8°.

Weak stimuli caused movements of the legs and facial muscles. Frequently the eyes were opened widely, the eyebrows were raised, and the pupils became dilated. The motor restlessness and facial grimaces increased with stronger stimuli. With increasing forcefulness of the stimuli, the responses became more jerky until finally a startle reaction extinguished all other movements. Afterward the infants kicked their legs violently and usually started to cry loudly. The respiratory rate increased with weaker stimuli of any kind. As in the adult, a cold stimulus of great intensity was followed by one deep inspiration.

CRUDDEN (1937) similarly tested the reaction of the newborn to heat stimuli by fastening the capsule to the back of the leg and filming the infant's reactions. There were no reactions at temperatures of 33 to 34 $\pm$ 5 °C. Cold stimulation of the leg near the region of the knee usually caused hip flexion and only in a third of the infants extension. Stimulation of the calf very frequently elicited general reactions, with strong movements restricted to the lower extremities and with extension more frequent than flexion. The appearance of the scratch reflex was a strong indication that the infant was able to localize the stimulus. Newborns seem to show a stronger response to the lowering of temperature than to the raising of it, as PRATT, NELSON, and SUN (1930) have also reported. Rapid adaptation was distinctly evident in all cases. Temperatures of 16° to 45 °C were applied; the reaction time varied from 2.5 to 11 seconds with a mean of 6.8 seconds.

Does the warmth of the mother's breast play a role in the infant's search for food?

In order to decide this question SIKORSKI (1908) brought a bottle filled with warm water near the infant's face and observed lively head and lip movements until the infant had turned his face completely toward the source of heat. By placing electric heat lamps at a distance of 1 m from both sides of the infant's head, I did not obtain any stimulus response. STIRNIMANN (1939) stimulated the cheeks of 50 newborns with radiating heat at a distance of 10 cm and he failed to obtain a response in only seven infants. The others showed strong movements; four turned their heads to the source of the heat, nine opened their mouths and at the same time turned toward the source of the heat, and five only opened their mouths. In other infants the responses were less distinct; in some a startle reaction appeared.

STIRNIMANN examined temperature reactions in 100 one-day-old infants. He touched the cheeks with glass tubes filled with water at temperatures of 43° to 45° or 15° to 17 °C. When hot water was used, some infants responded indistinctly and others turned their heads to the stimulated side. Some also opened their mouths and licked or moved their tongues. Thus, a rooting reflex was elicited although these newborns had not yet been breast fed. Of course, the tactile sense was also stimulated by these experiments. When stimulated with cold water, the infants became restless and many tried to withdraw the stimulated skin area from the stimulus. Some children attempted to push the tube away with their hands.

SCHENDEL (1944 [*15a*]) conducted similar experiments on the cheeks of 50 hungry newborns and infants (up to the age of two months). He tested their heat sensitivity with two copper boxes shaped like rubber stamps and filled with water. Prior to the experiment one of the boxes was filled with water at a temperature of 42 °C. He brought the boxes near the cheeks of the infant, to a distance of ½ cm, without touching them. Six of the 50 infants turned their mouths to the source of heat, the direction depending on its location. Four infants moved their lips, and one opened his mouth and thrust his lips forward snoutlike. After being fed only one infant turned to the source of heat.

The rooting reflex, which can easily be elicited at any time by touching the oral region of the hungry infant, can also be elicited by warming the same skin region. Of course, this reaction to warmth cannot be expected to occur with the same degree of certainty. The vital process enabling the newborn to find the mother's nipple is reinforced by this mechanism. When the infant is not hungry, we note a decrease of sensitivity to heat as well as to tactile stimuli.

Since the infant receives his nourishment at body temperature when breast fed, the influence of food temperature on the behavior of the infant has been repeatedly investigated. JENSEN (1932) and HERRMANN (1932) are two of the investigators. PRATT, NELSON, and SUN (1930) observed the facial reactions, particularly those of the mouth, when the infant was given distilled water at temperatures varying from 8° to 53°C. The reactions were minimal when water of body temperature was given.

JENSEN developed a procedure in which the eating reflex was used as an indicator of the reaction. The sucking reactions to various stimuli were compared with a control of a standard sucking reaction to milk at 40°C given for 10 seconds. After an interval of 20 seconds experimental solutions were given for 10 seconds (sterile water, 6% glucose solution, a saline solution, milk at a temperature above and below 40 °C). After another 20-second interval the infant again received the control nourishment. Since the control nourishment remained unchanged, each difference between an experimental and standard response was interpreted as a differentiating reaction. JENSEN examined 17 infants in the first twelve days of life with this procedure. Great differences in the stimulus threshold were found among the infants, but the threshold usually remained fairly constant in the same infant. Food at a temperature above 50° and below 20° was less acceptable than that at 40°.

According to HERRMANN, mother's milk during feeding usually has a temperature of 32° to 42°. Even with fluctuations within this range, some manifested preference for certain temperatures. If for any length of time the nourishment had a temperature of 20° to 30° many, though not all, infants drank much less than previously. Furthermore, alimentary disturbances occurred in many cases. On the other hand, GIBSON (1958) observed that 89% of 150 infants took ice-cold nourishment without any difficulty if they became used to it gradually. The bottles were taken from a refrigerator having a temperature of 5°, and the infants thrived on this diet.

STIRNIMANN (1939) tested the hands of 50 newborns with hot and cold tubes and obtained fairly inconclusive responses. Thirty-seven infants remained unresponsive. SCHENDEL demonstrated in 30 newborns that the palmar grasp reflex (closing of the hand on touch of the palm, see p. 158) is simultaneously elicited in this type of experiment. If cold water is used, the tactile stimulus has a propensity for closing and the temperature stimulus for opening of the hand. When water at a temperature of 28° was used, all 30 infants closed their hands; at 23°, only 15 did so; at 16°, 5; and at 10°, none. The opening of the hands was accompanied by a distinct expression of displeasure, some newborns even pushing the tube away.

One infant spread his fingers immediately when touching the tube and tried by determined movements to push it away. Nevertheless, the tube continued to touch his palm, but as soon as it began sliding away, as though by the infant's efforts, his fingers were flexed by a reflex action. The tube again elicited the palmar grasp reflex by means of the gliding movement. The infant was still holding the tube but started to cry.

According to STIRNIMANN, the skin regions of the infant most sensitive to heat are first the face and then the soles of the feet. Heat makes the whole foot turn toward the stimulus; some infants even grasp the tube with their flexed toes. On the other hand, cold stimuli caused distinct withdrawal of the foot and an expression of displeasure.

The labyrinth is also irritated by cold stimuli. This reaction is discussed on page 150.

Taste

(1, 3, 54)

The area of taste sensitivity of the oral mucosa is still more extensive in the infant than in the adult. According to VON SKRAMLIK (1926), who reviewed the available information, the following areas in the adult respond to taste: the anterior part of the dorsum linguae, the edges and the tip of the tongue, the region of the papillae vallatae and foliatae, the soft palate and the adjacent parts of the uvula, the anterior aspect of the arcus glossopalatinus, the posterior pharyngeal wall in the region of the root of the tongue, and the epiglottis and part of the inner larynx. In the infant the following additional regions are taste sensitive: the whole dorsum linguae, the inferior aspect of the tip of the tongue (KIESOW, 1894, 1904), the hard palate (KIESOW), and probably also the mucosa of the lips and cheeks (URBANTSCHITSCH, 1876; KIESOW). KIESOW regards this regression of taste sensitivity that occurs with increasing age as a repetition of the phylogenetic pattern.

During childhood the various types of taste papillae are subject to certain histological changes (WETZEL, 1938). In the first years of life they are fairly equally distributed over the dorsum of the tongue, but in the older child and in the adult they are concentrated on the edges and are sparse throughout the dorsum linguae (VON SKRAMLIK, 1926). GOLDSCHMIDT (1927) was able to prove no fixed relation between the child's age and the distribution of the taste papillae on the tongue from histological examinations and observations of living children one week to thirteen years of age, but WETZEL rejected his findings.

In 1826 JÖRG and MAGENDIE independently described the taste reactions of the newborn. These reports had been forgotten when KUSSMAUL rediscovered the same reactions in 1859. The newborn responds to the intake of differently tasting substances in different ways. The response to unpleasant substances is characterized by grimaces of the face, tight shutting of the eyes, opening or grimacing of the mouth, protruding of the lips and tongue, and increased secretion and flow of saliva carrying away part of the bad-tasting material. Frequently gagging or vomiting occurs. We see the same facial expression in the infant as in the adult, in the infant in an even purer form, on intake of such unpleasant substances. Figures 2 and 3 illustrate these reactions in a newborn, and Fig. 4 in an infant of two and a half months.

Sweet-tasting substances, however, cause sucking movements. Since the same sucking movements occur on mere touching of the tongue and lips and since it is not possible to stimulate the sense of taste independently from that of touch, it cannot be automatically deduced that these movements are caused by stimulation of the taste sense. The licking movements of the tongue that occur under the influence of sweet material may be caused by taste stimulation, as they do not occur otherwise.

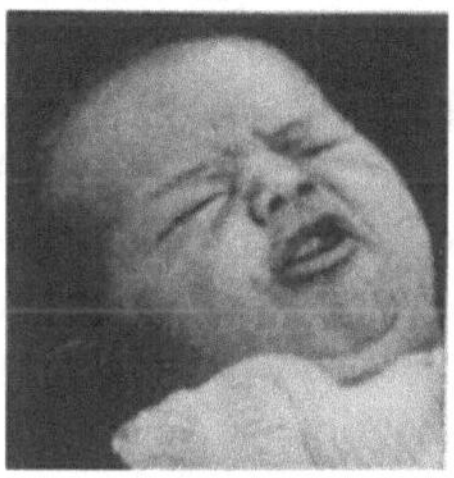

Fig. 2

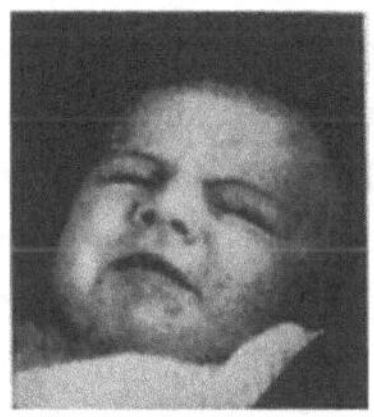

Fig. 3

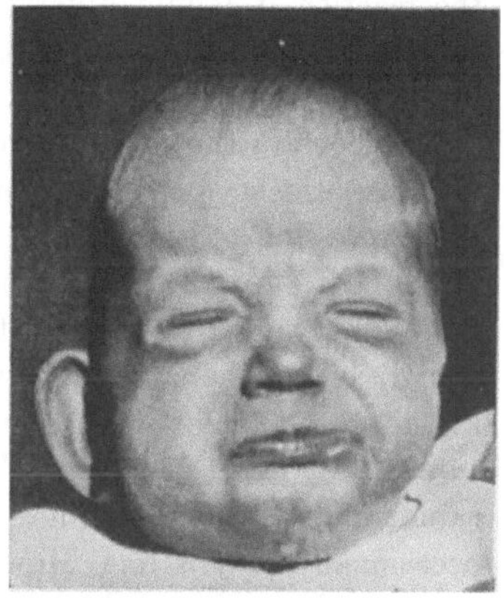

Fig. 4

Fig. 2. Reaction to intake of 3% quinine solution in a female newborn 3 weeks old.

Fig. 3. Reaction to intake of 3% quinine solution in a female newborn 4 weeks old.

Fig. 4. Facial reaction, spreading from mouth to eyes, caused by quinine in a female infant 2½ months old: defense position of mouth (slight lowering of corners, lower lip turned down, fold between lip and chin, nasolabial fold deepened, flow of saliva); defense position of eyes (tight shutting of eyelids, brows furrowed, slight horizontal fold on bridge).

KUSSMAUL observed facial grimaces in his subjects after consumption of quinine, salt, and vinegar, while dextrose was taken without resistance. His findings were confirmed by all other investigators. There is no doubt that these reactions are inherited and are not acquired through individual experience, as they can be seen prior to the first intake of food. Apparently the effect of the undesired stimulus is decreased by this response, and, on the other hand, the enjoyment of sweet substances is enhanced by sucking and licking.

According to KUSSMAUL, the various taste reactions appear even in the prematurely born (seventh and eighth months of pregnancy). I gave a few drops of a 3% quinine solution to a human fetus 23 cm long delivered by Cesarean section; thereupon swallowing movements occurred such as in response to water, without grimacing of the mouth. A taste reaction could not thus be proved at this stage of development. STIRNIMANN (1936) reported similar findings. He tested a prematurely born infant (at the beginning of the sixth month of pregnancy) ten minutes after delivery. Oil of anise and lactose, which are usually very well taken, caused a raising of the lower lip, while material of different taste failed to elicit any responses.

According to KUSSMAUL and PREYER, it is possible for the newborn to differentiate among bitter, sour, and salty substances. PREYER stated that a sour-tasting substance causes the corners of the mouth to be drawn upward and laterally, a reaction said not to occur with bitter stimuli. I was not successful in recognizing from the facial expression which type of unpleasant-tasting material was producing the reaction. I cannot differentiate the expression caused by salty, bitter, or sour substances and am only able to distinguish the facial expressions caused by "bad" or "good" tasting substances.

Occasionally the newborn reacts to the intake of bad-tasting material with the facial expression usually accompanying the intake of sweet substances and with the expression of displeasure to the intake of sweet-tasting material (KUSSMAUL, GENZMER, PREYER). According to KUSSMAUL, these paradoxical responses may in part be explained by the fact that the infants had received a bad-tasting substance shortly before the sweet substance and that the effect of this had not yet worn off at the start of the new experiment. Of course, there will be no reaction if the solution is too weak. The threshold varies in each newborn and may even vary in the same infant at different times. KUSSMAUL and STIRNIMANN think that surprise may occasionally play a role. However, I do not wish in this connection to relate the concept of surprise to the content of consciousness — what do we know of the consciousness of the newborn! I think that this reaction should be interpreted only as interference by an unaccustomed stimulus, a mechanism similar to the startle reaction.

STIRNIMANN (1938) tested the taste sensation of 100 newborns during the first 24 hours after birth prior to their receiving any nourishment. His conclusions were based not merely on the facial reaction and the force of the elicited sucking movements but on the total behavior of the infant. Some infants even used their hands to remove the bad-tasting material from the mouth. The degree of rejection was least for sour, more for salty, and greatest for bitter substances. According to STIRNIMANN, a facial expression that is absolutely characteristic for a specific simple taste sensation does not exist.

All 100 newborns reacted to lactose. One infant did not respond to distilled water, another not to citric acid, two not to salt, and two not to salt and quinine. For some infants the bitter material was more unpleasant than the salty; for others the opposite was the case. Comparison between salt and quinine solutions gave similar results. STIRNIMANN concluded that the newborn distinguishes among the four basic taste sensations. I cannot accept this conclusion in view of the variations from infant to infant and the marked fluctuations of the taste sensations at various times in the same infant. It could be that substances that impress the adult as having different tastes affect the newborn like different concentrations of the same substance. STIRNIMANN experimented with very strong solution concentrations, with which the threshold values were quickly reached.

KULAKOWSKAJA (1930) determined in the newborn the threshold values of solutions of varying strengths; she started with weak concentrations and judged the stimulus response from the facial expressions and the movements of the infant. Adults frequently were able to taste a 0.004% solution of quinine. However, only some newborns responded to a 0.01% quinine solution; all, including a prematurely born weighing 800 gm, responded to a 0.05% solution. The adults also reacted to weaker citric acid concentrations than did the newborns. Two infants, weighing 800 and 900 gm, reacted to a 0.25% solution. Six infants became accustomed to the sour taste, but all showed a longer lasting and stronger response to sour taste than to quinine. Adults were barely able to perceive the sweet taste of a 1% sugar solution. Infants drank this concentration quietly; at 1 to 2% they licked their lips; at 5% a satisfied facial expression appeared and they were smacking their tongues against their lips. The prematurely born showed fewer responses, some only at concentrations of 25 to 50%.

The responses of the infants to taste stimuli can be demonstrated by graphic recordings. CANESTRINI (1913) registered cerebral pulsations and respiratory rates. He found that a sweet taste sensation would cause relaxation; a salty sensation, slight restlessness and cessation of sucking movements; and sour and bitter sensations, marked motor restlessness. He was not able to demonstrate a difference between the effect of bitter and sour solutions. This corresponds to the results of simple observation.

It is more useful to record sucking movements. With this method ECKSTEIN (1927) observed that in most newborn, premature, and older infants the most pronounced sucking movements were elicited by saccharine and sugar; this reaction was diminished with lemon, although in some cases it was more pronounced; salt yielded the weakest reaction. In some cases sucking movements occurred even on feeding of quinine, but usually they were inhibited. Of course, the degree of concentration of the solution is an essential consideration in these experiments.

When comparing the effectiveness of distilled water on the sucking and general reactive movements of infants with that of "taste solutions," i.e., comparatively weak solutions of sugar, salt, quinine, and citric acid, PRATT, NELSON, and SUN (1930) did not find any great differences. During the first days of life, sucking movements were the most frequent responses to all types of stimuli. With increasing age, sucking movements not caused by sugar became less frequent in the newborn and all stimuli other than sugar produced increased grimacing.

JENSEN (1932) made a thorough study comparing taste solutions with human milk, in many experiments recording the sucking movements. He found differences between the infant's taking of human milk and his taking of a bottle filled with air or salt solutions of a certain strength, but not with sour milk or sugar and water. The sucking reaction was a more sensitive indicator than the grimacing of the face. Partially satiated infants were able to differentiate better than those who were very hungry or completely satiated. The threshold values differed considerably among the infants but had a tendency to remain unchanged in the same infant during the investigation.

The reported results contain many contradictions that can be explained by the marked fluctuations in excitability. Reactions depend on whether the newborn is healthy or sick; hungry or not; sleeping or awake; crying, quiet, or disturbed; and sucking on his fingers or not. Whether he is tested on the first day of life or later, once or repeatedly, further affects the response. Therefore, it is very difficult, in spite of all efforts, to conduct such investigations under well-controlled conditions. But even if such conditions could be achieved, the degree of excitability may still differ among infants.

The taste sense of the infant is far more delicate than can be concluded from these tests. It is sometimes very difficult to overcome the aversion of the infant to new and different-tasting food. There will always be a day when every infant must receive salty-tasting food (e.g., vegetables) for the first time. Until he becomes accustomed to the new taste sensation, the feeding of the new food usually requires great effort. These difficulties can be overcome; we, as physicians, must insist on it. A mother who gives her child only the food he desires exposes him to great dangers. Certainly the taste reaction is very useful, but it also may be injurious to the child.

According to PREYER (1895, p. 83), no child should be forced to eat food that is distasteful to him. "Does not the infant from the beginning resist — and rightly so — drinking sour milk, and at the critical time of weaning it is not the infant who deserves punishment, but the nursemaid who insists on weaning."

Modern pediatrics has not proved PREYER right. With his taste sense the infant cannot distinguish between food that will agree and food that will disagree with him. His reactions essentially depend on the sweetness of the food. Therefore, we transform rejected food into a "supranormal" substance by adding sugar or saccharine (see p. 607). Treatment with sour milk has proved to be one of the greatest advances in pediatrics.

The attempts of many mothers to adapt the food to the infant's taste should also be discouraged. If a child refuses a certain kind of food, the mother comes to the conclusion that this food disagrees with him and removes it from his diet. By this process of elimination a one-sided diet, with all its shortcomings, results.

Quite often the infant refuses even human milk after having received other food prior to the feeding. Even if his refusal for the most part depends on the taste of the food — there are many other reasons — it is mainly directed against the unaccustomed or nonsweet taste. In institutions we usually succeed without any difficulty in getting the infant accustomed to a kind of food refused at home.

It is sometimes surprising with how much sensitivity the infant may respond to differences in taste. For instance, frequently an infant who had received one of the usual feeding formulas containing cane sugar will refuse the less sweet-tasting human milk even when it is given by bottle. This resistance can easily be overcome by adding saccharine to the bottled human milk.

Febrile diseases and nutritional disturbances decrease appetite at all ages. This serves a very useful purpose, because in these types of disease a reduced intake of solid food, but not of fluid, is indicated. Not infrequently resistance to food intake poses great difficulty in the care of sick infants so that tube feeding may become necessary.

So far I have not succeeded in finding one healthy newborn who does not respond in the usual way to bad-tasting material if given to him in appropriate concentrations. Even in anencephalia the taste reflexes are similar to those of normal infants (STERNBERG, 1907; HEUBNER, 1909; BROUWER, 1916; GAMPER, 1926). During the early time of life these reflexes are really superfluous, as at this time human milk is the only food that the infant usually receives.

Severe taste disturbances are not rare after the neonatal period. According to LICHTENSTEIN (1894), taste reflexes exist only in some rachitic children; in others they are diminished or totally absent so that very bitter quinine solutions are taken without any resistance. Some feeble-minded behave similarly. LICHTENSTEIN recognized that the taste disturbance in rickets is not permanent, the symptom gradually disappearing after recovery from this illness.

These findings were confirmed soon afterward. BÜSSEM (1895) was able to demonstrate the frequent absence of taste reactions in rachitic infants. At the same time, he observed some nonrachitic infants in whom all taste reactions were also missing. NEUMANN (1896) reported on some rachitic as well as nonrachitic infants with absent taste reactions.

Based on our own experience, there is no doubt that the taste reaction is frequently disturbed in rickets. This disturbance is more likely to be present if the other clinical signs are very severe. As these signs disappear, the taste disorder also gradually disappears. As CZERNY (1921) stressed, taste disturbance is not of peripheral but of central origin, an indication of "cerebral" rachitis.

In addition, there are some forms of mental retardation characterized by the absence of taste reflexes. But we also saw some nonrachitic and nonretarded infants, suffering from severe atrophy, in whom every taste sensation was missing. With improvement in their general health, their taste sensations gradually, after many months, returned. Even in some infants who readily accept very bitter quinine solutions it is still possible to demonstrate a certain degree of taste sensibility by adding great amounts of quinine to their feeding formula. They may take a few sips but then will refuse to take the remainder. Even children of school age who were severely undernourished took very bitter quinine solutions without resistance.

Flow of Saliva and Taste Stimuli

KRASNOGORSKI (1931) examined the influence of taste stimuli on the flow of saliva in children beyond infancy. He did this by obtaining and measuring separate portions of saliva from the different glands. Table III shows the influence of certain kinds of food on the parotid gland. Lemon, mossberries, chocolate, cheese, and smoked sausage produced the greatest salivary flow; of the rejected foods, salt and citric acid did so. Alkaline solutions and spices had a minimal stimulating effect. The quantity of the saliva depended on the quantity of the stimulus and the duration of its effect. Other results of his investigations are discussed on page 519.

Smell

(1, 3, 6, 45)

The olfactory zone of the infant differs anatomically from that of the adult in that it does not contain pigment. This pigment can be first demonstrated at the age of six and a half years (PETER, 1929). It is doubtful that, as in the taste zone of the oral mucosa, the olfactory zone decreases in size after birth.

*Table III: Unconditioned Flow of Saliva on Feeding of Various Kinds of Food (Average Values in 6 Children of the Same Age)**

Food	Quantity of Stimulus (gm)	Median Value Parotid Salivation During 3-Minute Period (cc)	Food	Quantity of Stimulus (gm)	Median Value Parotid Salivation During 3-Minute Period (cc)
Chocolate	20.0	9.2	Carrot	20.0	6.5
Sugar	20.0	5.8	Red beets	20.0	6.2
Dark bread	20.0	6.1	Cucumber	20.0	3.5
White bread	20.0	5.3	Tomato	20.0	2.1
Smoked sausage	20.0	6.2	Lettuce	20.0	1.8
Liver	20.0	5.5	Fruits:		
Cheese	20.0	5.9	Lemon	20.0	9.0
Butter	20.0	2.3	Mossberries	20.0	7.8
Milk	20.0	0.4	Orange	20.0	7.8
Isotonic solutions:			Sour apple	20.0	7.1
Potassium bromide	100.0	3.2	Sweet apple	20.0	5.6
Sodium bromide	100.0	1.8	Grapes	20.0	6.4
Calcium bromide	100.0	1.4	Pear	20.0	5.7
Vegetables:			Melon	20.0	3.5
Cabbage stems	20.0	7.4	Watermelon	20.0	2.4
Cabbage leaves	20.0	6.0	Watermelon juice	100.0	0.4

The olfactory and trigeminal nerves take part in the olfactory sensation. The pure sensation of smell is mediated by the olfactory nerve while the stinging or scratching sensation occurring, for example, upon inhalation of ammonia or vinegar is transmitted by the nervus trigeminus. As many substances irritate both nerves at the same time, their function usually cannot be differentiated.

However, occasionally this differentiation is possible. STIRNIMANN (1936) observed an arhinencephalic infant in whom, as the autopsy later showed, both olfactory nerves and the lamina cribrosa were absent. This infant responded to Chenopodium oil and ammonium carbonate. As he had no olfactory nerves, the stimulus must have been transmitted by the trigeminal nerve.

It is easier to recognize trigeminal than olfactory irritation because stimuli received by the trigeminus more frequently cause disagreeable sensations and therefore defensive movements. These factors have to be considered in evaluating findings in infancy.

KUSSMAUL was (1859) able to demonstrate no definite groups of reactions to olfactory stimuli in the newborn, but he observed that sleeping children disliked strong smells (e.g., asafetida, Dippel's Oleum). The children squeezed their eyelids tightly together,

* According to KRASNOGORSKI-MACHTINGER (1931).

grimaced, became restless, moved their arms and legs, and finally woke up. But the olfactory sense became quickly dulled, and in later experiments no reactions were observed in the same infant.

KUSSMAUL's conclusions have often been confirmed. GENZMER's concurrence (1873), however, does not have sufficient validity inasmuch as he applied the various substances to the infant's lips and thus concurrently stimulated the tactile sense. KRONER's experiments (1882) are more convincing. He noted that hungry newborns refused to take a breast that had previously been painted with gasoline or liquid amber, while the other breast was immediately accepted. DARWIN (1877) reports that his child at the age of 32 days recognized the nipple of his mother at a distance of 75 to 100 mm. He believed that the baby was guided by his sense of smell or temperature. PEREZ (1902) and W. STERN (1923) reported that an infant's acceptance of the breast was guided by the breast's smell. It would indeed be difficult to prove that only the olfactory sense is responsible for this behavior. PREYER (1895) stated that the infant does not recognize his mother by her smell at night and is not guided by his sense of smell when searching for his mother's breast. VON ALLESCH (1921) observed a chimpanzee baby searching for its mother's breast: "Even when his mouth was at a distance of 1 cm from the nipple the baby was not at all directed to it." This proves the ineffectiveness of the sense of smell in this situation. That puppies behave differently was shown by the experiments of BIFFI (1902), who excised their olfactory lobes. Afterward the puppies, even if they had withstood the operation well, were not able to find the nipples of the mother and would have starved without assistance. Therefore, BURTON's statement (1957) that infants and most other young mammals find the nipples at first only with the aid of the olfactory sense seems without foundation as a generality.

KRATSCHMER (1870) elicited from the nasal mucosa of rabbits reflexes of the trigeminus affecting the respiratory and the circulatory systems. A transitory respiratory and cardiac arrest occurred after the blowing in of tobacco or similar substances. When CAMMANN and I (1932 [*45*]) examined the effect of carbon dioxide in respiratory disturbances of the normal and premature newborn we found a corresponding reaction. At the beginning of the CO_2 inhalation, a fright reaction, consisting of general movements and respiratory arrest as in KRATSCHMER's reflex or hyperventilation, occurred frequently. This reaction occurred in none of the 7 trials with pure oxygen, twice in 9 trials with atmospheric air, twice in 17 trials with a mixture of O_2 and 5% CO_2, and in 23 of 42 trials with pure CO_2. Figure 5 llustrates KRATSCHMER's trigeminal reflex in the infant.

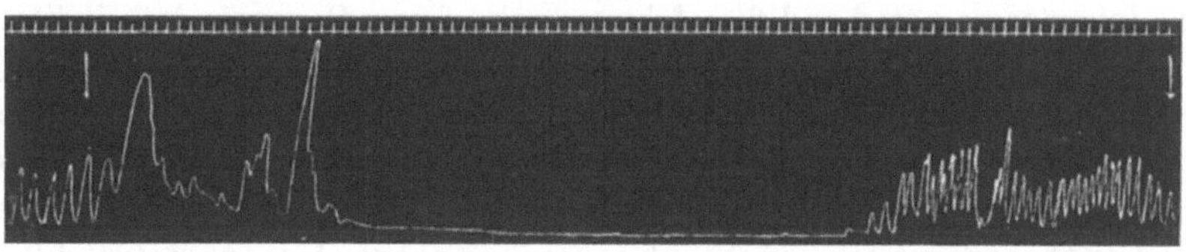

Fig. 5. KRATSCHMER's reflex: Respiratory arrest in a premature infant after inhalation of O_2 with 5% CO_2 (sec). Between the arrows: inhalation of O_2 with 5% CO_2.

CANESTRINI (1913) found no change in the respiration and cerebral pulsations of newborns in most of his experiments with smelling substances, even when strong odors were used. If the infants reacted at all, brain pulsations and breathing were accelerated and deepened and became irregular. Only in exceptional cases did respiratory arrest which could be interpreted as a KRATSCHMER reflex occur. The fact that we shortened the time of the exposure to the gases may perhaps explain why our findings differ from the results of others. Concurring with CANESTRINI, I found that the sense of smell in the newborn is more difficult to stimulate than the other senses.

CIURLO (1934) saw a distinct change in the respiratory recordings of the newborn after exposure to odorous substances that do not affect the trigeminal nerve (oranges, lavender, valerian, mint). It is possible to conclude from a newborn's behavior during breast feeding that he is already able to differentiate between agreeable and disagreeable smells.

In the experiments of PRATT, NELSON, and SUN (1930), the general movements of 48 newborns under the influence of odorous substances were continuously recorded. Fifty-nine per cent of the infants reacted to ammonia, 32% to acetic acid, 5% to valerian, 4% to oil of cloves, and only 1% to air (which had been used as a control). STIRNIMANN tested 100 newborns during the first 24 hours and before their first feeding with oil of anise, oil of chenopodium, and carbonate of ammonia. Two infants did not respond at all although they had responded to taste stimuli. Six infants did not respond to oil of anise, six not to oil of chenopodium, and eight not to carbonate of ammonia. In some children the reaction was first positive and later negative or vice versa. Contrary to CANESTRINI, STIRNIMANN (1936) found irritation of the olfactory nerve to be more effective than that of the trigeminus. The reactions, occurring in different combinations, consisted of sneezing, hyperventilation, sucking or licking movements, choking or flow of saliva, flaring of the nostrils, opening of the mouth, grimacing, closing of the lids, turning of the head, crying, and warding off with the hands. According to STIRNIMANN, the newborn can doubtlessly differentiate between oil of anise, oil of chenopodium, and ammonia. He certainly differentiates between agreeable and disagreeable smells. The same objections can be raised to the assumption of a still finer differentiating ability as were raised regarding taste.

DISHER (1934) examined the reactions of newborns to chemical stimuli applied to the nose. Pipettes were used as containers for the smelling substances. By mixing their contents with pure air in syringes she obtained three different levels of concentration. The smells of violets, asafetida, sassafras, lemon, turpentine, and pyridine were used in this experiment and compared with the effects of pure air. The reactions of the infant were noted and filmed by two observers. DISHER's data indicate that the newborn often reacted more distinctly to odors than to pure air. PRATT (1954), who reports on these experiments, interprets the findings differently. He believes that the reactions to air were not clearly enough differentiated from those to odorous stimuli. He believes that the percentage and the extent of the reactions seem to depend on the concentration of the substances and not on their qualitative differences.

KULAKOWSKAJA (1930) tried to determine the threshold values of smelling substances in the newborn. The infants began to react to a solution of 0.02%

tincture of asafetida, mature infants earlier than premature ones; adults were able to recognize the smell in a concentration as low as 0.002%. Infants responded with defensive motions to fruit extracts and artificial substances, but they did not respond to the odor of human milk. In these experiments sensitivity to odor also became rapidly dulled.

KRASNOGORSKI (1913) examined, utilizing conditioned reflexes, the differentiating ability of the olfactory sense in the older infant. Infants seven to nine months old were already able to differentiate rather easily between artificial smelling substances and camphor.

Vision

(1, 3, 14, 15, 16, 19, 54, 70, 9a, 17a)

> The eye would never have seen the sun if it were not like the sun.
>
> PLOTINUS (205–270 A.D.)
>
> Were not the eye so like the sun It ne'er the sun would see
>
> GOETHE, *Zahme Xenien*, 3. Reihe
>
> Were the sun not like the eye It never could light up the sky.
>
> VON UEXKÜLL (1940)

Anatomy

Compared with the whole body, the eye of the infant is very large. WEISS (1934) reports its average weight (in gm) as:

Newborn	2.20
1 year	4.05
2½ years	4.44
15 years	6.50
Adult	7.45

BADTKE's (1958) figures differ somewhat from those of WEISS. He indicated that there is little growth of the eye after the fifth year and that it reaches its ultimate size at 10 years. While the total weight of the adult is 21 times that of the newborn, the eye of the adult weighs only 3.25 times as much as that of the newborn (WEISS).

The size of the eyeball corresponds to its weight. In the newborn the average longitudinal axis is 16.4 mm (17.5–15.75), the average horizontal diameter 16 mm (17.75–14.5), and the average vertical diameter 15.4 mm (17–14.5); the corresponding measurements in the adult are not much greater (23.85, 24.42, and 23.70 mm, respectively) (WEISS). The cornea of the infant is almost as large as that of the adult.

The comparatively large size of the infant's eye and the wide opening of the lids make the eye the dominating part of the face (LANGE, 1937). The brightness of the eye, which depends on the pressure within the eyeball, the moisture of the conjunctiva, and the size of the pupil, is very striking in the healthy and happy infant. This luster decreases when the infant is in a bad mood and completely disappears in severe illness.

All parts of the eye necessary for sight are completely formed at birth; the only exception is the most important part of the retina, the fovea centralis, which in the newborn is less developed than the other parts of the retina. The cones in this area, which determine the strength of vision, are at first plumper, shorter, and less numerous than in later life. According to WOLFRUM (1908), the thickness of the cones in the fovea centralis is 5μ and its length 8μ (in the adult 15 and 60μ, respectively). The number of cones in a given area generally increases to four times the original number.

SEEFELDER (1927) found the fovea of a sixteen-week-old infant to be well developed; the central cones were already considerably higher and finer than the parafovea cones, and the granules of the cones were arranged in three or four layers while only one layer is usually found in the newborn and two layers in an eight-week-old infant.

This development is completed at the end of the first half year. Before that time, especially in the neonatal period, the fovea centralis is not predominant compared with other parts of the retina (WOLFRUM).

In premature infants the cornea appears slightly opaque so that the iris, which usually is small in these infants, can barely be seen. According to SCHMÖGER (1955 [*17a*]), the vascularized pupillary membrane (membrana iridopupillaris) may still cover the anterior surface of the stratum iridis, the pupil, and thus also the anterior surface of the iris. It is only in the last months of pregnancy that the membrane begins to dissolve, thus enabling the light to freely enter the eye. If the pupillary membrane is still completely intact, the whole cornea has a grayish-white and opaque appearance when seen from the front; but when it is seen from the side, its transparent nature can be recognized. According to SPECIALE-CIRINCIONE (1917), this deceptive impression is due to the fact that the cornea, pupillary membrane, and lens are still closely adjacent to each other, especially in the center; but according to SCHMÖGER, the pupillary membrane, which is vascularized, can by itself account for the opacity (see p. 555). The persistence of the intact pupillary membrane at birth according to the duration of pregnancy is given in Table IV.

*Table IV: Frequency of Persistence at Birth of Pupillary Membrane in 298 Newborns According to Duration of Pregnancy**

Length of Infant at Birth (cm)	Time of Birth (wk. of pregnancy)	No. Newborns	Newborns with Pupillary Membrane	%
under 35.3	before 28th	19	19	100
35.6–40.5	28th–32nd	51	42	82.4
40.6–45.8	32nd–36th	182	83	45.7
over 45.8	after 36th	46	7	15.2

* According to SCHMÖGER.

The pupillary membrane persists more frequently in the less developed premature infants. The "opaqueness" of the cornea caused by the membrane can easily be taken for a pathological phenomenon (see p. 555).

The iris of the newborn has a dark grayish-blue color; the epithelium of the iris is visible because of the meager development of its stroma (SEEFELDER, 1927). The final color of the iris develops only after several months.

Even in his time ARISTOTLE already knew that the eyes of the newborn are blue and that they often change their color later *(Creation and Development of Animals)*. The number of children with blue irises decreases gradually in the course of the first year. SCHINDLER (1919), in Munich, investigated the proportion of infants under one year old with blue irises. The results were as follows (in %):

Quarter of the First Year	I	II	III	IV
228 healthy infants	75.7	70.0	63	53.2
173 infants with malnutrition	38.8	38.8	27.3	23

These statistics indicate that the incidence of blue irises in healthy infants was approximately twice that of those suffering from malnutrition.

The eye of the newborn is at first farsighted (2 to 3 diopters) because cornea and lens are more convex at that time of life (SEEFELDER). ELSCHNIG (1904), using a skiascope, found a nearsightedness of 4 to 7 diopters in the first days of life. This is attributed to the involuntary accommodation during the first visual trials. If this condition is eliminated by a drop of atropine, the nearsightedness is replaced by a certain degree of farsightedness. According to GLEISS and PAU (1952), the hyperopia of the mature newborn develops only in the last months of pregnancy, as has been found by examining premature newborns.

From the anatomic standpoint the newborn's eye is functioning, although the incomplete development of the fovea centralis and the accomodative deficiency interfere with adequate focusing on objects.

The visual ability of the newborn is often disturbed by retinal hemorrhages; according to VON SICHERER (1907), these can be found in 20% of all infants. EDGERTON (1934) observed them in 41% of 458 newborns. These hemorrhages are sometimes sporadic, but sometimes they occur in such great numbers that only few areas of the eyegrounds remain free. The location proves that we are dealing with a birth effect: The right eye is more affected in the left occipital anterior position, the left eye in the right occipital anterior position. The hemorrhages subside after a few days. Lasting damage is unknown.

Pupillary Reactions

(14)

The median diameter of the adult pupil is 2 to 5 mm, depending on the degree of illumination. The pupils contract in sleep — the degree depending on the depth of sleep — and considerable dilatation occurs immediately on sudden awakening.

The pupils of the newborn are comparatively smaller than those of the adult. Their size is, according to PFISTER (1899), 1.5 mm in diameter under average conditions and increases during the first year of life to 2.5 mm. SCHUH (1937), who examined children's eyes with HAAB's pupillometer, gives the following average values for the size of the pupils (mm):

Age 0—10 days	1 mo	1/4 yr	1/2 yr	3/4 yr	1 yr	2 yr	3 yr
3.40	3.53	3.77	4.41	4.42	4.48	5.64	5.09

According to these figures the pupil enlarges until the second year and then decreases in size. GUERNSEY's figures (1929) are smaller: in newborns 0.7—1.1 mm with enlargement in the next six months to 1.3 mm.

BARTELS (1904) found the size of the pupils in sleeping newborns to be 1.5 mm; GUDDEN (1910) found them to be 2.2—2.5 mm. According to GUDDEN, the pupils slowly dilate on awakening to 3—3.5 mm. BARTELS reported the same findings. Starting with the third month the pupils are more distinctly contracted, but rapid dilatation on awakening is still absent; GUDDEN regards this process as a reflex of the cerebral hemispheres.

These findings do not agree. The reason for this may be found in the different methods of examination and in the varying degrees of illumination used in these investigations.

The pupil of the normal newborn contracts when reacting to light. The response is at first somewhat slower in young or premature infants. The consensual reaction exists from the start (PFISTER, 1899). MAGITOT (1909) tried to discover when the pupils react to light for the first time during fetal development. He found the reaction always present in fetuses at the sixth month of gestation and later, but it was found only in one of four fetuses at the fifth month — and this in the heaviest fetus (860 gm). MAGITOT stressed that central visual pathways are still absent at this stage of development and that myelinization of the optic nerve has not yet taken place. But in spite of this fact the reflex is not interrupted. BOLAFFIO and ARTOM (1924, 1926) found the pupillary light reflex in newborns born in the seventh month of gestation. MINKOWSKI (1928), however, found it absent in all examined fetuses of 10 to 34 cm in length, i.e., at the age of three to seven months of gestation.

The median value for the reaction differences in pupillary size is, according to PFISTER, 0.9 mm in the first month of life. This gradually increases with age and reaches a value of 1.1 mm in the fifth month and of 1.4 mm at the end of the first year. In the sixth to twelfth year this reaction difference amounts to 1.9 mm. DE RUDDER (1926, 1927) reached the same conclusion, i.e., that the ability of the pupil to react gradually increases with age. He determined the motor sensitivity of the pupil with HESS's differential pupilloscope. The size of the pupil rapidly increases in the first months of life; during the second half-year the curve begins to flatten. Adult values are reached at about the twelfth year. DE RUDDER did not succeed in using the pupilloscope in premature infants. According to my experience it is quite difficult to obtain valid values for infancy with the pupilloscope.

Generally both pupils of the infant have the same size. But occasionally I found in otherwise healthy infants unequal pupils with marked sensitivity to light. I was unable to find any reason for this phenomenon. In some infants PFISTER saw slow fluctuations in the size of the pupils without any external cause. I observed the same phenomenon in several premature newborns. There is no explanation for this phenomenon either.

In addition to reacting to light the pupil of the adult can also dilate in response to other types of external stimuli. PFISTER found that the dilatation of the pupils after skin stimulation did not occur before the end of the second month, and this reaction occurred in only 20% of the tested infants; in the second three months of life this reaction occurred in 87%; later it occurred in only 40% of the examined subjects. The dilatation of the pupils as a reaction to sound stimuli occurred only after the tenth week. It could be demonstrated in 64% of the cases examined in the seventh to ninth month. BARTELS (1904) reported different results; he saw the dilatation of the pupils on strong skin stimulation in all examined newborns. BACH (1908) also found the reaction occurring earlier than did PFISTER.

PEIPER (1926 [*14*]) examined infants with the HESS pupilloscope and did not find the pupillary reaction with regularity. On pinprick and sound stimuli it was sometimes present and sometimes absent in the same infant. With increasing age the probability of the reaction increased. Sometimes, however, it could be clearly demonstrated even in the smallest premature infants.

If the pupil of the adult is observed through a magnifying glass it will be noted that the rim of the iris is in a state of constant restlessness. Continuous minimal fibrillations occur at a rate of 30 to 120 times per minute. This "unrest" of the pupils is increased by external stimuli, affective reactions, and generally during mental efforts (BUMKE, 1914); it cannot be suppressed voluntarily. That pupillary unrest is absent in the young infant could be demonstrated with certainty in my investigations utilizing HESS's differential pupilloscope. The age at which pupillary unrest occurs for the first time cannot be precisely stated, since it appears gradually. At several months of age the pupil may at times be completely inactive, but as soon as the infant becomes excited the rim of the iris starts to fibrillate. Exterior stimuli also cause a longer lasting pupillary unrest. The undulations at first occur slowly and take a concentric course; at a later age the rim of the pupil fibrillates rapidly and irregularly. Continuous pupillary unrest can be demonstrated in infants four to five months old but not before that time.

Pupillary unrest is probably at first absent because it correlates only with the higher brain functions, possibly those of the cerebral hemispheres. Iridal fibrillations are independent of the dilatation of the pupils caused by external stimuli, which occurs earlier in life.

Blinking

(9a)

Blinking increases in frequency during infancy, as can be seen in Table V, which gives the findings of KÖDDING (1940 [*9a*]) and KNORR (1929).

Table V: Average Frequency of Blinking at Various Ages

Age	Average No. Blinks (per min) According to KÖDDING [9a]	Average No. Blinks (per min) According to KNORR (1929)
Newborn	2	1.3 (1–4)
Prematures and 1st mo	1 (0.8–1.7)	
2nd–12th mo	2.3 (1–4)	
2nd yr	1.8 (1–2.5)	
2nd–5th yr		3
3rd–6th yr	4.4 (3–6)	
6th–12th yr		8
7th–11th yr	6 (4–9)	
12th–16th yr		10
16th–20th yr		19
20th–60th yr		7–10

In infancy blinking becomes less frequent during severe illness. In four toxic infants KÖDDING found only 0.5 to 0.8 instead of the usual 1 to 3 blinks per minute. Six infants suffering from pylorospasm had an incidence of 1 to 1.5 blinks per minute; some weeks after surgery the number increased to 3 to 4 per minute (Fig. 6). The respiratory rate may also decrease simultaneously with the blinking rate.

The frequency of blinking is also considerably decreased in meningitis. Its progressive decrease in tuberculous meningitis can be seen in Fig. 7.

KÖDDING was unable to demonstrate a regular blinking rhythm corresponding to that of respiration. The relations between sucking and blinking are discussed later (see p. 446).

The lid movements occurring during opening and closing of the eyes are frequently not coordinated at first (RÄHLMANN and WITKOWSKI, 1877; PREYER, 1895). Quite often one eye is open and the other closed or both are not equally open. The infant's eye movements have already become conjugated in the first month. Sometimes during the second month the lids start to follow the bulbi, i.e., they are raised when looking upward and lowered when looking downward. From the beginning light causes a reflex closing of the lids. Even in sleep they are shut more tightly when light falls on them (KUSSMAUL, PREYER, CANESTRINI). This is one of the many protective reflexes with which we react to external stimuli in sleep without interrupting our sleep. If the eyelids are opened by force, the lids try to close again by reflex action. This defensive reflex is accompanied by thrashing "windmill" movements of the arms and legs and a "liberation" movement of the head (ESENTE, 1958).

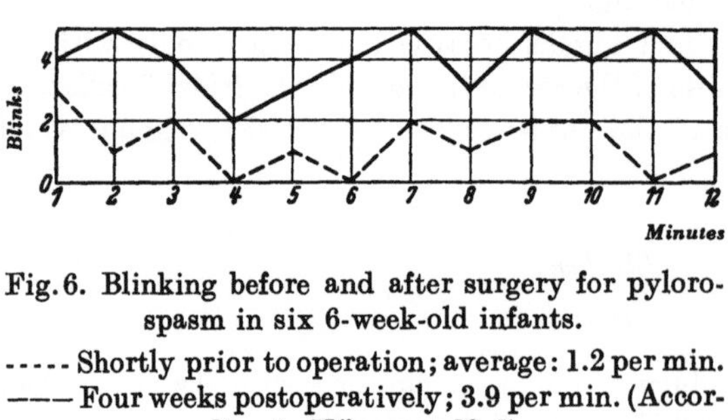

Fig. 6. Blinking before and after surgery for pylorospasm in six 6-week-old infants.
----- Shortly prior to operation; average: 1.2 per min.
—— Four weeks postoperatively; 3.9 per min. (According to KÖDDING [9a])

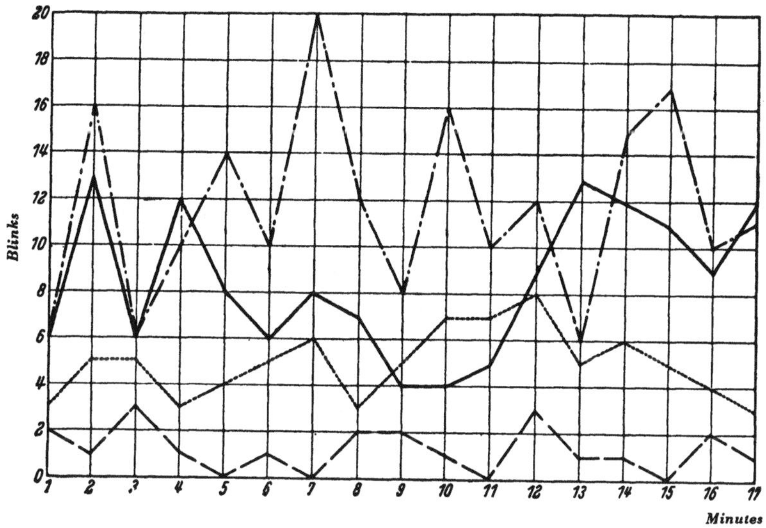

Fig. 7. Decrease of blinking frequency in the course of tuberculous meningitis in a 1-year-old child. Blinking frequency -.-.- average 11.6 per min June 6, ——— 8.3 per min June 9, ... 5 per min June 10, ----- 1.2 per min June 12th. Death: June 13. (According to KÖDDING [9a])

The adult closes his eyelids by reflex action if an object is suddenly moved toward his eyes (blink reflex). The newborn does not yet have this reflex; RÄHLMANN (1891), PREYER (1895), and SOLTMANN (1880) observed this reflex at about the eighth week of life. PFISTER (1899) noted in an extensive series of infants its first appearance between the second and fourth months. According to RÄHLMANN, at first it occurs only if the object approaches the visual direction of the infant's eyes, but it is absent until the fifth month for objects approached from the periphery.

KASAHARA and INAMATSU (1931) examined 2,293 Japanese infants, age two to twelve months, for the occurrence of this reflex. They found it at:

2 mo in 0 %	8 mo in 83.9%
3 mo in 2.0%	9 mo in 91.8%
4 mo in 19.8%	10 mo in 94.9%
5 mo in 58.8%	11 mo in 98.5%
6 mo in 80.1%	12 mo in 100.0%
7 mo in 79.1%	

An evasive movement of the head or trunk or a defensive movement of the arms on fast approach of an object is first observed after infancy (KASAHARA and INAMATSU).

The blinking reflex is a conditioned reflex in the PAVLOVian sense: The unconditioned closing of the lids on touch will finally, after frequent repetition, be conditionally connected with the approach of an object.

Sound stimuli as well as the touching of the eyes or other parts of the face may cause closing of the lids. These reflexes will be discussed elsewhere.

Eye-Neck Reflex

(15, 16, 17, 11a)

A reaction that can be called the eye-neck reflex (PEIPER, 1926 [*15*]) can be elicited in young infants, especially those prematurely born, by a sudden light stimulus. The infant is placed in an erect position (the head must not be supported). If the eyes are suddenly illuminated, the head is jerked backward so that the whole body assumes an opisthotonic posture. It becomes clear that a tonic state occurs when one fastens an electric bulb to the forehead of the infant so that the eyes remain illuminated when the current is switched on. The infant assumes the opisthotonic posture as long as his eyes are exposed to the light.

This reflex, which protects the eye just as the closing of the lids and the narrowing of the pupils, can be found in all normal and premature newborns. Premature newborns retain it for the first three months of life, while in normal newborns it becomes infrequent from the second week on. CATEL (1928) was at times able to demonstrate this reflex, though only feebly, in a feeble-minded eight-month-old infant.

LOTH (1952 [*11a*]) examined over a fairly long time the eye-neck reflexes in 220 infants, of whom 180 were born prematurely. It was absent in only two dying infants. Thirty-five prematurely born who died a few hours or days after birth temporarily showed this reflex, and at times it was even very pronounced. Again and again the movement was very impressive, frequently occurring after a certain latency period and causing the head to be jerked far back to the nape. The least mature newborn, weighing only 600 gm, was ten and a half hours old on admission; though it was delayed, the reflex could be unequivocally demonstrated in him. In older infants, after the third to fifth month, complete adaptation to the dark followed by sudden illumination were necessary to elicit the reflex. The degree of the resulting reflex depended generally on the degree of the dark adaptation prior to the test, the brightness of the light source, the degree of satiation, the sleeping or waking state, and the general state of health. In icterus neonatorum the movements often occurred more sluggishly than otherwise. WOLMUTH and PETRORINI (1958) could not demonstrate the reflex in two cases, in a prematurely born with icterus gravis and in another with cerebral hemorrhages.

Since the infants did not stay in the institution long enough, LOTH was able to observe the disappearance of the reflex, at the age of three to nine months, in only seven children. It was still present in two of the mature infants at the age of six months, in one at seven months, and in two at the ninth month of life.

The reflex was absent in a young chimpanzee that CATEL (1950, 1951) observed for ten months from the twelfth day of life.

The reflex depends on the brightness of the light and the state of adaptation of the eyes. Thus, with the aid of the threshold values of this reflex, the degree of adaptation to darkness and the relative effects of brightness of differently colored lights could be determined for the eye of the premature infant (see p. 71). Probably this reflex is related to the righting reflexes, which were investigated by R. MAGNUS (1924), as we have to assume that the reflex is finally suppressed in the growing infant by the progressive development of the cerebral hemispheres. But, as R. MAGNUS told me, he did not succeed in evoking head movements caused by light stimuli in thalamic preparations in the rabbit and cat. Either the conditions under which the eye-neck reflex occurs in animals are not yet known, or this is a phenomenon in which man has an advantage over the animal.

METZGER (1931) conducted intensive investigations on tonus changes to light. When one eye of an adult human or of a rabbit was subjected to a light stimulus an increase in tonus on the same side occurred; this was characterized by a tendency to fall, an increased tonus of the arm, and a backward stance. If the head was turned, a tendency to fall in the direction of the illuminated eye occurred. Vestibular turning is increased or decreased by the optic tonus. The increase of tonus depends on the strength of the light source and its wave length.

METZGER also examined newborns, but he found it more difficult to prove unequivocally the tonus effect in them, because of their general restlessness, than in the erect standing adult. He therefore could conclude from his examinations only that light is one of the conditions of the outside world that may cause muscle reactions. The eye-neck reflex, which is based on a light tonus of the neck muscles, was unknown to him at the time of his investigations and he did not discover it.

I have repeated METZGER's experiments in young infants by placing tightly fitting "light spectacles" on them, which permitted by appropriate light switches the illumination of each eye independently of the other. My attempts to use one-sided illumination failed to cause any one-sided light tonus as reflected in the asymmetrical neck righting reflexes of the limbs. When one eye or half of the retina of one eye was illuminated, the head was jerked backward in the same manner as if both eyes were exposed to light.

Sight (Fixating)

(54)

At first the newborn moves each eye independently of the other. It may even happen that one eye moves and the other does not. His eye movements are similar to those noted in older children during sleep. But, in addition, conjugate movements may already be observed on the first day of life (HERING, 1868; PREYER, 1895) and both forms of movement may appear in the same infant. However, according to PREYER, simultaneous movements of both eyes to the same side are not completely symmetrical; rather, they seem to occur at times only by chance among all other forms of eye movements. Gradually these purposeful movements occur more often while the purposeless ones become increasingly infrequent and finally disappear. STIRNIMANN (1940), who investigated this phenomenon in thousands of newborns, noted continuous squinting only in children who had suffered either a definite or a probable birth injury. He observed a quickly passing squint without apparent cause in only two infants. He therefore believes squinting to be a rare exception. A. STERN (1925) and GUILLAUME (1924) observed only conjugate movements in their children.

Exact vision is only possible for the adult when he is looking at an object (turning of the head), focusing, and accommodating. Looking at an object is a process in which both eyes are engaged in a common action. They adjust themselves in such a manner that the viewed object casts its picture on those areas of both retinas where it can be seen most distinctly. When the object moves, both eyes follow so that the picture on the retina remains in the same spot. The movements of both eyes of the adult are necessarily coupled to each other (conjugate). It takes great effort to interrupt this connection temporarily. Suddenly occurring paralysis of the eye muscles causes double vision.

As BUFFON (1773) already knew, newborns usually are not yet able to fixate. Anyone who observes them frequently can have no doubt about this fact. But under favorable conditions they may be induced to focus on an object.

According to HELMHOLTZ (1896, p. 599), an infant one to two weeks old confronted with a burning candle becomes restless and turns the eyes to and fro, evidently trying to bring the bright flame into focus. As soon as he has obtained the correct position of his eyes he follows the slow movements of the light. However, in the beginning he is not sure how to focus on a flame that is in the periphery of his visual fields. After two to three weeks he succeeds in doing so rather quickly, but it is only much later that he can reach out with his hand for a visualized object.

New and more detailed observations of this kind were made by ANDRÉ-THOMAS and SAINT-ANNE DARGASSIES (1952, p. 78; 1954) and later by ANDRÉ-THOMAS and AUTGAERDEN (1959, p. 23). SAINT-ANNE DARGASSIES brought the newborn with his head erect slowly to a window. In the beginning he only blinked, and after a week he started to open his eyes. When he was turned slowly away from this source of light, his head and eyes remained directed toward the light; when he was turned 180°, he tried to look again toward the window. If the rays of light met the pupil again, the child responded to this stimulus by turning first his eyes and then his head toward the source of light as if the light were causing a tropism. On the tenth day the newborns made weak efforts to follow a moving object with their eyes but usually without moving their heads.

ANDRÉ-THOMAS and AUTGAERDEN (1959, p. 96) observed during the second half of the first month of life a reflex caused by darkening: If a shade is brought in front of one eye without touching it, the head is turned to the opposite side to free the vision of the covered eye. If the shade is rapidly moved from one eye to the other, the head turns to and fro at the same rate.

WATSON (1924) induced 20 newborns to fixate on their first day of life; he placed them on their backs and in a darkened room he let one ray of light shine on them. Under these conditions the newborns focused after only a few seconds if the light was not moved more than 20° from the direction of their vision. BÁRÁNY (1921) reported that newborns were able to follow a weak light directed to the center of the pupils. In his color experiments on newborns, STIRNIMANN (1944) observed, quite unexpectedly, that the infants were frequently able to follow the movement of a colored card with their eyes and often also with their heads. WOLFF (1959) observed in one-day-old newborns movements following light stimuli. MCGINNIS (1930), however, saw successful movements of this type only in the third or fourth week.

GESELL, ILG, and BULLIS (1949) reported that fixation in the neonate is in general vague and very variable in its duration. The newborn does not yet move his head to look at an object. He stares vaguely and without interest, but occasionally he becomes active. Then his eyes follow over a short part or even over a whole quadrant of his visual field the movements of a dangling ring brought into his field of vision. He will look at this dangling ring if it is suspended at a distance of 4 to 10 in. and when focusing intensely he will even open his mouth.

LING (1942) correctly assumes that the great differences in opinion among various investigators on the earliest occurrence of focusing is caused by the great variations in the experimental settings. She therefore studied the onset of focusing in 25 infants whom she observed for a considerable period of time during their first six months of life under specific and constant conditions. She placed the bed of the infant in a cylinder that was transparent only on one side. The interior was uniformly white and served as a background for a movable black disk suspended in the visual field of the infant.

Under these conditions the newborn focused several hours after birth for the first time in a rudimentary way and reached full ability at four to five weeks of age. The first focusing attempts were accompanied by physical phenomena such as arrest or at least change of most movements, the stopping of crying or whimpering, slight raising of the upper lids and eyebrows, frowning and raising of the chin, and an immobility of one or both eyes. The infant focused better on the disk when it was close and this at first with only one eye which became fixed but still showed minimal deflections in horizontal or vertical directions. After this focusing with one eye, a gradual development of binocular focusing is noted.

In general, moving bright objects attract the infant's attention most easily. W. STERN (1932) and GUILLAUME (1924) found that their children could follow a moving object for the first time on the fifth day. PREYER's son did this on the twenty-third day (1895), the child observed by ESPINAS (1883) on the twenty-sixth day, and CUIGNET's child at three months (1871). However, these are only occasional observations and therefore must not be interpreted in the sense that the children were already able to fixate regularly at the particular age stated. At two months they are able to follow a slowly moving bright object with some regularity, and still later they succeed in searching for it. According to RÄHLMANN (1891) infants cannot intentionally look for an object before the middle of the fifth month, but this appears to me a rather late estimate. My own experiences make me share the point of view of HEUBNER (1919), who observed regular fixation usually in the third month; of course, considerable variations are possible.

S. S. and J. J. B. MORGAN (1944) made careful investigations on the onset of fixation under as uniform conditions as possible. They observed 18 infants continuously for 90 days in an institution. A red, blue, and white disk was shown to the infants. More than 75% were already able to focus during the second five-day period, and at the age of 50 days all were able to do so. Seventy-five percent could follow the disk for a short distance horizontally between the eleventh and fifteenth day and for the entire distance back and forth between the thirty-sixth and fortieth days. To follow an object vertically was a bit more difficult than to do so horizontally. Seventy-five percent followed the object vertically for a short distance between the twenty-first and twenty-fifth day and for the entire distance back and forth between the thirty-sixth and fortieth day. To follow a circular movement was still more difficult. Seventy-five percent followed for a short circular distance between the thirty-first and thirty-fifth day and for the entire circular movement between the fifty-first and fifty-fifth day. Between the thirty-sixth and fortieth day 75% followed a person moving next to the bed, and between the forty-sixth and fiftieth day the person was followed for a short distance back and forth; the person was followed continuously between the sixty-first and sixty-fifth day.

KESTENBAUM (1930) intensively examined the development of eye movements in 70 infants, age ten days to thirteen months. He reports a rather normal optic righting movement in the second or third week and at about the same time the ability to follow a moving object; this is, of course, not in the usual form of a directed movement. At this age the movements do not appear to be gliding in nature but rather jerky, as if composed of many small focusing movements. A gliding movement develops only gradually as these jerky movements subside. The directed movement has now developed from the combination of many small focusing movements. Sometimes this transition occurs after the third month, always at about the fifth month of life. Disturbances of directed movements that manifest themselves in jerking movements and occasionally occur in adults correspond to an earlier stage of development. McGINNIS (1930) stated that the first successful focusing movements of the eyes occur in the third or fourth week; they depend to a great extent on the experience and the age of the infant.

BÁRÁNY (1918) and McGINNIS observed nystagmus in the newborn. KESTENBAUM (1930) saw it in some cases in the fifth week in the form of "railroad nystagmus," i.e., later than the jerky directing and focusing movements. The nystagmus is always present at the age of three and one half months, i.e., approximately at the same time that the directed movement, with which it is closely connected, becomes gliding.

According to KESTENBAUM a "staying behind" of the eyes, or nystagmus, during passive turning of the head can be observed at the earliest age. If an object is shown that can be brought into focus, a compensatory eye movement maintains the eyes almost always in the direction of the object. If such an object is not shown, this deviation can be found only in very young infants while a nystagmus, i.e., a deviation interrupted again and again by jerks, instead of the compensatory simple deviation, can be seen in older children. KESTENBAUM thinks that the basis of this phenomenon is the predominant influence of the labyrinth, which after birth leads to such deviations in very small infants and in rabbits. Later the eye remains fixed on the focused object.

CARMICHAEL (1949, p. 137) electrically recorded ocular movements in young infants. These movements were elicited by placing the head of the infant in a striped turning drum. An optokinetic nystagmus was thus obtained. Similar experiments of TRINCKER (1955) are described on page 74.

The body righting reflex of the head (see p. 167) can be repressed when the infant looks at an object. This reflex can be elicited by placing the infant on his back, holding him by his legs, and turning his pelvis sideways. Usually the body and head follow this movement very rapidly. If a light is placed in front of the infant in a darkened room, the onset of the reflex is delayed although not completely suppressed;

the onset of the reflex will be markedly accelerated when the infant's eyes are blindfolded. When the infant learns to focus, the uncoordinated eye movements disappear but they are still present during sleep later on.

The direction of vision influences the frontalis muscle. Young infants who are able to focus lift their eyebrows and wrinkle their foreheads horizontally through contractions of the frontalis muscle when looking upward. This reflex may also occur on one side only. If an infant is made to look obliquely upward and to the left, the wrinkles of the forehead on the left side will become more distinct and the left eyebrow will be raised higher at the same time (Fig. 8). The visual field in the upward direction is thus enlarged. I term this reflex the "ocular frontalis reflex." PREYER (1895) already knew of this bilateral reflex and found it after the second three months of life. It can be easily observed in adults.

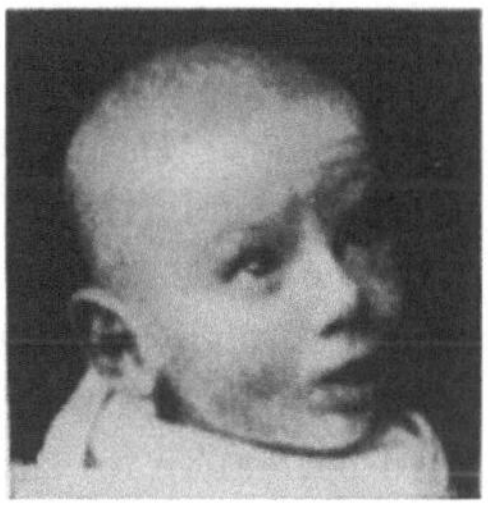

Fig. 8. Left ocular frontalis reflex in a 7½-month-old boy. Looking upward and to the left. The eyebrow is raised and the left side of the forehead shows horizontal wrinkles. Opening of the mouth when looking attentively.

The contraction of the frontalis muscle may be elicited in another manner. If the head of a premature infant is strongly bent forward, the eyes open and horizontal wrinkles appear on the forehead (neck-frontalis reflex). The reflex opening of the eyes was already described by BARTELS (1910). The forehead wrinkles vary greatly from infant to infant; sometimes they are only minimal, sometimes surprisingly distinct.

If the head of an infant who shows a good reaction is turned to the side, let us say to the left, the wrinkles of the forehead appear only on the left or are more pronounced on that side. The turning of the head does not seem to affect the frontalis directly but influences it by way of the compensatory eye position, which in turn causes the oculofrontalis reflex.

ORTEL (1944) describes a similar "lid opening reflex" caused by a change of position: The eyes of lightly sleeping newborns, premature newborns, and infants are opened if the infants are rapidly brought from a horizontal dorsal to a vertical position. Deep sleep represses this reflex, and it is less frequently elicited in the first days of life. It can be most frequently demonstrated during the first month. After the fifth month of life it is difficult to determine whether the opening of the lid is caused by the reflex or by awakening.

A higher cerebral center in which the visual impressions are digested is prerequisite for the ability to focus. Severely retarded children are not yet able to focus after the first year; some will never be able to do so. The eyes of these children wander around in a conjugate but purposeless manner; occasionally they are directed to an object but immediately lose it from their field of vision.

Sudden light may cause a fright reaction with or without the eye-neck reflex. This stimulus may also cause changes in respiration and brain pulsation, but CANESTRINI's tracings are not convincing since they are distorted by general movements of the infant.

Soft light attracts infants to such an extent that they turn their eyes and even their heads toward it very early. The turning of the eyes toward light has been described by BUFFON (1773). KUSSMAUL noted this turning to the light quite frequently in newborns, even in a two-day-old infant born two months prematurely. His observations have been confirmed by GENZMER (1873), KRONER (1882), and others. PREYER's child, whose facial expression changed during the day when his eyes were shaded, turned to the light for the first time on the sixth day of life. W. STERN (1932) even speaks of an orienting instinct that also responds to sound and touch stimuli.

It may even be possible that the use of the eyes changes the anatomic structure of the nerve tissue (FLECHSIG, 1876; BERNHEIMER, 1889; WESTPHAL, 1894, 1897). AMBRON and HELD (1896) kept one eye of newborn animals open for some length of time while the other eye remained closed. When examining the optic nerves of these animals, he found that distinct myelinization had occurred only in the optic nerve of the eye exposed to light. WESTPHAL found in prematurely born infants who had been alive for some time myelinization in excess of what could be expected considering their stage of development.

HYDÉN (1952) asked the question: "How does the time at which an adequate stimulus reaches the nerve cell affect the chemical composition and therewith the action of the neurons?" To come to a decision he examined the ganglion cells of the retinas of rabbits that had been kept for ten weeks after birth in complete darkness. He found that the number of ganglion cells in the animals raised in darkness was only one-sixth of that in the controls. Pentose nucleoproteids could not be found at all in the animals raised in darkness, in contrast to the animals raised in light.

ESENTE (1958) observed that a prematurely born infant will for some weeks after birth be temporarily more advanced than the mature newborn in his vitality and in his general visual behavior.

Light is an absolute prerequisite for the correct innervation of the eye. Under circumstances of long-lasting light deprivation, infants develop nystagmus and jerky head movements. This clinical entity, known as spasmus nutans, develops slowly and consequently does not appear before the second year of life. RAUDNITZ (1897, 1911, 1918) proved that it is a disease caused by light deprivation, which may occur because the infants stay in the darkest corners of dark rooms. If they are adequately exposed to light, they will recover in a few months. RAUDNITZ was able to produce spasmus nutans in puppies kept in the dark (see p. 631).

However, the cause of nystagmus of the newborn, described by RÖMER (1891), ZIPPERLING (1913), THORNVAL (1920/21), and SCHUR (1922), is uncertain. BERBERICH and WIECHERS (1924) tried to explain it as a sequela of cerebral hemorrhage due to birth trauma. STERN and SCHWARTZ (1924), who share this opinion, made a number of observations to prove this hypothesis. In their series they observed nystagmus in only one-third of the infants after normal delivery but found it present in all eleven infants born in abnormal delivery; it could not be demonstrated in those delivered by Cesarean section before labor had started. The nystagmus was always most pronounced in the first hours after birth. STERN and SCHWARTZ saw a relation between nystagmus and the duration and size of the caput succedaneum. From these findings they drew the conclusion that the common underlying cause was perinatal damage to skull and brain.

CATEL (1932) examined 16 prematures in lateral position for spontaneous nystagmus and found it to be frequently, but not always, present until the second and third month after birth, even in two-thirds of those infants whose births had been normal. The incidence of severe nystagmus in the child is greater after very difficult deliveries. CATEL considers, as does BARTH (1933), the ocular tremors of the newborn to be physiological and compares them to the nystagmus of miners, which is caused, according to OHM (1918), by specific external stimuli, for example, the bent position, the inadequate lighting of the mines, innate or acquired decrease of the light sense, and finally the hypersensitivity of the vestibular apparatus. That this apparatus is more sensitive immediately after birth was proved by SCHUR (1922) and CATEL. According to CATEL the prenatal fetus is in a situation similar to that of the dogs kept in darkness by RAUDNITZ.

BARTELS (1932) objected to this hypothesis: The nystagmus of the miners usually consists of a trembling of the eyes (pendular trembling) that remains more or less constant regardless of the direction of the gaze; it usually is slightly rotatory. The nystagmoid eye movements of the infant show quite a different pattern, one which conforms more to the eye movements of blind individuals. They consist of irregular nystagmoid jerks interrupted by fluttering movements, occasional very rapid tremors, and sometimes athetoid eye movements. Since BARTELS found the same eye movements to be also present in an infant with anophthalmus, other forces besides light must play a role in the origin of the nystagmus, i.e., stimuli from the lower centers of the eye. It must be that these nerve stimuli cause eye movements in the blind adults because they do not receive visual stimuli from the outside, and the same applies to infants, probably because they suffer, so to speak, from psychic blindness.

Labyrinthine nystagmus is discussed elsewhere in this volume (see p. 151).

If an adult looks at an object that is close to him, his lines of vision converge, the muscles of accommodation contract, and the pupils are constricted. According to GENZMER (1873), in the newborn every position of convergence caused by a strong light perception is accompanied by a strong contraction of the ciliary muscles. He noted this phenomenon when observing lenticular reflections and concluded from these findings that a connection pre-existed between convergence and accommodation. PREYER (1895) was able to produce convergence of the eyes in an infant two to six weeks old by bringing a bright object near; he did not make any experiments on accommodation. RÄHLMANN (1891), however, did not observe any narrowing of the pupils in the convergent movements which may occur by chance in the newborn during the first weeks of life; he found them only at about the end of the fifth week, i.e., during the time of the first focusing attempts. According to PREYER, the narrowing of the pupils in the newborn is not always associated with convergence. He could demonstrate these findings with certainty in the ninth week. LING (1942) observed the first convergence at about seven weeks under exactly the same conditions.

Reliable and thorough investigations on the development of accommodation apparently do not exist. So far we do not know whether the newborn is able to contract the ciliary muscles of both eyes equally or if this ability is acquired. Neither do we know when he starts the purposeful use of his ciliary muscles. As mentioned, ELSCHNIG (1904) explained that the nearsightedness of the infant in the first days of life is due to the involuntary contraction of the accommodation muscles in his first focusing attempts.

From the available investigations it therefore appears that constriction of the pupils and convergence and contraction of the ciliary muscles are not as closely associated at birth as later in life.

Color Perception

(16, 19)

Many different answers are given in the literature to the question of whether or not the infant has color perception. TRACY and STIMPFL (1912), SHINN (since 1893), and BERNFELD (1925) stated without ever conducting any investigations that the newborn is totally color-blind. VON JASCHKE (1917) stated that, in the consensus of all investigators, color perception of the newborn does not exist. According to SHINN, the color sense develops in the second half of the first year; according to UFFELMANN (1881) and GARBINI (1894) in the second year; and according to PREYER (1895), HEUBNER (1919), and SZULINA (1925) even as late as in the third year. GARBINI is of the opinion that color perception is fully developed in boys at the end of the fourth year and in girls between the fifth and sixth year; ROSELLI (1903) and SMITH (1936) believe that the color sense develops earlier in girls than in boys. DARWIN (1877) thought at first that his children were completely color-blind and discovered later that this fear was unfounded.

On the other hand, RÄHLMANN (1903) and STIRNIMANN (1944) attribute a completely developed sense of color to the newborn; K. BÜHLER (1922) feels the same, basing his opinion on WOOLEY's experiments (1909) with a six-month-old infant.

PREYER (1882) was probably the first to examine in more detail color perception of children; he tried to determine at what age his son could name colors correctly. This ability requires, of course, not only the sense of color but also memory, speech, and sufficient attention, i.e., qualities having nothing to do with color perception *per se*. One can therefore at best conclude from these experiments that the children could differentiate colors at a given time. It remains unknown whether or not they could have done this earlier.

To counter these objections a procedure was developed that did not rely on language and that therefore could be applied much earlier. A choice of different colors was presented to the infants, and the colors that they chose most frequently were noted (BALDWIN, 1898; HOLDEN and BOSSE, 1900; WOOLEY, 1909), or those that attracted their attention most frequently (VALENTINE, 1919; STIRNIMANN, 1944). Some investigators admit that only the preferred colors may be recognized in this manner; others presume that the favored color is at the same time the first that has led to color perception, an assumption lacking any proof from the known facts. The first question that has to be solved is whether one color is noted first or whether all colors are simultaneously perceived at birth.

Statements about the sequence of the favored and best recognized colors vary greatly, as the following summarization shows:

BALDWIN (1898): blue, red, white, green, brown.
BINET (1890): red, blue, orange, brown, pink, violet, green, white, yellow.
GARBINI (1894): red, green, yellow, orange, blue, violet.
HOLDEN and BOSSE (1900): red, orange, yellowish-green, blue, violet.
JASTROW: blue before all other colors.
LINDER: blue before all other colors.
MARSDEN (1903): newspaper, yellow, red, blue, white, green, black, brown.
MEUMANN (1911): white and black, red, blue, green, yellow, violet, orange.
MYERS (1909): red and yellow before all other colors.
OLTUSZEWSKI (1897): red and green, yellow, blue.
OSVALDO (1938): yellow, green, blue, red.
PREYER (1882): yellow, brown, red, violet, black, pink, orange, gray, green, blue.
STAPLES (1932): red, yellow, green, blue.
TYLER (as reported by Shinn): no preference for any color.
UFFELMANN (1881): red, green, blue, yellow.
VALENTINE (1913/14): yellow, white and pink, red, brown and black, blue and green, violet.
WINCH (1910): white and black, red, blue, green, yellow, violet, orange.
WOOLEY (1909): red, yellow, blue, green.

Undoubtedly much effort has been devoted to the determination of favored colors, but these experiments did not have tangible results. The color sequence of every investigator deviates from those of the others except for WINCH, who concurs with MEUMANN and has stated so. Even if red and yellow are more attractive to the infant than other colors, there is no tangible proof for the statement of COMPAYRÉ (1900) that color perception of the infant, starting with red and yellow, develops according to the spectrum because these colors correspond to the maximum wave lengths.

The hypothesis that the infant's color sense develops gradually may even have practical consequences. MAGNUS (1879) recommended a special color chart to develop the infant's color perception. Others, like UFFELMANN and GARBINI, supported his recommendation, but this procedure has long been forgotten.

Whoever interpreted these findings as proof of the incomplete development of the color sense in infants contradicts the known facts about adult color perception. Forms of incomplete color blindness as stipulated for the infant are unknown in the adult and incompatible with the present theories on color. Only complete color blindness, which also has been attributed to the infant, may occur in the adult.

There is a color theory (FICK, 1890, LADD-FRANKLIN, 1893; SCHENCK, 1907) founded on an ontogenetic hypothesis. This theory regards rods and cones of the retina as different ontogenetic stages and sees in the rods the lower stage that is supposed to contain the primordial visual substance. From this substance the somewhat more sensitive "white-perception" substance of the cones first develops, which is still present in the periphery of the adult's retina. This white-perception substance subdivides into the "yellow-and-blue" substance, which is found in the "red-green-blind" zone of the retina. Finally a "red-and-green" substance develops from the yellow substance. According to this theory, the color sense develops quite differently from what one would expect from the experiments that have been cited. The sequence of white and black, blue and yellow, and red and green has not been established at all. On the contrary, more recent research has led to the opinion that ontogenetically and phylogenetically the cones function earlier than the rods (see p. 74).

STIRNIMANN (1944) tested 200 newborns by means of a technique similar in principle and obtained a totally different result. He exhibited cards covered with paper of different colors from a distance of about 10 cm. He determined the duration of the gaze induced by the various colors. The mean values of the duration of gaze, expressed in seconds, gave a specific index for each color. Blue elicited the longest response; then in sequence came green, red and white, and finally yellow, a sequence, incidentally, differing from all those mentioned. STIRNIMANN concluded from his experiments that newborns reacted in general typically to different colors although no reaction consistent in every detail could be determined for a single color. Nevertheless, he felt the conclusion was justified that the newborn differentiates between colors and thus has color perception. Because of the wide scatter and overlap the evidence does not seem compelling to me. Color-blind persons react not only in general but also in individual cases differently from those who are not color-blind. Even if the newborn does not possess any color perception, the distribution and the mean values for the duration of the gaze in response to the various colors will vary because these colors differ in their relative brightness values (as discussed later) and because the whole behavior of the infants and the duration of gaze depend too much on other variables, such as mood, hunger, distraction, etc.

STIRNIMANN concluded from additional experiments conducted in a similar manner that the newborn responded differently to color contrasts and simple forms such as a circle or a horizontal line.

CHASE (1937) examined 24 infants fifteen to seventy days old at the start of the examinations; he projected two colors of equal light intensity on a screen opposite the face of the infant and this in such a manner that one color formed a small circle within the other color. By means of a special technique the impression was given that the inner color circle moved within the other color. With this arrangement it was possible to find out if the infant would follow this movement with his eyes. This experimental design proved the infant's ability to differentiate between red and yellowish-green, red and green, red and bluish-green, yellowish-green and green, yellowish-green and bluish-green, and green and bluish-green. A colorless dark spot inside a lighter spot was not followed with the eyes.

RÄHLMANN (1903), KRASNOGORSKI (1913), and KASSATKIN (1953) developed reflexes in infants conditioned to colored milk bottles and with this procedure were able to prove the ability to distinguish between colors at an early age (see p. 452).

One objection has to be raised against these and all other previously described experiments: Sufficient attention has not been paid to the fact that rays of different wave lengths cause brightness as well as color sensations. Both sensations are independent of each other. Only brightness reactions take place in the periphery of the visual fields. According to a well-known physiological law, the relative values of brightness of different colors change in unequal proportion with illumination and the corresponding condition of the observing eye. If, for example, the eye adapted to brightness looks at a very strong sun spectrum, it finds the brightest spot in the yellow portion (wave length of about 590 $\mu\mu$) and from this spot the brightness of the spectral colors decreases in both directions. If, on the other hand, the observing eye is adapted to darkness, a fairly weak spectrum appears to be colorless. The brightest spot is now found in the violet zone (wave length of 530 $\mu\mu$). This corresponds to PURKINJE's phenomenon. A red and a blue object may appear equally bright if they are fully illuminated, but after the illumination has been greatly diminished and corresponding adaptation to the dark has occurred the blue appears to be distinctly brighter than the red.

The relative brightness values of colors seen during twilight differ completely from theire relative brightness values during the day; this applies not only to the normal retina, but to the color-blind as well (VON KRIES, 1909). The eye of color-blind individuals adapted to light or to darkness functions exactly as the dark-adapted eye of the color-sensitive person; it lacks color sensitivity and the brightest spot of the spectrum appears to him in the violet zone even when the eye is adapted to light. Those who are completely color-blind see the colors in a brightness different from that seen by color-sensitive persons; they can therefore distinguish between colors having equal brightness for color-sensitive persons because the brightness has different degrees for their eyes.

Should the infant have this form of inborn complete color blindness — such color blindness has been attributed to him — different colors that the adult would see as equally bright must appear to him as having different brightness. He could distinguish them without having color perception. My experiments, which are described in another place, have shown that the infant's eyes can distinguish between different degrees brightness. Even the experiments of STIRNIMANN, RÄHLMANN, and KRASNOGORSKI still allow the possibility that the infant may be color-blind. Similar objections have been raised — and rightly so — against many experiments on the color sense of animals where similar conditions exist (NAGEL, 1894, 1901). One can avoid these difficulties as, for example, VON FRISCH has shown in his famous experiments with bees. If a certain color, for example blue, is compared with a series of different grays, one of the shades of gray will be equal to blue for the completely color-blind person.

As a basis for similar experiments, it is important to examine the degree of brightness of the different colors seen by the infant. This would furnish certain points of reference if we accept the simple hypothesis that during the infant's development only such deviations of color perception may occur as are known to us in the adult.

I therefore determined (1927 [*18*]) with white and colored lights in light- and dark-adapted premature infants the stimulus threshold for the eye-neck reflex (see p. 60), which depends on the degree of brightness. It was shown that the sensitivity to light of different wave lengths is increased considerably after dark and in subdued illimination. For example, the sensitivity to gray increased a hundred times in dark adaptation when compared with light adaptation. A comparison also showed that in the premature infant and in the adult the relative brightness values of red, yellow, green, and blue were equal to each other during light adaptation and that in dark adaptation and decreased illumination the values changed in the same way, i.e., that green and blue became brighter when compared to red and yellow (Fig. 9). Under these circumstances the relative brightness of the colors moves in the premature as well as in the adult to the short wave (ultraviolet) end of the spectrum, i.e., PURKINJE's phenomenon appears.

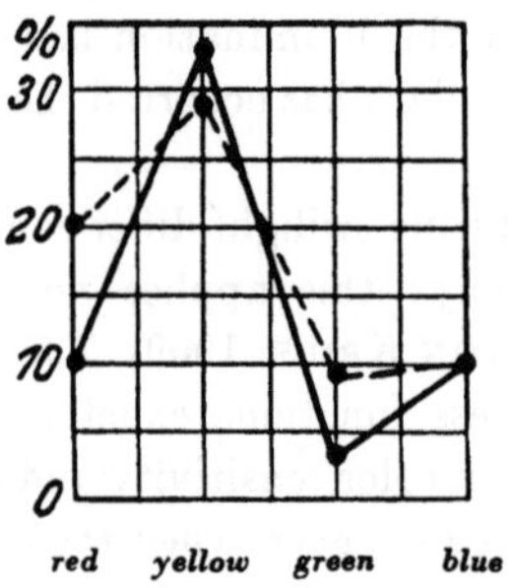

Fig. 9 a

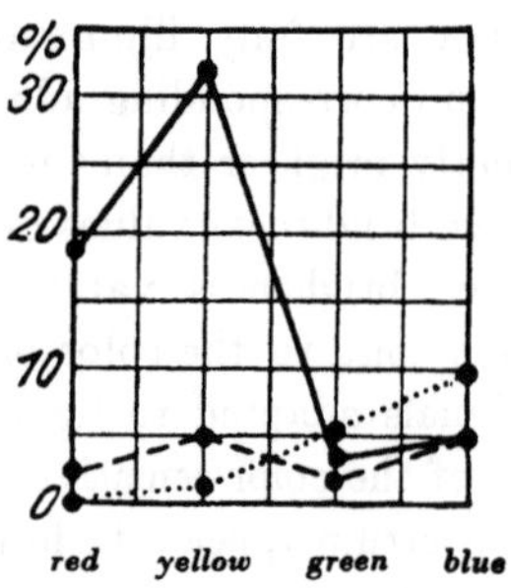

Fig. 9 b

Fig. 9a. ——— Relative brightness in a premature infant during light adaptation.
----- Relative brightness in an adult during light adaptation.

Fig. 9b. ——— Relative brightness in another premature infant during light adaptation.
----- Relative brightness in the same infant during dark adaptation.
.... Relative brightness in an adult during dark adaptation.

J. M. SMITH (1936) doubts that the evidence of my experiments is convincing, and MUNN (1938) agrees with her. She found the greatest brightness after dark adaptation in the red zone when using the graphs made by her according to my instructions. This statement taken by itself is correct, but J. M. SMITH relates the relative brightness to white light ("percentage of brightness relative to white light") while I use the hypothesis that in PURKINJE's phenomenon the relative values of brightness of both ends of the spectrum are related to each other and not to white light. As an example, the graph of the prematurely born infant (Fig. 9b) will be discussed in detail here: In light adaptation yellow is more than six times as bright as blue; in dark adaptation it is just as bright. In light adaptation red is about four times as bright as blue, but in dark adaptation blue is twice as bright as red.

J. M. SMITH (1937) answered my objection correctly in stating that we have a different concept of PURKINJE's phenomenon. To decide this difference of opinion, it is best to quote PURKINJE (1825) himself. He writes:

"Objectively the degree of illumination has a great influence on the intensity of the color quality. To convince oneself one has only to look at the colors before daybreak when dawn just starts to break. First one sees only black and gray. Especially the brightest colors, red and green, appear 'darkest.' For a long time yellow cannot be distinguished from rosy red. I can recognize the color blue first. The different shades of red that burn brightest in daylight, like crimson, cinnabar, and orange, remain darkest for a long time, altogether disproportionate to their usual brightness."

Accordingly, the increase of the relative brightness values of colors *at the ultraviolet end of the spectrum as compared with the red end in dark adaptation* are described as PURKINJE's phenomenon in all reports of physiologists known to me.

I have concluded from my experiments that brightness sensitivity in man does not develop after birth but that the ability to see in daylight and at dawn is present in the premature infant in the same way as is characteristic for adults. The twilight vision permits us to draw a conclusion as to the functional state of the rods in the retina, which are the only functioning elements in congenital total color blindness. But, in addition, the other light-sensitive visual cells, i.e., the cones, in the premature newborn must begin some of their future functioning although we cannot decide with certainty whether they are already fully functioning, i.e., whether they serve to transmit color.

My experiments on prematurely born infants were continued and enlarged by TRINCKER (1954); he based his experiments on the findings of BIRUKOW (1949), who studied the development of daylight and twilight vision in the larvae of frogs. BIRUKOW found at the lowest stage a maximum of the brightness values of twilight vision in the yellow zone, which is characteristic of pure cone vision. This indicates that at that stage the cones are fully functioning in the transmission of the brightness values of daylight vision. These values coincide relatively and absolutely with those of adult animals. A weak influence of the rods may be noted even at the lowest stage, but it is still far below that of the brightness distribution in twilight vision. This brightness becomes distinct only in later developmental stages and reaches absolute and relative values for all colors in adult animals after metamorphosis. The twilight values of colors characteristic of the completed development do not appear simultaneously for all colors during the development of the individual, but appear in succession. This made BIRUKOW assume that three different forms of rods exist.

TRINCKER examined premature and mature newborns and infants in the first ten weeks of life by determining the threshold value of the eye-neck reflex for red, yellow, yellowish-green, bluish-green, and blue during light and dark adaptation. While I used lights of different strength, he worked with monochromatic light stimuli of equal intensity. He found in each mature and premature infant a difference in the relative brightness values during light and dark adaptation according to PURKINJE's phenomenon. TRINCKER proved also — what I had missed — that the degree of the phenomenon depends on the stage of the infant's development.

While the graphs of relative brightness values show significant differences during dark adaptation in the various stages of development, this is not the case during light adaptation. As in frogs, brightness values for the cones corresponding to those of the adult are found in all examined stages of development. The function of the rods undergoes developmental changes, but the portion of this development that can be experimentally studied is more limited in man than in the frog. The maturity pattern of brightness values follows the same sequence of wave lengths in man and in the frog: first yellow and yellowish-green, then red, and finally blue. Complete adaptation to blue occurs in the ontogenesis of man at a stage when this experimental technique can no longer be used. Therefore, TRINCKER also wishes to apply BIRUKOW's hypothesis to man, i.e., that three forms of rods having different functions mature in succession. Incidentally, he could have deduced the possibility of such a development from my findings, since the values obtained in light adaptation coincide much better than those obtained in dark adaptation in the adult (the values for yellow and red coincide better than those for blue).

In summary, it may be said that TRINCKER finds the sensitivity to brightness of the cones to be already fully developed in mature and premature newborns, while color differentiation appears during the first few months of life. The sensitivity of the rods to various spectral hues is reached later in the ontogenetic and phylogenetic development than the same function of the cones. *The function of the cones therefore begins earlier than that of the rods.*

This concept contradicts the ontogenetic theory of color vision of FICK, LADD-FRANKLIN, and SCHENCK (see p. 69), which stipulates the opposite and regards the rod as the ontogenetically primary visual cell form. The anatomic investigations compiled by D. and J. TRINCKER also indicate that the development of the cones precedes that of the rods: SEEFELDER (1910) found as early as the third embryonic month definite cone cells in the centralis area while the development of the rods in the paracentral zone does not begin before the fifth month (according to CHIEVITZ, 1887, and FALCHI, 1888).

BIRUKOW and TRINCKER agree with WALLS (1934), who regards the cones as the phylogenetically primary visual cell form; he based his opinion on histological examinations of reptile retinas. The rods develop only when the transition to night living occurs. If the mode of living changes again, cones may develop from the rods.

D. and J. TRINCKER (1955) based their additional experiments on the hypothesis that the matching of brightness of colored and gray papers done with peripheral vision by light-adapted adults should also be possible in the central vision of newborns since the values of color-brightness of the cones in all examined developmental stages coincided with those of the adults. If the infants have only cone vision without color discrimination, they ought to confuse colors and grays of equal brightness in their central vision.

The experiment proceeded as follows: Using a rotating drum, ribbons made of alternating colors and gray stripes were moved in front of the infant's eyes. In more than 100 normal newborns and in most of the viable premature infants these tests triggered unequivocal optokinetic nystagmus in the first weeks of life.

In the first two weeks no color differentiation occurred; the reaction was absent if gray and color were equal in brightness. The first color that caused reactions in the third and fourth week was blue when paired with gray of equal brightness; red and yellow followed in the fourth to eighth week; green, which was last to be differentiated (in the seventh to eleventh week), completed the trichromatic vision. The fact that blue initiates the sequence agrees with BIRUKOW's findings in the frog.

These experiments allow us to draw a conclusion only on the relation of color sensations to each other, for we have no knowledge of their quality and we cannot have such knowledge. What NAGEL (1894, 1901) said about the animal applies also to the infant:

"It would be useless, even senseless, to puzzle over the quality of the color sensations of the animals, i.e., to discuss the absolute quality of their sensations. But it is not senseless or useless to investigate whether the different colors, in other words rays of different wave lengths, affect the visual organs of the animals specifically in a different manner, i.e., whether colors can be distinguished by animals."

According to PREYER (1895), CUIGNET (1871), and CANESTRINI (1913), the newborn infant is supposed to be photophobic, but TIEDEMANN (1787), ESPINAS (1883), and COMPAYRÉ (1900) have reported just the opposite. GUTMANN (1924) conducted his own investigations on this problem and even claims that he can differentiate newborns who are photophobic, who are indifferent to light, and who like it. There is nothing easier than to change an infant who enjoys light into a photophobic baby and vice versa. One has only to bring the child from a lighted into a darkened room and keep him there for some time and then examine him with the same light source. The infant adapts to darkness as well as the adult. As mentioned before, this sensitivity to colorless light increases a hundredfold after complete adaptation to darkness.

It is obvious that the infant's pupillary reactions could be used for the testing of his color perception in the same manner as VON HESS (1916) did in the adult and the animal. DE RUDDER (1926, 1927) felt that these experiments are only possible in the second half of the first year, when the pupil has become sufficiently light sensitive. Of course, the results of these observations are distorted by the restlessness of the infant, as I can confirm from my own experience.

Color blindness ought to be recognized most distinctly in the pupillary response values for red. The values determined by DE RUDDER coincide quite satisfactorily with those which were theoretically calculated. Thus, we cannot support the assumption that the color sense of the infant differs from that of the adult.

Another series of experiments on the influence of light on the behavior of the newborn was made at the IOWA CHILD WELFARE RESEARCH STATION. The investigators used the stabilimeter polygraph described by PRATT, NELSON, and SUN (1930). During the examination the infant rests on a base that is easily movable (the stabilimeter) and that is sensitive to each movement of the child. Its deflections vertically and horizontally are transmitted onto a paper strip by two pens. The strength of the stimuli to which the infant is exposed for some length of time, at least for several minutes, can be exactly graded. The mobility of the infant is measured by the number of deflections of the polygraph per time unit and is compared with the mobility under other conditions.

With this methodology IRWIN and WEISS (1934) found that the mobility of the newborn after exposure to light for several minutes usually decreases with increasing light intensity, in other words, that the effect of light is calming. However, mobility was not greatest in complete darkness.

In her studies of color vision in the newborn, J. M. SMITH (1936) used this methodology. She observed 20 newborns, age seven to nine days. Every infant was exposed for five minutes to a blue, green, or red light following a period of five minutes in a dark room. The average values of mobility in a dark room and on exposure to colored light and the average values of the time of crying were determined. From these determinations the relative brightness values of the three colors were calculated.

The infants moved and cried most when in a dark room. Fifty percent of the movements occurring in the dark room were suppressed by blue light, 30% by green light, and 23% by red light. Surprisingly great were the sex differences in color sensitivity. In general, the boys moved more than the girls. In the boys blue had a marked inhibiting effect, green was less effective, and red not at all. In the girls all three colors had a suppressing effect; the most marked was that of blue and the least that of green. The relative brightness values of the three colors determined by the percentage of inhibition – taking 100 as the highest value – were for boys: blue 100, green 50, and red 14; for girls: blue 100, green 69, and red 77. J. M. SMITH concludes from this that newborn boys behave similarly to completely color-blind adults and that the girl's behavior is similar to protanopes. Boys thus possess a monochromatic and girls a dichromatic color system at birth. The author believes that her findings can serve as a basis for a phylogenetic color theory: The development of color vision progresses gradually, and more rapidly in girls than in boys.

These results differ greatly from my findings. It can be demonstrated at any time that the eye-neck reflex in the normal infant depends only on the brightness of the light stimulus, while the mobility depends on many other internal and external factors in addition to the color of the light. J. M. SMITH compared the mobility of the infant during exposure to colored lights with that during the preceding darkness. If, instead, the succeeding period of darkness would have been compared with the preceding period of illumination, the results might have been different. The degree of mobility fluctuated quite considerably in the different periods of darkness. Finally, the reported figures are only the average values of ten boys and ten girls. The extent of the scattering is not quite clear. As stated previously, color-blind persons generally and individually behave differently from color-sensitive persons. The previously mentioned investigations of IRWIN and WEISS, indicating that the mobility of the newborn decreases with increasing light strength but is not at its peak in complete darkness, make the evaluation of these factors still more uncertain.

D. and J. TRINCKER (1955) stress the brief adaptation periods used by J. M. SMITH and write:

"Despite these shortcomings in the methodology, a comparison of the values of equal intensity in Figs. 5 and 7 of SMITH's paper proves that PURKINJE's phenomenon applies here also (exactly as in PEIPER's work based on the higher brightness values of red in light adaptation). SMITH's rationale for rejecting PEIPER's explanation of his findings is therefore contradicted by her own material."

These authors reject her assumption of different forms of color blindness in male and female newborns.

As soon as children are old enough to choose intelligently, additional experiments pertaining to their color sense can be made. I (1927 [*19*]) followed the method used by W. KÖHLER (1915–1918) in his experiments on chimpanzees, in which the animals were conditioned to certain colors, using for this purpose the "two-box" method: The desired object (chocolate, a toy, etc.) was in one of two cardboard boxes identical in appearance. On the front of one box was the color to be matched; on the other, one of the 30 shades of gray in ZIMMERMANN's color scale, which was successively changed. The child had to compare the colors red, yellow, green, and blue with each of the various shades of gray. Should the child be completely color blind, the colors would appear to him identical to the shades of gray. The experiments showed that two neighboring shades of gray could not be used for conditioning. Since the child was able to differentiate with great accuracy between colors and gray at the beginning of the third year of life, he could not have been completely color-blind at this stage of development.

In additional experiments the child had to differentiate between a greenish-yellow and a red light; the addition of glass of various shades of gray in front of the light served to grade the brightness of the red light. As he could differentiate these lights, red-green blindness could be excluded. Similarly, he differentiated between a bluish-purple and a bluish-green light which was gradually darkened similarly to the red light. Thus, blue-yellow blindness could be excluded.

These experiments — together with my described investigations on prematurely born infants — indicate that at the beginning of the third year of life no known forms of incomplete or complete color blindness existed in the examined children and that their ability to differentiate between colors surpassed that of completely color-blind or bichromatic persons. If one does not wish to make the improbable assumption that infants during their first years of life have a form of incomplete color blindness that is so far unknown or that they are anomalous trichromats, the only possible conclusion is that they have color vision identical to that of color-sensitive adults.

TRENDELENBURG (1943) arrived at a corresponding result when he examined a three-year-old girl with his "color-spot procedure." The child had the task of finding among small freely moving paper disks those corresponding to a shown color. With these experiments, bichromatic and anomalous color perception could be excluded. The child was directed to draw with differently colored pencils; the absence of a red-green blindness could thus be demonstrated at the age of two although the child could not name the colors. At age five — this test was not done earlier — she could trace the numbers, letters, or winding lines of STILLING's and ISHIHARA's plates with a pencil. So far it has not been possible to prove known forms of complete or incomplete color-blindness in children. On the contrary, these forms can be excluded after a certain age.

The statement that the small number of names for colors in the ancient languages proves that the ancient Greeks and Romans were color-blind had a similar fate. The magnificent frescoes in Pompeii and Herculaneum imply a highly developed sense of color (TRENDELENBURG). TRENDELENBURG and I. SCHMIDT proved by corresponding experiments of alternate choices that the lower apes (Java and Rhesus monkeys) have the same normal trichromatic color system as man. According to the experiments of GRETHER (1940), however, the color vision of the chimpanzees is said to range phylogenetically below that of man. TRENDELENBURG doubts the validity of such a conclusion.

With the previously described two-box method (PEIPER, 1927 [*19*]), we can prove that the infant is able to choose within a simply structured situation, in this particular situation according to the relative brightness of two toneless colors. The child first learns to choose the brighter of two specific brightness grades. If he is then presented with two grades of the same difference in brightness values but with the formerly brighter area reversed to become comparatively darker, he then rejects the brightness degree chosen until now and turns to the still brighter shade. He thus differentiates not according to the absolute brightness quality but according to the relation of the two brightness degrees. KÖHLER (1915—18) successfully conducted these experiments on chimpanzees, chickens, and finally, in a single experiment, on a three-year-old child. The youngest child with whom I had success with this type of experiment was one year and ten months old, but some children who were a year older could not yet be trained. If their choice could be at all influenced, it always occurred according to the selfsame pattern (i.e., contrary to the absolute quality) if one disregards occasional mistakes. WEIGL (1930) obtained the same results. However, to train children to differentiate the size or amount of geometric structures (colored round disks, up to three colors per disk; see p. 528) was more difficult and uncertain. In these choices the developing intelligence played an important part: The child learns surprisingly fast and can be easily trained for a new type of differentiation while a chicken cannot be similarly retrained. KÖHLER found this in his experiments of long duration, which finally proved to be unsuccessful.

Learning to See

Although the newborn reacts to light, real vision has to be preceded by experience and practice: Seeing has to be learned. The newborn is not yet able to make conjugate eye movements, focus on an object, or follow a moving object with his eyes, and he probably cannot even accommodate.

ZETTERSTRÖM (1951) examined the origin and development of the electroretinogram during the first year of life. It is either not present at all at birth or it consists in a weak rise of the base line. During the first year of life a distinct rise of the beta potential takes place. The electroretinogram of the infant becomes identical to that of the adult at the end of the first year of life.

There is an old dispute on the question of whether the ability to recognize space is innate or acquired through experience. According to BERKELEY's theory (1709), only a part of the knowledge of the external world is received and transmitted by our eyes; the rest is based on experience and acquired knowledge. The sense of vision informs us only on brightness, color, and a certain arrangement of lines and points, but not on real form, position, and size. We are informed about these factors by the tactile sense.

However, KANT established the theory of the apriority of spatial sensations. VON KRIES (1909, 1924) understands this to mean that all our sensory perceptions are spatial and that this is caused and can be explained by our psychologic and physiological characteristics. The empiricist theory founded by BERKELEY (1709) and CONDILLAC (1754) was represented particularly by HELMHOLTZ (1867). He interprets sensory perceptions only as signs; it is our intellect's responsibility to learn to understand their meaning.

These two opinions differ sharply from each other. VON KRIES takes a middle of the road approach: Fully developed vision is the result of a learning experience prepared or perhaps started by events following developmental laws. By developmental laws VON KRIES means the laws governing the totality of events and originating in the particular species and in heredity. On the other hand, there is acquired knowledge. VON KRIES shares completely HERING's opinion (1868) that the spatial sense is a necessary and direct component of optic sensations, just as are the qualities of light and color: It is not a question of an either/or in the dispute between nativism and empiricism, but of the evaluation of the roles played in certain actions by disposition and by learning. According to LÖHLEIN (1931), the infant starts to use an already existing disposition when learning to see; he practices, so to speak, his piano playing on an established keyboard.

As the picture of the outside world projected by the lens is upside down on the retina, the question of why in spite of this we see the objects of the outside world upright has to be answered. BUFFON believes that newborns at first see objects upside down. MAGENDIE (1826) objected to this hypothesis. In 1939 this opinion is still presented in ELSENHANS' Textbook of Psychology. By a process of "rethinking" we are supposed to subsequently become used to seeing correctly.

However, in 1754 the sensationalist CONDILLAC gave a better explanation, which in principle may still be regarded as correct:

> "The eyes perceive light and color where they ought to be according to the findings of the tactile sense. They see on top what the tactile sense recognizes as being on top, at the bottom what it recognizes as being at the bottom; in brief, they see objects in the same position as the sense of touch represents them to be. That the picture is upside down causes no interference since there exists no 'up' or 'down' for the eyes as long as they have not yet been thus instructed. Only the sense of touch, which alone can discover these relations, can also teach the eye to evaluate them correctly."

One of the chapters of CONDILLAC's treatise, therefore, has the heading: "How the Sense of Touch Teaches the Other Senses About the Outside World." CONDILLAC thus arrived at a hypothesis which has again and again been confirmed by observations on infants.

WUNDT (1910), on the other hand, calls this theory a mere prejudice of the eighteenth and to some extent also of the nineteenth century. He believes in the opposite development of the seeing mechanism: Tactile impressions can only be localized with the aid of the spatial perceptions of the visual sense. In general, spatial perceptions originate through the integration of tactile stimuli into perceived visual images.

However, this concept is contradicted by the behavior of the infant at the time his visual perception develops and is combined with the existing space perceptions. From birth – that is, before he can focus – certain reflexes, such as the rooting reflex and the protective skin reflexes, enable him to adjust to space to the degree that is important for him. Later on, when his eyes detect the new space picture, the visual sense is directed by the existing "space picture," which has developed through the action of the skin senses and the proprioceptors of the muscles, tendons, and joints. Objects reaching his hands are moved back and forth and are soon brought to the mouth where they are further examined by his lips and tongue. At that time the hands come into his visual field and are again and again attentively observed. Infants in their first months of life will quite often place their empty hands in front of their faces and will observe the play of their fingers for quite some time and in a peculiar way.

HELMHOLTZ (1896) recognized the meaning of this behavior:

"We can easily find out how this knowledge of the importance of visual images is first acquired by infants by observing them when they are occupied with their toys: how they touch them, observe them for hours from all angles, turn them, put them in their mouth, etc., and finally try to drop or to break them. We observe that they repeat this behavior day after day. There is no doubt that this is the school in which they learn to understand the nature of the objects that surround them and at the same time the perspective relations of these images. In this manner they also learn to use their hands. The observation of younger infants also teaches us that they do not yet have this knowledge during the first weeks of life."

HELMHOLTZ draws the conclusion
"that the interpretation of some of the simplest and for the infant most important visual images has to be learned. It is not given to him by any innate organization from the beginning and without prior experience. We do not have to decide to what degree a similar conclusion may by extended to newborn animals. The mental activities of animals are perhaps more limited by their instincts, which enable them to move more securely in a narrower field than would be suitable for the later development of man, who can choose freely."

PALAGYI (1924) also believes that the tactile sense must be regarded as our original sense or, so to speak, as the root of our whole sensitivity. The other senses would be unable to create a basic knowledge of the outside world without the aid of the tactile sense. The sense of touch is the only complete sense; all others only complement it. With the aid of visual perception we transfer the optic image of an object to the space where we can feel it with our tactile perception.

To prove this hypothesis PALAGYI cites his own observations, which correspond to those already mentioned: Infants age four to six months observe their hands and feet attentively when touching any objects or themselves. "Through these experiments in the first months of life we learn to coordinate our visual sensations with the sense of touch." PALAGYI refers to similar observations that PREYER made on his child when the child was eighteen weeks and again when he was twenty-three weeks old (1895, p. 162).

The rooting reflex is a good example of this hypothesis: When we touch the region of the infant's mouth, reflexes are elicited that turn the head and move the lips so that the touching object is drawn into the mouth (Fig. 167, see p. 405). This life-preserving function, i.e., to find the nipple of the mother in space and to grasp it with the mouth by means of this turning, is innate, but the ability to turn to the nipple or the bottle when it appears in the field of vision without touching the mouth is not innate. This is rapidly learned (Figs. 196, p. 450, and 197, p. 451). From the rooting reflex there develops a conditioned reflex that at the sight of the breast or bottle turns the head to the correct position in space.

HELMHOLTZ (1867, 1896, 1912) called the interpretations of the visual images on the basis of experiences "unconscious conclusions" (see p. 511). According to PAVLOV (1909; III, 1, p. 83 and IV, p. 120) they have to be considered as conditioned reflexes: When the physiologist recognizes that in order to form a concept of the real size of an object a certain size of the picture on the retina and simultaneously a certain action of the external and internal eye muscles are needed, he defines this pattern as the mechanism of the conditioned reflex. If a certain combination of stimuli from the retina and from these muscles coincides several times with the tactile stimuli of an object of a certain size, this becomes a signal, a conditioned stimulus, of the actual size of the object. According to this point of view, the basic facts about the psychologic part of the physiological reactions are nothing else than a series of conditioned reflexes.

According to JOHANNES MÜLLER (1837), everything, even our body, is seen upside down. Also, the image of our touching hand is seen in this way so that there is no disharmony between seeing upside down and feeling upright. We therefore call the objects upright, for this is how they appear to us. It is, according to LOTZE (1852), a prejudice to assume that in the real position of the image on the retina there exists a motive for the psyche to perceive it in the same aspect. The position of the image on the retina makes no difference to the psyche. To achieve harmony between the visual localization and that obtained by proprioceptive and tactile sensations, the direction of the eye muscle movements has to agree with that of a touching limb. This can only be accomplished by an upside-down retinal picture on the concave eyeground or by an upright picture on the eye's convex anterior surface. HELMHOLTZ believes that the perception of the image on the retina does not originally effect any definite localization on the body; this is only gradually acquired by experience whereby visual perceptions are guided by skin sensations.

The findings of EWERT (1930) also favor this concept. He repeated the previous experiments of STRATTON, and for weeks he inverted the retinal images of adults with lenses placed in front of the eyes. In the beginning the ability to localize by touch, hearing, and seeing was considerably disturbed, but after fourteen days of practice the disturbance was totally overcome in some fields, in some, however, only in part. The eye easily learned to adjust to the new conditions. Never during the entire period of observation, however, did there occur a reversion of the perceived surroundings which would have compensated for the inversion caused by the lenses.

KRÜGER (1939) made similar experiments. He arranged mirrors in the form of spectacles in such a way that the entire surroundings would be seen upside down. He wore these spectacles several hours a day for four weeks. After some practice, using the tactile sense as a guide, he was able to read, write, and draw. He concluded from this experience that the empiricist theory can be applied to a major portion of the space perceptions and that in order to perceive correctly it does not make any difference whether the retinal image is seen upright or upside down.

We definitely have no reason to believe that the infant ever sees the surroundings upside down because of the upside-down image on the retina.

Originally the mouth serves as the most important sensory organ and at the same time as a tool. It feels, tastes, grasps, and sucks. The hands, which touch and turn objects again and again, also serve as a sensory organ and as a tool by guiding everything to the mouth for further examination. Therefore, mouth and hands are at first of much greater importance for the behavior of the infant than the eyes. However, in the course of further development the use of the mouth as a tool and of the hands as a sensory organ gradually decreases in importance while the eyes begin their real mission. The gradually growing importance of the eyes can be demonstrated by an everyday example: While the rooting reflex can first be elicited only through tactile stimuli, the hungry infant, when he is able to recognize the bottle, will after several months make the same "following" movements with his eyes and head as soon as a bottle is moved in his field of vision. The unconditioned searching for the breast (rooting) has been conditionally connected with the sight of the bottle (see p. 451).

To what degree the space concept of the infant has been enlarged by his learning to see has so far not been examined. It is certain that light stimuli from a great distance, for example, sun rays, can elicit reactions, but does the infant see things two-dimensionally or actually in space? Perhaps a study of accommodation in infants could help here. The additional space acquired by the infant through vision is probably at first not larger than 1 or 2 m in diameter and has approximately the form of half of a sphere surrounding his head. By the later acquisition of the skills of turning his head and body and finally by moving around on his own, the surrounding space is tremendously enlarged.

Gradually the infant learns to recognize objects that can be grasped and to judge distances. He has, so to speak, "to see them out of the surrounding." This is made easier for him if the objects are moved in his field of vision and, particularly, if they are presented to him. However, the child does not reach for the "unreachable moon," as W. STERN (1932) has correctly stressed. He may at times reach with his hands for distant desired objects as adults also may do. This may have caused the wrong interpretation that he wants to take the moon into his hands. The proverbial "reaching for the moon" is just a symbolic poetic expression for the so often unrealizable wishes of the child and it has been taken literally without any justification. It would indeed be strange if the infants were always reaching just for the moon and not for other distant objects that are more conspicuous and enter their fields of vision more frequently.

WEIGT (1955) examined the development of the visual fields in infants from the age of one day to eighteen months by moving a source of light or a shining object from the periphery to the center and by determining the point at which the infants became distracted and followed the object with their eyes. There was a distinct development of the visual field which started in the third month and ended at about the twelfth month. This development does not progress evenly, but certain developmental stages can be clearly distinguished. When the infant starts to walk, his visual field is approximately as large as that of the adult.

The infant's learning to see has been compared to that of those born blind and who as adults acquire vision by means of operations. For example, the nineteen-year-old boy operated on by RÄHLMANN (1891) first had to learn how to focus: The first movements of his eyes were not conjugated; it was difficult for him to detect an object visually and to keep it in focus. In the beginning objects in the lateral field of vision were not noticed at all, and it was equally impossible to elicit a "threat reflex" in him from the side. Furthermore, he was unable to estimate distances correctly or to recognize visually round and angular objects that were known to him from his tactile experiences. Indeed, he learned to use his eyes in exactly the same sequence as does the newborn.

The direction in which we visualize an object with both eyes and through simple perception does not proceed from one of the eyes but from a place between them, i.e., the "cyclops," or "middle eye." That this is the original sensation of the infant was shown by VELHAGEN (1935) through the following experiments in children age fifteen months: The adult holds a napkin ring like a monocle in front of one eye. When trying to imitate this the child will always hold the ring between his eyes in front of the bridge of the nose, i.e., where the line of vision would place the "cyclops" eye. Thus, the child at this age has not yet had the experience of seeing with two eyes.

Stereoscopic vision can, according to SACHSENWEGER (1959), be proved in infancy shortly after binocular conjugate movements have been established. While the development of general visual acuity is almost completed in the first years of life, the development of visual acuity in depth takes many years. In early childhood only the forward and backward position are roughly defined; vision in depth is still weakly developed at school age; development is completed at age nine, although some children have already reached this stage at age six.

Hearing

(1, 3, 6, 10, 12, 51)

According to LANGE (1929), the newborn has only an eardrum ring in lieu of the auditory canal, and the mastoids are completely absent. The narrow and short membranous auditory canal is filled with the secretion of the epidermis, the vernix caseosa. The eardrum of the newborn has almost reached its final size (diameter 7½ to 9 mm, as compared with 9 to 11 mm in the adult). The tympanum and the middle ear bones also grow only very little after birth. LANGE states that the cavity of the tympanum, which until birth contains an embryonic cushion of mucus, fills with air during the first respiratory and swallowing movements so that the noninflamed middle ear always contains air. The ductus cochlearis of the newborn is anatomically complete and functioning (KOLMER, 1924).

Certain histological findings in the inner ear (adhesion of the membrana tectoria to the papilla spiralis) led NEUBERT (1954) to assume that hearing analysis of the ear is not possible until ten to twelve days after birth, at the earliest.

Sound conduction is probably impaired by the condition of the external auditory canal and perhaps also by that of the tympanum. According to ALEXANDER (1911), the hearing acuity of the adult under those conditions would decrease to a distance of 1 to 2 m for the volume of the usual speaking voice; the hearing acuity of the newborn should be estimated accordingly. According to LANGE, the mere obstruction of the external auditory canal can cause a certain degree of deafness. After removal of this obstruction the condition of the tympanum and the inner ear will not prevent normal hearing; on the contrary, the ear of the newborn is physically adapted to hearing. HEUBNER (1919) pointed out that through bone conduction the sound stimulus reaches the acoustic nerve when an obstruction of the external auditory canal exists. All these facts refute the statements of PORTAL (1818), PREYER (1895, p. 49), REMPLEIN (1952, p. 79), and TUMLIRZ (1955, p. 106), who maintained that the fetus and the newborn are deaf because there is no air in the tympanum before breathing has started. COMPAYRÉ (1900) stated that intermediate stages exist between the complete deafness of the first days and the time of complete hearing; hearing develops gradually and this serves as a protection for the infant.

"What would happen if a piercing cry, shrill whistle, or a tumultuous noise would irritate the hearing of the newborn? The premature irritation would endanger the infant so that he might lose his hearing forever."

For a long while these views were presented in the literature, although from the beginning there were dissenting statements. Today only those who have no direct experience maintain that the newborn is deaf.

These differences of opinion arose from the fact that it is by far more difficult to demonstrate the effect of sound stimuli in the infant than in the adult, whom one has only to question about his perceptions. As in the corresponding examinations of the other senses, it is necessary to observe the infant's responses to sound stimuli. Although nobody would think that an adult is deaf because he does not react to a sound stimulus in an obvious way, this wrong conclusion has been drawn again and again for the newborn.

When other sensory organs are stimulated, local reactions appear in the organ itself. However, in man neither the inner, middle nor external ear responds visibly to sound stimuli. On the other hand, these local reactions are known in animals. According to PREYER (1885), the external ear of guinea pig embryos twitched after loud whistling. KURODA (1939) noted in a newborn monkey *(Macaca zynomolgus)* a retraction of both external ears and scalp after the same kind of stimulus. The skin of the forehead was drawn upward, the eyeballs also turned upward, and the eyes were shut. The head usually turned in the direction of the source of the stimulus. According to W. N. and L. A. KELLOGG (1933), the young chimpanzee retracts its external ears after a shrill sound stimulus. A similar reflex has occasionally been found in the adult man (CEMACH, 1920). The muscles of the external ear may thus still be a phylogenetic vestige. H. S. and H. B. FORBES (1927) in a few instances apparently noted a similar reflex in newborns.

The reactions to sound consist generally of the following phenomena: When the eyes are open the aural-palpebral reflex of MÜLLER-BECHTEREV, which is also called the cochleopalpebral reflex (closing of the lid on sound stimulus), is elicited. When the eyes are closed, there is a tighter shutting of the eyes; furthermore, the fright reaction, change of respiratory rhythm, motor restlessness, facial grimacing, and turning of the head to the sound source, and sometimes the calming of a crying infant, take place. BRONSCHTEIN and PETROWA (1952) observed that sucking movements in young infants stopped on sound stimuli.

Besides these phenomena, HETZER and TUDOR-HART (1927) described many additional reactions in the newborn: blinking, opening of the eyes, placing the hands in front of the face, making a fist, wide opening of the eyes, restless movements, the startle reaction, wide opening of the mouth, twitching of the eyes, cessation of movements, wrinkling of the forehead, cessation of crying, turning of the eyes in the direction of the sound, extension of the arms, spreading of the fingers, pursing of the mouth, tense facial expression, expression of comfort, sucking movements, sounds of discomfort as soon as the stimulus has stopped, noisy crying, sounds of discomfort while the stimulus lasts, smiling, protruding the tongue, smiling after the stimulus has subsided, turning of the head, and sounds of comfort after the subsidence of the stimulus. The further development of the responses to sounds and noises in infancy is described by LÖWENFELD (1927).

The external testing conditions are especially important in determining the response of infants to sound. It is apparent that the surroundings have to be completely noiseless for a certain length of time prior to the introduction of the sound stimulus. Important, too, is the excitability of the infant. If he is occupied, sucking, moving, lively, or crying, or if he has just been taken out of bed, we can hardly expect recognizable reactions, whereas he may be more easily stimulated during complete rest and immobility shortly before falling asleep or during sleep. After all, in the first weeks of life we are dealing mostly with fright effects, which also occur in the adult only under specific circumstances. An important source of error lies in the fact that the excitability diminishes when the stimulus is repeated often. A few experiments without results do not permit us to judge the hearing ability of the infant.

All these difficulties explain the wide differences in opinion on the question when the newborn begins to hear. Probably the first statements about this problem were made by JÖRG (1826): Only shrill sound stimuli, irritating the acoustic nerves, can cause fright and startle reactions of the newborn whether he is awake or asleep. Sensitivity for everything we may call either sound or tone develops in the first days after birth. On the other hand, a contemporary of JÖRG, the French physiologist MAGENDIE, asserted in the same year that the newborn does not seem to listen even to the strongest sounds. Since that time, this difference of opinion has persisted in the literature.

KUSSMAUL (1859) could not obtain a response to sound stimuli in the first days of life. He cites FELDBAUSCH, who succeeded in obtaining such responses in three-day-old infants. POLI (1856) noted that the first responses occur 10 minutes after birth; KUTVIRT (1911, 1912) after half an hour; MOLDENHAUER (1881), KRONER (1882), and KÖLLREUTER (1907) after a few hours; DEMETRIADES (1924) after 2 hours; SACHS (1893) after 5 hours; GENZMER (1873) from the first or second day; PREYER (1895) from the first to third day; SIGISMUND (1856), SIKORSKI (1908), and MEYER (1912) at the age of several weeks. BLANTON (1917) stated that a reaction to sound immediately after birth was uncommon, but he observed it during the first days of life. CANESTRINI's tracings (1913) of respiration and brain pulsations in newborns always showed changes under the influence of sound stimuli. Among 70 infants, age 6 hours to fourteen days, he found none who did not respond in any way to sound stimuli. FRÖSCHELS and BEEBE (1946) observed in 31 of 33 newborns, age one-half to nine days, distinct reactions to whistling, while stimulation with tuning forks remained ineffective.

MUZIO (1934), who examined newborns with tuning forks and observed their facial expressions, found them sensitive to sound during the first minutes of life, especially when he used high-pitched sudden sounds. RICHMOND, GROSSMAN, and LUSTMAN (1953) obtained similar results. They examined the hearing ability of newborns, age one-half hour to eight days, and of prematurely born, age one to fifteen days, with a cowbell; they did this in infants who were awake, drinking, or in light or deep sleep. Lid closing and the fright reaction were interpreted as evidence of hearing. Infants who did not respond were re-examined shortly afterward. Under these conditions all infants could be stimulated by sound. The eyelid reaction occurred in 70% and the fright reaction in 47% of the observations. It was easiest to stimulate infants in light sleep. A reaction can only be expected with certainty when the newborn is not drinking or crying or when he is in deep sleep.

I (1924 [*10*]) have observed a change in the respiratory tracing under the influence of sound stimuli in four of six newborns during their first hour of life. The youngest of these infants was 25 minutes old when examined. The same experiment also succeeded with prematurely born infants. A prerequisite for this response is that the infants are breathing regularly before the start of the stimulus. Of the two newborns in whom I was not successful, one had not calmed down enough after birth to breathe rhythmically. The other infant did not react, although he was completely quiet.

SCHALTENBRAND and GIRNDT (1925) observed twitching and temporary irregular breathing on sound stimuli in thalamus preparations of cats. The reported tracings correspond exactly to those I obtained from infants who could be regarded as "pallidum preparations" (see p. 265). The changes in the respiratory rhythm after sound stimuli are applied to rabbits and frogs have been described (see p. 25).

That the excitability disappears quickly if the same stimulus is repeated at short intervals has been reported by various investigators: disappearance of the cochleopalpebral reflex (MOLDENHAUER, 1881), of the reaction of respiration and brain pulsations (CANESTRINI, 1913), and of the movement reaction (PRATT, 1934). For this reason WESSÉN (1928) stipulates an interval of ½ to 1 minute between tests.

I traced the breathing of the sleeping infant and could thus observe this phenomenon in more detail. It was necessary to obtain a stimulus strength that would cause a respiratory reaction but that would not interrupt the sleep. Because of the suppressing effect of sleep, the reaction was sometimes delayed by several seconds (Fig. 10). The stimulation of the respiratory center during sleep that is not completely interrupted sometimes results in temporarily faster, slower, or more shallow breathing and occasionally also in prolonged respiration (Fig. 1, p. 24). With a sufficiently strong sound stimulus another very typical reaction may appear in most instances: At first there is a very deep breath followed by respiratory arrest or, less frequently, by a few shallow breaths (Figs. 10

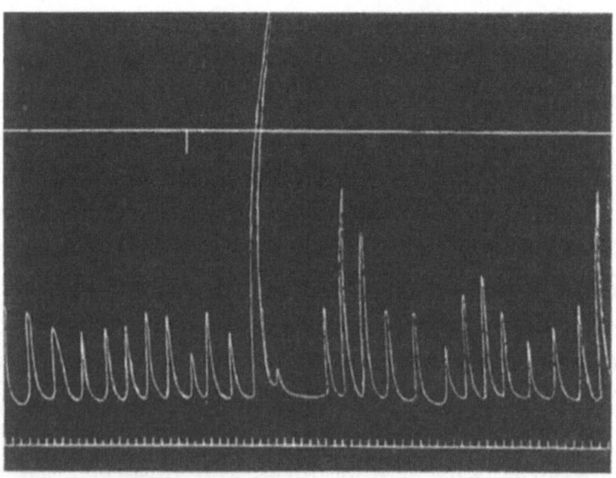

Fig. 10. After sound stimulus in prematurely born infants during sleep, delayed inspiration reaction and periodic breathing. Top: bell; middle: breathing; bottom: time (sec).

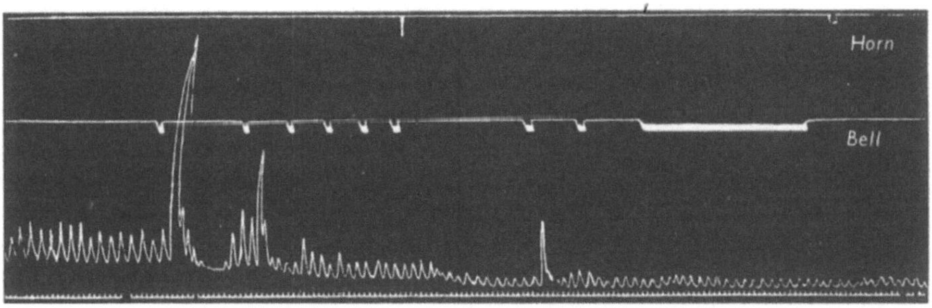

Fig. 11. Absence of the respiratory reaction after repeated sound stimuli in a sleeping prematurely born infant, age 9 wk, 2900 gm (sec).

and 11). After this reaction has subsided (about 5 to 10 seconds later) the previous respiratory pattern may be restored. But quite often the respiration remains irregular in depth; periodic breathing (Fig. 10) lasting for two to three periods in the same strength and then returning to its previous pattern will often occur after the first reaction. Under the influence of the stimulus there first may be one considerably deepened breath, i.e., strong excitement, followed by a prolonged respiratory arrest, i.e., strong respiratory suppression. Afterward the excitatory phase of the respiratory center may temporarily decrease to periodic breathing. All these happenings occur in the respiratory center successively. They can be explained by the fact that the nervous energy, i.e., the functioning ability of the nerves, is limited in quantity. If the existing energy of the repiratory center is depleted by the suddenly occurring increased excitement, its function is suppressed until the nervous energy is replenished.

If a stimulus is repeated for a second time, shortly after the first reaction has subsided and the breathing has calmed down, a second reaction may be completely absent even if the second stimulus was considerably stronger than the first (Fig. 11). Sometimes the state of nonexcitability is only evident after the second or third stimulation. Of course, this situation can occur only if the infant did not wake up from the preceding stimulus. The duration of the nonexcitability is quite short. If it is present at all, it generally lasts for 10 to 15 seconds after the first stimulation. If the two stimuli occur too soon after each other, they will have an accumulative effect and will wake up the infant. Sometimes a subthreshold stimulus may fail even with gradual increases of the stimulus strength and more prolonged intervals of application; at first the infant does not respond at all and then awakens suddenly when the strength is further increased.

The phenomenon of a fright stimulus followed temporarily by nonexcitability to the same stimulus can still be demonstrated in the adult. It is a salutary mechanism that adjusts the body to its environment by creating a new nervous equilibrium. Pavlov described a similar occurrence in the waking animal: If a new stimulus suddenly occurs in the environment, a general reaction, which can be characterized as an investigation reflex, takes place; the animal directs its receptive faculty to the new stimulus; it looks up, listens, etc. (the "What's the matter?" reflex).

If the same stimulus is repeated at short intervals or remains in effect for any length of time, the investigatory reflex becomes gradually weaker and finally disappears.

In both cases, in the sleeping infant and in the waking animal, the same stimulus in the same area causes first excitement and then inhibition. PAVLOV assumes that repeated stimulation of a certain area will exhaust the nervous energy, and he regards exhaustion and inhibition as identical phenomena. With our series of experiments we arrived at the same conclusion: One stimulus may exhaust the nervous energy of the sleeping infant in a certain area and may cause even stronger inhibition. There is no contradiction in the fact that an alert and therefore noninhibited brain cannot be as easily exhausted.

BERBERICH and WIECHERS (1924) examined the blinking reaction of newborns and prematures with noise makers and high-pitched tuning forks; they found only a few children who did not respond in the first days and weeks. VOSS (1927) regards the absence of such responses as a serious prognostic sign in premature infants, while CATEL (1928) stresses, and rightly so, that on the basis of his experiments the absence of such a reflex does not permit a judgment as to the infant's viability and hearing faculty.

WESSÉN (1928) was able to prove the presence of the cochleopalpebral reflex to sound stimuli in 15 of 23 newborns in the first 10 minutes after birth. Of 38 newborns examined within the first hour postpartum, the reflex could be found in 28. STIRNIMANN (1940) examined 100 newborns on their first day of life with a tuning whistle to determine their reaction to the notes F and A and treble F without finding any difference in their reactions to the various tones. Only six children reacted to all tones, 16 to none.

The IOWA CHILD WELFARE RESEARCH STATION thoroughly examined the effects of light and sound stimuli on the newborn under controlled conditions. They again used the described stabilimeter polygraph (see p. 75). PRATT, NELSON, and SUN (1930) applied 5836 sound stimuli of different kinds to 59 newborns and noted reactions in 46%. Two percent of the infant scalmed down under the effect of these stimuli. The reactions depended only in part on the stimulus strength. Movements of the body were found in 26%, of the limbs in 36%, and of the eyes in 34%. The relatively high number of eye movements was explained as the infants' initial search with their eyes for the source of sound, but we are not told how often they were actually looking toward the source of sound.

In this connection STUBBS (1934) investigated the question of whether sounds of different duration, strength, or pitch cause different reactions in the newborn. Indicators were the reactions recorded on the stabilimeter polygraph, the changes in the continuously recorded respiration, and the movements seen by the observer. For comparison the actual period of examination was preceded by a base-line period of double length. With this method it was found that the responses increased with increasing duration and strength of the sound stimuli, while the influence of the pitch of the tone was less evident. More sensitive than the stabilimeter tracings, which recorded the general movements, were the respiratory tracings, in which one could note about four times as many reactions to the stimulus, i.e., changes in the respiratory rhythm (slowing, accelerating, shallowing, deepening). Also in my experience, the observation of changes in the respiratory tracing is the most sensitive method for judging the effect of the stimulus.

As previously described (see p. 75), IRWIN and WEISS (1934) proved that an increasing light intensity has a calming effect on the movements of the newborn. WEISS (1934) similarly compared two sound stimuli of different strengths. Each experiment lasted 5 minutes. Two experiments in which the infants were kept in dimly lighted or darkened rooms served as controls. The mobility was determined with the aid of the stabilimeter. With this experimental design it was found that sound noticeably decreases the mobility of the newborn. If light and sound (both of which decrease mobility when used individually) were applied at the same time, the calming effect was noticeably increased.

KRASNOGORSKI (1913) states that the hearing analyzer, in the PAVLOVIAN sense, is only slightly developed in the infant. The infant cannot establish conditioned reflexes to specific sounds or chords, whereas dogs can differentiate between eighth notes. ALDRICH (1928) developed conditioned reflexes to sound stimuli in a three-month-old infant and thus proved the infant's ability to hear, which had been doubted before. He associated the sound of a small bell with scratching of the infant's sole. After 15 to 20 repetitions the child cried and retracted his foot on the sound of the bell. KASSATKIN and LEVIKOVA (1935) noted the earliest conditioned reflexes to sound stimuli in the first half of the second month. In 1953 KASSATKIN was successful at an even earlier age of the infant: He tried to establish a conditioned feeding reflex to whistling in an infant. He started his experiments on the sixteenth day of life and noted on the twenty-seventh and twenty-eighth days sucking movements on single sound stimuli in one to two combinations; in another infant he observed these responses only on the thirty-fourth and forty-first days under the same conditions. After the forty-third day the conditioned reflex was still weak but could be demonstrated every day; on the fifty-fifth day it was present in all or almost all combinations. MARQUIS (1931) even stated that she elicited conditioned sucking movements on sound stimuli in seven of eight newborns, a claim that WENGER (1936) could not confirm in his later investigation.

HETZER and TUDOR-HART (1927) reported that the human voice is far less exciting for the newborn than sounds and noises, but it increases its effect by the end of the first month. The voice can cause sucking movements in the third week and smiles in the second month. The question of whether or not the newborn recognizes the voice of his mother is discussed on page 609.

Every mother knows that crying infants can be calmed by words or musical sounds. CANESTRINI and HETZER and TUDOR-HART were able to confirm this effect experimentally.

There is no doubt that the newborn is not deaf. All differing statements can be explained by errors in the methodology. The question of whether the newborn hears less well at first cannot be solved because there are no adequate methods of investigation. The prematurely born newborn responds distinctly and exactly as a normal newborn to sound stimuli with fright reactions and changes in the respiratory rhythm months before his theoretical "day of birth." Therefore, we could infer that the infant has this faculty while in his mother's womb, although PREYER (1885) has stated that "with a probability near to certainty" the infant has no sense of hearing before being born.

It would probably be easiest to recognize the stimulus effect through observation of AHLFELD's intrauterine respiratory movements of the fetus (see p. 354), but these are too infrequent to serve as a basis for an extended series of observations. Therefore, I (1924 [*10*]) studied the fright reactions occurring in the fetus in the womb under the same conditions and in the same manner as after birth. One can see them by close observation of the body of the mother or feel them most distinctly by placing the hand on the mother's abdomen. Furthermore, the movements of the anterior abdominal wall can be recorded. Finally, the statement of the mother who feels the child's movements can be used. Since the sound is greatly weakened before it can reach the infant, I used a loud and shrill automobile horn which I sounded near the infant's head. An absolute prerequisite for the success of this experiment is immobility of mother and child and complete quiet before the start. Furthermore, the mother must not be frightened by the strong sound but must continue to breath calmly, which is not at all easy.

The demonstration of the fright reaction was successful under these conditions in about one-third or one-fourth of the infants. The movements caused by this stimulus vary. Sometimes the abdomen of the mother is only shaken by a kick followed by complete calm. There was probably a sudden startle reaction of the fetus. In other trials great restlessness was noted; several respiratory movements of the mother were recorded before the movement stopped. One could make the objection that the mother was frightened and not the fetus. But whoever has once felt the fetal movement with his own hand will never believe this. The statements of the mother who feels all the movements of the child also contradict this objection. Recording of the mother's respiration shows that the movements of the fetus also occur when the breathing of the mother remains unchanged.

If the stimulus is repeated several times and at short intervals, the fetus' responses disappear quickly; quite often only the first attempt is successful. Thus, the same phenomenon occurs in the child before and after birth. The effective stimulus leaves a mark in the unborn as well as the newborn child that prevents the formation of a new reaction.

H. S. and H. B. FORBES (1927) observed that the unborn child may respond with sudden movements to a loud sound stimulus outside his mother's body four to five weeks before birth. The pregnant mother was lying in a metal bathtub filled with water. The stimulus consisted of tapping on the bathtub with a metal object. The ensuing movements of the infant were felt by the mother and seen by the observer.

A detailed re-examination of my findings was made by SONTAG and WALLACE (1935), who observed reactions to sound stimuli in an extended series of unborn children. The following experiment of WALLACE, however, was not successful; he attempted to establish a conditioned reflex in the unborn child by associating the unconditioned movement reaction to a vibrator with the sound of a tuning fork. WALLACE concluded that either a conditioned reflex cannot yet be established in the unborn child or that the stimulus that was used was not strong enough. BERNARD and SONTAG (1947) examined the changes occurring in the heart rate of the unborn infant under the influence of sound stimuli. They found a distinct temporary increase of the heart rate shortly after application of the stimulation. The unborn infant showed sensitivity to a great variety of sounds.

Generally an unborn child is not exposed to such strong stimuli as, for example, the sound of a car horn. However, pregnant mothers repeatedly reported to me that they felt unpleasantly strong movements of the child during concerts. One mother told STIRNIMANN (1940) that she stopped attending concerts at the sixth month of pregnancy because she experienced very violent child movements during these concerts. H. S. and H. B. FORBES and, more recently and independently of these authors, MAYER (1953) reported similar incidents.

In a similar manner, FLEISCHER (1955) examined 78 unborn children with exactly uniform strong sound stimuli. He did not find a reaction of the fetus before the seventh month of gestation; it was strongest in the ninth month. FLEISCHER also observed the disappearance of the reaction after repeated stimulation. Based on his observations he did not believe that the hearing of the mother had any influence on the occurrence of the reaction. When he deafened the mother with the BÁRÁNY drum and, in addition, stopped up her ears and turned her head away from the sound source, so that she could not hear the sound at all or could perceive it only faintly, the reaction could nevertheless be obtained. The unborn child of a deaf-mute person also showed a distinct reaction to sound.

Of course, it is possible that the reaction could be mediated by the sense of touch and not by hearing. In order to clarify this problem, FLEISCHER in some cases used stimuli of a high pitch (1000 cps). The same reactions occurred under these conditions. It has been stated, however, that man may experience a tactile sensation when stimulated with 1000 cps, especially when the intensity of the tone is very great. Therefore, deaf-mute school children who were blindfolded were examined as controls. It was shown that totally deaf children could not perceive a sound stimulus of 520 cps at a distance of 20 cm from their heads and necks, a stimulus which would be more likely to cause a tactile sensation than a sound of 1000 cps.

"If one does not wish to assume that the threshold for tactile stimuli is lower in the fetus than later in life – and we have no arguments supporting such an assumption – this fact can be regarded as a confirmation of the hypothesis that the observed reactions are in fact the result of a stimulation of the fetus' ear."

Just as the newborn can occasionally look at the light source (see p. 62), he can sometimes turn his head or eyes to the source of sound (BLANTON, 1917; HETZER and TUDOR-HART, 1927; STIRNIMANN, 1940; and P. H. WOLFF, 1959). PREYER's child (1895), whom he observed continuously, turned his head in the direction of the sound for the first time in the eleventh week. In the twelfth week the turning of the head occurred rapidly, although the eyes did not immediately turn in the direction of the sound. MEYER (1912) regarded the newborn as deaf and described the different stages at which the infant learned to search for the source of sound. The infant responds first to loud sounds with movements of the limbs. Later he listens when, for example, the nurse calls him, but does not respond to unfamiliar sounds. Still later he tries to single out the voice of the nurse and also reacts to unfamiliar sounds. Then comes a stage when he can distinctly localize the mother's call, but is unable to localize unfamiliar sounds. Finally, he accurately localizes all noises and calls. Infants react differently to familiar and unfamiliar sounds. Some seven- to eight-week-old infants may be able to localize calls; older ones can localize calls and noises. Disease delays the ability to localize. LÖWENFELD (1927), who is also convinced of the deafness of the newborn, states that the ability to search for the source of sound with the eyes develops in the third month of life. However, according to ANDRÉ-THOMAS and AUTGAERDEN (1959), the infant turns his head in the second month or a little later when he hears his mother's voice.

According to my experiments, the reaction time to sound stimuli, as indicated by the start of the fright reaction, is about 0.25 second in waking infants age two to four months; the mean deviation is 0.067 second. Some of the shortest reaction times are below the median values for the adult, which have been determined by different methods. The reaction time of the infant is distinctly prolonged during sleep: There may be a duration of several seconds. In premature infants it is prolonged even when they are awake.

IRWIN (1932) found a value of 0.18 ± 0.03 second under similar conditions. The reaction time based on the respiratory response was determined by STUBBS and IRWIN (1934) as 0.09 ± 0.025 second.

Speech, which develops at the end of the first year, depends on hearing. Although the first words are acquired in a pattern similar to conditioned reflexes — one might better speak of conditioned connections — there is no question but that the further development of linguistic ability is in the neurological sphere. The increasingly better evaluation of sound perceptions, which during the development of speech makes hearing a "higher" sense, is the task of the brain and not of the ear.

GENERAL CAPABILITY OF THE SENSES

> Every animated body absolutely must have a tactile capacity... Every animal must possess this sense; it does not have to possess another.
>
> ARISTOTLE: *De Anima*

The question as to which is the most important sense was formerly answered without any reservations: For DEMOCRITUS all senses are special branches of the sense of touch. ARISTOTLE writes: "Taste and tactile senses are senses that are indispensable to living beings; evidently none could exist without the sense of touch." THOMAS AQUINAS describes this connection in a similar way: "The sense of touch, the most perfect of all senses, is more perfect in man than in any other being." According to the philosopher CARUS (1808), the sense of touch is the original sense. "It is the fundamental sense and as such it is the most necessary and most indispensable of all senses." Strangely enough, this physiological knowledge of the ancient philosophers was later more or less lost.

From the preceding presentation it is evident that after birth every sense is functioning. There is no proof whatsoever to the contrary, no matter how often such an opposing statement may have been made. We must concede only that their sensitivity has not yet been shown to be commensurate with that of later life; it is possible that with further improvement of the investigative techniques certain differences may be demonstrated. Certainly one sense mediates more reflexes in the infant than others; a greater variety of reflexes can be elicited through the eye than through the ear. However, the same differences exist in the adult. Although he is – in contrast to the infant – able to give exact information on his sound perceptions, it cannot be concluded from this fact that the function of the infant's ear is inferior. TUDOR-HART's statement (1957) that the human newborn possesses fewer reflexes to external stimuli than an animal contradicts our experience.

There are radical differences in the manner in which sensory impressions are used during ontogenetic development. As we know from the observation of anencephalics (GAMPER; 1926; CATEL and KRAUSPE, 1930), sensory reactions of normal newborns are possible without the function of their cerebral hemispheres; these sensory reactions are present even when only the medulla oblongata is the highest functioning part of the central nervous system (CATEL and KRAUSPE). In the course of continued maturation the "relay station" ascends to higher levels of the central nervous system so that sensory impressions can be progressively better evaluated.

According to BURDACH (1838), vision and feeling develop first, then hearing and taste, and finally the olfactory and tactile senses. LÖBISCH (1851) gives the same sequence. There is, however, no agreement on the sequence in which the various senses gain importance for the infant's behavior. The available statements are combined in Table VI and are compared with my own experiences.

CANESTRINI (1913) gives the following sequence: taste, hearing, vision, touch, and smell. This scale, which is conspicuous for its underestimation of the importance of the skin senses, can be explained by the fact that only the effect of sensory stimulation on brain pulsations and respiration had been examined. The results would have been different had CANESTRINI also observed, besides these general reactions, the local responses, which occur so frequently after stimulation of the skin senses and which are almost completely absent on sound stimuli.

Table VI: Order of Importance of the Senses According to Several Investigators*

CANESTRINI	W. STERN, KOFFKA	K. BÜHLER	REMPLEIN	PEIPER	
				Newborns	Adults
taste	touch	vision touch	touch taste smell	touch taste	vision hearing
hearing					
vision	taste smell vision	hearing		vision hearing	touch taste
touch		taste	vision		
smell	hearing	smell	hearing	smell	smell

* In newborns except for last column.

W. STERN (1923) gives another sequence, which is also accepted by KOFFKA (1923): The tactile sense develops first; then concurrently, taste, smell, and vision; and last, hearing. In this connection W. STERN refers to the brain research of the anatomist FLECHSIG; one may even assume that the ranking of the senses has been established under his influence.

According to FLECHSIG (1876), the nervous pathways at birth are in varying stages of maturation and function. Myelinization occurs gradually and in stages; thus, the individual systems can be identified by their state of myelinization. Maturation is not yet complete at birth; the various sensory organs are connected with the cerebral cortex according to an exactly predetermined sequence. At first the central connections that transmit tactile impressions and at the same time all pathways that transmit general sensations, such as pain, hunger, and thirst, as well as the nerves involved in the sense of position (originating from the muscles, tendons, and joints) develop. The tractus olfactorius is formed at approximately the same time. The optic pathways mature considerably later, but they are myelinated in the normal newborn; the acoustic pathways mature last.

However, the maturation of the cortical pathways is of no importance in relation to function since – according to our present knowledge – the reactions of the newborn are independent of his cerebral hemispheres.

The myelinization of the peripheral sensory nerves, which, contrary to the central pathways, are already functioning in the newborn, has been disregarded in setting up these sequences. If the sensory function were to be judged by this myelinization, another sequence would have to be established. Myelinization is least developed in the distal portions of the newborn's optic nerves (A. WESTPHAL 1894, 1897). At that time the orbital part is still completely nonmyelinated. The acoustic nerve, however, is most developed and this is precisely the nerve where one would not expect the most development on the basis of W. STERN's sequence. According to E. STERN (1950), however, hearing is the first sense to develop.

According to REMPLEIN (1952), the order of rank of the senses corresponds to the sequence in which myelinization occurs, first in the "nearsenses" and then in the "farsenses."* REMPLEIN stated that the newborn is "completely deaf."

From all this evidence we may conclude that we cannot judge the function of a nervous pathway by its anatomic structure. Independently of the degree of myelinization – the myelin by itself cannot transmit the stimulus – experiments show that each sensory organ can elicit reflexes and that each sensory nerve of the newborn is therefore able to transmit the stimulus to the brain. Even nonmyelinated nerves can do this, as shown by the functioning of the optic nerves in the newborn. Premature fetuses, whose central nervous system still lacks myelin, are already receptive to sensory stimuli. By this fact the assumption, which is quite often stated as being self-evident, that nonmyelinated nerve fibers cannot yet transmit stimuli is refuted. Many investigators have expressed opposing views (see p. 540).

TAINE (1870), who observed a girl from the time of birth, is sure that for this

* These are literal translations of newly coined words, "Nahsinne" and "Fernsinne" – words which do not otherwise exist in the German language – The translators.

infant the outside world consisted only of sounds and color spots that she could not localize. At the age of two and one half months she recognized the direction of certain sounds; at three months she turned her eyes and head to the object she wanted to see. With increasing experience she learned to differentiate five or six voices and as many faces. "At about the age of three months she started to make touching movements with her hands and to move her arms to reach objects and thus to associate tactile and muscular impressions of distance and form with the color spots."

According to K. BÜHLER (1922), from the start the visual and tactile senses have leading roles in mental development. Hearing assumes an important role when speech is learned, but taste and smell lag behind. However, we actually cannot speak of a leading role of vision during the first months of life. Simple observation shows that the infant is sensitive to pressure and pain and withdraws every skin area from the irritating pin point.

Food intake, the infant's most important neural activity next to respiration, is at first completely dependent on the so-called "lower" senses, i.e., the skin senses, especially the tactile and pressure sense, and taste. It is useless to establish an order of rank among these senses. They function whenever they are needed. The stimulation of taste causes strong sucking movements during hunger (see p. 472); after the meal there may occur at best a few weak sucking movements. Quite often additional food is absolutely rejected. The skin senses support the food intake in a similar way. Mechanical stimulation of the oral region causes the rooting reflex and sucking movements of the hungry infant but refusal of the proffered food in the satiated baby.

The "higher" senses, vision and hearing, can also be stimulated from the start, but they have only little influence on the infant's behavior at first and do not serve the food intake function. During the development of the child the spatial observations of the eye will be adjusted to the body and supervised by the skin senses (see p. 79). With the aid of the higher senses involuntary conditioned reflexes of food intake are gradually formed (see p. 450). The first understanding of language and the first spoken words are acquired in a similar way. Thus, progressive development furthers particularly the influence of vision and hearing, which more and more supersede the functions of the lower senses. It is just as difficult to establish an order of rank between eye and ear in the infant as it is in the adult. In man the olfactory sense remains in the background from the beginning.

Speculations on the relations between eye and ear were made long ago. LEONARDO DA VINCI (1452–1519) says:

"Who would not want to lose hearing, smell, and touch rather than vision? For he who loses vision is expelled from the world; for he cannot see the world anymore, and such a life is a sister to death."

GÜTTICH (1944) supplements LEONARDO:

"The eye is the leading sense for mature man as an individual being, but the ear is most important for the development of the individual as well as the whole human species. For the inner man with all his thoughts and feelings is based on human language. Language is the basis of the mental development of the human race. It is dependent on the ear and hearing. There are no concepts without words; language is the original basis of the human mind. We think in words that have become concepts. Without language there is no community, no state, no philosophy, no religion, and no art."

In the order of rank of the senses, the position and movement sense is missing because the recipients of its stimuli (the inner ear; the pressure sense organs of the skin; the proprioceptors of the muscles, tendons, and joints; and the eyes) are so varied that they cannot be compared to the other senses. One phenomenon can already be found by direct observation: The eyes, which are later effective in maintaining equilibrium, are at first of no importance for this function and accept this task only in the course of development. One of the consequences of the later development of the position and movement sense is the fact that the infant, who vomits so easily, does not get seasick.

The higher senses developed phylogenetically after the lower senses, i.e., in the same way as in the infant. Hearing developed last; in the vertebrates distinct hearing organs can only be found in increasing perfection ascending from the Amphibia (ECKERT-MÖBIUS, 1959).

The sequence in which the infant's senses affect the nervous system has a deeper meaning: for the baby sucking on the breast of his mother, only the events and objects in his immediate surroundings are vital; those farther away are of no concern. Thus, the lower senses develop first; they are, perhaps with the exception of the temperature sense, stimulated only by immediate touch and therefore transmit messages only from the immediate surroundings. The higher senses, which bring messages from far away, start to function later.

According to VON UEXKÜLL (1949), proximity is characterized by an increasingly dense protective wall of the senses. Touch, smell, hearing, and seeing enclose man like a gown the four layers of which become increasingly thinner toward the outside. We call this island of the senses his outside world. For man all distant objects are only seen; if they approach, they become heard, then smelled, and finally touched objects. Man can even take these objects, all of which have sensory qualities, to his mouth and thus make them tasted objects. The infant first conquers proximity with his lower senses and later distance with his higher senses.

As important as the physiological concept of the senses may be, this is not sufficient to explain the infant's behavior under the influence of sensory stimuli. The innate releasing mechanisms play a most important role (see p. 594).

SUMMARY

Sensory stimuli induce both uniform general reactions that are identical upon stimulation of different sensory organs and localized reactions in the stimulated sensory organs.

The pain sense is well developed in the newborn. The reactions have the purpose of withdrawing the irritated skin area from the stimulus. They are subsumed under the designation "protective skin reflexes." At first the face is the most sensitive area. The ability to scratch oneself is not innate but develops gradually during the first twelve months. Pain sensitivity to electric currents is remarkably low in the first years of life.

The newborn is very sensitive to tactile and pressure stimuli. A series of reactions characteristic of infancy, such as sucking movements and the tonic palmar and plantar grasp reflexes, are elicited in this manner.

The newborn is sensitive to warm and cold stimuli, even without being directly touched. Locally circumscribed stimulation will evoke turning of the stimulated body part toward or away from the stimulus and general responses, such as restlessness, fright reactions, grimacing, and crying. On local stimulation of the cheek with a warm stimulus the rooting reflex may occur (see p. 404); this vital reaction, usually elicited by touch, is also secured by warmth. It is found more often in the hungry than in the satiated infant.

The newborn can already clearly distinguish sweet from salty, bitter, and sour substances, as can be noted by his facial expression. There is no proof that he can differentiate between salty, sour, and bitter substances. Taste disturbances may occur in severe rachitis, atrophy, mental retardation, and long-lasting malnutrition.

Olfactory stimuli may irritate the olfactory nerve and the trigeminus. When the trigeminus is stimulated, KRATSCHMER's reflex (respiratory arrest) occurs in the infant. In addition, many newborns respond to odors by turning toward them but more often with defensive movements. It has not been proved that they can find the mother's nipple with the aid of the olfactory sense. At seven to nine months conditioned reflexes to certain odors can be obtained.

The comparatively large eyes of the newborn are already anatomically functional. The lids close on stimulation by light.

The pupil reacts to light from the beginning. It dilates on pain or sound stimuli even in the premature and newborn, but not with regularity. Pupillary fluctuation is still absent during the first months of life.

The average blinking frequency increases during childhood. The coordination of lid movements develops during the first months. The blink reflex appears as a conditioned reflex during the first year of life.

The eye-neck reflex, which depends on the degree of brightness of the light source, has been observed only in the young infant, not in the animal.

Under ordinary conditions the newborn is unable to fixate on objects or to follow them with his eyes. This ability develops gradually during the first months of life and is well established at about the third to fifth month. Moderate light attracts the eyes of the infant.

PURKINJE's phenomenon can be proved even in prematurely born infants. The retinal sensitivity is considerably increased during dark adaptation.

In all the developmental stages of normal and premature newborns during the first ten weeks of life the relative brightness values of monochromatic spectral lights during light adaptation correspond to those of the adult. During dark adaptation, however, they differ significantly from those in the adult, and they become equal only after a developmental period which may last to ten weeks in the prematurely born. The final twilight values are first noted in the spectral center (yellow and yellowish-green), then at the long-wave end (red), and finally at the short-wave end of the spectrum (bluish-green, blue, and bluish-purple). Phylogenetically and ontogenetically the function of the cones reaches maturity prior to that of the rods.

The newborns do not differentiate between colors in the first two weeks of life. In the third or fourth week gray can be distinguished from equally bright blue, in the fourth to eighth week red from yellow, and finally in the seventh to eleventh week green is recognized.

The presence of all known forms of incomplete and complete color blindness can be ruled out by the beginning of the third year of life.

In acquiring spatial perceptions in the early time of life, vision conforms to the other senses, especially those of touch and pressure. This explains the "erect" perception of the upside-down retinal image. Originally this image is unrelated to the body concept of the infant. Increasing experience coordinates the image with the other sensory impressions.

Normal and premature newborns are not deaf. Responses to suddenly occurring sound stimuli include blinking, change in respiratory rhythm, the fright reaction, and general motor restlessness. A temporary inhibition follows, making the infant insensitive to further sound stimuli. Even the unborn child responds to strong sound stimuli with fright reactions, which can be observed from the outside, and he also is temporarily insensitive to further sound stimuli.

Some newborns are already able to turn their eyes or heads toward the source of sound. In general, this ability develops at the age of several months.

It is doubtful whether newborns cán already form conditioned reflexes to sound stimuli. The infant can do this after several months. The first words are acquired as conditioned reflexes or, rather, as conditioned connections.

Reflexes can be elicited in the newborn from all senses without the mediation of the cerebral hemispheres.

The behavior of the newborn is determined by the lower senses, which respond to nearby stimuli, particularly the tactile sense and taste. Their influence is later superseded by the higher senses, which are stimulated from a distance, i.e., eye and ear, but they do not disappear.

REFERENCES

General Reactions

BAGLIONI, S.: Zbl. Physiol. *7*, 177 (1907); Erg. Physiol. *11*, 526 (1911).
BROWN, T. G.: Erg. Physiol. *15*, 480 (1916).
CANESTRINI, S.: Über das Sinnesleben des Neugeborenen (The Sensory Functions of the Newborn). Berlin, 1913.
CEMACH, A.: Beitr. Anat. usw. Ohr *14*, 1 (1920).
CLARKE, F. M.: J. Genet. Psychol., Worcester *55*, 415 (1939).
DRESEL, K.: Erkrankungen des vegetativen Nervensystems, im Handbuch der inneren Krankheiten von Kraus-Brugsch, 1921 (Diseases of the Vegetative Nervous System, in the Textbook of Internal Diseases by Kraus-Brugsch, 1921).
GILDEMEISTER, M.: Pflügers Arch. *162*, 489 (1915); *197*, 432 (1922).
HUNT, W. A.: J. Genet. Psychol., Worcester *55*, 215 (1939).
– F. M. CLARKE, and E. B. HUNT: J. Psychol. *2*, 201 (1936).
– and C. LANDIS: Psychol. Rev. *44*, 267 (1938).
JONES, H. E.: According to M. G. Jones, in Murchison. p. 277.

KELLAWAY, P., and B. J. FOX: J. Pediatr., St. Louis *41*, 262 (1952).
LANDIS, C., and W. A. HUNT: The Startle Pattern. New York, 1939; according to Pratt in Carmichael, p. 233.
MAYR, F.: Jb. Kinderhk., Vienna *5*, 118 (1862).
MORO, E.: Münch. med. Wschr. *1918*, 1147; *1920*, 300.
MÜLLER, K.: Über Reflexe im Säuglingsalter (Reflexes in Infancy). Diss. Munich, 1911.
SIEFERT, E.: Pflügers Arch. Physiol. *64*, 321 (1896).
STRAUB, W.: Biochem. Zschr. *54*, 156 (1913).
STRAUSS, H.: J. Psychol. *38*, 111 (1929).
STRÖDER, J., and B. SANDEN: Ann. Paediatr. *184*, 374 (1955).
WATSON, J. B.: Psychology from the Standpoint of a Behaviorist. 2nd ed. Philadelphia and London, 1924.
WIESER, S. T., and K. DOMANOVSKY: Arch. Psychiatr. Nervenkrankh. *198*, 257 and 267 (1959); Dtsch. Z. Nervenhk. *175*, 520 (1957).
– K. DOMANOVSKY, and G. HEINEN: Arch. Kinderheilk. *155*, 17 (1957).
– Wien. med. Wschr. *1958*, 1105.
WOLFF, P. H.: Psychosomatic Med. New Series *21*, p. 110 (1959).
YERKES: J. Comp. Psychol. *15* (1905), after Hempelmann.
YLPPÖ, A.: Zschr. Kinderhk. *20*, 283 (1919).

Skin Sensations

BECKER, H.: In Peter Wetzel u. Heiderich, Handbuch der Anatomie des Kindes (in Peter Wetzel's and Heiderich's Handbook of the Anatomy of the Child). Vol. 2., p. 221. Munich, 1922.
– in Pfaundler-Schloßmanns Handbuch der Kinderheilkunde (in Pfaundler-Schloßmann's Handbook of Pediatrics). Vol. 10, p. 22. Berlin, 1935.
– in Brocks Biologischen Daten für den Kinderarzt (in Brock's Biological Data for the Pediatrician). 2nd ed. *2*, 1067. Berlin, Göttingen, Heidelberg, 1954.
FREY, M. VON: Im Handbuch der normalen u. pathologischen Physiologie (in Handbook of Normal and Pathological Physiology). Vol. 11, p. 94. Berlin, 1926.
GOLDSCHEIDER, A.: ibid. pp. 131, 181.
HENSEL, H.: Erg. Physiol. *47*, 166 (1952); in H. PRECHT, J. CHRISTOPHERSEN and H. HENSEL, Temperatur und Leben (Temperature and Life). p. 375. Berlin, Göttingen, Heidelberg, 1955.
JALOWY, B.: Zschr. Anatomie *109*, 344 (1939).
THÖLE: Neurol. Zbl. *31*, 610 (1912).

Pain

ANDRÉ-THOMAS: Études neurologiques sur le nouveau-né (Neurologic Studies on the Newborn). pp. 57, 98. Paris, 1952.
BÜHLER, C., and L. SPIELMANN: Zschr. Psychol., Leipzig *107*, 3 (1928).
CZERNY, A.: Jb. Kinderhk. *33*, 1 (1892).
GENZMER, A.: Untersuchungen über das Seelenleben des neugeborenen Menschen (Investigations on the Psychic Functions of the Newborn). Diss. Halle, 1873.
GOLTZ, F.: Beiträge zur Lehre von den Funktionen der Nervenzentren des Frosches (Essay on the Theory of the Functions of the Nerve Centers of the Frog). Berlin, 1869.
HARTMANN-KARPLUS, D.: Jb. Kinderhk. *132*, 140 (1931).
JENNINGS, H. S.: Das Verhalten der niederen Organismen (Behavior of Lower Organisms). Leipzig and Berlin, 1910.
KRONER, F.: Breslauer ärztl. Zschr. *4*, 37 (1882).
PETROSELLI, F.: Ref. Zschr. Kinderforsch. *35*, 83 (1929).
REMPLEIN, H.: Die seelische Entwicklung in der Kindheit und Reifezeit (Psychological Development in Childhood and Adolescence). 3rd ed., p. 78. Munich and Basel, 1952.
SHERMANN, M., and I., and C. D. FLORY: Comp. Psychol. Monogr. (Am.) *12*, 4, 1 (1936).

SZYMANSKI, I. S.: Psychol. Forsch. *2*, 298 (1922).
TUMLIRZ, O.: Anthropologische Psychologie (Anthropological Psychology). 2nd ed., p. 105. Basel, 1955.
WOLOWIK, A. B.: Jb. Kinderhk. *115*, 185 (1927).

Touch and Pressure

ANDRÉ-THOMAS, CHESNI, and AUTGAERDEN: Presse méd. *1954*, 41.
CAMERER, W.: Zschr. Biol. *17*, 1 (1889).
CAREDDU, C.: Boll. Soc. ital. Pediatr. *4*, 238 (1935).
CZERMAK, J. N.: Ges. Schriften (Collected Writings). Vol. 1, Sect. 1, p. 303. Leipzig, 1879.
EBBECKE, U.: Mschr. Kinderheilk. *99*, 401 (1951).
ESCHERICH, T.: Münch. med. Wschr. *1888*, 687.
FREY, M. VON: Handbuch der norm. und path. Physiol. (Textbook of Normal and Pathological Physiology). Vol. 11, p. 126. Berlin, 1926.
FRIEDMANN, P.: Zschr. Physiol., Leipzig *103*, 185 (1927).
HARTMANN-KARPLUS, D.: Jb. Kinderhk. *132*, 140 (1931).
HOGG, J. D.: J. Comp. Neur. *75*, 371 (1941).
HOOKER, D.: Proc. Ass. Res. Nerv. Ment. Dis. *33*, 98 (1954).
MORO, E.: Wien. klin. Wschr. *1906*, 377.
NASSAU, E.: Jb. Kinderhk. *151*, 46 (1938).
PREYER, W.: Die Seele des Kindes (The Soul of the Child). 4th ed., p. 65. Leipzig, 1895.
ROSENBAUM, S.: Dtsch. med. Wschr. *1958*, 1128.
STIRNIMANN, F.: Rev. frç. Pédiatr. *13*, 496 (1937); Schweiz. Zschr. Psychol. *3*, 245 (1944); Annal. pädiatr. (Basel) *157*, 17 (1941); Das Kind und seine früheste Umwelt (The Child and his Earliest Environment). Psychol. Praxis. H. 6, p. 41. Basel, 1947.
STRÜMPELL, A. V.: Dtsch. med. Wschr. *1904*, 1411.
VIERORDT, K.: In Gerhardts Handbuch der Kinderkrankheiten (in Gerhardt's Handbook of Pediatrics). Vol. 1. Tübingen, 1877.

Temperature Sensation

BLANTON, M. G.: Psychol. Rev. *24*, 456 (1917).
CRUDDEN, C. H.: J. Exper. Psychol. *20*, 350 (1937).
HERRMANN, M.: Mschr. Kinderhk. *51*, 49 (1932).
JENSEN, K.: Genet. Psychol. Monogr. (Am). *12*, 361 (1932).
KAWURA: Münch. med. Wschr. *1944*, 40.
PRATT, K. C.: in Murchison. p. 163.
– A. K. NELSON and K. H. SUN: The Behavior of the Newborn Infant. Ohio State Univ. Studies, 1930.
PREYER, W.: Die Seele des Kindes (The Soul of the Child). 4th ed., p. 75. Leipzig, 1895.
SIKORSKI, I. A.: Die seelische Entwicklung des Kindes (The Psychological Development of the Child). 2nd ed. Leipzig, 1908.
STIRNIMANN, F.: Zschr. Kinderpsychiatr. *5*, 1 (1939).
– Psychologie des neugeborenen Kindes (Psychology of the Newborn). Zürich and Leipzig, 1940.

Taste

BROUWER, B.: Zschr. Neurol. *32*, 164 (1916).
BÜSSEM, H.: Jb. Kinderhk. *39*, 166 (1895).
CZERNY, A.: Rachitis, im Handbuch von Kraus-Brugsch ("Rickets" in Handbook of Kraus-Brugsch). Berlin and Vienna, 1921.
ECKSTEIN, A.: Zschr. Kinderhk. *45*, 1 (1927).
GAMPER, E.: Zschr. Neurol. *102*, 154; *104*, 49 (1926).
GIBSON, J. P.: Pediatrics *52*, 404 (1958).
GOLDSCHMIDT, H.: Zschr. Kinderhk. *45*, 28 (1927).

HEUBNER, O.: Charité-Ann. *33* (1909).
JENSEN, K.: Genet. Psychol. Monogr. (Am.) *12*, 5/6 (1932).
JÖRG, J. C. G.: Handbuch zum Erkennen und Heilen der Kinderkrankheiten (Handbook of Diagnosis and Treatment of Childhood Diseases). p. 125. Leipzig, 1826.
KIESOW, F.: Wundts philosoph. Studien *10*, 342 (1894).
– Zschr. Psychol., Physiol., Sinnesorg., Leipzig *36*, 90 (1904).
KRASNOGORSKI, N.: Erg. inn. Med. *39*, 613 (1931).
KULAKOWSKAJA, E.: Zbl. Kinderhk. *23*, 434 (1930).
KUSSMAUL, A.: Über das Seelenleben des neugeborenen Menschen (Psychology of the Newborn). Leipzig, 1859.
LICHTENSTEIN, A.: Jb. Kinderhk. *37*, 76 (1894).
MAGENDIE, F.: Lehrbuch der Physiologie (Textbook of Physiology). Vol. 1, p. 109. German translation. Tübingen, 1826.
NEUMANN, H.: Arch. Kinderhk. *41*, 155 (1896).
PREYER, W.: Die Seele des Kindes (The Soul of the Child). 4th ed., p. 80. Leipzig, 1895.
SKRAMLIK, E. VON: Handbuch der niederen Sinne (Handbook of the Lower Senses). Leipzig, 1926; Im Handbuch der norm. und patholog. Physiologie (in Handbook of Normal and Pathological Physiology). Vol. 11, p. 306. Berlin, 1926.
STERNBERG, W.: Zschr. Psychol., Physiol., Sinnesorg., Leipzig *27*, 77 (1907).
STIRNIMANN, F.: Jb. Kinderhk. *146*, 211 (1936); Rev. frç. Pédiatr. *12*, 453 (1936).
URBANTSCHITSCH, V.: Beobachtungen über Anomalien des Geschmackes (Observations on the Anomalies of Taste). p. 10. Stuttgart, 1876.
WETZEL, G.: In Heiderich-Peter-Wetzels Handbuch der Anatomie des Kindes (in Heiderich Peter-Wetzel's Handbook of the Anatomy of the Child). Vol. 1, p. 669. Munich, 1938.

Smell

(also see under *Taste*)

ALLESCH, G. H. VON: S. ber. Akad. Wiss. Berlin *1921*, II, 672; Naturw. *1921*, 774.
BIFFI: according to Canestrini. p. 87.
BURTON, M.: Die Kindheit der Tiere (The Childhood of Animals). p. 119. Hamburg, 1957.
CIURLO, L.: Valsalva (It.) *10*, 22 (1934).
DARWIN, C.: Kleinere Schriften (Lesser Writings) (1877). Published by E. Krause. Vol. 2., p. 134. Leipzig, 1886.
DISHER, D. R. (1934): according to K. C. Pratt, in Carmichael. p. 240. 1954.
KRASNOGORSKI, N.: Jb. Kinderhk. *78*, 373 (1913).
KRATSCHMER, F.: S. ber. Akad. Wiss. Wien, Math.-naturwiss. Kl. *62*, Part II, p. 147 (1870).
KRONER, F.: Breslauer ärztl. Zschr. *4*, 37 (1882).
PEREZ, G.: La psychologie de l'enfant (The Psychology of the Child). 6th ed., 1902.
PETER, K.: Heiderich-Peter-Wetzels Handbuch der Anatomie des Kindes (Heiderich-Peter-Wetzel's Handbook of the Anatomy of the Child). Vol. 2, p. 218. Munich, 1929.
STERN, W.: Psychologie der frühen Kindheit (Psychology of Early Childhood). 3rd ed., p. 43. Leipzig, 1923.

Vision

AMBRON, H., and H. HELD: Arch. Anat. u. Physiol. *1896*, 203.
ANDRÉ-THOMAS and SAINT-ANNE DARGASSIES: Études neurologiques (Neurologic Studies). p. 78. Paris, 1952.
– and S. AUTGAERDEN: Psycho-affectivité des premiers mois (Psychoaffectivity of the First Months of Life). pp. 23, 96. Paris, 1959.
BACH, L.: Pupillenlehre (The Pupil). p. 72. Berlin, 1908.
BADTKE, G.: In K. Velhagen, Der Augenarzt (in Velhagen's The Ophthalmologist). 1, 353. Leipzig, 1958.
BALDWIN, I. M.: Die Entwicklung des Geistes (The Development of the Mind). German translation. Berlin, 1898.

BÁRÁNY, R.: Acta otolaryng., Stockholm *1*, 97 (1918).
– Arch. Augenhk. *88*, 139 (1921).
BARTELS, N.: Zschr. Augenhk. *12*, 638 (1904).
– Arch. Ophth. *76*, 1 (1910).
– Dtsch. med. Wschr. *1932*, 1477.
BARTH, H.: Zschr. Hals-Nas.-Ohr.kh. *35*, 1 (1933).
BERBERICH, I., and A. WIECHERS: Zschr. Kinderhk. *38*, 59 (1924).
BERKELEY, G.: Works. I. Dublin, 1784. A New Theory of Vision; see also J. St. Mill: Dissertations and Discussions. *2*, 84. London, 1868. Bayley on Berkeley's Theory of Vision.
BERNFELD, S.: Psychologie des Säuglings (Psychology of the Infant). Vienna, 1925.
BERNHEIMER, S.: Über die Entwicklung und den Verlauf der Markfasern im Chiasma (The Development and Pathways of the Myelinated Fibers in the Chiasma). Habilit.-Schrift. Heidelberg, 1889.
BINET, A.: Rev. philosoph. *30*, 582 (1890).
BIRUKOW, G.: Zschr. vergl. Physiol. *31*, 322 (1949).
BOLAFFIO, G., and G. ARTOM: Ref. Zbl. *38*, 418 (1924); Zschr. ges. Neurol. *103*, 320 (1926).
BUFFON: Allgemeine Naturgeschichte (General Natural Science). German translation. Part 5. Berlin, 1773.
BÜHLER, K.: Die geistige Entwicklung des Kindes (The Mental Development of the Child). 3rd ed. Jena, 1922.
BUMKE, C.: Die Pupillenstörungen (Pupillary Disturbances). Jena, 1904.
CATEL, W.: Mschr. Kinderhk. *38*, 303 (1928).
– Dtsch. med. Wschr. *1932*, 997, 1478.
– Zoolog. Garten N. F. *16*, 244 (1950); *18*, 19 (1951).
CHASE, W. P.: J. Exper. Psychol. *20*, 205 (1937 I).
COMPAYRÉ, G.: Entwicklung der Kinderseele (Development of the Child's Psyche). German translation. Altenburg, 1900.
CONDILLAC, E. B.: Abhandlung über die Empfindungen (Treatise on the Sensations), (1754). German translation. Berlin, 1870.
CUIGNET, L.: Ann. d'oculist. *66*, 117 (1871).
DARWIN, C.: Kleinere Schriften (Lesser Writings) (1877). Published by E. Krause. Vol. 2, p. 134. Leipzig, 1886.
EDGERTON, A.: Arch. Ophth. (Chicago) *11*, 838 (1934).
ELSCHNIG, A.: Zschr. Augenhk. *11* (1904).
ELSENHANS, T.: Lehrbuch der Psychologie (Textbook of Psychology). 3rd ed. Tübingen, 1939.
ESENTE, I.: Physiologie de la vision chez le prématuré et le nourrisson normal (Physiology of Vision in the Premature and the Normal Infant). Paris, 1958.
ESPINAS, Observation sur un nouveau-né (Observation on a Newborn). Ann. de la faculté des lettres de Bordeaux, 1883.
EWERT, P. H.: Genet. Psychol. Monogr. (Am.) *7*, 177 (1930).
FICK, A.: Pflügers Arch. Physiol. *47* (1890).
FLECHSIG, P.: Die Leitungsbahnen im Gehirn und Rückenmark des Menschen (The Pathways in the Brain and Spinal Cord of Man). Leipzig, 1876.
FRISCH, K. v.: Die Sprache der Bienen (The Language of Bees). Berlin, 1924.
GARBINI, A.: Arcivio per l'Antropol. *24*, 71 (1894).
GESELL, A., F. ILG and G. BULLIS: Vision. Its Development in Infant and Child. p. 190. New York, 1949.
GLEISS, J., and H. PAU: Klin. Mbl. Augenhk. *121*, 440 (1952).
GRETHER, W. F.: J. Comp. Psychol. *29*, 167 and 187 (1940); J. Exper. Psychol. *26* I, 394 (1940 I).
GUDDEN, H.: Münch. med. Wschr. *1910*, 405.
GUERNSEY, M.: Psychol. Bull. *26*, 160 (1929).
GUILLAUME, P.: J. de Psychol. norm et pathol. *21*, 112 (1924).

GUTMANN, N. I.: Arch. ges. Psychol. *47*, 108 (1924).
HELMHOLTZ, H.: Handbuch der physiologischen Optik (Handbook of Physiological Optics). 1st ed. 1867, 2nd ed. Hamburg and Leipzig, 1896. p. 599, 3rd. ed. (with appendices by Gullstrand, von Kries and Nagel). 1909–1911.
HERING, E.: Die Lehre vom binokularen Sehen (Theory of Binocular Vision). Leipzig, 1868.
HESS, C. VON: Arch. Augenhk. *1916*, 80.
HEUBNER, O.: Erg. inn. Med. *16*, 1 (1919).
HOLDEN, W. A., and K. K. BOSSE: Arch. of Ophthalm. *29*, 261 (1900).
HYDÉN, H.: Die Chemie und der Stoffwechsel des Nervengewebes (The Chemistry and Metabolism of Nervous Tissue). 3. Kolloquium d. Gesellsch. für physiol. Chemie in Mosbach/Baden. Berlin, 1952.
IRWIN, O. C., and L. A. WEISS: Studies in Infant Behavior I. Univ. Iowa Studies Child Welfare *9*, 4, 139 and 165 (1934).
JASCHKE, R. T. VON: Physiologie, Pflege und Ernährung des Neugeborenen (Physiology, Care and Nutrition of the Newborn). p. 110. Wiesbaden, 1917.
JASTROW: according to Kahn.
KASAHARA, M., and St. INAMATSU: Arch. Kinderhk. *92*, 302 (1931).
KASSATKIN, N. I.: see p. 536.
KESTENBAUM, A.: Gräfes Arch. Ophth. *124*, 113 (1930).
KNORR, A.: Würzbg. Abh. N. F. *5* (1929).
KÖDDING, I.: Mschr. Kinderhk. *84*, 212 (1940).
KÖHLER, W.: Aus der Anthropoidenstation auf Teneriffa (The Anthropoid Station in Teneriffe). II.–IV. Abh. Preuß. Akad. Wiss., Berl., Physik.-math. Kl. 1915–1918.
KRASNOGORSKI, N.: Jb. Kinderhk. *78*, 373 (1913).
KRIES, G. v.: in Helmholtz.
– Zschr. Physiol. Sinnesorg. *15*, 247 (1897).
– Immanuel Kant. p. 12. Berlin, 1924.
KRONER, F.: Breslauer ärztl. Zschr. *4*, 37 (1882).
KRÜGER, U.: Über die Art der Wahrnehmung eines künstlich verkehrt gemachten Gesichtsfeldes (The Characteristics of Perception in Artificially Reversed Vision). Diss. Breslau, 1939.
LADD-FRANKLIN, C.: Zschr. Psychol. Physiol. *4*, 211 (1893).
LANGE, F.: Die Sprache des menschlichen Antlitzes (The Language of the Human Face). Munich, 1937.
LINDER: according to Kahn.
LING, B. C.: J. Genet. Psychol. *61*, 227 (1942).
LÖHLEIN, W.: Jenaer akadem. Reden. H. 13. Jena, 1931.
LOTZE, R. H.: Medizinische Psychologie (Medical Psychology). p. 362. Leipzig, 1852.
MCGINNIS, H. M.: Genet. Psychol. Monogr. (Am.) *8*, 4 (1930).
MAGITOT, A.: Ann. d'ocul. *141*, 161 (1909).
MAGNUS, H.: Tafeln zur Erziehung des Farbsinnes (Plates for the Training of the Color Sense). Breslau, 1879.
MAGENDIE, F.: Lehrbuch der Physiologie (Textbook of Physiology). Vol. 1, p. 76. German translation. Tübingen, 1826.
MARSDEN, R. E.: Rev. Psychol. *10*, 37 (1903).
METZGER, E.: Gräfes Arch. Ophth. *127*, 296 (1931).
MEUMANN, E.: Vorlesungen zur Einführung in die experimentelle Pädagogik (Lectures on the Introduction to the Experimental Pedagogic). 2nd ed. Leipzig, 1911.
MINKOVSKI, M.: In Abderhaldens Handbuch der biol. Arbeitsmethoden (in Abderhalden's Handbook of Biological Methods). V 5 B, Lief. 253, p. 558. Berlin and Vienna, 1928.
MORGAN, S. S., and J. J. B.: J. Pediatr. *25*, 168 (1944).
MÜLLER, J.: Handbuch der Physiologie des Menschen (Handbook of the Physiology of Man). Vol. 2, p. 357. Koblenz, 1837.
MUNN, M. L.: Psychological Development. Boston, 1938; according to Pratt, in Carmichael. p. 204. 1949.

MYERS, Ref. i. d. Jber. Ophth. *40*, 49 (1909).
NAGEL, W. A.: Bibliotheka zool. H. 18. Stuttgart, 1894.
– Über den Farbensinn der Tiere (The Color Sense of Animals). Wiesbaden, 1901.
NASSAU, E., and A. BORTNICK: Ann. paediatr. Basel *181*, 371 (1953).
OHM, J.: Jb. Kinderhk. *88*, 397 (1918).
OLTUSZEVSKI, W.: Die geistige und sprachliche Entwicklung des Kindes (The Mental and Language Development of the Child). Berlin, 1897.
ORTEL, D.: Über Lidöffnungsreflexe bei Säuglingen (The Lid Opening Reflex in Infants). Diss. Greifswald, 1944.
OSVALDO, S. G.: Ref. Zbl. Kinderheil. *34*, 2 (1938).
PALAGYI, M.: Naturphilosophische Vorlesungen (Lectures on Natural Philosophy). 2nd ed. Leipzig, 1924.
PAVLOV, I.: Works, see p. 568.
PFISTER, H.: Arch. Kinderhk. *26*, 11 (1899).
PREYER, W.: Die Seele des Kindes (The Mind of the Child). 4th ed. pp. 77, 162. Leipzig, 1895.
PURKINJE, J.: Beobachtungen und Versuche zur Physiologie der Sinne (Observations and Experiments on the Physiology of the Senses). 2, 109. Berlin, 1825.
RÄHLMANN, E.: Zschr. Sinnesphysiol. *2*, 53 (1891).
– Ophthth. Klinik. *7*, 321 (1903).
– and WITKOVSKI: Arch. Anat. u. Physiol. *1877*, 454.
RAUDNITZ, R. W.: Jb. Kinderhk. *45*, 145 (1897); *73*, 259 (1911); *87*, 15 (1918).
RÖMER, A.: Med. Korresp. bl. württembg. ärztl. Ld. ver. *61*, 265 (1891).
ROSELLI: Boll. tell. Osped-Oftalmico, Rome, i. 1903; Ref. Ophth. Rev. *22*, 210 (1930).
RUDDER, B. DE: Zschr. Kinderhk. *41*, 555 (1926); *43*, 323 (1927).
SACHSENWEGER, R.: Wissensch. Zschr. Univ. Leipzig *8*, 233 (1959).
SCHENCK, F.: Pflügers Arch. Physiol. *118*, 129 (1907).
SCHINDLER, E.: Zschr. Kinderhk. *19*, 153 (1919).
SCHUH, R.: Über die normalen Pupillenweiten von Kleinkindern und Säuglingen (The Normal Pupillary Diameters in Young Children and Infants). Diss. Erlangen, 1937.
SCHUR, E.: Zschr. Kinderhk. *32*, 227 (1922).
SEEFELDER, R.: In Heiderich-Peter-Wetzels Handbuch der Anatomie des Kindes (in Heiderich-Peter-Wetzel's Handbook of the Anatomy of the Child). Vol. 2, Lief. 1. Munich, 1927.
SHINN, M. W.: Univ. California Publ. *1* (1893–1899) and *4* (1907).
SICHERER, V.: Dtsch. med. Wschr. *1907*, 1564.
SMITH, J. M.: Studies in Infant Behavior III. Univ. Iowa Studies Child Welfare *12*, 1, 93 (1936).
– Child Development *8*, 301 (1937).
SOLTMANN, O.: In Gerhardts Handbuch der Kinderkrankheiten (in Gerhardt's Handbook of Pediatrics). Vol. 5, p. 1. Tübingen, 1880.
SPECIALE-CIRINCIONE: Ann. ottalm. (Pavia) *40* (1917).
STAPLES, R.: J. Exper. Psychol. *15*, 119 (1932).
STERN, A.: Arch. Gyn. *124*, 689 (1925).
– and P. SCHWARTZ: Klin. Wschr. *1924*, 931.
STERN, W.: Psychologie der frühen Kindheit (Psychology of Early Childhood). 3rd ed. Leipzig, 1932.
STIRNIMANN, F.: Ann. paediatr., Basel *163*, 1 (1944); Psychologie des neugeborenen Kindes (Psychology of the Newborn). Zürich and Leipzig, 1940.
SZULINA, A. I.: Ref. Zbl. Ophth. *13*, 460 (1925).
TAINE, H.: De l'intelligence (Intelligence). 2nd ed. *2*, 121. Paris, 1870.
THIELE, W. H.: Zbl. Gynäk. *1947*, 536.
THORNVAL, A.: Acta oto-laryng. Stockholm *2*, 451 (1920/21).

TIEDEMANN, D.: Beobachtungen über die Entwicklung der Seelenfähigkeit bei Kindern (Observations on the Development of Psychological Ability in Children) (1787). Altenburg, 1897.
TRACY, F., and J. STIMPFL: Psychologie der Kindheit (Psychology of Childhood). 4th ed. Leipzig, 1912.
TRENDELENBURG, W.: Der Gesichtssinn (Vision). Berlin, 1943.
TRINCKER, D.: Naturwiss. *41*, 310 (1954); Forsch. and Fortschr. *29*, 12 (1955).
TRINCKER, D. and J.: Graefes Arch. Ophthalm. *156*, 519 (1955).
TYLER, E. T.: according to Shinn.
UEXKÜLL, J. v.: Bedeutungslehre (Theory of Meaning). Bios. Vol. 10, p. 47. Leipzig, 1940.
UFFELMANN: Hygiene des Kindes (Hygiene of the Child). Leipzig, 1881.
VALENTINE, C. W.: Brit. J. Psychol. *6*, 636 (1913/14).
VELHAGEN, K.: Klin. Mbl. Augenhk. *94* (1935).
WALLS, G.: Amer. J. Ophth. *17* (1934).
WEIGL, E.: Ber. Ges. Experim. Psychol. 11. Kongr. p. 182. Jena, 1930.
WEIGT, G.: Die Entwicklung des Gesichtsfeldes beim Säugling (The Development of the Visual Field in the Infant). Diss. Leipzig, 1955.
WEISS, L. A.: Studies in Infant Behavior I. Univ. Iowa Studies Child Welfare *9*, 4, 11 (1934).
WESTPHAL, A.: Arch. Psychiatr. *26*, 1 (1894); *29*, 474 (1897).
WINCH, W. H.: Amer. J. Psychol. *21* (1910).
WOHLMUTH, G., and R. PETRORINI: Mschr. Kinderhk. *106*, 334 (1958).
WOLFF, P. H.: Psychosomat. Med. *21*, 110 (1959).
WOLFRUM, M.: Berl. dtsch. Ophthalm. Ges. 35. Versamml. Heidelberg, 1908. p. 206.
WOOLEY, H. T.: Psychol. Rev. (Am.) *16*, 963 (1909).
WUNDT, W.: Grundzüge der physiologischen Psychologie (Principles of Physiological Psychology). 6th ed. *2*, 734. Leipzig, 1910.
– Grundriß der Psychologie (Principles of Psychology). 12th ed., p. 126. Leipzig, 1914.
ZETTERSTRÖM, B.: Acta paediatr. (Stockh.) *40*, 569 (1951).
ZIPPERLING, W.: Zschr. Kinderhk. *5*, 31 (1913).

Hearing

ALDRICH, C. A.: Amer. J. Childr. *35*, 36 (1928).
ALEXANDER, G.: Zschr. Sinnesphysiol. Abt. 2, *45*, 153 (1911).
– Ohrenkrankheiten, in Pfaundler-Schloßmanns Handbuch der Kinderheilkunde (Diseases of the Ear, in Pfaundler-Schlossmann's Handbook of Pediatrics). Vol. 6, Leipzig, 1912.
ANDRÉ-THOMAS and AUTGAERDEN: Psycho-Affectivité des premiers mois (Psycho-Affectivity of the First Months of Life). p. 102. Paris, 1953.
BERBERICH, I., and A. WIECHERS: Zschr. Kinderhk. *38*, 59 (1924).
BERNARD, J., and L. W. SONTAG: J. genet. Psychol. *70*, 205 (1947).
BLANTON, M. G.: Psychol. Rev. (Am.) *24*, 456 (1917).
BRONSCHTEIN, A. J., and J. P. PETROVA: Pavlov's Zschr. höh. Nerventät. Dtsch. *2*, 441 (1952).
CATEL, W.: Mschr. Kinderhk. *38*, 303 (1928).
CEMACH, A. I.: Beitr. Anat. usw. Ohr. *14*, 1 (1920).
COMPAYRÉ, G.: Entwicklung der Kinderseele (Development of the Child's Psyche). German translation. Altenburg, 1900.
DEMETRIADES, T.: Ref. Zbl. Kinderhk. *16*, 282 (1924).
FORBES, H. S. and H. B.: J. comp. Psychol. *7*, 353 (1927).
FLEISCHER, K.: Zschr. Laryng. *34*, 733 (1955).
FROESCHELS, E., and H. BEEBE: Arch. Otolaryng. (Chicago) *44*, 710 (1946).
– according to Pratt, in Carmichael. p. 211. 1949.
HETZER, H., and B. H. TUDOR-HART: Quellen und Studien zur Jugendkunde (Sources and Studies on Youth) No. 5. Jena 1927.

HEUBNER, O.: Erg. inn. Med. *16*, 1 (1919).
IRWIN, O. C.: J. comp. Psychol. *32*, 233 (1941).
– and L. WEISS: Studies in Infant Behavior I. Univ. Iowa Studies Child Welfare *9*, 4, 139, 165 (1934).
JÖRG, J. C. G.: Handbuch der Kinderkrankheiten (Handbook of Pediatrics). p. 124. Leipzig, 1826.
KASSATKIN, N. J.: Frühe bedingte Reflexe beim Kinde. Aus dem medizinischenSchrifttum der SU (Early Conditioned Reflexes in the Child. In the medical writings of SU). No. 16. Berlin, 1953.
– and A. M. LEVIKOVA: J. genet. Psychol. *12*, 416 (1935); J. Exper. Psychol. *18*, 1 (1935).
KELLOGG, W. N., and L. A.: The Ape and the Child. p. 37. New York and London, 1933.
KÖLLREUTTER: Zschr. Ohrhk. *53*, 123 (1907).
KOLMER, W.: Alexander-Marburg-Brunners Handbuch der Neurologie des Ohres (in Alexander-Marburg-Brunner's Handbook of the Neurology of the Ear). Vol. 1/1. Berlin and Vienna, 1924.
KRASNOGORSKI, N. J.: Jb. Kinderhk. *78*, 373 (1913).
KURODA, R.: Acta Psychol. Kujo (Japan) *1*, 3 (1939).
KUTVIRT, O.: Beitr. Anat. usw. Ohr *4*, 166 (1911); *5*, 247 (1912).
LANGE, W.: Heiderich-Peter-Wetzels Handbuch der Anatomie des Kindes (Heiderich-Peter-Wetzel's Handbook of the Anatomy of the Child). Vol. 2, Lief. 2. Munich, 1929.
LÖWENFELD, B.: Zschr. Psychol., Leipzig *104*, 62 (1927).
MAGENDIE, F.: Lehrbuch der Physiologie (Textbook of Physiology). Vol. 1, p. 95. German translation. Tübingen, 1826.
MAYER, A.: Münch. med. Wschr. *1953*, p. 900.
MEYER, I.: Mschr. Ohrenhk. *46*, 449 (1912).
MOLDENHAUER: Festschr. Credé. Leipzig, 1881.
MÜLLER, J.: Handbuch der Physiologie des Menschen (Handbook of the Physiology of Man). Vol. 2, *438*. Koblenz, 1840.
MUZIO, O.: Ref. Zbl. Kinderhk. *29*, 134 (1934).
NEUBERT, K.: In F. Zöllner, Audiologie (in F. Zöllner's Audiology). p. 31. Stuttgart, 1954.
POLI: Ref. Arch. Ohrenhk. *41*, 82 (1896).
PORTAL, Dtsch. Arch. Physiol. *4*, 640 (1818).
PRATT, C.: J. genet. Psychol. *44*, 96 (1934).
PREYER, W.: Spezielle Physiologie des Embryo (Special Physiology of the Embryo). p. 481. Leipzig, 1885; Die Seele des Kindes (The Soul of the Child). 4th ed., p. 56. Leipzig 1895.
RICHMOND, J. B., H. J. GROSSMAN, and S. L. LUSTMAN: Pediatrics *11*, 634 (1953).
SCHALTENBRAND, G., and O. GIRNDT: Pflügers Arch. Physiol. *209*, 333 (1925).
SACHS, R.: Arch. Ohrenhk. *35*, 28 (1893).
SIGISMUND, B.: Kind und Welt (Child and World) (1856). 2nd ed. Braunschweig, 1897.
SIKORSKI, I. A.: Die seelische Entwicklung des Kindes (The Psychological Development of the Child). 2nd ed. Leipzig, 1908.
SONTAG, L. W.: Amer. J. Dis. Childr. *60*, 621 (1940).
– and R. F. WALLACE: Child Development *6*, 253 (1935); not available to me, cited according to Kellogg and Wenger.
STIRNIMANN, E.: Psychologie des neugeborenen Kindes (Psychology of the Newborn). Zürich and Leipzig, 1940.
STUBBS, E. M.: Studies in Infant Behavior I. Univ. Iowa Studies Child Welfare *9*, 4, 77 (1934).
– and O. C. IRWIN: according to Pratt, in Carmichael. p. 211. 1949.
TUMLIRZ, O.: Anthropologische Psychologie (Anthropological Psychology). 2nd ed., p. 106. Munich, 1955.
VOSS, O.: Acta otolaryng. Stockholm *11*, 73 (1927).
WEISS, L.: Studies in Infant Behavior I. Univ. Iowa Studies Child Welfare *9*, 4, 1 (1934).

WENGER, M. A.: Studies in Infant Behavior III. Ibid., 12, 1, 11 (1936).
WESSÉN, E.: Acta obstetr. *7*, 136 (1928).

General Function of the Senses

ARISTOTLE: Über die Seele (On the Soul). German translation by A. Busse. 2nd ed. 434b., 92. Leipzig, 1922.
– Von der Zeugung und Entwicklung der Tiere (The Procreation and Development of Animals). German translation by H. Aubert and F. Wimmer. Book 5, p. 361. Leipzig, 1860.
BÜHLER, K.: Die geistige Entwicklung des Kindes (The Mental Development of the Child). 3rd ed. Jena, 1922.
BURDACH, K. F.: Die Physiologie als Erfahrungswissenschaft (Physiology as an Empirical Science). 2nd ed. 2, 198. Leipzig, 1838.
CATEL, W., and C. A. KRAUSPE: Jb. Kinderhk. *129*, 1 (1930).
CARUS, F. A.: Nachgelassene Werke (Posthumous Works). Part 1. Psychology. Vol. 1, p. 151. Leipzig, 1808.
DEMOCRITUS, according to W. Durant: Die Geschichte der Zivilisation (The History of Civilization). Vol. 2. Das Leben Griechenlands (Greece). p. 413. Bern, 1947.
ECKERT-MÖBIUS, A.: Zschr. ärztl. Fortbild. *53*, 1254 (1959).
FLECHSIG, P.: Die Leitungsbahnen im Gehirn und Rückenmark des Menschen (The Pathways in the Brain and Spinal Cord of Man). Leipzig, 1876.
GAMPER, E.: Zschr. Neurol. *102*, 154; *104*, 49 (1926).
GÜTTICH, A.: Neurologie des Ohrlabyrinths (Neurology of the Labyrinth). Leipzig, 1944.
KOFFKA, K.: Die Grundlagen der geistigen Entwicklung (The Basis of Mental Development). 2nd ed. p. 91. Osterwieck, 1925.
LEONARDO DA VINCI: Das Buch von der Malerei (The Book on Painting). p. 12, German translation. Vienna, 1882.
LÖBISCH, I. E.: Entwicklungsgeschichte der Seele des Kindes (Ontogenesis of the Soul of the Child). p. 96. Vienna, 1851.
REMPLEIN, H.: Die seelische Entwicklung in der Kindheit (Psychological Development in Childhood). p. 79. Munich, Basel, 1952.
STERN, W.: Psychologie der frühen Kindheit (Psychology of Early Childhood). 3rd ed. p. 45. Leipzig, 1923.
THOMAS AQUINAS: Summa theologica. Vol. 6 (91, 3). Munich and Heidelberg, 1941.
TUDOR-HART, B.: Zschr. ärztl. Fortbild. *51*, 937 (1957).
UEXKÜLL, J. v.: Nie geschaute Welten (Worlds Never Seen). 9th–12th ed. p. 10. Berlin and Frankfort on the Main, 1949.
WESTPHAL, A.: Arch. Psychiatr. *26*, 1 (1894); *29*, 474 (1897).

3. DEVELOPMENT OF THE FACIAL EXPRESSION FROM THE SENSORY FUNCTIONS

(1, 54, 55, 60)

THEORIES OF EXPRESSION

The most important investigations on the facial expression of the adult are those of BELL (1872), PIDERIT (1919), DARWIN (1872), WUNDT (1877), KRUKENBERG (1920), and LERSCH (1932). The facial expression of the sick was described by KILLIAN (1934) and FERVERS (1935). DARWIN, KRUKENBERG, and FERVERS also studied the facial expression of children; in KRUKENBERG's book particularly there are many good illustrations. PREYER (1895) described the facial expressions of his own child during the first year of life in great detail; MAYR (1858), SOLTMANN (1887), PFAUNDLER (1906), FEER (1921), and RACH (1932) described the facial expressions of sick children. In KILLIAN's monograph on the *Facies dolorosa* there are some illustrations of the facial expressions of children. The anatomy of the facial muscles was first investigated by DUCHENNE by means of electric stimulation of the various muscles; additional anatomic studies were done by VIRCHOV (1908) and BRAUNS (1921). Histories of the various theories on facial expression were written by KRUKENBERG, K. BÜHLER (1933), and SOMBART (1938).

It would seem appropriate to use for comparison the facial expressions of apes or monkeys and other animals, as has already been done by DARWIN. However, so far there are no detailed and thorough investigations taking into account the immediate sensory reactions. In HEMPELMANN's book (1926) there are several quite good illustrations of expressive movements of chimpanzees.

As far as I know, the first attempt to explain the development of facial expressions was made by CARUS (1831). He did so on the basis of a "philosophical anatomy" which attributes a symbolic significance to the various parts of the human body. For example, the lacrimal glands are facsimiles of the gall-producing liver: severe distress causes a flow of gall, but moderate suffering such as grief and sorrow cause lacrimation. CARUS similarly explained the expressive movements of laughter, anger, and other emotions. In this part of CARUS' psychology, which can be only briefly mentioned here, we can easily recognize the philosophy of romanticism, which by that time had penetrated the natural sciences and consequently medicine. It is instructive to compare this theory with a statement made at the same time by JOHANNES MÜLLER (1840) in his famous textbook: "The bases of the relations of facial muscles to specific emotions are completely unknown."

Still, since the time this question was raised, it has not come to rest. PIDERIT (1867) was the first to recognize the positive connections between facial expressions and sensory impressions. According to him, the facial expressions evoked by ideas relate to "imaginary sensory impressions," which have the same characteristics as the sensory impressions caused by external sensory stimuli. The facial expressions accompanying pleasant or unpleasant ideas facilitate the reception of harmonious (pleasant) sensory effects and render the reception of nonharmonious (unpleasant) sensory impressions more difficult. The effect of an idea depends on the degree of its pleasantness or unpleasantness and on the suddenness with which it appears. With a knowledge of these mechanisms PIDERIT created a basis for further research even if he himself did not implement his ideas in all details. Shortly afterward GRATIOLET (1873) presented a similar point of view.

SPENCER (1855) tried to explain the remarkable fact that emotions are specifically reflected in the facial musculature. He did this in the following manner: When a weak wave of nervous irritation passes through the nervous system, it acts with the most strength where the muscles have to overcome the least resistance. Since the facial muscles are relatively small and are attached to easily movable parts, the effect on the face becomes very clear and distinct. Nowadays, of course, we may give another explanation: The most important sense organs are located in the face, and their reactions cause facial expressions.

DARWIN's work (1872) excels in a great many individual observations, but he does not do justice to PIDERIT and views facial expressive movements as remnants of reactions that originally had been purposeful in the course of evolution. He states that three laws are significant in the development of the most important facial expressive movements:

1. The effect of purposefully associated habits. Movements suitable for satisfying a desire or facilitating a sensation are frequently repeated and thus become habits. Eventually they are performed, whether useful or not, as soon as the same desire or the same sensation is even weakly felt.

2. The effect of contrast. According to the first law, certain psychological states cause habitually fixed or predetermined actions. In an opposite psychological state there appears a strong spontaneous inclination to perform movements of the opposite type, which are independent of their usefulness.

3. The immediate effect of the stimulated nervous system on the body. The reactions in this group are dependent on the condition of the nervous system, but are independent of free will and usually also of habit (e.g., effect on the heart action).

According to WUNDT (1877), similar emotions and sensations are combined with each other, for example, the sensation of esthetic or moral satisfaction with that which is sensually pleasant. The inner psychological states such as fear, sorrow, joy, etc. are linked not only to faint reflections of moral sensibility but also at the same time to movements and sensations of movements corresponding to the natural reactions of our sensory organs to external perceptions. Thus, the facial expressions originally indicating only the reaction of the perceiving organ to the sensory stimulus become a general form of expression of our sensations and emotions. For example, if we feel "bitter" anger, our mouth is spontaneously contorted as if a bitter substance had touched our tongue.

BÜRGI (1934) follows a train of thought similar to that of WUNDT. He stresses the near kinship between the effect of narcotics and that of works of art on man, the common factors being the stimulation of pleasant sensations with an immediacy of effect undisturbed by thought processes.

According to WUNDT, other expressive movements come about by the following process: Ideas are imitated by gestures and thus make the accompanying words plainer (for example, affirmation, negation, shrugging of the shoulders in case of doubt, etc.).

Finally, a third group of expressive movements is explained by WUNDT as the result of direct stimulation of the nervous system by strong emotions and the reaction of the muscles to these stimuli. This group corresponds in principle to that cited in the third basic law of DARWIN.

Later investigators for the most part based their work on that of PIDERIT, DARWIN, and WUNDT by modifying and implementing the basic laws, but without essentially augmenting them in any way. For example, KRUKENBERG (1920) believes that sensations and sensory perceptions direct the first facial movements after birth into certain pathways and that these movements become progressively smoother by frequent repetition. In addition to real sensory stimuli, in the course of further development there appear through sensory conceptions stimuli of pleasant or unpleasant ideas that cause the same movements in these smoothed pathways. Therefore, pleasant or disagreeable esthetic sensations and conceptions evoke the same responses in facial expression as pleasant or unpleasant sensory perceptions.

GRUHLE (1930) presents a completely different view. He is of the opinion that the newborn possesses certain inborn motor-secretory mechanisms. Occasionally these mechanisms are initially brought into action, so to speak, by chance without any underlying specific emotion. However, from the beginning they instinctively become related to the feelings even if at first they are restricted to gross pleasure-displeasure states. Gradually the emotions become more distinct and separated from each other, and at the same time certain motor relations become clearer. Expressive movements without real emotions become more and more infrequent; the same applies to emotions without expression. Thus, the connection between emotions and their means of expression becomes increasingly more intimate. GRUHLE considers this relation primary and "alien to the senses" (not related to the sensory organs). Thus, he is opposed to PIDERIT's view.

According to VON WYSS (1929, 1931), who expresses a point of view similar to that of GRUHLE, from the start expressive movements are bound to very specific affects; an emotion and the expression belonging to this emotion are an indivisible entity from the very beginning. Expressive movements are not, as DARWIN believed, remnants of originally purposeful actions that had their significance in the phylogenesis of man and animals. Rather, their meaning can be seen as pure expressive phenomena of the emotional inner life, starting with that stage of the spiritual life in which the specific experience of feelings and emotions appears. Accordingly, VON WYSS quotes PICARD's phrase: "Originally there was the smile of the mind, and for this smile the muscles were created so that it could be expressed through them." But how shall we conceive this process biologically?

GRUHLE and VON WYSS did not take issue with PIDERIT's views, to which they placed themselves in total opposition. They do not consider the problem in question, namely, why certain facial expressions became the expressive vehicle for specific emotional states. If I understand them correctly, they feel that this question should not even be discussed. However, the reasons for such a neglect of PIDERIT's theory should be given in detail; after all, a great part of this question can be answered through PIDERIT's theory.

The facial expressions of the child are, with the exception of the first period of life, much livelier and more expressive than those of the adult. This progressive age-linked limitation is explained by the general decrease of movements extending into old age and by education and social restrictions, the latter quite frequently making it appear undesirable to show one's emotions too openly. In order to study man's facial expression in pure form, the observation of children is particularly valuable, especially since their skin is thinner and consequently more transparent (VIRCHOV, 1908). Therefore, the excellent illustrations in KRUKENBERG's book (1920) mostly depict school-age children. It is not easy to obtain such photograpsh; even a three-year-old child behaves differently when he is aware that he is being observed.

The facial expressions of the adult are intelligible only if their development from the newborn period has been observed, for some expressive movements of the adult that otherwise cannot be explained have their origin in some peculiarity of the first years of life. They are preserved for a long time after the original cause has vanished, and later they may even develop further. Therefore, it is necessary to follow from birth the most important forms of human facial expression and by this means establish why a certain facial expression reflects a specific inner process and to follow the basic processes causing this development.

The numerous theories concerning the phylogenesis of facial expression are often incompatible. Therefore, they have to be compared with the events that take place during the development of the child's facial expression. Contradictions thus found speak against the correctness of these theories because we have to assume that a phylogenetic acquisition as comparatively recent as the facial expression cannot originate in an essentially different way in the young child from the way it did in the ancestors of our species.

Photographs are only an expedient, since the movements of expression do not constitute a specific state, but are a continuous process accompanying emotions. We are used to catching the "fruitful moment" (LESSING), i.e., to obtaining photographs that appear characteristic to the eye. It should be stressed, however, that quite often photographs are obtained that do not reflect exactly the desired expression or that are otherwise unsatisfactory. It is a moot question whether such photographs, which no doubt reflect a certain state, should be used. In studying children we need not consider the numerous subtle nuances that impart a characteristic charm to the facial expression of the adult; these expressions are described in great detail by PIDERIT, KRUKENBERG, and LERSCH. In children, particularly infants, one deals, compared with the adult, only with coarse expressive movements.

SENSORY REACTIONS OF THE FACE, "SPREADING REACTIONS"

The movements of facial muscles are the basis of facial expression; they form radiating and circular "frames" of the mouth, nose and, eyes, i.e., of the seats of three senses whose original mission is increasing the perceptive ability for welcome stimuli and decreasing it for unwelcome ones. This is accomplished by a widening of the openings by the muscles arranged in a radiating fashion and a narrowing of them by those circularly arranged. Furthermore, the reaction may spread to a nonstimulated sensory organ that also will be opened or closed. I term this process the "spreading reaction." It is also possible that an unrelated stimulus, particularly a tactile one, may change the perceptive faculty of a sensory organ in the same manner as does a specific sensory stimulus.

Facial expressions have their origins partly in local sensory reactions, described in detail in Chapter 2. Briefly summarized, the following processes take place:

A vivid facial expression already occurs around the mouth of the newborn, particularly with sucking in reaction to sweet-tasting food and with the very characteristic grimacing of the face in reaction to disagreeably tasting substances (Figs. 2—4). This may occur in various degrees. The grimacing is also seen in the adult when he is nauseated by disagreeable taste sensations or repulsive ideas. However, in the infant the facial expression is purer and more unrestrained than in the adult, who has become used to controlling his facial expression.

When the infant is subjected to a disagreeable taste, the lips are opened and the corners of the mouth are drawn apart so that the copiously secreted saliva will flow out. Frequently, choking occurs. The lower lip is turned down as in pouting. A transverse fold appears above the chin, and the nasolabial folds become more accentuated. Sometimes the corners of the mouth are turned down in a characteristic manner while the mouth itself remains closed. When the nose participates in the defense reaction, the alae are elevated and flare and the nose is wrinkled. The eyes are shut, oblique forehead wrinkles appear above the bridge of the nose (an effect of the eyebrow wrinkler), and sometimes a horizontal fold separates the bridge of the nose from the forehead, a phenomenon alien to the adult's facial expression (Fig. 12). Horizontal wrinkles of the forehead appear when the eyes are opened and are therefore not part of the expression denoting a bad taste sensation.

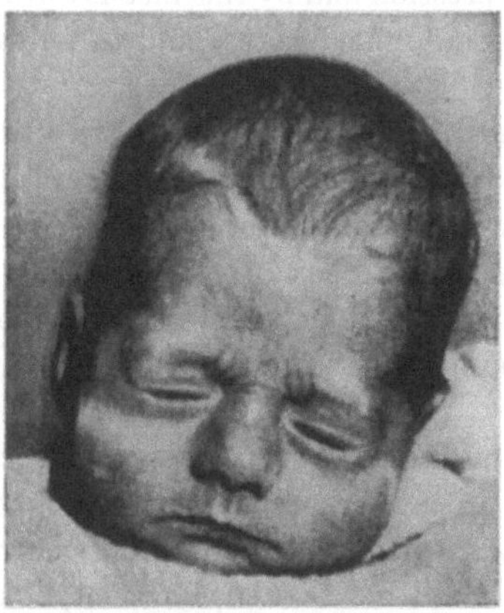

Fig. 12. Behavior of a 7½-week-old girl (1,200 gm) on exposure to blinding photic stimuli. Wrinkles in the medial upper corner of eyebrows and across dorsum nasi. Fold between chin and lower lip. Same folds and wrinkles appear during crying. (Admission weight 890 gm.)

Tactile stimuli of the newborn's mouth cause reactions similar to those of taste stimuli. Stroking of the oral region causes a rooting reflex and sucking movements. In the same manner an infant starts sucking if a rubber nipple is brought to his lips or placed in his mouth. However, if the pharynx is touched with a tongue depressor, the infant starts to gag and to grimace, as if to an unpleasant taste sensation.

A sharp taste sensation causes lacrimal flow in somewhat older children, as in the adult, but sometimes it causes only increased blinking.

Sucking movements increase the taste stimulus, whereas the distortion of the mouth, the increased secretion and flow of saliva, and the gag reflex decrease the contact between the bad-tasting substance and the taste buds. As this defense spreads to the eyes, we note the following phenomena: lid closing, oblique folds of the forehead, a horizontal fold on the back of the nose, and lacrimal flow ("spreading reaction").

Toward the end of the first year of life the infant, in response to a very bad-tasting substance, shakes himself almost as a dog does when coming out of the water. In a one-year-old boy it was possible to stop this unconditioned reflex: Cod liver oil was described to him as something particularly good so that he was elated with joy prior to taking the oil. After taking it he shook himself, but then continued to shout with joy. Later on the "shaking" disappeared and the joyful elation before and after the intake of cod liver oil increased. A happy affect was produced as a conditioned reflex that inhibited the original sensory reaction.

PAVLOV described a similar experiment in the dog: He combined strong, very painful electric stimuli with the feeding of the animals. Even though the skin of the animals was severely damaged by the electric current, there occurred no defensive reactions, but only a vivid feeding reaction, i.e., the dog turned to the side where the food was usually given, licked himself, and secreted saliva.

The facial expression denoting a bad taste sensation later becomes the general expression of disgust; the turning down of the corners of the mouth in itself indicates bad humor. Furthermore, the spitting that is supposed to indicate contempt – "you gall me" – reminds us of expelling disagreeable-tasting substances from the mouth, and the sticking out of the tongue makes us think of the tongue expelling the nipple at the end of the feeding.

The defense reaction against bad odors consists of respiratory arrest (KRATSCHMER's reflex, see p. 51) and sneezing. Also, turning up of the nose repulses the irritating stimulus, particularly when air is being expelled. "Phew" becomes an expression of contempt: "How it smells in here!" or "You stink." Breathing by movements of the alae nasi, an important sign of pneumonia, is similarly connected with expiration in this reaction. The nostrils are enlarged at the end of each expiration. The adult knows that irritation of the trigeminal nerve by stinging substances such as ammonia causes irritation of the nasal mucosa and of the conjunctiva and consequent lacrimation. The defense spreads from the nose to the eyes. In the newborn it would, of course, be better to omit this test.

The eyes of the newborn often have a peculiarly pinched expression caused by the CREDÉ instillation of silver nitrate solution (done immediately after birth), even if true conjunctivitis does not always develop. A certain degree of photophobia, which also is often found in the newborn, may be connected with this procedure.

According to REUSS (1927), in some infants we may see a transverse wrinkle on the dorsum nasi caused by a contraction of the dorsal nose muscle. Probably this wrinkle also indicates only an irritation of the eyes caused by the silver nitrate instillation.

Photic stimuli cause shutting of the eyes, a response that spreads to the mouth, as seen in Fig. 12. As a result of simultaneous contraction of the eyebrow wrinkler, vertical wrinkles appear on the forehead above the bridge of the nose. The action of the dorsal nose muscle causes a distinct transverse wrinkle or several wrinkles of the bridge of the nose (Fig. 12). In the adult the eyebrow wrinkler and the dorsal nose muscle are also activated when the eyes are shut tightly. Both muscles together form the only existing part of a second external muscle ring, which protects the eye through contractions in the same manner as does the eye ring muscle proper (orbicularis oculi). Only in older children may we observe the secretion of tears after exposure to a blinding light. This lacrimation, being also a protective reflex, usually results from tactile stimuli to the eyelids and originally serves the purpose of washing away foreign bodies from the conjunctiva.

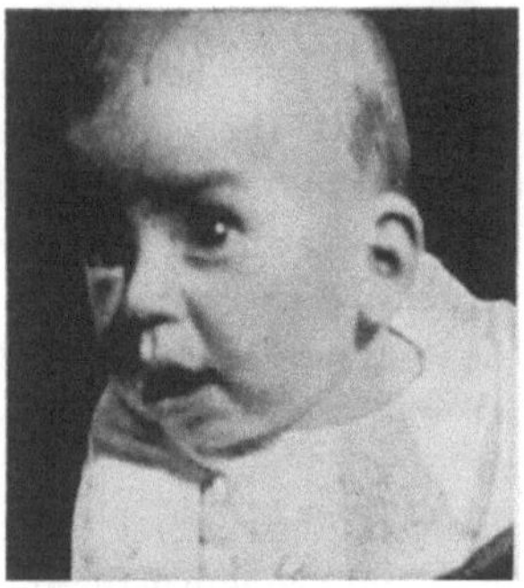

Fig. 13. Spreading reaction in a 7-month-old boy. Opening of eyes and mouth during attentive observation.

Early in life the infant turns his eyes and head toward a soft light. In somewhat older infants the activity of opening the eyes spreads to the mouth (spreading reaction) so that the mouth is opened in attentive observation or astonishment (Fig. 8, p. 65 and Fig. 13).

The ear does not have any muscles that participate in facial expression; nevertheless, it participates in the formation of a spreading reaction. After infancy the mouth is opened when the child is listening. This fact has been observed for a long time, but a satisfying explanation has not been found. KRUKENBERG (1920) was of the opinion that in this manner, by passage through the mouth, sound conduction is also favorably influenced. DARWIN (1872) first thought that the direction of sound could perhaps be better recognized if the tone were to enter the ear by still another route, but later he rejected this idea because the hearing was not improved by opening of the mouth. If a watch is placed in the mouth without touching the buccal mucosa, the tick-tock sound is heard less distinctly than if the watch were to be held outside the mouth. DARWIN did not find another explanation. We can account for this phenomenon by the spreading reaction, which at the same time explains why the mouth is also opened during attentive observation.

The aural-palpebral reflex can also be explained by the spreading reaction: Sound does not cause a stimulus response of the ear itself, but by the defensive spreading to the eyes the eyes are shut and occasionally frowning occurs (oblique frontal folds or wrinkles).

In addition to the corresponding sensory stimulus, an inappropriate stimulus, particularly touch, may elicit a sensory reaction. It has been previously mentioned that the eyes shut and lacrimate when touched and that when the mouth is touched the rooting reflex and sucking movements appear or the nipple is ejected. Stabbing pain, for example, the drilling of the dentist, may cause tears without crying. Intake of food that is too hot may likewise cause tears as a spreading reaction from stimulation of the pain or temperature sense.

Spreading reactions originate from involuntary sensory reactions and they also accompany the voluntary opening and closing of sensory organs; they even extend to the hands and feet, which cannot really be considered as sensory organs. However, they frequently have the same mission; this is particularly evident in blind individuals, for in them these organs may also be opened and closed. Such accompanying reflex movements can be seen in children of preschool and school age. Of course, many children do not manifest these reactions; sometimes these reactions are barely noticeable. Not infrequently, however, they are surprisingly distinct. In adults they are probably very rare, if they occur at all. So far no investigations have been done in this area.

If we ask a child who shows this reaction to open his eyes very wide, he also opens his mouth; and, vice versa, if he voluntarily opens his mouth very wide, the eyes are also opened wide and the eyebrows are raised. In voluntary tight shutting of the eyes the lips are pressed together. It cannot be decided whether or not these spreading reactions are favorably influenced by the child's incomplete understanding of the command. Testing various reactions of this type in short succession in the same child is not recommended.

Reactions may spread from voluntary facial movements to the hands and — to a minor degree — even to the feet. This accompanying activity has been described by DRÄSECKE (1921), GÖTT (1922), BECHER (1932), SCÄCHTER (1935), and ALBERTS (1955). If a child is asked to open his mouth very wide, his hands are opened and his fingers are spread at the same time. This spreading reaction is found in varying degrees in most, though not all, children. It is not restricted to the phenomenon just described, but corresponds to a general mode of reaction that has been less frequently noted and has been overlooked by the cited investigators: If a child is asked to open his eyes very wide, the fingers are spread just as when he is opening his mouth. If one asks him to shut his eyes tightly, he will make a tight fist. Children usually have more difficulty in closing their mouths tightly. But even then we may note a hint of fist closing. Occasionally we may find similar weak accompanying movements of the toes. Testing these various reactions in short intervals in the same children is also not recommended. The reaction may also spread from the fingers to the face. If a child is asked to squeeze the examiner's hand, or even better, his finger, the child will close his eyes if he exerts great effort. THOMAS (1943) thoroughly studied the frequency of this spreading reaction in 649 children age three to nine years. He also found paradoxical reactions. K. MÜLLER (1957) noticed less frequent finger spreading after mouth opening in blind children and adolescents than in children and adolescents with normal vision.

In a nine-year-old girl, RENNERT (1956) observed an abnormally pursed opening of the mouth and spreading of the fingers, which occurred tic-like and corresponded to the previously described accompanying movements. She was an intelligent girl who lived in constant fear of a temperamental and impatient mother. RENNERT interpreted this phenomenon as a remnant of a complicated instinctive pattern corresponding to the animal sequence of surprise, fear, startle, wish to attack, grabbing, eating.

The palmar-mental reflex and BABKIN's reflex (Fig. 14; also see p. 417) reveal a hand-mouth coordination even in the newborn. In the adult these reflexes can be manifested in pathological cerebral conditions. They are phylogenetically explained. BABKIN's reflex is described by I. LESNY (1960) as a palmar-mandibular sign.

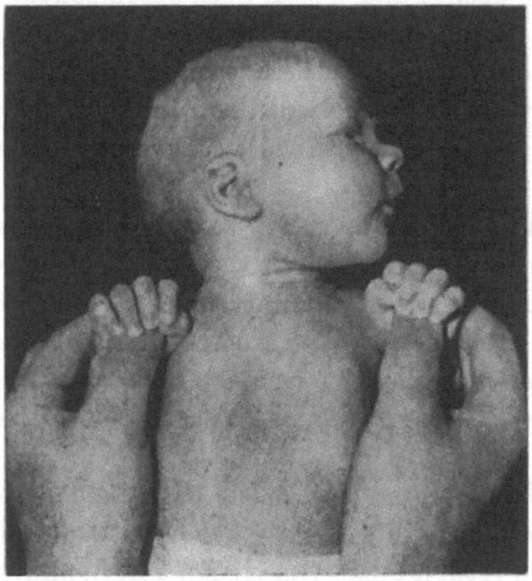

Fig. 14 a

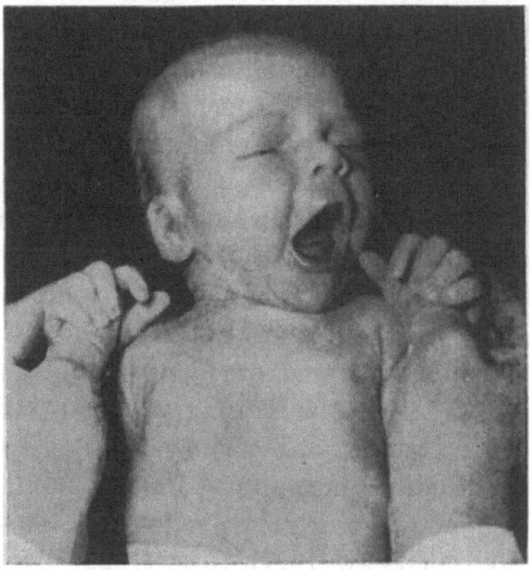

Fig. 14 b

Fig. 14. Babkin's reflex in the newborn: pressure on the palm causes opening of the mouth. [K. LIPPMANN, 1958 *(22a)*]

STRÖDER and SANDEN (1944) observed that fanning of the toes occurred in normal and premature infants up to the ninth to eleventh week of life as a spreading reaction after the introduction of a nasal catheter (see p. 27).

Active raising and lowering of the tip of the foot can cause corresponding involuntary accompanying movements of both hands. This spreading reaction is particularly evident in children whose foot muscles have become paretic as a result of poliomyelitis. The increased effort needed for the activation of the paretic muscles apparently favors the occurrence of the spreading reaction.

EXPRESSIVE MOVEMENTS AND THE RESPIRATORY CENTER

Of the general reactions that have been described, the change in respiratory rhythm caused by sensory stimuli considerably influences facial expression. Many expressive movements originate from the combination of respiratory with sensory and spreading reactions.

Thus, the position of the mouth does not depend only on sensory stimulation; it is also influenced in various ways by the respiratory center, i.e., in laughing, sighing, groaning, panting, moaning, crying, screaming, etc. In crying and laughing particularly, the effects of the respiratory center combine with the sensory reactions to form a uniform expressive movement.

CHARLES BELL (1872) had already postulated a special nervous "respiratory system" whose combined function included directing the respiratory muscles, the muscles of speech, and the expressive movements of the face. BELL's ideas were not accepted in his time, but today they are not completely forgotten. The "faciorespiratory movement form" (WEINGREW, 1934) is based on his theories; this phenomenon is said to become more pronounced phylogenetically with the higher development of cerebral function. It is manifested in man even in the newborn stage and is of greatest importance for his first breath.

The explosive expirations during coughing and sneezing are combined with opening of the mouth, contraction of the orbicularis oculi, and lacrimation. The narrowing of the lid opening and the lacrimal flow again represent a defensive spreading reaction, which in this case has its origin in the respiratory system.

During loud screaming the thoracic movements of the infant are markedly increased, while during rest the respiration is almost exclusively diaphragmatic with very little thoracic expansion. Furthermore, the ratio between inspiration and expiration is changed: brief, deep inspirations are followed by long, at times very long, expirations. This can become very annoying to the physician who attempts auscultation of a screaming child. At the peak of the screaming there usually are rapid inspiratory and expiratory movements, but in the beginning there are only several single diaphragmatic contractions. The sobbing noise originates during inspiration, but the real crying, screaming, moaning, panting, and whining occur during expiration. The alae nasi flare at the end of an expiration. Even if the various sounds appear very much alike to a stranger, the experienced mother knows the screaming and crying patterns and can immediately recognize the cause, i. e., whether the infant cries because he is hungry, afraid, or cold, or he has wet his diapers, or something unusual has occurred.

When the infant or young child stops screaming or crying, the respiratory rhythm changes immediately in a very characteristic way: Although the infant's mood has already changed and has become more pleasant and the respiration in itself has become calm and regular, the respiratory pattern is interrupted at irregular intervals by single deep sobbing inspirations. The infant may even start to smile weakly. The frequent occurrence of interrupted sobbing is caused by a struggle of the centers for the regulation and control of the respiration. When the usually inhibited part-center on which screaming and sobbing depend stops being active, it does not lose its increased state of irritation at once, but in stages. From time to time nervous energy sufficient to break through the inhibitions and to commandeer the respiration temporarily collects in this center. After the discharge of this energy, this center is again suppressed. A very similar struggle for the control or regulation of respiration occurs in central respiratory disturbances of infants (see p. 318). It is noteworthy that in this condition the expressive movements of the face, from which we usually draw conclusions about the infant's mood, do not conform to the respiratory movements.

Of all expressive movements, screaming causes the most pronounced and protracted change of the respiratory rhythm. It is difficult to judge the effect of smiling, but in laughing there occur, with the aid of the abdominal muscles, sharp expirations of shorter or longer duration. Astonishment causes deep inspiration followed by a period during which the thorax remains immobile. It was previously described (see pp. 24 and 86) how fright, through a sudden sensory stimulus, changes the infant's respiration.

EXPRESSIVE MOVEMENTS OF THE FACE

According to Fischl (1922), it is incorrect to state that at rest newborns and young infants have no facial expression. Rather, characteristic crease formations of the face of these infants are not infrequently seen. Three groups can be distinguished. The most frequent is the formation described by Reuss (1927), for which Fischl uses the term "eyeglass furrow" and which is similar to the pressure mark that eyeglasses leave on the bridge of the nose. This is a transverse wrinkling at the bridge of the nose of greater or lesser depth, occurring as single or multiple furrows. It was found as a solitary formation in 16.8% of the children examined. Fischl termed another type of facial wrinkle "glabella furrows": fine lines, etched on the skin as with an etching needle, radiating from the inner eye corner or from slightly above the eye to the forehead. This form was found as the only formation in 18.4% of the children examined. The rarest type of crease formation, observed in only 2.8%, was horizontal wrinkling of the forehead occurring as one or several furrows and extending across part of or all the forehead. The facial expression formed by the facial muscles moves mostly in the marked pathways of these furrows.

Crying and laughing change the respiratory rhythm and cause the uttering of sounds and facial grimacing. These two opposite phenomena have such a similar appearance at the start of the emotion that at first we hardly know whether a child will cry or laugh. Also, when the mood changes from crying to laughing, it is difficult to recognize the moment of transition. Probably this corresponds to the phylogenetic development in that screaming and crying developed prior to smiling and laughing.

Rapid mood changes are characteristic of the infant. It is easy to make him laugh, cry, and laugh again successively at short intervals. The adult maintains the same mood for a much longer period of time. Compared with the more complicated mood variations in later life, it is sufficient to differentiate a good and a bad mood in the infant.

Screaming and Crying

Alsbald ein neues Kind
Die erste Lufft entfindt,
So hebt es an zu weinen;
Die Sonne muß ihm scheinen
Den viermal zehnden Tag,
Eh als es lachen mag.
O Welt, bey deinen Sachen
Ist weinen mehr als lachen!

FRIEDRICH VON LOGAU (1604–1655)

As my co-worker L. SCHMIDT [1950 *(16a)*, see p. 361] has proved, respiration starts immediately after birth with one or several gasping breaths. A newborn never cried during the first gasps, but always a few seconds later. Sometimes after the first cry a transitory gasping respiration occurred again. In one newborn the first cry was observed to occur 26 seconds after the first breath.

PEIPER, HEMPEL, and THOMAS [1951 *(114)*] filmed the beginning of the newborn's respiration; on the film frames (Fig. 143, p. 363) the first gasping breath is easily distinguishable from the first cry, which is accompanied by activity of the facial nerve.

Various philosophical and psychological explanations (see p. 7) have been given for the crying that occurs immediately after birth, but a physiological explanation is the most logical: The long duration of the birth act, with the pressure extending from all sides on the body and with the continuously increasing respiratory embarrassment, is no doubt very unpleasant. As soon as possible the infant therefore makes the same use of his inborn ability to cry as he does later in life in all unpleasant situations.

Whining and crying are almost the only sounds the newborn is able to utter. The infant cries and screams more frequently than the small child and the latter, in turn, more than the school child, whereas in the adult, particularly the male, crying occurs less and less frequently, at least in our time. This was different in the sentimental eighteenth century.

ALDRICH, SUNG, and KNOP (1945) and ALDRICH, NORVAL, KNOP, and VENEGAS (1946) observed the frequency, duration, and causes of crying in infants up to the age of seven weeks during their institutional stay and after discharge. In the institutional setting, 50 infants were continuously watched by observers who alternated with each other. After an infant's discharge the mother reported the observations she had made at home. The 50 newborns in the institution cried an average of 117 minutes per day. The shortest average crying time of a newborn was 48.2 minutes per day; the longest, 243 minutes, that is, five times as much. The crying time was thus shown to vary considerably from infant to infant. Surprisingly enough, it was quite difficult to establish a clear-cut relation between food intake and weight gain on the one hand and crying on the other. Another relation, however, became quite evident: The newborns in the institution cried less if they were better cared for by the nursing personnel. The newborn in the institution cried an average of

11.9 times per day for a duration in excess of 3 minutes each time; the somewhat older infant at home cried only four times a day. This shows that the newborns' reactions of discontent diminish when their care is adapted to their individual needs. This became particularly evident when the duration of care in the institution was increased from 0.7 hours to 1.9 hours per infant. With this change the duration of crying decreased by more than half.

In addition to unknown reasons, crying is mainly caused by hunger, vomiting, and soiled and wet diapers. According to ALDRICH and his co-workers, an infant should not be allowed to cry longer than 3 minutes before he receives satisfaction.

According to TISCHLER (1957), the infant's crying distinctly depends on the feeding times. Toward the end of the first month the infant nearly always cries when he awakens between meals. He becomes extremely excited if the meal is delayed beyond the usual time of four hours. The crying stops if the nurse touches his mouth, but is resumed shortly afterward and is even more violent. Infants rarely cry shortly after feeding. With extended time intervals from the last feeding the average duration of crying increases; it reaches its maximum shortly before the next feeding. After the third month, when babbling starts, crying is distinctly diminished.

As the child grows older the reasons for crying or screaming change. The infant cries because of hunger, wet diapers, and strange impressions or when left alone, in short because of all kinds of internal and external unpleasant sensory impressions. However, once excited he cries much longer than the unpleasant sensory impressions last. It is nearly always possible to calm him by rocking.

The small child still lives only for the present moment and therefore also cries when irritated by unpleasant sensory stimuli, but he can control himself better. He cries during unpleasant experiences, for example, when he cannot do what he wants to do, when he is denied a wish, when he is ashamed of himself, when he is afraid, when he is homesick, etc., but still only for immediate reasons.

Later in life sad ideas or concepts are of increasing importance as immediate causes for crying, while unpleasant sensory perceptions become of less significance. However, strong physical pain can cause crying or screaming at any age, but with advancing years man learns to control himself better and better.

Moodiness or irritability of long duration, not infrequently combined with lack of movement, are found in children as an expression of physical illness, such as rickets, infantile cachexia, and abdominal disorders. Their disposition is similarly affected when they are in pain or are afraid that they will be subjected to painful manipulations. They particularly give vent to their bad mood when approached; when they are left alone, their faces are expressionless and apathetic. Only slightly downward turned corners of the mouth are sometimes observed, as noted in some photographs of children in KILLIAN's monograph (1934). The pitiful whimpering of severely ill infants is very different from the crying and screaming of healthy children.

According to AMES (1941), the crying infant moves his limbs violently, the legs more than the arms, unilaterally more frequently than bilaterally, and with flexion more pronounced than extension. Within only half a second a noncrying infant can change to a loudly crying one. From the first recognizable signs of crying it usually takes about 3 to 4 seconds for the crying to become fully developed.

Very excited children are inaccessible to external stimuli. Friendly talk or stroking is useless, having no effect at the peak of excitement. The available nervous energy is fully occupied by the excitement and is not available for other purposes.

The small child is generally incapable of any other action during a crying spell. At the beginning of the bad mood the muscles relax. The elevated arm drops (Fig. 15).

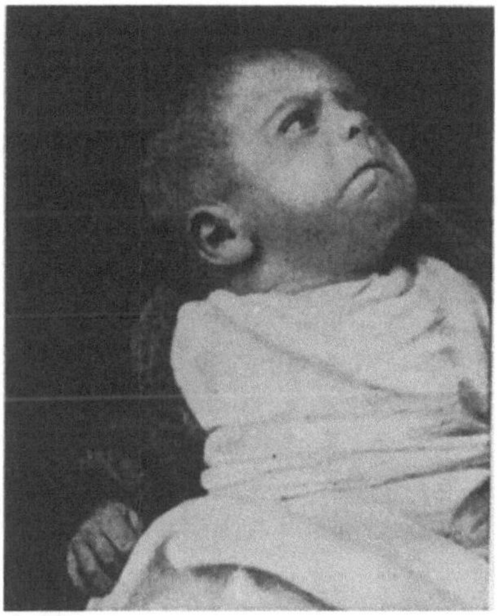

Fig. 15

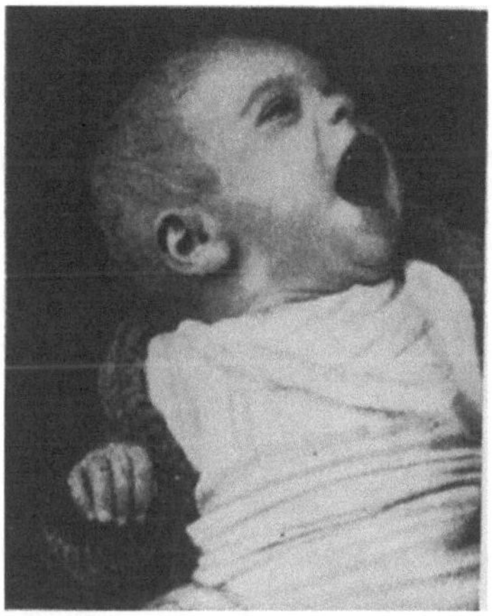

Fig. 16

Fig. 15. Bad mood in a 4-month-old girl. Corners of the mouth turned downward, dimple on chin due to contraction of the chin muscle, hint of vertical forehead wrinkles due to contraction of the eyebrow wrinkler, relaxation of the arm muscles.

Fig. 16. Laughing in a 4-month-old-girl. Mouth wide open, distinct lower chin fold, arms elevated.

If the child has an object in his hand, he may let it drop; however, during great excitement he may display significant physical strength, be it only kicking with his arms and legs or trying to reach a certain goal. Thus, he will cling desperately to his mother and will resist an examination violently, at times very deftly through snake-like sliding and gliding movements. In the most violent temper tantrums many children stamp their feet alternately or throw themselves on the ground, screaming loudly and thrashing about. From these patterns a conditioned reflex may develop.

After a long period of crying the face, ears, and scalp become reddened; frequently irregular, indistinctly circumscribed red blotches appear on the upper part of the chest. The tendency to flush during crying varies in degree in different children. DARWIN (1872) already noted an increase of blood flow to the naked scalp as the first sign of an impending attack of rage in an infant.

The development of each child teaches us again and again that crying and screaming are originally elicited only by unpleasant sensory perceptions, the only effective cause at all ages. From this fact we can postulate that the facial expression of a bad mood had its origin in the defensive sensory reactions, which diminish the undesired sensory stimuli as much as possible and thus make them innocuous. This can still be seen today in the expressive movements of a bad mood in the infant and in the adult.

At the onset of a bad mood the corners of the mouth are turned down by contraction of the musculus triangularis with the mouth at first remaining closed (Figs. 15 and 17). At the same time dimples appear above the chin where the chin muscle is attached to the skin. The nasolabial fold becomes more pronounced, the eyes may

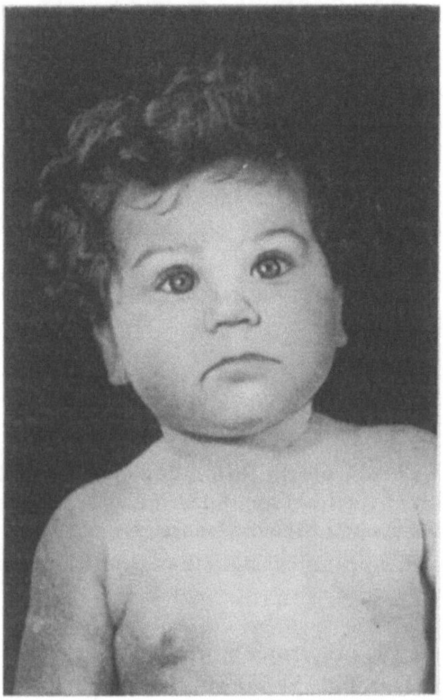

Fig. 17 a

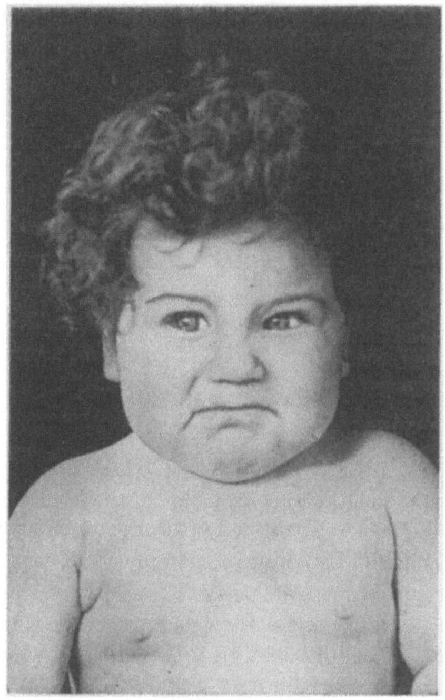

Fig. 17 b

at first stay open but they lose their sparkle, and vertical wrinkles appear on the forehead.

DARWIN observed that the downward turning of the corners of the mouth begins at the sixth week, but PREYER first observed it at the eighteenth week of life. HOLLIS (1913), however, described and illustrated it in a three-week-old infant and BLANTON (1917) found it in a newborn. A corresponding facial grimacing in a newborn before the actual crying was described by WOLFF (1959). In any event, the downward turning of the corners of the mouth becomes characteristic of the onset of bad humor only at the age of several months. Man retains at least a hint of this mode of expression all his life: the corners of the mouth are still turned down in the adult when he is in bad humor, and they remain so when he suffers a painful sickness of long duration.

DARWIN explains this facial expression as follows: When infants cry the muscles around the eyes contract tightly. As a result of this contraction the upper lip is raised; this is counteracted by muscles that draw downward and that move the corners of the mouth so that the mouth will not be opened, despite the upward pull. It is simpler, however, to assume that the reaction to a sensory stimulus is the original basis of the expressive movement. A disagreeable taste sensation, for example, causes the same turning down of the corners of the mouth, as previously described (Fig. 4, p. 45). The defensive reaction of the mouth to a very bad and acrid taste sensation spreads to the eyes. They are closed; vertical forehead wrinkles and, to a lesser degree, a horizontal fold over the bridge of the nose appear; there also may be tears. Thus, it can be stated that the facial expression of bad taste, as shown in Fig. 4, is unmistakably similar to the expression of bad humor.

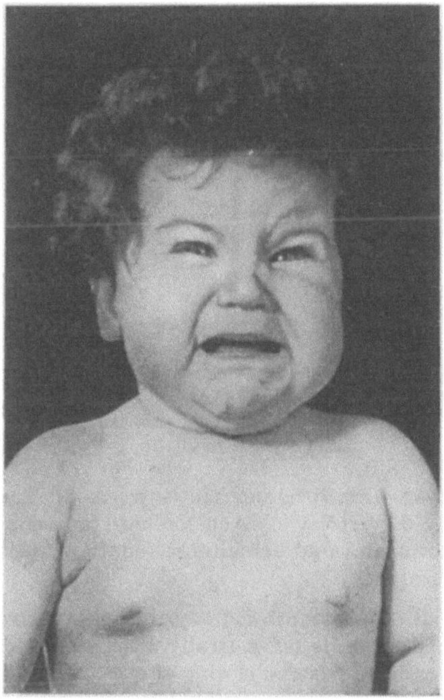

Fig. 17 c

Fig. 17. Transition to an increasingly bad humor in a 10½-month-old girl. (a) Eyes wide open, eyebrows raised, corners of mouth turned down. (b) Lid openings narrowed, eyebrows lowered, region above nose bridge puffed by contraction of eyebrow wrinklers, corners of mouth turned downward further, nasolabial folds, dimples at chin due to contraction of the chin muscle. (c) Lid openings still more narrowed, eyebrows turned down, folds above nose bridge due to stronger contraction of eyebrow wrinkler, nasolabial folds, square opening of mouth for crying, dimples at chin increased.

When the young infant turns the corners of the mouth down in anger, the movement derives from an old phylogenetic reflex and not from personal experience; for even the breast-fed infant, who has not yet tasted bad-tasting food, turns the corners of his mouth down in the same way.

Pouting is another preliminary to crying. The infant distorts his mouth, corners wide apart, and pushes his lower lip forward and outward so that the oral opening forms a small square. The lower lip is separated from the chin by a distinct furrow, and the chin pad moves upward as a result of the contraction of both chin muscles. Vertical forehead wrinkles appear, and the eyes are half closed (defensive spreading reaction). A similar facial expression occurs in infants who do not want to swallow the milk that has entered their mouths. They pout their lower lips in quite a similar way and let the milk flow out again. As the infant's mood is always bad when this occurs, it is possible to understand that pouting develops from the refusal of food. This expressive movement is evidently very old, for it can be found even in the ape. DARWIN shows a picture of a chimpanzee whose big lips are pushed farther forward than would be possible for the infant.

The mouth is opened as far as possible for loud crying and screaming (Fig. 18). The nasolabial fold becomes markedly accentuated and is even extended somewhat downward beyond the corners of the mouth. The eyes are more or less closed (Figs. 17 b,

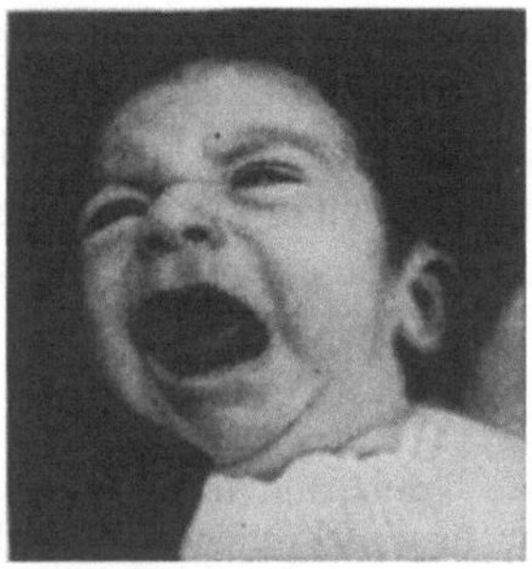

Fig. 18

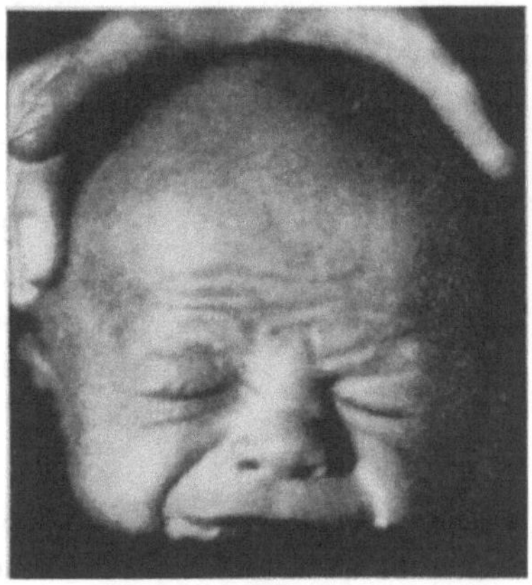

Fig. 19

Fig. 18. Loud screaming in a 4-week-old girl. Mouth wide open and fold between chin and neck. Eyes in a defensive position: lids half closed, vertical folds on forehead (musculus corrugator supercilii, eyebrow wrinkler) and horizontal fold across nose (dorsal portion of musculus nasalis).

Fig. 19. Horizontal and vertical folds on forehead of a 2-month-old boy. Strong forward bending of head causes horizontal folds as a neck-frontalis reflex (really also opening of eyes). Crying at the same time causes vertical folds and closing of eyes.

18, 20b), and the vertical forehead wrinkles become very pronounced. Through contraction of the muscle on the bridge of the nose (musculus nasalis) a horizontal wrinkle appears across the nose (spreading reaction). Another fold may be formed right below the chin by the opening of the mouth during crying (Fig. 17), screaming (Fig. 18), and laughing (Figs. 16 and 20a). Fat infants show this fold even at rest (double chin).

Horizontal folds on the forehead are not rare in infants; they appear independent even of the infant's mood and direction of his gaze. When these horizontal folds above the bridge of the nose are absent, the infant's face does not show a worried expression. FEER (1921) has drawn our attention to the frequent presence of such horizontal folds in infants with pylorospasm. I would rather not attribute this phenomenon to the sickness itself, but to the accompanying weight loss, for these folds occur particularly in cachectic infants and disappear again, just as in pylorospasm, when the general state of health has improved.

In adults physical or mental suffering causes vertical folds above the bridge of the nose and, somewhat higher and limited to the middle portion, horizontal folds on the forehead (contraction of the musculus frontalis), so that the entire picture of these folds resembles a Greek π.

Darwin (1872) described in detail the "grief wrinkles" that often persist in adults as fixed facial features. A classic example is the expression of Laocoön, whose facial muscles are simultaneously exposed to two opposite influences: the eyebrow wrinkler (musculus corrugator supercilii) tries to narrow the lid opening while the musculus frontalis attempts to widen it. We can thus understand Darwin's observation that the same forehead folds are formed when we try to focus on an object in the face of blinding light. A sentiment such as painful astonishment causes a similar picture.

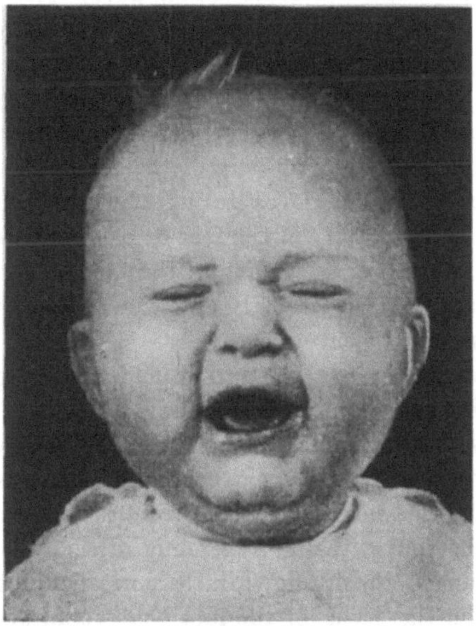

Fig. 20 a

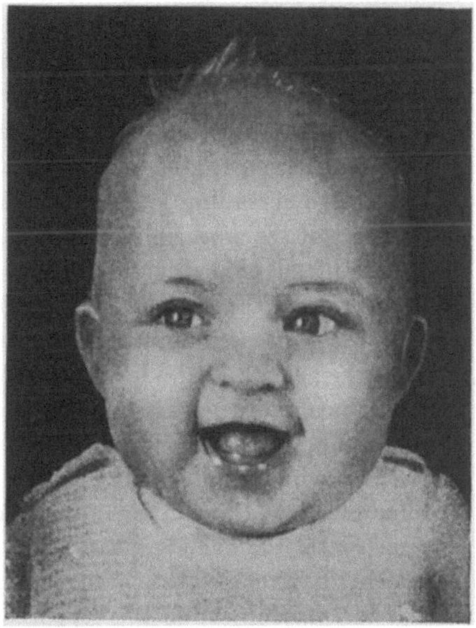

Fig. 20 b

Fig. 20. A 5-month-old girl. (a) Crying: Eyes closed, tops of eyebrows lowered, pad and vertical folds above bridge of nose due to contraction of eyebrow wrinklers; mouth open, nasolabial folds, dimple on chin. (b) Laughing: Eyes and mouth open, nasolabial folds, no dimple on chin.

Occasionally similar forehead wrinkles develop even in quite young infants, especially during illness. Such a child (observed by me for a considerable period) lost this expression with improvement of his state of health. The folds shown in Fig. 19 were artificially caused by eliciting the neck-musculus frontalis reflex through the strong forward bending of the head during crying. Quite often only the middle portion of the forehead is horizontally wrinkled, just as in the adult. Krukenberg

(1920) shows the picture of a scared boy (his Fig. 235) in whom the same facial expression is quite distinct. Nevertheless, the facial expression is not that of a child; it really belongs to a later age.

According to BRAUS (1921), the joint action of the supercilii corrugator and the frontalis muscles occurs only when the body is erect. However, we see this joint action in the infant long before he is able to assume the erect position.

The contraction of the muscles surrounding the eyes during crying is supposed to have as its purpose the protection of the blood vessels of the eyeball, which are in danger of rupturing because of the increased blood pressure (BELL, 1872). Correspondingly, lid closing is said to occur on other occasions when the blood pressure rises, such as during laughing, coughing, sneezing, contraction of the abdominal muscles, and strenuous activity (DARWIN). It is possible that this phenomenon may play a certain protective role during a considerable rise in blood pressure, but to me the real reason for the closing of the lids and the wrinkling of the forehead is the defensive spreading reaction, which extends to the eyes during crying. Other movements of the crying child, such as the turning away of the face or of the entire body, the lifting of the hands in front of the eyes, or the hiding of the face behind the hands or in the skirt of the mother, are defensive too. BELL's theory cannot explain these often simultaneously occurring movements.

After the neonatal period tears are shed during crying, but during loud screaming the eyes have no tears. The first tears during crying appear, according to DARWIN, between the twentieth and one hundred and thirty-ninth day of life. The ability to shed tears on tactile stimuli, though, is present from the very beginning. DARWIN attributes the shedding of tears during crying to a reflex that arises from the spastic contraction of the eyelids, possibly in connection with the hyperemia of the eyes. Associated habituation supposedly then connects the crying with the corresponding mood. We may object to this hypothesis on the basis of the fact that voluntary forcible pressing together of the lids does not cause a flow of tears. In my opinion we are dealing here with a spreading reaction to the eyes from disagreeable external stimuli, as in the closing of the lids. Tears are caused to flow by touch (foreign body!), glaring light, biting taste, stinging smell, and sharp pain stimuli, i.e., by the most varied disagreeable sensory stimuli. BECHTEREV (1913) already gave a similar explanation: The shedding of tears during crying corresponds to the reflex lacrimation on the touching of the conjunctiva or cornea.

During crying the flow of the nasal mucus is also increased; this can be explained without difficulty as a spreading reaction, but cannot be explained by DARWIN's hypothesis.

The screaming and crying of the infant, disturbing as it may be, is sometimes of vital importance. It is true that to an extent innate reflexes protect the newborn from damage, but beyond these mechanisms he is helpless; he cannot free himself from many disagreeable or dangerous situations. That is his mother's task. Crying and screaming call for her help long before the child can speak and at the same time indicate to her where he is. The crying and screaming of healthy infants can be stopped at any time by lifting them up, by carrying them around, and by rocking movements. It does not matter who does this, but it must be done skillfully. Older infants who can distinguish between their mothers and strangers can be calmed only by their mothers.

What even a young chimpanzee can achieve by crying is shown in a report of TEMBROCK (1949) from the Munich zoo: A young chimpanzee was lying alone in its cage at night. As the blanket was not sufficient, it started to cry; the result was that a chimpanzee in the neighboring cage handed it a blanket through the grill. However, it did not stop crying. Now the chimpanzee in the cage next to the infant's neighbor handed a blanket to the neighboring chimpanzee, which in turn handed it to the infant.

Smiling and Laughing

Incipe, parve puer, risu cognoscere matrem!
(Learn, little boy, to recognize your mother by her laughter!)

Virgil, *Eclogue* IV, 60

Laughing and its preliminary stage, smiling, develop only after the neonatal period, but no data as to the exact time can be given. A mother who lovingly observes her child will be more inclined to define a minor movement in the face as a smile than would a stranger. In general, a true and distinct smile can at the earliest be recognized at the end of the first month. PREYER (1895) saw his sleeping child move his mouth as in a smile for the first time on the tenth day. On the twelfth day a similar expression was noted when the child was awake, but the "consciousness necessary for its perfection" was still absent. Only on the twenty-sixth day did the smiling become an established facial expression. Audible and visible laughing was observed by PREYER for the first time when the child was 23 days old. Contrary to most other authors, BLANTON (1917) observed the first smile even in a newborn, age four days. DARWIN (1877) saw it appear in one infant on the forty-fifth day and in another on the forty-sixth day, and BECHTEREV (1913) observed it during the seventh to tenth week. Of the 18 infants that S. S. and J. J. B. MORGAN (1944) observed for a considerable period of time under constant conditions in an institution, 75% smiled for the first time between the fifty-sixth and sixtieth day and "crowed" between the sixty-sixth and seventieth day. According to ANDRÉ-THOMAS and AUTGAERDEN (1959), the first smile appears at the age of five to six weeks.

O. KOEHLER (1954), on the other hand, showed pictures of a newborn smiling unilaterally on his fourth and seventh day of life. The infant was also able to smile bilaterally; it took little more than 1 second from the first movement of the corners of the mouth until the return to the neutral position. Observers were convinced that the smile followed the expression of discomfort exactly at the moment when the discomfort ended.

P. H. WOLFF (1959) observed smiling not induced by an external cause as a slow, weak sideward or upward drawing of the mouth without rhythmic crying movements or contraction of the other facial muscles during irregular sleep or drowsiness, but never during normal sleep.

The infant does not learn to smile by imitation; otherwise he would not smile at all. The sequence from smiling to laughing to shouting with joy is not learned; no imitable example for shouting with joy exists.

Quite often "Stäupchen,"* i.e., spastic movements of the face (see p. 249) may have been taken for smiling, for occasionally these movements resemble smiling very much. They appear unilaterally or bilaterally and also during sleep. "The infant is playing with the angels" is the popular saying.

This phenomenon has been known for a long time. "We occasionally observe that an infant smiles in his sleep during the first months of life... However this is nothing but a spastic movement of the facial muscles" (LÖBISCH, 1851).

"Sometimes infants are seen to smile and cry in their sleep immediately after birth. However, when awake they do not laugh and cry until 40 days have passed" (HIPPOCRATES). "During the first 40 days after birth infants do not laugh and cry in the daytime, but they sometimes do both at night. They usually do not feel when they are tickled, but they sleep most of the time" (ARISTOTLE). PLINY THE ELDER and LOGAU (see p. 119) also state that the infant does not laugh before the fortieth day.

Before the infant is able to smile, his good humor becomes evident in his facial expression only by the disappearance of all signs of discomfort. He may be brought to a state of happy excitement only by direct sensory stimuli. He smiles and laughs, particularly when other persons occupy themselves with him, when he is given his bottle or a toy, or when somebody makes noises, and he responds early in life to the smile of an adult. Tickling produces loud laughing easily and early in life; the areas on both sides of the mouth are the most ticklish.

Only healthy infants smile or laugh. This ability disappears as soon as the general state of health deteriorates significantly because of an infection, an alimentary disturbance, or any other illness. In this situation infants reject otherwise welcome stimuli. The return of smiling is always a sign of recovery.

Although the young infant is not able to laugh, laughing is a cerebral function of a lower order. The feeble-minded laugh often without sufficient reason.

DARWIN (1872) described the beginning of laughing in anthropoid apes. If, for example, the axillae of chimpanzees were tickled, they made giggling or laughing sounds and their eyes sparkled and the corners of their mouths were retracted so that the lower eyelids were slightly wrinkled. According to W. KÖHLER (1915), anthropoid apes do not cry and laugh like man, but, when they are tickled, their laughter is somewhat similar to the human in its rhythmic panting and grunting. R. M. and A. W. YERKES (1929) find impatience, anger, and rage to be quite distinctly present in all primates, but only in the anthropoids do these expressions of emotion resemble those of human infants. Joyous excitement becomes more and more human-like from the lemur to the ape. Smiling and laughing are absent in the lemur. The descriptions of these signs of emotion in the monkey are not convincing, while similarity to corresponding human emotional expressions is quite distinct in the

* German popular expression for which no medical or literary equivalent exists; "little distempers" would be an English approximation — The translators.

anthropoid. The baby gorilla "Goma" in Basel occasionally opened its mouth during the first weeks of life and drew the lips back from the gums when it was stroked on the vertex or back. SCHENKEL (1960) views this phenomenon, variations of which are encountered in all anthropoids, as the beginning of laughter. No subhuman being can laugh or appreciate humor as man does. Dogs are able to grin by retracting their ears and the corners of their mouths (DARWIN; BECHTEREV, 1913).

The musculus zygomaticus,.to a much greater extent than the musculus risorius, is the true laughing muscle; the zygomatic muscle also causes the expression of joy. The first smile of the infant consists of a fleeting broadening of the mouth, which at first remains closed. The lips usually move, but not always bilaterally. The eye muscles do not yet take part in smiling; only the brightness of the eyes is increased during joyous excitement so that the corneal reflexes become more pronounced.

From this beginning real laughing develops in many stages. The joyful expression of the face persists for a longer period. The mouth becomes broader so that the nasolabial folds become more distinct. The upper lip is slightly raised, and the teeth of the upper jaw become visible. The mouth is more or less open (Fig. 20b), causing a transverse fold below the chin (Fig. 16). The eyes usually remain open; quite often the transverse fold, which separates the lower eyelid from the cheek and becomes most distinct at the nasal side, is deepened. A respiratory change is associated with the grimacing of the face: a short inhalation is followed by several exhalations accompanied by crowing sounds, and from this continuous laughter gradually develops.

A form of joyous laughter foreign to later years is characteristic of infancy (Fig. 16): the infant opens his mouth wide and utters crowing or jubilant sounds. The eyes are wide open, shine, and look at the object of joy. BRANDES (1939) shows a picture of an adult female orangutan laughing in this manner. According to HEMPELMANN (1926), the chimpanzee also opens its mouth and eyes in joyous expectation. The baby gorilla "Goma" in Basel started to laugh at exactly three weeks of age, just like the infant in Fig. 16 (HAUSER, 1960). The expressive movement leads in the infant, as well as in the anthropoid, to an especially distinct opening of both sensory organs. AHRENS (1955) compares this form of laughter with the spreading of the beak of young blackbirds.

Joyous excitement seizes the whole body of the infant much faster and more intensively than it does in the adult. The infant joyfully kicks his arms and legs. When he is able to sit, he jumps on his seat; and when he can stand and run, he dances exuberantly with joy.

The adult who laughs continuously and violently may show additional phenomena alien to the infant. LESSING already mentioned that "bodily crying is, so to speak, the highest degree of bodily laughing." Looking at the corresponding illustrations in the reports of PIDERIT, WUNDT, and others, we must acknowledge the similarity of intensive laughing and crying. It is the defensive reflexes of the eyes that finally act when man laughs "until the tears come." However, the infant cannot reach this stage. He can scream for a long time, but he cannot laugh continuously. He does not experience threatening respiratory embarrassment, nor does his face redden as does that of the adult.

At present no accepted physiological or psychological theory of laughter exists. We can understand the described opening of the mouth and eyes and the defensive reflexes of adults — but why is the mouth broadened during smiling or laughing? The opinion that this facial expression is based on a sensory reaction has repeatedly been expressed. GRATIOLET's (1873) view of smiling as the expression of an uninhibited "happy" respiration is not too convincing. WUNDT (1877) and KRUKENBERG (1920) claim to see in smiling the facial reaction to sweet taste. KRUKENBERG shows the picture of a child of school age whose expression after experiencing a sweet taste at least approaches that of a smile. However, the child may have really smiled at this moment. In any event, a sweet-tasting substance causes a completely different response in the infant: he does not smile, but rather starts to suck and swallow. According to BRAUS (1921), the widening of the nostrils is most important for the formation of a smile because it favors agreeable odors while closing of the nares characterizes sadness. I cannot share this opinion, at least as it pertains to infancy, for this is just the time when laughing develops. As much as the infant's alae nasi take part in respiration during crying and respiratory embarrassment, they stay rather immobile during smiling and laughing. After the neonatal period, when infants are able to smile, this expression can most easily be elicited, as previously mentioned, by stroking both sides of the face near the corners of the mouth. One could, so to speak, interpret smiling as a bilateral rooting reflex of the lips.

Although smiling is the most distinct expression of a friendly attitude, phylogenetically it has been repeatedly regarded as an originally hostile expressive movement. EBBINGHAUS (1913), for example, adheres to older theories that attribute the showing of teeth during laughter to the hostile expressions of our phylogenetic ancestors. There is said to exist a kind of cruel laughter that is the expression of a hostile and malicious sentiment. In the laughing of disdain and scorn, too, hostile intentions can be recognized. The friendly, comfortable, and good-natured laughing conforms to the hostile type insofar as it is based on a more or less pleasure-accentuated feeling of superiority, of preparedness, and of readiness for battle. Similarly, HOCART (1929) interprets the "smiling" of the dog as an intention to tear something with his teeth (a good-natured snapping). Correspondingly, smiling is said to have originally contained the intention of biting. Laughing, again speaking of the dog, is to be interpreted as barking mellowed by tenderness.

The objection to this opinion is that it is not very probable that a friendly expression developed from a hostile sentiment, and, if we may at all draw a conclusion on the phylogenetic history of man from the development of the infant, another contradiction occurs. The infant who smiles or laughs for the first time usually responds to the smile of another person; it is not possible to ascribe to him a hostile or even a defensive effort.

Perhaps smiling and laughing cannot, as most other expressions, be traced to the characteristics of infancy. Although this expression is not absolutely restricted to man (we hear of grinning apes and dogs), we no doubt find its specific development only in man. The joyous, in contrast to the sad, expression becomes even more varied and meaningful after infancy. The adult cries only rarely, but smiles and laughs more frequently. Smiling has become a form of social greeting; in the infant it first appears as a reaction to the people in his environment. According to RIPIN and HETZER (1930), the infant does not smile before the fifth month upon seeing his bottle, but smiles considerably earlier upon seeing a human face or hearing the sound of a human voice. On a phylogenetically lower stage in the distant past a smile could have been the key to the vital distinction between the other man's friendly or hostile intentions; that this expression may have had a certain usefulness therefore cannot be disputed. It was DARWIN who pointed out that expressive movements not only have biological importance but that they also serve to further mutual understanding and thus the welfare of mankind. HAYWORTH (1929) similarly traces the origin of laughter to a sound signal primitive man gave to his ancestral brother when danger had passed and relaxation and the feeling of security could take over. HAYWORTH interprets smiling as a later, visible form of this signal.

Attention, Astonishment, Fear

When a stimulus of minor strength affects man at rest, his senses "open up" and focus in the direction of the stimulus so that its perception is facilitated. It is the same phenomenon that PAVLOV describes in the animal as the "What-is-the-matter?" reflex. Even the young infant has such comprehensive attention that no other animal except the ape approaches or equals him (THORNDIKE, 1922).

In attentive seeing (Figs. 8 and 13, pp. 65 and 114) and hearing (see p. 114) the mouth and eyes are opened simultaneously. Accordingly, the expression of astonishment consists of an opening of the mouth and eyes and a contraction of the forehead muscle, the latter opening the eyes still more and effecting horizontal forehead wrinkles. In young infants this expression is rare; it develops at approximately the beginning of the second half year of life, when the infant can lift his head freely and turn it to the object of his attention. Then he opens his eyes and mouth wide (Fig. 13), and the mobility characteristic of his waking hours disappears completely. In the infant the raising of the eyebrows and the horizontal wrinkles of the forehead do not yet belong to the expression of astonishment; according to DARWIN (1872), with whom I agree on this, in infancy there is only a trace of this reflex as an expression of astonishment. Adults, on the contrary, react mostly with their eyes, often only with twitching of the eyebrows, and without opening their mouths.

Evidently the visual field is enlarged by opening the eyes wide and raising the eyebrows. Even the young infant raises his eyebrows and wrinkles his forehead horizontally when looking upward (Fig. 8, eye musculus frontalis reflex). We "look up" or "lift" our eyes if something attracts our attention. Thus, a muscle movement that originally raises the gaze develops into an expression of astonishment.

The question why the mouth is opened to express astonishment is answered in various ways. According to PIDERIT (1919), hearing is improved by opening of the mouth, a hypothesis disputed by DARWIN by reason of his previously mentioned experiences (see p. 114). DARWIN believes that we prepare for every effort by inhaling deeply and that this inhalation is facilitated by opening the mouth. Furthermore, GRATIOLET, DARWIN, and DE SANCTIS (1906) state that stimulation of one cerebral area as a result of attention inhibits another area. However, the spreading reaction is a much simpler explanation. The mouth is opened during astonishment, just as during attentive seeing or hearing. Quite often the mouth is a little pouted during intensive attention, as described by PREYER, and if this is combined with a smacking or inspiratory noise the similarity to a sucking movement becomes still greater. According to LERSCH (1932), the tongue often protrudes between the lips when infants are astonished, an expressive movement sometimes persisting until adult life.

In disagreeable astonishment resembling fright the adult opens his mouth and inhales deeply. He does not exhale immediately afterward, but for a short time maintains his thorax in expansion. At the same time he raises his flexed arms, palms forward. DARWIN gave an illustration of this posture in his Table 7, Fig. 1. In LEONARDO DA VINCI's *Last Supper* the third apostle from the left is pictured with his arms in this posture. The painting represents the moment when the disciples are startled by the words of Christ: "One of you shall betray me."

DARWIN explains the arm position of the astonished adult by his basic rule of contrast: the sudden lifting of the arms is in complete contrast to the position maintained in a state of indifference; therefore, an astonished person unconsciously executes this movement. Of course, the question should be raised why just this and not another position is assumed. However, there is still another explanation: the described posture of the arms corresponds exactly to the infant's posture in sleep (see p. 170). Thus, it could be postulated that great surprise consumes nervous energy and causes a temporary regression of brain function to a lower stage, as in other situations.

No expressive movement of fear or horror exists during infancy. Evidently a stage of cerebral development such as that attained beyond the first year is needed to enable man to be afraid. Accordingly, we find the expressive movements of horror only in the highest animals who show certain traces of reason, for example, the chimpanzees (W. KÖHLER 1915–18). LANGE-EICHBAUM (1928) even sees the germ of a prereligious mystic experience in this phenomenon. The infant has not yet reached this stage; instead he starts to cry during all disagreeable experiences without further thought and thus skips, so to speak, the stage of fear and horror.

MAYR (1858) gave an excellent description of the expression of fear in a child in danger of suffocation because of laryngeal stenosis. He mentions the wide opening of the eyes and nostrils, the broadening of the mouth, and the formation of horizontal forehead wrinkles.

The small child stares with wide open eyes, quite often with his mouth open, at the frightening object. At the same time his face turns pale and his heart beat increases. I cannot say with certainty whether or not his pupils also dilate, as J. MÜLLER (1840) and GRATIOLET (1873) reported in the adult. In any case, in a state of fear there appear reflexes which increase the perceptive ability of the eyes as well as the mouth, although the sensory impressions are of a disagreeable nature. Fearful trembling and chattering of the teeth hardly ever occur in the young child. In his fear, however, he will cling frantically to anything, preferably his mother. In contrast to later years, the expression of fear does not persist very long in the young child; rather, he soon starts to cry with all his strength.

Concentrated rumination is foreign to the child and he therefore also lacks the facial expression for thinking. For example, a young child in the posture of RODIN's *Thinker* would not look at all like a child. The expression accompanying an adult's thinking consists of vertical forehead wrinkles caused by contraction of the eyebrow wrinkler. It reflects, to a certain extent, a strained and introspective regard (DE SANCTIS, 1906), and thus may also be characterized as resulting from direct sensory reactions.

Negation and Affirmation

DARWIN (1872) – incidentally, CHARMA (1846) noted this before him – traced the expressions of negation and affirmation to the infant's behavior during food intake.

"The first act of negation in infants consists of refusing nourishment; I repeatedly observed in my own children that they express this by drawing their heads sideways away from the breast or from anything that is offered to them on a spoon. When accepting nourishment and taking it into their mouths they bend their heads forward... When refusing food, especially when it is urged upon them, they often move their heads several times to and fro as we do when we shake our heads to express negation."

The shaking of the head as an expression of negation results from the satiated infant's refusal of additional nourishment. The satiated infant (see p. 473) pushes the nipple away with his tongue and turns his head sideways; quite often his head and the upper part of his body sway to and fro several times, so that a movement ensues that conforms to the later expression of negation. However, according to PREYER (1895), the infant also makes shaking movements with his head at the beginning of the feeding when grasping with his mouth (see p. 473), but these movements soon disappear; the turning away from the breast or bottle, on the other hand, is gradually supplanted by the negative shaking of the head characteristic of the adult. According to DARWIN, the same expression of negation is found in the different races of man, although not without exception. The present-day Greeks and Turks, as well as certain other peoples, throw their heads back as a sign of negation and click their tongues. The throwing back of the head, already known to HOMER, is also traced by DARWIN to the refusal of nourishment. GARNER (1900) saw adult apes shake their heads as a sign of negation.

Nodding as an expression of affirmation is explained by DARWIN as the head movement of the infant grasping the offered nipple of his mother with his mouth and bending his head forward for this purpose. It is true that the hungry infant at times shows other and opposite movements that are caused by the nipple's position in relation to the infant's mouth (see p. 405).

SPITZ (1959) likewise traces the expression of negation to the infant's intake of nourishment; he relates it to the rooting reflex (which he terms "searching behavior") after a lateral touching of the mouth. However, it is simpler, according to DARWIN, to trace the expression of negation to a negative action than to an affirmative action. The expression of affirmation originates, according to SPITZ, in the movement with which the (still hungry) infant of several months follows the evading nipple, lifting his head while doing so. This forward movement of the head in the sagittal direction – it is shown as a conditioned reflex in Fig. 196 (p. 450) – is not yet identical with nodding, but it is, according to SPITZ, tiresome and therefore often ends in a nodding movement in older infants who are in a half-seated position. The expression of affirmation can more easily be traced to the innate nodding of the head occurring as a part of the rooting reflex in the newborn when his lower lip is touched (Fig. 167, p. 405). Of course, maintains SPITZ, this reflex does not exist: "In any case there is no nodding of the head among the innate behavioral patterns that come with the infant to this world."

Man first learns to negate and subsequently to affirm. PREYER's child shook his head at the age of 13 months when somebody said "no, no," but did not nod before the fourteenth month when somebody said "yes, yes." And, the affirmative nodding was not done with the same regularity as the negative head shaking. In the various human races, DARWIN found more exceptions to the usual pattern of affirmation than to that of negation. It seems that observation and education are more essential to create the affirmative than the negative expression. In the Teneriffe chimpanzees of W. KÖHLER, lively nodding of the head was a sign of joyous excitement.

These gestures do not begin to exhaust the infant's expressive possibilities when he wishes to negate and to reject or to affirm and to consent joyfully. Where the adult would be satisfied with a simple shaking of the head, the infant will partially close his lids (PREYER's child did this in the eighth month). The small child withdraws his hand, runs away, throws himself onto the floor and hides his face behind his hands or in his mother's skirt, shakes, and starts to cry or to scream furiously. On the other hand, if something welcome is offered to him, he shouts with joy, waves his hands, dances joyfully around the room, points with his fingers to the desired object, and tries to grasp it. It takes considerable time until, with increasing age, these varied expressive movements gradually diminish and, finally, the head movements remain as almost the only sign of affirmation.

Although the spreading reaction was unknown to him, RACH (1932) described how the sensory organs open during affirmation and close during negation:

> When a child is ready for a medical examination "...he stretches out his open hands. The head is pushed forward, the eyes are open, the forehead is smooth, and the corners of the mouth are turned upward. However, when the child is resisting, the face is turned away, he shrinks back, the forehead is wrinkled, the corners of the mouth are turned down, the lower lip protrudes, the lid openings are narrow, and the hands are kept near the body with the fists clenched."

EBBECKE (1955) gives an explanation for shrugging of the shoulders. He describes a three-year-old girl who comes home alone from a walk and, when asked where her sister is, answers, "gone." In so saying she supinates her open hands, spreads her fingers, and shrugs her shoulders, holding her head obliquely as if she wanted to say: "I have empty hands and what they carried fell out. I have empty shoulders and what they carried is thrown down." DARWIN observed two sisters who shrugged their shoulders at the age of sixteen and eighteen months. PREYER's child did this for the first time at fifteen months.

A tightly closed mouth and thin lips are signs of suppressed mental activity or emotional excitement (KRUKENBERG, 1920). SPENCER (1855) explained the pressing together of the lips occurring at the onset of anger as a desire to prevent the drawing back of the lips and the showing of the teeth that occurs in fully developed wrath. I see in the pressing together of the lips and the closing of the teeth a memory of the refusal of food during the first years of life. Anyone who observes infants or small children who are fed against their will will recognize the origin of this facial expression. As soon as the spoon comes near their mouths, the children turn their faces away; if they are prevented from doing this, they press their lips and jaws together, thus preventing the intake of food or making it more difficult. Hence, the expression "Verbissenheit"* originated from the rejection of food in childhood.

BASIC DETERMINATIVE PHENOMENA

As can be seen from the following tabulation, sensory reactions are based upon distinct laws that explain many otherwise incomprehensible expressive movements, but, of course, not every characteristic of expression and gesture.

I. Reactions of Sensory Organs that Change the Receptive Ability

Facial Organ	*Decreasing Receptive Ability*	*Increasing Receptive Ability*
Mouth	Moving of lips (downward turning of corners of mouth, pouting of lower lip)	Opening of mouth
		Rooting reflex
	Saliva flow, spitting, tight closing of jaws	Sucking movements
	Ejecting of nipple from mouth with tongue	
Eyes	Blinking, lid closing	Wide opening of lids
	Narrowing of pupils	Dilatation of pupils
	Eye-neck reflex	Contraction of frontalis muscle (horizontal forehead wrinkles)
	Contraction of orbicularis oculi, contraction of corrugator supercilii (vertical forehead wrinkles)	Raising of eyebrows
	Contraction of musculus nasalis (dorsal portion) (horizontal nasal wrinkle)	
	Secretion of tears	
Nose	Sneezing	Flaring of nostrils

II. Participation of Nonstimulated Sensory Organs (Spreading Reactions)

Spreading from:	
Mouth to eyes	Blinking or lid closing on biting taste (sour or bitter)
	Tears on biting taste
	Tears after hot fluids
Eye to mouth	Opening of mouth during attentive observation
Nose to eyes	Tears and blinking on stinging odors
Eyes to nose	Sneezing on light stimuli
Ear to eyes	Blinking and forehead wrinkling on sound stimuli
Ear to mouth	Opening of mouth during attentive listening
Pain sense to eyes	Tears (without crying)
Tactile and pressure sense to eyes (foreign body)	Lid closing, tears
Tactile and pressure sense to mouth	Mouth opening and lip movement (rooting reflex), retching

* Literally, "bitingness"; the term means "doggedness" or "stubbornness" and is derived from the clenching of the teeth – The translators.

Briefly stated, the following are the conclusions that can be derived from the preceding observations: The sensory reactions, from which the facial expressions have developed, originally decrease or improve the perceptive ability of the sensory organs of the face by narrowing the openings of the sensory organs through the circular muscles or by enlarging them through the radiating muscles. The secretions of the mucous membranes (saliva, tears, nasal mucus) likewise serve to ward off the stimulus. Stimuli alien to the senses, especially touch, may have the same effect as specific sensory stimuli.

Often the reaction spreads from a stimulated to a nonstimulated organ, which will then likewise assume a better receptive or defensive position. This spreading reaction may occur even when a sensory organ is voluntarily opened or closed, whereby the hands and — to a minor degree — even the feet may act as sensory organs do.

The spreading reaction reminds us of the "displacement activities" in man and animals described in 1940 by TINBERGEN. In this phenomenon built-up pressure is frequently released by means of nervous coordinations which were often operative. Thus, during intense general excitement unexpected movements which in themselves have a clearly defined species-preserving role but do not fit into the given situation at all will often occur. The excitement, so to speak, leaps over into the wrong track and thus leads to a senseless instinctive action (LORENZ, 1943). In this category belong the eating movements of the threatened stickleback and the cleansing movements of pigeons and ducks during sexual excitement. The chimpanzee who is not given the usual reward after completion of its task starts to scratch itself all over its body.

Examples of displacement activities in man, according to TINBERGEN, are the scratching behind the ear in embarrassment and the kiss — the infantile sucking movement — during sexual excitement. Surely custom also plays a certain role in kissing. The small child, though he is not far removed from infancy, has to learn how to kiss (second year) and is not able to do it *a priori* (PREYER). Kissing is by no means a characteristic of all human races, as DARWIN has shown. In some races other signs of affection are known; for example, the New Zealand aborigines and the Laplanders express affection by mutual rubbing of noses. In the baby chimpanzee that W. N. and L. A. KELLOGG (1933) raised, kissing was a frequently occurring expressive movement. TEMBROCK, too, found kissing to be an expression of sympathy in chimpanzees.

Expressive movements develop from unconditioned sensory reactions in the same way as do conditioned reflexes. The nature of the sensations that at first accompany the newborn's sensory stimulations cannot be described. WUNDT (1914) called them "dark" sensations. We may thus attribute to the newborn "dark" sentiments which arise at the same time as the sensations but which last longer. We know from later life that the same sentiments may also occur without sensory stimulation. They are associated with the same reflexes that are caused by sensory stimulation (PIDERIT, WUNDT).

WATSON (1930) showed that affects can be conditioned just as well as simple reflexes; he demonstrated this by combining disagreeable fright stimuli with others which in themselves were harmless. For example, he showed an infant of eleven months a white rat with which the infant had played for weeks and at the same time made a sound with an iron bar behind the child's back. The child had a violent startle reaction, falling forward and hiding its face in the mattress. After this experiment had been repeated several times the child was shown the rat without the noise. It immediately started to scream, and at the same time turned sideways, fell, and crawled hurriedly away.

The infant greets feeding with great joy. This phenomenon is just as much conditioned as the reflexes that develop in connection with food intake (see p. 450). Soon the joyous greeting is destined for the mother even if she does not bring food but just approaches the infant. In this way the infant's tie to his mother gradually develops.

There are a number of expressive movements that are innate and that therefore must have originated otherwise. They develop through processes governed by genetic laws that remain effective later, as VON KRIES postulates (see p. 78).

It is thus evident that expressive movements may develop under the influences of inborn tendencies, imitation, education, and social customs, and it is not usually possible to differentiate among the various influences. DARWIN was able to demonstrate that the basic forms of expression are the same in all human races, and from this fact he concluded that the human species is thus a genetic entity.

An adult seeing another in a happy or sad mood is easily affected by this mood. In the same manner as a tuning fork vibrates with its appropriate tone, "compassion," which is the exact translation of the word "sympathy," originates. The small child shows the same tendency. Without knowing the reason, he laughs and cries with others with all his heart. Even the infant soon mirrors the facial expression of his counterpart. But it can never be ascertained whether or not this is a "conscious" imitation. It is difficult, too, to estimate the role of imitation in the development of the infant's facial expression. The available results of investigation (BÜHLER and HETZER, 1928; GUERNSEY, 1928; KAILA, 1934) do not give a clear answer to this question.

GUERNSEY discovered a smile in response to another smile even in a two-month-old infant. According to BÜHLER and HETZER, an infant three to four months old cannot yet comprehend facial expression, but only the human presence, to which he responds. At the age of five to eight months, however, he mirrors the joyful or rejecting expression, and at the age of eight months he comprehends facial expressions. At first he may imitate an angry look, but then he suddenly breaks out in a bright smile. He thereby frees himself from the compulsory reflex transmission and evidently regards the situation as a game. A fast change of mood, however, is quite a natural occurrence in infants and should not be judged according to adult standards.

KAILA states that in the second three months of life the infant responds with an expression of displeasure to facial grimacing in the area of the eyes, whereas grimaces in the area of the mouth stimulate and amuse him. KAILA found an imitation of facial movements in only approximately 15% of infants, age two to four months, i.e., less frequently than one would expect.

AHRENS (1954, 1956) re-examined the conditions under which the infant smiles for the first time in response to visual stimuli. He examined the effect of devices showing the outline of a face with two small black circles as "eyes."

At the age of one to two months individual marks, best in the form of dots of medium size, will elicit a smile. Small angular marks are less effective. The form of the device is in itself not important. Starting with the second month the effectiveness of the individual marks decreases and is replaced by the expression of the human face. However, the human face cannot yet cause varying reactions; even a very angry look will cause smiles. At the age of two to three months the upper half of the face of the device is sufficient; at four months the lower half is sure to be missed. At five months the appeal of the device decreases, and by six months it has disappeared. Now the body outline of the size of an adult has to be added. At six months and with increasing experience the infant recognizes the adult as nurse and provider of food.

At the age of five months an angry face and a strange expression evoke negative reactions. Pieces of cardboard in front of the eyes and horizontal forehead wrinkles have the same effect. From the fifth to the seventh month the infant responds with a smile to the broadened mouth of the examiner.

AHRENS interprets his findings as do those engaged in animal behavioral research (LORENZ, TINBERGEN) (see p. 607). He sees key stimuli in the individual marks described that fit the successively maturing release mechanisms of smiling or laughing. The hypothesis that the first smile or laugh may be an instinctive movement is supported by its first occurrence as a "vacuum activity" (without an external stimulus). There may be a developmental stage in infants, similar to that in animal babies, where they are especially open to impressions (see p. 609). AHRENS stressed — and rightly so — the biological meaning of smiling: by smiling the child attempts to gain the love and protection of his mother.

There is no doubt that an infant can most easily be made to smile, without being touched, by the sight of his smiling mother. She enjoys just seeing her child and is happy over each sign that she can interpret as a response to her feelings, i.e., when he smiles at her, stretches his arms toward her, or even utters the syllables "ma, ma." How much of his actions is instinctive and how much is an understanding of the sight of his mother remains unknown; and, after all, it does not matter, for sooner or later the understanding of his mother will quite certainly come.

SPITZ (1946) examined the effect of a person's face on 145 infants three to six months old. With few exceptions the infants smiled when the face smiled or nodded, when it assumed a threatening expression to frighten the infant, when the nodding face was covered with a mask, or when a nodding crude rag doll served as stimulus. At this age the infant cannot yet understand the expression of a human face. The effective stimulus consists of parts of the human face combined with movements. During the first half year the smiling response indicates the emotional maturity of the infant.

Since the newborn, who surely cannot observe his environment, can already cry, this expressive movement is certainly no imitation. However, are there also other expressive movements of the individual that develop without any imitation and that are based only on an innate anlage resulting from genetic laws? These questions can be answered through the observation of the blind and especially of the deaf-mute blind. Although the facial expressions of such children have so far not been thoroughly examined, some data can be gathered from the existing sparse reports in the literature.

According to DUMAS (1932), the blind person is completely unaware of his facial expression; he notices it only when his attention is drawn to it. As BIRCH-HIRSCHFELD (1880), DE SANCTIS (1906), KRUKENBERG (1920), and MISTSCHENKO (1934) stated, blind persons who become blind sometime after birth and those who have still retained partial vision possess facial expressions that do not essentially differ from those of seeing persons. In those who become completely blind shortly after birth, the ocular area (orbicularis oculi, corrugator supercilii, and frontalis muscles) is not involved in the facial expression; these muscles remain rigid and motionless, but the mouth has complete movement, albeit, perhaps, in a somewhat cruder form. The explanation is simple: since the blind do not experience an immediate sensory reaction through their eyes, they do not have the prerequisite for the formation of the expressive movements originating from the eye.

Blind children also differ in other respects from their seeing contemporaries. Perusal of the pedagogic literature on the blind shows that "eye-boring" has been known for a long time; FRANCESCHETTI (1947) described it as a digital-ocular sign. According to K. MÜLLER (1955, 1957) and MACKENSEN (1956), blind infants usually press their fists or the knuckles of their index fingers to their eyes. Older children bore with their thumbs or index fingers between the eyeballs and the orbital rims or press several fingers or their whole hands on the closed lids of their eyes. In most blind children the eye boring begins during the first year of life and disappears during the fifth and sixth year. Sometimes it is found in retarded children even without ocular disorders (K. MÜLLER).

The eye-boring may be combined with other uniform rhythmic movements. These include to-and-fro movements of the head while lying down, similar movements of the upper part of the body while sitting, rocking movements while standing or walking, tapping with the hands, and similar movements. Frequently blind children may show several of these patterns at the same time (K. MÜLLER, MACKENSEN). Similar movements are quite often found in childhood as an expression of institutionalization (see p. 630).

Blind deaf-mutes are still less accessible to environmental impressions than are the blind. Their expressive movements evidently are mostly on an innate basis and are not individually acquired. GOODENOUGH (1932) describes a ten-year-old girl who was congenitally deaf-mute and blind, cause unknown. This child's laughter was clear and musical and could not in any way be distinguished from that of a normal child. It is thus evident that laughing can be learned without imitation. One of the illustrations shows this child in a state of puzzled attention because a small doll had been placed between her neck and dress. The mouth is half open. The blind eyes, too, are open as far as possible and the eyebrows are raised (spreading reaction). The same expression of astonishment as that in a child having all his senses has been produced. Thus, it appears that the expressive movement of astonishment is probably innate.

Further research on suitable human beings will no doubt be very rewarding. For the time being we do not know the exact degree to which imitation may influence the development of the facial expression. That imitation does have such influence may be seen in the response of each infant to the smile of his mother. Perhaps imitation does not influence the quality of the expressive movement so much as the child's age at the first appearance of this movement.

The stronger the influencing stimulus and the younger the child, the more extended and distinct the expressive movements. Stronger stimuli and irritations break through, so to speak, the separating layers of the individual nerve pathways. The younger the child, the less effective are the separating layers from the start and the wider the spread of the excitement. The infant still cries and exults with his whole body; he kicks happily with his arms and legs. The small child throws himself on the floor screaming with rage and thrashing about in all directions. In the adult, though, the corresponding expressive movements are much more inhibited; only the most powerful and painful sensory irritations and the strongest excitements affect his whole body. The chimpanzee's behavior seems to be similar to that of the infant (ROTHMANN and TEUBNER, 1915).

The younger the child, the more is his cerebral function governed by affects. In the completely awake infant whose nervous energy does not happen to be occupied with food intake, the periods of tranquillity are considerably shorter than those in the adult. Only comparatively late and under the inhibiting influence of the cerebral hemispheres does the mood become more balanced. In early childhood, emotions still immediately follow internal or external stimulation. Past events may perhaps influence the mood by the formation of conditioned reflexes or in this case conditioned affects, but the contemplation of past or the anticipation of future joy or suffering is the more alien to the child the younger he is.

The expressive movements of emotions, affects, and conceptions originate from the sensory reactions, and these in turn developed from the central phenomena accompanying sensory reactions. Only sensory reactions depend directly on the stimulation of the senses. Emotions and affects arising at the same time outlast the actual stimulation. At a higher stage of cerebral development the same emotions and affects may arise without external stimulation, and they too are accompanied by the respective expressive movements. However, it will never be possible to determine to what degree these emotions and affects can be perceived as such by the newborn.

Involuntary expressive movements are regulated by the striatum and pallidum. HESS (1949) was able to elicit rage and affective defenses in an intact cat by electrically stimulating certain areas of the diencephalon, where he had placed very fine electrodes. VON HOLST (1957, 1958) placed extremely fine electrodes in the brain stem of freely roving birds and by stimulation elicited natural instinctive behavior, such as flight, evasion, and attack, toward animals of different and the same species.

The lower the phylogenetic and ontogenetic stage, the more predominant the true sensory reactions, which are transmitted by the brain stem. The presence of these reactions in newborns and anencephali shows that these reactions and also certain affective expressive movements are possible without the action of the cerebral hemispheres. Sensory reactions are preserved in the higher developmental stages, but they are repressed by the higher cerebral functions, which developed from these sensory reactions.

Language, the basis of the highest human brain function and culture, is possible only through the senses. Also, thinking did not appear suddenly and abruptly;

distinct beginning stages also dependent on simple stimulation of the senses can be traced. The merely "practical" intelligence (STERN, 1928) forms the lowest step in the development of intelligence. It acts under the influence of internal and external stimuli for the present only and, for example, assists by indirect means in reaching desired visible objects that cannot be simply grasped (W. KÖHLER, 1915–1918). Such thinking, which can be found in the young child as well as in the chimpanzee, consists of the expedient use of existing sensory impressions. "Theoretical" intelligence (STERN), which has reflective thought as its goal, develops later. Since an affect may occur in the adult without any immediate external stimulation, thinking processes that are also independent of immediate external influences finally occur in him. Expressions such as "sensing," "grasping," "comprehension" and "conception," "perceiving" and "insight" may be explained by the fact that these terms have been transferred from the sensory sphere to that of thought. Thus "Scharfsinn"* and "Schwachsinn"* do not denote the sharpness of the senses, but rather the intellectual endowment.

That a lower sense, such as taste, also participates to a great extent can be seen in the numerous expressive movements of refusal depending on taste, as has been described. We surely cannot weigh the influence of the individual senses, but I cannot agree with K. BÜHLER (1922) when he attributes scant importance to the sense of taste for the individual's psychological development. We can see it in etymology: the Romans formed the word "sapientia" from the word "sapere," which means "to taste," and we speak of "sweet love," "bitter disappointment," or "sour expression." These figures of speech are not just simple similes or metaphors; neither are the expressive movements of sadness, which reproduce the sensory reactions of bad taste by the turning down of the corners of the mouth. Even though these designations for thinking processes develop through language usage, their original kinship with sensory processes is not thereby eliminated.

Following a similar train of thought, LORENZ calls to our attention expressions directly bound to the central representation of space that are still being used today for the highest and most abstract performances of the human mind:

> "We gain 'insight' into an 'entangled' 'connection' – like an ape in an entanglement of branches – but we really have 'grasped' an 'object' only if we 'comprehend' it completely. In the last three expressions the age-old primacy of the haptic over the optic is brought out in a beautiful manner" (see p. 79).

To prove that all thinking is spatial in origin, LORENZ quotes PORZIG (1950):

> "Language translates all nonvisual relations into spatial relations. Not just one language or one group of languages does this, but all without exception do it. This characteristic is one of the unchangeable traits (invariables) of human language. Time relations are expressed in space: before or after Christmas, within a space of two years. In mental processes we speak not only of 'outside' or 'inside' but also 'above and below the threshold' of consciousness, of the 'subconscious,' of the 'foreground' or 'background,' of 'depths and layers' of the mind. Space serves as a model for all nonvisual relations: besides working, he teaches;

* Scharfsinn, literally "sharp sense," means sagacity or keen mind; Schwachsinn, literally "feeble sense," means feeble-mindedness. The English expressions "good sense," "common sense," or "sensible" are analogous — Translators and Editor.

ambition was bigger than love; behind this measure was the intention – it is superfluous to amass examples that can be collected in any desired number in any piece of written or spoken language. This phenomenon gains importance from its general extent and from the role it plays in the history of language; it is not only in the use of prepositions, which originally designate everything related to space, that we can demonstrate this, but also in verbs and adjectives."

We have already discussed in some detail how philosophy and psychology, as well as the natural sciences, also trace the higher life of the psyche to the activity of the senses (see p. 15).

PHYSIOGNOMY

There's no art
To find the mind's construction in the face.

SHAKESPEARE, *Macbeth* I, 4

Physiognomy, according to LICHTENBERG (1778), is the art of finding the construction of the mind and heart from the form and condition of the external parts of the human body, especially of the face (except for all temporary emotional signs). Efforts of this kind have also enjoyed great popularity in nonscientific circles because great significance for daily life was attributed to this art. It would be of the utmost importance if the facial expression were to betray the mental qualities and if we could discern a person's intellect, mental endowment, and character from his face. Other methods of achieving this goal have also been tried, such as interpreting handwriting, lines on the palm, and, more recently, capillaries of the finger tips. However, in spite of all such efforts no reliable standard has so far been found.

Thus, the history of physiognomy, which dates back to ARISTOTLE and has been continued by LAVATER, GALL, and many others, is a history of errors. Even the most recent literature on the subject does not reveal an attempt to find scientific proofs for the physiognomic theories. To be a good physiognomist one has only to possess the right kind of imagination, as KASSNER (1922, 1932), one such author, boasts of himself.

LAVATER, who had even GOETHE temporarily under his spell, was violently attacked by LICHTENBERG (1778), who disproved his theories. LICHTENBERG had already recognized the relations between transient facial expressions and the constant expression of the face: "Pathognomonic signs (expressions of passions) do not always disappear completely but leave physiognomic impressions." SCHILLER (1793) expressed it similarly: "The constant features were originally nothing more than movements which finally, with frequent repetition, became habitual and imprinted lasting traces."

These words of SCHILLER's explain why no lasting facial expressions remain in the infant in spite of the lively play of his features: the same facial expression has to be repeated frequently before it will leave lasting traces. The formation of wrinkles is also an outcome of the age-related, gradually progressive loss of fat of the subcutaneous tissue, a process which is culminated in old age. In contrast, the faces of healthy infants remain free of wrinkles, so that infants at rest are virtually expressionless. Since aging causes additional differences, children's faces seem much more similar to each other than those of adults. Thus, physiognomy can be applied to the child even less than to the adult.

Only in exceptional cases did the painters of the Middle Ages succeed in reproducing the facial expression, or rather the lack of a constant facial expression, of infancy. In paintings of Madonnas, the Christ child does not have the appearance of a child at all; for example, when sitting on the lap of his mother he blesses the believers with his raised hand. The infants look too old, not so much because of their body size, but because of the proportions of the body parts (for example, the heads are comparatively too small). MANTEGNA, though, pictured a Madonna with a real newborn (as illustrated in the books of FISCHL and KRUKENBERG).

In infancy, illnesses of long duration, especially nutritional disturbances, cause a premature loss of fat of the subcutaneous tissue so that wrinkles form and the face has a senile appearance, even more so because the infant is toothless.

SUMMARY

During phylogenesis the sensory reactions from which the facial expression has developed decrease or increase the receptive ability of the sensory organs in the face. The circular muscles narrow and the radiating muscles distort or enlarge the openings of the sensory organs. Fluid secretions (saliva, nasal mucus, tears) serve as a defense against irritation. Stimuli remote from the sensory organs, especially of touch, may have the same effect as do the specific sensory stimuli.

Frequently the reaction spreads from a stimulated to a nonstimulated organ, which in turn assumes an accepting or defensive position. This spreading reaction may occur even when a sensory organ is voluntarily opened or closed, whereby the hands and, to a minor degree, even the feet may behave as sensory organs. Some characteristics of the facial expression can thus be explained as spreading reactions.

The most important determinants of the facial expressions are the eye and mouth and to a lesser degree the nose; however, even the ear may be at least the source of a spreading reaction.

From the various associations and sequences of the reactions of the facial muscles and their combinations with the activity of the respiratory center originate the most important expressive movements of the infant. Often we are not able to recognize precisely the role that developmental laws, imitation, and education play in this process.

The mouth and eyes assume a defensive position at the beginning of a bad mood as well as in crying and screaming. The facial expressions of defiance, negation, and stubbornness correspond to those signifying refusal of food. The laughing infant opens his mouth and eyes, but a certain defensive position of the eyes may also occur. That smiling is connected with the reactions to food intake is doubtful. Affirmation, attention, astonishment, and fear open the sensory organs.

Facial expressions, gestures, emotions, and affects are more lively, less restrained, and less inhibited in infants than in adults.

The higher brain function of man has developed from the senses, as shown by their accompanying expressive movements.

REFERENCES

AHRENS, R.: Zschr. exper. u. angew. Psychol. *2*, 412 and 599 (1954); Nervenarzt *27*, 108 (1956).
ALBERTS, M.: Nervenarzt *26*, 228 (1955).
ALDRICH, C. A., M. NORVAL, C. KNOP, and F. VENEGAS: J. Pediatr. *28*, 665 (1946).
–, C. SUNG, and C. KNOP: J. Pediatr. *26*, 313 (1945); *27*, 89 and 428 (1945).
AMES, L. B.: J. Genet. Psychol. *59*, 239 (1941).
ANDRÉ-THOMAS and AUTGAERDEN: Psycho-Affectivité des premiers mois (Psycho-affectivity of the First Months). p. 99. Paris, 1959.
ARISTOTLE: Naturgeschichte der Tiere (Natural History of Animals). Book 7, Ch. 9, 3, 166 of the German translation by A. Karsch. Stuttgart, 1866.
BECHER, H.: Wien. klin. Wschr. *1932*, II, 1155.
BECHTEREV, H. v.: Objektive Psychologie (Objective Psychology). Leipzig and Berlin, 1913.
BELL, C.: Anatomy and Philosophy of Expression. 6th ed. London, 1872.
BIRCH-HIRSCHFELD, F. V.: Dtsch. Rundschau *6*, 4, 41 (1880).
BLAKE, J. R., and E. C. KUNKLE: Arch. Neurol. (Am.) *56*, 337 (1951).
BLANTON, M. E.: Psychol. Rev. *24*, 456 (1917).
BRANDES, G.: Buschi. p. 41. Leipzig, 1939.
BRAUS, H.: Anatomie des Menschen (Human Anatomy). Vol. 1. Berlin, 1921.
BÜHLER, C.: Zschr. Psychol., Leipzig *132*, 1 (1934).
–, and H. HETZER: Zschr. Psychol., Leipzig *107*, 50 (1928).
BÜHLER, K.: Ausdruckstheorie (Theory of Expression). Jena, 1933.
–, Die geistige Entwicklung des Kindes (The Mental Development of the Child). 3rd ed. Jena, 1922.
BÜRGI, E.: Münch. med. Wschr. *1934*, 1137.
CARUS, C. G.: Vorlesungen über Psychologie (Lectures on Psychology) (1831). New edition. Leipzig, 1931.
CASTRODE, A. D.: Rev. neurol. *1*, 15 (1926).
CHARMA: according to DARWIN.
DARWIN, C.: Biographische Skizze eines kleinen Kindes (Biographic Sketch of a Small Child) (1877). Collected short writings translated into German by E. Krause. pp. 2, 139. Leipzig, 1886.
–, Der Ausdruck der Gemütsbewegungen bei Menschen und Tieren (The Expression of Emotions in Man and Animals). German translation, Stuttgart 1872.
DRÄSECKE, J.: Zschr. Nervenhk. *68/69*, 344 (1921).
DUMAS, G.: Bull. Acad. Méd. Paris *107*, 17, 607 (1932).
EBBECKE, U.: Dtsch. med. Wschr. 1955, 45 and 88.
EBBINGHAUS: Grundzüge der Psychologie (Fundamentals of Psychology). Vol. 2, p. 633. Leipzig, 1913.
FEER, E.: Diagnostik der Kinderkrankheiten (Diagnosis of Children's Diseases). Berlin, 1921.
FERVERS, C.: Der Ausdruck des Kranken (The Expression of the Sick). Munich, 1935.
FISCHL, R.: Mschr. Kinderhk. *43*, 411 (1929).
FRANCESCHETTI, A.: Ophthalmologica (Basel) *114*, 332 (1947).
GARNER, L. R.: Die Sprache der Affen (The Language of the Ape). Leipzig, 1900.
GOODENOUGH, F. L.: J. abnorm a. soc. Psychol. (Am.) *27/3*, 328 (1932).
GÖTT, T.: Zschr. Neurol. *66*, 92 (1922).
GRATIOLET, P.: De la physiognomie et des mouvements d'expression (Physiognomy and Expressive Movements). 3rd ed. Paris, 1873.
GRUHLE, W. H.: K. Birnbaums Handwörterbuch der medizinischen Psychologie (Dictionary of Medical Psychology). Leipzig, 1930. Ausdruckspsychologie, Mimik, Kindheit (Expressive Psychology, Facial Expression, Childhood).
GUERNSEY, H.: Zschr. Psychol., Leipzig *107*, 105 (1928).

HAUSER, F.: Documenta Geigy. Goma, das Basler Gorillakind (Goma, The Basel Gorilla Baby). Bull. 4. Fig. 6. Basel, 1960.

HAYWORTH, D.: Ref. Zbl. Neurol., Leipzig *53*, 567 (1929).

HERSCOVITZI, J.: Zbl. ges. Neurol. *31*, 90 (1922).

HEMPELMANN, F.: Tierpsychologie (Animal Psychology). Leipzig, 1926.

HESS, W. R.: Das Zwischenhirn (The Diencephalon). Basel, 1949.

HIPPOCRATES: Sämtliche Werke (Complete Writings). 3, 648. German translation by R. Fuchs. Munich, 1900.

HOCART, A. M.: Ref. Zbl. Neurol., Leipzig *51*, 405 (1929).

HOLLIS, W. A.: Lancet *1913*, I, 23.

HOLST, E. VON: Naturwissensch. *44*, 549 (1957); *45*, 579 (1958).

KAILA, E.: Zschr. Psychol., Leipzig *132*, 1 (1934).

KASSNER, R.: Physiognomik (Physiognomy). Munich, 1932.

—, Die Grundlage der Physiognomik (Fundamentals of Physiognomy). Leipzig, 1922.

KELLOGG, W. N. and L. A.: The Ape and the Child. New York and London, 1933.

KILLIAN, H.: Facies dolorosa (The Painful Facial Expression). Leipzig, 1934.

KOEHLER, O.: Umschau *54*, 321 (1954); Zschr. menschl. Vererb. u. Konstit.lehre *32*, 390 (1954).

KRETSCHMER, E.: Medizinische Psychologie (Medical Psychology). 4th ed. Leipzig, 1930.

KRUKENBERG, H.: Der Gesichtsausdruck des Menschen (The Facial Expression of Man). 2nd ed. Stuttgart, 1920.

LANGE-EICHBAUM, W.: Genie, Irrsinn und Ruhm (Genius, Madness and Glory). p. 282. Munich, 1928.

LERSCH, P.: Gesicht und Seele (Face and Soul). Munich, 1932.

LESNY, I.: Zentralbl. Kinderheilk. *77*, 227 (1961).

LESSING, G. E.: Brief an Mendelssohn (Letter to Mendelssohn). November 18, 1756.

LICHTENBERG, G. C.: Über Physiognomik wider die Physiognomen (On Physiognomy against the Physiognomists) (1778), in Lichtenberg's Miscellaneous Writings. Vol. 4, p. 1. Göttingen, 1844.

LÖBISCH, I. E.: Entwicklungsgeschichte der Seele des Kindes (Ontogenesis of the Soul of the Child). p. 41. Vienna, 1851.

LOGAU, F. v.: Sämtliche Sinngedichte (Collected Epigrams). Bibliothek des Stuttgarter literarischen Vereins. Vol. 113, p. 309 (2 Thousand, 4 Hundred 90). Tübingen, 1872.

LORENZ, K.: In G. Heberer. Die Evolution der Organismen (The Evolution of Organisms). p. 115. Jena, 1943. 2nd ed., p. 153. Stuttgart, 1954.

—, Z. Tierpsychol. *5*, 235 (1943).

MACKENSEN, G.: Die Psychomotorik blinder Kinder (Psychomotor Activity of Blind Children). Beiheft 26 d. Klin. Mbl. Augenhk. Stuttgart, 1956.

MAYR, F.: Jb. Kinderhk. *2*, 24. Vienna, 1858.

MISTSCHENKO, M. N.: Ref. Nervenarzt *7*, 522 (1934).

MORGAN, S. S. and J. J. B.: J. Pediatr. *25*, 168 (1944).

MÜLLER, J.: Zur vergleichenden Physiologie des Gesichtssinnes der Menschen und Tiere (Comparative Physiology of the Vision of Man and Animals). p. 289. Leipzig, 1826.

—, Physiologie des Menschen (Physiology of Man). Vol. 2, p. 92. Koblenz, 1840.

MÜLLER, K.: Mschr. Kinderhk. *103*, 219 (1955); Dtsch. Zschr. Nervenhk. *177*, 150 (1957).

PFAUNDLER, M. v.: Semiotik der Kinderkrankheiten, in Pfaundler-Schlossmanns Handbuch der Kinderkrankheiten (Semiotics of Children's Diseases, in Pfaundler-Schlossmann's Handbook of Children's Diseases). 1st. ed., p. 37. Leipzig, 1906.

PIDERIT, T.: Mimik und Physiognomik (Facial Expression and Physiognomy) (1867). 3rd ed. Detmold, 1919.

PLINY THE ELDER: Naturgeschichte (Natural History). 2, 2 (Book 7, Ch. 1) of the German translation by Wittstein. Leipzig, 1881.

PORZIG, W.: Das Wunder der Sprache (The Miracle of Language). Dalp Collection. 2nd ed. 1957, pp. 24, 209. Munich (Bern), 1950.

RACH, E.: Wien. klin. Wschr. *1932*, 39, 40.

REUSS, A.: Physiologie des Neugeborenen, in Halban-Seitz, Biologie und Pathologie des Weibes ("Physiology of the Newborn," in Halban-Seitz' Biology and Pathology of the Woman). Vol. 8, p. 2. Berlin and Vienna, 1927.
RIPIN, R., and H. HETZER: Zschr. Psychol., Leipzig *118*, 82 (1930).
ROTHMANN, M., and E. TEUBNER: Abh. preuß. Akad. Wiss., Berl., Physik.-math. Kl. 1915.
RUSINOVA, J. G.: Zbl. Neurol. *74*, 611 (1935).
SANCTIS, S. DE: Die Mimik des Denkens (The Facial Expression of Thinking). German translation. Halle, 1906.
SCÄCHTER, M.: Clin. pediatr. *17*, 433 (1935).
SCHENKEL, R.: in Goma, das Basler Gorillakind (in Goma, the Basel Gorilla Baby). Bull. No. 2. Basel, 1960.
SCHILLER, F.: Über Anmut und Würde (On Grace and Dignity). 1793.
SOLTMANN, O.: Jb. Kinderhk. *26*, 206 (1887).
SOMBART, W.: Vom Menschen (On Man). Berlin, 1938.
SPENCER, H.: Die Prinzipien der Psychologie (The Principles of Psychology). Vol. 2 of German translation. Stuttgart, 1855.
SPITZ, R. A.: Genet. Psychol. Monogr. *34*, 57 (1946).
–, Nein und Ja (No and Yes) (Supplement to "Psyche"). Stuttgart, 1959.
STERN, W.: Die Intelligenz der Kinder und Jugendlichen (The Intelligence of Children and Adolescents). 4th ed. Leipzig, 1928.
STRÖDER, J., and B. SANDEN: Ann. Paediatr. *184*, 373 (1955).
TEMBROCK, G.: Grundzüge der Schimpansenpsychologie (Fundamentals of the Psychology of the Chimpanzee). Berlin, 1949.
THOMAS, H.: Unzweckmäßige Mitbewegungen (Purposeless Associated Movements). Diss. Leipzig, 1943.
THORNDIKE, E. L.: Psychologie der Erziehung (Psychology of Education). German translation. Jena, 1922.
TINBERGEN, N.: Zschr. Tierpsychol. *4* (1940).
TISCHLER, H.: Zschr. Psychol. *160*, 209 (1957).
VIRCHOV, H.: Arch. Anat. *1908*, 371.
VERGIL: Ländliche Gedichte (Bucolics). *1*, 128 of German translation by J. H. Voss. Altona, 1830.
WATSON, J. B.: Der Behaviorismus (Behaviorism). Berlin and Leipzig, 1930.
WEINGREW, S. M.: Arch. Pediatr. (Am.) *51*, 501–515, 567–588, 624–643 (1934).
WOLFF, P. H.: Psychosomat. Med. (Am.) *21*, 100 (1959).
WUNDT, W.: Dtsch. Rundschau *3*, 7, 120 (1877).
–, Völkerpsychologie (Ethnologic Psychology, Vol. 1: Language). 3rd ed. Leipzig, 1911. Grundriß der Psychologie (Fundamentals of Psychology). 12th ed. Leipzig, 1914.
WYSS, W. H. VON: Nervenarzt *2*, 534 (1929).
–, Körperlich-seelische Zusammenhänge von Gesundheit und Krankheit (Body-Mind Relations in Health and Sickness). Leipzig, 1931.
YERKES, R. M. and A. W.: The Great Apes. p. 569. New Haven, 1929.

4. REFLEXES OF POSITION AND MOVEMENT

(1, 17, 36, 41, 59, 95, 140, 141)

From birth the body, with all its parts, is subject to the force of gravity. Many movements and reflexes are either facilitated or impeded by, but are otherwise independent of, the force of gravity. In this category belong, for example, the protective skin reflexes (see p. 30), the rooting reflex (see p. 404), and the eye-neck reflex (see p. 60). The newborn is already able to right the position of head, trunk, arms, and legs in relation to each other so that standard positions are repeatedly assumed. The body-head righting reflex and the neck-body reflex serve this purpose (see p. 166).

However, the newborn almost completely lacks the ability to make a movement whose goal it is to oppose gravity. He cannot yet place his head in the right position in space. It takes a year until the child can walk freely and erect. Quite considerable efforts of his own are needed to enable him to accomplish successively the following: hold his head up, sit, sit up, stand, creep, crawl, stand up, and finally take the first free steps. Each small progress succeeds only after numerous failures.

During the first year, reflexes of position and movement develop that aid the infant in his efforts and that later lose importance as the development of higher centers makes them superfluous. They do not disappear completely and may be reactivated when pathological loss of the function of the inhibitory higher centers occurs.

Equilibrium is assisted by the receptors of the sense of position and movement, which control the position of the individual limbs in relation to each other and the body position in relation to the force of gravity. Knowledge of these connections began with the basic experiments of FLOURENS, who in 1824 destroyed the semicircular canals of pigeons and afterward observed striking disturbances in movement.

The eyes, the pressure organs of the skin, and the proprioceptors of the muscles, tendons and joints, and, most of all, the inner ear serve as receptors for the sense of position and movement. Often several receptors of different sensory organs act simultaneously to complete a certain task so that many reactions are secured by many means and, therefore, persist even if a sensory organ or a brain portion is not functioning. These stimulations, originating from the sensory organs, are not assembled in one area. There is no "static" cerebral cortex that could be compared to the optic or the acoustic cortex (MAGNUS, 1924, p. 5). This makes an understanding of these fundamental nervous phenomena difficult, especially since they occur almost unconsciously.

Our knowledge has been much enlarged by MAGNUS (1924) and his co-workers, most of all RADEMAKER (1926, 1931). They experimentally found a great number of position and movement reflexes in animals by sections through the brain stem, excising certain cerebral areas or inactivating them by means of toxic substances, or by eliminating sensory organs. Only after these operations was it possible to observe previously undetected groups of reflexes.

According to MAGNUS, the tonic neck and trunk reflexes, whose proprioceptive afferent pathways originate in the neck and otoliths of the labyrinth, pass through the rhombencephalon and the upper cervical cord, and the righting reflexes are transmitted by centers of the pons and mesencephalon. From their animal experiments MAGNUS and RADEMAKER had assumed that the righting reflexes depend on the red nucleus. However, MUSSEN (1927) found that in animals with intact cerebral hemispheres the electric destruction of both red nuclei caused only ataxia of several hours duration but not a loss of the righting reflexes. BARD (according to SCHALTENBRAND, 1956) saw the righting reflexes reappear after a prolonged interruption in deeply decerebrated animals kept alive for a long time.

JUNG and HASSLER (1960) therefore indicated that MAGNUS's and RADEMAKER's views, the core of which is that body position depends on reflex balance, were too limited and have to be supplemented by HESS's (1954, 1956) observations of direction-specific motion responses in cats after electric stimulation of the mesencephalon and diencephalon, by comparison of the behavior of different animal species, and by clinical observations of extrapyramidal movements in man. MAGNUS overestimated the importance of the peripheral regulation originating in the labyrinth and in the muscle receptors and underestimated the innate coordinated activity of the central nervous system occurring without external influences. Only both together insure body posture. The balance of different forces, from which the normal body posture results, does not come about through a balance of reflexes, but through a central regulation by certain brain stem centers that can be influenced by receptors and reflexes. However, it is possible to confirm that body position is regulated in centers above the pons. Some formations in the tegmentum and tectum mesencephali and in the diencephalic subthalamic nuclei are most important for this purpose, and HESS's experiments permit us to relate this function to certain nuclei and pathways of the brain stem.

In spite of the great progress initiated by the electric investigations of HESS, the fact remains that MAGNUS discovered a series of reflexes that participate in the regulation of body posture. These reflexes can hardly be specifically recognized and distinguished in the adult because his body position is in various ways secured and determined by the cerebral hemispheres. In the adult too, however, the deeper reflexes can be demonstrated when pathological changes eliminate the influence of higher centers.

Animal experiments always leave open the question of how far they can be applied to man. JUNG and HASSLER stress that most of MAGNUS's experiments were made on rabbits, RADEMAKER's on dogs, and HESS's on cats. These animal species differ even from each other in their extrapyramidal movement systems. The difference between them and man is still greater, for it is much easier for the quadruped to maintain equilibrium than it is for the erect and merely biped man. Man and animal also differ basically in their mode of independent locomotion.

In the development of the static abilities during the first period of life we can distinctly follow the basic law, which generally governs the slow awakening of the human brain function and which, for example, can be recognized equally well in the maturing respiratory center (see p. 313): The individual brain parts start to function in the same sequence in which they have been formed phylogenetically and ontogenetically. Thus, from the reflex groups, which appear during the first year of life and which correspond mostly to those found by MAGNUS and RADEMAKER, we may draw conclusions on the pattern of human brain function. However, investigations on normal children do not permit us to draw conclusions on the areas in which the special centers are located and which regulate the especially frail equilibrium of man.

At birth the various brain portions do not function equally well; rather, their function starts from the midbrain and continues to the cerebral hemispheres, which at birth are still neurologically inactive. Therefore, they have no inhibitory influence on the deeper cerebral portion and as a consequence the reflexes of the midbrain are not suppressed and become manifest. Many static reflexes can be more easily shown in the infant than in the adult.

Although infants vomit easily, they do not become seasick and are not disturbed by a rocking motion. The reflexes of position and movement must reach a certain stage of development for their coordination to be disturbed.

The young infant at the beginning is in a state that the physiologists try to reach artificially in animal experiments by eliminating higher brain parts in order to ascertain their influence on brain function. Their experiments are grossly contaminated by the operative procedure so that it quite often is not possible to differentiate between the elimination of a brain part and the consequences of the surgical intervention. On the other hand, in the infant we have a being that lacks the function of higher cerebral regions and is not in postoperative shock. Since the animal experiments of the physiologists permit only limited conclusions pertaining to man, they fortunately can be supplemented by observations of the young infant.

The inner ear contains the sensory organs that are sensitive to progressive forward or turning movements as well as to changes of direction in relation to gravity.

Each of the three semicircular canals originates in an ampulla membranacea from the utriculus. The cristae ampullares project into the three ampullae as comb-like, semilunar ridges. Their sensory epithelium, which is innervated by the vestibular nerve, contains bulky cells with very fine hairs covered by a jelly-like mass, the cupula cristae ampullaris, which on cross section has the appearance of a broad lancet. This cupula rises to the roof of the ampulla so that it forms a separating wall between the endolymph on both sides at rest and on rotation. According to the MACHE—BREUER theory (1874—1875), the sensory epithelium of the crista ampullaris is stimulated when shifting of the endolymph, which is caused by progressively accelerated or decelerated rotation in the plane of the semicircular canal, deflects the cupula. STEINHAUSEN (1927, 1931, 1932, 1933, 1934) succeeded in directly observing the cupula of a living pike and was able to confirm that the cupula is deflected by heat or by progressively accelerated or decelerated rotation, as postulated by the MACH–BREUER theory.

The tracing of the action potentials of vestibular nerve fibers (ADRIAN, 1943) proved that most fibers are constantly firing in a regular rhythm. These potentials decrease or disappear completely during rotation in the direction of the ampulla; after stopping they temporarily reappear in greater numbers. Their frequency is at first increased as the cupula rotates away from the ampulla, but rapidly diminishes as the deflection decelerates. At a sudden stop the impulses cease for a while, as in the deflection of the cupula to the other side.

In the vestibular part of the labyrinth the fibers of the vestibular nerve innervate the columnar epithelium, which forms the maculae of the sacculus and utriculus. This epithelium is covered with hairs that have at their ends a jelly-like mass. The otoliths, very fine prismatic crystals of calcium carbonate, are imbedded in this mass. Because gravity continuously attracts the otolithic membrane, the otoliths continuously irritate the epithelial hairs by pulling, pressing, or shearing them, depending on the position of the head in relation to them. In addition, progressive movements (acceleration or deceleration) in a straight direction affect the same area (HASEGAWA, 1932). The different effects of the pressure exerted by the otoliths and their changes due to shifts of the head position relative to the direction of gravity comprise the stimuli for the sensory epithelium. Thus, the relation of the head to the vertical axis is continuously checked and if necessary changed by reflex action.

REFLEXES OF POSITION AND MOVEMENT INDEPENDENT OF GRAVITY

VESTIBULAR REACTIONS

Caloric Stimulation

In caloric examination the auditory canal is irrigated with cold or warm water. In the adult, warm water causes nystagmus to the irrigated side and cold water, nystagmus to the opposite side. THORNVAL (1920/21) examined 74 newborns and always obtained a caloric reaction. SCHUR (1922) found the same reaction in response to irrigation as to rotation; he was able to elicit a distinct nystagmus in two newborns. BERBERICH and WIECHERS (1924) irrigated with cold water and noted deviations of the eyes toward the irrigated side in normal and premature newborns. Nystagmus occurred only rarely. Children just beginning to run about, according to SCHUR, were calorically very sensitive. Adult values were obtained only in the third year. VOSS (1927) irrigated with ice-cold water and observed nystagmus or deviation of the eyes in normal infants; when he irrigated with warm water, he noted deviations in the opposite direction. Also, according to GALEBSKY (1927), all three semicircular canals react to caloric stimulation immediately after birth. According to CATEL (1928), a weak cold stimulation is below the threshold for the vestibularis of fully mature infants and older children, but is often a threshold value for prematures. The absence of the reaction does not prove, as VOSS and BERBERICH and WIECHERS thought, damage of the vestibularis from birth trauma. However, the frequently noted reaction in the prematurely born infant points to a higher reflex excitability. CATEL found nystagmus on cold stimulation in only 17% of normal newborns, small children, and school children.

Electrical Stimulation

For electrical stimulation of the vestibularis in adults a current of 3 ma is necessary; to obtain the same result in children of sitting and walking age SCHUR needed 5 to 7 ma and in newborns 8 to 10 ma. SCHUR explains this by stating that it is a known fact that the younger the child, the more difficult it is to stimulate the nerves electrically.

The report of BALDENWECK and GUY-ARNAUD (1940), that on electrical vestibularis stimulation they found the same values in normal and premature newborns as in adults cannot be reconciled with SCHUR's findings. As they found only a few lower values in the caloric and rotatory testing, they considered the vestibular apparatus of the newborn to be completely developed.

Stimulation by Progressive Movements, Mono Reflex

The usual examination of the labyrinth on a rotating chair is executed in adults as follows: The subject is turned ten times with increasing speed, and then the rotation is abruptly stopped. Nystagmus to the left follows rotation to the right, and nystagmus to the right follows rotation to the left. Eye deviation and head turning can be easily prevented in the adult by focusing and other muscle influences.

Corresponding experiments in infants had the following results: In many trials BARTELS (1910) was always able to elicit nystagmus immediately after birth; it could best be shown after the seventh day of life. If the infants were rotated on a horizontal plane, their heads made distinct countermovements. In premature newborns nystagmus could only be elicited several days after birth. ALEXANDER (1911), who rotated the newborns in a dorsal position, found this reaction in 92 of 118 trials. In prematurely born infants he noted it only after several days. During rotation the heads deviated mostly in the direction of the rotation and afterward in the opposite direction. Often the eyes deviated also. Children who at first did not react normally later showed the usual reactions.

In BÁRÁNY's (1918) experiments on newborns the head movement always followed the plane and direction of the slow component of the nystagmus. Reactions of the extremities could not be demonstrated. SCHUR, who first rotated the infants while keeping them in the erect position, found nystagmus in only 5 of 20 trials. During rotation the head deviated in the opposite direction and afterward in the same direction. If the child was rotated in the lying position, as in ALEXANDER's experiments the same deviations of the head were found. BERBERICH and WIECHERS examined numerous normal and premature newborns. Their results depended on the infant's position (erect, dorsal, or lateral). During rotation the head and eyes moved in the opposite direction and afterward in the same direction. If immediately after stopping the movement nystagmus occurred, it was opposite to the direction of rotation. When deviations of the eyes occurred, they moved in the direction of the turning.

When examining infants (probably mostly newborns) VOSS (1927) noted that during rotation the head and eyes trailed in the opposite direction and afterward in the direction of the rotation and that the nystagmus which occurred during rotation was in the direction of the rotation and afterward in the opposite direction. Head and eye deviations always occurred exactly in the plane and direction of the slow component of the nystagmus. The opposite observation in infants in the dorsal position made by ALEXANDER and SCHUR can be explained by the fact that the infant's head was not, as in VOSS's examinations, lying on the outside, but on the inside. LAWRENCE and FEIND (1953) rotated normal newborns three hours to ten days of age in the dorsal position, with their heads above the turning point, on electrically movable disks ten times in 20 seconds under controlled conditions. When awake all children showed head and eye deviations during acceleration opposite to the direction of rotation and during deceleration in the direction of rotation. All children showed normal nystagmus during acceleration and during some rotations at constant speed. Sleeping newborns, although they reacted while awake, could not be stimulated.

Furthermore, SCHUR found that older infants responded with nystagmus after only four or five rotations, while adults needed ten rotations for this effect to occur. If the child could walk, he would deviate considerably from the direction of the rotation after the test. SCHUR concludes from his observations that the infant's labyrinth, compared with that of the adult, is oversensitive.

If during rotation with the subject in the lateral position a downward vertical nystagmus is elicited, the eyes are opened and forehead wrinkles appear (BARTELS, 1910; BARANY, 1918; VOSS, 1927). In connection with these findings, MYGIND observed a pursing of the upper lip, a reaction that VOSS also occasionally saw. In horizontal nystagmus VOSS sometimes found a facialis sign of the cheek corresponding to the slow component of the nystagmus and also a widening of the lid opening.

According to DE KLEIJN and MAGNUS (1921) and DE KLEIJN (1932), the reflexes of progressive movement in a straight direction are also mediated by the semicircular canals. On the basis of these investigations ISBERT and PEIPER [1927 *(17)*] termed corresponding reflexes in young infants "semicircular canal reactions of the limbs." HASEGAWA (1932) demonstrated in the guinea pig that the reaction to progressive motion in a straight direction can no longer be produced after the otolithic membranes have been removed by centrifugation of the animals. He therefore states that in the animal in the normal position the two sacculi are responsible for the reaction of straightforward motion in the direction of the dorsoventral axis and the two utriculi for the motion in the direction of the longitudinal body axis. HASEGAWA's concept has gained more and more ground although it has not yet been definitely proved. In the literature the term "MORO reflex" has been accepted.

Acceleration of motion in the forward-backward direction can be easily carried out in the infant. MAGNUS and DE KLEIJN (1912) noticed a spreading apart of the arms when the body was changed from the sitting to the lying position or when the sharply forward bent head was bent backward. They described this reflex as a labyrinthine reaction of the limbs. Similarly, BLANTON (1917) noticed an upward motion of the arms as reactions to "dropping" of the infant.

MORO (1918) elicited this reflex by hitting the infant's pillow with both hands. Thereupon both arms, half extended at the elbows, are spread apart and the fingers are also spread. The legs move somewhat more weakly and less distinctly. Afterward the limbs return to the original position. MORO did not express any opinion about the receptors involved in this reflex. As can easily be demonstrated, this is a vestibular reaction caused by the head being shaken, i.e., a brief acceleration and deceleration in the sagittal direction. If the infant is placed on a hard base, for example, a table, the reflex does not occur because the head is not shaken when this type of base is hit.

The MORO reflex can also be elicited by shaking the head, i.e., through progressive movement in a sagittal direction with the young infant's body placed on the forearm of the examiner so that the infant's head rests on the palm face upward; if the hand, which protects the skull, is now bumped against the edge of a table, the reaction occurs, incidentally without leaving a transitory refractory period. This experiment contradicts the opinion of ANDRÉ-THOMAS (1954) and SAINT-ANNE DARGASSIES (1960) that we are not dealing with a vestibular reaction but with a deep neck reflex elicited by the movement of the head. Neck reflexes of this type are so far not known. Also, it would be impossible to understand which receptors play a role in this phenomenon. The neck reflexes of the limbs described and named by MAGNUS and DE KLEIJN are, in contrast to the MORO reflex, tonic.

The reaction occurs if the infant is suddenly moved upward or downward ("lift reaction"), forward, backward, or sideward. To what extent the head is shaken depends on the manner in which the child's neck muscles are innervated at the moment of the experiment. The "readiness-to-jump" (MAGNUS), which consists of the same reaction pattern, is tested in the infant by holding him in a horizontal position free in space and face downward, and then suddenly moving him downward. SCHALTENBRAND (1925) found that this jumping readiness is present only in weak form in the infant; occasionally it was absent in the second half of the first year of life, but in older children it was usually clearly demonstrated.

Frequently the MORO reflex is elicited by the infant himself when he shakes his head. The usual kicking movements as well as coughing and sneezing may easily have this effect. WOLFF (1959) observed the same form of motion in infants who were sleeping and not moving prior to the occurrence of this phenomenon. Therefore, an irritation of the vestibularis nerve was out of the question.

The MORO reflex, consisting of an extensor movement of the arms, is the opposite of the startle or fright reaction (see p. 25), which is a flexor movement. The two reactions differ from each other also in the fact that the MORO reflex can be elicited repeatedly, in constant series, i.e., it does not leave a transitory stimulus-refractory period, while a certain amount of time must elapse before another fright or startle reaction can be elicited. Finally, the MORO reflex can be elicited only by shaking of the head, but the startle reaction always occurs when a strong stimulus affects the infant who is at complete rest. A sound stimulus is as effective as light or any other type of stimulus. More detailed investigations in this field are not available.

HOOKER (1942) noted the MORO reflex in two human fetuses (nine and a half weeks of gestation) when they were placed below a camera to be photographed. KRAUSS (1951) observed the same reflex in a fetus in the third month of gestation (see p. 548). The MORO reflex is always present in the first three months of life. In the first week of life we note clonic movements of the arms during the reflex in approximately two thirds of all newborns; this phenomenon can be only rarely demonstrated in the second week, and later it disappears. The MORO reflex becomes less frequent in the subsequent six months of life (fourth to ninth) and then completely disappears. MORO states that it persists longer in the mentally defective.

In the second half year of life the MORO reflex becomes by far less frequent in the form of the "tilt and lift reaction." However, some remnants can still be easily demonstrated in older children: If we take a child on our lap and let him suddenly fall a short way between our legs, he will spread his arms automatically. According to SCHALTENBRAND, the jumping readiness is still maintained in the adult and is the cause of the well-known radius fracture (Colles' fracture).

The MORO reflex has nothing to do with holding tight or clinging to the mother's body (MORO). An infant would never be able to get hold of an object with this quickly disappearing movement. STIRNIMANN came to the same conclusion. When he placed his arm between the two approaching hands, the infant did not grasp the arm but evaded it. The reaction can also be found in young mammals (rabbits) who, being "nest squatters" (see p. 586), are not carried around by the mother. Rather, ROBINSON's tonic palmar grasp reflex serves the purpose of holding tight to the mother. This reflex is not too well known in the literature and frequently is confused with the clasping reflex, although it is elicited in a different way and represents a different reaction (see p. 158). DOFLEIN (1914), to whom MORO refers, has interpreted the palmar grasp reflex as the clasping reflex in man and monkey.

Asymmetrical Movement Reflex of the Limbs

(17)

If the infant's head is sharply turned to the side, a transitory asymmetrical movement of both arms in the form of a fencing position will not infrequently occur. This head movement probably elicits the MORO reflex, but its appearance is altered by the influence of the asymmetrical tonic neck reflexes of the limbs (SCHALTENBRAND, ISBERT and PEIPER). However, to solve this question elimination of the labyrinths would be necessary.

This reflex is more common than the asymmetrical tonic neck reflex themselves. It is regularly found in premature and normal newborns, becoming less frequent until it is completely absent in the second half year of life.

If the infant's head is shaken while in a lateral position, only the MORO reflex will always appear and not the asymmetrical reaction.

POSITION REACTIONS

According to MAGNUS the position and equilibrium of the whole body during rest depend on the static reflexes, i.e., continuing reflexes having tonic properties. The pressure sense of the skin and the proprioceptors of the tendons, muscles and joints, and the eyes, in addition to the organs of the antrum, serve as receptors.

Compensatory Eye Movements

The compensatory eye movements are elicited by the brain-directed ascending stimuli coming from the proprioceptors of the neck muscles, the eyes, and the otolithic membranes. In lower mammals the purpose of these movements is to avoid, as much as possible, changes of the retinal pictures during head movements. In higher animals they are replaced by voluntary eye movements.

Tonic neck reflexes of the eyes, i.e., horizontal deviations, were first demonstrated in newborns by BÁRÁNY (1921): If he fixed the head and rotated the body by 90° in the longitudinal axis, both eyes turned to the same side. This reflex was observed only in the first two days of life. VOSS (1927), who confirmed these findings, also tested the vertical downward and upward deviation of the eyes by moving the body backward toward the occiput (or forward toward the face), keeping the head in a fixed position. In both situations compensatory eye movements occurred. VOSS observed rotary eye movements when he moved the body from the side toward the face while the head was held in a fixed position.

BARTELS (1910) described "doll eye movements" in normal and particularly in premature newborns: If the infant's head is flexed in the transverse axis, the eyes move in the opposite direction, i.e., in forward bending of the head the eyes move upward and vice versa. Of course, in such a procedure the labyrinth is stimulated in addition to the proprioceptors of the neck muscles, so that doubt remains as to where the primary stimulus strikes.

I was able to demonstrate these "doll eye movements" easily; they were also described by ESENTE (1958).

WILLI (1950, 1952) described a reaction that is difficult to classify, namely, the reflex-like "phenomenon of the setting sun." It is elicited by an rapid change of position (transition from the lying to the sitting position). The eyeballs are turned, inward and downward for a few seconds, as is the case in the gaze of hydrocephali, so that the iris is partially covered by the lower lid and the white sclera above remains visible in the wide open eye. After a few seconds the bulbi return to their original position. The phenomenon may also appear without external cause and can also be elicited by sound, heat, and tactile stimuli. STILLHART (1954) observed it in 5.7% of prematurely born infants and in 3.4% of normal newborns, usually in the first weeks of life. The sign has no clinical significance in premature infants, in whom it can be demonstrated during the first three months of life. Occasionally it is seen for several weeks in absolutely healthy normal newborns; it also accompanies kernicterus or elevated cerebral pressure. If it persists very distinctly for more than three months, or at the minimum four weeks, severe permanent brain damage exists.

Willi interprets this phenomenon as a sign of diencephalic damage. In this connection he refers to the well-defined "raising and lowering centers" demonstrated by Hess in the cat in the region of the posterior commissure, where electrical stimulation causes a lowering or raising of the eyes and the head.

Voss did not note tonic labyrinthine reflexes of the eyes in newborns.

Posture Reflexes

The posture reflexes guard the body position during rest. The relative position of the body parts depends on a regulated tonus distribution of the total body musculature and on a tonic fixation of the joints.

Tonic Neck Reflexes of the Limbs

The proprioceptors of the neck muscles, which are stretched by movements of the head, serve as stimulus receptors. According to Magnus (1924, p. 51), there is a general rule, applicable to man and all animal species, that after a sideward rotation of the head the tonus of the extensor muscles of the fore and hind limbs is increased on the side to which the face is turned and is diminished on the other side. To simplify the terminology he introduced the term "jaw-arm and -leg" for the limbs of the side to which the face is turned and "skull-arm and -leg" for the limbs of the opposite side to which the occiput is turned. Magnus made his experiments in decerebrated animals whose labyrinths were destroyed. It should be taken into consideration that in the human infant the labyrinth in particular could influence these reflexes. If the infant's head is turned to one side, vestibular reflexes usually appear; we have to wait until they subside to observe the tonic neck reflexes. In the normal young infant they were not found by Magnus; but later they were described by Landau (1923, 1925), Schaltenbrand (1925), Isbert and Peiper [1927 *(17)*], Byers (1938), and Gesell (1938).

Asymmetrical tonic neck reflexes of the limbs cause the "fencing position" (Fig. 21). The skull-arm is raised at the shoulder and flexed at the elbow, rotated inward, and adducted; the jaw-arm is lowered at the shoulder, extended at the elbow, and rotated outward. Thus, the hand of the skull-arm is nearer to the head than that of the jaw-arm. In the legs there is, though less frequently, a corresponding flexion of the skull-leg and extension of the jaw-leg. Whether the infant's head is turned actively or passively is of no importance.

Frequently this position is not maintained "in tonus" because the voluntary movements of the infant interfere with this position. When we observe the infant who is awake and uncovered, we note that the arms and legs are in continuous motion; however, on prolonged observation we will see that there is always a tendency to flex the skull-limbs and to extend the jaw-limbs.

The so-called fencing position of the arms has been known for a long time as a clinical sign of toxic states in infants. I consider this to be nothing more than an asymmetrical tonic neck reflex of the arms that occurs in toxic states because higher inhibitory brain centers are eliminated by the disease. A similar explanation can be given for the occurrence of this phenomenon during sleep when children turn their heads sideward (SCHALTENBRAND, ISBERT and PEIPER).

The younger the child, the more frequent the asymmetrical tonic neck reflexes of the limbs. ISBERT and PEIPER found these reflexes to be most frequent in premature infants; they occurred in approximately half of the normal newborns during the first week and only occasionally later. They were absent in the second half year of life. Their disappearance with progressing maturation is interpreted as an indication of the fact that the cerebral cortex is gaining control over the lower reflex

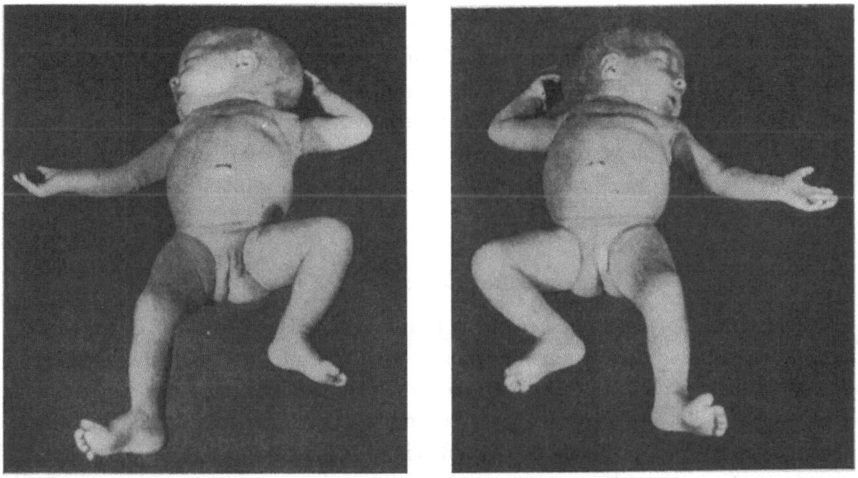

Fig. 21. Asymmetrical tonic reflexes of the limbs (jaw-arm and -leg extended, skull-arm and -leg flexed) in a prematurely born girl at birth (1550 gm); age: 2 months, 9 days (2920 gm).

centers (RIESEN and KINDER, 1952). According to MINKOWSKI (1928), it is not possible to speak of somewhat constant and regular tonic neck reflexes in fetuses; however, it occasionally is possible to observe a hint of these reflexes in fetuses, particularly in older ones. These reflexes manifest themselves by contralateral reactions of the limbs of both sides if the head is turned sideward.

In arrested cerebral development the reflexes can persist beyond the first six months of life (ISBERT and PEIPER, BYERS, GESELL). Their reappearance in meningitic states is described elsewhere (see p. 296).

ISBERT and PEIPER did not find a reaction in healthy children when the head was turned at the sagittal axis. In two dying premature infants and in one flaccid newborn it was noted that in this type of head turning the leg on the side to which the head was turned was extended at the knee and hip joints while the other leg was flexed or remained in a flexed position. In these observations we cannot exclude with certainty the possibility that the labyrinth may also play a role.

BRUDZINSKI's neck sign is elicited by the forward bending of the head and consists of flexion of both legs at the hip and knee joints. According to FREUDEN-

BERG (1921), it is occasionally observed in healthy young infants; however, in general it is considered to be an important meningitic sign and therefore is discussed elsewhere (see p. 293). The interpretation of this neck sign as a tonic neck reflex is not firmly established (see also WARTENBERG, 1952). According to MAGNUS this may be the underlying mechanism of this sign, but it may also have another explanation.

Tonic neck reflexes of the supporting tone of the extremities are discussed on page 198.

Tonic Skin Reflexes

(3, 36)

I use the term "tonic skin reflexes" for a group of reflexes that are elicited by stimulation of the skin and that cause tonic flexion or extension of the limbs and tonic flexion of the spine. The leg and, to a lesser degree, the arm usually react uniformly so that hip flexion is accompanied by flexion of the knee and vice versa. The same usually applies to extension of hip and knee joints. Frequently the vertebral column participates in this reaction.

Tonic Skin Reflexes of the Limbs

At rest the infant usually keeps his limbs in a flexed position. If by chance the limbs are extended, the infant responds with flexion, particularly leg flexion, to various kinds of stimuli. This reflex is most distinctly elicited if the skin of the flexor surface of a joint is stimulated by stroking with a pin point or the handle of the reflex hammer or if pressure is applied with the handle of the reflex hammer. The reverse reaction, namely, extension of the joints in response to skin stimulation of the outer surface, is far less frequent. If the reaction is not disturbed by spontaneous movements of the infant, it persists for the duration of the stimulus, i.e., has tonic qualities.

If the inner surface of the axilla is stimulated, the arm is slightly rotated inward and adducted, thus "closing" the axilla. On stimulation of the inguinal region or the inner surface of the extended thighs the stimulated leg is flexed at the knee and hip joints. On one-sided stimulation both legs are often simultaneously flexed, while the crossed flexor reflex of the legs is elicited by flexion of one leg (see p. 294). Stimulation of the flexor surfaces of elbow and knee does not regularly cause similar reactions.

So far the tonic skin reflexes of the limbs of animals have not been studied in detail. Neither is it known how long they persist in children. If later in life because of arthritis or bone fractures the affected extremity remains in flexion, the skin as well as the more deeply situated organs may mediate the same reaction.

Tonic Palmar Grasp and Plantar Grasp Reflexes

(17, 36, 115)

Palmar and plantar grasp reflexes in the infant correspond to each other. The palmar grasp reflex is elicited by touching the palm; a pressure stimulus has an even better effect. The stimulus leads to a tonic flexion of the second to fifth fingers

(Fig. 22); by means of this response the touching or pressing object is grasped. The plantar reflex is elicited in the same manner. Most effective is strong pressure applied to the anterior end of the sole, whereupon the five toes are flexed; this flexion is tonic. The toes can also grasp a slender object (Fig. 23). The plantar grasp reflex is markedly weaker than the palmar grasp reflex; however, it can always and with absolute certainty be elicited, while the attempt to elicit the BABINSKI sign in the first months of life causes a variety of toe movements that are unpredictable (see p. 285).

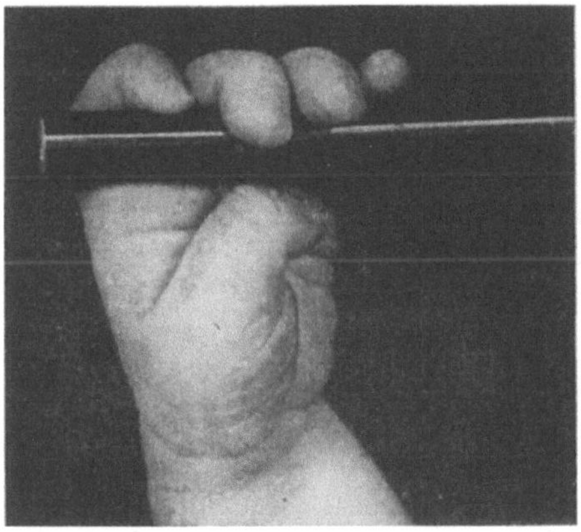

Fig. 22

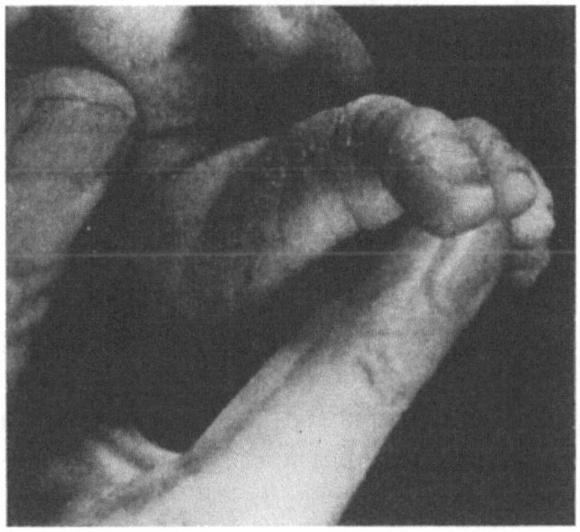

Fig. 23

Fig. 22. Palmar grasp reflex of the newborn. Touching the palm causes flexion of the fingers and grasping of the touching object.

Fig. 23. Plantar grasp reflex of the newborn. Pressure on the anterior end of the sole causes flexion of the toes.

According to MINKOWSKI (1928), stroking of the palms of fetuses frequently causes flexion of the fingers and flexion, abduction, and sometimes also opposition of the thumb; a distinct grasp reflex is the result, and sometimes the hand makes a fist. This reflex has even been found in fetuses 65 mm long, a size which corresponds to the eleventh week of gestation. The reflex has been observed as early as the sixth month of gestation. HOOKER (1938) also noted that the grasp reflex in human fetuses develops in two succeeding phases: the closing of the fingers and the grasping. The closing of the fingers already appears at approximately the eleventh week of pregnancy as a quick but incomplete flexion of the fingers with the exception of the thumb. The reflex becomes increasingly more complete so that by sixteen weeks of gestation a fist closure occurs on tactile stimulus of the hand. At this stage of development the fingers and hand do not yet grasp the object. In the twentieth week of pregnancy the closed fingers appear to hold weakly a thin glass stick placed between them. At 22 weeks the grip is so strong that the observer can move the whole arm with the glass stick. Contrary to MINKOWSKI's observations, HOOKER was unable to elicit any thumb movements by means of tactile stimuli. The tonic palmar grasp reflex of the newborn constitutes a continuation of this reflex.

If a slender object, for example, a stick, is placed in the hands of a newborn, his hands grasp it. He can cling to the stick in this manner so that for a short time his hands can carry the weight of his freely suspended body (Fig. 24). It sometimes takes a minute or more before he lets go. If this experiment is repeated at short intervals, the grasping periods gradually become shorter. The infant's grip is considerably stronger before than after feeding. Infants who suffer from pylorospasm show a particularly firm grip. This impressive reflex was first described by ROBINSON (1891) as a "clinging reaction."

SCHNEEBERGER (1958) noted that the suspended infant is excited; he cries and moves strongly, and the peripheral vessels are dilated.

According to PRECHTL (1953), the fingers successively close around the stimulus object after the palm has been touched and in the following sequence: middle finger, ring and little fingers, index finger, and thumb. Since originally infants clung to the fur of their mothers with the aid of this reflex, it is no coincidence that this reflex can even now best be elicited by hairs.

According to ANDRÉ-THOMAS (1952, p. 115), the thumbs of idiots, imbeciles, and other mental retardates are often in continuous flexion. This hand position is said to be a continuation of the fetal position.

Healthy normal and premature newborns always show the palmar grasp reflex. RICHTER (1931) measured the clinging time of 27 newborns for long periods and noted that the strength of the reflex varied from day to day and from infant to infant. The reflex appeared to be strongest after the twelfth day of life. In my experience it gradually becomes more infrequent in the fourth to sixth month and disappears at the end of the first year of life. According to SITTIG (1932), it disappears on the average at six months.

GIORDANO (1953) examined 282 infants and found that up to the fifth month of life the palmar grasp reflex was always present in healthy infants. Starting with the sixth month it became more infrequent and weaker. In older infants it is difficult to distinguish this reflex from spontaneous grasping.

ISBERT and PEIPER *(17)* noticed a certain relation between head and leg position in approximately half of the newborns when the infants were clinging in the pull-up position: If the head is bent backward, the legs are adducted to the body with pronounced flexion of the knee joints. If the head is bent forward, the legs are extended at the hip and knee joints and hang down flaccidly. Equilibrium appears to be thereby improved. The newborn does not yet possess the pull-up reaction of the arms and the labyrinthine righting reflex of the head. RADEMAKER (1931, p. 334) compares the lack of these reactions in the human newborn with the lack of the pull-up reaction in decerebrated and newborn dogs (see p. 205). These reactions were present in intact animals and in those whose cerebellum or cerebral hemispheres were removed. In animals whose labyrinths were removed these reactions were present only when the eyes were not closed.

As ANDRÉ-THOMAS and his co-workers (1954) have reported, it is possible to inhibit the MORO reflex by means of the palmar grasp reflex: When the MORO reflex is elicited by shaking of the head, the arms, hands, and fingers are extended. However if this reflex is evoked in a newborn in whom a tonic palmar grasp reflex has been unilaterally elicited by placing an object in one hand prior to this, the palmar grasp reflex persists unilaterally while the other arm moves according to the pattern of the MORO reflex. If the palmar grasp reflex has been elicited bilaterally, it is also maintained bilaterally.

In the grasp reflex HALVERSON (1937) distinguishes between the automatic hand closing and the object holding that follows. Both stages can be separated by means of the stimulation, i.e., by cautiously touching the palm or the anterior surface of the fingers with a soft object. Thereupon the fingers are slowly and incompletely bent. As soon as the fingers meet resistance when touching a hard object during this flexion, the relative stretching of the flexor tendons causes a proprioceptive flexion.

WIESER and DOMANOWSKY (1957) were unable to demonstrate the influence of the asymmetrical and symmetrical tonic neck reflex and of the otolith reflex on the tonic palmar grasp reflex of the newborn. The fright reaction could reinforce the grasp reflex.

According to RICHTER (1931), the reflex is even stronger in the newborn monkey. Such an animal suspended itself with only one hand for 33 minutes. KURODA (1930) reported on the duration of suspension of a newborn monkey *(Macacus cynomolgus)* five days old weighing 420 gm: right hand 0, left hand 10 seconds, right foot 15 seconds, left foot 5 seconds, both hands and feet simultaneously 45 seconds. A corresponding reflex can also be noted in the thalamus preparation of a monkey (KARPLUS and KREIDL, 1914; MAGNUS). As soon as the animal has recovered from the operation it is able to grasp a finger that is placed in its hand; held by both hands they bring head and body back to normal positions by vigorous pull-ups, and they maintain this position.

The cerebral hemispheres are not needed for this reflex; on the contrary, the reflex is inhibited by the function of higher cerebral parts. WATSON (1924) and PEIPER found this reflex to be very distinct in cases of anencephalia in man, but CATEL and KRAUSPE (1930) and SCHENK (1932) noted only a weak indication of the reflex in these cases.

The reflex represents a phylogenetic memory of the time when the young animal was clinging to the fur of its mother's abdomen, as young monkeys still do today. Thus, the palmar grasp reflex really serves the purpose that was erroneously attributed by MORO to the "clasping" reflex.

The tonic plantar grasp reflex (Fig. 23) was first described by VAN WOERKOM (1912). Only rarely can the infant maintain himself in arm and leg suspension for a short time (Fig. 24b).

F. and W. STIRNIMANN (1940) tested the reflex with an instrument on which the applied pressure could be measured. Of 1000 newborns, 984 responded with tonic plantar grasp reflexes by flexion of all toes. This flexion becomes a real grasping by the action of the adductor hallucis muscle. Threshold values of 40 to 70 gm of pressure are certainly normal; values of 30 to 80 gm are borderline. Higher and lower values suggest a disturbance; the probability of this disturbance is in direct relation to the deviation from the mean values. Ten infants responded only with fanning of the toes, and six failed to respond even at a pressure of more than 100 gm. When the reflex was absent or when the toes were only spread apart, diseases or previous birth injuries were present almost without exception.

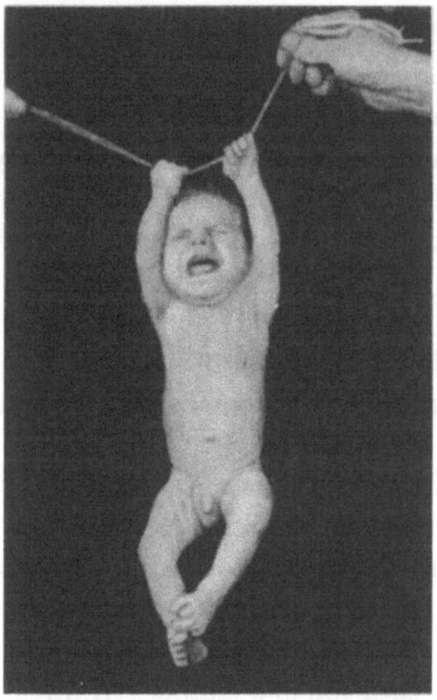

a

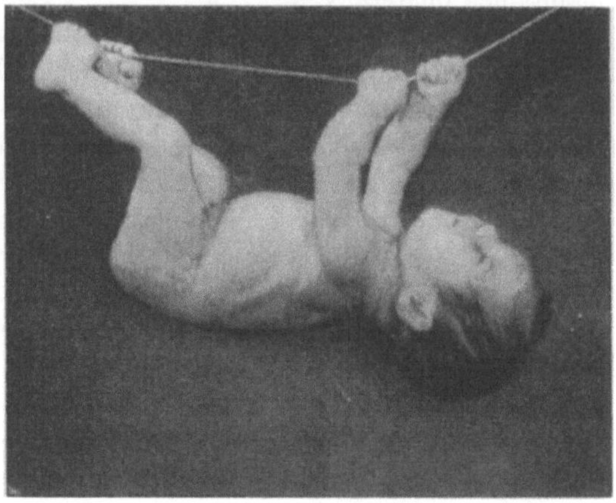

b

c

Fig.24. Grasp reflex in a prematurely born 4-week-old twin (weight 2300 gm). (a) Arm suspension from the palmar grasp reflex. (b) Arm suspension with simultaneous leg suspension from palmar and plantar grasp reflexes [According to PEIPER (*115*)]. (c) Sloth (*Choloepus didactylus* Lin.) suspended from a tree (from Brehm's Tierleben, 4th ed. Leipzig and Vienna, 1912, Vol. 10, p. 543).

The reflex disappears, according to SITTIG (1932), at the age of one year and, according to DIETRICH (1957), between the sixth and twelfth month of life, i.e., when the infant begins to stand. The human infant is not able to maintain himself in free suspension by his foot alone, as the newborn monkey reported on by KURODA was able to do.

Until now little attention has been paid to the tonic plantar grasp reflex, although it is observed in the infant with far greater regularity than the BABINSKI sign.

Spinal Reflex (Incurvation of the Trunk, GALANT Reflex)

(1, 3, 17, 36)

In 1904 BERTOLOTTI described a "réflexe dorsolombaire" in the child: If the dorsal skin in the lumbar region is stimulated at the level of the first lumbar vertebra by pinching or tickling, a quick defense movement with contraction of the dorsal muscles appears. This reflex is more common in children than in adults and most lively in the first two years of life. In 1912 the reflex was rediscovered in children by NOICA and given the term "réflexe de la masse musculaire sacrolombaire." According to NOICA, it is found in almost all children, but rarely in adults and old people.

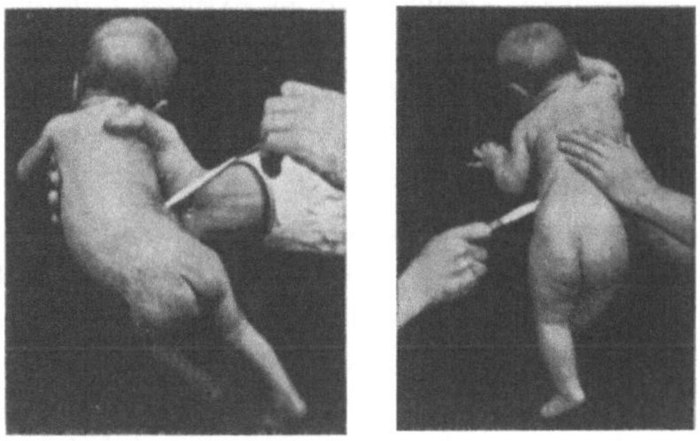

Fig. 25. (a) Spine reflex. (b) Spine reflex with simultaneous tonic neck reflex of the limbs.

In 1917 VERAGUTH described a corresponding reflex in healthy infants and sick adults, and in the same year GALANT also described it in normal infants and named it the spinal reflex (Fig. 25a). This reflex consists (according to GALANT) in the following phenomenon:

> "When the dorsal skin near and along the vertebral column is stroked, the infant forms an arch with his body; the concavity of the arch is directed toward the stimulated area, and by arching in the opposite direction the infant evades the stimulus."

According to GALANT, this reflex is most distinct in the first months of life and later becomes weaker.

Augmenting GALANT's findings, ISBERT and PEIPER [1927 *(17)*] found that on application of this stimulus the pelvis is flexed backward and the ipsilateral leg is extended at the knee and hip joints and the other is flexed (Fig. 25a and b). This reaction does not occur in each instance and also may be easily disturbed by spontaneous movements of the infant. It becomes apparent that this reflex has tonic qualities when a reflex hammer is pressed against the lumbar region. Children assume this position as long as the stimulus lasts if no spontaneous movements interfere. The tonic qualities of the reflex cannot be demonstrated by mere stroking.

The same reaction can also be elicited from the chest and abdomen. The infant retracts the stimulated area so that the vertebral column forms an arch, which is open to the stimulated area, while the ipsilateral leg is frequently extended at the knee and hip joints. The most sensitive areas are the lateral parts of the trunk; the reflex becomes weaker when elicited anteriorly or posteriorly near the midline. Frequently the head position can be changed by stimulation of the anterior surface of the trunk: Stimulation of one side causes turning of the face to the stimulated side. By alternately stimulating both sides it is sometimes possible to turn the infant's head to and fro at will. This reflex can also be elicited from the axillae, but not from the thighs, and in sensitive infants it can also be elicited from the proximity of the midline. Rotation of the head can effect tonic neck reflexes of the limbs.

ISBERT and PEIPER saw this reflex in almost all infants. The leg position described occurred in the great majority of all prematurely born infants. It was present during the first half year of life in half of the infants born at term; later it occurred only occasionally. Apparently an external stimulus is not always needed. A prematurely born infant suffering from unilateral purulent pleuropneumonia and an infant suffering from a gluteal abscess lay in positions corresponding to that of the spinal reflex during their illnesses.

I. and P. DE PULGAR MARX (1955) described the following reflexes in newborns and prematurely born infants: The infant is placed in the abdominal position on a hard base or on the hand of the examiner. With his finger the examiner then strokes the infant along the spine from the pelvis to the nape. Thereupon flexion of the legs occurs and with it lordosis of the spine, elevation of the pelvis, flexion of the arms, lifting of the head, loud crying culminating in apnea and cyanosis, emptying of the bladder, and relaxation and bulging of the rectum with bowel movement; after the reflex has fully developed there is general hypertonia lasting for a few seconds. This reflex can be evoked for a few weeks and then gradually disappears. VOLLMER (1958) tested this reflex in 123 infants, age 4 to 46 days, and found arm flexion in 99%, crying in 98%, leg flexion in 97%, lordosis in 85%, head lifting in 78%, relaxation of the rectum in 1%, and bladder emptying in 1%. The reflex becomes weaker in the second month and disappears in the third. It is regarded as evidence of immaturity.

It is doubtful whether LEINER (1926) observed the same reflex in the phenomenon that he describes:

> "Upon stroking the skin between the shoulder blades and the spine downward to the sacrum of a child lying in the appropriate position (horizontal abdominal position without a pillow, arms held parallel to the trunk) with the handle of a reflex hammer, a distinct longitudinal wave-like contraction corresponding to the attachments of the dorsal muscles to the vertebral column occurs."

LEINER rarely or never found this reflex in infants; this militates against this reflex being identical to the spinal reflex, which was unknown to him. He found the reflex described by him with regularity only after the first year of life.

HINSCHE (1939, 1942, 1943) made a comprehensive study of the subject based on extensive investigations of his own. He distinguishes between two absolutely different groups of reflexes in the dorsal region, i.e., tonic skin reflexes and the muscle reflexes proper as defined by HOFFMANN. Since both groups occur in the same region, they frequently combine, causing contradictory phenomena.

The phylogenetically very ancient spinal reflex is a pure skin reflex: Touching the skin at the spinal column causes an increase of tonus, apnea or dyspnea, lumbar lordosis with backward bending of the pelvis, kyphosis in the region of the thorax, forward bending of the head, extension and spreading of both legs, and slight extension of the arms. Lateral stimulation causes scoliosis, rotation of the spine, extension of the leg near the stimulus and flexion of the other leg, and turning of the head with a corresponding arm movement. All such signs are very rarely found in the same person. Th s comprehensive reaction, which represents a tonic reflex of the spine, is termed by HINSCHE "dorsal posture reaction"; GALANT's spinal reflex represents its essence. Particularly sensitive persons contract their muscles jerkily on the slightest touch so that their bodies become hard as a board in an opisthotonic extension tonus. This reaction may even occur when a person is approached or is expecting the stimulus, thus forming a conditioned reflex.

Since each muscle may contract in response to a pull, there are as many individual reflexes of the back as there are back muscles. HINSCHE describes the individual reflexes of the trapezius muscle, latissimus dorsi, teres major, erector trunci, and gluteus maximus. Where several muscle layers cover others, we have to observe the reactions of the deeper through the superficial layers.

Only the responses of the gluteal muscles will be discussed here. According to HINSCHE, the dorsal posture reaction is triggered by touching the buttock region. In addition, he differentiates between two individual reflexes that can best be elicited in a subject in the knee-hand position, but that are mostly absent in the prone position. One reflex consists of the contraction of all buttock muscles, a process that spreads to the other side (nates reflex). It is triggered by tapping the muscle attachments of the fascia lumbodorsalis, crista iliaca, sacrum and coccyx, trochanter major, and tractus iliotibialis. The particular reflex of the musculus gluteus maximus is triggered by tapping along the crista iliaca and in the nearby region of the iliac bones and results in individual contractions of the coarse fasciculi of the gluteus maximus muscle.

In addition, I have found that there is still another reflex of the same muscle elicited in the prone position and triggered from the skin. Pricking or stroking the buttock with the tip of a pin causes unilateral contraction of the muscle. This reflex of the gluteus maximus muscle permits easy recognition of paralysis of the muscle, a not infrequent state in poliomyelitis.

MINKOWSKI (1928) found spinal reflexes in fetuses even at three months of gestation; this reflex consisted of a curving of the spinal column toward the side next to or distant from the stimulus. DIJKSTRA (1960) could nearly always demonstrate the reflex in newborns after the fifth day of life. The reflex is always present in premature infants and in children during the first years of life and is not uncommon in the fifth to sixth year. According to HINSCHE, the dorsal posture reaction is still common during school age, but becomes less common with increasing age. This reaction particularly persists in those who are physically and mentally retarded.

HINSCHE states that the reappearance of the dorsal posture reaction after it has subsided is always a pathological sign (foci in the brain or spinal cord, glomerulonephritis in children, kidney and gallstone attacks, bleeding ulcers, etc.). An asymmetrical response of the two sides is not a pathological sign.

HINSCHE and PRECHTL (1956) consider the dorsal posture reaction to be a remnant of an ancient form of behavior, having originated phylogenetically as a tonic skin reflex of the spine at the stage of the amphibia and reptiles. In quadruped locomotion the dorsal skin is an important sensory organ that regulates the movements of the vertebral column. After the transition to the erect body posture this task and mission are taken away from the dorsal skin and transferred to the front of the body, particularly to the now "liberated" arms (see p. 243).

The righting reflexes (MAGNUS, 1924) enable the body to obtain the normal posture among the greatest variety of possible positions.

Neck Righting Reflex of the Body

(1, 17)

According to SCHALTENBRAND (1925), the neck righting reflex of the body is elicited in the infant by placing him on his back and turning his head sideward. The whole body then follows this rotation. If the thorax is kept in a fixed position, the pelvis swings in the opposite direction. Frequently the whole body follows this rotation at once so that the infant rolls completely over to one side. If the reflex is weak, it will take some time until the body follows. As in the reverse reflex, namely, the body righting reflex of the head, a general restlessness first appears. The infant makes purposeful and purposeless movements until he has reached the normal body position, in which he will then finally remain (Fig. 26).

It may be that this reflex has a certain importance in the birth act. When the body of the infant follows the rotation of the preceding head, it may be that the neck righting reflexes of the body, which are so distinct in the newborn, are playing an important role. When the head rotates around the sagittal axis, the pelvis some-

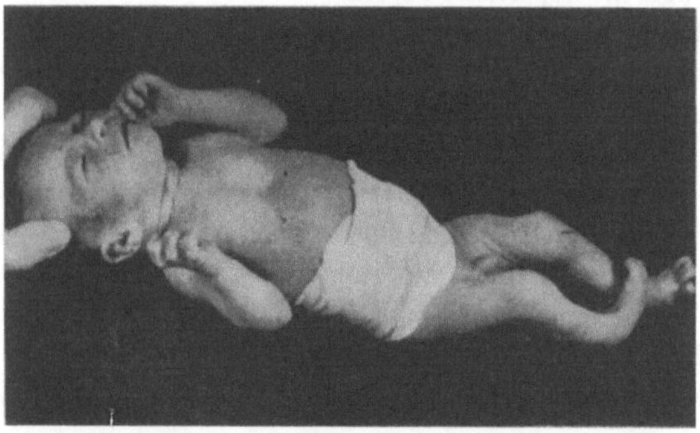

a

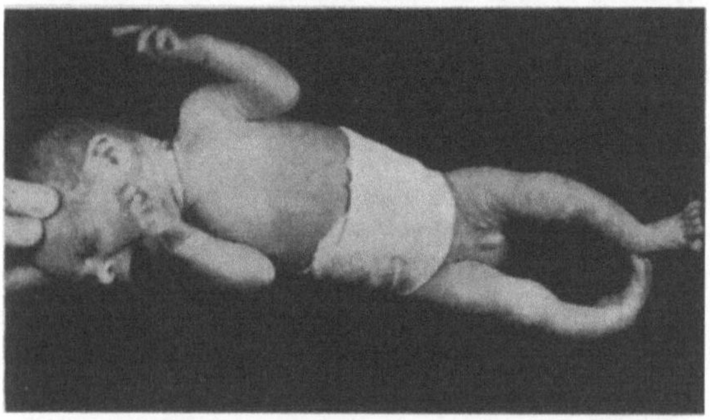

b

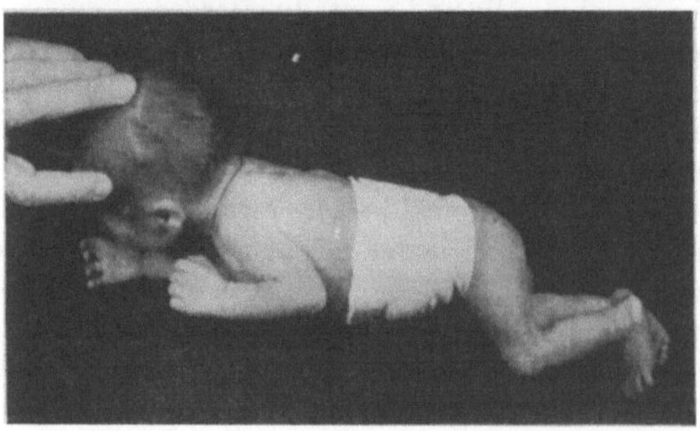

c

Fig. 26. Neck righting reflex of the body. (a) Initial position. (b) Head turned to the right. (c) Trunk and legs have followed the turning of the head.

times rotates to the same side. In the first year of life it is usually possible to demonstrate this reflex. Later it gradually disappears. Isbert and Peiper were able to confirm the findings of Schaltenbrand, but Spitzer, Kramer, and Rabkin (1958) usually found the reflex to be absent.

Body Righting Reflex of the Head

(17)

The body righting reflex of the head is the counterpart to the previously described righting reflex: If an infant is placed in the lateral position on a table and his legs are then grasped, thus turning the pelvis to the other side, the trunk and then the head roll over to this side too [Isbert and Peiper, 1927 *(17)*]. The rotation may take place so rapidly that pelvis, trunk, and head are moved simultaneously (Fig. 27).

Optic righting reflexes can easily be eliminated by covering the eyes. Any influence of the labyrinths (destroyed in Magnus's animal experiments) is quite improbable in the infant with normal labyrinths, as direct observation proves.

If the infant's chest is held and his pelvis is swung sideways, the head is turned to the opposite side. This corresponds to the above-described neck righting reflex of the body with the chest held (Fig. 28).

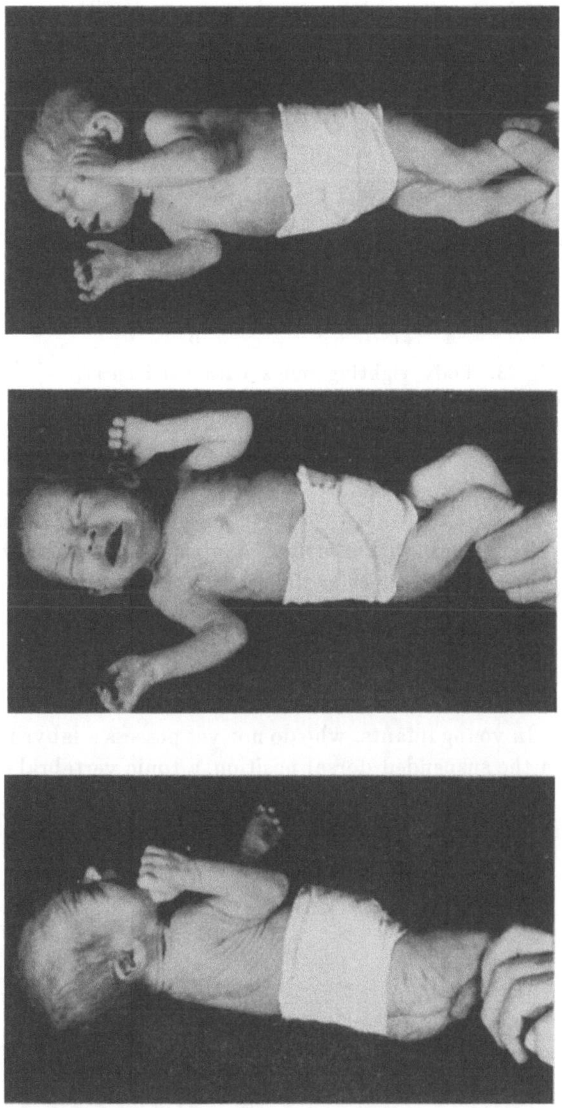

Fig. 27. Body righting reflex of the head in the newborn. On passive rotation of both legs at the vertical body axis the trunk and head follow the rotation.

RADEMAKER (1931, p. 310) demonstrated the same reflex in a dog without a cerebellum:

"If the pelvis of a dog lying in the right lateral position is first turned to the dorsal and then returned to the lateral position, the trunk and head will also turn and will assume a left dorsal position; the head does it even when the anterior part of the body is held by the front legs."

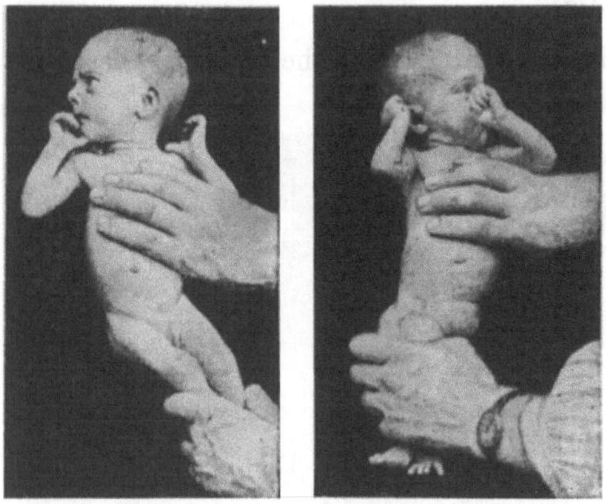

a b

Fig. 28. Body righting reflex with fixed chest. (a) Pelvis is passively swung to the left; head is actively turned to the right. (b) Pelvis is passively swung to the right; head is actively turned to the left.

The reflexes in man are fundamentally different from the body righting reflexes of the head in the animal as described by MAGNUS. In his experiments the animal's head was brought to a normal position in space, whereas in the infant's reflexes the head maintains a normal position in relation to the body. STENVERS (1918) described a corresponding reflex in a sick woman. When she was lying in the dorsal position a corresponding turning of the head could automatically be elicited by rotating the pelvis; by this rotation the head again assumed its normal position in relation to the pelvis. If the head was kept in a fixed position, only the eyes moved in the direction of the pelvis rotation.

The reflex with the free-moving chest is nearly always found during the first year, but with the fixed chest it is much rarer.

Tonic Vertebral Column Reflex of the Legs

(36)

In young infants, who do not yet possess a labyrinthine righting reflex of the head in the suspended dorsal position, a tonic vertebral column reflex of the legs [PEIPER, 1931 *(36)*] can be demonstrated in this position. The production of a passive lordosis of the lower spine is the eliciting stim-

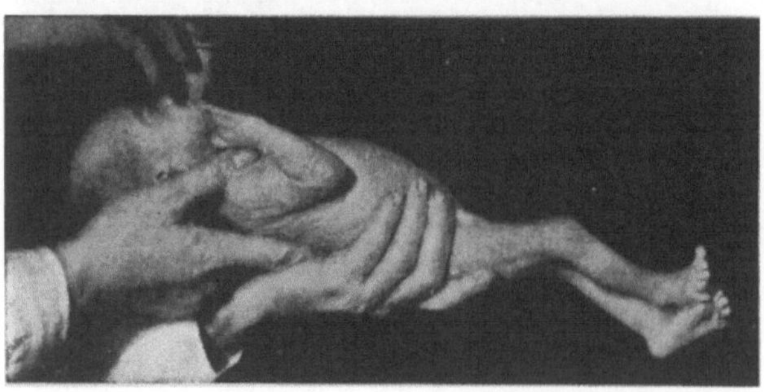

Fig. 29. Tonic vertebral column reflex of the legs; lordosis of the spine effecting extension of both legs.

ulus, the result being the extension of both legs at the knees and hips, which is the opposite of the usual position of flexion. The reverse phenomenon, flexion of both legs in kyphotic posture, also often occurs, though not with the same certainty.

To elicit the tonic vertebral column reflex of the legs it is best that the child be freely suspended by supporting the sacrum so that the spine is overextended by the weight of the lower half of the body and assumes a lordotic position (Fig. 29). Subsequently the infant extends both legs at the knees and hips.

However, if the spine is pressed into a kyphotic position by the hand of the examiner while the head of the infant is somewhat lowered, both legs are flexed at the knees and hips. It is unlikely that tonic labyrinthine reflexes of the limbs are involved in this behavior, but it is hard to refute their participation in an infant with a normal labyrinth.

The reflex can be inhibited on one side if a small clamp is fastened to a toe, thus eliciting a flight reflex by constant irritation (Fig. 30).

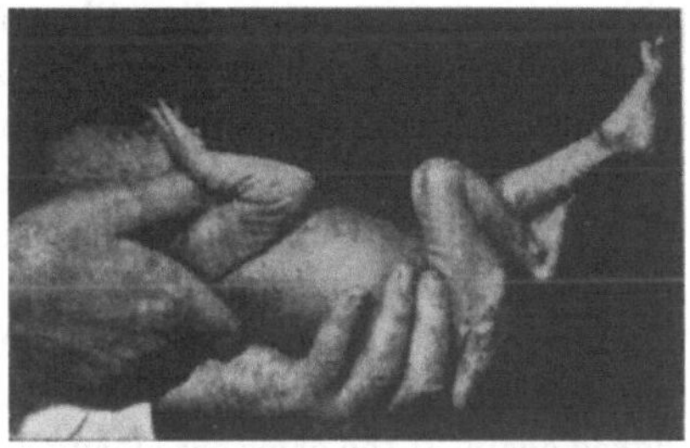

Fig. 30. Tonic vertebral column reflex of the legs inhibited on one side by a clamp on the toe (flight reflex).

Preferred Posture of the Infant
(1, 17, 19a, 23a)

When the infant is not moving, he prefers a definite, generally symmetrical, dorsal posture (Fig. 31). The arms are very much flexed at the elbows, and the clenched fists and forearms rest on both sides of the head with their dorsal surfaces on the supporting base. The legs are rotated outward and drawn to the body, so that the big toes are sometimes abducted and somewhat opposed. Sometimes the elbows are flexed so much that the hands lie behind the neck.

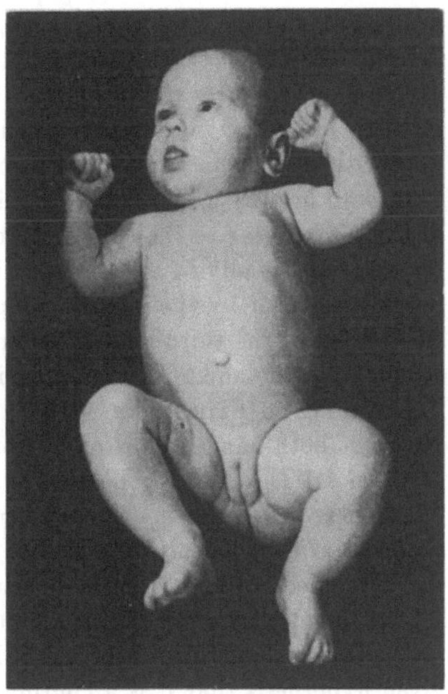

Fig. 31. Preferred posture of the infant shown in a 3-week-old girl.

WENGRAF (1959) does not believe that this posture in the alert infant is a resting position. The anatomists understand this term to mean the position of the joints to each other caused by gravity when the muscles are completely relaxed. The preferred posture of the awake infant is conditioned by a tonic and sometimes also a phasic innervation. This may explain why the awake infant will occasionally maintain this posture if he is held in a sitting position.

Quite similar to this, but according to the roentgenographic examinations of DREXLER and WENGRAF (1959), nonetheless distinctive is the sleeping posture of the infant. The difference is that in the transition from the waking to the sleeping state the humerus rotates outward by 90°, so that during sleep the whole length of the dorsal surface of the forearm and hand rest on the base while the upper arm is only moderately abducted. According to DREXLER and WENGRAF, the arm of the sleeping infant is no longer innervated and therefore sinks down in response to gravity.

The sleeping posture may be absent in healthy infants; for example, some sleep only if they lie on their stomachs.

R. M. and W. A. YERKES (1929) described the same sleeping posture in a newborn chimpanzee; a corresponding illustration is found in the publication of C. F. and M. M. JACOBSEN and YOSHIOKA (1932). The baby gorilla "Goma," raised without his mother in Basel, assumed the same sleeping posture (SCHENKEL, 1960).

SAINT-ANNE DARGASSIES (1954) described the reflex return of the arms to the flexor position: If one tries to stretch the flexed forearms in the sleeping posture, resistance will be met. If one insists, the extension will be incomplete. As soon as the forearms are released, they again assume their starting or an even more pronounced position. The more quickly the arms are released, the more quickly they will be flexed again. If the arms are slowly and gradually released, they momentarily remain in extension.

THUROW [1956 *(19a)*] and NAUMANN [1958 *(23a)*] observed the sleeping posture of infants and small children and distinguished three postures: the infant's sleeping posture, the transition sleeping posture (in which the flexion of the extremities decreases), and the extension sleeping posture. Until the sixth or eighth month the infant's sleeping posture predominates; it may still occur sporadically in healthy and sick infants after the third year. Sick infants prefer this posture during the first months. In healthy infants the transition sleeping posture appears gradually beginning with the second month and is predominant after the eighth month.

In both healthy and sick children it is most frequently found from the beginning of the second year to the end of the third year. The extension sleeping posture first appears occasionally in healthy and sick infants in the second month, but is always quickly replaced by the infant's sleeping posture or the transition sleeping posture. At the end of the third year approximately 80% of children assume the transition sleeping posture and 20% the extension sleeping posture.

According to STIRNIMANN (1947), the newborn has his own sleeping posture. He holds both hands below or in front of the mouth and, less frequently, in front of the nose. This sleeping posture is said to have its origin in the prenatal period.

Under certain conditions the preferred posture of infancy may persist longer or even reappear after it has already ceased. In developmental disturbances such as severe feeble-mindedness and cerebral rachitis, it is still found in the second and third year. I have found it as a meningitic sign (see p. 290) even in the eighth year of life.

The preferred posture of the infant has no satisfactory explanation. The previously given hypothesis that it is a continuation of the intrauterine posture is refuted by the fact that the child assumes a totally different posture in his mother's womb because he is exposed to the pressure of the uterus from all sides.

The possibility that the flexed posture of the infant could simply be attributed to the predominance of the flexor over the extensor muscles was considered. To come to a decision RITTER (1924) ascertained the weight of the flexors and extensors in the newborn, assuming that the greater strength would correspond to the greater bulk. However, it was shown that the proportions were just the reverse of the findings in the adult: The bulk of the extensors surpassed that of the flexors, a ratio which was previously noted by THEILE (1884). Nevertheless, the flexor posture is a characteristic of the infant and must be explained by central and not peripheral influences.

DOXIADES (1931) made tracings of the contractions of the muscle antagonists in infants; he sees an innate resting tonus of the flexors as the cause of the flexor posture. He considers the flexor posture to be a reflex, and believes that it is elicited by brain-directed stimuli that keep the movement centers in higher tension. This is supported by the clinical observation that a constant tonus in the flexor muscles can frequently be demonstrated. The forearms of the sleeping infant do not always lie on the base, but often are entirely free in the air, supported only by the elbows; thus their position does not depend simply on gravity.

The flexor posture of the young infant resembles a forced posture in adults, i.e., the squatting or climbing posture that occurs through the loss of the striatum function (FOERSTER, 1913). The absence of the striate inhibitions gives predominance to the pallidum, which phylogenetically is said to have served the climbing function (see p. 219). A posture that today is still characteristic of the monkey is said to result in this way. Thus, the sleeping posture of infants is traced to the same cause as the forced position of adults with striatum lesions, namely, the absence of the striatum function. DREXLER and WENGRAF, however, explain the phylogenetic similarity of the sleeping posture of infants to the fundamental posture of other primates as a result of identical skeleton formation rather than of a process of innervation when falling asleep.

Some otherwise healthy infants exhibit a distinct forced opisthotonic posture from birth. They apparently persist in a posture that they had already assumed before birth and that must have been a severe obstacle to delivery, since those infants are frequently born in abnormal positions. According to this explanation, the forced posture is the cause and not the result of the abnormal delivery. A healthy nervous system is also the prerequisite of a normal birth; infants with paralysis of the lower part of the body due to spina bifida cystica are therefore often born in breech presentation.

Other forced postures are found in some feeble-minded infants. Sometimes they are the expression of asymmetric tonic neck reflexes of the limbs which appear because of the absence of higher inhibitory centers. For other cases we have no explanation.

GAMPER's "Bowing" Reflex

(140)

GAMPER (1926) reported his observations of a human male anencephalus with arhinencephalia and encephalocele in whom only the mesencephalon was developed. He observed this anencephalus for a few days during the infant's third month of life until his death. The following was noted: If pressure was applied to both lower legs while the infant was lying on his back, first the head and then the trunk were raised from the base so that he finally reached a sitting position with his back extremely curved. This movement did not proceed uninterrupted, but was often accompanied by sideward turns of the head and trunk. Even without external stimulation the infant had the tendency when lying on his back to bend his head forward and thus to raise it from the mattress.

MAGNUS, to whom GAMPER had shown a film of the infant, interpreted the reaction as follows:

"This infant, freely suspended in air in the dorsal position, behaves like a mesencephalic preparation; he brings his head into normal position by a combination of labyrinthine and neck righting reflexes through ventral flexion of the neck with successive ventral flexion of the trunk; the reaction often begins by the infant's turning the head sideward and a corresponding turning of the body. We can observe that we can also elicit the same reaction in the animal by an accessory stimulus – as here, by pressure on the lower leg – if the reaction does not occur spontaneously."

GAMPER's reflex is not mentioned in later descriptions of anencephali (GLATZEL, 1927; CATEL and KRAUSPE, 1930; SOKOLANSKY, 1930; SCHENK, 1932; ANDRÉ-THOMAS, 1952, 1954; WILLI, 1952). Only MONNIER and WILLI (1947) observed a reflex somewhat similar to that described by GAMPER in an anencephalus in whom only the rhombencephalon was developed: During the attempt to overcome the flexor hypertonia of the legs by passive stretching of the head, the trunk and arms were raised as in the first phase of the MORO reflex.

PEIPER, HEMPEL, and WÜNSCHER [1959 *(140)*] observed GAMPER's reflex in all seven living anencephalic infants they examined.

As an example, case 2 is here reported: A female anencephalus S. B., birth weight 2800 gm, was examined for the first time on the second day and was clinically observed until her death at the age of eight weeks. In the beginning the infant was doing fairly well. She was alternately excited or completely unexcitable. She cried, yawned, sneezed, and coughed and was awake and slept at regular intervals just like a normal baby. GAMPER's reflex could best be elicited by slow and firm, but not jerky, extension of the hip joints. Following this stimulation the child gradually sat up with increased tonus of the abdominal muscles. She did not fall forward, but remained sitting, almost erect, and started to make licking movements of the mouth and tongue (Fig. 32). Repeated testing of the reflex, one stimulation shortly after the other, resulted twice more in a sitting position, then only in a nearly vertical raising of the trunk. The child did not immediately fall back after sitting up, but sank back slowly, making slight pendular lateral to-and-fro movements and showing gradually increasing trembling of the trunk muscles. The reflex lasted up to 25 seconds.

GAMPER did not name the reflex first described by him, but spoke only of the "process of lifting oneself up." However, if the reflex is elicited in a lateral position the same chain reflex occurs without the child's sitting up (Fig. 33). The examiner has to support the sacrum in such a manner that the body cannot move backward.

Fig. 32. GAMPER's reflex with sideward turning, elicited in a female anencephalus in the dorsal position by extension of the thighs at the hip joints.

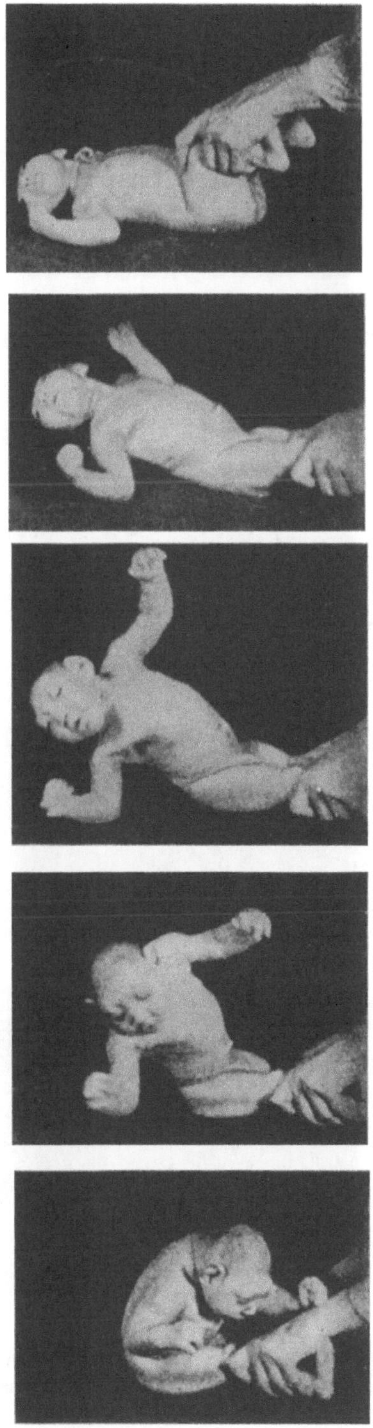

Since a bowing movement of body and head develops from each starting position, the term "bowing reflex" has been chosen.

In unpublished investigations, H. THOMAS (1950) has demonstrated that GAMPER's bowing reflex can also occasionally be elicited in healthy premature infants. After passive extension of the legs at the knees and hips the child sat up for a moment (Fig. 34).

Since GAMPER's bowing reflex can also be elicited in the lateral position and always brings the head to an erect position (vertex up and mouth horizontal), it cannot be a labyrinthine righting reflex of the head, as MAGNUS assumed.

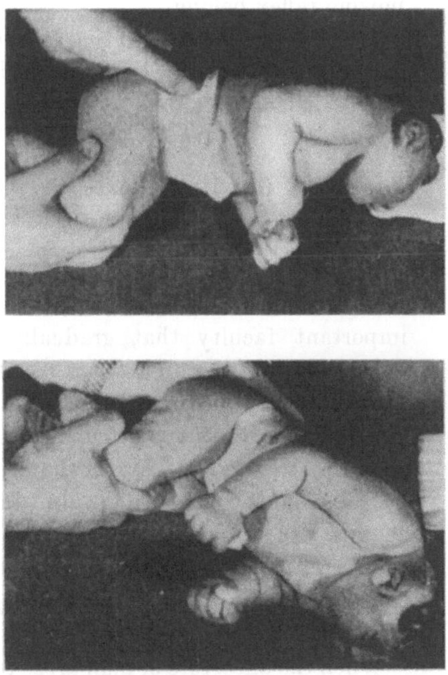

Fig. 33. GAMPER's reflex in an anencephalus in the lateral position. On extension of the thighs at the hip joints the body bows forward.

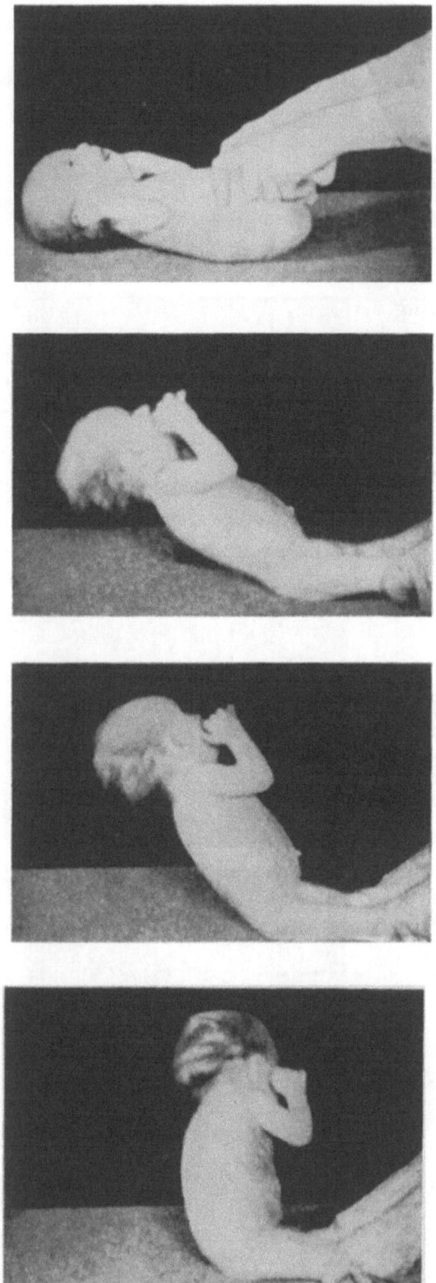

Fig. 34
GAMPER's reflex elicited in a prematurely born 2-month-old infant by extension of the legs at the hip joints.

To elicit GAMPER's reflex, external stimuli are important. It is not a simple righting reflex, but a mass or chain reflex. Perhaps it is also a protective reflex, as we can interpret from its progression in the lateral position.

This chain reflex, which starts at the head and spreads to the trunk, is directed by the lower half of the body. In view of the anatomic findings in anencephali in whom only the bulbopontine part of the brain has developed and the importance of the formatio reticularis for tonus regulation, as especially established by MAGOUN and co-workers (1952), it is suggested that the formatio reticularis, liberated from cortical, cerebellar, and striate-pallidal influences, is the highest structure of the central nervous system that could be responsible for these mass reflexes, to which GAMPER's bowing reflex belongs.

It should be mentioned that this reflex can only occasionally be found in healthy premature infants.

Gravity Reflexes

(139)

Besides the reflexes of position and movement mentioned thus far, the most important faculty that gradually develops during the first year of life is that of reflex adjustment of the positions and movements of head, trunk, and extremities to gravity. This faculty makes it possible for the infant to finally acquire the upright body posture and the free gait. I have named the reflexes that make this function possible "gravity reflexes."

When the ancestors of man arose to an upright body posture, the center of gravity moved upward considerably, while the number of his supports was reduced from four to two. How this change endangered man's equilibrium can still be seen in each child who — repeating the phylogenetic pattern — tries to get up and assume an upright body posture. Generally the adult is no longer conscious of the difficulty of this task.

The upright posture frees the arms from their tasks of body support and locomotion. The hands thus become free for the many tasks that have made man man. The child is still very much in need of his arms to gain the upright posture. As a rope walker uses a horizontal bar to keep his equilibrium, the child stretches out both arms or only one — as the need arises — to restore the threatened equilibrium. When sitting or suspended he also uses his legs for this purpose. With the aid of his arms he draws himself up onto objects within reach; he holds onto them and wanders along objects that are suitable for this purpose.

Labyrinthine Righting Reflex of the Head

(17, 41, 95, 139)

The labyrinthine righting reflex of the head is decisive for the ability to bring the body into upright position, i. e., to adjust it to gravity and move it forward. This reflex has the task of adjusting the head to gravity: vertex up and mouth horizontal. It was demonstrated by MAGNUS (1924) in experiments on animals and was examined in the child by SCHALTENBRAND (1925), PEIPER and ISBERT (1927), and PEIPER (1932, 1959).

The reflex is only faintly demonstrable in the neonate, but during the first year of life it becomes an important prerequisite for the infant's ability to lift his head and later also the upper part of his body while in the prone position, thus reaching the starting position for locomotion in the abdominal position. With the help of this reflex the infant learns to sit and stand, a prerequisite for the upright and free gait. For this purpose the chain reflexes, which depend on the labyrinthine righting reflex of the head, are very important.

MAGNUS demonstrated this reflex in the animal as follows: The animal is lifted up in the air, blindfolded, and supported by its pelvis, thus avoiding bodily contact with the base. If it is now bent forward, backward, or sideward, the head will try to keep its normal position in relation to gravity, righted by tonic reflexes.

With the body in the lateral position, these labyrinthine righting reflexes occur through the concerted action of the sacculus and utriculus otoliths of both ears (see p. 149). By reflex action each otolith supports the effort to turn the head in such a way that it itself is on top. Therefore, the head turns to the side of the operative procedure when one labyrinth has been removed. If both labyrinths are functioning, a compromise is reached that places the head in the normal position. According to RADEMAKER (1926), the labyrinthine righting reflex passes through the red nucleus and Monakow's tract, but not through the cerebral hemispheres and the cerebellum.

The labyrinthine righting reflex does not depend on the position the body assumes in relation to the vertical axis and thus to the head. The chain reflexes, which depend on the labyrinthine righting reflex, enable the body to turn freely around the head, which is stationary in space (Figs. 35—37, 42c, 45—49, 62). As in MAGNUS's animal experiments (1924, p. 17), the infant's head "is also, so to speak, fixed in space by a magic power: an extraordinarily obvious experiment."

Dogs, cats, and monkeys whose labyrinths have been removed are for some time after the operation again able to bring their heads into the normal position in relation to gravity when they are held suspended in air in different positions. This ability

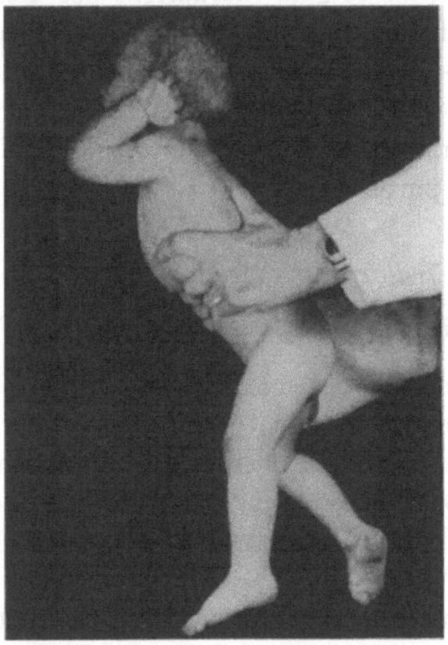

Fig. 35. Labyrinthine righting reflex of the head in a 12-month-old child whose body is suspended and bent obliquely forward.

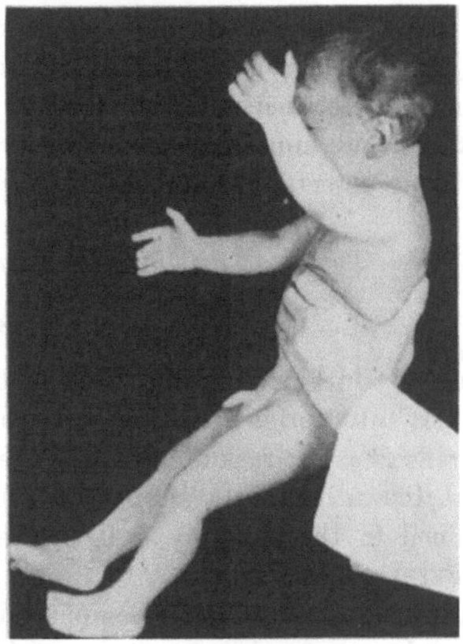

Fig. 36. Labyrinthine righting reflex of the head in a 12-month-old child whose body is suspended and bent obliquely backward.

disappears again when they are blindfolded, thus proving that their heads are also kept in the upright position by optic reflexes. However, in rabbits and guinea pigs deprived of their labyrinths, the inability to right themselves in relation to gravity also remains when they are held freely suspended in air with their eyes open. This indicates that these animals do not possess optic righting reflexes (MAGNUS and DE KLEIJN, 1930, p. 68). The optic righting reflexes are not effective under water in a dog deprived of its labyrinths. Without help the dog will drown when its labyrinths are absent (MAGNUS, 1924, pp. 225, 446).

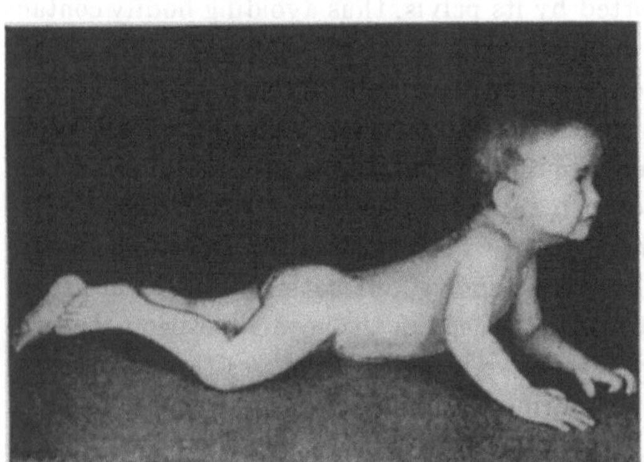

Fig. 37. Labyrinthine righting reflex of the head in a 12-month-old child in the abdominal position whose body is supported by the arms.

Usually the body position of

a man whose eyes are open also depends on optic righting reflexes. In older infants these optic righting reflexes can be made ineffective by means of opaque spectacles. Naturally the infants then become quite restless. That is why they are not wearing glasses in the photographs reproduced here. Optic righting reflexes are not yet present in neonates and very young infants.

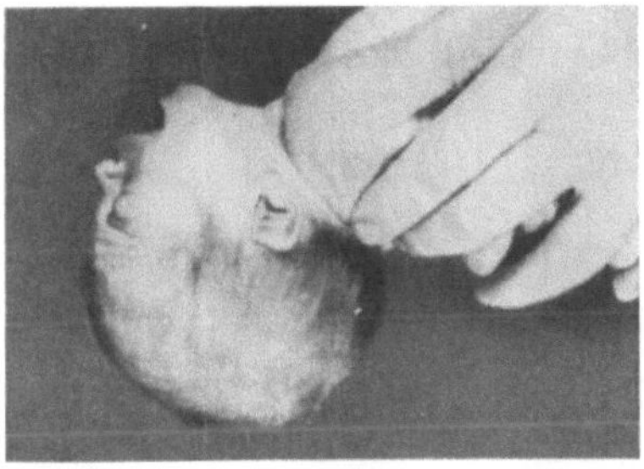

Fig. 38. Absence of the labyrinthine righting reflex of the head in a 5-day-old newborn whose head is bent backward.

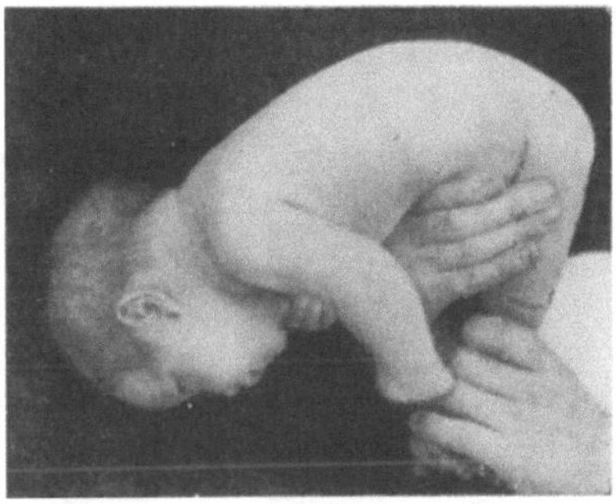

Fig. 39. Weak labyrinthine righting reflex of the head in an 11-day-old boy suspended in the abdominal position.

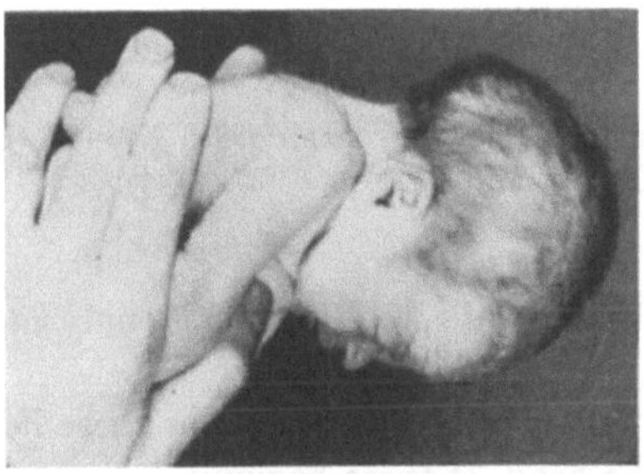

Fig. 40. Absence of the labyrinthine righting reflex of the head in a 5-day-old newborn whose head is bent forward.

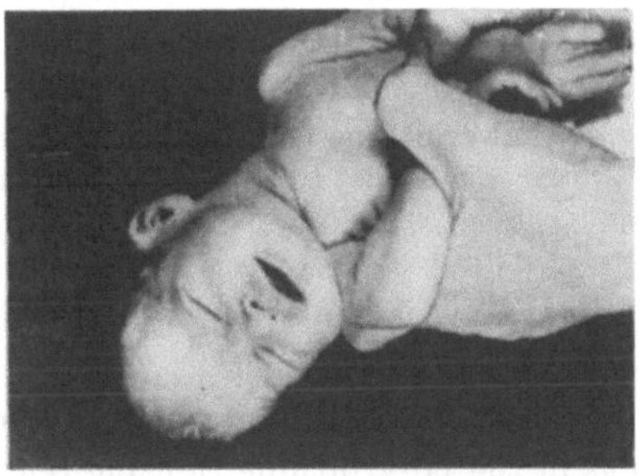

Fig. 41. Absence of labyrinthine righting reflex of the head in a 5-day-old newborn in the suspended lateral position.

During the first years of life the infant passes through the stage of the quadruped and thus repeats the phylogenetic pattern of man. During this time gravity reflexes corresponding to those demonstrated by MAGNUS and co-workers in quadrupeds can be demonstrated in him. On the other hand, the gravity reflexes, which aid the infant to achieve and maintain an upright body posture, do not play a role in the quadruped. Its body axis is horizontal, while in the erect man the axis is vertical. The animal that under natural conditions never stands in an upright position does not need a reflex balance for this particularly difficult task. Man and animal, therefore, cannot be compared in this respect.

SCHALTENBRAND (1925) states that with the body in the upright position the head of the newborn usually sinks downward in response to gravity.

If the body axis is bent backward, the head falls backward (Fig. 38). Almost no movements occur to lift it. Sometimes such movements become more distinct when the body axis, and with it the head, is slowly raised. If the body axis is bent obliquely forward so that the head falls forward (Fig. 40), weak, repeated countermovements which temporarily lift the head into the normal position may appear (Fig. 39). If

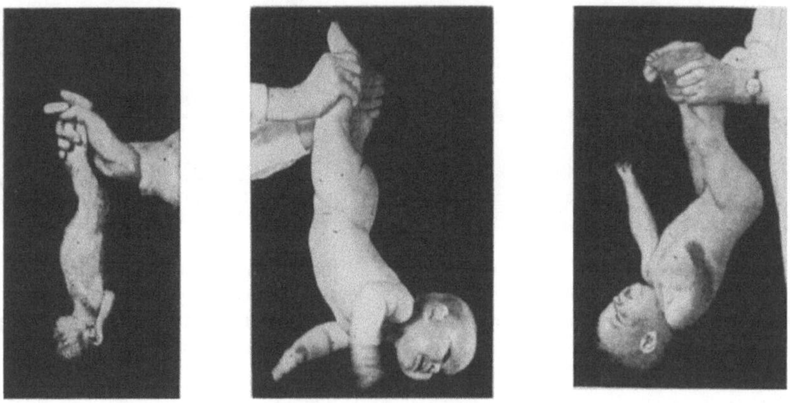

a b c

Fig. 42. Labyrinthine righting reflex of the head in vertical suspension. (a) 1 week; (b) 10 months; (c) 1 year. [According to ISBERT and PEIPER, 1927 *(17)*]

the body axis is bent sideways, the head falls sideways, too (Fig. 41). Countermovements are very indistinct.

The more flaccid the newborn is, the less tendency he has to exhibit these beginnings of the labyrinthine righting reflex of the head. If these beginnings are absent, the face is often turned to the lower shoulder during lateral bending, as described by SCHALTENBRAND.

The findings of RADEMAKER (1926, p. 136) in the suspended mammal can be compared with the following results: The intact rabbit, if suspended in the dorsal position, can bend its head forward considerably and thus bring it into a normal position in space (Fig. 69b, p. 192). The rabbit deprived of labyrinths, on the other hand, lets its head fall backward in response to gravity when placed in the same position (Fig. 69a).

In the suspended dorsal position the newborn acts just like the rabbit deprived of labyrinths (Fig. 68). He does not yet possess the labyrinthine righting reflex of the head and the chain reflexes based on this reflex. Figures 70 and 71 (p. 193) show what the infant is able to do when both reflexes have developed. In Fig. 68 the older infant is suspended in the dorsal position with his pelvis supported. The body forms an obtuse angle with the extended legs so that his weight is distributed, though not quite equally, between the lower and upper halves of the body. Of course this is not a completely balanced position.

With the body in the suspended lateral position (Fig. 66a) the rabbit deprived of labyrinths lets its head fall just as the newborn does (Fig. 68, p. 192), while the intact rabbit (Fig. 66b) lifts its head into the normal position (RADEMAKER, 1926, p. 137). Thus, the labyrinthine righting reflex of the head is absent in both the human newborn and the rabbit deprived of labyrinths.

The beginnings of this reflex can be clearly demonstrated in the newborn: If placed on a table in the abdominal position, some will repeatedly lift their head with a jerk from the base, i.e., against gravity, but let it fall again immediately.

Even in the first week of life the baby gorilla "Goma," in Basel, could lift its head from the base and hold it for several seconds while in the abdominal position. The effort to bring the head into an upright position was conspicuous (SCHENKEL, 1960).

Young animals often behave differently. MAGNUS (1924) demonstrated the labyrinthine righting reflexes very distinctly even in newborn rabbits, dogs, and cats.

Fig. 43. Head down suspension of a thalamus preparation of a rabbit. The head does not follow gravity, but is brought into normal position in space by the labyrinthine righting reflex. (Sketch from photograph by MAGNUS, 1924, p. 218)

If a newborn is vertically suspended with the head down (Fig. 42a), his head hangs downward in response to gravity. However, in some cases it is strongly bent backward even in the first week, and with regularity in the second three months of life, so that the body assumes a distinct opisthotonic position (Fig. 42b). The thalamus preparation of the rabbit will also overextend its head backward strongly in a corresponding position (Fig. 43).

On the other hand, older infants, starting approximately in the seventh month, try to get up partly by bending their heads and partly by flexing their hips (Fig. 44). In the oblique lateral position with the head down, they turn their faces to their lower shoulders.

SCHALTENBRAND (1925) observed the labyrinthine righting reflex of the head in all older infants. Of course it was often very much inhibited, came late, or disappeared after a short time. Infants who trusted the examiner often did not show the reflex at all.

As previously described, the labyrinthine righting reflex of the head occurs very distinctly when the older infant is in the abdominal position, but it is only weak and transitory in the newborn. If the body is resting with the entire back on a base, the labyrinthine righting reflex is absent in all ages; otherwise the preferred posture of the infant and the sleeping posture of the adult would not be possible.

After the neonatal period the infant soon learns to bend his head forward when in the dorsal position and thus to lift it from the base when he wants to grasp a seen object, the breast, or the bottle with his mouth (Fig. 196, p. 450).

Chain Reflexes
Dependent on the Labyrinthine Righting Reflex of the Head
(17, 41, 95, 139)

If the labyrinthine righting reflex raises the head of an older infant to the normal position and if the body is fixed in an unsuitable position at the same time, chain reactions will occur. These reactions maintain the threatened equilibrium by righting the upper and lower parts of the body and the arms and legs in relation to the head. The individual chain reflexes depend on the infant's position: whether he lies, sits, or stands; whether his body axis is horizontal, oblique, or vertical; whether his

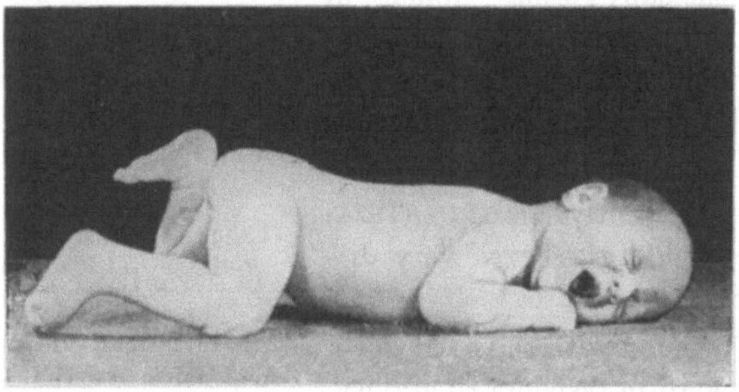

Fig. 44. Labyrinthine righting reflex still absent in a 20-day old girl. In lieu of this reflex the head is turned sideways (protection against suffocation).

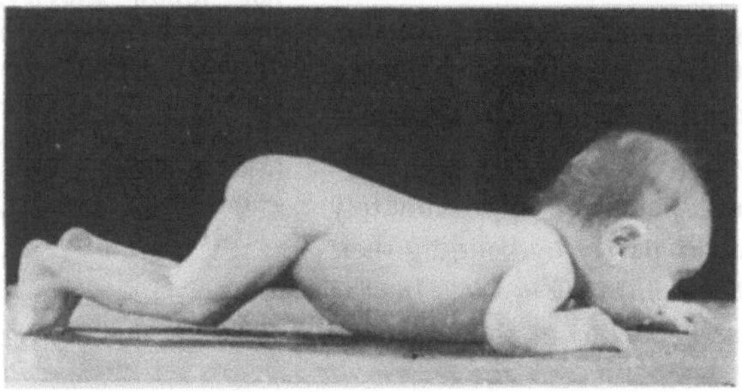

Fig. 45. Beginning of the labyrinthine righting reflex of the head and of supporting reactions in a 2-month, 10-day old girl in the abdominal position. Brief raising of the head, gaze obliquely downward, horizontal forehead wrinkles due to contractions of the frontalis muscle serve the effort to lift the head and gaze.

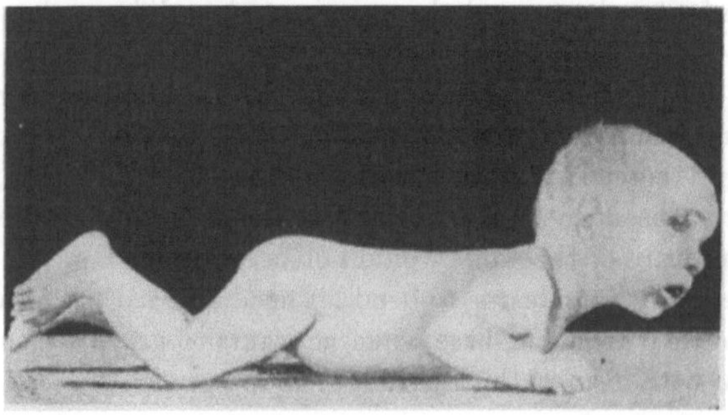

Fig. 46. Labyrinthine righting reflex of the head in a 3-month, 7-day-old boy in the abdominal position. Raising of the head last longer but head is not brought completely into normal position. Support reactions of the arms, flexed at an acute angle, start to lift the upper part of the body from the base. Gaze is raised higher.

body hangs freely or is supported in some manner. In the horizontal position there is a substantial difference, depending on whether the child is in the abdominal, dorsal, or lateral position. In "suspension" there is a difference, depending on whether the shoulders, the center of the body, the pelvis, or the legs are supported. Whether the infant can support himself with one or both hands or can hold onto something, which the older infant always tries to do if balance is difficult, are also important factors.

As confusing as all these possibilities seem at first sight, all the ensuing reflexes and the postures dependent on them follow strict laws; these bring the child, through

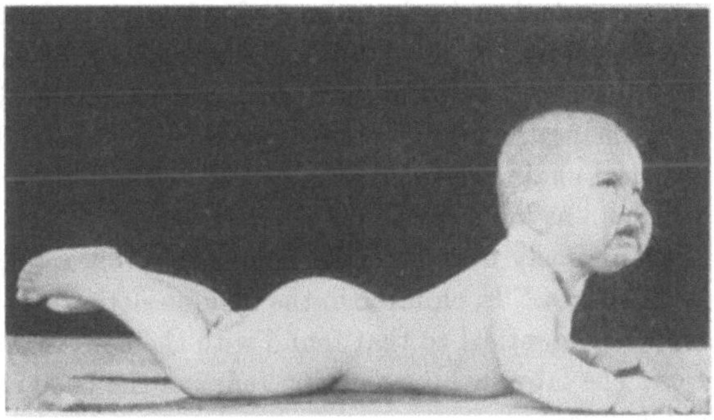

Fig. 47. The labyrinthine righting reflex of the head in an 8-month-old girl brings the head into the normal position in space: vertex up and mouth horizontal. The gaze is straight ahead and upward. Supporting reactions of the arms, flexed in a right angle, lift the upper part of the body farther from the base.

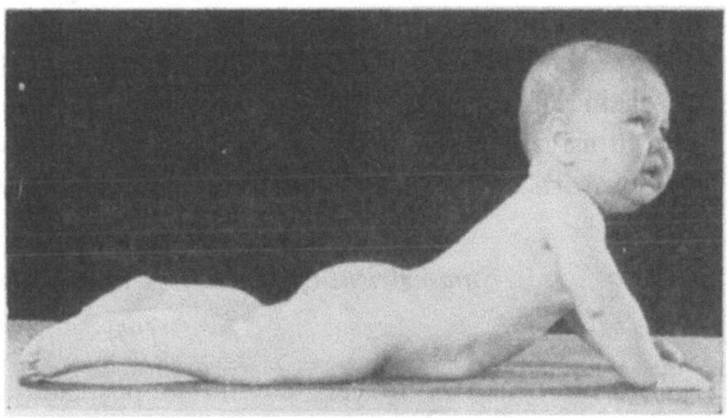

Fig. 48. Supporting reactions of the extended arms lift the upper part of the body still farther from the base in an 8-month-old girl.

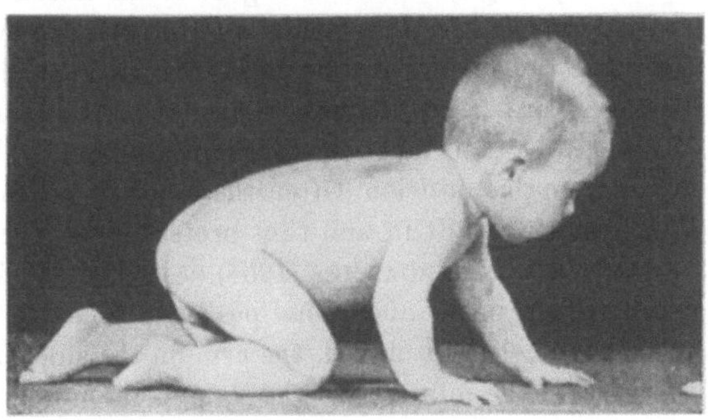

Fig. 49. Beginning of crawling in a 257-day-old girl: supporting reactions of the extended arms and flexed knees; cross coordination (left arm and right leg forward).

reflexes of which he is not conscious, into the exact position under the given conditions to maintain the equilibrium which would otherwise be lost.

According to ANDRÉ-THOMAS (1952, p. 178), however, in man and in most animals the static forces of the body are supposed to determine the position of the head. Actually, the opposite is the case: First the head rights itself in space; thereby it acquires the ability to determine through reflexes the position of the other body parts. As soon as the head is shifted from its normal position the chain reflexes that have righted the body in relation to the head are extinguished (p. 188).

If the infant's body is in an upright or an oblique lateral position with the head up, the labyrinthine righting reflex of the head becomes more and more distinct with increasing age. As soon as the infant succeeds in righting his head in space, his whole body tries by means of reflexes to adjust to this position. Chain reflexes starting from the head regulate the position of the neck, trunk, arms, pelvis, and legs down to the tips of the toes in order to maintain equilibrium. Especially important among the many possible positions are the abdominal position, in which a symmetrical chain reflex occurs, and the oblique lateral position with the head up, in which an asymmetrical chain reflex occurs. When one examines children by lifting them up freely from the base, some chain reflexes appear as especially pure "suspension reflexes." In children in the oblique sitting position a reflex occurs that is fundamentally identical to that found with the body suspended in the oblique lateral position. The labyrinthine righting reflex of the head is always the starting point for the chain reflexes.

It is hard to tell to what extent body-righting reflexes and tonic neck and labyrinthine righting reflexes of the rest of the body — in addition to the labyrinthine righting reflex of the head and the neck righting reflex — also take part in the formation of the chain reflexes in the different positions. In any case the result is that positions that adjust the whole body to the head position — and this in the oblique lateral position with the head up — occur in a surprisingly useful way.

Symmetrical Chain Reflex in the Abdominal Position

(95)

Vigorous neonates in the abdominal position repeatedly and jerkily raise their heads and let them drop again. It is at first impossible for them to hold their heads up for any length of time. The gaze of the raised head is directed downward in the early period of life. This reflex is of importance even at this stage of development. Through the raising of the head the nasal orifices, i.e., the infant's respiratory passages, are freed. It has happened that weak infants lying in the abdominal position with their faces on soft pillows suffocated because they could not free themselves from this position. Often the infant is at least able to turn his head sideways if he cannot yet lift it, and thus protect himself from suffocation (Fig. 44).

SAINT-ANNE DARGASSIES (1954) described the following rather constant reaction of a newborn in the abdominal position: When the head is turned sideways and the arms have not been moved, the arm on the side to which the head has been turned is raised to the infant's face as though the infant wished to turn the head back. If the head is turned to the other side, the other arm makes the same movement. The reaction becomes more pronounced if a steady pressure is applied to the right or left buttock. The side of the arm movement is determined by the direction of the face. The movement of the arm is facilitated when the shoulder is slightly raised by means of the position of the head. After the second month the infant is able to lift his head and shoulders somewhat. This movement facilitates the forward motion of his forearms before he can use them for support.

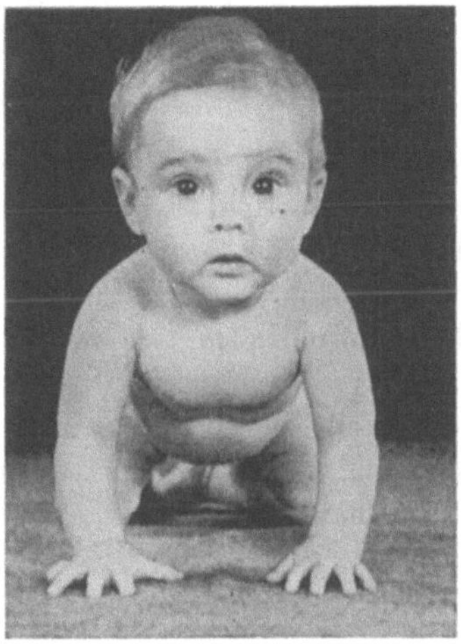

Fig. 50

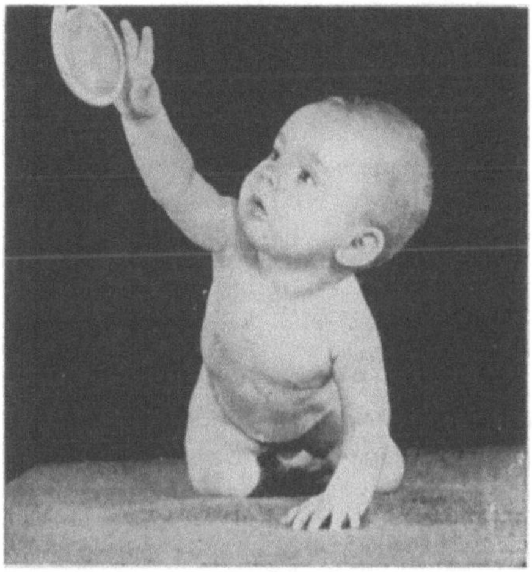

Fig. 51

Fig. 50. Anterior view of crawling position in a 257-day-old girl showing supporting reactions of the extended arms and flexed legs.

Fig. 51. Crawling position in a 257-day-old girl showing lifting of one arm for grasping.

In the course of development the infant is progressively more successful in lifting first his head and then the upper half of the body when in the abdominal position (Figs. 44—51). The gaze is directed forward more and more, i.e., to the future direction of movement.

The position of the infant's head, as seen in Figs. 46—48, increasingly elicits the neck righting reflex, which by reflex action brings the body into the correct relation to the head. The muscles of the trunk and legs, too, come under the influence of the tonic labyrinthine and neck righting reflexes; the effect of these reflexes cannot be distinguished in infants as clearly as in the animals of MAGNUS's experiment. As Figs. 45—48 show, the infant becomes more and more successful in raising his head and this leads to more and more pronounced lordosis and finally (Figs. 47 and 48) to a vigorous extension at the hips, knees, and ankles. Often the reflex stretching of the lower half of the body becomes so pronounced that the legs do not touch the base even in the ordinary abdominal position and are even a little elevated at the lower end, as can be recognized on close observation.

The reflex also extends to the arms (Figs. 45–51); at first the forearms lie flat on the base. The arms have to stiffen in the shoulders and elbows so that the child does not collapse. In this process first the upper arms and later the extended arms (Figs. 48–50) are transformed into supporting pillars. In the last analysis these are supporting reactions, which MAGNUS (1924) and RADEMAKER (1931) described in the animal (see p. 194). However, there is a difference between the infant and RADEMAKER's experimental animals: In the animals the legs are transformed into straight supporting pillars, while the arms of the infant are at first still flexed at the elbows.

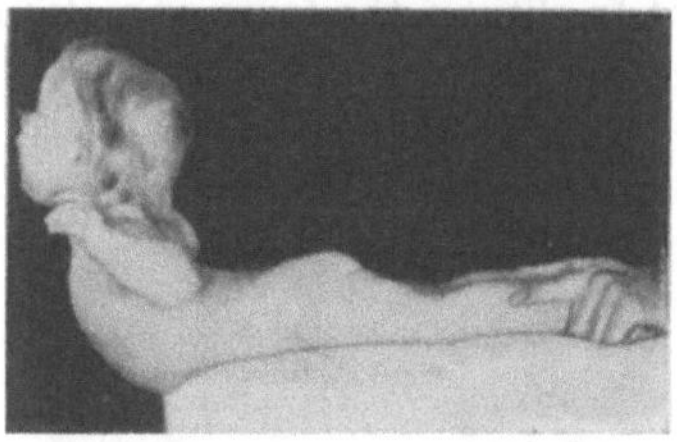

Fig. 52. Symmetrical chain reflex in a 10-month-old girl in the abdominal position. Upper half of body is not supported. Both arms stretched sideways.

In the infant shown in Figs. 47 and 48 a distinct chain reflex starting at the head and influencing the neck, trunk, arms, and legs to the tips of the toes occurs. I term this reflex a "symmetrical chain reflex in the abdominal position."

The reflex becomes even more distinct when the infant has to exert himself to assume the position. When older infants in the abdominal, lateral, or dorsal position are moved head forward over the edge of a table so that only the pelvis remains

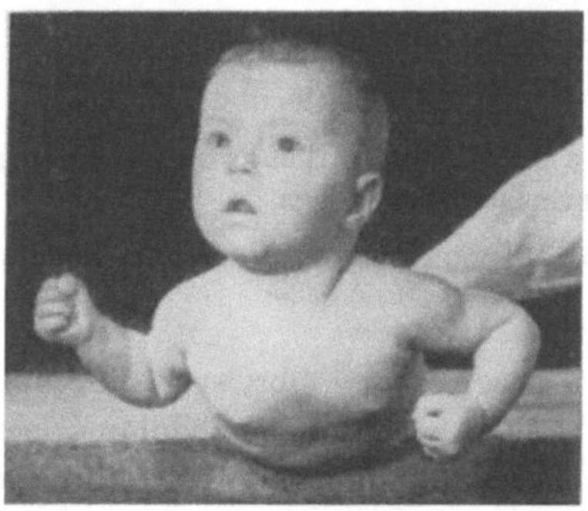

Fig. 53

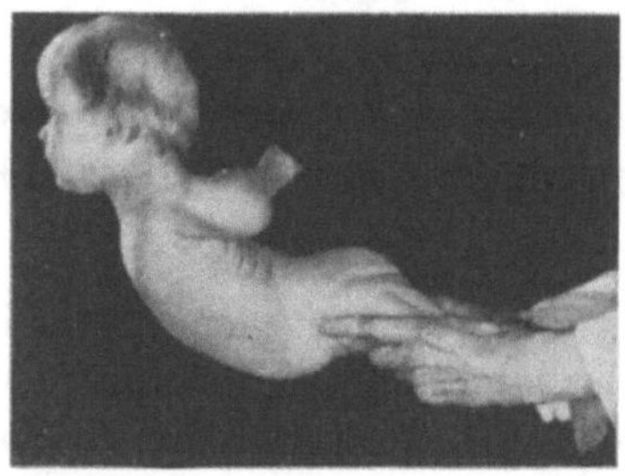

Fig. 54

Figs. 53 and 54. Symmetrical chain reflex in a 10-month-old girl in the suspended abdominal position. Both arms spread horizontally.

supported, many of them are still able to avoid letting the upper half of their bodies sink down.

In Figs. 52 and 53 the infant is in the abdominal position and moved head forward over the edge of the table as far as the umbilical region. In this critical position the head and upper half of the body are steeply raised so that a still more marked lordosis occurs than is seen in Fig. 48. At the same time the arms are raised horizontally,

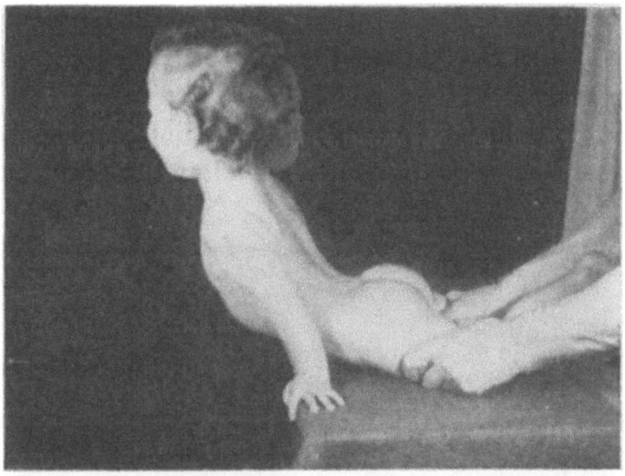

Fig. 55. Infant 11 months old in abdominal position, moved far out from table, head forward with only the pelvis supported. Labyrinthine righting reflex of head. Upper half of body, supported by arms, steeply raised.

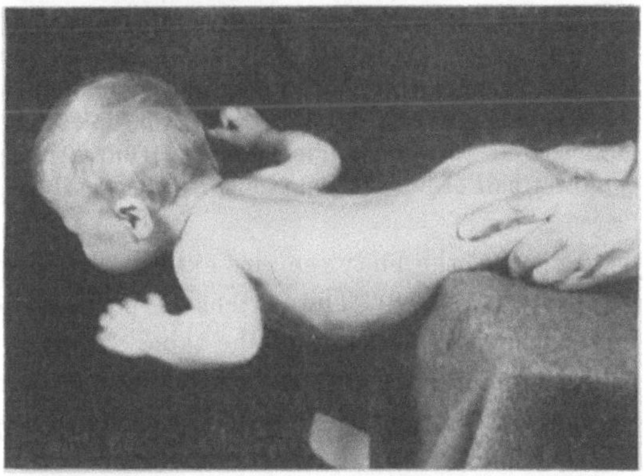

Fig. 56. Infant $11\frac{1}{2}$ months old in abdominal position, moved away from table, head forward. Labyrinthine righting reflex of head. Upper half of body and arms horizontal.

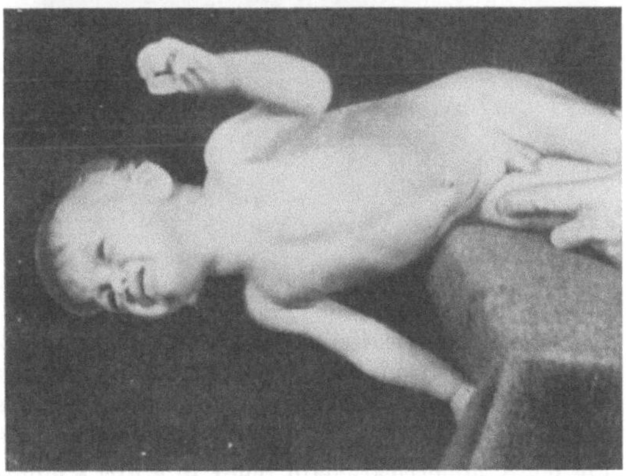

Fig. 57. Infant $11\frac{1}{2}$ months old in lateral position, moved from table as far as region of pelvis. No labyrinthine righting reflex of head. Child supports himself with one arm against table.

sideways, and a little to the back, and are flexed at the elbows. The center of gravity of the upper half of the body is hereby moved backward, making the difficult position easier for the infant.

Some infants find other solutions. They place their arms on the table in back of them and in this way lift themselves up (Fig. 55), or with heads raised they hold their bodies in the horizontal position (Fig. 56). In the lateral position another child gave up the labyrinthine righting reflex and supported himself with one arm on the table so that the head and upper half of the body remained in the horizontal position (Fig. 57).

It is understandable that the infant cannot endure such an effort for very long; he soon tires and lets his head and the upper half of his body sink. However, he is really able to do more. If he is grasped by his thighs and held in the abdominal position without further support, a quite similar posture requiring much more strength occurs (Fig. 54).

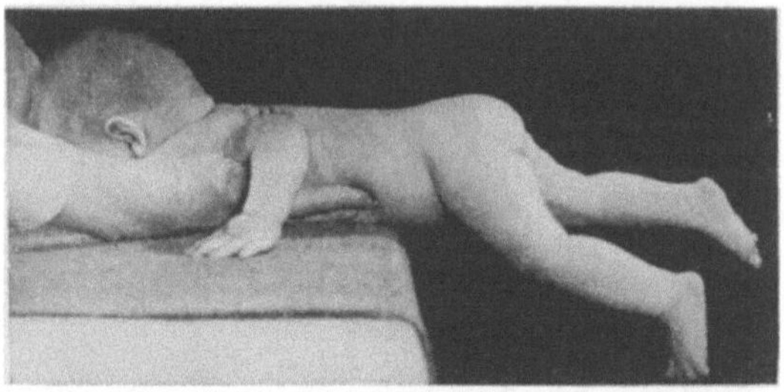

Fig. 58. Symmetrical chain reflex in a girl 5 months, 13 days old in the abdominal position. She holds pelvis and legs in an almost horizontal position although lower half of her body no longer rests on table.

To show that we are dealing with a chain reflex in Fig. 48, the demands on the child may be increased in another direction. If he is held at the chest right below the neck and moved so that his legs extend more and more beyond the table, the lower half of the body remains extended even when the legs and pelvis are no longer supported (Fig. 58). This extension position does not change even if the child is moved beyond the edge of the table as far as the umbilical region or still farther. There are children whose whole bodies can be moved from the supporting base like rigid sticks without their losing the extension position if they are adequately held.

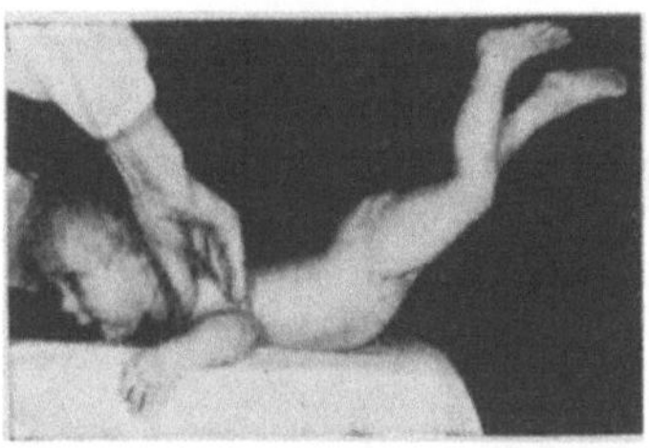

Fig. 59

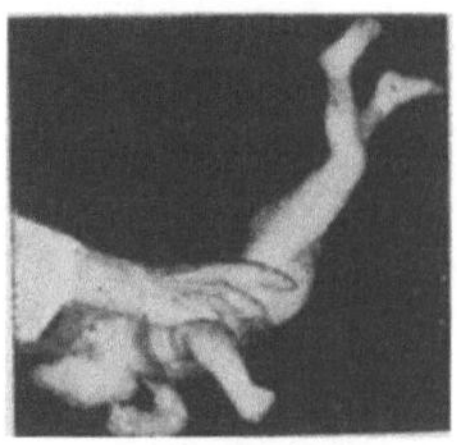

Fig. 60

Fig. 59. Symmetrical chain reflex in a girl 8 months old (same child as in Fig. 54) in the abdominal position. Through pressure on chest near neck, chest is lowered and lower half of body is raised in a lordotic body posture. Head is, as much as possible, brought into normal position in space.

Fig. 60. Symmetrical chain reflex in a girl 8 months old (same child as in Fig. 54) in the oblique position with the head down. Procedure in suspended child similar to that in Fig. 59 also causes extension and raising of lower half of body. Head is, as much as possible, brought into normal position in space.

The result is a symmetrical suspension reflex with the body in the abdominal position; the child acts as if he were still resting on the table, as in Fig. 48. The demands on the child can be increased even further. One can, starting with the child in the abdominal position, lower — again like a stick — the upper part of the chest and at the same time cause the lower half of the body to rise, whereby a pronounced lordosis results (Fig. 59). The corresponding reaction with the body in oblique suspension, and the head down, is a symmetrical chain reflex where a marked lordosis also occurs (Fig. 60). In all these reactions (Figs. 52—60) there is clear evidence of the child's attempts to bring the head as far as possible into the normal position in space and to keep it there, corresponding to the labyrinthine righting reflex of the head. The reactions disappear or cannot be elicited if the head is pushed down. It is thus clear that we are dealing with obvious chain reflexes starting at the head.

The behavior of the older infant in the suspended abdominal position supported at the abdomen was described in detail by LANDAU (1923). In this position the infant lifts his head and pelvis somewhat (corresponding to the labyrinthine righting reflex) so that his body forms an upward open arch; at the same time the legs are extended. If the head of the suspended infant is now pushed down and the labyrinthine righting reflex is thus extinguished, the muscles of the pelvis become flaccid and, conforming to gravity, permit the pelvis and legs to sink down. This shows that the chain reflex depends on the labyrinthine righting reflex of the head.

We can easily understand that the positions shown in Figs. 52—60 soon become uncomfortable for the infant, for these positions require much exertion; most adults could not accomplish such performances. Therefore, the infant soon collapses and lets the body drop in accordance with gravity. With frequent repetition of the experiment, infants learn, as LANDAU has already mentioned, to avoid the effort so that the reflex becomes more or less indistinct. Thus, practice decreases the reaction. In the second year of life the reflex disappears completely. If children of that age are brought into the same position, they immediately let their muscles relax and let the free body parts hang down. It is very rare to find the reflexes as distinct as in our illustrations. However, those who know about these reflexes are often able to demonstrate them at least to a minor degree. In art, flying angels have been represented in similar positions (especially as in Figs. 53 and 54).

In younger premature infants resting in the abdominal position, the symmetrical chain reflex does not occur at all; it appears gradually in normal newborns during the first month of life. After the fourth week it can be demonstrated in almost all infants when they are lying on a table. It is present in suspension in almost a third of the infants age two to six months, and in an overwhelming majority in the second year of life. If it is present at all in suspension, it can always be extinguished by pushing the head down.

Asymmetrical Chain Reflex in the Suspended Oblique Lateral Position, Head Up (1, 17)

If an infant is supported in the region of his loins and is lifted in a suspended oblique lateral position with his head up, his body assumes a peculiar position (Fig. 61). The head is again raised by the labyrinthine righting reflex into the normal position in space and the upper half of the body follows this movement so that he is half upright. The pelvis is also lifted somewhat sideways so that the body forms an upward open arch. Arms and legs also show specific positions. For example, if the body is lowered to the right, the left arm and leg are laterally extended and raised horizontally to the left. The right arm is brought back to the left as far as possible. The right leg is flexed at the hip and rotated outward somewhat and the foot is

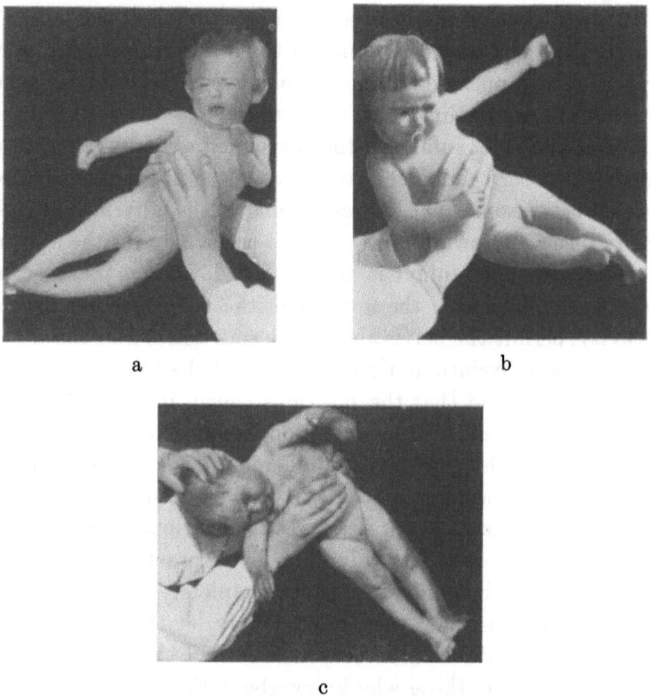

Fig. 61. Asymmetrical chain reflex with the body in suspended oblique lateral position and head up. (a) Left lateral position. (b) Right lateral position. (c) Extinction of suspension reflex in right lateral position by passive lowering of head.

extended with a distinct raising of its inner edge. In brief: In the oblique lateral position the upper extremities are extended and the lower limbs are flexed.

If the body of the infant is moved from one oblique position to the other, the head keeps its position in space while the extremities change their positions. The experiment can be repeated as often as desired, but it can be disturbed by the voluntary movements of the infant. The position of the lower half of the body is most easily disturbed. It does not matter whether or not the infant is blindfolded. The peculiar arm posture is frequently absent in younger infants.

Thus, a labyrinthine righting reflex of the head with a neck righting reflex followed by the whole body also occurred in these infants. How far the tonic neck and labyrinthine reflexes are responsible for the arm and leg positions cannot be stated in detail. In any case, in all parts of the body this reflex presents a picture exactly opposite to that described for the symmetrical chain reflex in the abdominal position, which also depends on the labyrinthine righting reflex of the head.

This reflex, too, can be extinguished by passively moving the head. The position of the arms and legs remains unchanged when the head is turned sideways. However, if the head is rotated to the lower shoulder in the sagittal axis, the upper half of the body and the pelvis are suddenly lowered and the arms and legs drop flaccidly (Fig. 61c).

The child maintains the threatened equilibrium by means of the posture he assumes when suspended in the oblique lateral position. The reflex can be most clearly observed in this position, but suspension does not play a role in daily life. However, corresponding reflexes appear if the child is in danger of falling to the side when sitting or walking. In Fig. 62., for example, the lower part of the abdomen has been brought into an oblique position by the observer. The child would have lost her equilibrium had not countermovements of the body and of the arm and leg on the opposite side occurred. Comparison of Fig. 62 with Fig. 63, where the child held at the chest is suspended in the oblique lateral position, shows that in principle we are dealing with identical chain reflexes both when the child is in suspension and when he sits in the oblique lateral position.

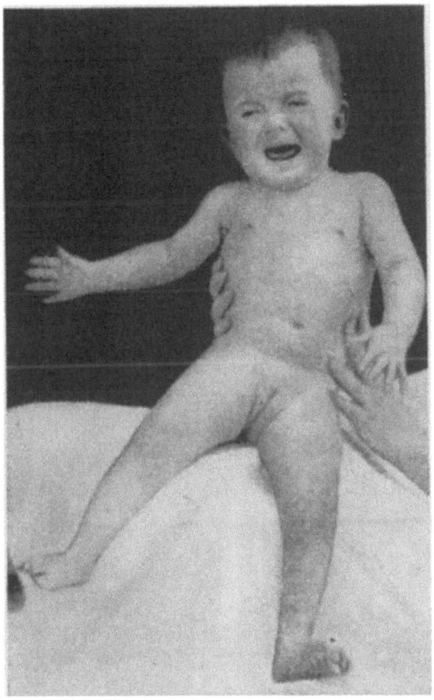

Fig. 62. Counterposition of right arm and right leg in a girl 11½ months old sitting in the oblique lateral position.

Figure 64 shows another position in which the infant maintains his equilibrium by means of the position of his arms.

Slight deviations of equilibrium, which are always immediately compensated for by such reflexes, occur quite often. The body of the unsupported sitting infant constantly sways to and fro in a much coarser way than that of the adult, who cannot remain immobile in a sitting position either.

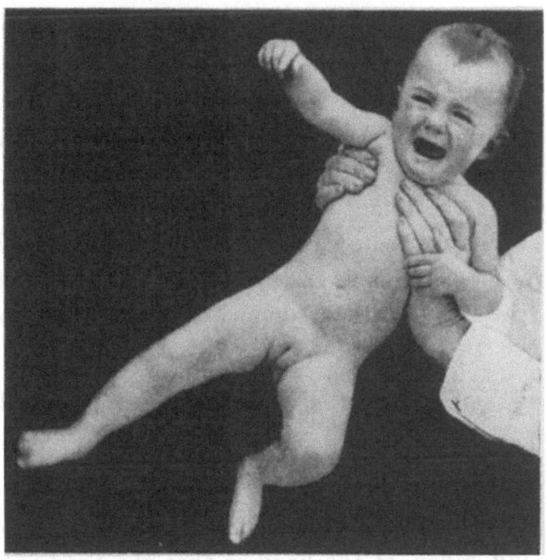

Fig.63. Counterposition of right arm and right leg in a girl 11½ months old in suspension in the oblique lateral position.

The asymmetrical chain reflex in the oblique lateral position is rarely found as a suspension reflex in the pure form described above. The position of arms and legs is often not completely attained. A bad mood or strong voluntary movements will have an inhibitory effect. The raising of the head with the body in the lateral position, which LANDAU also mentions, is regularly found in infants at this stage of development.

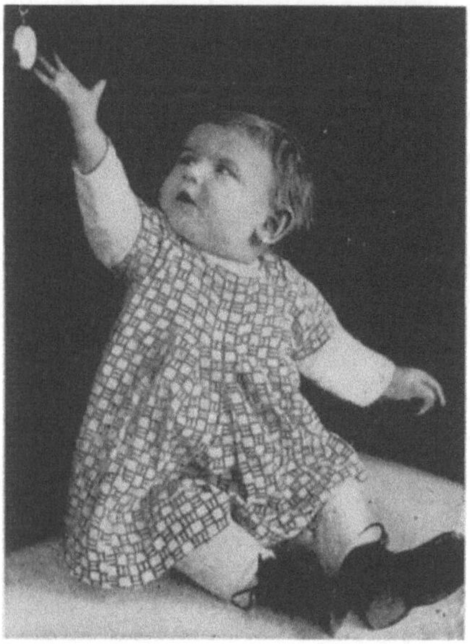

Fig. 64. Maintenance of equilibrium by means of arm position in a 10-month-old boy.

The reflex appears in the fourth to sixth month of life; it is absent in the newborn (Fig. 65). The rabbit deprived of labyrinths, suspended in the lateral position, reacts in the same way, while the intact animal lifts its head and upper half of the body into the normal position in space (Fig. 66). If present, the reflex can always be extinguished by passive head movements.

In chain reflexes with the body in the oblique lateral position, the compensatory position of the extremities is changed when the body position is changed in relation to the head, and at the same time the labyrinthine righting reflex fixes the head in space. The chain reflexes hereby differ basically from the tonic labyrinthine reflexes of the limbs that have been demonstrated in decerebrated animals (see p. 191). In these reflexes the extensor tonus of the extremities is changed when the head is moved in the direction of gravity while the position of the trunk remains unchanged, i.e., all four extremities are uniformly flexed and extended.

The compensatory reflexes described above, which restore the disturbed equilibrium, have been known for a long time and have been represented in many statues. In art theory this is known as counterpoising.

If the infant's body in the lateral position is rotated downward 90° so that the head is directed obliquely downward, the body and extremities will maintain the same position as in the oblique upright position of the upper half of the body. Only the face is turned somewhat to the lower shoulder.

Symmetrical Chain Reflex in the Suspended Oblique Dorsal Position, Head Up (17)

If the infant is lifted by the upper half of his body and is bent obliquely backward, he will move his head forward in a manner corresponding to the labyrinthine righting reflex of the head so that the head will again assume its normal position. The pelvis is raised forward and the arms and legs are extended forward (Figs. 36, p. 176, and 67). It is hard to determine the degree to which the position of the extremities is determined by tonic influences coming from the labyrinth or neck. The reflex is independent of the eyes. It cannot be completely extinguished by passive head movements and is somewhat less frequently found than in the oblique lateral position.

Symmetrical Chain Reflex in the Suspended Horizontal Dorsal Position

(1, 17, 36, 95, 139)

The labyrinthine righting reflex of the head, as previously mentioned, is completely absent when the body rests on a base in the usual dorsal position. If a newborn is held in a suspended dorsal position by supporting his sacrum (Fig. 68), he is unable to lift his head to the normal position in space.

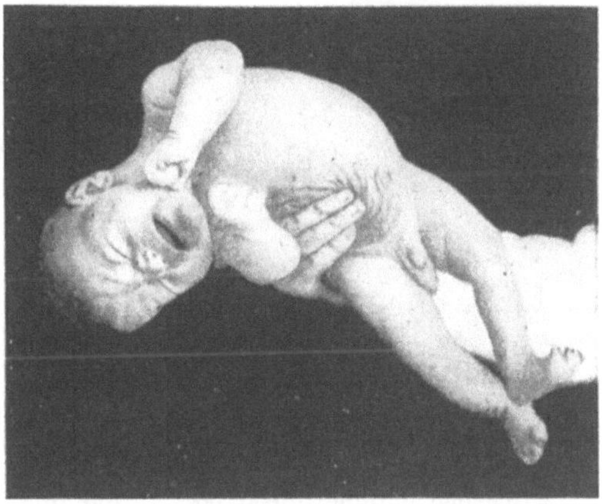

Fig. 65. Absence of the labyrinthine righting reflex of the head in the newborn, an 11-day-old boy, in the lateral position. The body responds to gravity.

In the same position the rabbit without labyrinths acts in the same manner, whereas in the intact animal the labyrinthine righting reflex raises the head and upper half of the body to the normal position in space (Fig. 69).

Fig. 66. (a) Rabbit without labyrinths; (b) intact rabbit in suspended lateral position. In the animal without labyrinths the head responds to gravity; in the intact animal the labyrinthine righting reflex raises the head to normal position. (Sketch after RADEMAKER, 1926, p. 137)

The reflex can be very distinctly demonstrated in the second half year of life. Again asymmetrical chain reflex starts from the head and attempts to maintain equilibrium in this difficult position (Figs. 70 and 71): The supported sacral area forms, so to speak, the vertex of an upward open angle; one of its sides is formed by the head and upper half of the body and the other by the extended legs (Fig. 71). The arms are extended more or less forward.

Tonic Labyrinthine Reflexes of the Limbs

According to the experiments on animals by MAGNUS and DE KLEIJN, tonic stimulations which effect the extension and flexion of the extremities start from the labyrinth (sacculus and utriculus). These stimulations are characterized by the fact that there is one position in space in which the extensor tonus of the limbs is strongest and another position in which it is weakest. The two head positions form an angle of 180° to each other. MAGNUS and DE KLEIJN demonstrated these reflexes in a child with brain damage and described in detail the method of examination that should

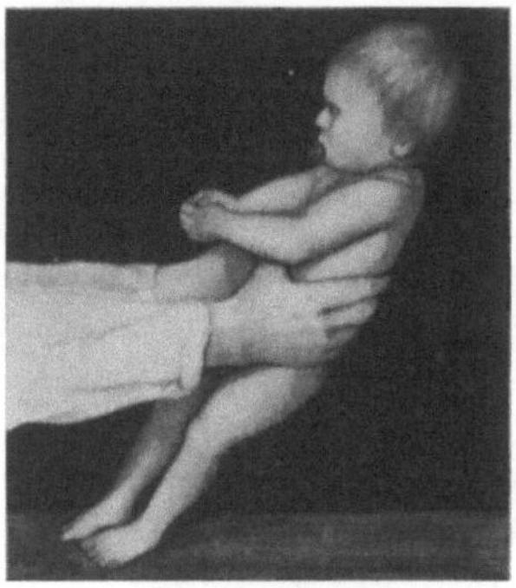

Fig. 67
Symmetrical chain reflex in suspended oblique dorsal position, head up.

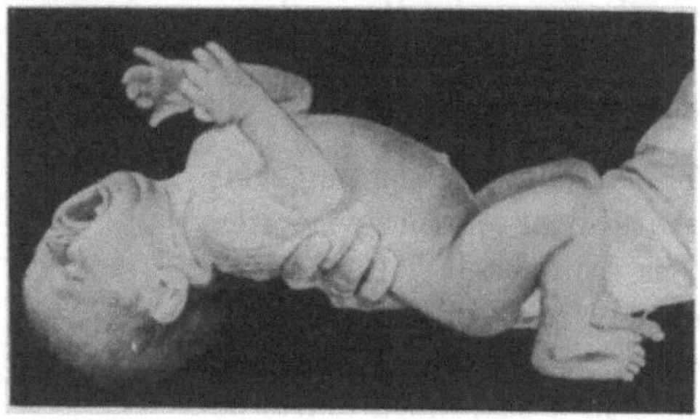

Fig. 68. Absence of the labyrinthine righting reflex of the head in a newborn boy 11 days old in the dorsal position. Body responds to gravity.

be used in man. So far no one has succeeded in demonstrating such reflexes in healthy infants. VOSS (1927) stresses the possibility that vestibular reflexes following motion may be confused with these reflexes. The same could happen in the tonic vertebral column reflex of the legs (see p. 168).

Tonic labyrinthine reflexes of the supporting tonus of the extremities are discussed on page 198.

Fig. 69. (a) Rabbit without labyrinths; (b) intact rabbit in suspended dorsal position. In the animal without labyrinths the head responds to gravity. In the intact animal the labyrinthine righting reflex maintains the head in the normal relation to gravity. (Sketch after RADEMAKER, 1926, p. 136)

Reflexes of Station and Gait Dependent in Part on Gravity

(41)

In standing and walking the body has to be moved and held opposite to the pull of gravity. This task is facilitated by suitable reflexes that have been examined, following MAGNUS's investigations (1924, 1930), by RADEMAKER (1926, 1931) in extended experiments on mammals, especially dogs, without cerebral hemispheres or a cerebellum. PEIPER [1932 *(41)*] and BALDUZZI (1932) demonstrated corresponding reflexes in infants. These reflexes gradually disappear when they are "covered" by the action of the cerebral hemispheres or become superfluous in the arms because of the upright body posture.

Only healthy infants can be used for these investigations. Poor development due to different causes, such as alimentary disorders, infections, or rachitis, considerably

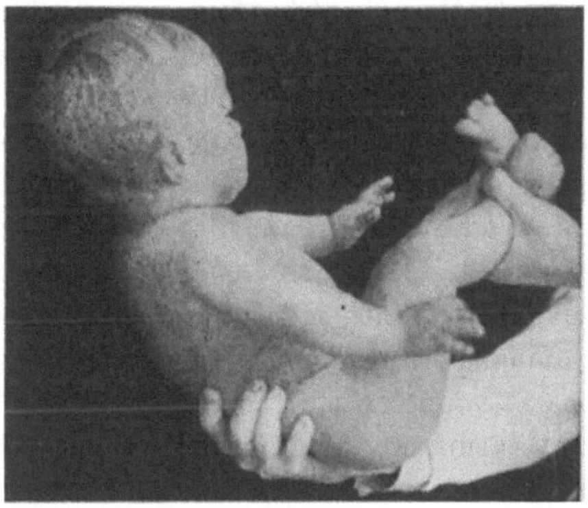

Fig. 70. Suspension reflex in an 11-month-old boy in the dorsal position.

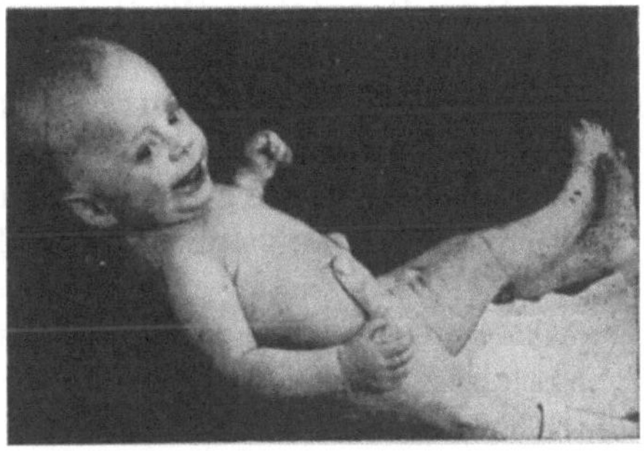

Fig. 71. Suspension reflex in an 8-month-old boy in the dorsal position.

retard the learning of standing and walking if the disturbance lasts for any length of time. These diseases also impair the mood of the infant. This further inhibits station and gait. In general a child who knows the examiner and who enjoys his own performance will achieve considerably more when in a good mood than when in a bad mood. Experiments that are carried out only once by an observer unfamiliar to the child will not elicit the child's optimal response.

Standing Readiness

(41)

RADEMAKER terms the ability to bring the extremities into the position necessary for standing "standing readiness." For this function stimuli arising in different areas of the body surface are needed. Standing readiness implicates the reflexes of the cerebral hemispheres, can be understood as a conditioned reflex, and is found in intact animals and in those without a cerebellum, but is absent in animals without cerebral hemispheres and in decorticated spinal cord and thalamus preparations.

According to RADEMAKER, standing readiness is also said to be absent in infants during their first year of life, corresponding to similar observations on newborn puppies. If one holds infants under the axillae and moves them downward until their legs touch the ground, this contact does not make them put their feet down; on the contrary, the legs are drawn up and lifted from the floor. My experiences contradict this statement. It is possible to elicit stepping movements even in newborns when they are put down (see p. 211).

Infants in the upright position have a tendency to extend their legs when the soles touch the ground. At first only the ability for supporting reactions, not a distinct standing readiness, exists when the legs are brought into the appropriate position. A readiness to stand appears only in the fourth to sixth month of life when the dorsal aspect of the feet is touched: The child is held in the upright position so that he can touch the edge of a table from below with the dorsal aspect of his feet. He flexes first one leg and then the other, placing each with the sole of his foot on the table, and then extending it again until he has climbed onto the table with both legs. SAINT-ANNE DARGASSIES (1954) describes a similar reflex as "réflexe d'enjambement." In the dorsal position only very weak corresponding reflexes can be demonstrated; in the first three months they are barely noticeable.

In the abdominal position with the arms propped up, the infant lifts his abdomen from the base by kneeling to a certain extent with adducted thighs at a very early age (fourth to sixth month). Only in the seventh to ninth month does he acquire the ability to lift himself up from the abdominal position onto the extended legs, an achievement that is possible with his arms at a much earlier age.

A readiness to stand exists also for the infant's arms, which support the upper half of the body in the abdominal position. The ability to raise the upper half of the body develops in approximately the fourth to sixth month; the infant at first supports himself on his forearms and later on his extended arms. In suspension with the head down, a position corresponding to that in RADEMAKER's experiments on animals, a standing readiness of the arms usually does not occur when the table is touched; grasp reflexes appear in lieu of this phenomenon.

The optic readiness to stand was demonstrated for the forelegs by RADEMAKER in his experiments on animals. In infancy it also can be tested on the arms because the child is accustomed to the abdominal position and at first can recognize where to put his arms much more easily. If a six-month-old infant is held in the abdominal position and lowered in this manner to a base, he will stretch both arms out to stand on it (Fig. 72). If the infant thus supported sees his mother in front of him, he will walk toward her on his hands (wheelbarrow experiment of ANDRÉ-THOMAS and AUTGAERDEN, 1959, p. 103). A readiness of the arms also occurs if the child is held at the trunk and is slowly inclined forward from the standing position.

The optic standing readiness of the legs develops approximately in the ninth to twelfth month of life. This can be demonstrated by holding the infant upright and suspended so that he has to flex his legs at the hips in the forward direction in order to reach the supporting plane with his soles.

The disappearance of these reactions in the dark shows that all these stimuli are transmitted by the eyes.

Supporting Reactions

(41)

The supporting reactions described by MAGNUS (1926) and RADEMAKER (1931) cause a supporting tonus in the muscles of the extremities which makes standing possible. These reactions are elicited through the effect of the base on the correctly placed hands and feet; the result is an extension of the extremities that is needed for static functions. Through simultaneous contractions of the flexors and extensors the limbs are transformed into solid pillars and thus support the body.

Usually the supporting reactions are elicited by the weight of the body, which presses the limbs against the base. In the stepping movements of the newborn (see p. 211), however, they can be elicited independent of the direction of gravity. This is done in the following manner: The examiner holds the infant in the horizontal position and presses him against a vertical wall or holds him in a vertical position with the head down and presses him against a horizontal wall from below.

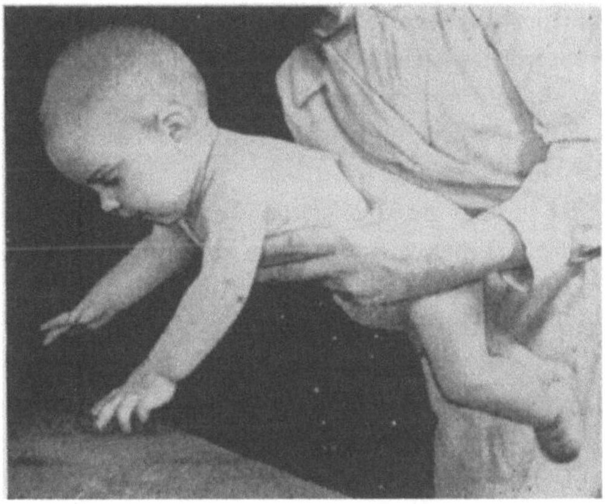

Fig. 72. Optic standing readiness of the arms in a 257-day-old girl. In the abdominal position and in suspension lowered to a supporting plane the infant stretches her arms out even before touching the plane in order to support herself on it.

The flight reflex must be regarded as a negative supporting reaction. It is closely related to the protective skin reflexes (see p. 30). A slight stimulation of the sole, such as a pin prick, causes flexion of the leg at the knee and hip joints. BABINSKI (1915, 1922) described this "réflexe de défense" as due to damage to the pyramidal tracts, the same as the toe sign named after him. In the German literature this sign is known as a shortening reaction of the legs or flight reflex. It is physiological in early infancy. The time of its disappearance has not yet been investigated in detail. This sign also possesses tonic qualities. If a small clamp that does not pinch too much is fastened to the sole, the leg remains flexed as long as the clamp pinches.

According to RADEMAKER (1931), supporting reactions and supporting tonus are absent in infants during the first half year of life; the legs immediately bend when one tries to have the infant stand. The absence of these phenomena cannot be a consequence of a weakness of the leg muscles, since infants execute kicking and defense movements with considerable strength. However, the newborn shows positive supporting reactions. If one is placed in an upright standing position, his legs are frequently transformed into rigid pillars that carry the weight of the body for a short time. It sometimes takes several seconds before he topples over from this position. We can observe that the infant does not completely extend his legs at the knees and hips as the adult does, but stands in a manner similar to that of the ape, his knees and hips slightly flexed. The positive supporting reaction of the legs is a prerequisite for the stepping movements of the newborn. ANDRÉ-THOMAS (1952) describes reactions that are essentially supporting reactions with the title "redressement statique," with no reference to RADEMAKER.

In the following months the supporting reactions of the legs become stronger; it takes longer and longer before the infant's legs give way. The supporting reactions are aided by the slow disappearance of the original stepping movements. The reactions now also appear in the oblique body position, where formerly stepping movements would have occurred. It is possible to elicit supporting reactions on an oblique or vertical wall if the infant's soles are firmly placed on it. The supporting reactions occurring during crawling movements of the newborn in the abdominal position (see p. 217) are so strong that the infant pushes himself off the hand of the examiner and moves ahead.

RADEMAKER analyzed the individual stages of the supporting reactions in the animal and also investigated the conditions under which they occurred. The positive supporting reactions depend on the touching of the soles (magnetic or exteroceptive reaction), on the position of the distal parts of the extremities (hands, fingers, and toes), and on the pressure on the soles.

In the experiments on animals certain differences between the magnetic reactions of the front and hind legs exist. In dogs without a cerebellum the hind legs react best in the dorsal position. Intact dogs have changing reactions; the strength of the reactions may differ even in the same animal. In intact dogs they are easily inhibited by the experimental conditions, such as the dorsal position; they are nonexistent in cats, rabbits, and apes.

RADEMAKER described the magnetic reaction of the hind legs in the dog without a cerebellum as follows: At the slightest touch of the toe tips an extension of the leg by which the touching finger is pushed away occurs. If the finger is gradually retracted, the hind leg follows, as though attracted by a magnetic force, until the leg is extended. The extension persists as long as the contact lasts, even 45 minutes or more. All joints are fixed and the leg is as rigid as a pillar. When the contact stops, the leg immediately moves again.

RADEMAKER (1931) and PEIPER [1932 *(41)*] were unable to demonstrate magnetic reactions in the legs of the infant, but BALDUZZI (1932) described them as follows: When the foot of an infant who is resting in the dorsal position with the legs adducted is touched in front, it is sometimes possible to cause extension of the leg as long as the contact lasts. If it is touched for any length of time, the leg will eventually be flexed. If the contact is nevertheless continued, the leg will immediately be extended again. The reaction occurs in the abdominal and dorsal positions and is most distinct when the infant is supported by the axillae and held in suspension. The reaction becomes more distinct with increasing age; it can be demonstrated in 20% of newborns and becomes constant in the fifth month. Later it is less frequent. It disappears in approximately the twelfth month of life, as soon as the child can stand erect.

As far as the duration of the magnetic reaction is concerned, there is considerable difference between the findings of RADEMAKER in the animal and the observations of BALDUZZI in the human infant. The supporting reactions may persist for some time after the first year in diseases and disturbances of development.

According to RADEMAKER, magnetic reactions of the front legs of intact dogs and those without a cerebellum are found in the abdominal position, but not in the dorsal position. Apes respond to the touch of the palms with grasping movements; they flex their fingers and adduct their arms. The tonic palmar grasp reflex (see p. 158) of the human newborn corresponds to this reaction with the exception that the arm is at first not adducted. I could not find any magnetic reactions of the arms in infancy.

Pressure on the soles, according to RADEMAKER, causes extension of the legs in intact dogs and in those without a cerebellum when in the dorsal position. BALDUZZI had difficulty in demonstrating a similar reaction in the newborn. However, if he placed the infant in the abdominal position as though to elicit crawling movements and held a little plank against the soles, stronger or weaker pressure caused stronger or weaker extension of the leg.

According to RADEMAKER, passive extension of a proximal joint of the front or hind legs causes an extension and a more or less marked fixation of the other joints in the intact dog and in those without cerebral hemispheres or a cerebellum, while passive flexing of a proximal joint is accompanied by flexion and decrease of the fixation of the other joints. BALDUZZI placed an infant in the abdominal position and passively flexed his soles and toes. The infant extended his legs so forcefully that he succeeded in pushing himself forward. The same occurred in the dorsal position with the head up. Often the stimulus had to be repeated several times in order to elicit the reaction.

According to BALDUZZI, the reactions described by him as positive support reactions become more marked until the sixth month of life, are always present at that age, and later become indistinct again.

Generally the adult does not use his arms for support, but rather for grasping; therefore, supporting reactions of the arms are not easily demonstrated in the adult. It is different in the infant. In his first creeping and crawling attempts he uses his arms more than his legs for the support of his body. During these movements we find supporting reactions of the arms that are very similar to the supporting reactions of the front and hind legs of animals. Later, when walking in the erect posture is achieved, these reactions are no longer needed. However, they appear immediately when the adult tries to walk on all four limbs.

No supporting reaction of the arms is found in the newborn. However, a standing readiness of the arms flexed at the elbows soon develops, as has been described. In approximately the fourth to sixth month the infant acquires the ability to support himself in the abdominal position with his hands only, while wrists, elbows, and shoulders are fixed (Fig. 75). The arms are only slightly flexed at the elbows.

Further supporting reactions of the arms occur when the sitting infant protects himself from falling. He supports himself with one arm or with both at the base, as can be observed in every infant who is insecure in sitting.

Strong supporting reactions can also be elicited in the arms with the body in the standing position if the child is on his tip toes and the body is bent forward (Fig. 84, p. 207). The same reaction to forward bending also occurs in the standing adult. RADEMAKER sees these reactions as myotatic reflexes of the stretched muscles of the lower legs.

The supporting reactions are of clinical importance. It has long been known that a severely rachitic child brought to a standing position adducts his legs to his body and cannot be persuaded to stand (Fig. 73). He is as little ready to assume the beginning position of his arms for creeping or crawling (Fig. 74). Readiness to stand and supporting reactions of the arms and legs are absent in such a child.

Compared with the animals investigated so far, the human newborn occupies a special position, since the supporting reactions are absent in newborn puppies and rabbit. Nothing is known about the behavior of newborn apes. However, supporting reactions are not a characteristic of infancy. RADEMAKER reports observations of reactions in adults that correspond more or less to those in dogs. These reactions have clinical significance in diseases of the cerebellum.

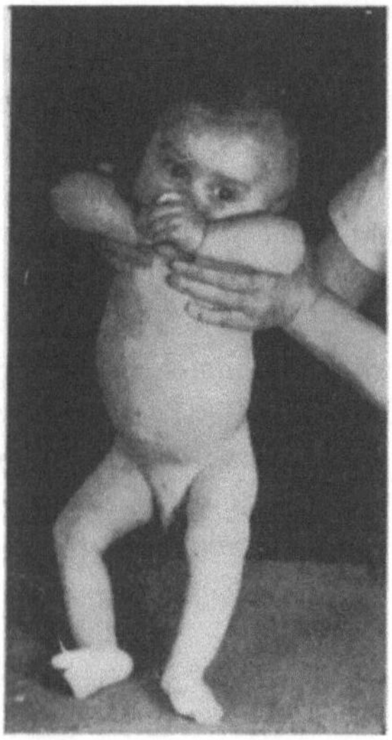

Fig. 73. Absence of standing readiness and supporting reactions of the legs in severe rachitis with brittleness of the bones in a 1-year, 8-month-old infant (weight 6700 gm, length 66.4 cm).

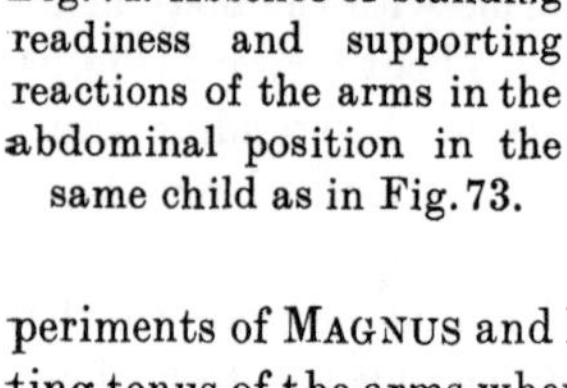

Fig. 74. Absence of standing readiness and supporting reactions of the arms in the abdominal position in the same child as in Fig. 73.

Changes of Supporting Tonus Through Head Movements (41)

MAGNUS and DE KLEIJN demonstrated in their experiments on animals that the tonus of the extensors of arms and legs is influenced by tonic neck and labyrinthine reflexes. In the young infant these asymmetrical tonic neck reflexes of the limbs can frequently be demonstrated (see p. 156), but no details are known about the influence of tonic labyrinthine reflexes. RADEMAKER showed that the supporting tonus in the animal likewise depends on the head position. This is also the case in the infant.

According to MAGNUS and DE KLEIJN, the raising of the head in intact animals and in those without a cerebellum causes an increase of extensor tonus in the front legs and a decrease in the hind legs. RADEMAKER expanded these findings by proving that the raising and lowering of the head also alters the supporting tonus and the position of the weight-carrying legs; this is particularly caused by the tonic neck reflexes. According to BALDUZZI, the supporting tonus in the human newborn is only rarely influenced by the raising of the head, but if the change occurs it is quite obvious. In the dorsal position the raising of the head causes an increase of the supporting tonus of the legs, and the opposite head movement makes the supporting tonus disappear.

When I placed a seven-month-old infant in the abdominal position so that he supported himself on his extended arms (Fig. 75), as was done in the animal experiments of MAGNUS and RADEMAKER, I could not find any alteration of the supporting tonus of the arms when I raised or lowered the infant's head. In the standing position the supporting tonus of the legs is not changed through the same head movements either. Only in bending down, which is a complicated reflex process, is the lowering of the head connected with a decrease of the extensor tonus of the legs (Fig. 96, p. 231).

In the young infant, extension of the jaw-limbs and flexion of the skull-limbs occur upon turning of the head (see p. 156) under the influence of tonic neck reflexes. BALDUZZI (1932) was able to demonstrate in the newborn lying in the dorsal or

Fig. 75

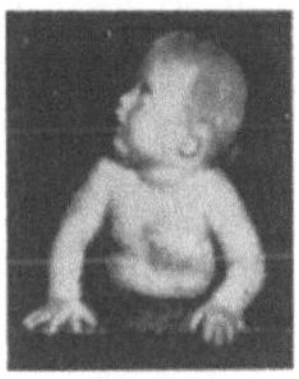

Fig. 76 a

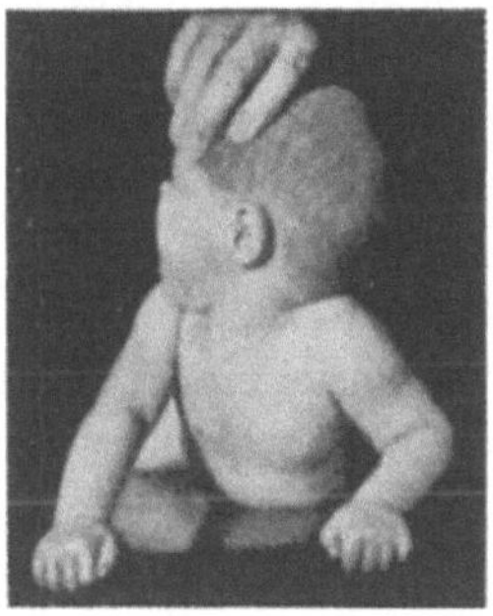

Fig. 76 b

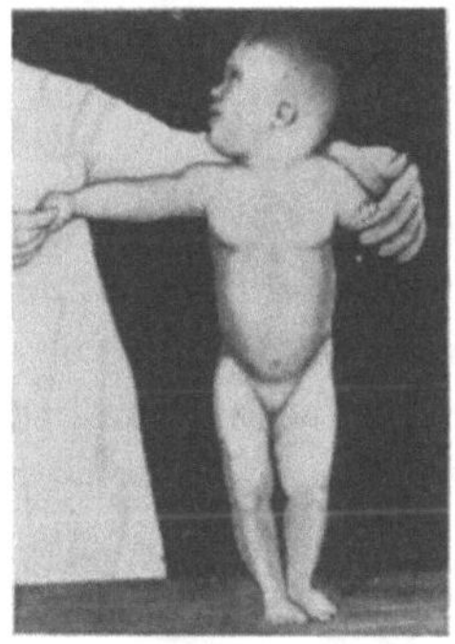

Fig. 77

Fig. 75. Supporting reactions of the arms in a girl 6½ months old in the abdominal position.

Fig. 76. Influence of the tonic neck reflexes on the supporting tonus of the arms in active (a) and passive (b) turning of the head of a 7-month-old girl. The jaw-arm is extended and the skull-arm is flexed.

Fig. 77. Influence of the tonic neck reflexes of the supporting tonus of the arms in a 7-month-old girl in the standing position during active sideways rotation of the head. Extension of the jaw-arm and flexion of the skull-arm. The child holds fast to the hand of the examiner (palmar grasp reflex) and is not held by him.

abdominal position that head turning increases the supporting tonus in the jaw-limbs and decreases it in the skull-limbs.

If an infant seven months old is placed in the abdominal position with the arms propped up, active or passive turning of the head causes extension of the jaw-arm and flexion of the skull-arm, corresponding to MAGNUS's rule (Fig. 76). The same influences of head turning on the supporting tonus of the arms can also be demonstrated in the standing position (Fig. 77). However, the influence of head turning on the supporting tonus of the legs was, at the most, barely noticeable.

One of the many experiments RADEMAKER conducted on the influences of the tonic labyrinthine reflexes on the supporting reactions in dogs without a cerebellum was imitated in the human newborn by BALDUZZI. RADEMAKER found a distinct supporting tonus in the hind legs when the dog was in the abdominal, dorsal, or vertical position and its mouth was directed upward, but if the mouth was directed downward the tonus was absent in all positions and could be elicited neither by touching nor by static stimulation of the soles of the hind legs. To demonstrate this phenomenon in newborns, BALDUZZI fastened the infant to a small board so that chest and abdomen became immobile and the head could not change its position in relation to the body. If the infant was in the dorsal position, lowering of the head (by lowering the head end of the board) caused the disappearance, and raising of the head an increase, of the supporting tonus. This phenomenon could be shown immediately after birth and became even more distinct in the second month of life. With the body in the abdominal position the effect of this experiment was uncertain because the infant became restless.

According to Balduzzi, the crossed flexor reflex of the legs accompanying flexion of the other leg (see p. 294) is inhibited in the infant by the magnetic reaction.

Vertebral Column Reflex

(139)

If we try to bring a newborn into the sitting position or if we hold him up in oblique suspension, he will collapse. The dorsal muscles are not yet strong enough to maintain the vertebral column in extension under the stress of weight. For the same reason it is necessary to support the newborn's body during his stepping movements (see p. 211). The infant acquires the ability to keep his vertebral column in extension only in close association with the development of the labyrinthine righting reflex, which places the head in the right position in space. Not only the cervical but also the thoracic spine turns with the head (Fig. 76). In the course of the infant's development the raising of the head in the abdominal position leads to a strengthening of the thoracic spine so that finally in the sixth to ninth month the child can raise one arm without falling over (Fig. 51, p. 183).

Fig. 78. "Erecting" vertebral column reflex in a dog without a cerebellum. The dog is kept in suspension by its horizontally held thighs, head down. On passive flexion of the thighs the upper part of the body becomes erect and the mouth turns upward. (Sketch after Rademaker, 1931, p. 199)

In the older infant (Fig. 79) and in the dog (Fig. 78) we can find the same conspicuous reflex, which completely raises the body in the suspended lateral position with the head down. Rademaker (1931, p. 199) described it as a vertebral column reflex. If a dog is lifted from the floor by the horizontally held thighs, the spine of the vertically suspended trunk is usually so relaxed that during passive movements the upper part of the body will swing to and fro. If the thighs are then passively flexed at the hips, the muscles of the spine suddenly contract and bring the upper part of the body into an erect position. If the thighs are directed forward, the upper part of the body remains in an almost horizontal position. The neck muscles contract at the same time so that the animal's mouth is directed upward. Rademaker found this reflex in intact (?) dogs and in those without a cerebellum, cerebral hemispheres, or labyrinths. It was absent during deep narcotic states and in newborn puppies during the first week of life.

A corresponding reflex can be demonstrated in the older infant (Fig. 79): The infant is held in the suspended abdominal position by his thighs, which are horizontal; the trunk is directed downward because of gravity. If the thighs are now flexed at the hips the head and trunk are suddenly raised until they are completely erect. Sometimes they are really hurled upward to an erect position against gravity and reach a by far more upward erect position than RADEMAKER's dog, which raises the spine to the horizontal, i.e., normal, position and only turns the mouth upward.

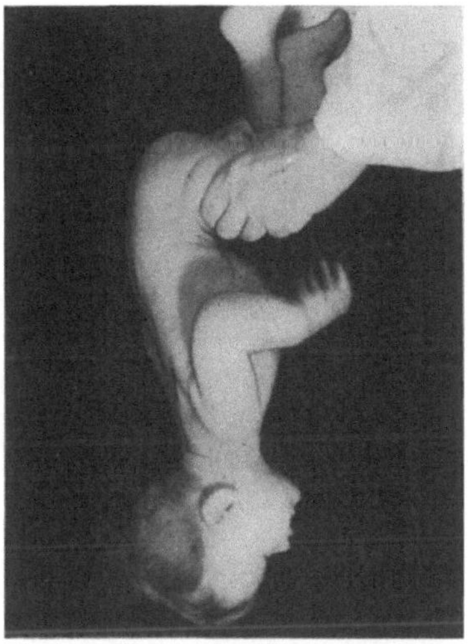

Fig. 79 a

This reflex can also be elicited in the suspended lateral position: If an infant ten to twelve months old is seized by his thighs and suspended in the lateral position, he can hold this position for a short time. The head remains in an approximately lateral position and shows little tendency to be raised. Perhaps it is mainly righted in respect to the body by the body-righting reflex of the head. However, if the thighs are grasped and the pelvis then rotated around its sagittal axis (Fig. 80), the vertebral column reflex will bring the child into the erect position in the same way.

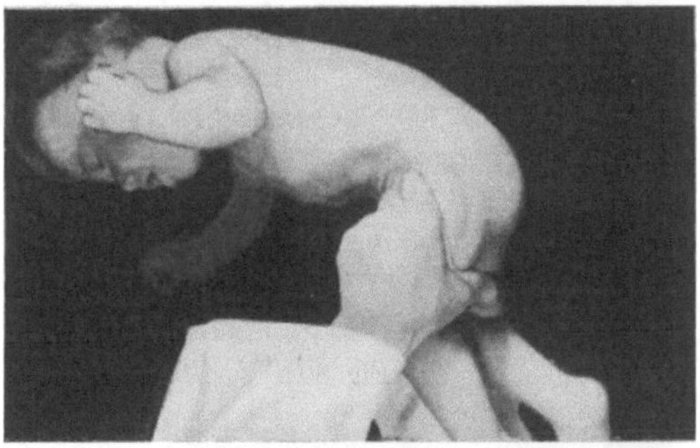

Fig. 79 b

Which receptors are active in eliciting the vertebral column reflex is uncertain. It could be that a chain reflex from the labyrinthine righting reflex of the head effects the raising of the trunk, for we can clearly recognize (Figs. 79 and 80) that the head precedes the body in the erecting motion.

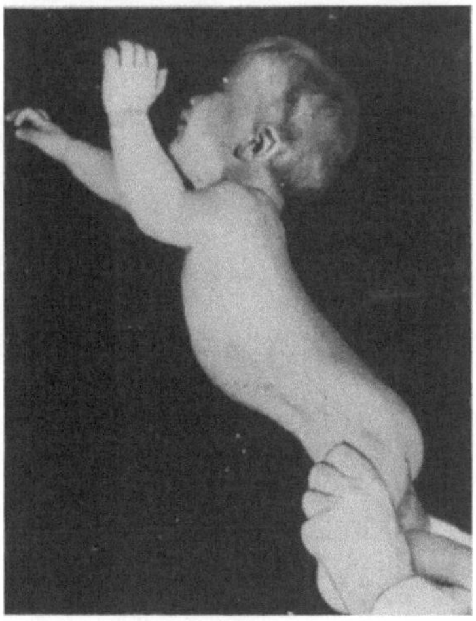

Fig. 79 c

Fig. 79. "Erecting" vertebral column reflex in an infant 11½ months old. The thighs are held horizontally; the head is in a downward direction. On passive flexion of the thighs the head and upper part of the body are raised. The head leads the movement.

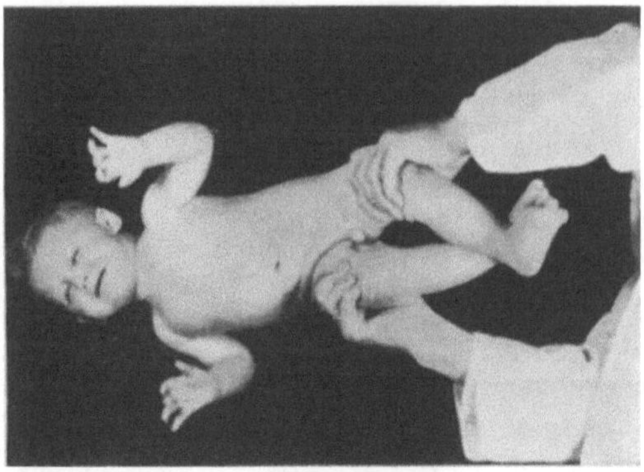

Fig. 80 a

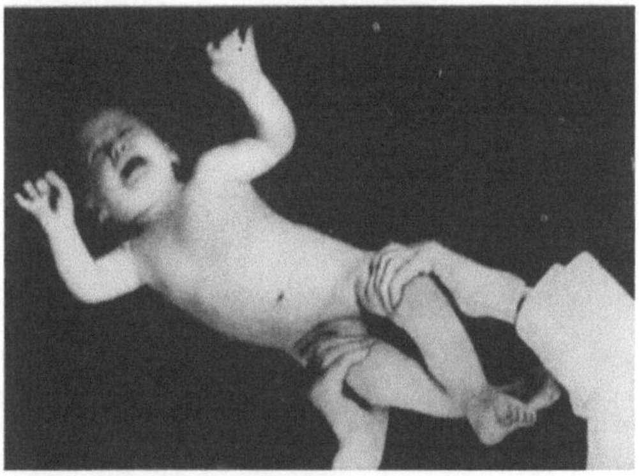

Fig. 80 b

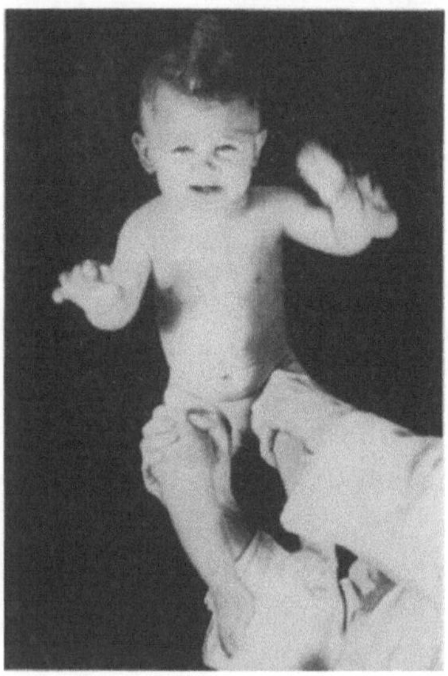

Fig. 80 c

If the child is resting on the table in the dorsal or lateral position instead of being suspended, the corresponding flexion of the thighs in the hips is not successful. Optic righting reflexes are not involved, since the experiment also succeeds in the dark.

A reflex raising of the trunk can be elicited, according to SAINT-ANNE DARGASSIES (1954), in the following manner: The child is held upright with his back toward the examiner, who holds his right forearm in front of the infant's pelvis. In this position the trunk and head will bend forward. If the examiner now repeatedly presses the plantar region of the child with his free hand the tibiotarsal joints move and the extensors are activated and extend the legs; thus, the trunk is again raised. When the head is bent forward and lowered, breathing is made difficult. To enable the infant to breathe and cry better, the extensors of the neck contract strongly, so that the head is raised. The infant also raises his body axis with repeated pushes and leans against the observer.

A tonic vertebral column reflex of the leg is described in another context (see p. 168).

Rocking Reaction
(41)

If a forepaw, for example, the right one, of a standing dog is passively lifted, a strong

Fig. 80. "Erecting" vertebral column reflex in suspended horizontal lateral position (same child as in Fig. 79). The child can keep this position for a short time. If the pelvis is rotated around its sagittal axis by the thighs, the head and upper part of the body are raised.

extension and abduction of the raised leg is felt with each passive movement of the trunk to the right ("rocking reaction" according to RADEMAKER). The reaction of the lifted leg is caused by the position change of the opposite leg, which rests on the ground.

This rocking reaction is absent in decerebrated dogs, in spinal cord preparations, and in newborn puppies. RADEMAKER did not find this reaction in young infants; he could demonstrate it in children only when they started to stand and walk.

I was also unable to demonstrate this reaction during the first months of life, even when distinct stepping movements were present. Later, in approximately the second half of the first year, the reaction is very clearly seen, but is easily overshadowed by irregular interfering movements that appear when the child starts to develop great interest in standing and walking. With this reservation rocking reactions can sometimes be demonstrated even in the arms, when the infants prop themselves up on their arms in the abdominal position.

Resisting and Limping Reactions

(41)

Passive movement of a statically involved leg toward the median position evokes a resisting reaction, and if the movement is directed away from the median position, a limping reaction will result (RADEMAKER). For example, if the body of a standing dog is pushed in a certain direction, let us say to the right, the right front and hind legs at first will resist the pressure (resisting reaction). If the body is moved still further to the right, the animal does not fall, but the right legs are lifted and put down farther to the right (limping reaction). Both reactions are elicited by shifting of the body in various directions: forward, backward, or sideward. They serve to preserve equilibrium, by counteracting the toppling of the animal.

RADEMAKER found the resisting reaction in intact dogs and in dogs without cerebral hemispheres or a cerebellum, but it was absent in decerebrated animals and in spinal cord preparations.

The deeper the point of gravity is located in the body and the better it is supported, the more effective is the reaction. It therefore is most distinct in the dog standing on its four legs (RADEMAKER).

Such a position is rare in man even in infancy. The child starts to crawl before he starts to walk. However, when he has acquired the ability to move forward on his extended arms and at the same time (and this is more difficult and happens less frequently) to move on his legs when they are extended at the knees, he usually can also stand up and walk and he makes use of these skills. Still, in the abdominal position with propped-up arms, the starting position for creeping, a distinct resisting reaction of the arms may be recognized (Fig. 81). If in this position the upper half of the body is pushed to the side, it is possible to feel and see a short resisting reaction of one arm. As soon as the resistance is overcome, the arm is lifted and set down farther to the side. In the standing position the passive lateral shifting of the body causes a limping reaction of the leg, but a resisting reaction of the leg cannot be felt. This will have little effect, since the center of gravity is very high.

The limping reaction of the leg is elicited by a shift of the animal's body against the supporting base or, vice versa, by a shifting of the base in relation to the body, whereupon the supporting tonus decreases in the propped-up leg and the leg is lifted up and placed in another position. Intact dogs show the most distinct limping reactions in the front legs in the outward and forward directions; reactions in other directions are not elicited with the same regularity. In animals without cerebral hemispheres or a cerebellum the reactions are associated with certain interferences; they are absent in decerebrated dogs, in spinal cord preparations after the posterior

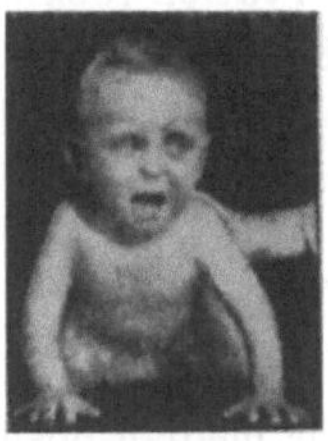

Fig. 81

Fig. 82 a

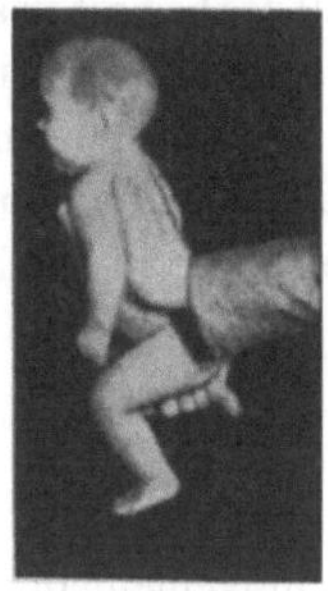

Fig. 82 b

Fig. 81. Resisting reaction of the right arm on pushing of the body to the right in a 13-month-old girl.

Fig. 82. Limping reaction of the leg in a 7-month-old girl. The right leg is passively flexed at the knee and hip. Upon touching the floor with the left sole the left leg is extended. Now the upper part of the body is passively moved forward. (a) Following this procedure the left foot is placed in tip-toe position. (b) A moment later the left leg is suddenly flexed at the hip, knee, and ankle joints and placed forward (limping reaction of leg). The next stage is again extension of the leg.

roots have been cut, and in newborn puppies during the first weeks of life (Rademaker).

In infancy, limping reactions can be elicited easily and distinctly when the procedure shown in Fig. 82 is used: One leg, for example, the right, is passively flexed in the standing position. Then the sole of the left leg, which is still flexed, is placed on the base, whereupon it will be extended. If the upper half of the body is now moved farther and farther forward, the left foot increasingly goes into the tip-toe position. Suddenly the whole left leg collapses, makes a limping step forward, and is again extended.

Limping reactions of the legs can be elicited by shifting the body forward, backward, and sideward, but they do not occur with the same distinctness in all directions. These reactions can best be obtained in forward-directed movements. They are found in newborns and even in premature infants weighing as little as 1200 gm if the infants are at all capable of supporting reactions of the legs. The presence of stepping movements in the newborn (see p. 211), though, is not necessarily associated with the elicitation of limping reactions.

"Limping" reactions identical to those in the front legs of animals can be demonstrated in the arms of infants. In the abdominal position with the arms propped up, distinct reactions are seen when the infant's body or the supporting base of the arms is shifted. For example, if a piece of cloth on which one arm is resting is gradually pulled forward, the arm at first follows and the upper half of the body, which is supported by the arms, is somewhat lowered. Suddenly the arm is flexed, placed backward, and then again extended. If the support on which both arms of the infant are propped up is removed, both arms show corresponding movements, at times simultaneously and at times in succession.

In contrast to the limping reactions of the leg, the limping reactions of the arms cannot be elicited with the same certainty. Of course, the child can easily see the mat that moves directly before his eyes and tries to hold onto it; this naturally interferes with the process. The reactions are also most distinct in forward movements, but occasionally occur in sideward and backward displacements.

In infancy, limping reactions of the legs do not depend exclusively on the extension of the statically involved extremity, as they did in the animals in RADEMAKER's experiments. The infant shows similar reactions when he extends his elbows and flexes his knees in the crawling position. In addition, I observed a corresponding reaction in a sitting infant 18 months old who could not yet stand or walk. He was sliding through the room (see p. 230) by alternately shifting his buttocks. The child was set down so that only one buttock, for example, the right, rested on a movable piece of cloth and the other directly on the table. If the cloth was then moved forward and at the same time the left buttock was held to the table, a distinct limping reaction of the right buttock occurred.

Pull-Up Reactions of the Arms

(41)

In intact cats and in those without a cerebellum, RADEMAKER found a very complicated balance reaction in which the grasp reflex and the pull-up or prop-up reaction of the front legs and standing readiness of the hind legs participated. This reaction was demonstrated by letting the lower half of the cat's suspended body drop. Dogs have no grasp reflex, but corresponding pull-up or prop-up reactions of the leg were present in intact animals and in those without a cerebellum or cerebral hemispheres; they were absent in decerebrated and newborn animals.

Intact apes have a tonic palmar grasp reflex (see p. 158). As FOERSTER (1913) has shown, they possess an automatic flexor synergy of the arms that enables them to hang for hours in a climbing position without moving. GIERLICH (1913) found a reciprocal relation between the posture of the arms and legs in the climbing leap of the apes. He distinguishes between two stages; in the first stage the arms are lengthened as much as possible and the legs shortened, and in the second stage the opposite takes place — the arms are shortened and the legs are lengthened.

Similar reflexes can be seen in the human infant. The newborn possesses only the tonic palmar grasp reflex; the pull-up reaction of the arms appears only after several months.

The quadruped animal needs the front legs for the support of its body. The infant at first uses his arms in the same way, but during the first year of life he has already attained the human ability to sit upright and to stand. The palmar grasp reflex and the pull-up reaction are of great importance for the first sitting and standing attempts. In the erect position the infant at first needs to protect himself from falling by propping himself up with his arms on the base when sitting, or by grasping a suitable object when sitting or standing. Thus he manages to maintain his equilibrium, which he loses more easily in the erect than in quadruped posture. Later, when the infant has become secure in sitting without support, standing, and walking, this pull-up reaction of the arms decreases in importance. The reaction is tonic, as is the palmar grasp reflex.

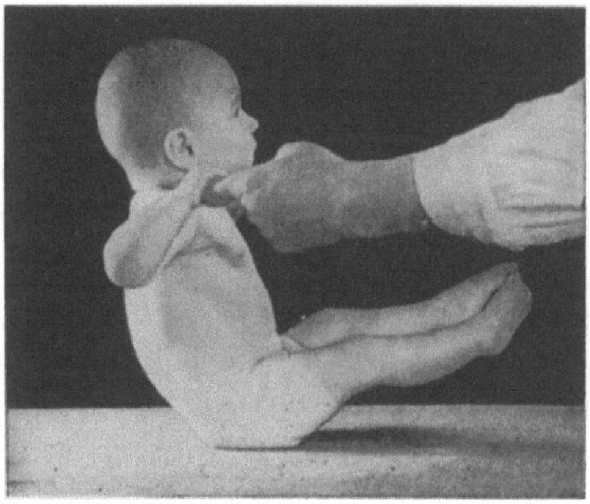

Fig. 83. Tonic pull-up reaction of the arms and extension reaction of the legs in backward-tilted sitting position in a female infant 18 months, 13 days old. The child holds onto the examiner's fingers herself and is not held. The weight of the extended and raised legs, fixed at the hips, levers the trunk forward.

An especially impressive chain reflex that depends on the pull-up reaction of the arms is shown in Fig. 83. The infant sits in an obliquely backward-tilted position and with flexed arms holds onto the fingers of the examiner. Thus, with the aid of the labyrinthine righting reflex the head and then the upper half of the body barely attain the erect posture while the lower half of the body is still directed backward. To compensate for this, the extended legs, fixed at the hips, are lifted from the table. Thus, the rigid body forms a right angle with the supported buttocks. The weight of the legs no longer rests on the supporting base, but pushes the upper half of the body forward in order to restore the lost equilibrium. When the adult gets up from a dorsal position without using his arms, his body passes through similar stages.

The same pull-up reaction of the arms occurs in the standing position when the body is tilted backward. Figure 85 shows the opposite of Fig. 84; in the latter, strong supporting reactions of the arms appear when the body is tilted forward. In both positions the body tries to approach the normal posture with the aid of the arms. Both illustrations explain the action of the arms when the body is tilted obliquely sideways (Fig. 86). A supporting reaction occurs on the side to which the infant is tilted, and on the other side there is a pull-up reaction. In addition, the labyrinthine righting reflex of the head can be recognized in the three illustrations. In Figs. 84 and 85 the legs act in a manner similar to the arms: in Fig. 84 they are extended and the infant stands on tip toes; in Fig. 85 they are slightly flexed at the hips and knees. All these reactions are distinct only as long as the infant opposes the shifting solely with his arms. Sooner or later he will restore the lost equilibrium by one step forward, backward, or sideward, and this will make these reflexes unnecessary.

It is not a rare occurrence in daily life that only one arm gets a chance to hold on and execute a supporting or pull-up reaction.

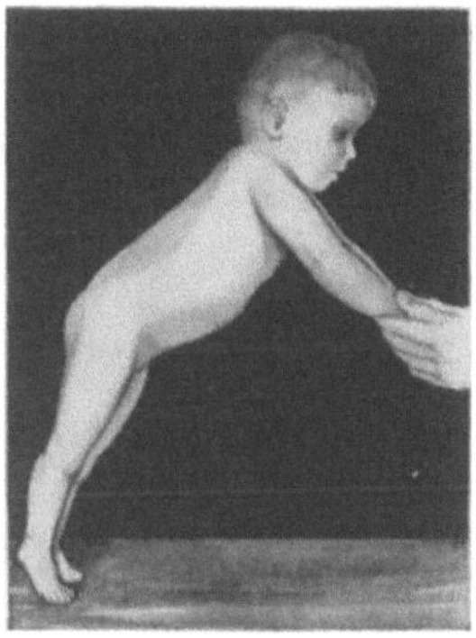

Fig. 84

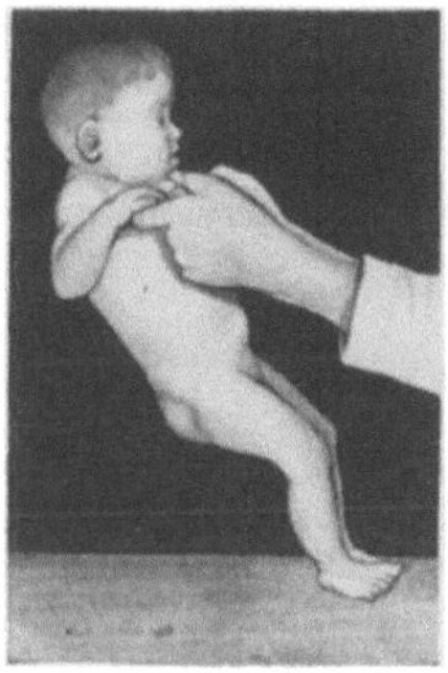

Fig. 85

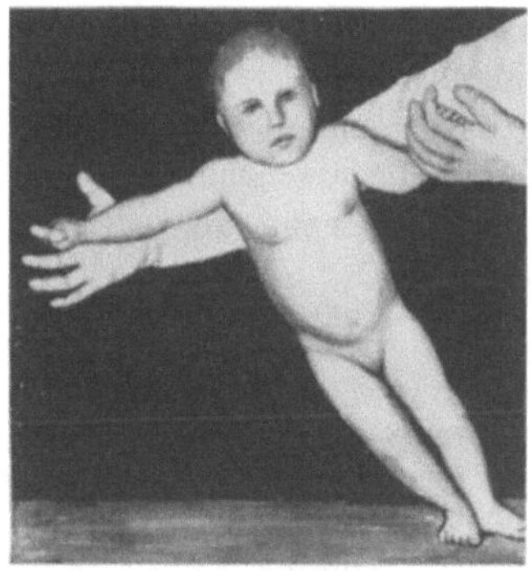

Fig. 86

Fig. 84. Supporting reactions of arms and legs and standing on tip toes in forward-tilted standing position in a girl 6½ months old.

Fig. 85. Tonic pull-up reaction of the arms in backward-tilted standing position in a 7-month-old girl. The child holds on and is not held.

Fig. 86. Supporting reaction (right) and pull-up reaction of the arm in a girl 6½ months old in the standing position while the body is obliquely tilted to the right. The child holds on and is not held.

SUMMARY

During the development of the infant's static functions, groups of reflexes like those examined by Magnus and his co-workers in animals occur temporarily. These investigators inactivated higher brain portions that otherwise would have inhibited these reflexes. When in the course of human development the corresponding higher centers are formed, these reflexes are also inhibited in the infant, but they may reappear later in illnesses affecting the inhibiting centers.

Receptors for the reflexes of position and movement are the sensory organs of the inner ear (semicircular canals, sacculus, and utriculus), the sensory pressure organs of the skin, the proprioceptors of the muscles, tendons, and joints, and the eyes. Quite often several receptors serve the same task so that these reflexes are secured in many ways.

The position reactions, for the most part innate, determine the position of the individual limbs in respect to each other by means of tonic reflexes.

Ordinarily the gravity reflexes are not consciously perceived even by the adult; these reflexes have the task of righting the body in relation to the vertical axis and thus of supporting the attainment and maintenance of the correct posture and the unsupported gait. They begin weakly in the newborn and develop during the first year of life. The leading reflex is the labyrinthine righting reflex of the head, which rights the head in relation to space. Chain reflexes originating from this reflex bring trunk and extremities into the best possible balance under given conditions.

Other reflexes depend only in part on gravity. With the body in the abdominal position, the same supporting reactions as are seen in the front and hind legs of quadrupeds occur in the arms and legs of the infant. With the body in the erect position, the supporting reactions of the arms disappear and those of the legs are maintained.

The fundamental resemblance of the motor and postural reflexes in man and animal shows that ontogenetically and phylogenetically the human and animal brain have the same structure.

REFERENCES

ADRIAN, E. D.: J. Physiol. *101*, 389 (1943).
ALEXANDER, G.: Zschr. Sinnespsychol. Abt. 2, *45*, 153 (1911).
—, Ohrenkrankheiten, in Pfaundler-Schloßmanns Handbuch der Kinderheilkunde (Diseases of the Ear, in Pfaundler-Schlossmann's Handbook of Pediatrics). Vol. 6. Leipzig, 1912.
ANDRÉ-THOMAS et al.: Presse méd. *1954*, 146, 885.
—, and SAINT-ANNE DARGASSIES: Études neurologiques sur le nouveau-né et le jeune nourrisson (Neurological Studies on the Newborn and Young Infant). p. 207. Paris, 1952.
—, and S. AUTGAERDEN: Psycho-Affectivité des premiers mois (Psycho-affectivity of the First Months). p. 103. Paris, 1959.
BABINSKI, I.: Rev. neurol. (Fr.) *28*, 2145 (1915); *1922*, 1049.
BALDENWECK, L., and GUY-ARNAUD: Presse méd. *1940* I, 47.
BALDUZZI, O.: Zschr. Neurol. *141*, 1 (1932).
BÁRÁNY, R.: Acta oto-laryng. (Stockholm) *1*, 97 (1918).
—, Arch. Augenhk. *88*, 139 (1921).
BARTELS, M.: Arch. Ophth. *76*, 1 (1910).
BERBERICH, I., and A. WIECHERS: Zschr. Kinderhk. *38*, 59 (1924).
BERTOLOTTI, M.: Rev. neurol. (Fr.) *12*, 1160 (1904); *20*, 242 (1912).
BLANTON, M. G.: Psychol. Rev. (Am.) *24*, 456 (1917)
BREUER, J.: Med. Jb. (Vienna) *1874*, 72; *1875*, 87; Pflügers Arch. *48*, 195 (1890).
BYERS, R. K.: Amer. J. Dis. Childr. *55*, 696 (1938).
CATEL, W.: Mschr. Kinderhk. *38*, 303 (1928).
—, and C. A. KRAUSPE: Jb. Kinderhk. *129*, 1 (1930).
DIETRICH, H. F.: Amer. J. Dis. Childr. *94*, 265 (1957).
DIJKSTRA, J.: An Investigation of the Galant Reflex. The Little Club Summer Meeting. Groningen, 1960 (in press).
DOFLEIN, s. R. HESSE and F. DOFLEIN: Tierbau und Tierleben (Animal Structure and Animal Life). Vol. 2, p. 661. Leipzig and Berlin, 1914.
DOXIADES, L.: Zschr. Kinderhk. *52*, 85 (1931).
DREXLER, L., and F. WENGRAF: N. österreich. Z. Kinderheilk. *4*, 27 (1959).
ESENTE, L.: Physiologie de la vision chez le prématuré et le nourrisson normal (Physiology of Vision in the Premature and Normal Infant). p. 22. Paris, 1958.
FOERSTER, O.: Berl. klin. Wschr. *1913*, 1217 and 1255.
FREUDENBERG, E.: Münch. med. Wschr. *1921*, 1646.
GALANT, S.: Der Rückgratreflex (The Spinal Reflex). Diss. Basel, 1917.
GALEBSKY, A.: Acta oto-laryng. (Stockholm) *11*, 409 (1927).
GAMPER, E.: Zschr. Neurol. *102*, 154; *104*, 49 (1926).
GESELL, A.: J. Pediatr. (Am.) *13*, 455 (1938).
GIERLICH, N.: Über Symptomatologie, Wesen und Therapie der hemiplegischen Lähmung (Symptomatology, Nature, and Therapy of Hemiplegic Paralysis). p. 57. Wiesbaden, 1913.

GIORDANO, G. G.: Acta neurol. (Neapel). Feb. 1953. III. Quaderno. p. 313.
GLATZEL, H.: Zschr. Neurol. *111*, 529 (1927).
HASEGAWA, T.: Pflügers Arch. Physiol. *229*, 225 (1932).
HALVERSON, H. M.: J. genet. Psychol. *51*, 371 (1937).
HINSCHE, G.: Roux' Arch. Entw.mech. *139*, 124 (1939).
–, Palaeontolog. Zschr. *23*, 1 (1942).
–, Psychiatr.-neurol. Wschr. *45*, 281 (1943).
HESS. W. R.: Das Zwischenhirn (The Diencephalon). 2nd ed. Basel, 1954; Thalamus und Hypothalamus. Stuttgart, 1956.
HOOKER, D.: Proc. Am. Philosoph. Soc. *79*, 597 (1938); Psychol. somat. Medicine *4*, 199 (1942).
JACOBSEN, C. F., M. M. JACOBSEN, and J. G. YOSHIOKA: Compar. Psychol. Monogr. *9*, 1 (1932).
JUNG, R., and R. HASSLER: in J. Field, H. W. Magoun and V. E. Hall, Handbook of Physiology. Sect. 1. Neurophysiology. Vol. 2. Washington, 1960.
KARPLUS, J. P., and A. KREIDL: Arch. Anat. and Physiol. (Berlin and Leipzig) *1914*, 155.
DE KLEIJN, A., and R. MAGNUS: Pflügers Arch. Physiol. *145*, 537 (1912); *186*, 1 and 61 (1921).
–, and C. VERSTEEGH: Pflügers Arch. Physiol. *232*, 465 (1932).
KÖHLER, W.: Aus der Anthropoidenstation auf Teneriffa (The Anthropoid Station in Teneriffe). II. -IV. Abh. preuß. Akad. Wissensch. Berlin. Physik.-mathemat. Abt. 1915. 1918.
KURODA, R.: Acta Psychol. Keijo (Japan) *1*, 3 (1930).
LANDAU, A.: Klin. Wschr. *1923*, 1253.
–, Mschr. Kinderhk. *29*, 555 (1925).
LAWRENCE, M. M., and C. R. FEIND: Pediatr. *12*, 300 (1953).
LEINER, C.: Med. Klin. *1926*, 1882.
MACH, E.: Grundlinie der Lehre von den Bewegungsempfindungen (Fundamentals of the Theory on Sensations of Movements). Leipzig, 1875.
MAGNUS, R.: Körperstellung (Body Posture). Berlin, 1924.
–, Pflügers Arch. Physiol. *145*, 537 (1912).
–, and A. DE KLEIJN: Zbl. ges. Neurol. u. Physiol. *44*, 780 (1926); Körperstellung, Gleichgewicht und Bewegung bei Säugern. Haltung und Stellung bei Säugern. Handbuch der Physiologie (Body Position, Equilibrium and Movement in Mammals. Posture and Position in Mammals; Handbook of Physiology) *15*, 1, 29, and 55. Berlin, 1930.
MAGOUN, H. W.: Arch. Neur. *67*, 145 (1952).
MONNIER, M., and H. WILLI: Ann. Paediatr. (Basel) *168*, 289 (1947).
MORO, E.: Münch. med. Wschr. *1918*, 1147; *1920*, 360.
MUSSEN, A. T.: Brain *50*, 313 (1927).
MYGIND: according to Voß.
NOICA: Rev. Neurol. *20*, I, 134 (1912).
PRECHTL, H. F. R.: Klin. Wschr. 1956, 281.
–, Umschau *53*, 656 (1953).
PULGAR MARX, I., and P. DE: Rev. espan. pediatr. *11*, 317 (1955); also see Zentralbl. Kinderheilk. *58*, 220 (1957).
RADEMAKER, G. G. J.: Das Stehen (Standing). Berlin, 1931.
–, Die Bedeutung der roten Kerne und des übrigen Mittelhirns für Muskeltonus, Körperstellung und Labyrinthreflexe (The Importance of the Red Nucleus and the Other Parts of the Mid-brain for the Muscle Tonus, Body Position, and Labyrinthine Reflexes). Monograph. d. Gesamtgeb. Neurol. u. Psychiatr. No. 44, Berlin, 1926.
RICHTER, C. P.: Arch. Neurol. (Am.) *26*, 784 (1931).
RIESEN, A. H., and E. F. KINDER: Postural Development of Infant Chimpanzees. Yale Univ. Press. New Haven, 1952.
RITTER, C.: Jb. Kinderhk. *104*, 293 (1924).
ROBINSON, R.: The Nineteenth Century *30*, 831 (1891).

SCHALTENBRAND, G.: Zschr. Nervenhk. *87*, 23 (1925); *175*, 109 (1956).
SCHENK, V. W. D.: Zschr. Neur. *142*, 469 (1932).
SCHENKEL, R.: In Goma, das Basler Gorillakind (in Goma, the Basel Baby Gorilla). Bulletin No. 2. Basel, 1960.
SCHNEEBERGER, A.: Zschr. Kinderpsychiatr. *25*, 254 (1958).
SCHUR, E.: Zschr. Kinderhk. *32*, 227 (1922).
SITTIG, O.: Med. Klin. *1932*, 934.
SOKOLANSKY, G.: Zschr. Neur. *145*, 576 (1933).
SPITZER, R. L., Y. KRAMER, and R. RABKIN: J. Pediatr. *52*, 149 (1958).
STEINHAUSEN, W.: Pflügers Arch. *217*, 747 (1927); *228*, 322 (1931); *229*, 439 (1932); *232*, 500 (1933).
SAINT-ANNE DARGASSIES: Études néonatales (Studies on the Newborn). *3*, 101 (1954); World Neurol. *1*, 71 (1960).
STENVERS, H. W.: Arch. néerld. Physiol. *2*, 669 (1918).
STILLHART, H.: Helvet. paediatr. acta. S. C. *9*, 298 (1954).
–, Über die klinische Bedeutung des sog. reflexartigen Phänomens der untergehenden Sonne beim Neugeborenen (On the Clinical Importance of the So-Called Reflex-Like Phenomenon of the Setting Sun in the Newborn). Diss. Zürich, 1954.
STIRNIMANN, F.: Das erste Erleben des Kindes (The First Experience of the Infant). Frauenfeld and Leipzig, 1933.
–, Ann. paediatr. *160*, 1 (19 3).
–, Das Kind und seine früheste Umwelt (The Child and His Earliest Environment). p. 43. Basel, 1947.
–, and W.: Ann. paediatr. (Basel) *154*, 249 (1940).
THEILE, F. W.: Nova Acta physico-med. Acad. Leopold.-Carol. (D.) *46*, No. 3, 133 (1884).
THORNVAL, A.: Acta oto-laryng. (Stockholm) *2*, 451 (1920/21).
VERAGUTH: Neurol. Zbl. *1918*, No. 7.
VOLLMER, H.: Amer. J. Dis. Childr. *95*, 481 (1958).
VOSS, O.: Acta oto-laryng. (Stockholm) *11*, 73 (1927).
WARTENBERG, R.: Die Untersuchung der Reflexe (The Examination of Reflexes). p. 166 of German translation. Stuttgart, 1952.
WENGRAF, F.: Briefl. Mitteilung, 1959.
WIESER, S. and DOMANOWSKY: Dtsch. Zschr. Nervenhk. *175*, 520 (1957).
WILLI, H.: Ann. paediatr. (Basel) *174*, 87 (1950); *178*, 321 (1952).
–, Études néonat. *1*, 25 (1952).
VAN WOERKOM, W.: Rev. neurol. Paris *20*, II, 285 (1912).
YERKES, R. M., and A. W.: The Great Apes. pp. 264 and 267. New Haven, 1929.

5. LOCOMOTION

PRELIMINARY STAGES OF LOCOMOTION

(1, 30, 95, 115, 119, 130)

The movement rhythms that make the locomotion of the vertebrates possible, such as walking, running, swimming, and flying, though under the influence of higher brain centers still depend on reflex centers in the spinal cord. These movement rhythms are therefore maintained when the spinal cord is separated from the brain. Thus, sharks can still swim after their heads are cut off, and spinal cord preparations of frogs are still able to creep, crawl, jump, and swim. Ducks still make swimming movements after sectioning of their cervical cords if artificial respiration is supplied. Decapitated birds still run or fly for a distance. von Holst (1951) reported that an isolated section of the abdominal spinal cord of the rainworm placed across two electrodes in a humid chamber could for a long time still create pulsating discharges conforming to the normal crawling rhythm of this animal.

The brain has the task of regulating the spinal cord action, which is in itself independent, to the existing conditions and to accelerate it or to slow it down.

In the young infant, whose cerebral hemispheres are still nonfunctioning, rhythmic stepping, mounting, crawling, and swimming reflex movements can also be elicited under favorable conditions. They occur even in the newborn although the child begins the independent use of these movements in his second year of life at the earliest. There is no temporal connection; the infant learns independent locomotion only some time after the preliminary stages have disappeared.

Stepping and Ascending Movements

(1, 30, 119, 125)

To elicit stepping movements [Peiper, 1928 *(1)*, p. 47] (Fig. 87), the trunk of the newborn is held by the examiner's hands and the infant is placed on a table as for ordinary standing. Tilting the body slightly forward is recommended because the experiment fails if the newborn is in an oblique backward position. An infant several months old usually pulls both legs up as soon as his feet touch the table.

The newborn reacts differently. Both legs are extended, and by simultaneous contraction of the flexors and extensors they become rigid supports more or less capable of carrying the body (supporting reactions, as the term is used by MAGNUS and RADEMAKER) (see p. 194). The standing infant flexes one leg at the knee and hip. If the examiner follows this movement with the infant's body, the child places his foot down just beyond the other foot; he then makes the same movement with his other leg, which has not yet moved. While the foot is being moved forward, it may leave the ground completely or be dragged over the surface. Thus, regular and completely coordinated stepping movements, in which the position of one leg always depends on the other, originate. The infant is in no way dragged over the table surface by the examiner, but strides with real steps. Of course, the infant cannot walk for long distances because he soon tires.

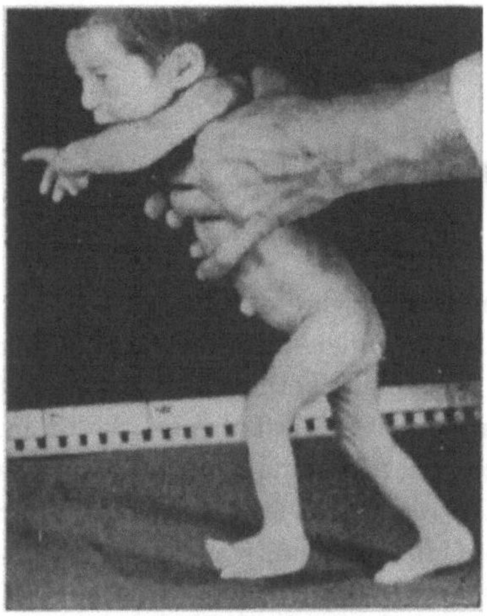

Fig. 87. Stepping movements of a newborn (frame of a film).

The distance covered by each pair of steps amounts to approximately 20 cm. If the child is made to stride on soot-blackened glossy paper, a pattern of the foot tracks is obtained. In these tracings we can see that the toes are definitely kept pointed out. SHIRLEY (1933) observed the same foot position in infants who were just learning to walk; later, when the children could walk well, the feet no longer formed an angle to the midline.

Not infrequently the newborn strides with his legs crossed. He places the preceding foot directly in front of the other rather than lateral to it, and at times even puts it in a somewhat more medial position, so that the other foot if brought forward would become entangled with the heel of the standing leg. Usually the foot is dragged forward in spite of that, but occasionally one has to remove the obstacle if further stepping movements are to follow. Crossing of the legs, which is caused by increased innervation of the adductors, is quite characteristic of the newborn, who not infrequently also lies with the legs crossed. The resulting gait is quite similar to that seen in Little's disease except that the newborn does not walk on his tiptoes.

The stepping movements of the newborn differ from real walking in that only the legs are used; the other parts of the body do not participate at all. However, the legs can carry the weight of the body only for a few moments, and the trunk cannot remain erect by itself. If one were to let go, the infant would immediately collapse in most cases and in some after only a few moments of standing. The newborn strides with slightly flexed knees and hips, just as the small child does.

According to ANDRÉ-THOMAS and SAINT-ANNE DARGASSIES (1952), stepping movements can also be elicited in the newborn by holding him suspended a few centimeters above the ground, tilting him forward and moving him so that he does not touch the ground.

Stirnimann (1938) examined the stepping movements of 75 newborns on their first day of life. He found real steps in 12 and no sign of them in only 30 infants; in the others they were not definite. At the age of 9 to 14 days the incidence increased: Of 100 infants tested, 35 were really stepping, an additional 18 infants did this with crossing of the legs, and 5 achieved only one step. Thus, stepping movements were found at the end of the neonatal period in 58% of the infants tested, while they occurred in only 16% on the first day of life. Hungry or slightly excited infants do best in these experiments.

The stepping movements of young infants have no temporal relation to the child's mastery of walking. According to Stirnimann, these phenomena disappear in the fourth or fifth month of life while real walking starts only in the ninth to tenth month. Thus, independent walking appears only several months after the disappearance of the movement rhythm evoked solely by reflex action. McGraw (1940) even speaks of an inhibiting phase in which the feet act as if they were sticking to the ground. André-Thomas and Saint-Anne Dargassies (1952) practiced the stepping movements with newborns continuously two to three times a day in order to maintain the movements for some length of time, but could not prevent their disappearance. After several months the posture of the infant becomes progressively worse and he is less and less inclined to stride. His legs flex and no longer support the weight of the body (disappearance of the supporting reaction), and he no longer moves his feet but drags them over the floor; in brief, the infant can no longer stride. André-Thomas and Autgaerden (1953) persuaded a mother to practice stepping movements with her child from his first day of life. After two months the movements became hesitant, slow, and arrhythmic, but they were maintained. Afterward the child again made rapid progress and walked without support at the end of the ninth month. In some infants whose brain function did not develop, the stepping movements persisted for some months and then disappeared permanently without the subsequent appearance of an unsupported gait. In others who were checked until the age of fifteen months or two years the persistence of the stepping movements of the neonatal period could be demonstrated.

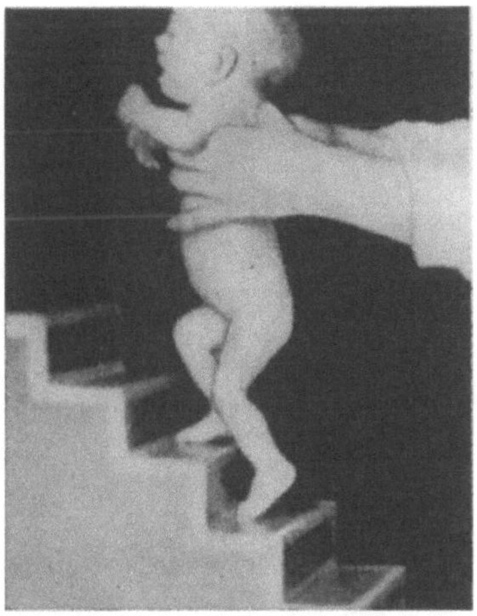

Fig. 88. Ascending movements of a newborn (frame of a film).

The stepping newborn can also ascend steps [Peiper, 1953 *(125)*]. To demonstrate this activity it is best to use a small staircase with steps approximately 8 cm in height (Fig. 88). If one holds the infant with the upper half of his body tilted slightly forward, lets him step as usual on a plane, and after a few steps lets him ascend the stairs, he can do this with no difficulty at all. He first places one foot on the higher step, extends his leg through a supporting reaction elicited by the contact of the sole with the step, then lifts himself up if the movement is followed with the supported body, and passes the standing leg with the other foot going to the next step, where the maneuver is then repeated. Thus, the neonate is able to ascend several steps in succession in a fluid and strictly rhythmic movement sequence just as the adult does. Sometimes the rhythmic course is slightly disrupted by the fact that the ascending leg does not pass the standing leg but remains standing on the same step, whereupon the standing leg is again lifted.

However, it is most difficult to have a newborn descend stairs in a similar way. It is necessary each time to flex the standing leg, which tries to stretch when the sole touches the step. (This flexion at the knee and hip is contrary to the supporting

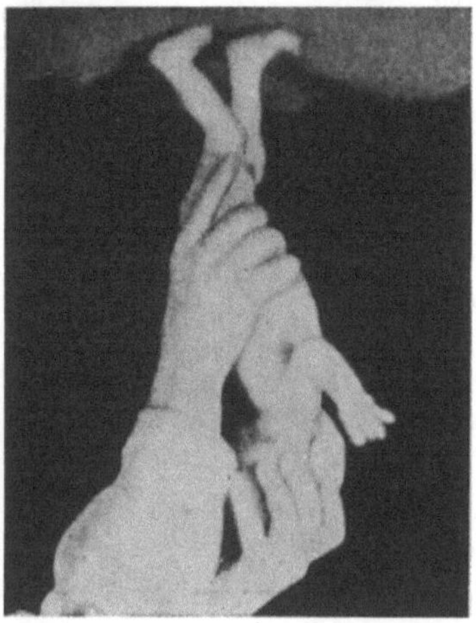

Fig. 89. Stepping movements of a newborn, head down (frame of a film).

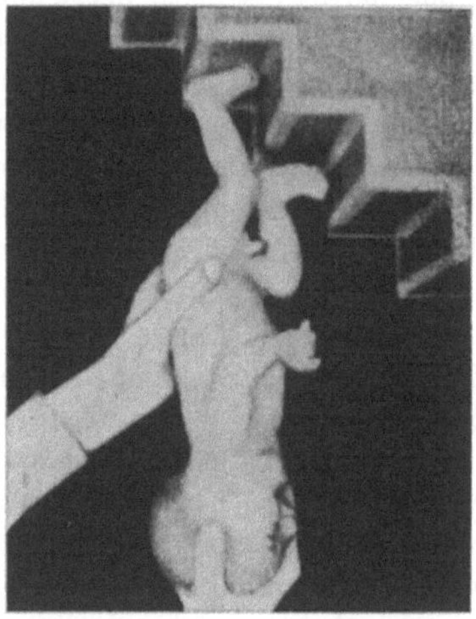

Fig. 90. Ascending movements of a newborn, head down (frame of a film).

extensor reflex reaction.) The tightly held body of the infant has to be pressed downward. The descent that thus ensues is disharmonious, shows little rhythm, and cannot be compared with the fluid ascent. Stirnimann (1938) states that in the stepping movements of the newborn the vestibular apparatus is stimulated by the unaccustomed (erect) position of the head whereby certain obscure muscle and joint sensations are said to play a role. However, it can be shown that the vestibular apparatus most likely does not play any part in the neonate's stepping and ascending movements.

It is possible to let infants whose bodies are in a horizontal position walk upon and "ascend" a vertical wall and then make them do the same thing upside down with their feet on the inferior surface of a horizontal wall, such as a low ceiling. A somewhat rough surface, such as a woolen blanket, serves best as a supporting plane for the sole. If an infant is tested in the horizontal position with the face up, occiput lightly supported, and his soles touching the vertical wall, he is able to ascend this wall with definite steps. If he is rotated 180° and his face is directed downward and carefully supported, he will descend the wall. If he is rotated 90° to a lateral position and the lower half of his face is supported, he will stride on the wall even in this position. The contact of the soles with the vertical wall causes supporting reactions of the legs as in the stepping movements in the erect position. These reactions always precede the stepping movements. If a small staircase is fastened to the vertical wall so that the steps are in a vertical downward direction, the infant can ascend the steps with his body in a horizontal lateral position if his body is supported and helped along.

In the vertical position with the head down, the young infant is able to stride on the underside of a horizontal plane (Fig. 89). In this position he can even "ascend" stairs that are fastened to the ceiling and are directed downward, as shown in Fig. 90. One must consider that standing on his head is not as unnatural for the newborn as it is for the adult, since the child was in this position during the last months of gestation when his mother was standing erect. Therefore, no particular discomfort can be noted in the newborn during these experiments.

Understandably, the more difficult the infant's task, the more easily he gives up. Thus the striding and ascending of the newborn in the horizontal and vertical position with the head down are not always as rhythmic as in the erect position. Since the infant has to be held more tightly than in the regular position this may also interfere with the motion.

The fact that the newborn steps and mounts with his body axis in any position in relation to the vertical axis proves that the basic reflexes are independent of gravity. STIRNIMANN's hypothesis that the vestibular apparatus participates in these movements is hereby refuted. From the beginning I have assumed that we are dealing with deep-seated reflexes that are later inhibited by the progressive maturation of the brain.

As is true of so many other reflexes of position and motion, the stepping movements of the infant are also inhibited by the progressive maturation of the cerebral hemispheres. In animal experiments it is just the reverse: Similar movements are observed in animals only after the higher brain portions have been cut off. It is an old experience that decerebrated animals do not become motionless but may run until they collapse. We must be dealing here with very deep-seated reflexes, since the running movements are maintained even after the lumbar region of the spinal cord has been cut (GOLTZ, 1892; FREUSBERG, 1874; PHILIPPSON, 1905; SHERRINGTON, 1910; T. G. BROWN, 1916; MAGNUS, 1924; and others). When the brain stem is cut, the running movements are sometimes present and sometimes absent (MAGNUS; BERITOFF, 1923; MELLA, 1923; LAUGHTON, 1924). Only stepping movements, such as are seen in the human newborn, occur in spinal cord preparations; real walking is impossible since the legs cannot carry the weight of the body and the supporting reactions are consequently absent. LANDÉ (1924) could elicit stepping movements even in spinal cord preparations of frogs by dragging them across the table.

SHERRINGTON (1906) applied faradic stimulation to the hind leg of a spinal cord preparation of a dog and evoked rhythmic stepping movements in the other hind leg. The frequency of this movement rhythm was half that of scratching movements.

Brown (1916) observed coordinated walking movements in all extremities of newborn kittens after severance of the umbilical cord. Kabat, Dennis, and Baker (1941) interrupted the cerebral blood circulation of dogs for several minutes and after restoring the circulation they observed rapid coordinated rhythmic movements of all extremities in the unconscious animal lying on its side. Ebbecke (1956) observed a stage of narcosis in the cat in which the animal executed rhythmic reflex stepping movements when held obliquely by the upper part of its body and was completely motionless and flaccid when this stimulus was not applied.

If the spinal cord of the adult is separated from the brain, the same movements may occur. Böhme (1917) found that in patients with transverse spinal cord lesions flexion and extension of one leg occurred if the other leg was extended or flexed. In one of these patients he could even elicit completely regular stepping movements by constant electric stimulation of the skin. None of these faculties can be found in the healthy adult, since healthy adults are inhibited by the activity of the cerebral hemispheres. According to Böhme, however, they may be used in the voluntary gait.

In the course of evolution and of development one can often recognize the phenomenon described by Steiner in 1885, viz., that the action-regulating centers ascend in the direction of the proximal part of the medulla (cervicorostral), may frequently be observed. A task ordinarily performed by a certain region is taken over by a higher placed area and is then executed in an improved manner. A sequence of different superimposed layers is thereby established.

It therefore makes a considerable difference whether an already active higher brain portion is eliminated in a higher animal or whether at a lower stage of cerebral development only this lower part is present. The lower the animal species, the smaller the residual extrapyramidal motor system need be to maintain the normal tonus distribution, the normal body position, and the method of locomotion (Spatz, 1927). Therefore, the injury of a young animal can much more easily be compensated for than that of an old animal. Newborn rabbits whose dorsal cords had been severed by Spatz moved with the aid of "spontaneous movements," which Freusberg had already described in 1874. They started to move so soon after the operation that it was repeatedly thought that the operative procedure had been a failure. In the adult animal the hind legs remained paralyzed after the same operative procedure.

Deficits in the adult resulting from a pathological condition of higher brain regions are therefore not linked to the lack of maturation of these higher centers when found in infancy. Rather, the anencephalus of Gamper (1926), in whom only the midbrain had developed, proved what a multitude of tasks can be taken over by the lower brain portions in man. For this reason caution must be used when applying the results of animal experiments to man.

The rule of the ascending direction of development in the brain described here is the counterpart of the so-called law of the cervicocaudal (descending) direction of development in the spinal cord. This law is described elsewhere (see p. 237).

Crawling Movements

"If a newborn is placed in the abdominal position on a flat table and his soles are supported by the hands of the examiner or by an object, he starts to crawl; he makes crawling movements sideways with his spine, adducts one leg after the other to the trunk, and by alternating movements pushes himself away from the hand of the examiner if the hand follows or does not release him. At the same time he alternately flexes his arms and, supported by his hands, lifts the upper half of his body from the table. It is possible to make the child crawl over the whole length of the table if the examiner follows the infant's sole with his hand touching it, thus giving him an opportunity to 'push off'" (BAUER, 1926).

As in stepping and ascending movements, the same supporting reactions that MAGNUS and RADEMAKER observed in experimental animals occur in the legs. Through these reactions the body is pushed forward against the unmovable resistance. According to BLANTON (1917), STIRNIMANN (1938), and my own observations, crawling movements may occur even without stimulation of the soles and are caused only by the abdominal position. They cannot be elicited with the body in the lateral or dorsal positions. When the crawling movements described by BAUER were still unknown, BLANTON reported that a newborn seven days old crawled 6 in. in 20 minutes. By stimulating the soles of human fetuses, MINKOWSKI (1928) elicited diagonal arm reflexes that he believed to represent an ontogenetic preliminary pattern of locomotion. I was not able to observe any distinct coordination of arm movements with those of the legs or even paired coordination of arm movements or of leg movements, as seen in the crawling movements of older infants.

I suspect that crawling movements, which are more distinct in hunger, have a connection with the searching for the maternal breast. In our phylogenetic ancestors the infant hung from the maternal body, as baby monkeys still do today (see p. 161).

BAUER was able to elicit crawling movements in all the young healthy infants he examined. The movements gradually disappear in the fourth month (according to STIRNIMANN as early as the third month), but persist in the prematurely born infant until the fifth month of life. In 9 of 75 newborns STIRNIMANN found crawling movements the first day of life caused solely by the abdominal position and in 12 other infants in this group when the soles were supported. 9 to 14 days after birth 18 of 100 infants crawled when in the abdominal position and for 41 other babies in this group pressure on the soles was needed to elicit these phenomena. They thus became more frequent in the neonatal period, even though the infants did not practice these movements. BAUER maintains that their subsequent disappearance cannot be prevented by practice.

According to BAUER's observation, nonsupported crawling in order to reach an object (my term: "creeping"; see p. 224) starts at the earliest at the age of seven and a half months. There is a distinct time interval between the reflex crawling movements and the nonsupported creeping. The crawling movements, just as the stepping movements, can therefore not be regarded as a kind of preliminary practice. BAUER also states that crawling movements may persist in infants with cerebral pathological conditions beyond the fifth month of life; they were found in different forms of mental retardation up to the age of thirteen months. Similarly, other deep reflexes of the position and movement senses, such as the MORO reflex, are observed beyond the usual age if the inhibitory centers are not developing.

Swimming Movements

(12a)

Swimming movements of the young infant were described by MUMFORD as early as 1897, independently by AMES (1937), and in great detail by McGRAW (1939, 1941, 1949).

As can be seen in the illustrations, MUMFORD holds the infant, with the head supported and in the abdominal position, over a base and does not immerse him in water. If the infant starts to crawl in this position, it is as though he were moving through a liquid or semiliquid substance, i.e., sometimes paddle movements occur three times successively, then there is a pause, and then again three paddle movements occur. The similarity to swimming movements even increases on closer observation. If the fingers are spread apart, the wrist is rotated so that the ulnar side and the knuckle of the little finger are turned upward as though a cup is being lifted to the mouth, although the elbow is extended rather than flexed. The hand is moved backward and upward as in swimming. According to MUMFORD, this movement — he evidently means the stretching — continues to be associated with yawning for the rest of life.

McGRAW observed the swimming movements of 42 children from the age of 11 days to two and a half years in 445 experiments and has filmed their movements in part. In some cases the observations extended over eighteen to twenty months. If an infant a few weeks old supported in the horizontal position is placed in water, distinct coordinated extension and flexion movements of the arms and legs occur; these movements suffice to propel the infant forward for a short distance. Compared with the stepping and crawling movements, the swimming movements of the newborn are more simultaneous and more regular. The movements are not so distinct in air as they are in water. If the newborn is ducked, his respiration is inhibited by reflex action; older children cough and show other disturbances afterward. At the age of four months the swimming movements become less harmonious, and at the beginning of the second year voluntary movements of a different type appear. The child is never able to lift his head above the water to breathe. He therefore has to be supported under the chin during these experiments. The early disappearance of the swimming movements after the first months of life corresponds to the pattern of many other reflexes of position and movement.

My co-worker MAYERHOFER [1953 *(12a)*] confirmed the data of McGRAW and was also able to demonstrate swimming movements of cross-coordination patterns (see p. 225) in the majority of 113 infants age 50 to 150 days who were held in the abdominal position with chin supported. These swimming movements are more rhythmic than the crawling movements, become irregular after the fifth month and are finally uncoordinated (Fig. 91). In contrast, chimpanzees shy away from water and cannot swim at any age (YERKES, 1945). According to HEDIGER (1952), no anthropoids are capable of swimming. McGRAW (1939) found similar swimming movements in baby mammals (opossum, rat, cat, rabbit, guinea pig, and monkey). They were described earlier in the newborn guinea pig by AVERY (1928): The animals were able to keep above water by coordinated swimming movements and went down when they did not move, but the adult guinea pig avoids water if possible.

Naturally, real swimming has to be learned considerably later. In this connection the good experiences ERSTLING (1955) had when giving swimming lessons to three- to six-year-old children should be mentioned. They were supposed to swim a certain distance without support and in good form. Six children succeeded swimming 5 to 6 m without any support. Their ages are not given. The lessons met with external difficulties; otherwise the results might have been even better.

So-Called Climbing Movements

Climbing movements have been attributed even to the young infant. According to KLAATSCH (1922), the entire posture of the newborn reminds us of a climbing animal. The newborn's hands are comparatively strong (palmar grasp reflex) and the plantar surfaces of his feet are held against each other as if clasping a tree from both sides, an observation which, incidentally, is valid only for some infants. The big toes are said to approach each other to facilitate tree climbing.

According to FOERSTER (1921), the newborn's posture and movements are very similar to the climbing posture and the climbing movements of the apes because of phylogenetic correlations. The newborn corresponds to a thalamus-pallidum preparation, and by his posture and his movements he reveals the descent of man from a root from which present-

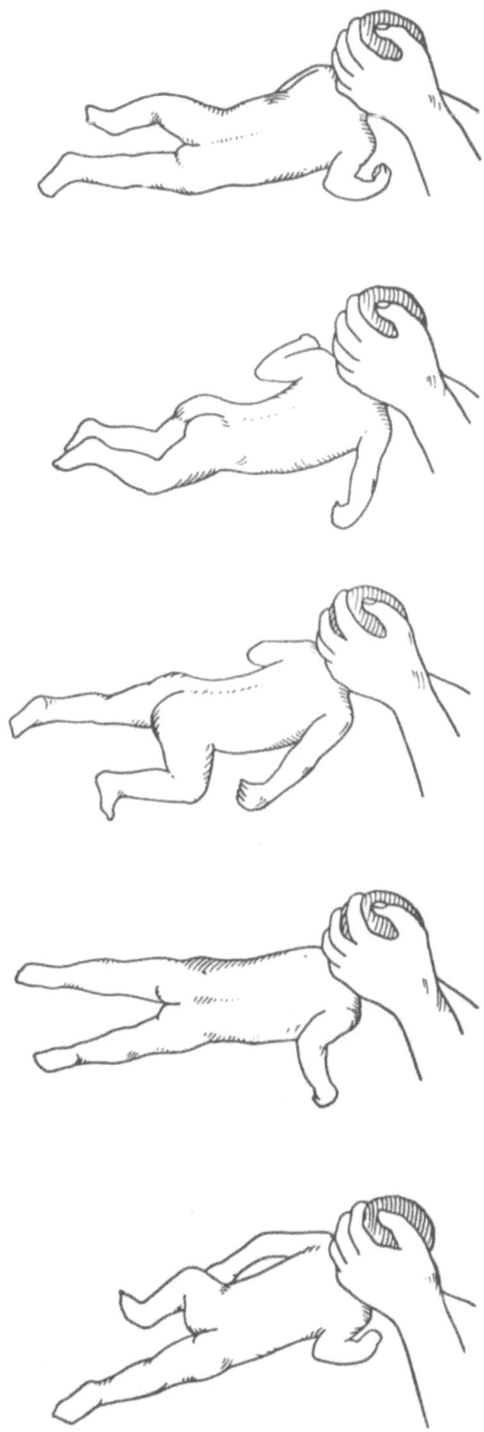

Fig. 91. Swimming movements of an infant: crossed coordination. [Sketch after MAYERHOFER's original photographs in his thesis *(12a)*]

day apes and man originated. Only by the maturation of the striatum does the original task of the pallidum, i.e., the regulation of climbing, decrease in importance and become inhibited so that the human achievements of standing and walking become possible. If this inhibitory influence ceases because of a pathological condition in the striatum, the climbing movements (climbing and the climbing jump) are again released. Thus, more or less coordinated mass movements of the trunk and the limbs, described in detail by FOERSTER and compared by him with the climbing movements of the apes, occur in congenital generalized athetosis.

WATSON (1924, 1930) also speaks of climbing movements in the newborn. If an infant holds onto something with his right or left hand, distinct "climbing movements" of the trunk and hips can be seen: Contraction waves – drawing the trunk and legs upward – seem to alternate with states of relaxation.

Because of these statements, which show little agreement with each other, I tried to evoke climbing movements in the newborn. For example, I tried to let him climp up a small ladder placed in a more or less vertical position while I supported his trunk. The experiments were as unsuccessful as were my efforts to induce the infant to climb when in the abdominal position. Certainly, there sometimes were almost coordinated-appearing movements, especially in the legs. But it happened so infrequently that chance could not be excluded.

PRECHTL (1953) recognizes coordinated and constantly recurring components in the mass movements of hungry infants: Raising of the arms and simultaneous opening of the hands alternate with downward movements and simultaneous closing of the hands while the legs remain immobile. The head is frequently bent backward. This sequence of movements often occurs in hungry infants in a "vacuum activity" i.e., without immediate external cause. Premature infants weighing between 1800 and 2500 gm and sometimes newborns born at term make these movements. PRECHTL does not wish to explain these movements as phylogenetic memories of life in trees; instead, he compares them with the corresponding behavior of the chimpanzee baby that, drawing himself up by alternating movements with both hands, climbs from the lap of its mother to the breast, where it holds onto the fur in order to drink until it returns to its position on the flexed thighs of its mother. During the first weeks of life the baby's legs do not participate in this motion.

PILLERI (1960) was able to elicit corresponding climbing movements of the arms with spreading and closing of the hands in a seventy-two-year-old woman suffering from Alzheimer's disease. By touching her with an object and having her merely look at it sucking movements could be elicited, and by removing the object a distinct persisting groping for it could be evoked.

The motherless chimpanzee Eva, raised in the Leipzig Pediatric Clinic, was able to stand up at the age of three and a half months; less than two weeks later she clim bedover the bedrail to the windowsill. She left her bed by climbing up before she could leave it when in the horizontal position (SCHNEIDER, 1950).

Statements on the development of the human baby's climbing ability differ; evidently various observers mean different things when talking about "climbing." PREYER (1895) observed his son's first climbing attempts when the child was in his twenty sixth month; the climbing efforts were successful at the twenty-eighth month. HETZER and KOLLER (1930) mention "climbing" among tasks for the third quarter of the second year of life and "climbing up on a chair" for the fourth quarter. On the other hand, DIX (1911) states: "SHINN and others observed, just as I did, that the instinctive climbing movements of the infant appear before walking." He reports that his child started to climb at the age of eight months. One day the infant sat unattended on his highchair and got up on the seat by drawing his legs up and pulling himself up to a standing position with the help of the back of the chair. SCHNEIDER concludes from this that climbing precedes walking in the chimpanzee and man. As far as man is concerned, we cannot agree with this statement.

The human child learns real climbing considerably later than he does horizontal locomotion, as PREYER has already stated. What DIX and others have observed are only the beginnings of climbing. This develops through many intermediate stages. DIX's child was able to climb to the first branch of a sloping or almost horizontal tree at the age of one year and eight months.

WARTENBERG (1952) states that reflex movements and leg positions that occur in climbing may also occur in pathological conditions of the pyramidal tracts. He includes in these phenomena STRÜMPELL's tibialis signs (raising of the inner edge of the foot when flexing the extended leg at the knee and hip, see p. 265) and the abduction of the thigh that accompanies the leg flexion in the hemiplegic adult. WARTENBERG regards the position of the extremities in spastic hemiplegia (arm flexion and leg extension) as a "frozen phase of climbing" and the opposite pattern (arm extension and leg flexion) as the second phase in Little's disease. He thus approaches GIERLICH's hypothesis (see p. 205) on the climbing jump of the apes. In WARTENBERG's opinion, BABINSKI's fanning sign (spreading of the last four toes on stroking of the sole) serves to enlarge the friction surface in climbing. WARTENBERG sees in the BABINSKI reflex part of an ipsilateral mass flexor reflex that is the opposite of a mass extensor reflex. Both reflexes are said to be vestiges of the climbing process. However, this conclusion would have to be based on the assumption that the dorsal flexion of the big toes really replaces an abduction that previously existed but must have been lost in the course of phylogenetic development.

> "As the human foot gradually assumed in the course of evolution a function in running and standing, the metatarsal-phalangeal joints gradually lost their ability to move the big toes sideways, as they had changed and were essentially executing only upward and downward movements of the toes."

To me this does not seem a valid argument, since young children are still able to abduct their big toes (Fig. 106, page 252). This faculty persists in infants and also in adults of primitive races. That we have lost it can be explained by the interference of shoes. What practice can accomplish is shown by the achievements of armless people who "handle" tasks with their feet almost as well as others do with their hands and fingers.

STAGES OF INDEPENDENT LOCOMOTION

(95, 119)

Man alone among all living beings walks erect, because his nature and his being are divine.

ARISTOTLE, *Parts of Animals* IV, 10

Primum temporis munus quadrupedi similem facit.

PLINY, *Natural History*, Vol. VII, Proemium

Gait Patterns

Certain patterns of gait serve the independent locomotion of man and animal on a level plane, i.e., strictly coordinated groups of movements that are regularly repeated. Each species has only a few types of gait, but there are innumerable possibilities for the use of extremities for locomotion. Thus, the horse generally moves by step, trot, or gallop although it can be specially trained in completely different types of gait; but when the external compulsion ceases, even trained horses fall back into their own gait patterns, which are evidently determined by their physique and their neurological constitution.

After infancy man, too, has several types of gait, i.e., walking, running, and hopping, at his disposal; dancing might be also included. But whereas the different gait types were investigated scientifically a long time ago, the independent locomotion of man has not been explored much since VIERORDT's first paper in 1881. BIRKMAYER and GOLL (1940) studied the conditions in childhood after infancy. Correct statements on the infant's independent locomotion were already made by SIGISMUND (1856) and PREYER (1895); detailed investigations on the subject have been reported, particularly by BURNSIDE (1927), and additional studies were carried out by TRETTIEN (1900), VARIOT (1926), BERSOT (1926), SHIRLEY (1933), AMES (1937), PEIPER [1942 *(95)*], and MCGRAW (1949).

The erect gait is the important final goal of the entire static development, which takes place in different ingenious interlocking stages. These stages of development enable the child to execute tasks of increasing complexity. Progress is made possible by the strong urge for movement and activity, which continuously drives the child to new experiments and which is not inhibited by many failures. To further this development by special aides who are trained in "infant gymnastics" seems to me needless, at least for the healthy child.

The described stepping, ascending, crawling, and swimming reflex movements of the infant already lead to more or less coordinated movements of the extremities, but have no temporal connection with the later appearance of independent locomotion. The kicking movements of the infant in the dorsal position show little coordination. These are strong flexion and extension movements of the legs (to a lesser degree of the arms) occurring in the big joints, i.e., in the hips, knees, shoulders, and elbows, and are not reflex in origin. They are frequently accompanied by strong lateral displacements of the pelvis. They are extraordinarily characteristic for the first months of life, in contrast to later development, but seem to show too little rhythm and coordination to permit a comparison with a running quadruped or man.

However, I distinctly observed coordinated leg movements during epileptic seizures. For example, in a six-month-old infant who suffered from purulent meningitis the following movement disturbances were seen during seizures: Alternately one leg was extended at the knee and hip joints and the other was simultaneously flexed, and this alternating pattern occurred many times; the arms did not participate in these movements. These movements gave the distinct impression of walking movements; naturally, it remains uncertain whether such a relation actually existed.

Such movements, just as the preliminary stages of locomotion, depend on the function of deeper brain centers. The ability of independent motion is acquired only with progressive maturation of the brain. In the infant, just as in the animal, only a few gait patterns are seen. To describe these patterns up to the fully developed crawling movements, McGraw distinguished 9 and Gesell and Ames as many as 19 different stages of development. But essentially there are only 4 gait patterns which can be clearly differentiated: creeping, crawling, sliding on the buttocks (here referred to as sliding), and the free erect gait. No two children will move in exactly the same way; sometimes the variations in the same patterns of gait may be quite considerable.

Clearly recognizable ideal forms, which are, so to speak, the goals of development, seem to exist only for crawling and the nonsupported free gait. In creeping a definite ideal form cannot be observed because the movements are not repeated with the same regularity as in the later forms of movement. The children may be able to slide with the same skill in totally different ways.

Even though the unsupported gait always has to be learned, the three other gait patterns do not necessarily have to precede it. Rather, external influences frequently determine whether one of these patterns is learned and is preferred or appears at the same time with others. Important and often deciding influences are the guidance of the mother and the example of other children, especially in institutions. For example, I saw that nearly all infants in a day room of a children's home used sliding for locomotion; this movement pattern, which is generally rare, had evidently spread there by example. Creeping is generally the preliminary stage of crawling; but infants quite often will completely skip these two gait patterns, which seem so natural, if they do not have a good opportunity to use them. In an experimental series, Variot observed that 42 of 80 infants skipped the crawling stage. For understandable reasons there is some reluctance to allow infants to crawl on the floor, especially in institutions. Therefore, the erect gait may be the first gait pattern seen in the child.

Marked fluctuations in the ages at which these various stages appear are explained by internal and external causes. More detailed data are given by McGraw, who examined 82 infants: Phase D, which approximately corresponds to creeping, was well developed some time between the one hundred and twenty-fifth to the four hundred and fifth day (average, two hundred and forty-fifth day). The infants could crawl fairly well some time between the two hundred and seventy-fifth to the six hundred and sixty-fifth day (average, three hundred and fortieth day). The coordination of the crawling movements was further improved afterward. Sliding usually occurs between the ninth and fifteenth month of life. Whereas these movement patterns appear gradually, the day on which the first free step is taken is a distinct turning point; it also receives a great deal of attention from the parents. In healthy infants this event occurs at approximately the end of the first year, but deviations of several months to an earlier or especially to a later time are not infrequent. CZERNY and KELLER (1925) saw an infant walk freely at nine months and SIGISMUND (1856), even at eight months. Some of the earlier authors give a later age for the start of walking. They must have observed more or less rachitic infants; for example, PREYER (1895) states that the first attempts at unsupported walking occur between the ninth and twenty-first month of life.

Creeping

Although the infant usually rests on his back, he creeps and crawls in the abdominal position. If he is placed on his abdomen by his mother before he can turn himself (see p. 235), the labyrinthine righting reflex early in life causes him to lift his head first and later also the upper part of his body. LASCH and FISCHER (1924), as well as K. S. HOLT (1960), compared the static development of two groups of infants. During the day the infants of one group lay on their stomachs and those of the other group on their backs. The infants in the abdominal position groupes raised their heads and propped themselves on their arms definitely earlier than those in the other group. The abdominal position is favorable for the development of the neuromuscular faculties and therefore should be frequently employed.

With the infant's strong urge for movement, minor changes in location must automatically occur in the unimpeded abdominal position even without intended locomotion. The raised upper half of the body may fall to the side or may turn to one side, while the lower part will simultaneously turn to the other side. Sometimes the infant rolls or turns over. If he tries to turn over when in the dorsal position, slight locomotion also occurs. However, all these uncoordinated movements cause only small changes in location and are not rhythmic.

The lowest pattern of rhythmic change of location, creeping, was clearly described and differentiated for the first time by BURNSIDE (1927). The starting position (Fig. 47, p. 181) is instigated in the abdominal position by the previously described symmetrical chain reflex, which originates in the labyrinthine righting reflex of the head. The head is in a normal position in relation to space and lifts the upper half of the body, which is propped up on the forearms. Strong supporting reactions in the shoulders and in the elbows, which are flexed at right angles, bring these joints into a fixed position. The infant's gaze is directed forward.

When the creeping infant tries to move forward, he at first does not want a change of location but is trying to reach an object that is visible in front of him but beyond the reach of his arms. The best way to make him creep is to place a desirable object as a goal in front of him. He immediately reaches for it, at first with both hands simultaneously, just as he extended both hands toward the desired object when he first learned to grasp objects while in the dorsal position. However, when both arms are extended forward, he falls forward to a certain extent and props his arms up again. Now he has merely to pull up his body. Possibly the forward urge is still further supported by the desire to grasp with his mouth, even if the object in question may later not be taken into the mouth. In creeping the abdomen is dragged over the base and, except for some spontaneous movements, the legs remain extended and do not support the locomotion; these two factors are of importance in distinguishing creeping from crawling.

If the infant does not reach his goal on his first attempts, a second and third attempt will be made, especially when the goal is moved farther back. Thus, small distances can be covered but, especially at first, there is little perseverance. The infant soon becomes tired and is easily distracted. A rhythmic repetition of the movement, which always occurs in the later phases of locomotion, exists for a brief period at best. In most cases the movements are clumsily executed with by far too great effort, as is the case with every new skill later in life. The infant often falls to the side in these attempts; some move forward only when lying on one side, whereby the arms are appropriately placed in an asymmetrical position. In other cases only one arm is flexed at the elbow joint, and the other extended as for crawling. Sometimes the movements are so clumsy, in spite of the infant's great effort to move forward, that he really moves backward more than forward. When the arms are moved forward alternately more than simultaneously, a preliminary stage of crawling is reached.

Crawling

Crawling clearly develops from creeping. It can be observed best in somewhat older infants who still like to crawl fast even though they are already able to walk freely; these children move much more uniformly than younger infants who have just learned to crawl.

The infant crawls with his head and body raised (Figs. 49, 50, and 92), but the head is not in as completely normal a position as in creeping (Fig. 47). The abdomen is no longer dragged over the base but is raised by the four extremities working in a coordinated fashion. Supporting reactions, which fix shoulders, elbows, and wrists, occur in the arms, which are now, unlike their position in creeping, extended at the elbows and overextended at the wrists. These supporting reactions of the extended extremities (Fig. 49, p. 181) evidently represent a higher stage of development than those of the flexed limbs during creeping (Fig. 47). They transform the arms into strong straight pillars that do not collapse but support the weight of the body. In propping up, the hand is overextended at a right angle against the forearm so that the entire palm touches the base. The hip and knee joints are fixed in a flexed position by supporting reactions. The ankle joints are extended as in creeping, so that the dorsal surfaces of the feet touch the floor provided no disturbing movements interfere. Occasionally it appears as though the backs of the toes support the locomotion by pushing themselves off from an uneven base.

Infants who do well in their crawling (Figs. 49 and 92) always move the contralateral extremities at the same time, the left arm together with the right leg and then the right arm with the left leg (SIGISMUND, 1856). This sequence of movements — I call it "crossed coordination" — does not correspond to the stepping but to the trotting of the horse. I did not see any other type of gait, such as the amble (simultaneous

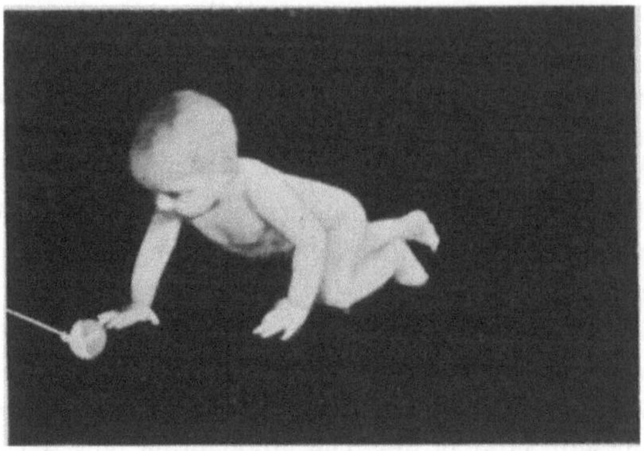

Fig. 92a

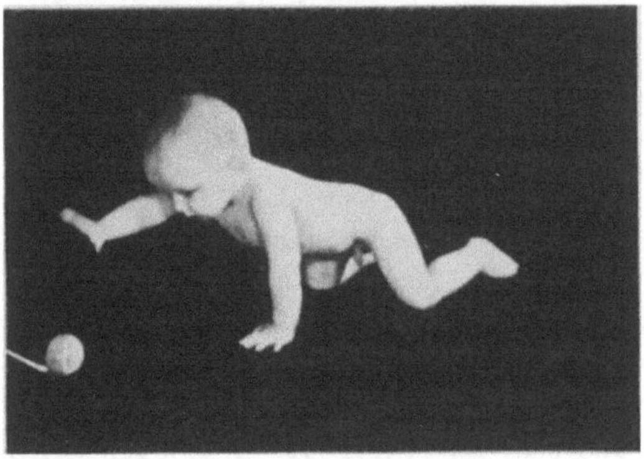

Fig. 92b

Fig. 92a and b. Crawling with crossed coordination in order to reach an object (frames of a film).

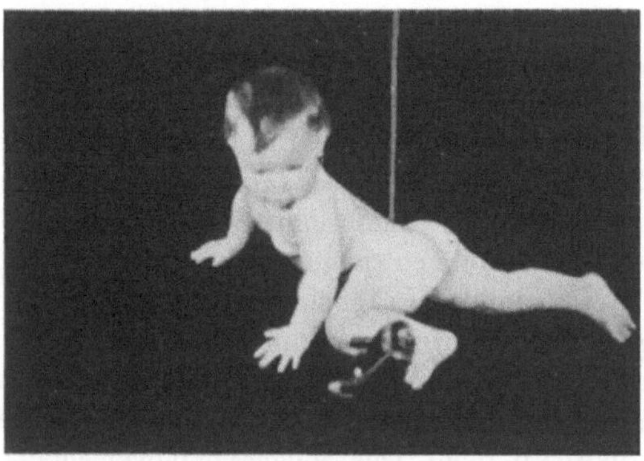

Fig. 93. Rotary crawling in order to reach an object (frame of a film).

moving of the right arm and leg alternating with the left side); neither did SIGISMUND or BURNSIDE (1927). However, TRETTIEN (1900) claims to have observed this gait pattern in 20% of the infants he studied, if I understand him correctly.

It has previously been mentioned that completely developed crawling occurs only in older infants. Younger babies, who are still unsteady, move only one limb at a time so that the body will be supported on the three other extremities. Since the arms are moved ahead shortly before the legs, the following sequence results: right arm, brief pause, left leg, longer pause, left arm, brief pause, right leg, longer pause. This much can be recognized by mere observation. AMES examined the differences between slow and fast crawling infants with the aid of films. He found that the arms are moved at almost the same speed but the legs of the slower infant are moved more slowly and the intervals between movements are longer.

Sometimes an infant who cannot yet crawl in the forward direction may, while crawling, turn in a circle to reach a desired object (Fig. 93).

According to ANDRÉ-THOMAS and his co-workers (1954), unexpected capacities for locomotion are evoked when an infant sees his mother. If, for example, a five-month-old infant is held and carried in the horizontal position toward a base until he can place his hands on it (standing readiness, see p. 193), he "marches" toward his mother when she calls him by alternately moving his arms in the forward direction. An infant three and a half months old who could not yet prop himself up on his extended arms "marched" on his flexed forearms toward his mother.

While the infant is learning to crawl, the following sequence of movements may develop temporarily: After some creeping movements the child lifts himself up from the starting position of creeping on his extended arms and flexed legs to the crawling position but soon sinks back to the creeping position; evidently the supporting reactions of the limbs are not yet sufficiently strong. Thus, the "game" is repeated several times.

Occasionally the infant crawls with only one leg while the other remains extended as in creeping. Some children crawl by setting a knee down on one side and a foot on the other at the same time while that leg is almost completely extended at the knee joint.

Generally the first two types of locomotion, creeping to a greater degree than crawling, cannot be recognized very distinctly at the beginning because the movements are still clumsy, i.e., they are executed with imperfect coordination and faulty rhythm.

Although the infant initially creeps and crawls in order to reach a desired object (Fig. 92b), a change of location becomes a more definite goal later on. The older infant often crawls for the sheer fun of it even when he has already mastered free walking.

In crawling on extended arms and flexed legs a slanting upright posture of the body results, although at the end of the first year the arms are somewhat shorter than the legs. In rare cases the infant moves forward with his elbows and knees in extension. Here, contrary to the pattern in real crawling, only the tips of the toes touch the floor, not the thighs or the backs of the feet. The gaze is still directed more downward than forward, unlike the situation in crawling. VARIOT saw this type of locomotion in only 3 of 80 infants. In this obviously awkward and uncomfortable position the chest is held lower than the pelvis because of the shortness of the arms, so that the whole body seems to be under an unnatural strain; this is in contrast to creeping and crawling, which adapt themselves so naturally to the body configuration. Unlike crawling, this position does not allow the labyrinthine righting reflex of the head to bring the head into the right spatial relations. Evidently the backward flexion of the cervical spine, necessary for this reflex, would be too unnatural. However, quadrupeds, such as the horse and the dog, naturally look straight ahead because of their totally different body structure.

The experiments BETHE conducted on animals to ascertain the adaptability (plasticity) of the nervous system can be repeated to a certain degree in the crawling infant. BETHE (1891) upset the classic center theory, according to which the reflexes and automatic movements are bound to the function of certain areas in the central nervous system, i.e., the centers. GOLTZ (1869) had already traced the goal-directed wiping movements of the decerebrate frog to a "central adaptability." BETHE showed that in man and animal coordinated movements reappear, quite often without any practice, after mutilations, injuries, or interventions involving crossing of muscles and nerves, transplantation of extremities to unnatural locations, etc. Therefore, we have to assume the existence of countless numbers of preformed centers if we wish to explain adaptability under these different conditions. The central nervous system does not act as a composite of reflex centers independent of each other, but functions as a whole. The reaction of the intact animal does not depend only on a certain center, but also on the total condition of its nervous system.

When BETHE experimented with dogs, six-legged insects, eight-legged spiders, or ten-legged crabs and removed any number of legs from various locations, the faculty of coordinated locomotion remained. A dog with one injured leg limps on the three intact legs almost as fast as before by virtue of an internal reorganization. When BETHE removed both fore or both hind legs, the dogs were still able to move forward.

It is also possible to compel the infant to execute a different pattern of locomotion if ordinary crawling is prevented. For example, I extended one leg at the knee by applying a splint or, even simpler, by binding both legs tightly together with a bandage but without impairing the ability to flex. Only rarely were the infants so bothered by this unaccustomed procedure that they gave up all locomotion. Sooner or later most children continued their crawling attempts if we succeeded in keeping them in good spirits.

The plasticity of the nervous system was clearly demonstrated, as the children were able to move forward in different ways in spite of their impediments. As far as possible they continued their former rhythms of movement, supplemented them with others, and thus achieved real forward movements. An infant whose legs were bound together in the manner described got up on his knees and moved forward by pushing first one and then the other knee a little forward by means of lateral rotations of his spine. A second infant made corresponding movements with his extended legs by alternately shifting his pelvis from one side to the other. Another moved forward by forward extension of both arms in the starting position for crawling, propping them up against the floor, and drawing himself forward until the previously flexed hips were again extended. Then the hips were flexed again. This "game" was repeated several times but was repeatedly interrupted by a sideways shift of the lumbar spine, as has been described. I did not see a return to creeping, which after all would have been possible. There is one difference from BETHE's animal experiments in that the infants solved their problems in so many different ways.

According to BÜSCHELBERGER (1958), infants who lose part of a limb or the entire extremity adjust much more easily and much more quickly to the loss and to an artificial limb than do adults who have suffered a similar loss. The child has no inhibitions; every means is right if he can only satisfy his curiosity and his urge to learn. In congenital absence of an extremity the child does not know any other condition and does not miss anything during the first years of life. By increased practice with the remaining limbs he may become unbelievably skillful, as the vaudeville artist Unthan and the Canadian lawyer Watsen have demonstrated.

Unthan was born without arms and learned to do everything we do with our hands with his feet. He used his feet for dressing and feeding himself and for writing and is said to have achieved the stature of a concert artist in playing the cello.

Watsen had poliomyelitis as a small child; both arms remained paralyzed. He became a "foot artist" too. He studied law and wrote his thesis with his feet, not requiring much more time than others do for this task.

BÜSCHELBERGER treated an armless boy in a home for the crippled. This boy also learned to write with his feet by daily hour-long practice, and he was able to guide the bookbinding knife very skillfully with his feet when working with cardboard and pasting. A sixteen-year-old student had lost his left leg up to the middle of the thigh through a streetcar accident when he was a small child. When BÜSCHELBERGER met him, he was the goalkeeper of his class's soccer team. When playing soccer he used an artificial leg, jumped in the middle of the crowd as any other goalkeeper would do, and kicked the ball high from the goal into the middle of the field.

In spite of BETHE's objections, I speak of centers because this mode of expression, which MAGNUS and RADEMAKER also employ, is in general use and is clear, while the theory of plasticity cannot supply vivid and distinct images.

VON HOLST (1951) believed that:

"The reflex theory, the center theory, the theory of plasticity, and their variations may apply to limited spheres, but they are certainly wrong when they claim totality. In reality, the nervous system consists of a number of distinct and quite differently functioning apparatuses although the histological findings seem to contradict this hypothesis."

According to JUNG (1953), the theories of centers and plasticity are not mutually exclusive but supplement each other as do the concepts of localization and "irradiation." The opponents of the center theory grasped this concept in much too narrow a sense. Organ activities are coordinated by certain parts of the central nervous system called "centers." Without these centers certain achievements are impossible. In the last analysis the theory of plasticity abandons the hypothesis of coordinating centers.

W. R. HESS (1954) does not want to abandon the center theory either. A spatial idea is connected with the concept of the center. The local accumulation of nervous elements is not the decisive factor; rather, the association of representatives of certain receptors and effectors and their combination for certain common functions are the significant aspects of this theory. The common goal is coordinated participation in a certain achievement; the local arrangement of the connecting and connected elements is of no significance.

Sliding

The locomotion of infants by sliding on their buttocks was described by SIGISMUND (1856) and VIERORDT (1877); it was reported later and in more detail by TRETTIEN (1900), BURNSIDE (1927), VARIOT (1926), and MCGRAW (1941). In order to slide, the infant must be able to sit unsupported. In this position he slides forward by digging his heels against the floor and by pulling himself forward with his feet. One child may not use his arms for support at all, another may lean to the side and push himself away from the floor with one hand, another may lean forward to push himself with both arms simultaneously and thus aid the motion, and still another may spread his arms horizontally to keep his equilibrium. Often only one leg is extended and the other flexed backward under the buttocks (Fig. 94). This movement pattern is quite often very asymmetrical. It cannot be decided whether or not this task is more easily accomplished with the aid of the arms. In any case the infant is able to slide in many different ways. By rhythmically repeating one of the movements described he moves about the room skillfully and rather rapidly.

Sliding is a frequently bypassed intermediate stage before walking and is abandoned in the latter's favor. On occasion I have seen an anthropoid and a young macaque move in this manner. Some crippled persons who had lost their legs or who were born without, or certain paraplegics, used to move about in a similar way before orthopedics was known.

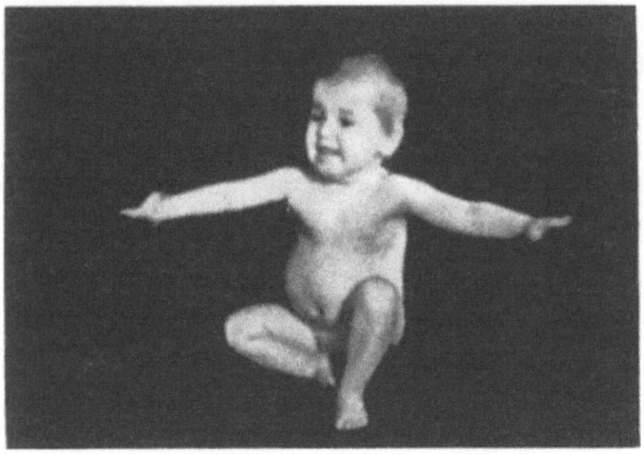

Fig. 94. Sliding:
The arms are spread and serve to maintain equilibrium (frames of a film).

Erect Gait

In the first independent attempts at standing, walking, and getting up from the floor, the arms, with which the infant holds onto things, are no less important than the legs (Fig. 95a and b). The mother places the infant near some supporting object before he is able to get up unaided. Only when he feels halfway safe does he dare to let loose with one hand and to grasp for something. Bending down is especially difficult (Fig. 96). Gradually the infant develops the faculty of moving along a suitable object, such as a railing, by repeatedly changing the position of the hands (Fig. 97). Thus the time of discoveries in the room starts. The room appears quite different to the infant, especially if he cannot crawl, than it does to the adult. He does not see the table, but only the tablecloth hanging down. If he grasps a corner of the tablecloth for his support it slides down and everything on the table tumbles after it. Chairs, seats, and sofas are not meant for sitting, but to serve as railings along which he can "handle." There are coasts, islands, and the still unattainable high sea (time of "coastal navigation," Fig. 98). The closed door is an unsurpassable obstacle for some time. A chair or a baby carriage is preferred and used as a movable support in walking.

In the beginning the infant anxiously avoids walking even if he is able to do so. If while holding his hand his mother suddenly lets go, he sits down or at most remains standing, but cannot be induced to walk on. The same happens when an only apparent support, such as a loose leading string, is withdrawn. Some infants who would love to walk unaided move so clumsily in their excitement that they fall down as soon as one lets go. If an infant has once been hurt very much by a fall, he will be frightened for some time, but sooner or later regains his courage and ventures his first unaided step, walking with outstretched arms to his mother. In this position the open distance is shortened by the length of the arms (Fig. 99). When the infant is walking for a considerable distance, his arms are likely to be extended horizontally to both sides just as if "rowing" in the air in order to maintain his equilibrium, which at first is always threatened. Later the arms are flexed at the elbows; the regular arm swinging during walking appears much later.

At first the infant walks straddle legged, as a sailor does when on the high seas, in order not to fall so easily. He usually stands and walks with slightly flexed hips and knees, but he is already able to extend his legs completely. At first he always tries to support himself with his hands in front of him. His lumbar spine is slightly kyphotic; however, when he stands completely free, his spine shows the same slight lordosis seen in the standing adult. The kyphosis of the dorsal vertebral column is compensated for by a lordosis of the cervical spine. The infant thus passes through a stage of development that is consistently found in the ape; a weak indication of the pattern is also found in primitive human races (SPITZY) (also see p. 242). The tips of the toes are turned outward at the onset of walking (BURNSIDE); they show a pattern identical to that seen in the stepping movements of the newborn (see p. 211). Only later, when the infant can walk well, are the feet moved parallel to each other (SHIRLEY).

It is quite evident that progress is favored by the always

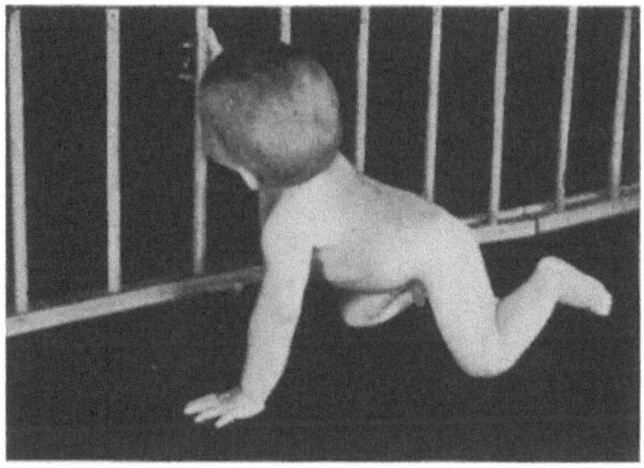

Fig. 95 a

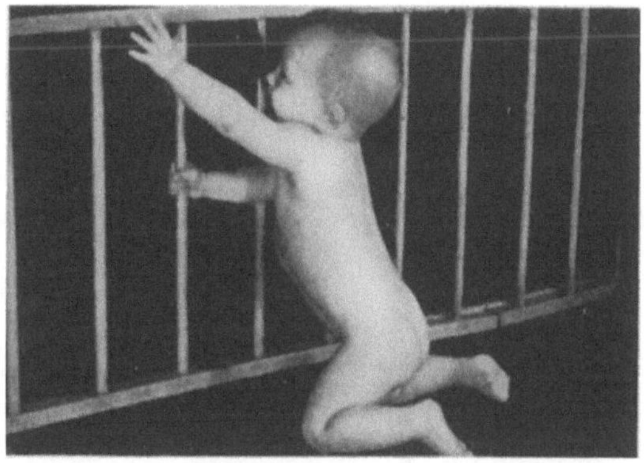

Fig. 95 b

Fig. 95a and b. Getting up with the aid of the arms (frames of a film).

Fig. 96. Stooping in order to grasp an object (frame of a film).

Fig. 97. Following an object along a railing (frame of a film).

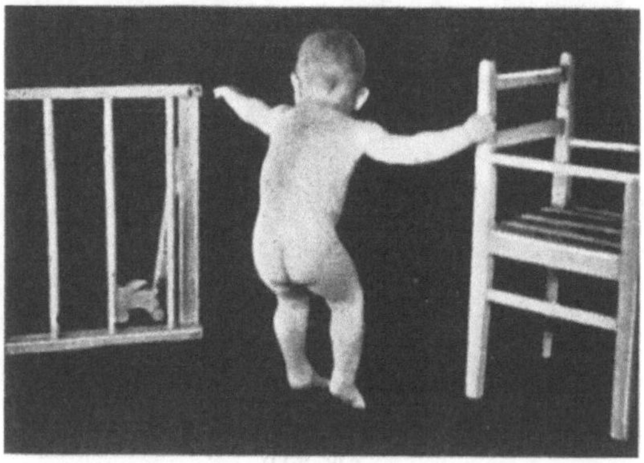

Fig. 98. "Coastal navigation" (frame of a film).

Fig. 99. The first free steps, arms extended to the goal (frame of a film).

available assistance of adults. Infants who cannot yet get up alone are often capable of standing and moving along a railing, an ability due solely to the support of adults. But an infant who is left completely to himself also eventually learns to walk without aid.

The question arises: To what extent does this progression depend on the maturation of the brain and how much is due to increasing practice and experience? A series of experiments on animals was made to study this question. CARMICHAEL (1926 and later) divided embryos of amblystomas into two groups, control and experimental. The experimental animals were immersed in water containing the anesthetic chloretone; the control group was placed in plain water. In the control group the usual strong reactions started at the usual time. The experimental group developed morphologically but remained completely inert because of the anesthetic. When placed in fresh water the poisoned embryos started to swim well within a short time.

We know from experiments on animals that even the ability to fly may occur solely through "maturation of the functions." In his experiments, GROHMANN (1939) deprived young pigeons of any opportunity to practice flying. When the pigeons were finally permitted to fly, from the start they could do so as well as their contemporaries with previous practice. Experiments to establish premature flying ability by forcing the pigeons to practice were unsuccessful. In these experiments, the ability to fly developed as a maturation process related to external development and could not be accelerated by practice and experience. Similarly POULSSON (1951) — as SPALDING had done as early as 1873 — placed young chimney swallows, which as nestlings flutter about a lot, in a narrow fenced-in box and fed them through the fence. When their siblings, which were left in the nest, were ready to fly out, he allowed the experimental group to be in the open also. They flew just as well as the control group. Ducks flying for the first time flew in an accomplished manner from the beginning and avoided obstacles. THOMAS and SCHALLER (1954) demonstrated that newborn optically isolated "Kaspar Hauser" cats focused, watched, attacked, struck, and clutched the fake prey just as well as their normally brought up siblings did. The innate catching automatisms were so charged that they emptied even as "vacuum-activity" reactions.

According to O. KOEHLER (1953):

"One does not learn movement patterns characteristic for the species but which are not known right after birth; rather, one has to wait until the needed structures are fully developed, and then the full ability is suddenly there."

That animals copy ("learn by seeing") different movements of the members of their own species or even of other species, i.e., that they learn by imitating what they see, is, according to KOEHLER, a widely held but entirely wrong opinion. In reality we deal with maturation processes that occur even if no model exists or with innate triggering mechanisms, for example, when a bird starting or only preparing to fly "infects" the other birds. There is an exception in birds: the imitation of heard patterns.

MCGRAW (1935) made comparative studies in a pair of young male twins. She practiced crawling and walking with one twin (Johnny) while preventing the other (Jimmy) from performing these motions. In spite of this, Jimmy crawled one week earlier than his twin brother. There was no time difference between their beginning to walk; only the "style" of walking was influenced by prolonged exercise.

GESELL and THOMPSON (1929) examined identical twins to ascertain their ability to ascend steps and to build towers. These skills began in the forty-sixth month of life. From this point on, one of the twins had a daily practice period for six weeks; then the other twin was trained in the same skills for two weeks. The latter climbed the four experimental steps immediately on her first attempt. After two weeks she possessed the same skill her sister had acquired after six weeks.

"One cannot prove conclusively that practice and exercise will accelerate the appearance of reaction patterns such as climbing steps or building towers. The time of their appearance is basically determined by the maturation of the neuronic structure."

I can report on a similar observation: A six-month-old girl was placed in a plaster cast because of a bilateral congenital hip dislocation and remained in the cast for a year. She was thus prevented from standing and walking. When she was one and a half years old, the cast was replaced by a half cast. One day after having been placed in the half cast she freed herself from it, fell from the bed, and appeared, walking erect and unaided, in the next room. Thus, she could suddenly walk at the age of one and a half years without ever having practiced standing or walking.

All these experiments and observations show how important cerebral maturation is for the development of new skills. The opinion expressed by JACQUES LOEB in 1899 is thereby confirmed: According to the empiricists, the infant must "learn to walk." The fact that the chick can walk immediately should be sufficient to refute the erroneous opinion of the empiricists. The difference between the chick and the human infant is that at the time of birth the structural development of the chicken is further advanced than that of the child. The child does not learn to walk, but he walks "automatically" as soon as his nerves, muscles, etc. have reached the stage of development necessary for this achievement.

Pediatrics formerly taught that one should not let an infant stand up too early but should restrain him at the beginning when he tries to get up so that he does not become bowlegged. This may have been justified at a time when rickets could not yet be sufficiently combated. Today we have such effective weapons against rachitis that we do not have to be afraid. We therefore welcome the infant's attempts and support them although we should take care not to overdo it. Often we have to contradict parents who, following antiquated ideas, try to restrain their children.

No less a man than KANT (1803) had already recommended that infants be allowed to crawl until they start to walk by themselves. Creeping and crawling are natural stages of development that evidently for some length of time correspond to the body structure and the nervous adjustment. Therefore, they should not be omitted even if in principle the erect gait can be learned without their assistance. It is easier for the child to creep and crawl at home than in an institution because the mother can spend more time with her child than the nurse with her patient. Since the child likes to put everything in his mouth, the floor on which he crawls must be meticulously clean. Tuberculosis may be easily acquired in the crawling stage.

I cannot share the opinion that the crawling period should not be shortened by early standing and walking attempts. One should give a child the opportunity to practice the skills appropriate to his age but without encouraging too rapid progress.

Every infant has a strong urge to move and to act, a drive which leads him again and again to undertake new experiments that he greatly enjoys even if they fail quite often. New tasks, for example, ascending steps, which at first is so difficult, are undertaken with great zeal until the new skill is mastered. Each bit of progress is paid for with innumerable fruitless attempts, many painful falls, and many blue marks. However, serious injuries are quite rare. Thus, the acquisition of unaided locomotion is a source of joy for the child and his mother. She supports the child's attempts by a series of assisting devices so self-understood that they do not have to be learned. The formerly widely used walking chairs, "protective fall hats," walking ropes, walking baskets, etc., are superfluous today and as good as unknown.

The ages at which erect body posture and free gait appear in the infant and the monkey baby are compared on page 591.

Stages of Standing Up

(141)

Static development starts from the dorsal resting position; the final goal, reached during the first year of life, is the erect unsupported gait. This requires the ability to get up unaided and without pulling oneself up by holding onto nearby objects. This highest stage is reached after several intermediate stages, which bring the infant nearer and nearer to his goal and which are closely connected with the different stages of unaided locomotion. Each new body posture to which the child progresses has been made known to him by his mother, who turns him on his stomach, sets him up, and stands him up long before he is able to turn, sit, and stand up by himself. Not discouraged by frequent failures, he untiringly strives for new bits of progress that bring him to a greater and greater distance from the floor and make the task of maintaining equilibrium increasingly more difficult.

The young infant is not yet able to turn from the dorsal to the abdominal position. On the other hand, the dorsal position is preferred at the beginning. According to BLANTON (1917) and WATSON (1924), many newborns placed in the abdominal position on a hard plane are gradually able by repeated and finally successful movements to turn to the dorsal position.

Even the newborn is successful in turning from the dorsal position to the side. In the third or fourth month of life the infant gradually learns to turn over completely onto his abdomen; first the head is turned, then the upper and finally the lower part of the body (SCHALTENBRAND, 1925). The skill of getting up from the dorsal position develops through many intermediate stages in accordance with definite laws. First, turning around has to be learned. According to MCGRAW's observations (1941), the newborn cannot yet turn over unaided, but occasionally his whole body rolls sideways with the aid of gravity and some of the infant's own movements. After some weeks the spinal extension phase begins: The infant simultaneously turns his head sideways and backward, whereby the shoulder on the side to which the back of the head is turned is lifted. Gradually the backward turning involves the entire vertebral column. During this progressive development the arm and leg on the side of the skull are raised and brought to the side of the chin. Usually the shoulders are turned first and then the hips until the infant is in the lateral position. Sometimes he rolls over onto the abdominal position. The automatic rolling phase develops from these maneuvers; it is at its height on approximately the two hundred and fiftieth day. The individual movements, which at the beginning are still fairly uncoordinated, interlock more and more ingeniously so that the exact starting point of this phase cannot be determined. The deliberate phase, which appears between the two hundredth and four hundredth day, consists of voluntary movements with the goal of crawling or sitting up.

The head leads the movement from the start, corresponding to the cervicocaudal direction of development. The labyrinthine righting reflex of the head is decisive in this process. That the body follows the position of the leading head proves that in uninhibited movements the neck righting reflex of the body is stronger than the body righting reflex of the head. This is surely a very appropriate arrangement.

Once in the abdominal position, the infant has achieved the potentiality of creeping and later of crawling.

It is difficult to say at what time the infant begins to sit with support, because this depends entirely on the kind of support. SHIRLEY (1933) cites the fourth month as the starting time, but with adequate support on both sides infants may sit much earlier. According to TRETTIEN (1900), the infant sits without support in the seventh to ninth month; according to VARIOT (1926), at the age of six to eight months; and according to SHIRLEY, in the seventh month. PREYER's (1895) child was able to sit unsupported for some time in the thirty-ninth week, but not without a backrest. In the forty-second week he sat naked in his bath, erect and with a straight back, without support. Of course, he had to give full attention to this task. He did not fall sideways as long as he was not occupied with new impressions. I can confirm this observation. If a child has just learned to sit up, a call that attracts his attention may at times suffice to make him topple over. BÜHLER and HETZER (1932) made the same observation: The seven-month-old infant sits unsupported as long as he is not distracted, but he immediately topples over if anyone makes him reach for a rattle.

As LAUDER BRUNTIN and later PREYER have observed, small children who sit on the floor turn the soles of their feet toward each other, a habit that may have resulted from the leg position at birth.

In approximately the tenth or eleventh month the infant sits up from the dorsal position without assistance. Many sick infants cannot get beyond this stage for a long time. They can sit and sit up but cannot stand.

We have previously stated that strong supporting reactions occur in the legs of the newborn when someone stands him up. But after several months the infant pulls his legs up to his body when the same attempt is made. When supported under the arms the infant stands in the eighteenth to twentieth week according to SIGISMUND (1856) and in the eighth month according to SHIRLEY (1933). Soon afterward the infant learns to hold on to something when standing. According to TRETTIEN (1900), the infant can stand unsupported sometime between the seventh to sixteenth, usually between the tenth and twelfth months. SHIRLEY gives the sixteenth month as the starting time for standing; this seems rather late to me. The skill of getting up without aid is learned later than the unaided gait. In the thirteenth to fifteenth month the infant gets up from the dorsal position without holding on to nearby objects to pull himself up. In getting up from the dorsal position the infant at first progresses through the same sequence in which he learned the individual movements (SCHALTENBRAND): He first rolls into the abdominal position, squats, sits down, and finally gets up by vigorously pushing himself up from the floor with extended arms.

The pattern of getting up changes in the second or third year (SCHALTENBRAND). Children facilitate the transition from the dorsal to the sitting position by turning slightly; then, keeping the pelvis in contact with the floor on one side, they prop themselves up with their arms, mainly on the same side. From this stage the final pattern of getting up develops in the fourth or fifth year: The body is now symmetrically rolled from the floor into the sitting position, with the arms at the same time propped on the floor. From the sitting position the infant simply stands up on his legs.

The starting dates cited previously are only average values. Fairly considerable deviations are not pathological. Diseases, especially rachitis and long-lasting nutritional disturbances, may temporarily delay development. Any urge to progress in the static development is absent under these conditions. In the feeble-minded the development is considerably retarded, and in severe cases it does not occur at all.

Cervicocaudal and Cervicorostral Directions of Development

Briefly summarized, the following stages of development of locomotion can be distinguished: turning from the dorsal into the abdominal position with the head leading; in the abdominal position: lifting of the head (labyrinthine righting reflex of the head), raising of the upper half of the body first on flexed and later on extended arms (supporting reactions), chain reflex of the entire body, creeping, raising of the body on the extended arms and flexed legs, crawling, and finally standing without aid and walking on extended legs without the assistance of the arms. This development mirrors a rule, the so-called law of the cervicocaudal (descending) direction of development.

This "law" resulted from the observations of early developmental stages, first in the animal and later in man (TRETTIEN, 1900; CHILD, 1925, 1929; VARIOT, 1926; BURNSIDE, 1927; COGHILL, 1929; LANGWORTHY, 1933; CARMICHAEL, 1933; MCGRAW, 1939; GESELL and AMES, 1940; GESELL, 1954; HOOKER, 1954, 1958). KINGSBURY (1924, 1926) rejects, and rightly so, the designation "cephalocaudal" direction of development as confusing and speaks of "cervicocaudal" and, in the opposite direction, "cervicorostral" development.

Anatomically and physiologically the development progresses like a wave moving in the direction from head to tail. Anatomically the proximal segments near the head have progressed further in their development than those near the tail. Thus, the body parts also successively serve independent locomotion; the involvement of the limbs in the uncoordinated and coordinated movement sequence progresses from the head, over the proximal part of the trunk and the fore legs to the distal part of the trunk and the hind legs and finally to the tail.

I cite COGHILL's investigations on the amphibian amblystomum as an example. Before the limbs are formed, winding movements of the front part of the body occur; with progressing development these spread increasingly backward to the hind part of the trunk and the tail. At first all movements start at the front and progress in the direction of the tail. The movements of the limbs, which have developed in the meantime, also follow this rule. The original locomotion consists of a slow swimming movement of the head and neck in which first the fore legs and later the hind legs become involved. The latter at first do not move simultaneously with the front legs; they do so only when the movement has spread from front to back.

Long before the investigators cited, PREYER made the same observation in infants:

> "It should be mentioned that the bilateral-symmetrical movements of the facial muscles and of the arms occur in reflexes at a much earlier time and more distinctly than those of the legs. Also, the abductions, adductions, supinations, and rotations of the arms appear distinctly in all variations at an earlier time than those of the legs."

The same rule can be deduced for human fetuses from the observations of KELLOGG (1941). He traced the movements through the abdominal wall of the mother and transferred them to a drum. He observed a distinct increase of movements with progressing fetal age. From the one hundred and seventieth to the one hundred and ninety-ninth day of fetal life the movements of the anterior part of the body were more numerous than those of the posterior part in 81.3% of the observations;

from the two hundred and twentieth to the two hundred and fifty-sixth day this value decreased to 55.6%.

The same results were obtained by HOOKER (1954) when he performed double simultaneous stimulations of the face and hand or the hand and foot in viable human fetuses. If he stimulated the face and hands at the same time in two fetuses (ten and a half and thirteen weeks of gestation), he obtained a facial reaction but not one of the hand. If he stimulated the hand and foot at the same time in two fetuses of the same gestation period, only the fingers were flexed and the flexion usually occurring when only the toes are stimulated was not obtained. The rostral area thus dominated the caudal region, or in other words the phylogenetically older area dominated the newer one.

The different forms of locomotion confirm this rule. The head precedes the body in turning. In creeping, the supporting reactions occur first in the flexed arms while the legs are still completely passive. In well-developed creeping and still more distinctly in crawling, supporting reactions of the extended arms occur. In creeping the arms alone serve the locomotion. During crawling, supporting reactions of the legs appear for the first time; the legs are flexed during the motion. Later, in the erect-free gait, the arms are no longer used for locomotion; therefore, the supporting reactions disappear in the arms and now appear in the extended legs. Participation of the limbs in locomotion in general and the onset and improvement of the supporting reactions (first in the flexed and then in the extended limbs) thus always develop first in the arms and then spread to the legs, so that the arms always remain one developmental stage ahead of the legs.

One could make the objection that supporting reactions of the legs occur even in the stepping and crawling movements of the newborn and are therefore present in the extended limbs at a time when supporting reactions in the arms are still completely absent. But these reflexes have no relation to the independent locomotion of the infant since they have disappeared prior to the beginning of locomotion.

The cervicocaudal direction of development can be distinctly recognized in BRANDES's observations (1939) of a baby orangutan: When at the age of three months the baby was hanging from a branch with its arms, only the front limbs were active while the hind limbs were, so to speak, inactively glued to the trunk. SCHNEIDER (1950) similarly recognizes the cervicocaudal direction of development in the acquisition of climbing skill which he observed in the young chimpanzee: The activities of the arms and hands preceded those of the legs and feet.

As previously mentioned, the cervicocaudal (descending) direction of development from the brain to the spinal cord proceeds in a direction opposite to the cervicorostral (ascending) development within the brain, which has been described in the animal (see p. 216) and can also be demonstrated in man on many occasions, for example, in the development of the respiratory center (see p. 313) and of the centers of motility (see p. 264). The brain centers in the infant mature in the same sequence in which they have been formed ontogenetically and phylogenetically. Thus, the brain stem is the center from which the development spreads in both directions, upward to the neocortex and downward to the spinal cord.

Phylogenetic Memories

(141)

As a preliminary to real locomotion the infant learns to master the abdominal position and in this posture to right his head into the normal position in relation to space, i.e., vertex up and the mouth horizontal. As long as they are alive, fish assume the abdominal position, which depends on the normal position of their heads in space. Only in the abdominal position can they move forward.

According to ABEL (1912), the oldest terrestrial vertebrates known from prehistoric times dragged their bodies on the ground when moving; the lowest terrestrial vertebrates still existing today (Urodela or Batrachia) move in the same manner. The limbs of these animals are so short that the body touches the ground not only when they are at rest but also when they are in motion. The infant is in exactly the same situation when he is creeping, the first type of rhythmic locomotion he masters. But at the same time there is an essential difference: The oldest terrestrial vertebrates probably moved on all four extremities just as today's amphibia do, whereas the creeping infant uses only his arms and not his legs. Later, when the child crawls on his arms and legs, the body is no longer dragged over the floor. In this connection human development does not completely agree with the phylogenetic pattern.

Movements similar to the infant's creeping are found in young animals of species related to distant phylogenetic ancestors of man. HARTMAN (1920) observed in the opossum and GOERLING (1920) in the kangaroo that the young of these marsupials proceeded to the pouches of their mothers after birth without any aid. If they do not reach the nourishing teats, they are doomed to starvation. This kind of locomotion was investigated in detail by LANGWORTHY (1933). He placed young opossums on a piece of gauze to which their claws could cling. They grasped the threads with their front claws and rowed forward by alternating strokes of the front legs. They were capable of carrying the weight of their bodies on one front leg in a manner similar to the palmar grasp reflex of the infant. The hind legs did not participate in the locomotion. The resemblance to the creeping of the infant is therefore obvious. This reflex enables the young opossum to reach the pouch by grasping the fur of its mother. There it secures its position at the nipple.

In the crossed coordination of the crawling infant (see p. 225) a very old ontogenetic and phylogenetic coordination of movement finds its expression. ARISTOTLE already attributed it to quadruped and multiped animals: the simultaneous forward moving of the crossed limbs alternating with the two other crossed limbs. Crossed coordination can be found in many mammals provided they do not move forward by ambling. In the horse, crossed coordination appears in trotting but not in stepping. I observed it distinctly in Javanese monkeys (macaques), which moved horizontally on all four extremities in a quiet gait pattern.

The original form of locomotion in crossed coordination succeeds the undulating movements descending from the head along the body, demonstrated by PREYER (1885) in frog embryos and COGHILL (1929) in embryos of the batrachian amblystomum. AVERY (1928) saw them in embryos of guinea pigs delivered by Cesarean section, BROWN (1921) in fetuses of cats, and BRAUS (1921) and DU BOIS-REYMOND

(1930) in amphibia and reptiles. Corresponding head movements were observed by CORONIOS (1933) in fetuses of cats between the twenty-fifth and thirty-fourth day of gestation and by HARTMAN (1920) in newborn opossums which tried to reach the pouches of their mothers after birth.

I found undulating head movements only rarely in infants but they were occasionally quite distinct. When, for example, reflex "crawling" movements were elicited in an infant two and a half months old, strong movements of the head alternating to the right and to the left occurred. When the infant was in the abdominal position, these movements also occurred without external stimulation. Simultaneous coordinated arm movements could scarcely be observed since, on the whole, crossed coordination was absent in the infant's reflex crawling movements, which have nothing to do with the later independent crawling. McGRAW found undulating head movements, probably of a similar kind, in the swimming movements of the human newborn (see p. 218).

In his erect gait, adult man still preserves a vestige of the crossed coordination his ancestors once used when they moved on all four limbs in the abdominal position. The rhythm of his swinging arms and his simultaneously moving legs corresponds to crossed coordination: With the forward motion of the right leg the left arm swings forward and vice versa. It is hardly possible to make the arms swing voluntarily in any other way.

The oldest mammals known to us originated in the Triassic period. They remained insect eaters, as they had been originally, mounted trees by climbing and there achieved, as monkeys, the half erect posture characteristic of primates. The ancestors of man finally returned to the ground, where they became fully erect (see p. 571).

The body posture of the creeping and crawling infant (Figs. 50 and 51) is half erect; his head always assumes the highest plane of the body position. His arms are therefore free for grasping, a skill he previously learned in the supine position.

In creeping the arms alone serve the motion, but in crawling they cooperate with the legs; however, the extended arms are a developmental stage ahead of the flexed legs. Finally, in the erect gait locomotion depends entirely on the now extended legs. Thus, the rule of the cervicocaudal direction of development is again confirmed. During the entire developmental period, including that of crawling, the arms never lose their ability to grasp so that neither the infant nor his animal ancestors become quadrupeds whose limbs, as, for example, those of the dog or horse, serve only locomotion and never grasping. This is different in man. The anterior and posterior extremities of the infant have additional tasks during crawling: The arms placed in front pull the body and the legs push it forward (Fig. 51). Amphibians and reptiles move in the same manner.

The infant does not pass even a temporary stage of being a pure quadruped nor does he become temporarily quadrumanous; he mainly uses his hands and only in exceptional cases his feet for grasping, although he does the latter, of course, more frequently than the adult does. Therefore, one could sometimes call him a two-handed creature and at the same time a quadruped. In any case, during a certain period of his development his locomotion passes through a stage corresponding to that of the oldest terrestrial vertebrates, as described above.

The details of the kinship relations between man and the anthropoids still living today are disputable. I therefore wish to discuss only one question: Apes living in trees use their arms and legs simultaneously when climbing and jumping. On the ground they generally become quadrupeds, but they can also move in an erect position. However, if young anthropoids are taken away from their natural environment and grow up in human surroundings, they have to get more or less used to the erect gait. In animal acts one likes to demonstrate how human they really are. They appear completely dressed up and display their skills in smoking or bicycling. Other animals can also sometimes be trained in an astonishing way. The behavior in their natural environment, which is the only topic discussed here, has nothing to do with such achievements. Gorillas also walk in the erect position in their natural habitat, as a film taken by CORDIER shows.

When, in the course of evolution, the ancestors of man assumed the erect posture, the entire supporting skeleton of the trunk, including the vertebral column and the thorax, had to be transformed. The quadruped's foreleg, which serves only standing and running, becomes man's arm, which, liberated from the task of supporting the weight of the body, only grasps; from the grasping foot of the ape the standing foot of man developed. Comparative anatomy and ontogenesis show distinctly the different stages of this transformation.

The grasping ability of the infant's big toe still reminds us today of the climbing foot of the ape. This ability may have been lost in our civilization only because we always wear shoes and thus no longer have the opportunity to practice. We have previously mentioned to what degree armless adults may attain skill with their feet. Adults of primitive races still often grasp with their big toes. Australian aborigines climb at great speed with the aid of their spread toes on slippery eucalyptus trees (KLAATSCH).

Our small children love to climb, too, and are not afraid of danger. Stairs, railings, ladders, and trees have a magnetic attraction for them. Scratches, bumps, and holes in their clothing can only temporarily dampen their joy in climbing. But in this regard the children of some primitive races are far ahead. BUCKMAN (1894) sees in the infant's climbing pleasure a direct throwback to our ape-like ancestors who lived in trees.

I was never able to find genuine climbing movements in the infant (see p. 219). It is true that the palmar grasp reflex has been interpreted by KLAATSCH as a survival from our climbing ancestors. But, as previously described, it enables the baby ape to grasp the fur of its mother and therefore has another meaning.

KLAATSCH (1922) and FOERSTER (1913, 1921) linked the posture and the movements of the newborn phylogenetically with the squatting position and the climbing movements of the apes. The pallidum, which originally controlled the mass movement of climbing, is inhibited by the maturation of the dominating striatum and serves the adult's motility. If this inhibition ceases because of a disease of the striatum, the phylogenetically older function of the pallidum, the climbing movements, comes to light again as a release phenomenon. FOERSTER especially describes congenital general athetosis. Whereas the phylogenetically older pallidum serves climbing, the phylogenetically younger striatum directs the specific human faculties of sitting, standing, and walking. The formation of the pyramidal tracts, on which the voluntary individual movements depend, leads phylogenetically beyond this striate function.

FOERSTER speculates on the phylogenetic development of the voluntary individual movements in the following manner: The motion of a single limb and later that of a part of a limb were isolated from the climbing movements, which are mass movements. However, in order that these movements could appear as individual motions the pallidum, which is capable only of mass movements, had to be dominated by an inhibitory station, the striatum. The striatum is capable of blocking a portion of the pallidum and of permitting the functioning of another portion that regulates the individual movements. A special type of tract, the pyramidal tract, which can transmit individual impulses, developed at the same time as did the striatum, but these tracts still pass through the pallidum because of their phylogenetic development.

I cannot share FOERSTER's opinions in one respect, i.e., in that I cannot see, as mentioned before, any climbing movements in the movements of the infant. Apart from this reservation his presentation is very convincing. In his investigations of striatal motility disturbances, FOERSTER generally arrived at views on the structure of brain function similar to those I have developed by another approach, i.e., through the investigation of central respiratory disturbances. He says: "The phenomenon which we see in the striatum overruling the pallidum is repeated in the cortex. Frontal areas are superimposed, as checking organs, over central areas. The course of development is always the same: The higher function develops from the lower by taking over parts of the latter and simultaneously inhibiting the lower by a superimposed organ so that only that part of the lower organ needed for the higher task remains active. Correspondingly, the respiratory center is composed of different part-centers; of these the lower and phylogenetically older centers are inhibited by the activity of the higher and phylogenetically younger centers."

The free gait was probably attained by human ancestors in a way similar to what we can still observe today in each child: First occurs the erect stance with the aid of the arms and then the navigation with supporting devices. Then the first short steps are taken with the arms extended forward between two supports, and finally the independent gait with arms extended sideways to preserve the equilibrium.

When the small child gets up, he first tries to support himself with his hands in front of him. Thus, he first stands and walks with his body bent forward and with his hips and knees flexed, as apes and bears do in the erect position. Even newborns in their stepping movements do not completely extend their knees and hips (see p. 211). This posture has an anatomic but not a neurological explanation: A neonate's corpse shows a certain contracture in the knees and hips; the angle at the hip is approximately 35° and at the knee 25° (BÖHM, 1928). The contractures of the hip joints are effected in part by the iliopsoas muscle and in part by the form of the head and the neck of the femur and the acetabulum as well as by their positions in relation to each other. The contracture of the knee is, according to BÖHM (1928) and HASSELWANDER (1938), based on a backward bowing of the upper portion of the tibia.

However, when the infant stands really without support, he also has a lordosis of the lumbar spine and an extension at the knees and hips just as the adult has.

The importance of the acquisition of the upright gait, the last and highest type of locomotion that can be reached in the course of phylogenesis and ontogenesis, cannot be overestimated for the real evolution of man; locomotion, which was left to the arms in creeping and later to the arms and legs in crawling, was taken over completely by the legs in the erect gait. The foot, which until now had served to grasp and climb, was taken over for standing, walking, and running. The arms had previously been used occasionally for grasping, but at first were still needed for locomotion and later for at least holding onto something during erect gait, as we can observe in any child who has just learned to walk. Only with the unsupported gait did the arms become completely free for the many functions enabling man to progress beyond the animal stage.

ARISTOTLE already recognized an essential part of this development:

"The hand is not a tool but a tool to replace tools... It serves man as a talon, claw, and hoof; he can arm it with spear and sword and with any tool he wishes... It can serve him, because it can grasp and hold everything. The hand is formed to accord with its natural purpose."

Following ARISTOTLE, THOMAS AQUINAS also called the hand "the tool of tools." Man does not possess the weapons of certain animals, such as horns and claws. In their place he possesses reason and hands with which he can in different ways procure for himself weapons, clothing, and everything that is needed for life. Best of all, he can manufacture an infinite number of tools for unlimited tasks.

KANT (1782) expressed the same thought. He speaks of the defense of man, "for which nature has given him neither the horns of the steer, nor the claws of the lion, nor the fangs of the dog, but only hands."

And CARREL (1936) pays this tribute to the hand:

"The hand is a masterpiece. It can sense and act at the same time. It acts as if it had the gift of vision... The hand adjusts itself to the hardest as well as to the most sensitive work: with equal ease it has handled the stone knife of the prehistoric hunter, the hammer of the blacksmith, the ax of the woodcutter, the plow of the peasant, the sword of the medieval knight, the controls of an aeroplane, the brush of the artist, the pen of the writer, and the thread of the silk weaver. It can kill and bless, steal and bestow gifts, sow grain in the fields and throw hand grenades in the trenches."

This has been possible only because the form of the human hand was arrested phylogenetically at a lower and undeveloped stage corresponding to that of the salamander and lizard-like reptiles. On the other hand, the "fingers" of hoofed animals have been transformed to such an extent that they are not capable of human functions. The outstanding adaptability of the human hand is due to the fact that it is not the product of any one-sided formative evolutionary development.

Thus, a passing stage in the genetic development of the ape became a final stage for man. According to the "fetilization" hypothesis of BOLK (1926), other primates end their development with a final stage that is not part of man's development. Man is, so to speak, a primate fetus that has attained its sexual maturity. K. LORENZ (1954) describes this with a term used elsewhere in biology — neoteny — i.e., a developmental inhibition permanently retaining the characteristics of youth.

It is frequently found in domesticated animals — and man is domesticated too. Neoteny considerably assisted humanization; decisive factors were the achievement of the erect gait which allowed the liberation of the still primitive hand for grasping and the development of speech and of thinking in close interdependence. According to GEHLEN (1958) and LORENZ, boundless curiosity is another featurc of humanization. It is found in animals and even in anthropoids, but only during a short period of their youth; in man it is also found in childhood but is retained throughout most of his life. Neoteny marks man even physically as an "unspecialized creature of curiosity" (see p. 571).

Discoveries of the Stone Age allow us to observe how man gradually created his tools. In this respect the infant, surrounded from birth by all the artifacts of adults, goes through a course of development completely different from that of his phylogenetic ancestors.

SUMMARY

In the first months of life regular stepping, ascending, crawling, and swimming movements can be elicited if the infant receives adequate support. The stepping, ascending, and crawling movements disappear several months prior to the time the infant starts to walk and crawl independently. They are gradually inhibited because of the maturation of the brain. In children with brain damage the crawling movements persist up to the second year.

The infant learns independent locomotion in several stages during his first year of life. From the basic posture of the supine position he turns over onto his abdomen with his head leading this movement. From the abdominal position creeping develops first and then crawling; sometimes sliding on the buttocks is learned and the erect gait is always learned. Some stages may be skipped.

Body and limbs start a coordinated sequence of movements following the cervicocaudal direction of development. First the labyrinthine righting reflex of the head develops in the abdominal position, then the neck righting reflex of the body, and finally the symmetrical chain reflex in the abdominal position, which represents the starting position for creeping. For creeping only the arms are used, but for crawling the legs are also used. The cervicocaudal direction of development manifests itself in the initiation and improvement of the supporting reactions, first on the flexed and later on the extended arms and legs. In creeping, crawling, and walking the arms are always one developmental stage ahead of the legs.

The ability to rise without support to a standing posture from the supine position is achieved in close association with locomotion; this development passes through the intermediate stages of turning to the abdominal position, squatting, sitting up, and finally standing up, during which the upper half of the body is pushed up from the floor by the arms. In succeeding years this involved procedure is shortened.

Comparison with amphibia and reptiles shows that the development of locomotion occurs in the young child in the same way as it does in phylogenesis.

REFERENCES

Preliminary Stages of Locomotion

AMES, L. B.: Genet. Psychol. Monogr., Worcester (Am.) *19*, 411 (1937).
ANDRÉ-THOMAS and SAINT-ANNE DARGASSIES: Études neurologiques sur le nouveau-né (Neurologic Studies of the Newborn). p. 133. Paris, 1952; Encephale *45*, 101 (1956).
—, and AUTGAERDEN: Presse méd. *1953*, 582.
AVERY, G. T.: Genet. Psychol. Monogr., Worcester (Am.) *3*, 247 (1928).
BAUER, J.: Klin. Wschr. *1926*, 1468; Kinderärztl. Prax. *2*, 256 (1931).
BERITOFF, J. S.: Pflügers Arch. Physiol. *199*, 248 (1923).
BERSOT, H.: Le rôle de la musculature dans le développement du petit enfant (The Role of the Musculature in the Development of the Small Child). Vevey, Montreux, Berne, 1926.
BLANTON, M. G.: Psychol. Rev. (Am.) *24*, 456 (1917).
BÖHM, M.: Klin. Wschr. *1928*, 551.
BÖHME, A.: Dtsch. Zschr. Nervenhk. *56*, 217 (1917).
BOLK, L.: Das Problem der Menschwerdung (The Problem of Humanization). Jena, 1926.
BROWN, T. C.: Erg. Physiol. *15*, 480 (1916).
BÜSCHELBERGER, H.: Zschr. ärztl. Fortbild. *52*, 386 (1958).
BUCKMAN, S. S.: The Nineteenth Century *36*, 1894 II, p. 733.
CARREL, A.: Der Mensch, das unbekannte Wesen (Man the Unknown). Stuttgart, Berlin 1936.
DIX, K. W.: Körperliche u. geistige Entwicklung eines Kindes (Physical and Mental Development of a Child). No. 1, pp. 59, 74. Leipzig, 1911.
EBBECKE, U.: Dtsch. Zschr. Nervenhk. *175*, 116 (1956).
ERSTLING, O.: Theorie und Praxis der Körperkultur (Theory and Practice of Body Culture) *4*, 191 (1955).
FOERSTER, O.: Zschr. Neurol. *71*, 1 (1921).
FREUSBERG, A.: Pflügers Arch. Physiol. *9*, 358 (1874).
GOLTZ, F.: Pflügers Arch. Physiol. *51*, 570 (1892).
HEDIGER, H.: Acta trop., Basel *9*, 100 (1952).
HETZER, H., and L. KOLLER: Zschr. Psychol., Leipzig *117*, 257 (1930).
HOLST, E. VON: Klin. Wschr. *1951*, 97.
HOLT, K. S.: J. Pediatr. *57*, 571 (1960).
KABAT, H., C. DENNIS, and A. B. BAKER: Amer. J. Physiol. *132*, 737 (1941); J. Physiol. *34*, 1 (1906); *40*, 28 (1910).
KLAATSCH, H.: Der Werdegang der Menschheit und die Entstehung der Kultur (Evolution of Mankind and the Origin of Culture). 2nd ed. Berlin, 1922.
LANDÉ, E.: Zschr. Biol. *81*, 224 (1924).
LAUGHTON, N. B.: Amer. J. Physiol. *70*, 358 (1924).
MCGRAW, M. B.: J. Pediatr. (Am.) *15*, 485 (1939); *17*, 747 (1940); In Carmichael, Manual of Child Psychology. 4th ed., p. 332. New York and London, 1949.
MELLA, H.: Arch. Neurol. (Am.) *10*, 441 (1923).
MINKOWSKI, M.: In Abderhaldens Handbuch der biologischen Arbeitsmethoden (in Abderhalden's Handbook of Biological Methodology). V, 5B, p. 542. Berlin and Vienna, 1928.
MUMFORD, A. A.: Brain *20*, 290 (1897).
PHILIPPSON, M.: L'automatie et la centralisation dans les systèmes nerveux des animaux (Automatism and Centralization in the Nervous Systems of Animals). Brussels, 1905.
PILLERI, G.: Zschr. ges. Neurol. *200*, 455 (1960).
PRECHTL, H. F. R.: Umschau *53*, 656 (1953); Mschr. Kinderhk. *101*, 519 (1953).
SCHNEIDER, K. M.: Zschr. Tierpsychol. 7, 485 (1950).
SHERRINGTON, C. S.: J. Physiol. *34*, 1 (1906); *40*, 28 (1910).
SHIRLEY, M. M.: In Murchison's Handbook of Child Psychology. 2nd ed., p. 236. Worcester, Mass., 1933.
STIRNIMANN, F.: Schweiz. med. Wschr. *1938*, 1374.

WARTENBERG, R.: Die Untersuchung der Reflexe (The Examination of Reflexes). p. 156. Stuttgart, 1952.
WATSON, J. B.: Der Behaviorismus (Behaviorism). p. 152 of German translation. Berlin and Leipzig, 1930.
—, Psychology from the Standpoint of a Behaviorist. 2nd ed. Philadelphia and London, 1924.
YERKES, R. M.: Chimpanzees. p. 17. New Haven, 1945.

Stages of Independent Locomotion

ABEL, O.: Grundzüge der Paläobiologie der Wirbeltiere (Principles of the Palaeobiology of the Vertebrates). Stuttgart, 1912.
AMES, L. B.: Genet. Psychol. Monogr., Worcester *19*, 411 (1937).
ANDRÉ-THOMAS, CHESNI and AUTGAERDEN: Presse méd. *1954*, 41.
ARISTOTLE: Naturgeschichte der Tiere (Natural History of Animals). German translation by Karsch. Stuttgart, 1866.
AVERY, G. T.: Genet. Psychol. Monogr., Worcester *3*, 247 (1928).
BETHE, A.: Handbuch der Physiologie (Handbook of Physiology). Vol. 15 (2), p. 1175. Berlin, 1931.
BIRKMAYER, W., and H. GOLL: Dtsch. Zschr. Nervenhk. *151*, 237 (1940).
BOIS-REYMOND, R. DU: Handbuch der Physiologie (Handbook of Physiology). Vol. 15 (1), p. 236. Berlin, 1930.
BRANDES, G.: Buschi. p. 28. Leipzig, 1939.
BRAUS, H.: Anatomie des Menschen (Anatomy of Man). Vol. 1, p. 406. Berlin, 1921.
BROWN, T. G.: Erg. Physiol. *15*, 480 (1916); J. Physiol. *49*, 208 (1914/15).
BÜHLER, C., and H. HETZER: Kleinkindertests (Tests on Small Children). p. 110. Leipzig, 1932.
BURNSIDE, L. H.: Genet. Psychol. Monogr., Worcester *2*, 284 (1927).
CARMICHAEL, L.: In Murchison's Handbook of Child Psychology. 2nd ed., p. 31. Worcester, Mass., 1933.
—, Manual of Child Psychology. p. 54. New York, London, 1946.
CHILD, C. M.: Anat. Rec. (Am.) *31*, 369 (1925); Protoplasma, Berlin *5*, 447 (1929).
COGHILL, G. E.: Anatomy and the Problem of Behaviour. Cambridge, New York, 1929.
—, Arch. Neurol. (Am.) *21*, 989 (1929).
CORDIER, C.: Herrscher des Urwalds (Rulers of the Jungle). Cinemascope-Farbfilm.
CORONIOS, F. D.: Genet. Psychol. Monogr., Worcester *14*, 287 (1933).
CZERNY, A., and A. KELLER: Des Kindes Ernährung (Nutrition of the Child). 2nd ed., Vol. 1, p. 767. Leipzig and Vienna, 1925.
FOERSTER, O.: Berl. klin. Wschr. *1913*, 1217 and 1255; Zschr. Neurol. *73*, 1 (1921).
GEHLEN, A.: Der Mensch (Man). 6th ed. Bonn, 1958.
GESELL, A.: in L. Carmichael's Manual of Child Psychology. p. 399. New York, London, 1954.
—, Die Ontogenese des kindlichen Verhaltens (The Ontogenesis of the Child's Behavior), in L. Carmichael's Manual of Child Psychology. New York, London, 1946.
—, and H. THOMPSON: Genet. Psychol. Monogr., Worcester *6*, No. 1 (1929).
—, and L. B. AMES: J. Gen. Psychol. *56*, 247 (1940).
GOERLING: Anat. Rec. (Am.) *19*, 259 (1920).
GOLTZ, F.: Beiträge zur Lehre von den Funktionen der Nervenzentren des Frosches (Essays on the Theory of the Functions of the Nerve Centers of the Frog). p. 86. Berlin, 1869.
GROHMANN, J.: Zschr. Tierpsychol. *2*, 132 (1939).
HARTMAN, C. G.: Anat. Rec., Philadelphia *19*, 251 (1920).
HEBERER, G.: Die Evolution der Organismen (The Evolution of the Organism). 2nd ed. Stuttgart, 1954–59.
HESS, W. R.: Das Zwischenhirn (The Diencephalon). 2nd ed., p. 72. Basel, 1954.
HESSE, R.: Naturwiss. *6*, 322 (1918)

HOOKER, D.: Genetics and the Inheritance Integrated Neur. and Psychiatr. Patterns, *33*, Proc. Assoc. Res. in Nervous and Mental Disease, p. 108 (1954).

–, Evidence of Prenatal Function of the Central Nervous System in Man. James Arthur Lecture. p. 19. New York, 1958.

JUNG, R.: Handbuch der Inneren Medizin (Handbook of Internal Medicine). 4th ed., V, 1, p. 99. Berlin-Göttingen-Heidelberg, 1953.

KANT, I.: Über Pädagogik (1803). Taschenausgabe der Philosophischen Bibliothek (On Pedagogics – 1803. Pocketbook of the Philosophic Library). p. 25. Leipzig.

–, Idee zu einer allgemeinen Geschichte in weltbürgerlicher Absicht (1782). Ges. Schriften (Idea on a General History from World Citizen's Viewpoint – 1782). Collected Writings. 8, 19. Berlin, 1912.

KELLOGG, W. N. J.: J. Genet. Psychol. Monogr., Worcester *58*, 307 (1941).

KINGSBURY, B. F.: Anat. Rec., Philadelphia *27*, 305 (1924); *33*, 73 (1926).

KLAATSCH, H.: Der Werdegang der Menschheit und die Entstehung der Kultur (Evolution of Mankind and the Origin of Culture). 2nd ed. Berlin, 1922.

KOEHLER, O.: In H. Friedrich, Lebendiges Wissen. Sammlung Dieterich (In H. Friedrich, Living Knowledge – Collection Dieterich). Vol. 99 p. 251. Wiesbaden, 1953. Sprache und unbenanntes Denken, in Fondation Singer-Polignac, l'instinct dans le comportement des animaux et de l'homme (Language and Non-Verbal Thinking, in Fondation Singer-Polignac: The Instinct in the Behavior of Animal and Man). p. 647. Paris.

LANGWORTHY, O. R.: Contr. Embryol. (Am.) *24*, No. 139, Carnegie Inst. Washington, 1933.

LASCH, W., and M. FISCHER: Klin. Wschr. *1924* II, 1362.

LOEB, JACQUES: Einleitung in die vergleichende Gehirnphysiologie und vergleichende Psychologie der wirbellosen Tiere (Introduction to the Comparative Cerebral Physiology and Comparative Psychology of the Invertebrates). p. 156. Leipzig, 1899.

LORENZ, K.: In G. Heberer, Die Evolution der Organismen (in G. Heberer's The Evolution of the Organism). 2nd ed. p. 131. Stuttgart, 1954.

MCGRAW, M. B.: J. Pediatr. (Am.) *15*, 485 (1939); *18*, 385 (1941).

–, A Study of Johnny and Jimmy. New York, 1935. According to a reference in J. Educat. Research, Nov., 1936, by Pratt.

PREYER, W.: Spezielle Physiologie des Embryo (Special Physiology of the Embryo). Leipzig, 1885.

POULSSON, H.: Zschr. Tierpsychol. *8*, 394 (1951).

SCHALTENBRAND, G.: Dtsch. Zschr. Nervenhk. *87*, 24 and *89*, 92 (1925).

SHIRLEY, M. M.: In Murchison's Handbook of Child Psychology. 2nd ed., p. 236. Worcester, Mass., 1933.

SIGISMUND, B.: Kind und Welt (Child and World) (1856). 2nd ed. Braunschweig, 1897.

SPALDING, D. A.: according to Eibe-Eibesfeldt: Naturwiss. Rundschau *9*, 86 (1956).

SPATZ, H.: Hdb. d. normalen und patholog. Physiologie, Vol. 10, p. 416. Berlin, 1927.

SPITZY, H.: Die körperliche Entwicklung des Kindes (The Physical Development of the Child). 2nd ed. Vienna, 1926.

THOMAS, E., and F. SCHALLER: Naturwiss. *41*, 557 (1954).

THOMAS AQUINAS: Summa Theologica. Vol. 6, p. 28 (91,3). Munich-Heidelberg, 1941.

TRETTIEN, A. W.: Amer. J. Psychol. *12*, 1 (1900).

VARIOT, G.: Bull. Soc. Anthrop. Par. *7*, 128 (1926).

VIERORDT, H.: Der Gang des Menschen (The Gait of Man). Tübingen, 1881.

VIERORDT, K.: In Gerhardts Handbuch der Kinderkrankheiten (in Gerhardt's Handbook of Pediatrics). Vol. 1, p. 186. Tübingen, 1877.

YERKES, R. M.: Chimpanzees. p. 55. New Haven, 1945.

6. MOVEMENTS

(1, 3, 11, 17, 29, 37, 71)

Since a great number of reflexes can be elicited in the young infant while "higher" brain activity cannot yet be observed the conclusion has been drawn that the infant's brain activity consists only of reflexes.

This opinion does not at all do justice to the facts. At each stage of development there exist functions that occur without external stimulation. The stimulations on which these functions are based develop in the centers simply by virtue of the fact that the centers have remained inactive for a time. To use a comparison from the theory of electricity: During their inactivity the centers have been charged and finally press for a discharge "by themselves," i.e., without external stimulation.

Even the behavior of the unicellular animal is not regulated only by external stimuli (ALVERDES, 1938). If a central nervous system exists, the rhythmic centers especially act by themselves. Thus, the respiratory center is susceptible to a high degree to influences of external and internal nervous stimuli. But in principle it works automatically. It does not stop functioning when all afferent nervous pathways are cut (WINTERSTEIN, 1911). Its anatomic neighbor, the sucking center, functions similarly. It is true that sucking movements appear mostly after the introduction of the nipple and the instillation of a few drops of milk into the mouth, but they may also occur without these stimuli when the infant is hungry. As VON HOLST (1939) has shown in the example of the rhythmic fin movements of the bonefish, automatic rhythmic stimuli occur in the central nervous system that cause coordinated, salutary, and purposeful processes of stimulation without any sensory stimulation pathways. VON HOLST demonstrated such processes even in higher mammals. Similar phenomena are the "vacuum activity" reactions (see p. 595), a special form of instinctive action that LORENZ (1937, 1955) and his co-workers described in the animal and that they specifically associated with VON HOLST's findings: If the eliciting stimuli are absent for any length of time, the stimulation threshold is lowered to such an extent that the instinctive function finally is released even without stimulation and because of this cannot attain its goal. Thus, TINBERGEN and KUENEN (1939) described the movements occurring when young thrushes opened their mouths wide, i.e., a sudden jerking up of the head and opening of the beak elicited by the arrival of one of the parents in the nest. If the young are hungry, this movement may occur as a "vacuum activity" reaction, i.e., without external cause.

The centers of motion behave in a similar manner. Even if they are usually reflex stimulated, they do not always need an external cause to be excited but often function "by themselves" without such a stimulus.

The periods of mobility and immobility in the infant do not exactly coincide with the periods of waking and sleeping. Graphic tracings are best for the continuous observation of movements. IRWIN (1932) continuously demonstrated and counted the movements of newborns for days. He found great differences among infants. The mobility while awake was six times as great as in sleep. These movements are probably related to the "Stäupchen."

FORMS OF MOTION

"Twitches and Flutters" * (Involuntary Sleep Movements)

(29)

It has long been known from clinical observation that a certain motor restlessness, populary called "Stäupchen," is characteristic for sleeping prematurely born infants. ZIPPERLING (1913) described these states as follows:

> "At first the infant lies quietly; suddenly the eyes are rolled and turned completely inward and then again turned into every imaginable position; in between there may be short lid spasms, lightning-like twitches of the orbicularis oculi muscles, and in some infants a twitching of both corners of the mouth, lasting a few seconds. The attacks may occur sporadically or may frequently recur."

According to ZIPPERLING, who regards these as nonpathological phenomena at this developmental stage, the movements are found in almost half of all infants during the first months of life. If more sensitive investigating procedures are employed, they can always be observed. But even the mere observation contradicts YLPPÖs assertion (1931) that prematurely born infants lie motionless and quiet all day long.

The fluttering movements of the face may have the appearance of a facial expression, particularly that of laughing (see p. 128). Thus, the statement of ARISTOTLE can be understood, i.e., that infants cry and laugh at night.

ECKSTEIN and PAFFRATH (1928) recorded on a kymograph the trunk movements of prematurely born infants and of infants born at term. They found that young premature infants were very rarely completely motionless. Periods of comparative rest, in which movements occurred only from time to time, were more frequent. The periods of mobility were irregularly spaced, contrary to the findings of SZYMANSKI (see p. 495). Occasionally periodic movements were observed during periods of calm.

According to ZIPPERLING, the movements are supposed to occur almost exclusively in the face. He probably arrived at this conclusion because he could see only the faces of the premature infants, for they were completely wrapped in warm covers. I made the movements of other body parts visible with the aid of air-filled rubber

* These are the closest American colloquial approximations to the very local German colloquialism "Stäupchen"; the Bavarian equivalent is "Frieseln"—Ed.

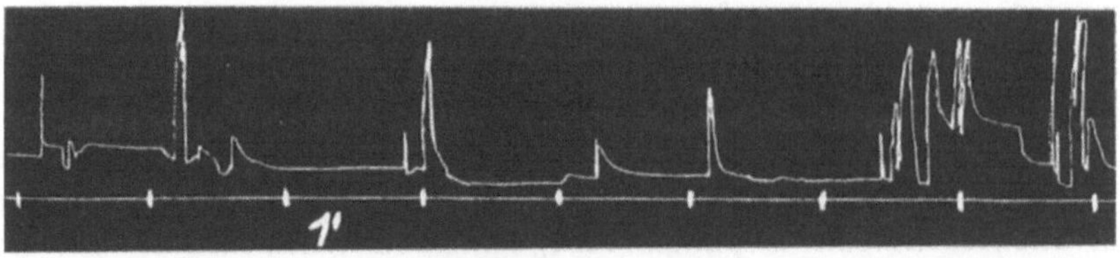

Fig. 100. Periodic movements of the feet of a sleeping prematurely born infant (weighing 1600 gm).

bags fastened to the body parts to be examined, for example, the chin, eye, palm of the hand, or sole of the foot. The most minute movements were thus transmitted to the rubber bags; their pressure oscillations were in turn transferred to a drum. In this manner premature infants and infants born at term were examined in their sleep.

It was found that the periods of complete rest were comparatively short. Sometimes a certain regularity in the change from rest to movement was undeniable (Figs. 100–102); at other times the movements were absent. The reproduced tracings (Figs. 100-104), which come from different body parts, do not differ from each other. From this it can be concluded that these movements do not occur only in the face. In clinically healthy premature infants they could always be demonstrated in the first months of life but in somewhat older infants they were not so frequent.

They were especially frequent in an impaired state of health. In an incubator the infants are lightly dressed and therefore can move unimpeded. Under these conditions a continuous motor restlessness can be shown during waking and sleeping; this corresponds to the tracings reproduced here.

To term all these movements "seizures" appears unwarranted. They have no relation at all to the epileptic seizures occurring later; rather, they permit us to give an absolutely favorable prognosis. It may be that the developing higher centers by which they are later inhibited are not yet completely functioning so that their inhibitory influence is effective at times and then disappears again.

Athetoid Movements

(11)

For the prematurely born, and to a lesser degree also for those born at term, athetoid movements of hands and feet, such as were described as signs of a striatal disease later in life, are characteristic (C. and O. Vogt, 1919, 1920; O. Foerster 1913, 1921). In the newborn they can be attributed to an immaturity of the corpus striatum, whose inhibitory function has not yet developed.

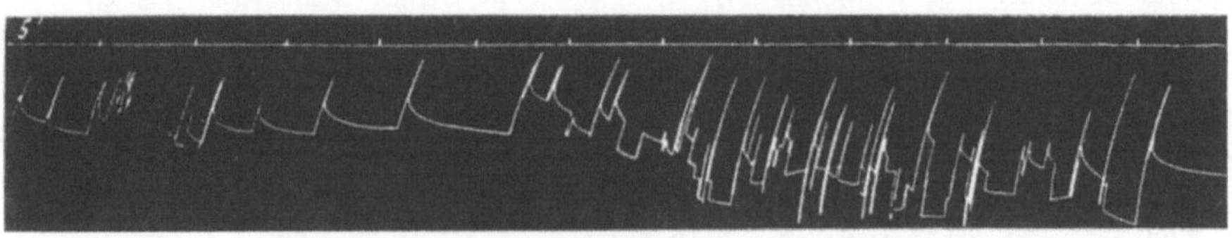

Fig. 101. Foot movements of a sleeping prematurely born infant (weighing 1200 gm).

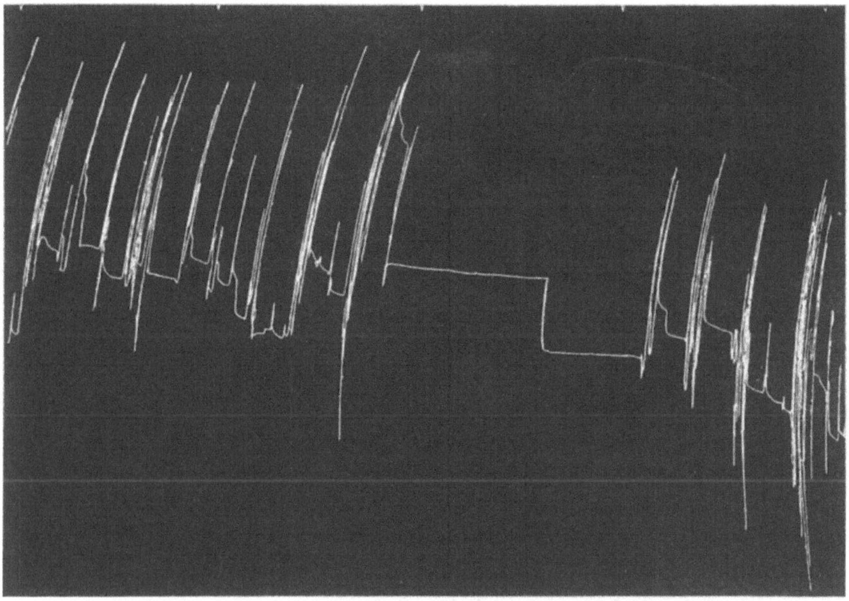

Fig. 102. Periodic hand movements of a sleeping prematurely born infant (weighing 1200 gm). (Time scale: 5-minute intervals.)

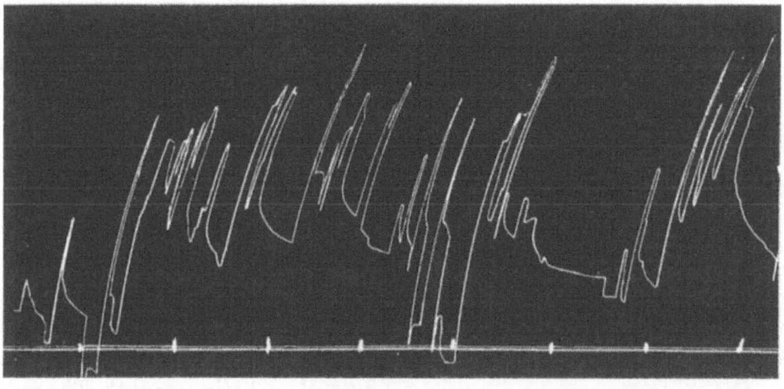

Fig. 103. Hand movements of a sleeping prematurely born infant (weighing 1200 gm). (Time scale: 1-minute intervals.)

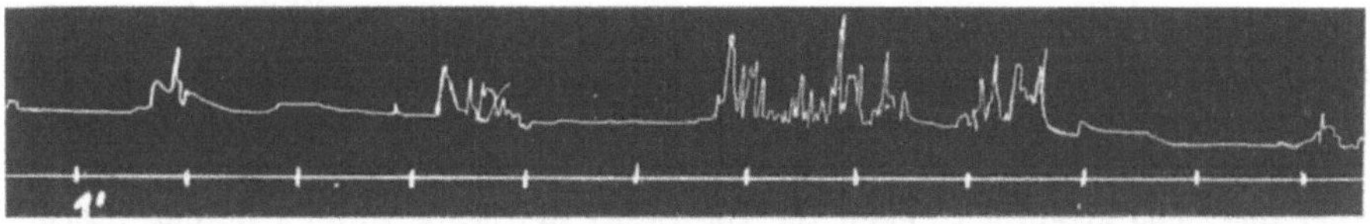

Fig. 104. Eye movements of a sleeping prematurely born 4-day-old infant (weighing 1200 gm). (Time scale: 1-minute intervals.)

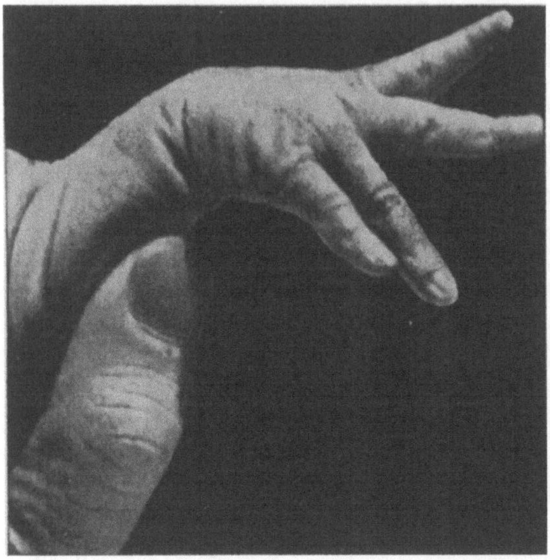

Fig.105. Athetoid extending, bending, and spreading movements of the hand of a prematurely born infant.

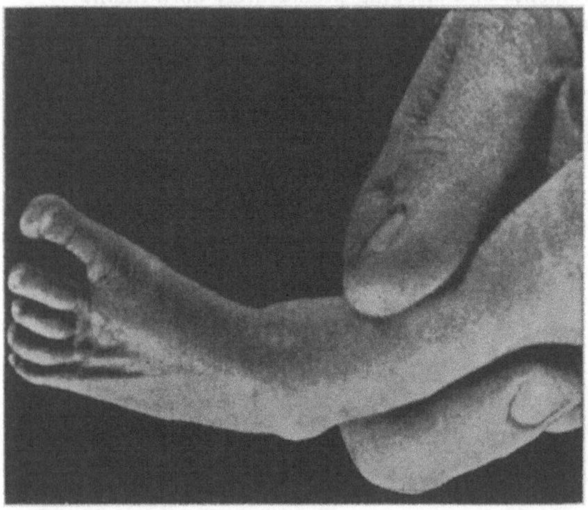

Fig. 106. Athetoid movements (abduction of the big toe) in a prematurely born infant.

If a healthy premature infant weighing approximately 1500 to 2000 gms is taken out of bed, an interplay of movements begins: The arms are flexed at the elbows almost at a right angle, and the forearms are markedly pronated so that the palms are turned outward. The wrists are extended. The individual fingers make peculiar extending, bending, and spreading movements. At the same time they are markedly overextended toward the back of the hand, as Fig. 105 shows. This interplay of movements attains its special characteristic by the movements of the fingers, which move independently of each other. Thus, the index finger is overextended toward the back of the hand, as seen in Fig. 105, whereas the little and ring fingers are flexed. Both hands are placed near the face and execute light wiping movements. In contrast to the later mass movements, the athetoid movements of the fingers are executed at strikingly slow speed and with only little strength. The individual positions are often maintained for some length of time and are frequently repeated.

Not quite so distinct, but still striking enough, are the movements that take place in the feet. Children are able to execute individual movements that as a rule the adult can no longer make. The ability to abduct the big toe as far out as is in Fig. 106 is later lost. The big toe can also be extended or overextended independently of the other toes. The attempt to elicit the Babinski sign leads to an athetoid interplay of the movements of all the toes, during which activity the big toe is only occasionally overextended (see p. 285).

The athetoid interplay of movements in the adult has been described in detail by FOERSTER. Some of the illustrations of congenital generalized athetosis that he includes correspond exactly to those reproduced here, only his patients often showed athetoid movements also in the knee, elbow and shoulder joints, trunk, etc., and these have no counterpart in premature infants. That FOERSTER's patients suffering from congenital generalized athetosis were prematurely born is not merely a coincidence.

Important conclusions about the function of certain brain regions can be drawn from the athetoid interplay of movements in the premature and in the congenital generalized athetosis of later life (FOERSTER, p. 265).

Even if it is not decisive, it should be stressed that, according to C. and O. VOGT, the pallidostriate connecting fibers are not myelinated in the newborn, while the structures of the pallidum show myelin sheaths early. These findings point to a different starting time for the functioning of the pallidum and for the striatum. MINKOWSKI states that the first movements of the fetus resemble athetoid movements (see p. 542).

Mass Movements

(1, 17)

The movements that best characterize the peculiar interplay of motion in the infant, because they are so different from those of the adult, were previously termed by me "reflex-like spontaneous movements," but now I would prefer the expression "mass movements," a term used in the American literature. They probably are approximately the same movements as those described by PREYER (1895) as "impulsive" movements. They are also called "choreatic" movements. The choreatic movements, though, are always restricted to individual muscle groups; voluntary, goal-directed individual movements that are always preserved in chorea, though twisted and distorted, are not present in the young infant.

The mass movements usually occur bilaterally but are often not completely symmetrical, since not all limbs happen to be at the same stage of the movement sequence. If, for example, one leg is extended at the knee and hip, the other will be flexed at the same time and vice versa. The kicking movements in the dorsal position, in which the arms also participate, are completely absent in the adult's interplay of movements. Furthermore, we have the characteristic phenomena of the symmetrical jerking movements of arms and legs that occur in the MORO reflex (see p. 153) and in the fright reaction; they are often elicited by the infant himself through a movement of the head that shakes the inner ear, but can also occur without apparent external stimulus. If the head is turned sideways, the movements become asymmetrical because of the influence of the tonic neck reflex of the limbs. The most striking movements occur in the big joints, i.e., the hip, knee, and elbow joints. The opening and closing of the hands are less distinct.

The head is frequently overextended backward. Crying children often stiffen and thus assume a distinctly opisthotonic posture. In the first months of life the head is bent forward only for the intake of food.

The phenomena described here as individual movements can naturally blend in the most varied ways, but on the whole the basic components always remain the same. Quite often they are also combined with true reflexes, as previously described. Thus, the infant's interplay of movement is comparatively monotonous, quite contrary to later life.

I cannot share FOERSTER's opinion that the mass movements of the infant are really climbing movements (see p. 219), just as I can see little in them reminiscent of the quadruped's locomotion.

Completely missing in the motor interplay of the infant is the voluntary, goal-directed, purposeful individual movement that occurs under the influence of the cortical centers and the pyramidal tract. The mass movements of the infant are awkward and abrupt and follow each other without connection; it makes no difference whether they occur because of an external stimulus or not. In contrast, the movements of the adult result one from the other in flowing transitions. An intended movement is supported by the muscle action of the whole body, always with the intention of expending as little strength as possible. The infant, on the contrary, is not at all economical with his strength; one could almost believe that the squandering of strength is the purpose of his movements.

The mass movements of the infant are regulated extrapyramidally, i.e., from the pallidum, which is not yet inhibited by the corpus striatum. Mass movements may still occur in the adult when he is suffering from extreme pain.

Voluntary Movements

The activity of the motor areas of the cortex and of the pyramidal tract becomes evident in the first voluntary and goal-directed movements, i.e., in the grasping movements toward a certain object within the visual field at approximately the fourth or fifth month of life. From this time on the extrapyramidal mass movements are gradually replaced by voluntary movements, which depend on the pyramidal tract and the cortex; many supporting movements of the other limbs or of the whole body are still retained. The infant does not yet grasp a visualized object with one hand only but with both hands together and may also even move his legs. His movements are explosive and clumsy. With the further development of the corpus striatum and the pyramidal tract during the first year of life the movements become increasingly economical, skillful, and thus similar to those of the adult. But even the small child still moves clumsily and awkwardly; later the individual movements become increasingly smoother and more balanced. The command over his own body is no longer difficult, and thus from the fourth year, according to HOMBURGER (1923), the movements appear to the eye of the adult as being soft, flowing, and graceful.

Grasping

Man shares with some animals, for example, apes, the faculty of grasping a seen object with his hands. This faculty is not yet present in the newborn, but develops through different intermediate stages in the course of the first year of life.

When touched in the oral region the infant grasps with his mouth in a goal-directed manner to seize the nipple. This reflex searching for the breast with the mouth (rooting reflex, see p. 404) is a phylogenetic vestige: Fish, amphibia, and reptiles grasp only with their mouths. The most delicate grasping movement that W. N. and L. A. KELLOGG (1933) observed in a female chimpanzee was done with the lips. She evidently could feel and handle small objects with her lips better than man can do. At first, when she still was walking on all four limbs, she carried objects in her mouth with greater ease than in her hands.

The neonate's hand grasps when the palm is merely touched by an elongated object (tonic palmar grasp reflex, see p. 158). The development of this reflex during the fetal period has been described by MINKOWSKI (1928) and HOOKER (1938) (see p. 544). Here a phylogenetic vestige is also involved, since newborn apes hold onto the fur of their mothers with this grasp.

The habit of taking objects held in the hand into the mouth for further examination is noted early in life. If the infant is used to having a rubber nipple in his mouth as a pacifier, he soon learns to put it in his mouth by himself.

The coordination between the eye and the hand, i.e., grasping with the hand an object that is merely seen but not touched, this goes beyond the more or less reflex-like movements. This is the first movement corresponding to the voluntary movements of the adult; it appears in approximately the fourth or fifth month. At first many unnecessary accompanying movements are made together with the grasping attempts. To grasp the object both arms and legs are moved simultaneously with lively muscle activity; often both hands do the grasping. The movement is clumsy and explosive so that several attempts are necessary to reach the goal. The unnecessary accompanying movements are suppressed only gradually.

BARUCK, LEROY, LAUNAY, and VALLANCIEN (1953) described the development of "voluntary" grasping in detail. They regard "spontaneity" (i.e., occurrence without apparent external cause), direction to a certain goal, and exact execution as signs of voluntary movement. The first stage consists of a platonic interest in the object seen, but without any action. In the third month jerky eruptions of affect occur as a "psychomotor" expression of joy when the object is seen. In the third or fourth month these affective eruptions of movement are directed toward the object, but they are still in the form of asymmetrical endeavors that do not reach the goal. At seven and a half months the first "really voluntary" movement occurs; thereby the desire (the affective part), the initiation of the movement toward the goal, and finally the exact realization, as an expression of affectivity, of reason (?), and of all psychomotor means of execution, are all connected and coordinated with each other.

The development of true grasping movements was investigated in detail by HALVERSON (1937) and CASTNER (1932) with the aid of motion pictures. They made the infant grasp for a ball or a die. Both investigators arrived at identical conclusions. According to HALVERSON the maturation of the grasping movements progresses from the larger muscles of the shoulders and the upper arm to the smaller muscles of the wrist, thumb, and index finger. CASTNER distinguished four stages:

1. Closing of the entire hand, a clumsy movement in the earliest stage: The infant succeeds in touching the object but only rarely in grasping it.

2. Grasping with the palm of the hand: The fingers are flexed downward so that they press the object toward the proximal part of the palm (thirty-second week).

3. Scissor-like closing: The thumb presses the object toward the index finger (forty-fourth week).

4. Pincer-like grasping: The object is held between the tips of the index finger and thumb (fifty-second week).

In the course of the infant's development the object gradually moves from the inner portion of the hand, with which it was at first grasped, to the tips of the fingers and at the same time in a radius-directed movement to the index finger and thumb.

Before the extremities are moved on one side only, the movements occur bilaterally; in most cases arms and legs are even moved simultaneously. According to CURSCHMANN (1906), in early childhood most voluntary movements of one side are accompanied by involuntary movements on the other side. These accompanying movements are reduced or suppressed in the course of development whenever they no longer serve any purpose but they may be released again later. Thus, CURSCHMANN usually found strictly isolated bilateral accompanying movements in the hands of all children under the age of ten years. They were more frequent and more common in younger than in older children. These accompanying movements of the opposite side appeared in younger children immediately during the first practice session, decreased after the stipulated and often unaccustomed movement had been learned, but increased again when the children became tired. Especially the spreading movements of the fingers and adductions of the thumb and of the other fingers caused accompanying movements on the other side. If the movements met resistance, the accompanying movements increased. The age limit for distinctly visible consensual movements on the opposite side surely goes beyond the twentieth year. The inclination to execute these accompanying movements is diminished or extinguished in children who play the piano, since they are accustomed to using their hands asymmetrically.

Right-Handedness

Right-handedness has been acquired rather late phylogenetically, for it cannot be proved that apes, including the chimpanzee, give preference to one hand (ROTHMANN and TEUBNER, 1915; LUDWIG, 1932). Only special and frequently repeated actions can lead to one-handedness in some animals (LUDWIG). FINCK (1941) had 30 chimpanzees reach through a wire mesh fence for pieces of fruit. Eighteen of them used the same hand for the 800 grasping movements which each animal had to perform in more than 90% of the trials, i. e, 9 used the right hand and 9 the left; 25 used the same hand in 80% of the trials, 11 the right and 14 the left hand; and 5 (among them two babies) showed only little preference or change in behavior. According to YERKES (1945), the chimpanzee often prefers the same side in actions that demand only one hand or one foot. But an animal which is predominantly right or left-handed can still consistently prefer one hand for certain purposes and the other for other purposes. A predominant right-handedness could not be observed.

Right- and left-handedness of man can be shown to have existed even in the early Stone Age; it may have occurred even earlier. There is no sure sign that left-handedness was considerably more frequent in the past than it is now (LUDWIG).

The question of right-handedness has repeatedly been examined even in the human newborn. Anatomic measurements of the circumference of the right and left wrist, the size of the palm, and the length of the forearm did not result in any tangible evidence of differences between right and left (WATSON, 1924). If one lets the neonate hang by his right or left hand with the aid of the palmar grasp reflex, the performances are equal on the right and left side (WATSON). M. and I. SHERMAN and FLORY (1936) measured the strength of the palmar grasp reflex in the newborn; it amounted to 1732 gm on the right and to 1765 gm on the left, a finding that does not point to a preference for the right hand. WATSON measured the entire motility of the right and the left hands with a special instrument without finding a significant difference. All these investigations show that right-handedness is not innate in man.

The same can be seen in electroencephalographic examinations. SCHÜTZ and MÜLLER (1951) and SCHÜTZ and SCHÖNENBERG (1951) report that EEG findings do not show any dominance of one hemisphere during the first year of life. After the first year a left dominance develops three times as often as does a right dominance.

BALDWIN (1898) reported on the first appearance of right-handedness. In the sixth to tenth month preference for one hand is not noted if grasping does not require great effort; rather, both hands are used at the same time more frequently than one alone. After the seventh to eighth month a preference for the right hand becomes noticeable if the object has to be grasped with greater effort. However, the left hand does not remain immobile but slowly follows the right. According to GESELL and AMES (1947), two pairs of opposite endeavors struggle with each other in the development toward right-handedness: bilaterality and unilaterality, right and left. The following intermediate stages occur: touching of the object unilaterally and with the left hand (sixteen to twenty weeks); bilateral touching (twenty-four weeks); unilateral touching, usually with the right hand (twenty-eight weeks); bilateral touching (thirty-two weeks); unilateral touching during the rest of the first year, at times the right, at times the left side preferred (fifty-two to fifty-six weeks); fast change with considerable use of both hands or the left hand alone (eighty weeks); right hand (two years); considerable ambidextrousness (from two and a half to four years); and from then on right-handedness.

VOELKEL (1913) examined 43 infants during the first year of life; in 50 experiments he tested each infant to determine the preferred hand for grasping. Table VII gives the results.

A preponderance of preference for the right hand after the seventh month and increasing from month to month can be noted in the table.

The conscious and successful education to right-handedness starts early, as observations in everyday life show. The child is repeatedly shown that he has to eat with his right hand and has to offer his right, his "nice" hand, for handshaking.

Table VII

Frequency of Grasping with the Hands During First Year of Life According to Age (%)

Month	4th	5th	6th	7th	8th	9th	10th	11th	12th
Right hand	38	42.4	46.5	52	65.6	77.4	80.8	82.7	88
Left hand	40	42.4	47.5	45.3	27.7	20.5	15.2	10.7	8
Both hands simultaneously	22	15.6	6	2.7	6.7	2.1	4	6.6	4

Accordingly, the use of the right hand gradually increases. BETHE (1925) examined 95 children at the age of one and a half to six years to ascertain their handedness; he tested the handshake, eating with a spoon, pointing to an object with the fingers, grasping for symmetrically arranged candies, lifting of objects from the floor, building with blocks, cutting with scissors, etc. The results are given in Table VIII.

Table VIII: Use of the Right and Left Hand for Grasping in the Second to Sixth Year of Life

Number of Children	Age (years)	Percentage		
		Left	Left = Right	Right
42	2–4	40.5	21.4	38.1
53	4–6	18.9	5.7	75.4

According to Table VIII, the number of originally right-handed children is small and does not surpass those who have a left-handed endowment. In succeeding years the percentage changes in favor of right-handedness. According to BETHE, a considerable portion of left-handed children and of those who have no preference for any particular side become right-handed under the influence of school, and afterward they are erroneously believed to be endowed with right-handedness. LUDWIG disputes the validity of BETHE's conclusions.

SCHAEFER's (1911) findings on school children constitute a continuation of BETHE's investigations to some extent. In a series of 17,083 children he found 692 left-handed children (4.06%), 37 ambidextrous children (0.21%), and 16,354 right-handed children (95.73%).

The number of left-handed children continuously decreased from the youngest to the oldest group. For every group of 100 left-handed children in the lower grades he found only 60 in the middle grades and 55 in the upper grades, a phenomenon that can probably be explained as resulting from education to right-handedness in school.

It has long been observed that certain disturbances, such as stuttering, epilepsy, and moral inferiority, can be found more frequently among left-handed than among right-handed children. For example, SCHAEFER found speech disturbances in 0.7% of the right-handed and in 2.3% of the left-handed population. However, in contrast to these statements, which could be easily augmented (LUDWIG). NADOLECZNY (1926) stated that among the stutterers he observed the percentage of left-handed persons was no greater than in the general population.

BETHE draws attention to the fact that statistically only the incorrigible left-handed adults are considered, but not the much greater number of those who have a disposition to left-handedness. The counterpart of this group, the confirmed right-handers, escape the evaluation altogether. There may be an increased number of inferior people in this group also.

That left-handedness is inherited is statistically proved (LUDWIG). But the question of why most people prefer the right side cannot yet be satisfactorily answered.

Convulsions*

(1, 29)

Convulsions are tonic-clonic movements that occur in attacks and are usually associated with other disturbances of brain function, such as a clouded sensorium and changes in reflex excitability and in the respiratory rhythm. According to FOERSTER (1926), who observed the effect of more than 100 epileptic attacks on the exposed brain, the brain blanches and contracts before the attack and then during the attack a venous stasis with bluish-purple discoloration immediately follows. According to the pathological-anatomic findings of SPIELMEYER (1926) and SCHOLZ (1938), circulatory disturbances in the form of angiospasms occur in the brain even before the onset of the epileptic seizure.

FOERSTER regards epileptic convulsions, i.e., the most important form of seizures, as a reaction to a stimulus and not as a release phenomenon in which the function of lower centers is set free by the loss of superimposed inhibitory centers. Therefore, no lower parts of the respiratory system are released during the severe respiratory disturbance accompanying each attack (see p. 382).

Recently it has been possible to elicit seizures in a number of epileptic subjects — and more frequently in patients with petit mal attacks — by means of intermittent, strong photic stimuli (flickering light, stroboscope). The eyes are exposed to a strong light source that flashes at rapid intervals at a controlled frequency. BICKFORD, DALY, and KEITH (1953) consider frequencies of 10 to 20 per second most effective.

Natural stimuli can also cause seizures at times. Thus, the rapid change of light and shade during fast driving on a tree-lined road may cause seizures as well as looking at the sun through windswept leaves or its reflection on a rough water surface (the breakers of the sea or the water of a glacier). One child succeeded in eliciting seizures in himself by rapidly moving his hand with fingers spread before his eyes when looking at a light source and another by moving his head to and fro in front of a lighted window with panes composed of colored glass (SCHAPER, 1953, 1954).

Chronological age influences the frequency of seizures: they are more frequent in the child than in the adult, and the incidence is greater in infants than in older children. The reasons for this prevalence are known only in part.

In childhood some seizures are associated only with certain diseases. Febrile seizures independent of the basic illness occur at the onset of a febrile infection of

* The electroencephalographic changes in convulsions will not be discussed.

any kind; the same applies to terminal seizures before death. Other diseases that are accompanied by seizures, such as tetany due to rachitis, are more frequent during childhood. The younger the children, the more often meningitis, which is often accompanied by seizures, occurs. The prevalence of epilepsy is less conspicuously related to specific age groups. The seizures often start early in life and may reappear even after an interval of many years. The many other factors that may occasionally cause seizures in children and adults will not be mentioned here in detail. As a rule the cause of the seizure cannot be determined solely by its clinical picture. The trigger point of an epileptic seizure may be any region of the central nervous system, i.e., the cortex, the subcortical centers, or the spinal cord (FOERSTER, 1926). Since we assume that the newborn still lacks functioning cerebral hemispheres, a deeper focus must be considered.

Frequent attempts have been made to connect the child's susceptibility to seizures with the characteristics of his brain function. The first to examine this problem in animals was SOLTMANN (1876–1880). After FRITSCH and HITZIG (1870) had discovered the motor centers of the cortex, SOLTMANN found that these centers were at first not responsive to electric stimulation in newborn dogs, cats, and rabbits; he concluded from this phenomenon that the cortical centers do not function shortly after birth and that they therefore cannot transmit any stimulations and inhibitions to the more deeply situated subcortical centers. Since a human infant in the sixth to eighteenth month of life is thought to correspond to a ten- to twelve-day-old animal when the state of myelinization is considered, the motor areas of the cortex do not function for a period after birth but only later are able to inhibit the more deeply situated centers. To a certain degree the absence of inhibitions is compensated for by the fact that the galvanic and faradic excitability of the peripheral nerves is much lower during the first period of life than later, a finding that C. WESTPHAL (1886) and A. WESTPHAL (1894, 1897) and many others have confirmed. Between the fifth and ninth month the infant must pass through his most unfavorable period of life, since the peripheral nerves are already excitable, but the inhibitory influence of the cortical areas is not yet sufficiently developed. This generates an "increased reflex disposition" or a "physiological spasmophilia"; the seizures are said to occur by reflex response to certain stimuli. The concept of a stimulus is so broad that even diseases of the intestinal or the respiratory tracts, inflammation, fever, and acute exanthemas can be included. The formerly widely held opinion that teething may also cause seizures has been abandoned today.

Later investigators, such as PANETH (1885), BARY (1898), and BECHTEREV (1888, 1889, 1898), were able to confirm SOLTMANN's findings in animals only in part and in part they found the opposite. WEED and LANGWORTHY (1926) repeated SOLTMANN's experiments with great care and exactness on young kittens several hours to 47 days old and agreed with him more than with later investigators. Electrical stimulation of the cortex caused nothing but movements of the opposite front leg in newborn kittens until the age of two weeks. Movements in the region of the muscles supplied by the facialis nerve and the third branch of the trigeminus and of the hind legs could be elicited only in older kittens in this manner. The cortical areas that proved by this process to be excitable corresponded absolutely to those of adult animals. Accordingly, LANGWORTHY (1927) found when he examined young kittens histologically that the cortical motor centers of the forelegs were matured more than the centers of the hind legs and those of the muscles supplied by the facialis nerve. However, neither were these results fully confirmed. HENRY (1943) found that in 28 newborn kittens up to 11 days of age localized reactions of all parts of the musculature (facialis, third branch of trigeminus, fore and hind legs) could be obtained by electrical stimulation of the appropriate cortical areas. The reactions are quickly exhausted and cannot be elicited as often as in older animals.

CRUIKSHANK (1954) summarizes the results of early and more recent investigations on electrical stimulation of the cortex in animals.

1. In general the fore limbs can be stimulated before the hind limbs.
2. The number of trigger points increases with increasing age.
3. The stimulation threshold is greater in younger animals and decreases with age.
4. At first the whole limb is moved; later individual segments respond.
5. It is difficult to find a relation between the development of the (nervous) function and the response to electrical stimulation.

However, the question of how far animal experiments can be applied to the human infant, whose cerebral development is completely different, always remains. C. WESTPHAL (1886), YLPPÖ, PEIPER [1928 *(1)*, p. 58], CATEL, and LIEBE (1940) tried to stimulate the cortical motor centers of young infants electrically through pre-existing cranial defects. The attempt was unsuccessful in all cases. Nor could it be demonstrated in any case that the stimulation was successful later. But these experiments cannot prove anything since the process that caused the cranial defects may also have involved the brain.

According to THIEMICH (1906), SOLTMANN'S theory is outdated. FLEISCHMANN (1877) had made the objection that strong stimuli do not produce any convulsions in otherwise healthy infants; even excessive stimuli, such as burns or peritonitis, cause convulsions only in exceptional cases. When it later became possible to identify clearly the clinical entity of the once so frequent tetany of infants, which then disposed of a great many of the cases of infantile convulsions, the need for further clarification of this issue became less important. Still, the infant's susceptibility to convulsions cannot be sufficiently explained by tetany.

SCHOLZ (1938) also rejected SOLTMANN's theory: The continuous disinhibition of the subcortical centers because of the immaturity of the cortex does not explain the tendency to convulsions in infants, because this tendency increases even after maturation of the brain. According to his pathological-anatomic findings in epileptic patients with brain defects and seizures starting in the first three months of life, cortical damage without involvement of the basal ganglia is by far more frequent so that the pathological condition of the basal ganglia cannot be made responsible for the convulsions. If one could construct a theory about the susceptibility to convulsions based on SOLTMANN's observations, one would arrive at the assumption that the infant has less opportunity to have generalized convulsions than the adult; the infant does not as yet have an excitable cortex, the damage of which so often causes convulsions in the adult. However, the facts do not agree with this theory.

In experiments on animals the immature brain responds more easily with convulsions to the same stimulus than does the mature brain. Thus, NACHTSHEIM (1932) bred epileptic and nonepileptic strains of rabbits and used them for pentylenetetrazol experiments. The spontaneous and the pentylenetetrazol convulsions were identical in their manifestations. A much greater susceptibility to convulsions was found in all young animals than in the old animals. It did not make any difference whether they were epileptic or nonepileptic. WEGMAN (1939) was able to demonstrate the susceptibility to convulsions of young animals in another manner: He warmed cats of different ages in special containers until their body temperatures had risen at least 3°. Young kittens were more easily subject to convulsions at a sudden rise of temperature than mature animals were. If the temperature was increased slowly, the convulsions became less frequent, even after a considerable increase of temperature. Convulsions could always be elicited in young kittens by rapid heating. Moreover, it is known that in infancy, too, accumulation of heat may cause convulsions.

FOERSTER (1956) and W. THEOPOLD and E. FOERSTER (1956) denied that these experiments yielded valid proof. Rabbits with a high threshold to pentylenetetrazol convulsions may show either a high or a low threshold to aminopyrine. Therefore, there is no susceptibility to convulsions per se, but there are only different types of susceptibility to convulsions on certain stimuli. When all sources for error had been avoided as much as possible, the seizure threshold for intravenous pentylenetetrazol injections was no higher in young rabbits than it was in the adult animals. The investigators mentioned therefore regard the theory of the increased susceptibility to convulsions of infants as lacking sufficient proof.

In any case, more convulsions can still be observed in pediatric hospitals today than in those for adults, although with the disappearance of rachitis, tetany of the infant has nearly disappeared. MÜLLER's (1957) investigations of 407 patients of all ages suffering from purulent meningitis also speak in favor of the hypothesis of an increased susceptibility to convulsions in infancy. A statistically significant increased number of convulsions in children, especially in infants, was found as compared with the incidence in adults, while hyperthermia and mortality, which could be taken as expressions of the severity of the illness, were equally distributed in all ages. According to BAMBERGER and MATTHES (1959), four times as many children under the age of four have convulsions as have older children with the same disease.

The susceptibility to convulsions in childhood may also be attributed to the sensitivity of the metabolism, the equilibrium of which is more easily disturbed in early childhood than later in life.

According to the results of experiments on animals conducted by ZIEHEN (1886, 1890), the cortex is involved in the clonic component of convulsions, while the tonic component is caused by an excitation of the more deeply situated centers. ZIEHEN's findings have been confirmed several times by physiologists. Thus, ROTHMANN's (1912) decorticated dog had only tonic convulsions in the limbs three years after the removal of the hemispheres; clonic convulsions occurred, though, in the facialis region. According to TRENDELENBURG, the clonic convulsions SCHOEN (1926) observed in decorticated rabbits had the characteristics of running and jumping movements and therefore cannot be regarded as clonic convulsions. TRENDELENBURG (1926) regarded it as a proven fact that the clonic components of convulsions have to be separated from the tonic components, not only insofar as their external appearance is concerned but also in regard to their origin, the clonic convulsions originating only in the cortex.

However, FOERSTER (1926) came to a different conclusion on the basis of the existing findings from experiments on animals. After elimination of the cortex or thalamus, clonic seizures may be caused by poisoning, although the cloni become slower and weaker under these circumstances. Tonic-clonic convulsions may occur even when the spinal cord is completely separated from the rest of the nervous system.

SPERANSKY (1950) came to a similar conclusion. He completely removed both cerebral hemispheres in a dog in several operative stages. The dog experienced a generalized increase of tonus and then myocloni, i.e., seizures such as are seen in epilepsy. Finally a status epilepticus developed; the dog died in coma 12 hours later. This experiment also proves that clonic convulsions may occur when there is no cortex.

The experience that convulsions in human newborns, whose cerebral hemispheres are not yet functioning, quite often have clonic qualities agrees with this hypothesis. In addition, we can often see a few distinct clonic contractions in the arms during the first days of life when the MORO reflex (see p. 153) is elicited.

WICHURA (1902) observed clonic twitching of the face and an arm in a human anencephalus (without cerebral hemispheres) on three occasions. In a similar infant whom I observed clinically for seven weeks from his birth to his death, a tendency to distinct convulsions of the hands and feet after elicitation of the MORO reflex existed during his whole life.

When judging this evidence we must say that the clinical experiences favor the possibility of clonic convulsions even without participation of the cerebral hemispheres.

According to BAMBERGER and MATTHES (1959):

"Because of the deficient maturation of the cortex and its topically arranged functions in infancy, the localized epileptogenic discharge may still be without its cortical 'keyboard,' and it makes its clinical appearance only after spreading into phylogenetically older subcortical regions of functioning."

It should still be stressed that convulsions always represent a pathological process and can never be explained merely by the immaturity of the brain. I wholeheartedly agree with CATEL (1951), who strictly represents this point of view. When GÖTT (1931) also mentions epileptiform attacks in the multiform circle of the signs of neuropathy, I cannot follow him. We always have to look for a serious illness behind the phenomenon of seizures.

Only the "flutters" of the prematurely born infant (see p. 249), movements that can easily be recognized as such, are not pathological signs; they may be associated with the immaturity of the motor centers. The emotional respiratory attack from the breath holding cannot be counted among the epileptiform seizures mentioned here because it consists essentially of a respiratory disturbance (see

p. 371). At the end of the respiratory arrest during the periodic respiration of the prematurely born a motor restlessness often occurs, but this does not present the appearance of an epileptiform attack. On the other hand, FOERSTER saw epileptic seizures in the respiratory pauses of CHEYNE-STOKES breathing.

Movement Centers of the Infant

(1)

The question of movement centers of the infant arose in approximately the middle of the last century, when the physiologists began to investigate the function of the spinal cord. PFLÜGER (1853), who designated the "mind" as the "sensorium," endeavored to prove the sensory function of the spinal cord. He started, for example, with the well-known fact that a decapitated frog whose thighs are dabbed with a drop of acid wipes the irritated area with the back of the toes of the same foot. If the lower part of the irritated leg was amputated in advance of the experiment, the dabbed spot was wiped with the sole of the other foot. From the purposefulness of these and similar movements, PFLÜGER concluded that they involved sensory and voluntary structures of the animal. He therefore declared that the hypothesis that only the brain is an organ of the sensorium is wrong; according to him, the sensory function is rooted in the whole cerebrospinal organ (that is, also in the spinal cord).

GOLTZ (1869, p. 52) avoids the expression "sensorium." "Should I choose the expression 'sensorium' or 'psychic functions' instead of the word 'mind,' these foreign words could not explain what I want to express any better." The attempt to explain the concept philosophically could not bring greater clarity either. Thus, GOLTZ omits altogether any explanations of concepts in his treatise "on the seat of mental function in the frog."

GOLTZ wanted to know which parts of the frog's body were necessary to achieve those functions that we assume are products of the mind. Extended experiments resulted in the following conclusions: "The brain seems to be the exclusive organ of the mind. All movements of the decerebrate animal can be traced to simple reflex phenomena." PFLÜGER's findings are not the result of any "mental capacity of the spinal cord," but are the result of a central adaptibility on a reflex basis. "It really seems funny to me if I have to state whether I believe in conscious sensations and conscious volition, cogitation, judgment, and memory in a decerebrated frog since the intact animal hardly has the rudiments of all these qualities." LOTZE (1853), who criticizes PFLÜGER's conclusions in detail, stresses that the purposeful movements of the decerebrate animals derive from a type of action dependent partially on reflexes and partially on actions practiced during the animal's lifetime. It therefore would seem completely unnecessary to incriminate the assistance as a vital intelligence, to regard the physical stimulation effect as sensations and sentiments and the reaction to a mechanical provocation as the result of cogitation and volition. Nothing could be found in the examples given by PFLÜGER that would exceed these limitations and that would show movements that the intact animal could not do equally well habitually and without specific consciousness.

VIRCHOW (1871) also opposes the concept of a special mental function in the spinal cord. The decerebrated frog possesses discriminating faculties as to agreeable and disagreeable states of his body that alllow him to remove the acid. However, it is impossible to postulate a special nonanatomic, immaterial agent that feels, thinks, and acts apart from the organic structure of the spinal cord.

PFLÜGER compared the movements of the decerebrated frog to corresponding movements of a sleeping three-year-old child. But from this similarity he does not wish to draw the conclusion that the sensory function of the brain is extinguished during sleep and that that of the spinal cord persists. Both actions are mediated by conscious activity and comprise the output of a sensorium reduced to a lower level.

According to VIRCHOW, the neonate does not in the least display any functions that allow us to conclude that conscious volition or actions exist. When he is hungry, he becomes restless, turns his mouth sideways, and moves his lips. "He searches for his mother's breast." He grasps it and starts sucking. If he is satiated, he lets go and falls asleep. If he cannot find the breast, his head becomes red and he becomes angry and starts to cry. "The newborn child is a magnificent example of an almost purely spinal cord creature; even his cerebral functions are still those of the spinal type."

SOLTMANN (1876—79; see p. 260) concluded that the cortical motor areas of the human neonate are not yet functional. This fact can be confirmed even if his arguments are no longer acceptable today. The extrapyramidal system, the centers of which are in the nuclei of the brain stem, i.e., the basal ganglia, is older than the system of the pyramidal tract and the motor areas of the cortex. Lower vertebrates possess only the extrapyramidal system; in the human infant the pyramidal system is not functioning at first so that he also has to move "extrapyramidally."

The pattern of the newborn's motions shows that the cortical motor areas anp the pyramidal tract are actually not yet functioning: Goal-directed voluntary individual movements, the centers of which are in the cortex, are still completely absent. In lieu of this, mass movements regulated by the pallidum are executed. Later the striatum has the task of inhibiting these mass movements. Since the striatum is not yet able to do this in the early time of life, FOERSTER (1913) named the newborn a "pallidum creature" after his highest functioning brain part. This expression has been fairly generally accepted in the literature. It is only MINKOWSKI who, without giving any detailed reasons, states that the cortex begins to influence the sequence of movements and reflexes shortly before birth.

According to STRÜMPELL (1901), certain muscle synergies, which appear in the adult only when the pyramidal tract is no longer functioning, are characteristic for the first month of life. STRÜMPELL's tibialis sign is one of these phenomena: When the leg is flexed at the knee and hip, the inner edge of the foot is raised through innervation of the tibialis anterior nerve; if the leg is extended, the edge of the foot is lowered toward the planta. Similar synergies can be evoked in the hands: When the fist is closed, the metacarpal part of the hand is overextended at the wrists; when the fist is opened, it is flexed. THIEMICH (1902) found these synergies in infants three to four months old and occasionally even earlier; they later disappear again.

When in their experiments on animals MAGNUS and RADEMAKER inactivated certain brain portions by section or extirpation, they saw reflex groups disappear; they saw others appear as a result of removal of the inhibitions. From these experiments the authors drew their conclusions on the functions of the inactivated brain regions. In newborn rabbits, cats, dogs, and guinea pigs MAGNUS (1923) was able to demonstrate the labyrinthine reflexes needed for the body posture, the righting reflexes of the head and of the body, the neck righting reflexes, and the tonic neck reflexes of the limbs; and he did this, in general, perfectly and completely. These are the same reflex groups that can be found in the human newborn and young infant but that disappear again in man and animal because they are inhibited by the function of higher centers.

Similarly WOLOCHOV (1951) found in newborn rabbits a number of characteristic reactions of the midbrain, such as wiping movements, the motor feeding reflex, scratching reflex, and licking reflex. All these reflexes are inhibited during growth by the influence of the cortex when it begins to function, but they can be elicited again after removal of the cortex.

The characteristics of the electroencephalogram (EEG) of children have been discussed in detail by GARSCHE (1954) in BROCK's *Biological Data for the Pediatrician.* GARSCHE is more cautious than BERGER, the inventer of the EEG (1932), in his statements on the relations of the EEG to the maturation of the brain. BERGER was able to demonstrate cortical potential differences only after the age of six weeks and concluded from this fact that only after the first weeks of life does the cortex begin to function in the newborn. Certainly, close relations between the morphological differentiation of the brain and the development of the function of the cortex may be assumed, according to GARSCHE, but so far it has not been possible to establish a dependence on anatomically demonstrable maturation processes in the brain. Repeated attempts have been made to relate the EEG patterns of the infant to specific maturation processes of the brain, but a final clarification of this question is still pending.

From the behavior of monsters without cerebral hemispheres, who are labeled mesencephali or anencephali and also, less appropriately, hemicephali, a certain insight into the functions of the cerebral hemispheres and of some deeper brain structures can be gained, for all the capabilities found in these monsters must have developed without the influence of the missing parts of the brain. Several such infants have previously been described, but most of the early and many of the more recent statements come from pathologists who placed the main emphasis on the pathological anatomic changes of the nervous system and who saw these infants only after their deaths. In the well-known case reported by EDINGER and FISCHER (1913), only the mother reported on the behavior of the child.

GAMPER (1926) was the first to examine in detail the total behavior of such an infant while the child was alive; he took MAGNUS's very important investigations into consideration and compared them with the anatomic findings. CATEL and KRAUSPE (1930) and SCHENK (1931, 1932) proceeded similarly. In GAMPER's case the maturation of the brain was arrested at the transition from the mesencephalon to the diencephalon. Thus, the mesencephalon was the highest point of development from the beginning and could develop its own capabilities according to its own laws and was not influenced by the function of the higher brain centers.

Of course, one has to make a reservation: These monsters without cerebral hemispheres show not only a developmental defect, i.e., absence of higher brain portions in an otherwise healthy brain, but also have severe brain disease, which has not only arrested the anatomic development but has quite often damaged the already formed and more or less functioning parts of the brain. In addition, some sensory organs or sensory nerves are frequently deformed. Furthermore, it must also be considered that these monsters described in detail by GAMPER, CATEL and KRAUSPE, and SCHENK were observed in detail by these authors only in the last week of their lives. The responsive capacities of such an infant may change considerably with time. This has been shown in such an infant whom I observed and followed clinically for seven weeks and on whom, unfortunately, an autopsy was not performed. Immediately after birth there was a period of great excitability that was greatly diminished with time. It is a common observation that sick infants are usually less excitable than healthy infants.

The infant described by GAMPER, an arhinencephalus with encephalocele, lived three months. He possessed all the systems of the spinal cord and of the medulla oblongata, the pons, and the mesencephalon, but only as far as these systems have their beginnings and ends within these segments. The other systems, which are connected to the higher brain parts, were however only minimally developed or completely absent. The cerebellum can be credited with some gross participation in the nervous functions; possibly some remnants of functioning nervous pathways were left in the thalamus.

Movements without external stimuli occurred mostly in the big joints and did not differ much from those of normal infants. All cranial nerves with the exception of the olfactorius and opticus could be stimulated. Fright reactions were easily obtained on pain, sound, motion, and pressure stimuli. Even the backward movement of the head was not absent. The child was usually lying in the sleeping posture of the infant. The muscle tonus, which was tested for its resistance to passive movements, was also appropriate.

The infant showed a distinct change between waking and sleeping, just as HEUBNER's (1909) two cases did. When he was awake, he kept his eyes open, occasionally showed quite lively facial expressions, and accepted food well. When he was asleep, his lids were closed, his eyes were turned upward in the sleeping position, his movements were sparse and limited, and the general posture was flaccid. At times the infant could be easily aroused from this state, but at other times even strong arousal stimuli had no effect. Thus, it seems that brain structures above the level of the mesencephalon are not necessary for the alternation between waking and sleeping, but this does not mean that higher brain structures do not participate in these functions in the higher stages of development. The same applied to the yawning-stretching reflex, which occurred very distinctly in the child when he was sleeping.

The infant was able to cough and sneeze but could not cry; he only whimpered. Occasionally he even showed a weak smile on his face. When bad-tasting food was offered, he grimaced just as the normal infant does. His ability to suck and swallow and the "oral righting automatism" (rooting reflex) were well developed.

The following position and movement reactions were present: the tonic neck reflexes of the limbs; the labyrinthine righting reflex of the head in the abdominal, dorsal, and lateral positions; the neck righting reflex; the reactions and delayed reactions of the head and eyes on turning; and the MORO reflex. A peculiar reflex should be especially mentioned: Pressure on the thigh induced gradual elevation of the head, followed by the upper part of the body, until the infant sat upright (see p. 171). Curiously enough, the palmar grasp reflex was scarcely noticeable.

The infant was so similar to a normal child in his general behavior that as long as the infant was alive GAMPER assumed that the thalamopallidum system, which is charged with the highest tasks in infancy, must be intact. Because of the early onset of the developmental disturbance (GAMPER), perhaps ancient phylogenetic functions could still develop in the infant's mesencephalon, which otherwise might have been inhibited by the development of higher brain structures.

The infant CATEL and KRAUSPE studied, who died on the seventh day after birth, was at a still lower stage of development. In this case only the spinal cord without the pyramidal tract and only those portions of the posterior cerebral structures influenced by the spinal cord and the cranial nerves were well developed, while the cerebral vesicle had failed to develop in the region of the mesencephalon, diencephalon, and telencephalon. These cerebral portions, the cerebellum and the pons, were replaced by degenerated vestiges. Accordingly, the red nucleus and the corpora quadrigemina were missing too; the optic nerves and the inner nose were deformed.

The infant was usually resting quietly in his bed. If he was aroused for feeding, he yawned frequently. He often yelled with a shrill voice like a cat, and was able to drink from the bottle. His general posture and muscle tonus corresponded to those of a normal newborn. Sometimes his fingers and toes performed spreading and extending movements. He did not respond to photic stimuli. An examination for caloric nystagmus was carried out with a weak stimulus and resulted in a response. If ice was brought to the sole of his foot, a wiping movement was performed with the other foot in a manner similar to that of the decerebrated frog. The infant responded to pain with a fright reaction. Of the reflexes of position and movement, the following were distinctly present: the body righting reflex of the head, the MORO reflex, and BRUDZINSKI's neck sign. The spinal reflex could also be demonstrated. The absence of some other reflexes may be explained by the general decrease of excitability before death; usually the excitability decreases in very sick infants.

Based on his experiments on animals, MAGNUS assumed that the centers for habitual posture are located in the region between the upper cervical cord and the mesencephalon. The body righting reflex of the head, the center of which in animals he thought to be in the red nucleus or somewhat lower, could be demonstrated in this infant. Thus, in man this center should be in the medulla oblongata. Of course, it may be possible that in the course of development higher centers take part in the function of originally lower brain parts.

The case of the yawning center may be similar. CATEL and KRAUSPE's observations show without question that the ability to yawn is still present even when the development has not gone beyond the medulla oblongata. The center for phonation also must be located in this region.

These comparative studies of monsters without cerebral hemispheres again prove the hypothesis that the newborn's cerebral hemispheres are not yet functioning neurologically. In contrast to the anencephalus, to man suffering from cerebral disease, and to the animal whose brain has been artificially mutilated, the newborn's cerebral function without benefit of his cerebral hemispheres is fused into an intact entity.

Physiologists remove the cerebral hemispheres in animals in order to recognize their function through the resulting deficits. But this operation causes shock, which may last for a long time and may easily not even be survived by the weakened animal. Thus, there exists the danger that deficits caused by shock are ascribed to the removal of the cerebral hemispheres. The extent to which these findings in animals can be applied to man is also questionable. In the newborn, nature gave us a creature with inactive cerebral hemispheres; the observation of the infant supplements animal experiments in a most fortunate manner.

Muscle Tonus

In general, the infant's extremities tend to assume a flexor position, which is most clearly expressed in his sleeping posture (see p. 169). This cannot be attributed to the fact that the flexor muscles outweigh the extensors, since THEILE and RITTER (see p. 171) have found that in the newborn the extensors are heavier than the flexors. The flexor posture of the infant, as well as the forced posture of the adult with a striate lesion, is explained by the absence of the striate function.

Newborns, especially premature infants, are very flaccid at first. But if the infant's muscle tonus is tested later by passive flexion and extension of the extremities, we have to overcome a greater resistance than in older children or adults. The resistance usually decreases or increases suddenly. We are therefore dealing with an extrapyramidal rigidity, since the pyramidal tract is not yet functioning.

It is obvious that to explain this physiological muscle hypertonia we should use a theory established by BRONDGEEST (1860) that explains the occurrence of spasms in adults with pyramidal tract deficits. In his experiments on animals BRONDGEEST found that the muscles became flaccid not only when the appropriate nerves were severed but also when the sensory posterior roots were eliminated. According to VON MONAKOW (1897), the pyramidal tract deficit disturbs the nervous homeostasis since the cerebral hemispheres constantly receive stimuli from the periphery; these stimuli cannot be transformed into orderly movements because of the interruption of the pyramidal tract. Thus, the entire stream of reflex-evoked stimuli flows to the lower motor centers and overloads them. It is this state of general irritation which induces the spastic paralysis.

A similar situation exists in early infancy before the pyramidal tract begins to function. In the early time of life, when the cortex is not yet functioning, the lower centers are more burdened by invading centripetal stimulation than later, when a part of the incoming stimuli is diverted to the cortex. The consequence is muscle hypertonia.

Saint-Anne Dargassies (1960)* distinguishes between two types of muscle tonus: (1) the resting tonus or sustained tonus with the attributes of elasticity and passivity, which can be tested; and (2) the tonus of action, which is responsible for the quality of the active movements of the infant, regardless of whether or not these movements have an external cause.

It is recommended that these different types of muscle tonus be distinguished since certain dissociated disturbances relate only to one form. Thus, elasticity may be pathological and the passivity may be normal, or vice versa. The sustained tonus may be decreased, but nevertheless the action tonus may be normal or even increased at the same time, as in mongolism, for example.

When this is known, the clinical examination is simple if one proceeds systematically, segment by segment, in order to evaluate the general muscle tonus and a possible difference between the two sides. For this examination the child lies on his back with all four extremities flexed as in the usual pediatric examination. The movements occurring without external cause and alternately are asymmetrical but equal in quality and frequency as far as the arms and legs are concerned. However, the trunk and neck show fewer movements. If the muscles of the neck, arms, legs, and trunk are successively tested, even the most sensitive infant is not unduly bothered.

Examination at the Cervical Level

The head is cautiously turned to bring the chin near the shoulders, and the resistance and motility of the two sides are compared. If the head is flexed, the extensors offer resistance; but if the head is overextended backward, no resistance of the flexors can be felt. This difference is physiological in the newborn. Pulling the infant up by his wrists is sufficient to prove this difference distinctly; the head remains overextended to the back but we can observe attempts of the infant to flex it. If the infant is brought into a sitting position, after an interval he should be helped to bring his head into an upright position.

Examination at the Level of the Arms

If one tries to extend the forearms, strong resistance will be felt. The arms will rapidly return to the flexed position as soon as the grip is released. This is the expression of hypertonia of the flexors characteristic for this age.

The elasticity of the extensors can be tested, segment by segment, if the motility and resistance are compared on both sides by: (1) bringing the hand near the opposite shoulder, (2) bringing the hand near the same shoulder and measuring the angle attained, and (3) flexing the wrist against the forearm and measuring the angle attained.

In order to examine the passivity tonus the arm is lifted from the base and then released. Normally it does not plop down; this would be pathological in a newborn born at term. Furthermore, the manner in which the hand can be made to

* I am indebted to Mme. Saint-Anne Dargassies (Paris) for permitting me to use the following statements.

dangle against the forearm is tested. If possible the dangling of the whole arm should be tested too. For this test the infant's body and head are held in an erect position. When the infant is repeatedly rocked sideways, his arms normally remain flexed and adducted to the thorax. In pathological passivity one or both arms are abnormally pendulous.

Examination at the Level of the Legs

Elasticity and passivity are examined in the legs in the same manner as in the arms. The feet are brought near the ears. In this test a limited motility may be noted. The angle in the knee joint is approximately 90° in the normal newborn. The passively flexed dorsum pedis should touch the lower leg lightly. The passively lifted and then released leg must not plop down.

The foot, one leg, or both legs, are swung with the infant in the erect posture, just as is done in testing the hands and arms. Whereas the arms are held adducted to the body, the legs are pendulous in the lumbosacral joint. If one leg is held, the free leg executes a passive movement of limited circumference. By comparing the two sides differences in flaccidity or passivity can be discovered.

Examination at the Level of the Trunk

Only the elasticity need be tested. We compare how far the trunk can be passively bent to the right and to the left sides; this is done by bending the pelvis sideways against the thorax. Both angles should be equal. The bending of the trunk is rather limited. If this movement is not possible and the pelvis cannot be brought near the abdomen, we can speak of truncal rigidity. If at the same time the trunk can be passively extended too far, a latent opisthotonus exists.

Trunk and neck tonus are significant indicators of the condition of the body axis. In all disturbances of the central nervous system this knowledge is of greater value than that of the tonus condition of the extremities.

By testing the ability to stand up, the importance of the action tonus can be recognized. The infant is held erect so that the soles of his feet are supported by the table; this stimulus can be further intensified by pressure on the shoulders. Through this maneuver the tonus of the legs is increased; the erect posture spreads from the legs to the trunk and finally to the neck. Thus, a sustained independent erect stance is achieved. The infant can also stand up by bracing himself against the body of the investigator. The trunk is thereby bent forward while the head droops; then the ankle joints are so strongly extended that the infant stands up actively.

This extremely important examination of the muscle tonus is not harmful if one follows the principle of excluding all infants whose conditions are critical.*

Very serious diseases in infancy are accompanied by a decrease in muscle tonus. This phenomenon is especially evident in newborns and premature infants who suffer, for example, from icterus or pneumonia. Nutritional disturbances in infancy also lower the muscle tonus; the decrease of tonus in the abdominal wall is an important early sign of these disorders. However, disturbances due to unbalanced carbohydrate feeding (CZERNY-KELLER disease) increase the muscle tonus.

---

* These remarks are based on SAINT-ANNE DARGASSIES' statements.

Opitz and Isbert (1925, 1926) determined the muscle firmness of older children by using Mangold's method. They found a rise with increasing age. Physical exertion, continued until exhaustion, caused increased muscle firmness. Daily muscle exercises increased the resting values.

Motility and Activity Drives

(73)

"The need to move is inborn in the infant" (Bersot, 1926). I term the inability to suppress all movement for any length of time while awake the "motility drive." This is a special and very characteristic quality of the child's cerebral function. The younger the child, the more distinct this motility drive. It is most pronounced during the first year of life. A healthy infant, awake and lying without covers, keeps moving practically without interruption.

Until recent decades it was the general custom in Germany to wrap the trunk and if possible every single extremity of the infant with long bands so that the infant's movements were almost completely suppressed. This procedure had already been recommended and was described in detail by Soranos of Ephesos in the second century A.D. and it is still practiced today in many countries as can be seen from the compilation of Ploss (1911). No reason for this custom is given. I would place this custom on the same level as the old Egyptian practice of wrapping their mummies in bandages and I suspect that this custom was originally based on ideas of magic. Later swaddling became so generally accepted — even contemporary philosophers think that the newborn is immobile (see p. 8) — that it became superfluous to give a reason. The expression "swaddling child" was for a long time more in popular use than the word "suckling." No one ever succeeded in finding a hygienic reason for the swaddling. But it is self-evident that skin care must suffer when the need for dry diapers cannot always be easily met and long bands have to be unwound before the diapers can be changed. Rousseau had already rejected swaddling — and rightly so. In Germany this custom was finally abandoned at the turn of the century, thanks to the development of pediatrics.

It is not difficult for the healthy adult to keep completely quiet for some length of time. I once tried to trace on a drum the pulsations of the anterior fontanelles of an infant. I succeeded only when the infant was asleep and was thus immobile. However, it was impossible to keep the head quiet by any type of arrangement because the infant became extremely agitated whenever this was tried.

This restlessness is utterly disturbing if one tries to determine the electric excitability of the nerves in the infant. But the task is easier if the child is examined during his meal. Not without reason do we have the popular expression "to calm"* the infant.

Besides the motility drive, an activity drive, which is closely connected with it, occurs even in infancy and causes the child to reach joyfully for objects, to touch

---

* "Stillen" — literally "to calm" or "to quiet" — means in German to breast feed the infant — The translators.

them, to turn them this way and that, to let them drop and to pick them up again, to rub or to rattle them, in short, to play. Evidently he enjoys this activity. A definite purpose for this activity cannot be recognized as long as the infant does not put objects in his mouth and suck on them. Later, when objects are placed one on top of the other and then noisily pushed down again or when they are arranged in different positions in relation to each other, a certain meaning may gradually be gleaned from these actions.

The younger the child, the less ability he has to stay at a particular task for any length of time. While the adult may persevere for hours at the same task, for example, writing, the young child will turn to a great variety of things in quick succession within several minutes' time. He is easily distracted and is not able to work industriously and with perseverance. From the beginning only the intake of food attracts his attention. Poorly endowed children of school age are just as easily distracted and as inattentive as infants are, and feeble-minded children show this behavior to a still greater extent.

The motility drive and the activity drive are still quite marked after the first year of life. As soon as the infant can walk his sphere of activity is enlarged as many new possibilities are accessible to him. Now his movements correspond roughly to those of the adult, only he is still clumsy and awkward. He puts much more effort into his standing and walking than the adult, who has long ago learned to economize his strength. The adult makes many movements unconsciously that the child attempts to do with the total effort of his personality and often even this in vain.

Thus, the child who has just learned to walk does not proceed with a light tread but plods forward with unnecessarily forceful use of his muscles and with close attention, either with courage and determination or with caution and fear, while his arms paddle through the air to maintain his threatened equilibrium.

When first learning a new skill the adult performs each manual movement in a similar way, with much greater effort and exertion than when he has mastered the skill by practice. In addition, the small child definitely enjoys moving about. He will ascend and descend the same steps again and again without becoming tired or being deterred or discouraged by a fall. He seems to be literally attracted by tasks that are still difficult for him. If he is in a happy mood, this motility drive is especially distinct. Whereas the adult would show only a pleasant and cheerful smile, the infant merrily jumps, frolics, and dances about the room.

The numerous hasty and clumsy movements cause many small misfortunes, such as the breaking of an object or the overturning of a filled cup. A classic example is the story of "Fidgety Phillip" in the book *Struwwelpeter*. It is indeed difficult for the child to suppress his urge to move when he first sits at mealtime in the company of adults.

Again and again the durability and cleanliness of the child's clothes are in danger. The mother is at first so proud of the bright white dress her daughter wears at parties or visits, but later the same dress will easily become the cause of punishment for the child. At a certain age boys often have scratches, bumps, and bruises on their knees.

The motility drive decreases only gradually with increasing age. In the fourth to sixth year of life it is still much more marked than in the adult, although the playful moving around becomes definitely smoother, more balanced, and more graceful. Thus, when taking a walk the child likes to cover two or three times as great a distance as the adult does by jumping over every rock, jumping from every ledge, running forward and backward, and making continuous detours.

Children start school at an age at which the motility drive has already noticeably decreased. Nevertheless, enforced sitting at first still creates considerable difficulty. This difficulty is easily underrated by the adult, who does not experience it himself. The long-lasting inhibition of movements during the school sessions is followed by a state of excitement, so that during recess the children rush down the stairs in order to run around in the playground. There, of course, they are forced by foolish teachers to stand in formation and to walk in a circle around the yard. This method of controlling the motility drive serves the needs of the orderly and methodical teacher more than those of the children. The urge to move again becomes evident when the children leave school. It is not necessary to describe the sight of children dashing out of the school building. Not without reason are schools in towns and in the country nowadays made recognizable to the motorist by special warning signs. Children are endangered not so much before school starts but after school ends because of the state of excitement caused by the long periods of sitting still.

PAVLOV (1932, p. 411) had similar experiences in his animal experiments. For some reason a dog was forced to stand for hours in a rack. The dog was fully awake, stood perfectly still and only moved its feet cautiously from time to time. As soon as the experiment was completed and the animal could leave the rack, a violent excitatory phase followed: The dog started to whine and to howl without calming down even after being called or beaten. After a period of rage lasting a few minutes, the dog calmed down and obeyed orders. In this excitement PAVLOV sees an "induction" that lasts for a short time after protracted and intensive inhibition.

Unfortunately today's school does not take the motility drive of the school child into account at all (HELLBRÜGGE, 1958). The excessively long and uninterrupted school sessions, homework, and riding to school force the child at an early age to sit still for long periods of time. To compensate for this, HELLBRÜGGE urges physical activities that periodically break up the whole school day instead of the daily period in the gymnasium desired by others. He finds it desirable, for example, to provide playgrounds with simple gymnastic equipment that can be used by the child himself for his physical development. HELLBRÜGGE and RUTENFRANZ's (1955) comparison of ten- to eleven-year-old school children and industrial workers as to their periods of work and recreation gives food for thought:

"The comparison with the workers proves the situation to be unfavorable for the school children, for the workers have almost three times as much free time at their disposal as the school children have when almost identical periods of stress are compared."

This unfavorable condition was found in Munich; but it also exists in many other places.

Quite often a state of excitement occurs before the child falls asleep. Be it a pillow fight or only the refusal to go to bed, we are again dealing with an "induction," which this time precedes the suppression of movements during sleep. If the small child awakes at night, he usually cries or calls for his mother. It seems that the regular alternation between excitement and inhibition is more easily disturbed in the child's brain than in that of the adult.

By comparison a single child plays quietly, but the play gets noisy as soon as several children join in. The greater the noise, the greater the fun for the children — unless they start fighting. Noise, which is inevitable during the play of normal happy children, may cause great difficulties. It is remarkable how many adults no longer remember their own youth and have no understanding at all of the behavior of children. The adults' demands are based on their own behavior and they take it for granted that children will behave in essentially the same manner as they do. If the children do not live up to the adult's expectations, the children are called "badly brought up." Old people, whose motility drive has been supplanted by a paucity of movement and whose need for rest has increased, are especially fastidious and exacting (see p. 278). Grandparents are in general more sensitive to children's noise than parents are; and if a grandmother has to take care of orphaned children, this alone may cause serious difficulties, which will affect the children even more than the grandmother. Unmarried old relatives, who have had little experience with children during their lifetime, are also very sensitive. If an unlucky star happens to bring them into contact with a family with many children, they soon flee and for a long time afterward they are still horrified by the "bad manners" they have witnessed.

Therefore, it is better for everyone concerned if older people do not live in the same house with small children. The children can become a nuisance, especially in winter, when they have to stay inside in crowded apartments instead of romping around outside. Many landlords are therefore a priori hesitant to take families with many children; the house and the garden suffer less and the other tenants are less inconvenienced when the families consist only of adults.

If a family with many children lives directly above an elderly childless couple, repeated complaints to the landlord and many encounters may easily occur. The fact that many adults are especially sensitive to noise aggravates things still more. They see in the harmless play of the children an intended challenge, against which they have to take corresponding measures. Children are often the cause of differences in opinion with neighbors who have no children of their own but are thus qualified to have very definite opinions about the raising of children; in spite of their good faith they do not have the slightest idea how much reality differs from their views. Since they have no experience of their own, they are convinced of the power of education and demand more from the educator than he can possibly achieve.

The urge to move — just as the pattern of movement — depends on the characteristics of the brain function of the child. We cannot expect children to display the poverty of movement seen in old people, just as we cannot shorten the length of the child's period of sleeping because an adult needs less sleep. Of course, this does not give children the right to romp around noisily and unchecked any time they please. Rather, we have to see to it that adults are inconvenienced as little as possible. The child should be guided to games that he enjoys and that will not cause a revolt of the whole neighborhood. Noisy games cannot and should not be completely prevented but do not have to take place just during the time of the afternoon nap of the adults. The children should, as much as possible, be sent outdoors, where they can run around unimpeded, so that they will come home tired. Unfortunately they are still too young for organized sports at the time when they are especially troublesome for adults.

Not only children, who get most of the blame, but also adults tend to demand too much from their environment. Fights cannot always be avoided. One of the many admonitions that the child continuously hears is: "Can't you ever sit still?" Those admonitions are certainly necessary, but the educator should know that sitting still is more difficult for the child than for him because of the child's nature. Whoever demands that a child always walk, so to speak, on his tiptoes does not do justice to the characteristics of the child. The urge to move is just as natural for the child as the need for rest is for the adult. Therefore, both have to be considerate of each other.

The justified demands that the children's motility drive makes are especially hard to satisfy in big cities. High, densely populated buildings line both sides of the street. The narrow, crowded sidewalk is not fit for play involving much moving around, the middle of the street is dangerous, and stepping on lawns and trespassing on gardens are forbidden and punishable. Small children who cannot be left unattended on the street have absolutely no opportunity to run around. Nearly half of the deaths at the age of three to twelve years are caused by accidents, mostly traffic accidents. At the beginning of our century traffic accidents were almost unknown.

One has only to observe how older children learn to help themselves. Thus, I saw a football game organized in the following manner in a Berlin street that has comparatively little traffic. The children had placed guards who had to announce each approaching car on both sides of the street. As soon as a car approached, the game was temporarily interrupted and everybody fled to the sidewalk. Not infrequently a child has to pay with his life for his motility drive. However, exaggerated care and supervision will deprive him of his greatest pleasures.

Municipal development plans, as well as designs of individual houses, should take into consideration the fact that children have to satisfy their urge to move. One-family houses with gardens are no doubt the best but also the most expensive solution. But even with more crowded living conditions larger or smaller playgrounds or play streets for children could be provided everywhere; they must — or should — be planned from the beginning when a new section is developed.

In Bavaria the child's right to a playground is now legally secured by a new building code: Each house having three apartments or more must provide a playground. These requirements are also in force for buildings already completed, "if it is indicated from the standpoint of health and for the protection of the children" (HARMSEN, 1960).

In addition, the children should be able to romp around extensively outdoors wherever they congregate. Each kindergarten, if it is living up to its name, should actually have a garden. The school playgrounds should be accessible to children in densely populated districts not only during but also after actual school hours. Pediatric hospitals absolutely need outdoor space for their children but today may still lack these facilities. Since nurses in general do not have enough time and knowledge, kindergarten teachers should direct the outdoor games, especially during long hospitalizations such as those for tuberculosis when bed rest is not required. The joy with which children play accelerates their recovery.

In spite of all efforts and in spite of gymnasiums and sports, the child cannot have the opportunities in a big city that he should have. Therefore, it has long since become the custom to leave the city for a time in the summer and to lead a more natural life in the country, in a resort, at the seashore, or in the mountains. Every child living in a big city should have such a period of recreation each year. Among other advantages this also gives him the opportunity to romp around without danger. The seashore is an unsurpassed playground for children.

The urge to move is a sign of health and *joie de vivre* and is therefore altered with the general change in the nervous system during disease. In a serious disturbance, for example, pneumonia, the child, just like the adult, asks to be put to bed, as a rule lies rather motionless, and economizes his strength. Already acquired static abilities are more or less lost, and the muscles become flaccid. When the happy mood disappears, the motility drive goes with it. Only during unpleasant events, for example, a medical examination, will the child desperately protest, fighting with his hands and feet, but he soon will fall back into his state of minimal motility. Some children become very excited during illness, thrash around restlessly, cry or scream, and are unable to sleep. There is no doubt that children who economize their strength so little are in a more unfavorable position and have more difficulties in overcoming their disease than those who sleep through their illnesses.

A temporary phenomenon in many small children characterized by a marked paucity of movements is the state of "animal hypnosis" (see p. 505). The paucity of movements and hypotonia of the rachitic child are also known. They are accompanied by a bad mood and a marked decrease of interest in the events in his environment. The children sit in an almost immobile position in their beds for hours, so that a little wooden block can remain unmoved on such a child's head for a long time, as CZERNY has demonstrated. In a normal child the block would slide down after a few seconds. With recovery the urge-to-move characteristic of the normal child returns.

The impulse to move and to be active is pathologically increased in the "agile" idiot, who occupies himself in a continual restless and more or less aimless manner.

Transformation of Movement Forms after Childhood

The permutation of motility patterns is by no means completed in childhood. As HOMBURGER (1923) has shown in detail in an important publication, each stage of human development, i.e., childhood, puberty, prime of life, declining years, and old age, has its own characteristic form of movement; this depends on the changing influence of the old ontogenetic and phylogenetic motor centers. The transformation does not occur uniformly. Periods of arrest in which the transformation is unnoticeably prepared for are followed by periods in which the progress becomes evident.

Adolescence causes marked disturbances in the harmony of the movements of children, more so in boys than in girls. Activities of everyday life, such as gait, work, and skilled performances, become ragged, distorted, clumsy, "sloppy"; the whole sequence of movements is again awkward and uncontrolled. Corresponding to the sudden mood changes to which children in pubescence are prone, facial expressions are also uncontrolled and exaggerated.

When this period has passed, a long-lasting uniform phase in which movements are executed without unnecessary exertion develops in the middle adult years. Gestures and facial expressions have become fixed characteristics of the personality. The facial expressions of many women remain more closely related to those of the child than the expressions of men do. Incidentally, women are greatly influenced by pregnancy, child birth, puerperium, and climacterium. HOMBURGER distinguishes two groups: virgin-like supple women and those who are matronly sedate and with considerable poise and strength in their bodies,

Old age shows a marked paucity of movement in both sexes; the impression of dignity which this age conveys is caused in no small degree by this paucity of movement. Many old people lose their ability to do several things at the same time, such as walking and talking.

OSERETZKY (1931, 1954) enumerated a great number of tasks for children age four to sixteen years; through these tasks it can be demonstrated – in a manner similar to that of intelligence tests – to what degree the child's motor ability corresponds to his chronological age. These tests have been continuously improved and have also been used in the examination of mentally ill and feeble-minded children. OSERETZKY stresses – and rightly so – that small children often lose their newly acquired faculties during severe illness; they can no longer walk, talk, or care for themselves. During recovery the static abilities reappear first; then – one after the other – the coordination of movements of the entire body and that of the arms and finally the accuracy of individual and chain movements reappear. The onset of stuttering and enuresis is often accompanied by disturbances of motility.

OSERETZKY's method soon became popular. This procedure was supplemented by KWINT (1931), who examined only the active motor components of facial expression. He proceeded from the idea that an especially highly developed play of movements is necessary in order for man to achieve facial expressions of the highest spiritual content. According to GÖLLNITZ (1954), the combined use of both methods yields good results in the diagnosis of early brain damage.

SUMMARY

The movement centers of the young infant become active even without external stimuli. His movements are fundamentally different from those of adults. He is a "pallidum creature." Through the infant's characteristic mass movements it becomes clear that the motor centers of the cerebral cortex, the pyramidal tract, and the corpus striatum are not yet functioning at this stage. They are controlled by the pallidum but are not yet inhibited by the striatum. A comparison of the infant's behavior with that of monsters without cerebral hemispheres proves that the infant's cortex behavior is not yet functioning. The characteristics of the muscle tonus of the young infant can be traced to the nonfunction of the cerebral hemispheres.

The prematurely born infant shows "Stäupchen" (uncoordinated twitching of different muscles, most prominent in sleep) and athetoid movements in addition to the mass movements.

In the older infant the mass movements of the pallidum are gradually inhibited by the striatum and are replaced by voluntary goal-directed movements, which depend on the centers of the cortex and the pyramidal tracts.

The newborn snaps for an object that touches his mouth. An object placed in his palm is tightly grasped. Grasping with the hand for a visible object appears in the fourth to fifth month. The grasping movements are noticeably improved during the first year of life; grasping with the fist is gradually replaced by a pincer-like grasping with the tips of the thumb and index finger.

Unilateral movement is preceded by the bilateral approach. Right-handedness is not innate; after the seventh month the right hand is progressively preferred for grasping. Education at home and at school changes original left-handedness to right-handedness. During the school years the percentage of left-handed children decreases.

The susceptibility to convulsions in the first years of life is caused by an as yet unexplained characteristic of the infant's brain function and has some connection with the immaturity of the brain and the sensitivity of the infant's metabolism.

In general, the younger a child, the greater his motility drive. What inexperienced adults often regard as naughtiness is the expression of a vital need. The child does not obtain what is due to him if sufficient consideration is not given to this need. In the planning of cities, as well as in the planning of houses, kindergartens, and pediatric hospitals, sufficient room for outdoor activities has to be considered. With advancing age the urge to move gradually decreases; this development is interrupted in the period of sexual maturation. Old people display a paucity of movements.

REFERENCES

ALVERDES, F.: Zschr. Tierpsychol. *2*, 258 (1938).

BAMBERGER, P., and A. MATTHES: Anfälle im Kindesalter (Convulsions in Childhood). pp. 16, 52, 458. Basel, New York, 1959.

BALDWIN, I. M.: Die Entwicklung des Geistes (The Development of the Mind). Berlin, 1898.

BARUCK, H., B. LEROY, J. LAUNAY, and B. VALLANCIEN: Arch. frç. Pédiatr. *10*, 425 (1953).

BARY, A.: Arch. Anat. u. Physiol. *1898*, 341.

BECHTEREV, W.: Neurol. Zbl. *1888*, 154; *1889*, 513; *1898*, 148.

BERBERICH, J., and A. WIECHERS: Zschr. Kinderhk. *38*, 59 (1924).

BERGER, H.: Münch. med. Wschr. *1932*, 1636; Arch. Psychiatr. (Ger.) *90*, 231 (1932).

BERSOT, H.: Le rôle de la musculature dans le développement du petit enfant (The Role of the Musculature in the Development of the Small Child). p. 116. Vevey, Montreux, Berne, 1926.

BETHE, A.: Dtsch. med. Wschr. *1925*, 681.

BICKFORD, R. G., D. DALY, and H. M. KEITH: Amer. J. Dis. Childr. *86*, 170 (1953).

BRONDGEEST, P. I.: Arch. Anat. *1860*, 703.
CASTNER, B. M.: Genet. Psychol. Monogr., Worcester *12*, 105 (1932).
CATEL, W.: Dtsch. med. Wschr. *1932*, No. 26; Mschr. Kinderhk. *52*, 1 (1932); *58*, 89 (1933).
—, Differentialdiagnostische Symptomatologie von Krankheiten des Kindesalters (Differential Diagnostic Symptomatology of the Diseases of Childhood). 2nd ed., p. 560. Stuttgart, 1951.
—, and C. A. KRAUSPE: Jb. Kinderhk. *129*, 1 (1930).
CRUIKSHANK, R. M.: In Carmichael, Child Psychology. 2nd ed., p. 195. New York, London, 1954.
CURSCHMANN, H.: Zschr. Nervenhk. *31*, 1 (1906).
ECKSTEIN, A., and H. PAFFRATH: Zschr. Kinderhk. *46*, 595 (1928).
EDINGER, L., and B. FISCHER: Neurol. Zbl. *1913*, 876.
FINCK, G.: Science *94*, 117 (1941) after Ciba-Zschr. *6*, No. 61, 2034 (1953).
FLEISCHMANN, L.: Klinik der Pädiatrik (Pediatrics Clinic). Vol. 2, 154. Vienna, 1877.
FOERSTER, E.: Dtsch. Z. Nervenhk. *174*, 140 (1956).
FOERSTER, O.: Berl. klin. Wschr. *1913*, 1217; Zschr. Neurol. *73*, 1 (1921); Dtsch. Zschr. Nervenhk. *94*, 15 (1926).
FRITSCH, G., and E. HITZIG: Arch. Physiol. *1870*, 300.
GAMPER, E.: Zschr. Neurol. *102*, 154; *104*, 49 (1926).
GARSCHE, R.: Klin. Wschr. *1953*, p. 118.
—, Electroencephalographie, in Brock, Biologische Daten für den Kinderarzt (Electroencephalography, in Brock's Biological Data for the Pediatrician). 2nd ed., Vol. 2, p. 856. Berlin, Göttingen, Heidelberg, 1954.
GESELL, A., and L. B. AMES: J. Genet. Psychol., Worcester *70*, 155 (1947).
GOLTZ, F.: Beiträge zur Lehre von den Funktionen der Nervenzentren des Frosches (Essays on the Theory of the Functions of the Nerve Centers of the Frog). Berlin, 1869.
GÖLLNITZ, G.: Die Bedeutung der frühkindlichen Hirnschädigung für die Kinderpsychiatrie (The Importance of Brain Damage in Early Infancy for Child Psychiatry). Leipzig, 1954.
GÖTT, T.: In Pfaundler-Schloßmanns Handbuch der Kinderkrankheiten (In Pfaundler-Schlossmann's Handbook of Children's Diseases). 4th ed., Vol. 4, p. 399. Berlin, 1931.
HALVERSON, H. M.: J. Genet. Psychol., Worcester *51*, 371 (1937).
HARMSEN, H.: In Der Mensch in der Großstadt. Das Heidelberger Studio (in "Man in the Metropolis," Heidelberg Studio). Suppl. 20, p. 157. Stuttgart, 1960.
HELLBRÜGGE, T.: Landarzt. *34*, 504 (1958).
—, and J. RUTENFRANZ: Dtsch. med. Wschr. *1955*, 1909.
HENRY, W.: Fed. Proc. (Washington) *2*, 21 (1943).
HEUBNER, O.: Charité-Ann. *33* (1909).
HOLST, E. VON: Erg. Physiol. *42*, 228 (1939).
HOLT, K. S.: Pediatr. *57*, 571 (1960).
HOMBURGER, A.: Zschr. Neurol. *85*, 274 (1923).
IRWIN, O. C.: J. Comp. Psychol. (Am.) *14*, 415 and 429 (1932).
KELLOGG, W. N. and L. A.: The Ape and the Child. p. 59. New York and London, 1933.
KWINT: Zschr. Kinderforsch. *38*, 143 (1931).
LANGWORTHY, O. R.: Contr. Embryol. Carnegie Inst. Wash. No. 104, p. 177 (1927).
LIEBE, S.: Mschr. Kinderhk. *83*, 1 (1940).
LORENZ, K.: Naturw. *25*, 289 (1937); Fol. Biotheoret. *2*, 17 (1937); in Heberer, Die Evolution der Organismen (in Heberer's The Evolution of the Organism). 2nd ed., p. 131. Stuttgart, 1955.
LOTZE, H.: Göttingische gelehrte Anzeigen. *3*, 1768 (1853).
LUDWIG, W.: Das Rechts-Linksproblem im Tierreich und beim Menschen (The Right-Left Problem in the Animal Kingdom and in Man). Berlin, 1932.
MAGNUS, R.: Körperstellung (Body Posture). Berlin, 1924.
1928.

MINKOWSKI, M.: In Abderhaldens Handbuch der biolog. Arbeitsmethoden (in Abderhalden's Handbook of Biological Working Methods). V, 5B, Lief. 253. p. 608. Berlin and Vienna, 1928.
MONAKOW, C. VON: Gehirnpathologie. Nothnagels spezielle Pathologie und Therapie (Cerebral Pathology, in Nothnagel's Special Pathology and Therapy). IX, 1. Vienna, 1897.
MÜLLER, KLAUS: Dtsch. Zschr. Nervenhk. *176*, 233 (1957).
NACHTSHEIM, H.: Dtsch. med. Wschr. *1932*, 168.
NADOLECZNY, M.: Die Sprach- und Stimmstörungen im Kindesalter. Hdb. der Kinderheilk. (Language and Phonation Disturbances in Infancy. In Handbook of Pediatrics) by M. von Pfaundler and A. Schlossmann. 2nd ed., Vol. 5, p. 87. Leipzig, 1926.
OPITZ, H., and H. ISBERT: Jb. Kinderhk. *108*, 1 (1925); *111*, 164 (1926).
OSERETZKY, N. J.: Psychomotorik. 57. Beih. Zschr. angew. Psychol. Leipzig, 1931.
—, Aus den Schriften der Sowjet-Union (from Writings of the Soviet Union). 1954. p. 561. Beih. z. Dtsch. Gesd.wes. *9*, No. 45 (1954).
PANETH: Pflügers Arch. Physiol. *37*, 202 (1885).
PFLÜGER, E.: Die sensorischen Funktionen des Rückenmarks der Wirbeltiere (The Sensory Functions of the Spinal Cord of the Vertebrates). Berlin, 1853.
PLOSS, H.: Das Kind in Brauch und Sitte der Völker (The Child in Popular Customs and Habits). 3rd ed., Vol. 1, p. 232. Leipzig, 1911.
ROTHMANN, M.: Neurol. Zbl. *31*, 1287 (1912).
—, and W. TEUBNER: Aus der Anthropoidenstation auf Teneriffa (From the Anthropoid Station in Teneriffe). Abh. preuß. Akad. Wiss., Physik.-math. Kl., Berlin, 1915.
SAINT-ANNE DARGASSIES: Briefliche Mitteilung für die 3. Auflage dieses Buches (Personal Communication for the 3rd Edition of this Book) (1960).
SCHAEFER, M.: Berl. klin. Wschr. *1911*, 295.
SCHAPER, G.: Z. Kinderheilk. *73*, 513 (1953); Mschr. Kinderheilk. *102*, 508 (1954).
SCHENK, V. W.: Zschr. Neurol. *142*, 469 (1932); Een Hemicephal. Proefschrift (A Hemicephalus). Utrecht, 1931.
SCHOEN, R.: Arch. exper. Path. (Ger.) *113*, 157 (1926).
SCHOLZ, W.: Mschr. Kinderheilk. *75*, 5 (1938).
SCHÜTZ, E., and H. W. MÜLLER: Klin. Wschr. *1951*, p. 20.
—, and H. SCHÖNENBERG: Zschr. ges. exper. Med. *117*, 157 (1951).
SHERMAN, M., I. SHERMAN, and C. D. FLORY: Infant Behavior. Comp. Psychol. Monogr. 12, No. 4 (1936).
SOLTMANN, O.: Jb. Kinderhk. *9*, 106 (1876); *11*, 101 (1877); *12*, 1 (1878); *14*, 308 (1879). In C. Gerhardts Handbuch der Kinderkrankheiten (in C. Gerhardt's Handbook of Children's Diseases). Vol. 5, p. 1. Tübingen, 1880.
SORANOS VON EPHESOS: Gynäkologie (Gynecology). German translation by Lüneburg. Munich, 1894.
SPERANSKY, A. D.: Grundlagen der Theorie der Medizin (Foundations of the Theory of Medicine). p. 22 of German translation. Berlin, 1950.
SPIELMEYER, W.: Dtsch. Zschr. Nervenhk. *94*, 54 (1926).
STIRNIMANN, F.: Ann. paediatr., Basel *157*, 17 (1941).
STRÜMPELL, A.: Zschr. Nervenhk. *20*, 436 (1901).
THEOPOLD, W., and E. FOERSTER: Z. Kinderhk. *78*, 218 (1956).
THIEMICH, M.: Zschr. klin. Med. *45*, 226 (1902); Pfaundler-Schloßmanns Handbuch der Kinderheilkunde (Pfaundler-Schloßmann's Handbook of Pediatrics). II, 2, Leipzig, 1906.
TINBERGEN, N., and E. KUENEN: Zschr. Tierpsychol. *3*, 37 (1939).
TRENDELENBURG, W.: Dtsch. Zschr. Nervenhk. *94*, 54 (1926).
VIRCHOW, R.: Über das Rückenmark (On the Spinal Cord). Samml. gemeinverständl. wissenschaftl. Vorträge. Hgb. Virchow u. von Holtzendorff. No. 120, p. 36. Berlin, 1871.
VOELKEL, E.: Zschr. Kinderhk. *8*, 351 (1913).
VOGT, C., and O.: J. Psychol. u. Neurol. *24*, 1 (1919); Ergänz. H. *25* (1920).
WEED, L. H., and O. R. LANGWORTHY: Contrib. Embryol. Carnegie Inst. Wash. No. 87, p. 89 (1926).

WEGMAN, M. E.: J. Paediatr. (Am.) *14*, 190 (1939).
WESTPHAL, A.: Arch. Psychiatr. *26*, 1 (1894); *29*, 474 (1897).
WESTPHAL, C.: Neurol. Zbl. *1886*, 361.
WICHURA, M.: Jb. Kinderhk. *56*, 131 (1902).
WINTERSTEIN, H.: Pflügers Arch. Physiol. *138*, 159 (1911).
WOLOCHOV, A. A. (1951), according to J. M. Kreps, S. D. Piragewa, D. A. Tschetwernikow and L. F. Pomasanskaja: Pawlow-Zschr. höhere Nerventätigkeit. Deutsch. *2*, 67 (1952).
YERKES, R. M.: Chimpanzees. p. 113. New Haven, 1945.
YLPPÖ, A.: Pfaundler-Schlossmanns Handbuch der Kinderheilkunde (Pfaundler-Schlossmann's Handbook of Pediatrics). 4th ed., Vol. 4. Berlin, 1931.
ZIEHEN, T.: Arch. Psychiatr. *17*, 99 (1886); *21*, 863 (1890).
ZIPPERLING, W.: Zschr. Kinderhk. *5*, 31 (1913).

7. REFLEXES OF CLINICAL SIGNIFICANCE

A number of reflexes that do not belong in the same category when judged by the way they receive stimuli and by the course of their reflex arcs are of significance for the diagnosis of pathology in the central nervous system. For practical reasons they will nevertheless be discussed together here. The significance of some reflexes extends far beyond infancy.

Testing of reflexes is more difficult in the infant than in the adult, since the infant does not relax his muscles on demand but becomes easily excited by the preparations for the examination and cries and moves violently. Therefore, a fairly quiet moment should be chosen for the examination. It is safest to examine the child during feeding, since his state of excitement completely subsides and he becomes immobile when he is sucking.

PATELLAR REFLEX, CROSSED ADDUCTOR REFLEX

(1, 1a)

The patellar reflex is always present in normal infants in the first year of life, as FARAGO (1887) and BYCHOWSKI (1908) have already stated. Other investigators, such as CATTANEO (1902), MÜLLER (1911), and BURR (1921), sometimes found it to be absent, but this may have been due to the difficulties in examination that have been described. According to PFAUNDLER (1902), the patellar reflex is often diminished or extinguished during pneumonia in children after the first year of life.

In the young infant the usual elicitation of the patellar reflex regularly causes a second reaction, i.e., the contraction of the adductors of the opposite leg, which is thus moved inward. Both reactions are independent of each other. If for some reason the patellar reflex is absent, the adductor reflex appears alone. Extension at the knee joint, which is characteristic of the patellar reflex, does not occur on the opposite side.

MINKOWSKI (1928) found the crossed adductor reflex even in a fetus only 6.5 cm long. My co-worker BOLL [1944 *(1a)*] observed it in 9 of 17 underweight infants weighing between 1700 and 3000 gm. He attributes absence of this reflex in the other infants to the difficulty in the examination.

SCHOO (1934) examined 112 infants and obtained the following results:

Quarter of first year of life	1st		2nd		3rd		4th	
	20+	39−	12+	24−	2+	4−	4+	7−

BOLL examined 417 infants during their first year of life to determine the presence of the crossed reflex. His results are given in Table IX.

Table IX: Presence of the Crossed Reflex in 417 Infants

Age (months)	Number of infants			
	Total	Positive	Negative	% Positive
1– 6	136	130	6	95.6
7	102	61	42	59.7
8–12	179	0	179	0

The crossed reflex was questionable in 9 of 186 children two to fifteen years old.

According to these findings the crossed adductor reflex is almost always present up to the seventh month; later it is extinguished. In postdiphtheric paralysis the patellar reflex may disappear, whereas the adductor response on the opposite side is preserved.

STRÜMPELL (1879) was the first to describe the crossed reflex in adults. He was able to demonstrate this reflex only very rarely. According to WEGENER (1927), who examined it in detail, the reflex is present in 10 to 50% of healthy persons. The presence of this reflex on both sides is not pathological, but its unilateral presence does not occur in healthy persons. In unilateral lesions of the pyramidal tract the reflex is frequently a concomitant symptom of spastic reflexes. WEGENER regards this phenomenon as a true crossed reflex closely connected with the individual reflexes. This hypothesis has yet to be proved. The reflex is thought to be an expression of a lateral connection and to be always present in latency. WARTENBERG (1952) maintained that the crossed adductor reflex in the adult is completely normal. He was able to elicit it from many points of the opposite leg and usually much more easily than the ipsilateral adductor reflex. The reflex is exaggerated in pyramidal tract lesions.

The presence of the reflex during the first period of infancy and its disappearance in the seventh month suggest the possibility of its being inhibited by newly developing cerebral centers. If it is found still later in healthy persons it must be due to faulty development of the inhibitory mechanisms.

If the skin of the newborn is cautiously pinched at the inner aspect of the thigh, the opposite thigh is suddenly adducted. This "crossed adductor reflex of the thigh" disappears at the age of six to eight weeks (ANDRÉ-THOMAS and AUTGAERDEN, 1959).

ACHILLES TENDON REFLEX

It is difficult to evaluate the Achilles tendon reflex in the young infant because of the smallness of the proportions; nevertheless, it should be possible to elicit the reflex regularly. SCHOO (1934) observed it in 34 of 46 infants up to three months of age and in 20 of 22 infants four to six months of age. W. and I. SHERMANN and FLORY (1936) found ankle clonus in half of the newborns. They attribute these findings to the nonmaturation of the pyramidal tract.

PLANTAR REACTIONS

BABINSKI's Sign

(3)

BABINSKI's sign (1898) consists in an overextension of the big toe when the sole of the foot is stroked. This diagnostically very important sign occurs in the adult in pyramidal tract lesions. According to WARTENBERG (1952), this overextension of the big toe is actually a flexion and as such is the most sensitive component of the flexor flight reflex, which is intended to shorten the leg just as the flexion at the hip and knee joints does. The neighboring muscles react most easily. The reaction becomes weaker in direct relation to the distance of the muscles from the stimulus. Therefore, the entire mass flexor reflex will not always occur, but only its most sensitive component, the toe movement. WARTENBERG's interpretation of BABINSKI's sign as a remnant of the climbing process is discussed on page 221.

HOFF and OSLER (1957) regard BABINSKI's sign as the result of posture and flight reflexes that act in opposite directions: If the sole of a patient with a spastic extensor contraction is stimulated, the flight reflex suppresses the extensor reflex at a certain stimulus strength. The results are: flexion of knee and hip joints, dorsal flexion of the foot, spreading of the toes, and overextension of the big toe. If the stimulus strength is decreased, the reaction also decreases. If the stimulus is decreased still further, BABINSKI's sign occurs.

According to BABINSKI, stimulation of the newborn's sole effects an extension of the toes because the pyramidal tract is not yet functioning. After BABINSKI's report the plantar reaction of the young child received so much attention from neurologists and pediatricians that for a long time it appeared that the behavior of the big toe was the only neurological sign characteristic of infancy. In RICHARDS and IRWIN's (1935) detailed and comprehensive compilation of the literature on this subject, 117 papers in this category were enumerated. Unfortunately it cannot be said that the scientific results are in proportion to the tremendous effort. If the findings of different investigators are compared, contradictions are encountered. These contradictions may be mainly due to the fact that the plantar stimulation will evoke various toe movements in the infant that are difficult to evaluate and that, incidentally, will even change when the stimulus is repeated. If the plantar reaction of an infant is tested, the result is not predictable as it is in the examination of other reflexes. The problem of toe movements after plantar stimulation differs absolutely from that of the patellar reflex or BABINSKI's sign in the adult. These latter reflexes are either present, difficult to elicit, or totally absent. These clear descriptions are not at all sufficient for the evaluation of BABINSKI's sign in infancy. For example, RICHARDS and IRWIN found seven reactions:

1. Flexion of all toes.
2. Extension of the big toe only.
3. Extension of one or of all four little toes.
4. Flexion of one or of all toes.
5. Mixed response to the stimulus (flexion followed by extension or vice versa).
6. No stimulus response.
7. Uncertain responses.

Only the behavior of the toes is considered in this compilation. If further reactions, such as ipsilateral or contralateral flexion at the hip, knee, and ankle joints, are observed, the result becomes even more complicated.

Therefore, we can understand that RICHARDS and IRWIN agreed in only 77% of their evaluations when judging the same reaction. For that reason they spoke of the "personal factor" of the examiner. It is self-evident that the contradictions in the reports of different investigators who have not collaborated will be even greater.

Attempts have been made to reduce the differences by equalizing the stimulus as much as possible. PRATT, NELSON, and SUN (1935) and RICHARDS and IRWIN (1930) devised special instruments for this purpose. However, most examinations were made in the customary manner.

MINKOWSKI (1928) followed the reaction patterns of the toes in human embryos obtained by Cesarean section from the earliest to later stages of development. On stimulation of the soles of embryos (two months of gestation) he found flexion and — two to three minutes later — extension of the toes and at times also of the foot alone, or this was combined with the drawing up of the legs and spreading of the toes; in the fourth to sixth month of gestation, extension, spreading of the toes, and transition from extension to flexion were seen; in premature newborns and newborns born at term extension and flexion were noted. HOOKER (1958) was able to elicit the first reflexes on plantar stimulation at ten to ten and a half weeks of gestation; the response consisted of plantar flexion of all toes. Mixed responses (flexion or extension) were first observed at eleven and a half weeks of gestation; they were often accompanied by flexion and subsequent extension of the knee. From twelve and a half weeks of gestation until the time of normal birth and beyond that time the BABINSKI position of the big toe was noted; later an upward (dorsal) directed and fanning position of the other toes was observed.

According to ENGSTLER (1905) and FELDMAN (1922), the pattern in the premature infant during the first week is similar to that in the infant born at term. Later a certain tendency to extension of the toes seems to exist in the prematurely born.

LANTUEJOUL and HARTMANN (1923) found flexion in 29 newborn infants immediately after birth. These authors examined 131 newborns 1 to 45 minutes postpartum. According to their findings flexion usually replaced extension after a few minutes.

Even at rest the young infant shows athetoid movements of the toes and also of the fingers; they have been previously described and illustrated (see p. 252). By plantar stimulation these movements can be increased or elicited if they were previously absent. The reaction does not consist of an individual movement of the big toe; rather, the toes make several movements successively until they return to the rest position. They are irregularly flexed, extended, or spread during this process;

the big and little toes may be abducted also. No one who observes young infants would ever for a moment think that the individual movements of the big toe have any significance different from that of the movements of the other toes. Therefore, it seems pointless to me to count the frequency with which a certain toe movement may be found at different ages. The existing contradictory statements (such as those of CATTANEO, 1902; ENGSTLER, 1905; MÜLLER, 1911; GALANT, 1917; BURR, 1921; BERSOT, 1920; MINKOWSKI, 1928; WOLPERT, 1926; DE BRUIN, 1928; ZADOR, 1927; PRATT, NELSON, and SUN, 1930; RICHARDS and IRWIN, 1935; W. and I. SHERMAN and FLORY, 1936; DIETRICH, 1957; and many others) do not permit any established rule. The various investigators agreed on only one finding, i.e., that the frequency of reflex extension of the toes gradually diminishes with increasing age.

The cautious conclusion that RICHARDS and IRWIN have reached, based on the existing literature and on extensive investigations of their own, agrees with my own views: The plantar reactions fluctuate and are inconsistent at birth and probably also during fetal life. A distinct toe reaction, at first having a slight tendency to extension and later (at the age of six to eighteen months) perhaps also to flexion, develops step by step from the mass movements of the extremities. The excitability gradually diminishes, until at the age of five years the only significant result of plantar stimulation is the complete absence of any response.

The changing behavior of the big toe on plantar stimulation cannot be explained simply by the nonmaturation of the pyramidal tract. The athetoid movements, to which during the first months of life the reactive movements of the big toe also belong, are caused by the nonmaturation of the striatum. As can be seen from the example of the respiratory center (see p. 313), the maturing nervous centers do not influence the nervous system suddenly; there is a period of transition during which their influence is only fluctuating and uncertain. Periods of functioning alternate with others during which their influence is again extinguished. If similar conditions exist in the motor centers, the fluctuating patterns of the toe reactions can be explained. A regular toe reaction, namely, the flexion of the big toe on stroking of the sole, is possible only when the higher centers predominate permanently.

VAN WOERKOM (1912) regards the BABINSKI sign as a vestige of phylogenetic development. According to SCHOCH (1948, 1949), this sign is a "release reflex," a counterpart of the plantar grasp reflex (see p. 158). Both reflexes are supposed to be carryovers from our phylogenetic ancestors who still grasped with their feet. Should this be correct, BABINSKI's sign should be found in their surviving kin, whose stage of development was arrested at the same level. But actually this is not the case; the reflex is absent in them. FULTON and KELLER (1932) examined more than 100 apes of different species, among them anthropoids, without ever finding extension of the big toe. Either they did not obtain a response at all or they obtained a flexor response on stimulation. Neither L. F. and M. M. JACOBSEN and YOSHIOKA (1932) nor CATEL (1950) were able to demonstrate BABINSKI's sign in newborn chimpanzees. FULTON and KELLER observed extension after lesions in the central nervous system and found that the level of the lesion had to be in proportion to the phylogenetic distance of the animal from man: The nearer the animal to man (phylogenetically), the higher the localization of the lesion.

Below the stage of mammals the pyramidal tract has little importance. This difference is not expressed in the plantar reactions since the anatomic differences of the feet do not allow for such a comparison (RICHARDS and IRWIN, 1935). Another plantar reflex of the young infant, though, represents a phylogenetic memory of the time of the life in trees. This is the tonic plantar grasp reflex (see p. 158). It can be found in the infant as well as in the ape and — contrary to BABINSKI's sign — can always be elicited by pressure on the flexor surface of the toes.

In children suffering from meningitis extension of the big toe can replace flexion of the leg at the knee and hip joints: If a meningitic sign that flexes the leg is elicited (for example, the neck sign or the inguinal flexor reflex) and this flexion is prevented by simultaneous pressure on the knee joint, the big toe is often extended (see p. 295).

ROSSOLIMO's and MENDEL-BECHTEREW's Signs

If a light, short tap applied to the inferior surface of the toes causes extension, the toes will execute a flexion, abduction, or adduction movement after a fairly brief interval (ROSSOLIMO). If the lateral aspect of the dorsum pedis is tapped on its proximal half, flexion of the second to fifth toes will occur when the sign is positive; otherwise, extension will occur (MENDEL-BECHTEREW). In the adult, both signs indicate a pyramidal tract lesion. In infants, DE BRUIN (1928) usually found analogous responses to the two reflexes. W. and I. SHERMAN and FLORY (1936) were able to demonstrate ROSSOLIMO's sign in 72.5% of all newborns.

Flight Reflex

(Withdrawal or Defense Reflex)

Slight stimulation of the sole, such as a pinprick, usually causes leg flexion at the knee and hip joints. BABINSKI (1915, 1922) considered this "réflexe de défense," as well as the big-toe sign named after him, to be the results of pyramidal tract lesions. In the German literature this reflex is called "shortening reaction of the leg" or "withdrawal reflex." It is physiologically present in early infancy. The time of its disappearance has not yet been investigated in detail.

The reflex also has tonic qualities. If a small, not excessively tight clamp is fastened to the sole, the leg continues to be flexed at the knee and hip joints as long as the clamp remains (see Fig. 30, p. 169).

ARM REFLEXES

In order to elicit LÉRI's forearm sign, the arm is allowed to relax as much as possible and is supported at the wrist by the examiner's hand. With his right hand the examiner bends the fingers of the patient against the palm and then the patient's hand against the forearm. The hand is rolled up, so to speak. As a consequence, the forearm is progressively and increasingly flexed as if controlled by a spring or an elastic band.

MAYER's finger-basal joint reflex is tested in the following manner: The hand of the person to be examined is brought into supination. Then the basal phalanx of a finger, usually the middle finger, is forcefully pressed against the palm. In a healthy person, flexion of the thumb in the metacarpophalangeal joint and opponens position will occur. Frequently flexion in the proximal joint and extension in the distal joint of the thumb will also take place.

The absence of LÉRI's and MAYER's reflexes in adults points to an organic lesion in the pyramidal tract, just as the appearance of the above-mentioned plantar reflexes does. According to GOLDSTEIN (1920), TIEFENSEE (1925), DE BRUIN (1928), et al., both reflexes are absent during the first two years of life and do not appear before the third or fourth year. According to GOLDSTEIN their appearance depends on the development of the central nervous system and therefore belongs in the same category as the disappearance of BABINSKI's sign. GOLDSTEIN relates it to the flexor synergy of the arms that enables the ape to grasp and to climb.

CREMASTERIC REFLEX

The cremasteric reflex is not so easily elicited in infancy as it is later in life and is therefore often missed (FARAGO, 1887; CATTANEO, 1902; MÜLLER, 1911; ARONOVITSCH, 1926; DVORAK, 1949). According to BRESA, it can always be elicited even in the newborn if a sustained contraction of the tunica dartos, which otherwise counteracts the reflex, is eliminated by warm baths.

A type of cremasteric reflex can also be observed in thin, easily excitable girls (SECKEL, 1932): If the inner aspect of the thigh is stroked, sporadic oblique twitching appears directly above the ipsilateral inguinal ligament; it originates in the lowest fibers of the musculus obliquus internus abdominis that go to the ligamentum uteri. DVORAK (1949) reported a corresponding reaction; he regards it as the lowest abdominal reflex.

ABDOMINAL REFLEX

(1)

According to MINKOWSKI (1928), the abdominal reflex can be distinctly elicited even in human fetuses of the fifth to sixth month of gestation. It is quite often accompanied by movements of the legs or lifting of the arms. The reflex is also usually present in the neonatal period and in infancy (HARLEM and LÖNNUM, 1957), but it is easily disturbed by the restlessness of the infant (BYCHOWSKI, 1908; FARAGO, 1887; CATTANEO, 1903; MÜLLER, 1911; BURR, 1921; and SCHOO, 1934).

HAVELKA (1955) was able to elicit the reflexes in some newborns and also in the prematurely born as early as the first day of life; however, the reflex usually appeared in the course of the first month of life. At eight months the abdominal reflexes could be elicited in all infants.

If the handle of a reflex hammer is pressed against the abdomen, the pelvis swings sideways, as in the spinal reflex (see p. 163). The reflex zone extends to the proximity of the midline, but the reflex cannot be elicited from the immediate proximity of the midline.

MENINGITIC SYMPTOMS AND SIGNS

(17, 90, 105, 108)

The suspicion of meningism or meningitis may be aroused early by a series of genera nervous phenomena, such as change of behavior, restlessness, ill humor, anxiety, apathy, vacant stare, sighing and loud screams, gnashing of the teeth, vomiting, complaints of headache and abdominal pain, hypersensitivity, slightly or markedly clouded sensorium, spasms, tremors and pareses, motor disturbances (which may reach the extent of convulsions), loss of bowel and bladder control, and transitory skin erythema. Although all these phenomena may also occasionally occur in other clinical entities, specific pathological signs exist that occur only in meningitic states and therefore have particular importance for the diagnosis of this disease.

Meningitic Posture

The previously described preferred position of the infant (see p. 169) is frequently found in meningitic children much beyond the first year of life; this is an expression of the fact that the cerebral function has in this respect regressed to a lower developmental stage. At the same time the children may be alert and able to give correct information. The oldest child with this posture whom I was able to observe was eight years old.

A similar but not identical meningitic posture was already described by RILLIET and BARTHEZ (1856):

> "The position is not constant; it is rare that the patients constantly lie on their backs. In other cases they lie now on their sides and now on their backs. The children like to change their position and rarely remain immobile on the same side. If they lie on their sides, they often pull their thighs up to the lower abdomen and the legs to the thighs. They remain lying in this position and complain if you try to turn them on their backs."

ECKSTEIN's (1931) description, when compared with this clear and lucid account, should be regarded as a backward step; he described the posture in tuberculous meningitis in the following manner: "Some children prefer a lateral position, turning their faces away from the light and toward the wall." ECKSTEIN speaks of a "hunting dog position," although I cannot imagine why hunting dogs particularly (why not other dogs?) should prefer this position. The most important feature of this position, the pulling up of the legs, is not mentioned. But why "hunting dog position"? The French, for example TAILLENS (1934), recognize a *position en chien de fusil,** i. e., a rifle cock position, corresponding exactly to that described by RILLIET and BARTHEZ and occurring under the same conditions. GLANZMANN (1943) described this position and reproduced illustrations of it. The "hunting dog position" is only a wrong and meaningless translation of the French expression. Not satisfied with this, ECKSTEIN knows also of a "hunting dog respiration" as a sign of tuberculous meningitis; here again we have to ask the question of why just hunting dogs should have a special kind of breathing.

The rifle cock position, which can be more simply termed "flexion position," is not limited to tuberculous meningitis. I distinctly noted this position to be pres-

* The French word *chien* means either *dog* or *(rifle) cock*—Ed.

ent in a twelve-year-old child with meningism and locally circumscribed tuberculous peritonitis and in a five-month-old infant with purulent meningitis. SCHAEFER (1939) described it in poliomyelitis. This position is a sign of the meningitic state but not of a specific disease.

A distinct flexed position without opisthotonus in which the child lies in a crouched lateral position occurs after infancy during prolonged severe states of hunger. After a period of many months there is even the danger of ankylosis in this position. By this flexed position the body surface, and consequently heat loss, is decreased. It is the logical outdoor sleeping position.

Nuchal Rigidity, "Kiss-the-Knee" Test

Healthy children are able to bend their heads to an extreme forward position. Sometimes they may so stiffen up while screaming that passive movements of the head become difficult. SAINT-ANNE DARGASSIES (1954) describes tonic reactions of the neck muscles in the newborn: If repeated pressure is exerted on the occiput in order to bend the neck, strong resistance of the extensor muscles is noted. However, on pressure on the frontal region, a weak contraction of the antagonists occurs. ROMBERG (1834) already recognized nuchal rigidity, i.e., resistance to forced bending of the head, as a meningitic sign.

In severe cases, particularly in epidemic meningitis, the children are in a constant position characterized by marked backward overextending of the neck. This backward arching may extend to the dorsal and even lumbar spine, so that lumbar puncture is very difficult to execute. Any attempt to compensate for this position is very painful. In milder cases the opisthotonus is noted only when an attempt is made to bend the child's head. In the absence of the opisthotonus the meningism may easily be overlooked because it may be thought unnecessary to look for other meningitic signs.

Children who are already able to sit up may lose this ability because of the stiffness of the vertebral column. An older child who is carried into the office lying in a horizontal position should immediately be suspected of meningitis. Nuchal rigidity of a lesser degree may still permit bending of the head and sitting up but will prevent further forward bending of the upper half of the body and can be recognized by the inability of the children to kiss their knees. In order to understand what they are expected to do the children must be at least three or four years old, and they must be in a cooperative mood to perform this task. If they are very resistive, which is quite often the case, it will not be possible to test this sign. When the meningitic state decreases in severity, it is possible to measure directly the decrease of the stiffness of the neck by the diminished distance between mouth and knees. The inability to kiss the knees corresponds to the "spinal sign" in the English literature.

BROCK (1925) and LANGE (1939) explain the stiffness of the neck on the basis of increased spinal fluid pressure. The pressure is increased when the child's head is bent forward (see p. 297). According to FULTON (1952), nuchal rigidity is caused by the irritation of the pain-carrying fibers of the trigeminal nerve; this irritation is more effective in causing a backward movement of the head when it occurs in the dura of the base of the skull and along the tentorium than when the irritation affects the cerebral hemispheres. Of course, slight drawing back of the head is noted in the decerebrated animal in response to irritation of practically every sensory nerve of the body, but the backward motion is most distinct and clear and lasts longest when the trigeminal nerve is stimulated.

Flexor Contraction of the Knee Joints (KERNIG's Sign)

KERNIG's sign (1884) is incorrectly described even in textbooks, and we therefore note many errors and contradictions. As an example, I quote the description given by ECKSTEIN in his handbook (1931) and in his textbook of pediatrics (1933): "KERNIG's symptom consists of the following: The patient flexes his leg like lightning when an attempt is made to lift his extended leg passively at the hip joint while he is sitting." Others consider the essential features of this sign to be that the patient experiences pain when a forceful attempt is made to lift the extended leg at the hip joint. This concept also differs from KERNIG's statement.

The discoverer's description is the final authority. The flexor contraction at the knee joint – this is the appropriate designation given by KERNIG –

"consists in the phenomenon that the legs of the patient who is sitting or placed in a sitting position cannot be extended at the knee joints, whereas the knees are not contracted when in a lying position; or... in the phenomenon that the mentioned contraction in the knee joints will occur only if the thighs of the meningitic patient are in a certain degree of flexion (at approximately a right angle) to the trunk."

The angle between thigh and trunk should be 90° but should never be acute.

The patient can be examined for the presence of the contraction either lying on his back or in the sitting position. If he is lying down, he is asked to sit up or he is placed in the sitting position. As soon as the trunk forms a right angle with the thighs, sometimes even earlier, an irresistible flexor contraction at the knees occurs. If the patient's knees are kept extended by force, his trunk cannot be raised far enough to form a right angle with the legs. If the patient is sitting with his legs hanging down, the legs cannot be completely extended at the knee joints and a flexor contraction occurs at the knees during this attempt. KERNIG states that it is immaterial whether the test is done with the patient lying down or sitting up. This is in accordance with my own experience.

KERNIG's sign is not a motor reaction, as KERNIG himself stressed. When the patient is brought to a sitting position it is the appearance of a noncompensable contraction at the knee joint that is important and not the pulling up of the legs. KERNIG has also stressed that pain is not a characteristic of his sign, although pain in the region of the sciatic nerve may occur during the testing and this pain may inhibit the flexion at the knee and hip joints. This phenomenon, known as LASÈGUE's sign, also occurs in meningitis. KERNIG's and LASÈGUE's signs are independent of each other. The testing of KERNIG's sign is in itself painless.

When WARTENBERG (1952) attributes KERNIG's as well as LASÈGUE's sign to a stretching of the sciatic nerve, he thereby places himself in opposition to KERNIG.

It is better not to speak of KERNIG-LASÈGUE's sign, as WARTENBERG does, because this obliterates the difference between the two signs.

KERNIG explains the way his sign comes about by ROGLET's hypothesis. ROGLET regards the comparatively short flexors of the lower leg, in addition to the influence of the increased pressure and the direct irritation of the lumbar cord, as the cause of KERNIG's sign. The flexors have reached the limit of their elasticity when the leg, extended at the knee, forms a right angle. If they become hypertonic because of meningitis and are thus shortened, the flexor contraction occurs.

BRUDZINSKI's Neck Sign

BRUDZINSKI's neck sign (1909) consists in flexion of both legs at the knees and hips on forced (passive) forward bowing of the head. The great value of this sign was soon confirmed universally, so that when we speak today of BRUDZINSKI's sign we always mean the neck sign, although BRUDZINSKI described a great number of other meningitic signs before and after he described the neck sign.

It is not generally known that BRUDZINSKI further amplified his statements on the neck sign in 1916: The enforced bending of the head may — and I can confirm this — also cause flexion of the arms at the elbows. What he meant by "a raising of the arms," which is supposed to occur at the same time, has never become clear to me. The nuchal rigidity may be so great that it is not possible to bend the child's head forward. In that case the neck sign is of course absent until it again becomes possible to bend the head forward.

According to WARTENBERG (1952), the spinal cord is stretched to its lowest roots when BRUDZINSKI's sign is elicited. In order to neutralize the pain the patient assumes a position that inhibits this stretching. Therefore, he overextends his head backward, bends the spinal column to a lordotic curvature, and flexes his legs.

MARANON (1927) described a not infrequent variation of this reflex under the term "nucleoplantar reflex": With one hand the examiner presses the head of the supine patient forcefully toward the chest and keeps the knees extended with his other hand. (It is recommended that an assistant keep the knees in extension since both hands are needed just to bend the head. With one hand the head of the patient is flexed and with the other the chest is held, opposing this pressure.) Instead of the pulling up of the legs, which has thus become impossible, both big toes are overextended, as in BABINSKI's sign. Unilateral appearance of this phenomenon is not infrequent. The neck sign and substitute reactions may appear simultaneously in the same child, but they may also occur apart from each other.

The interpretation of the neck sign is still disputed today (see p. 157). I was able to demonstrate it in early infancy only in exceptional cases.

Inguinal Reflex

(61, 90, 105, 108)

The inguinal reflex [PEIPER, 1936 *(61)*] is elicited by short energetic pressure on both inguinal regions; the reaction consists in flexion of both legs at the knees and hips, just as in BRUDZINSKI's neck sign. The elicitation may be painful. A very limited reaction to the unexpected stimulus, such as a brief twitching at the hips, may occasionally be found in healthy persons and therefore cannot be regarded as a pathological sign. In a lively reaction both legs are pulled up to the body with a loud expression of pain. If, as occasionally happens, there is a weak reaction it can be reinforced if both legs are placed in an abducted position prior to testing the reaction.

If the reaction is prevented by keeping the legs extended by pressing on the knees with the examiner's free hand, both big toes (or sometimes only one) may be overextended as substitute reactions in a manner similar to BABINSKI's sign. For this test strong pressure on the inguinal region is required. A reaction in the big toes may still be elicited by strong pressure when weaker pressure has been ineffective. This reaction of the big toes occurs only rarely when the legs have remained extended after pressure on the inguinal region. Both forms of elicitation of the inguinal reflex are independent of each other.

Unilateral elicitation of the inguinal reflex generally causes bilateral or — in rare cases, when the reaction is weak — unilateral flexion of the legs.

The elicitation of the inguinal reflex occasionally leads to arm flexion at the elbows in a manner similar to that seen in the neck sign.

Crossed Homologous and Nonhomologous Flexor and Extensor Reflexes of the Legs

(1, 18a)

The first meningitic signs described by BRUDZINSKI (1908) were the crossed ("contralateral") flexor and extensor reflexes of the legs. If one leg of the child is flexed at the knee and hip joints by the examiner, the other leg is simultaneously flexed by reflex action (crossed homologous flexion) or is extended (crossed nonhomologous extension). According to my experience, only the crossed homologous flexion reflex of the legs is of importance, and it alone will be considered in the following discussion. The younger the children, the more frequently have I found it in meningitis. I no longer observed this reflex after the fourth year of life.

Surprisingly enough, the reflex was not found by BRUDZINSKI in healthy persons although he examined young children, too. Actually the reflex can be demonstrated in every healthy infant during his first months of life. Infants of that age usually have both legs pulled up to their abdomen. But when they are uncovered, they occasionally extend their legs. When one extended leg is now suddenly flexed at the knee and hip joints, the other leg makes the same movement almost simultaneously. The decisive factor is the hip movement, since the reaction may also occur with the knee extended, whereas flexion of the knee alone without hip movement will remain ineffective. Flexion of the other leg does not depend on the disturbance, for it is absent when the other leg is tapped.

STENGEL [1945 *(18a)*] observed more than 1000 healthy infants in order to find out at what age the crossed flexor reflex of the legs disappears. He found that the reflex is present up to the seventh month, that it then gradually becomes less frequent. and that it is completely extinguished by the end of the first year. In feeble-minded children it may persist up to the fifth year, but later it can no longer be demonstrated. Children who suffered from severe, long-lasting alimentary disturbances showed the reflex up to the third year.

The crossed homologous flexor reflex of the legs cannot be elicited successively as often as desired; after a few tests it may become temporarily weaker or disappear, sometimes even after a single test. Not infrequently the leg reacts with a distinct delay.

In experiments on animals BRUDZINSKI was able to elicit the crossed homologous flexor reflex of the legs under various conditions, for example, in the frog after its spinal cord had been cut.

In explaining the appearance of this reflex in meningitic children, we must start from the fact that the reflex is always present in early infancy. When it reappears later in life, it evidently is released because of the elimination of higher centers.

As in the neck sign and in the inguinal reflex, an overextension of the big toe can at times be effected as a substitute reaction when the flexion of the legs is prevented by pressure on the knees. The substitute reaction may also occur when the flexion fails to appear. The reflex cannot be found after the fourth year of life. It does not depend on the usual form of elicitation.

I did not find corresponding crossed flexion and extension reflexes in the arms of children, nor did BÖHME and WEILAND (1918) find them in adults.

BABINSKI's Sign

BABINSKI's toe sign (see p. 285) can frequently be elicited in meningitis. In addition, there are, as has been described, other meningitic signs that likewise cause overextension of the big toe; they are the neck sign, inguinal reflex, and crossed flexor reflex of the legs if a flexion of the opposite leg is prevented. When KERNIG's knee contraction is tested, the reaction of the big toe may occur ipsilaterally or contralaterally. All these phenomena are independent of each other to a certain extent. Their frequent simultaneous occurrence indicates that they have as a common basis a higher entity, which I term the "meningitic system." The usual forms of the neck sign, the inguinal reflex, and the crossed homologous flexor reflex of the legs consist in the same movement, i.e., flexion of the legs at the knee and hip joints, and therefore can very well be classified together along with the described withdrawal reflex (see p. 288), which, according to BABINSKI, often accompanies the toe sign.

LASÈGUE's Sign

LASÈGUE's sign, which is frequently confused with KERNIG's sign but can always be distinctly differentiated from it (see p. 292), consists in the occurrence of pain along the sciatic nerve when the hip is flexed while the leg is extended. The youngest meningitic child in whom I could observe this sign was two and a half years old.

Tonic Neck Reflexes of the Limbs

(17, 106)

The asymmetrical tonic neck reflexes of the limbs after rotation of the head can be found in premature infants (see p. 156). They were demonstrated in tuberculous meningitis by WECHSLER (1925) and ISBERT and PEIPER [1927 *(17)*]. The youngest infant who showed these reflexes was four and a half months old; thus, he was just at an age at which these reflexes usually disappear. The oldest child was two and a half years old. The reflexes may occur unilaterally or may be noted in only one arm or one leg. Evidently the appearance of these reflexes can also be explained by the elimination of higher, usually inhibitory, centers.

Symphysis Sign

The symphysis sign (BRUDZINSKI, 1916) is elicited by pressure on the symphysis pubis and consists in rapid flexion of both legs at the knees and hips accompanied by abduction. The pressure is usually painful, but at times it is painless in spite of a lively reaction. Occasionally overextension of the big toe occurs, usually unilaterally but in rare cases bilaterally.

The point of elicitation is near the inguinal region, and the reaction is identical to that in the inguinal reflex; even the same substitute reaction occurs if flexion of the knees is prevented. But as a whole the inguinal reflex is considerably more frequent and easier to demonstrate.

Cheek Sign

The cheek sign (BRUDZINSKI) is elicited by pressing strongly with abducted thumb and index finger on the upper jaw just below the cheek bones on both sides. The reaction consists in rapid raising of both arms flexed at the elbows. The pressure may be either very painful or painless. I was able to demonstrate the cheek sign only rarely.

Tripod Sign

The tripod sign (AMOSS's sign) was first described in the English literature: If the child is brought from the dorsal into the sitting position, he braces himself on the base with both arms extended at the elbows and with the palms at both sides of his trunk, supposedly to relieve the spinal column. However, the reaction is not conscious, for the sign also occurs when children are completely unconscious and it may not even appear distinctly until then. Whenever the child is sitting up, his arms are extended and when he is lying down they are flexed again. This ought to be a pure reflex phenomenon. Even if the infants are held freely suspended in the air and the identical movements are executed with the upper half of their bodies, the same arm position occurs although such movements now make no sense. If the child is fully conscious, this phenomenon can be easily obscured because of the children's hypersensitivity and resistance.

If the tripod sign is very pronounced, the child, who is bracing himself with his arms, may be prompted to certain movements: When the upper half of the body is pressed backward, both arms will be placed farther backward. If the test is repeated, the arms will move back still farther. These movements are appropriate, for the arms can thus still support the upper half of the body in spite of its backward-shifted center of gravity. The tripod sign is always bilateral.

The diagnostic value of this sign, which is found often in older meningitic children, is limited by the fact that it may also occasionally occur in healthy children.

SIGNIFICANCE OF THE MENINGITIC SIGNS

(108)

After the first year of life the most frequent, and thus for the diagnosis of meningitis the most important, signs at the onset of the disease are: nuchal rigidity, BRUDZINSKI's neck sign, the inguinal reflex, the contraction of the knee joints (KERNIG), and BABINSKI's sign. Despite the frequency of their occurrence, the preferred position of the infant and the tripod sign are less useful since they may also occur occasionally in healthy children. The homologous crossed flexor reflex of the legs is always present in the early period of life and has not been observed in sick children after the fourth year of life. During this period the reflex is a useful diagnostic sign. The incidence of this sign, of the neck sign, and of the inguinal reflex is even increased if leg flexion is prevented during testing. Compared with these signs the other meningitic signs are of minor importance because they are less frequent and hence of less diagnostic value.

None of these signs is especially frequent in the early stages of the disease. The contraction of the knee joint persists longest, as KERNIG has already stated; it may outlast the other meningitic signs by many months. Nuchal rigidity may also have a rather long duration; the other signs are considerably more transient.

During the first year of life nuchal rigidity and the neck sign are conspicuously infrequent. The homologous crossed flexor reflex of the legs, BABINSKI's sign, the preferred position of the infant, and the asymmetrical tonic neck reflexes of the limbs can be more or less frequently demonstrated in healthy infants of that age too, and are therefore not useful for diagnostic conclusions.

The meningitic signs are completely independent of the state of consciousness and fever. After a lumbar puncture they may temporarily be present, may become more distinct, or may disappear.

Sometimes the better known signs are absent in a meningitic condition and only the more infrequent signs are present. A sign that can always be found in meningitis does not exist. These signs are never all present at the same time, and no two of them are absolutely linked with each other. Thus, the meningitic condition is characterized by an always recurring, at times more and at times less complete series of signs that complement each other, thus forming the meningitic system.

According to LANGE (1939), the increase in the spinal fluid pressure causes the nuchal rigidity and contraction of the knee joint; these reactions are supposed to decrease the stress on the posterior roots. All forced movements counteracting the relaxation of the stress are thought to cause resisting movements. HILLER (1936) states that similarly the response to each stretching of nerves or roots is a reflex defense tension causing the meningitic signs. This opinion is supported by BROCK's (1925) findings, i.e., the fluid pressure is increased by flexing of the head. But this does not explain the flexed position of the legs and the contraction at the knee joints, for the spinal fluid pressure does not increase when the legs are extended at the knees and are simultaneously flexed at the hips (BROCK). Neither can the posterior roots be relaxed by the phenomenon that occurs when bending of the head causes overextension of the big toe if leg flexion is prevented or absent. In the inguinal reflex and its substitute reaction, the eliciting stimulus does not cause any strain and the reaction does not abolish a strain, although we obtain the same response as in the neck sign.

The fact that the most important meningitic signs may also occur unilaterally likewise contradicts the hypothesis that they may be caused solely by increased intracranial pressure. If this pressure plays a role in the origin of the meningitic signs, they ought to disappear when spinal fluid is drained off. However, this it not the case, at least not regularly. On the other hand, markedly increased intracranial pressure from other causes, such as that from brain tumor or turricephaly, does not cause any meningitic signs at all.

Various meningitic signs, such as the neck sign, the inguinal reflex, the crossed flexor reflex of the legs, and the symphysis sign, have the same reaction in common: the shortening of the legs by flexion at the knees and hips. The parallel is even greater: When the flexion is prevented by pressure on the knees, the big toe may be overextended, though not always, as a substitute for the inhibited flexion; it does not make any difference whether or not BABINSKI's sign is otherwise present. At times the flexion of the legs does not even occur automatically and is replaced by the toe movement. During the attempt to prevent the contraction of the knee joint the identical position of the big toe is occasionally evoked ipsilaterally or contralaterally. In this connection BABINSKI's sign must also be regarded as a meningitic sign, although it may occur otherwise too. A corresponding agreement can be found in the flexor reaction of the arms, which can likewise be caused by elicitation of the neck sign and the inguinal reflex.

A similar relation can be found when unilateral or bilateral occurrence is observed. Usually the meningitic reactions occur bilaterally, but it may happen that they appear only on one side or to a more pronounced degree on one side than on the other. The same side may be favored on different days. A certain correlation between the different reactions may be found in this connection.

Overextension of the big toe together with flexion of the legs at the knees and hips is also seen in the flexor synergy occurring in spastically paralyzed extremities during voluntary and involuntary movements. FOERSTER (1913) states that in the so-called flexor reflex of the legs the thigh and lower leg are flexed, the foot is flexed backward and is supinated, the big toe is overextended, sometimes with pronounced supination, and the other toes are flexed or extended. In spastic diplegia the synergy elicited by the voluntary movements of one extremity may extend over the whole body. Thus, FOERSTER describes a flexion synergy of all four extremities that also involves the trunk. Likewise, all four extremities may be flexed simultaneously during the elicitation of the neck sign and the inguinal reflex.

FOERSTER attributes the flexion synergies to the fact that because of the pyramidal tract deficit the movements depend on subcortical centers. It is obvious that the same mechanisms can also be assumed for the meningitic centers. This does not mean that the pyramidal tract function has to be permanently eliminated; rather, it suffices that at the moment of the stimulation the function of the subcortical centers is predominant.

The meningitic signs appearing in an otherwise healthy young infant, i.e., the asymmetrical tonic neck reflexes of the limbs, the preferred posture of the infant, the crossed flexor reflex of the legs, and BABINSKI's sign, can also be explained as a regression of the cerebral function to a lower developmental state, as described above. An additional meningitic sign is the periodically interrupted respiration (see p. 323) which appears in premature infants as well as in meningitic children. These phenomena are characteristic of premature infants because the higher centers, which later inhibit these phenomena, are still immature and therefore not functioning. The meningitic condition can cause disintegration of the inhibitory centers, so that the brain function regresses to lower stages. This does not cause complete decerebration. In spite of many neurological phenomena of a lower order, the children may still be at the height of their mental functioning in other aspects. For example, they frequently can still talk. Meningitic phenomena are independent of the stage of consciousness.

The location of the meningitic centers is not known in detail. MAGNUS assumes the center of the asymmetrical tonic neck reflexes of the limbs in man and animal to be in the upper cervical cord. Similarly, the other centers of the meningitic system can be assumed to be in the brain stem.

Encephalitis, which could be considered a possible cause of meningitic signs, has not been proved a contributing factor by the pathologists. One cannot clinically recognize to what extent the stretching of the ventricular walls because of increased spinal fluid secretion or disturbances in the spinal fluid circulation or in the cerebral blood flow may play a role.

Symptomatology of Tetany

(3, 4, 25, 32, 13a, 14a)

Although the tetany of the young child can be easily diagnosed today, the characteristics and the significance of the signs of tetany have so far not been clearly recognized. The most important signs were originally discovered in the tetany of the adult, which has a completely different cause; and later they were rediscovered in the tetany of the infant. Besides the classical signs, which will be discussed later, other signs associated with tetany also appear: "tetany face" (UFFENHEIMER, 1905), which is caused by a contraction of the circular muscles of the eyes and mouth that results in a "cunning, tricky" facial expression; bowel and bladder disturbances due to sphincter spasms; intestinal spasms; bronchotetany (LEDERER, 1913, 1919); and sudden cardiac death, which is caused by a tetany of the heart muscle (IBRAHIM).

Signs of latent tetany are mechanical or electrical hyperirritability; in manifest tetany we see in addition generalized convulsions, laryngospasms, and carpopedal spasms.

Signs of Latent Tetany

Mechanical Hyperirritability

(24, 32, 13a, 14a)

Muscle twitches can be elicited in the tetanic child, best by tapping, in areas where the motor nerves are directly under the skin; the healthy child shows no reaction to this stimulation. CZERNY and KELLER (1928) rejected the expression "hyper-irritability of the peripheral nerves" for this phenomenon. They recognize only a hyperirritability of the central nervous system and its peripheral end stations. This question is still unsolved.

The facialis sign, originally described by CHVOSTEK (1876) in tetany of adults, consists in a spasm of the muscles served by the facialis nerve and is elicited by tapping the cheek between the lower edge of the cheek bone and the corner of the mouth. In severe hyperirritability it is sufficient to stroke the cheek (SCHULTZE's sign). According to MORO (1906), spasms may occur bilaterally in a newborn upon unilateral stimulation. The facialis sign must be distinguished from the "mouth phenomenon" (ESCHERICH, 1909) and the identical "lip phenomenon" (THIEMICH, see p. 407).

In a manner similar to that in the facialis sign, muscle spasms can be evoked in a tetanic child from other nerves also. The hyperirritability is "best tested in the radialis on its turning point at the lower third of the upper arm, whereupon a rapid dorsal flexion of the hand occurs, or in the outer portion of the peroneus at the capitulum fibulae; the stimulation of the latter causes an abduction of the foot with slight dorsal flexion." With this description BIEDERT and FISCHL (1902) anticipated the radialis sign, later described by HOFFA (1920), and the peroneus sign, described by IBRAHIM (1911, 1920) and LUST (1911). Both signs can also be tested in a crying infant; this is not possible when testing for the facialis sign.

The mechanical hyperirritability of many premature infants leads to a number of other phenomena apart from the signs mentioned. For example, if the nerves at the medial aspect of the upper arm are tapped, jerk-like flexion at the elbow, wrist, and fingers occurs. Depending on which nerves are struck, the movements described may appear simultaneously or may be restricted to individual muscle groups. The same phenomenon occurs when the plexus brachialis is tapped in the axilla. Occasionally movements of the individual fingers may be evoked from this region. In some children a distinct reaction in both arms occurs on unilateral plexus stimulation, i.e., a crossed reflex. Similarly, extensions at the knee, ankle, and toes can be elicited by tapping the anterior nerves of the thigh, and corresponding flexion may be elicited by tapping the posterior surface of the thigh. Thus, the peroneus sign is one of the phenomena of general mechanical hyperirritability and not even one of the most conspicuous signs. Whereas the right and left halves of the body are homologous, the upper and lower halves may react differently from each other. It frequently happens that hyperirritability is very pronounced in the arms, while unequivocal findings cannot be obtained in the legs. Also, the facialis sign proves to be rather independent of the irritability of other body parts. Similar observations can be made in tetany of the infant.

The muscles per se are also mechanically hyperirritable in tetany. THIEMICH, however, rightly stressed that this phenomenon is of no diagnostic value since it may also occur in emaciated and cachectic infants and adults. In many premature infants the hyperirritability of the muscles on tapping can be clearly recognized even if the stimulation of the nerves proves to be unsuccessful. If, for example, the musculus pectoralis major is tapped, a strong spasm occurs. Also, tapping of the muscles on the flexor or extensor aspect of the upper arms and forearms as well as on the thighs and legs leads to twitches in the stimulated muscles.

It is customary to call the signs of mechanical hyperirritability "phenomena" and not reflexes since their nature is debatable. BEHRENDT and FREUDENBERG (1923) conducted conclusive investigations on the respiratory tetany of adults. They injected local anesthetics into the proximal area of a nerve. If that part became insensitive and paralyzed, the distal hyperexcitability could nevertheless be increased to such a degree by hyperventilation that tapping of the nerves below the blocked area could cause muscle twitching.

From experiments on animals, McCOLLUM (1913) came to the same conclusion, i. e., that the hyperirritability is not connected with a change in the entire neuron including the ganglion cells of the anterior horns. He connected the arteria and vena femoralis of a healthy dog with the carotis and jugularis of another dog with fully developed tetany. Thus, the blood of a tetanic animal flowed through the peripheral portion of the nerves and muscles and the spinal cord remained unaffected. After an hour the periphery became electrically hyperirritable as in tetany. When the femoral vessels were reconnected with their own stumps and the legs were thus again supplied with their own blood, the level of irritability became normal.

Of course, even after these investigations, there always remains the possibility that in addition to the peripheral irritation, tapping of the nerve trunk may also cause a reflex that achieves the same end result. Some observations support such a hypothesis. Thus, the previously mentioned bilateral occurrence of the facialis and plexus brachialis signs after unilateral stimulation is only possible through an intermediate central relay station. Studies of the muscle action potentials could solve this question; to my knowledge investigations of this kind have not yet been made.

VAN DER ELST (1929) differentiates between the true CHVOSTEK sign and the facialis sign. Only the true CHVOSTEK sign is supposed to be the end result of mechanical hyperirritability of the nerves. The facialis sign, however, elicited in the usual fashion, represents a reflex; its reflexogenic zone is located below the cheek bone and its responding elements are the muscles of facial expression. This facialis sign disappears in the intact facialis after a procaine injection into the trigeminal nerve.

Of all the signs of tetany, the sign described by TROUSSEAU (1861) has been known the longest. According to JESSERER (1958), it has so far been observed only in tetany. It is elicited by tying the upper arm with an elastic bandage for several minutes or by compressing the vessels and nerves in the biceps groove with the finger. The result is a tonic spasm of the hand muscles that brings the fingers into the "position d'accoucheur." The sign cannot be as distinctly elicited in infancy as it can later in life; however, I saw it occasionally in newborn infants.

SCHLESINGER (1892), SCHÄFFER (1920), and NOTHMANN (1937) consider TROUSSEAU's sign to be a reflex. FÜNFGELD (1943) is also of the opinion that TROUSSEAU's sign will not occur without special proneness toward it of the subcortical centers, since the same position may also occur in extrapyramidal diseases. However, THIEMICH (1900) and MANN (1900) see in it the expression of an increased mechanical hyperirritability. BEHRENDT and FREUDENBERG (1923) found a marked TROUSSEAU sign in the adult after hyperventilation even after elimination of the sensory tracts in the plexus brachialis by local anesthesia; it could be elicited even more easily and more distinctly here than in the other arm.

Although, according to these findings, centers are not needed to produce TROUSSEAU's sign, FREUDENBERG still regards this sign as a reflex occurring as a tetanic spasm on the basis of profound peripheral changes.

TROUSSEAU's sign lies on the border between latent and manifest tetany.

Electrical Hyperirritability
(32)

Electrical hyperirritability was first found in the tetany of adults by ERB and later in the tetany of infants by ESCHERICH (1909), MANN (1900), and THIEMICH (1900). Of the greatest clinical importance is the fact that in the galvanic testing of the median nerve of tetanic children the cathodal opening contraction (COC) begins at below 5 ma, whereas this value is higher in healthy children. The value is lowered even in latent tetany, so that an early diagnosis can be made. PIRQUET (1906, 1907), who tested the peroneus, furthermore differentiates an anodic hyperirritability preceding the cathodic hyperirritability with values for the anodal opening contraction (AOC) below 5 ma. In higher stages the AOC sinks below the value of the anodal closing contraction (ACC). While MANN and THIEMICH found higher threshold values in healthy infants under the age of eight weeks than in older children, PIRQUET could find no difference based on age. In tetany the irritability to faradic stimulation is also increased; this examination, though, is of little value since comparisons cannot be made between findings obtained with different instruments and since no absolute standard values exist.

Serious objections from the standpoint of physics have been raised against the customary procedure of galvanic testing, i.e., mainly that immediately after the current is stopped the meter shows a higher potential that gradually sinks to a lower sustained value. Only this lower value can be read on the measuring device, since it cannot follow the first short current peak because of its slow response. But the stimulation occurs just in this first millisecond. These objections have long been recognized as valid. To overcome this difficulty, MANN (1900, 1904) and SALGE (1919) employed condenser discharges. FREISE and SCHIMMELPFENG (1925) followed GILDEMEISTER's example and inserted an especially high resistor to neutralize this initial rise.

A procedure that similarly avoids the sources of error mentioned is the determination of the chronaxie, which was introduced by LAPICQUE (1926) as a standard measurement of neuromuscular irritability. In this procedure the smallest current that just produces a closing contraction is termed a rheobase. It thus corresponds to the cathodal closing contraction (CCC) in the customary procedure. The chronaxie is the shortest time interval for which the current has to act to produce a threshold contraction at a strength of current double the rheobase. LAUGIER (1929) gives full information on this investigative procedure; BOURGIGNON (1923) summarizes it.

According to my investigations [1930 *(32)*] premature infants and newborns born at term show such great fluctuations in the chronaxie of the nerves that no comparable values can be obtained. The same can be seen in the findings of BANU (1922) (nerve chronaxie) and ROTHE (1929) (muscle chronaxie). The values seemed to change even during the same trial. With increasing age the spread of the fluctuation decreases. Mechanically hyperirritable premature infants were characterized by comparatively low values and a small amount of fluctuation in the first days of life.

In latent tetany of the infant, BOURGIGNON (1923, 1925) found the muscle chronaxie to be so inconsistent that he regards this variability as a characteristic of this disorder. The rheobase also fluctuated very irregularly and independently of the chronaxie. In manifest tetany low rheobase values and high chronaxie values were found. TURPIN (1925), GYÖRGY and STEIN (1928), ROTHE (1929), and PEIPER [1930 *(32)*] obtained the same results.

Although there are no objections to the chronaxie method from the standpoint of physics, the procedure has so far not become popular because it has as a prerequisite a cumbersome and expensive device and in the last analysis achieves no more than the old and customary procedure. The inconsistency of values obtained in early infancy makes the electrical testing procedures in general appear quite useless for that age.

Signs of Manifest Tetany

The signs of manifest tetany comprise carpopedal spasms, laryngospasms, and generalized seizures.

In carpopedal spasms the arms and — in adults less frequently than in children — the legs are in a state of tonic spasm of long duration, at times lasting for days. The upper arms are adducted, the elbows and wrists are flexed, and the fingers are in the "position d'accoucheur" (like "paws"). When the lower half of the body is involved in the reaction, the legs assume an extended position, the feet are in varus or equinovarus position, and the toes are in pronounced flexion. If the spasm is of long duration, edema may appear on the dorsal aspects of the hands and feet. According to BEHRENDT and FREUDENBERG (1923), TROUSSEAU's sign and carpopedal spasms are essentially the same processes.

Laryngospasm causes a severe respiratory disturbance. During inspiration the narrowing of the larynx causes a crowing or high-pitched sound that is accompanied by a drawing in of the jugulum and the upper abdomen; this phenomenon is very characteristic of this stage. In very severe cases the respiration stops completely and the child becomes cyanotic, restless, and unconscious. Clonic twitches may occur.

Formerly many a child died during the attack. If the child recovers, a long drawn out, harsh-sounding inspiration occurs and is repeated several times with diminishing stridor and finally the usual respiration returns.

According to GARSCHE (1950), the respiratory inhibition during the laryngospasm is a consequence of the irritation of the center for swallowing. Minor, usually subthreshold, stimuli cause the hypersensitive swallowing center to send motor impulses to the laryngeal muscles. Because of the physiologically intimate relation between respiratory and swallowing centers, first the respiration and then the swallowing movements, which had already caused a rise of the larynx, are arrested. The now beginning spasms of the laryngeal levators prevent the larynx from gliding down again. The respiratory arrest lasts as long as the elevation of the larynx. As soon as the spasms recede, the outside air streams into the larynx with a harsh sound (stridor).

Tetanic and epileptic seizures cannot be clinically distinguished from each other.

Significance of the Signs of Tetany

(24, 32, 13a, 14a)

The symptomatology of tetany appears in the newborn, in the infant after the first months of life, and in the child after infancy. At each of these three stages çertain signs are present and others are absent.

Tetany of the infant and small child, which develops at the earliest in the third or fourth month of life and is particularly prevalent from the seventh to eighteenth month, is the most important form and has been known longest. In this disease all the tetanic signs described may be present. In addition to the neurological phenomena of peripheral hyperirritability (increased mechanical and electrical irritability, TROUSSEAU's sign, carpopedal spasm), in the young child laryngospasm and generalized seizures are present as expressions of an increased central state of excitation. A complete laryngospasm can be found almost exclusively in tetany. FÜNFGELD (1943) did not see it at all in 184 adults with idiopathic tetany, but he, as well as BEHRENDT and FREUDENBERG (1923), observed laryngospasm in adults after hyperventilation.

Tetany of the child occurs when, in a rather mild case of rickets, recovery processes, which withdraw calcium from the serum and deposit it in the bones, suddenly begin. The serum calcium level is decreased and the serum phosphates are increased.

Table X: Frequency of the Facialis and Peroneus Signs in Mature and Premature Newborns [according to Rosenbaum, 1931 (13)]

Newborn category	Facialis sign + Peroneus sign +	Facialis sign + Peroneus sign −	Facialis sign − Peroneus sign +	Facialis sign − Peroneus sign −
Premature, 115	22	17	18	58
Mature, 115	8	12	16	79

KEHRER (1913) was the first to describe tetany of the newborn. A mechanical hyperirritability is more frequent in premature newborns than in those born at term, as shown by ROSENBAUM's [1931 *(13)*] comparative study (Table X). CATEL (1928) found a facialis sign in one third of the premature newborns he examined; the peroneus sign was less frequent. SCHAAF [1945 *(14a)*] tested the mechanical hyperirritability in 137 newborns continuously from their first day of life. Immediately after birth none of the infants was hyperirritable; the first symptoms appeared only after five to six hours, and frequently the onset was on the second or third day. In one case a facialis sign appeared only on the fifth day and in another a peroneus sign on the sixth day. The duration of the hyperirritability showed considerable fluctuation; it lasted longest when it had already appeared a few hours after birth. If it first appeared on the second or third day it usually lasted only one day. The longest period was ten days; usually the signs disappeared after four or five days and the facialis sign occasionally after only two or three hours. SCHAAF did not see any other tetanic phenomena in any of these newborns.

I found the determination of electrical irritability to be valueless in the recognition of tetany of the newborn, even if the physically objectionable customary procedure was replaced by a determination of the chronaxie.

WILLI (1939) distinguishes latent, manifest, and complex types of tetany in newborns. Manifest tetany causes fright reactions, restlessness, crying spells, tremulous movements, sometimes a coarse tremor of the hands and feet, and hypertonia; carpopedal spasms occurred twice, convulsions and twitching twice, and inspiratory stridor in one case. According to WILLI, the physiological drop in serum calcium values after birth is pathologically increased in tetany of the newborn. He found low blood calcium values in a great number of the tetanic newborns he observed.

VON REUSS (1914, 1918, 1942) and WILLI consider the possibility of a certain relation between tetany of the newborn and perinatal intracranial hemorrhages. However, proof is difficult because of the great frequency of both diseases and thus far has not been established. SCHNYDER (1952) observed marked mechanical hyperirritability during the first three days of life in 374 of 1534 prematurely born infants when the peripheral nerves were stimulated. This was always accompanied by a hypocalcemia.

FANCONI (1955) and JESSERER (1958) distinguish between two forms of tetany of the newborn: a benign form which appears during the first three days of life, is transitory, and is related to the rapid fluctuations in the blood calcium or blood phosphate levels in infancy or to other metabolic processes; and a severe form which lasts for several weeks or months and has various causes, i.e., true hypoparathyroidism when the mother suffers from this disease, an overfeeding of cow's milk, or malformation or severe trauma of the regulatory centers.

Mechanical hyperirritability, particularly a facialis or peroneus sign, is very frequent in older children. The significance of this phenomenon is still undecided, but one thing is certain: No manifest tetany originates from it as it does in infancy.

The investigations available are concerned mainly with the facialis sign, which has been known longest. Its average frequency cannot be established, since the findings vary greatly. The sign may appear or disappear in a child before our eyes while we remain incapable of influencing its course by our intervening. Whereas tetany of the infant is more common in spring, a connection between the facialis sign and the seasons is rejected by most investigators of older children. Many authors, for example, THIEMICH and FINKELSTEIN (1921), regard it as a sign of latent tetany. Others, such as HOCHSINGER (1911), NEUMANN (1912), and KLEINSCHMIDT (1918), interpret it as one of the phenomena of a neuropathy: according to UHSE (1951), it belongs in the group of the vegetative dystonias. Still others, for example, RAUDNITZ (1914), do not see anything of pathological significance in the facialis sign.

The latent and manifest forms of tetany are rarely observed today. No doubt this has some connection with the prevention of rachitis by vitamin D. Furthermore, I see the facialis and the peroneus phenomena of the older child far less frequently today than I did forty years ago. Particularly in older children, mechanical hyperirritability is frequently restricted to a single nerve distribution.

BORRINO (1908), GOETT (1918), HAMANN (1918), and SECKEL (1933) frequently observed peripheral mechanical hyperirritability, especially the facialis sign, in diphtheria and interpret it as an early sign of postdiphtheric paralysis. According to THIESS (1940), however, hyperirritability does not depend on the severity of the diphtheria and cannot be regarded as a forerunner of postdiphtheric paralysis. Since it can often be eliminated by vitamin D and as it may also occur in connection with other diseases, such as throat infections or scarlet fever, THIESS, following CZERNY's views, regards it as the expression of a tetany associated with the infection.

Incidentally, RAVE (1937) noted just the opposite: He found that as a rule an existing positive facialis sign disappears in the first 6 to 13 days of diphtheria and then reappears.

SUMMARY

Some clinically important reflexes are different in early childhood from those seen later in life. The elicitation of the patellar reflex in the young infant causes an adductor contraction of the opposite side. The Achilles tendon, the cremasteric, and the abdominal reflexes are more difficult to assess in the beginning than later in life, but they are usually present. On stroking the sole of the young infant's foot to elicit BABINSKI's sign, athetoid movements of the toes occur; the characteristic presence or absence of BABINSKI's sign develops from these movements only in subsequent years. LÉRI's and MAYER's signs appear only in the third or fourth year of life.

Among the meningitic signs the most significant for diagnosis are: nuchal rigidity, contraction of the knee joints, the neck sign, and the inguinal reflex — with their substitute reactions — and BABINSKI's sign. Occasionally the more frequently observed signs may be absent and only the rare phenomena are present. The most important meningitic signs are to some extent mutually dependent on each other and together form the meningitic system, which is again related to the flexion synergy of all four extremities, as described by FOERSTER.

The signs of latent tetany, i.e., mechanical and electrical hyperirritability and TROUSSEAU's sign, result from peripheral hyperirritability. The carpopedal spasms of manifest tetany are also connected with the peripheral hyperirritability, whereas the generalized convulsions and the laryngospasm are caused by central excitation.

REFERENCES

ANDRÉ-THOMAS and S. AUTGAERDEN: Psycho-Affectivité des premiers mois (Psycho-affectivity of the First Months). p. 27. Paris, 1959.
ARONOVITSCH, G.: J. Nerv. Ment. Dis. (Amer.) *64*, 235 (1926).
BABINSKI, I.: Sem. méd. (Fr.) *18*, 321 (1898).
–, Rev. neurol. (Fr.) *28*, 2, 145 (1915); *1922*, 1049.
BERSOT, H.: Schweiz. Arch. Neur. *7*, 212 (1920); *8*, 47 (1921).
DE BRUIN, M.: Acta paediatr. (Stockholm) *8*, 277 (1928).
BURR, C.: Amer. J. Dis. Childr. *21*, 529 (1921).
BYCHOWSKI, Z.: Dtsch. Zschr. Nervenhk. *34* (1908).
CATEL, W.: Zoolog. Garten N. F. *16*, 258 (1950).
CATTANEO, C.: Jb. Kinderhk. *55*, 458 (1902).
DIETRICH, H. F.: Amer. J. Dis. Childr. *94*, 265 (1957).
DVORAK, J.: Ars medici *39*, No. 2 (1949).
ENGSTLER, G.: Wien. klin. Wschr. *1905*, 567.
FANCONI, G.: G. Fanconi and A. Wallgren: Lehrbuch der Pädiatrie (Textbook of Pediatrics) 3rd ed., p. 710. Basel, 1954.
FARAGO, J.: Arch. Kinderhk. *8*, 385 (1887).
FELDMAN, W. M.: Amer. J. Dis. Childr. *23*, 1 (1922).
FULTON, J. F.: Physiol. d. Nervensystems (Physiology of the Nervous System). p. 108 of German translation. Stuttgart, 1952.
–, and A. D. KELLER: According to Richards and Irwin.
GALANT, S.: Der Rückgratreflex (The Spine Reflex, Thesis). Basel, 1917.
GOLDSTEIN, M.: Zschr. Neurolog. *61*, 43 (1920).
HARLEM, O. K., and A. LÖNNUM: Arch. Dis. Childh. *32*, 127 (1957).
HOFF, H., and G. OSLER: Neurologie auf den Grundlagen der Physiologie (Neurology Based on Physiology). p. 125. Vienna and Bonn, 1957.
HAVELKA, S.: Csl. pediatr. *10*, 458 (1955).
JACOBSEN, L. F., M. M. JACOBSEN, and F. G. YOSHIOKA: Comp. Psychol. Monogr. *9* (1932).
LANTUEJOUL, P., and E. HARTMANN: Rev. Neurol. *39*, 387 (1923).
MÜLLER, K.: Über Reflexe im Säuglingsalter (On Reflexes in Infancy, Thesis). Munich, 1911.
PFAUNDLER, M.: Münch. med. Wschr. *1902*, 1211.
RICHARDS, T. W., and O. C. IRWIN: Iowa Studies in Child Welfare *11*, No. 1. Iowa, 1935.
SCHOCH, O.: Grenzgeb. Med. *1*, 111 (1948); Homo *1*, No. 2 (1949).
SCHOO, A. G.: Acta paediatr. (Stockholm) *15*, 181 (1934).
SECKEL, H.: Jb. Kinderhk. *134*, 339 (1932).
SHERMAN, W., I. SHERMAN, and C. D. FLORY: Comp. Psychol. Monogr. *12*, No. 4, 1 (1936).
STRÜMPELL, A.: Arch. klin. Med. *24*, 175 (1879).
–, Dtsch. Zschr. Nervenhk. *15*, 254 (1899).
TIEFENSEE, K.: Arch. Psychiatr. (Ger.) *74*, 52 (1925).
WARTENBERG, R.: Die Untersuchung der Reflexe (The Examination of Reflexes). p. 163 of German translation. Stuttgart, 1952.
WEGENER, H.: Mschr. Psychiatr. *66*, 242 (1927).
VAN WOERKOM, W.: Rev. Neurol. *20*, II, 285 (1912).
WOLPERT, I.: Dtsch. Zschr. Nervenhk. *89*, 98 (1926).
ZADOR, J.: Mschr. Psychiatr. *64*, 336 (1927).

Meningitic Symptoms and Signs

BABINSKI, I.: Sem. méd. Paris. *16*, 321 (1898); Rev. neurol. Paris *28*, 2, 143 (1917) and *1922*, 1049.
BÖHME, A., and W. WEILAND: Zschr. Neurol. *44*, 94 (1918).

BROCK, J.: Zschr. Kinderhk. *40*, 552 (1925).
BRUDZINSKI, J.: Wien. klin. Wschr. *1908*, 255; *1911*, 1795.
—, Arch. Méd. Enf. *12*, 745 (1909).
—, Jb. Kinderhk. *71*, 242 (1910), Ref.
—, Berl. klin. Wschr. *1916*, 686 and 916.
ECKSTEIN, A.: Pfaundler-Schloßmanns Handbuch der Kinderheilkunde (Pfaundler Schlossmann's Handbook of Pediatrics). 4th ed., Vol. 4, p. 536. Berlin, 1931.
FOERSTER, O.: Berl. klin. Wschr. *1913*, 1217 and 1255.
GLANZMANN, E.: Einführung in die Kinderheilkunde (Introduction to Pediatrics). p.215. Berlin, 1943.
HILLER, F.: Lehrbuch der inneren Medizin (Textbook of Internal Medicine). 3rd ed., Vol. 2, p. 431. Berlin, 1936.
IBRAHIM, J.: Feers Lehrbuch der Kinderheilkunde (Feer's Textbook of Pediatrics). 15th ed., p. 428. Jena, 1944.
KERNIG, W.: Berl. klin. Wschr. *1884*, 829; Z. klin. Med. *64*, 19 (1907).
LANGE, J.: Mohr-Stähelins Handbuch der inneren Medizin (Mohr-Stähelin's Handbook of Internal Medicine). 3rd ed., Vol. 5, 1, p. 488. Berlin, 1939.
MARANON, G., Ref. Zbl. Kinderhk. *20*, 242 (1927).
RILLIET, F., and E. BARTHEZ: Handbuch der Kinderkrankheiten (Handbook of Children's Diseases). Vol. 3, p. 612 of German translation. Leipzig, 1856.
ROMBERG: Wschr. ges. Heilk. *1834*, No. 30, p. 479.
SCHAEFER, K. H.: In H. Kleinschmidt, Die übertragbare Kinderlähmung (in H. Kleinschmidt's Poliomyelitis). p. 132. Leipzig, 1939.
SAINT-ANNE DARGASSIES: Étud. néonat. *3*, 101 (1954).
TAILLENS, J.: In P. Nobécourt and L. Babonneux, Traité de méd. des enfants (Treatise on Pediatrics). V, 141. Paris, 1934.
WECHSLER, I. S.: Med. Clin. North Amercia *10*, 989 (1927); Arch. Neurol. *14*, 758 (1925).

Symptomatology of Tetany

BANU, G.: Recherches physiol. sur le développement neuromusculaire chez l'homme (Physiological Investigations on the Neuromuscular Development of Man). Thèse de Paris, 1922.
BEHRENDT, H., and E. FREUDENBERG: Klin. Wschr. *1923*, 866.
BIEDERT, P., and R. FISCHL: Lehrbuch der Kinderkrankheiten (Textbook of Pediatrics). 12th ed., p. 559. Stuttgart, 1902.
BORRINO: Ref. Jb. Kinderhk. *68*, 485 (1908).
BOURGIGNON, G.: La chronaxie chez l'homme (The Chronaxie in Man). Paris, 1923.
—, J. Médecine frç. *14*, 388 (1925).
CATEL, W.: Mschr. Kinderhk. *38*, 303 (1928).
CHVOSTEK, F.: Wien. med. Presse *1876*, 1201.
CZERNY, A.: Pädiatrie meiner Zeit (Pediatrics in My Time). p. 110. Berlin, 1939.
—, and A. KELLER: Des Kindes Ernährung (The Nutrition of the Child). 2nd ed., Vol. 2, p. 397. Leipzig and Vienna, 1928.
VAN DER ELST, P. R.: Jb. Kinderhk. *123*, 288 (1929).
ERB, W.: Arch. Psychiatr. (Berlin) *4*, 271 (1874).
ESCHERICH, T.: Die Tetanie der Kinder (The Tetany of Children). Vienna and Leipzig, 1909.
FANCONI, G.: Nebenschilddrüsen, Hdb. d. Inn. Med. (The Parathyroids, Handbook of Internal Medicine). 4th ed., VII, 1, p. 924. Berlin-Göttingen-Heidelberg, 1955.
FINKELSTEIN, H.: Lehrbuch der Säuglingskrankheiten (Textbook of the Diseases of Infancy). 2nd ed., p. 541. Berlin, 1921.
FREISE, R., and P. SCHIMMELPFENG: Mschr. Kinderhk. *30*, 450 (1925).
FREUDENBERG, E.: Pfaundler-Schloßmanns Handbuch der Kinderheilkunde (Pfaundler-Schlossmann's Handbook of Pediatrics). 4th ed., Vol. 1, p. 803. Berlin, 1931.

FÜNFGELD, E.: Die tetanischen Erkrankungen der Erwachsenen (The Tetanic Diseases of the Adult). Leipzig, 1943.
GARSCHE, R.: Arch. Kinderhk. *140*, 65 (1950).
GOETT, T.: Münch. med. Wschr. *1918*, 669.
GYÖRGY, P., and H. STEIN: Klin. Wschr. *1928*, 2424.
HAMANN, M.: Zschr. Kinderhk. *17*, 209 (1918).
HOCHSINGER, K.: Wien. klin. Wschr. *1911*, 1487.
HOFFA, T.: Dtsch. med. Wschr. *1920*, 1144.
IBRAHIM, J.: In: Feer's Lehrbuch der Kinderheilkunde (in Feer's Textbook of Pediatrics). Jena, 1911; Dtsch. med. Wschr. *1920*, 1359.
JESSERER, H.: Tetanie (Tetany). Stuttgart, 1958.
KEHRER, E.: Arch. Gynäk. *99*, 372 (1913); Jb. Kinderhk. *77*, 629 (1913).
KLEINSCHMIDT, H.: Berl. klin. Wschr. *1918*, 1017.
LABICQUE, L.: L'excitabilité en fonction du temps (Excitability and Temporal Factors). Paris, 1926.
LAUGIER, G.: Abderhaldens Handbuch der biologischen Arbeitsmethoden (in Abderhalden's Handbook of Biological Methods). Abt. V, Lief. 285, Teil 5, A, H. 5, Berlin, 1929.
LEDERER, R.: Zschr. Kinderhk. *7*, 1 (1913); *23*, 79 (1919).
LUST, F.: Münch. med. Wschr. *1911*, 1709.
MCCOLLUM, W. G.: Erg. inn. Med. *11*, 569 (1913).
MANN, L.: Mschr. Psychiatr. *7*, 14 (1900); Berl. klin. Wschr. *1904*, 872.
MORO, E.: Wien. klin. Wschr. *1906*, 377.
NEUMANN, H.: Dtsch. med. Wschr. *1912*, 813.
NOTHMANN, W.: Tetanie, im Handbuch der Neurologie von Bumke und Foerster (Tetany, in the Handbook of Neurology by Bumke and Foerster). Vol. 15, p. 173 (1937).
PIRQUET, C.: Verh. Ges. Kinderhk. *13*, 229 (1906); Wien. med. Wschr. *1907*, 1.
RAUDNITZ, W.: Verh. Ges. Kinderhk. *30*, 62 (1914).
RAVE, H.: Mschr. Kinderhk. *69*, 73 (1937).
VON REUSS, A.: Krankheiten des Neugeborenen (Diseases of the Newborn). Berlin, 1914; Halban-Seitz, Biologie and Pathologie des Weibes (Halban-Seitz, Biology and Pathology of the Female). 8. Vienna and Berlin, 1928; Pfaundlers Handbuch der Kinderheilkunde (Pfaundlers' Handbook of Pediatrics). Vol. 1, p. 74. Berlin, 1942.
ROSENBAUM, J.: Frühgeburt und soziale Lage (Premature Birth and Social Conditions). Thesis. Berlin, 1931.
ROTHE, H.: Jb. Kinderhk. *125*, 285 (1929).
SALGE, B.: Zschr. Kinderhk. *19*, 74 (1919).
SCHAAF, R.: Übererregbarkeitserscheinungen am Neugeborenen (Phenomena of Hyperirritability in the Newborn). Thesis. Greifswald, 1945.
SCHÄFFER, H.: Dtsch. med. Wschr. *1920*, 1073.
SCHLESINGER, H.: Neurol. Zbl. *1892*, 66.
SCHNYDER: Thesis 1952, according to H. Willi, Ann. paediatr. (Basel) *178*, 297 (1952).
SECKEL, H.: Dtsch. med. Wschr. *1933*, 1918.
THIEMICH, M.: Pfaundler-Schloßmanns Handbuch der Kinderheilkunde (Pfaundler-Schlossmann's Handbook of Pediatrics). II, 2. Leipzig, 1906; Jb. Kinderhk. *60* (1900); Mschr. Kinderhk. *1* (1902).
THIESS, A.: Mschr. Kinderhk. *84*, 224 (1940).
TROUSSEAU, A.: Medizinische Klinik des Hôtel Dieu in Paris (1861) (The Medical Clinic of the Hotel Dieu in Paris). Vol. 2, p. 162 of German translation. Würzburg, 1866.
TURPIN: Tetanie infantile (Infantile Tetany). Paris, 1925.
UFFENHEIMER, A.: Verh. Ges. Kinderhk. *22*, 240 (1905).
UHSE, H.: Dtsch. Gesd.wes. *6*, 1413 (1951).
WILLI, H.: Mschr. Kinderhk. *80*, 309 (1939).

8. THE NEUROLOGY OF RESPIRATION

MECHANICS OF RESPIRATION

At rest the infant breathes through his nose and therefore easily experiences difficulties in sucking when his airways are blocked by an inflammation of the posterior portion of the pharynx.

The mechanics of the thoracic movements in infancy were studied in detail by GREGOR (1903). In infancy, lung expansion is effected principally by the lowering of the diaphragm. Therefore, the respiratory movements of the abdominal wall are extensive, whereas the thoracic excursions are only minimal. In the erect position the ribs are at right angles to the vertebral column in the same manner as in quadrupeds; compared with that of the adult, the infant's chest is therefore in the inspiratory position. Respiration must overcome the resistance of the abdominal contents, which is greater in the lying than in the standing position. Breathing is shallow, and respiratory frequency is high. After the sixth month of life the thorax becomes lower anteriorly so that the ribs are directed obliquely downward. It may appear that this development is related to the attainment of the erect posture, but feeble-minded infants who constantly lie on their backs develop in the same manner. With the descent of the thorax the predominantly diaphragmatic respiration is gradually replaced by a mixed respiration in which diaphragm and thorax jointly participate. Finally, at approximately the seventh year of life, thoracic respiration is predominant. By the sixth month of life the breathing has already become deeper and slower with the descent of the thorax.

Diaphragmatic respiration is also phylogenetically older than thoracic respiration. According to HASSE (1893), it was the original breathing pattern of mammals. Only during the course of phylogenesis has thoracic respiration become progressively associated with diaphragmatic respiration.

During the first months of life breathing becomes considerably faster when oxygen demand is increased. But to ascribe to the infant a "physiological respiratory insufficiency" caused by the inability to expand the chest would mean that we are overlooking the fact that the "toxic" breathing of the infant, the most important sign of toxic states, is a pure thoracic respiration. The chest is also distinctly expanded during crying.

RESPIRATORY FREQUENCY

(3a)

With increasing age the frequency of respiration, i.e., the number of breaths per minute, diminishes. Fluctuations in the young child are so great that reliable mean values could scarcely be found in the following data VORMITTAG (1933) compiled from the existing literature:

Respirations per minute

Mature newborn infants, 16—93 breaths (DEMMING and HANNER, 1935, 1936)

Healthy infants, 37—49 breaths, extreme values 30—70 (ECKSTEIN and ROMINGER, 1921, 1922)

Prematurely born infants, 36—87 breaths (SHAW and HOPKINS, 1931). On the first day an average of 58 in those with a birth weight up to 1500 gm and 52 in those with a birth weight up to 2300 gm (MARX, 1959).

VORMITTAG, who traced the respiration on a drum, obtained the following values:

	Breaths per Minute	Mean
1st to 7th day	22–72	35.0
8th to 28th day	21–71.5	37.7
1st to 6th month	21–58	33.3
½ to 2 years	19.5–45	27.8
2 to 5 years	21–40	26.5
5 to 10 years	15–31	20.4
10 to 15 years	14–31	19.1

When the infant is awake, he breathes so irregularly that no comparative figures can be obtained. It is different if the respiratory frequency is investigated during sleep, when the breathing is somewhat slower than in the waking state. For many years I have ordered the routine determination of the respiration of all infants by the mere observation of them while they were asleep and had the results recorded on the temperature charts. Especially in young infants the respiratory graph is of great value in the evaluation of the general state of their health.

DIETEL [1954 *(3a)*] determined the respiratory frequency in newborns asleep after feeding in two ways: He counted the breaths by holding a loosely suspended cellulose sheet in front of the infant's nose or traced the respirations on a drum.

When he counted the breaths by the cellulose sheet method he found the following values in 115 newborns:

Days after Birth	Lowest	Highest	Mean
1	28	78	45.8
2	32	72	46
3	32	68	44.5
4	28	66	42.5
5	28	64	39.9
6	26	62	38.8
7	26	62	37.7

It can be seen that the mean value decreases gradually from the first to the seventh day.

The respiratory frequency was studied in 19 children before and after birth. The intrauterine values were between 30 to 72 respirations, with an average of 50.4 per minute; and on the first day of life the average was 49.9, with a spread between 32 and 72. If the intrauterine figures were high, they were also high after birth.

High respiratory frequency in newborns, especially in premature newborns, makes us suspicious of aspiration of amniotic fluid and dystelectasis (partial atelectasis) as well as of hyaline membranes [FRIEDRICH, 1958 *(25a)*], especially when this is accompanied by "Knorksen"* (loud, sharp, deep exhalation), inspiratory contractions of the thorax, and preinspiratory widening of the nostrils. The presence of amniotic fluid aspiration can be shown roentgenographically. If the newborn recovers from this state, the increased respiratory rate soon decreases to normal values and the aforementioned accompanying phenomena disappear.

Healthy sleeping infants, whether mature or immature, breathe approximately 35 to 40 times per minute and the prematurely born 35 to 45 per minute. The rate increases during infections, most frequently and most markedly in pneumonia. Values as high as 70 to 80 breaths per minute are often found in this disease. They often are the first sign to make us suspicious of the presence of pneumonia. Occasionally the respiratory rate may even be markedly higher. I repeatedly observed values of 120 per minute, and in one case I even found a frequency of 132 per minute on the recording drum; shortly afterward 150 breaths per minute were counted in this infant. This is a maximal value that was also seen by DEBRÉ, MARIE, LAMY, and DE FONT-REAULX (1939) in a prematurely born infant.

In pneumonia of the prematurely born infant, fever may often be absent and it may also be difficult to obtain the physical findings. Therefore, the tracing of the respiration is a valuable method. In interstitial pneumonia the respiratory graph shows an increase of frequency long before other signs appear and is the last sign to disappear when recovery from this disease occurs.

Besides pneumonia, meningitis also causes an increase in the respiratory rate, but the values are not so high. The respiratory rate may be permanently increased in congenital heart disease.

Not only the highest but also the lowest respiratory frequency values encountered throughout life occur in infancy. In the gasping respiration of premature infants the breathing slows down so much that the single breaths succeed each other after intervals of minutes. The longest respiratory pause that I was able to measure was 2 minutes. In a hunger state of long duration due to pylorospasm or insufficient food intake the respiratory rate may decrease to 20 or fewer per minute.

---

* German onomatopoetic colloquial or slang expression — The translators.

STRUCTURE OF THE RESPIRATORY CENTER IN HIGHER MAMMALS

(39)

In warm-blooded animals, a lesion of the medulla oblongata leads to death from respiratory arrest. FLOURENS concluded from this fact that a center, the so-called "life node," is located in this area and that this center regulates the entire respiration. Considerable objections were raised against this hypothesis, which postulated the presence of a respiratory center in one single area, for it was possible to demonstrate successfully in experiments on animals that respiratory movements could still persist under appropriate conditions after removal of the medulla oblongata. BROWN-SÉQUARD (1858, 1860) was even able to remove the "life node" without killing the animal by this operation. ROKITANSKY (1874) still observed respiratory movements in rabbits after he had severed the spinal cord at the level of the medulla oblongata if he increased the nervous excitability by strychnine. LANGENDORFF (1880) and WERTHEIMER (1886, 1887) conducted additional experiments on animals to prove the presence of spinal respiratory centers. If artificial breathing was administered for a time after removal of the medulla oblongata, respiratory movements gradually returned and frequently lasted for hours.

LANGENDORFF therefore considers the respiratory center to be merely a physiological concept and not an anatomic entity. This principle was accepted by the majority of investigators after a long struggle. According to LANGENDORFF, the respiratory center unites all segmental centers of the spinal cord and the medulla oblongata that have a respiratory function. It is therefore possible under certain conditions to release individual partial centers, of which the entire respiratory center is composed, from their association and to let them become independent. These findings may be explained by the fact that after decerebration respiratory centers of the spinal cord not previously recognizable begin to function and thus maintain respiration, albeit imperfectly.

The findings of comparative physiology are in agreement with these concepts. In many arthropods each breathing segment of the body possesses an autonomous respiratory center in the proper ganglion. Only during later phylogenetic development do the partial centers combine into a uniformly functioning respiratory center. The individual components, however, are still recognizable in its structure. LOEB (1903) therefore regards the respiratory center in the medulla oblongata as a segmental ganglion of a respiratory organ that exists only temporarily in the embryonic life of man, viz., the gills.

LANGENDORFF's experiments have been frequently repeated, but confirmed only in part. Thus, TRENDELENBURG's (1910) cooling experiments demonstrated that the controlling respiratory and circulatory center was located in the medulla oblongata. The unequivocally isolated respiratory centers of the spinal cord could not maintain rhythmic respiratory movements. BETHE (1925) explains TRENDELENBURG's findings by stating that the necessary conditions for the activation of the lower respiratory centers (pronounced asphyxation) were not sufficiently established by the nonirritating elimination through cold application.

BABAK (1921) and BETHE maintain that even though the conclusiveness of some of LANGENDORFF's experiments could be disputed the correctness of the LANGENDORFF-LOEB theory cannot be doubted. According to this theory the lower centers gradually became more dependent during the evolution of the species so that a specific area, usually anteriorly located, increasingly gained supremacy (BETHE). The central respiratory apparatus may be compared to a federated republic whose single members, independent to a certain degree, are united by an internal bond of communal life (BABAK).

There is no doubt that the respiratory center responds easily to external stimuli. But experiments on the isolated medulla have shown that it possesses automatism, i.e., an excitability originating in itself. According to OBERHOLZER and TOFANI (1960), the respiratory substratum in the medulla must be credited with a faculty of automatic rhythmic functioning. The intact living creature regulates its breathing by superimposed pontine to diencephalic and occasionally also to cortical "respiratory centers." However, as far as we know today, the bulbar centers are the minimal substratum that can maintain regular respiration.

Thus, the respiratory center becomes very similar to another automatically excitable organ, i.e., the conduction system of the heart. Such a comparison has already been made by many authors (LOEB, 1899, 1903; WENCKEBACH; LANGENDORFF; TRENDELENBURG; and others). In the heart the sinus node possesses the greatest automatism and enforces its rhythm upon the lower and less excitable segments of the conduction system. If the conduction system is interrupted, the automatism appears in the other segments.

KIRCHHOFF (1958) also finds similarities between the electrocardiographic pattern of infants dying of cerebral failure and the pattern of loss of the respiratory center: The basic rhythm slows down and the sinus node surrenders control to the subordinate centers. Finally, the pacemaker centers give way under various pictures and activity sequences.

The same applies to the individual partial centers of the respiratory center. Here also the highest component can be the most excitable, while the excitability threshold of the lower centers is not reached (LOEB).

At present the rhythmic function of the heart has been investigated more thoroughly than that of the respiratory center because the examination methods for the heart have been well established for many decades. The less well-developed investigation of the respiratory center can therefore gain much from comparison with investigations of heart function.

Otherwise we depend on animal experiments for the time being. Since the respiratory rhythm of man is changed in some diseases in a specific manner, animal experiments, in which the same changes of the respiratory rhythm can be successfully produced by eliminating higher brain regions, are of great importance. The findings suggest that human and animal respiratory centers function according to the same laws.

LUMSDEN (1922, 1924) made serial sections through the brain stem of cats and observed different types of breathing related to the level of the section. The gasping center is located in the lowest region near the point of the calamus scriptorius, immediately above is the exhalation center, at the level of the striae acusticae is the apnea center for inhalation, and finally in the upper half of the pons is the pneumotaxic center, which effects the highest type of mammalian respiration.

The lower the center is located in the brain stem, the older it is from the phylogenetic and ontogenetic standpoint. According to LUMSDEN, the gasping center is the remnant of an earlier respiratory device. It usually has no influence on respiration because it is inhibited by higher centers. However, if these inhibitory centers become ineffective (because of degeneration, poisoning, or section), the gasping center functions again. The gasping breath is shorter than the usual respiration; between every two gasping breaths prolonged intervals of variable duration occur. Gasping respiration can revive an animal in which the higher and more sensitive respiratory centers have been arrested, and thus can save its life. LUMSDEN compares it to the gasping respiration of a fish taken out of water. We shall show that a similar gasping respiration frequently occurs in the prematurely born infant.

The exhalation center probably participates in the tonus of the exhalation muscles during quiet exhalation; during active exhalation it is inhibited by the pneumotaxic center.

If the apnea center is still functioning, a definite pattern of periodic respiration emerges. First a deep inhalation occurs, and the inhalation position is maintained for some minutes. During the last period of this phase the muscles relax gradually and finally yield completely; this is followed by two or three fairly irregular breaths, and then again a prolonged inhalation takes place, and so on.

TEREGULOW (1929) and V. E. HENDERSON and SWEET (1929) repeated and checked LUMSDEN's animal experiments and ascribed his findings to a brain stem lesion or a section through the vagus but not to the elimination of higher brain regions. W. R. HESS (1931) also rejected these findings.

PITTS, MAGOUN, and RANSON (1939) and PITTS (1941) found a well-circumscribed inhalation center in the ventral formatio reticularis of the medulla oblongata of the cat. On minute electrical stimulation the thorax and diaphragm remained in an extreme tonic inhalation phase and rhythmic breathing was suspended; finally even death could occur. A corresponding region of exhalation was found dorsally and somewhat proximally and enveloping the proximal end of the inhalation region. If the stimulation was continued, the extreme exhalation position was maintained for 2 to 3 minutes, but death did not occur because the inhalation returned gradually. It was possible to achieve coordinated breathing by rhythmic stimulation of both centers. Activity of one center inhibited the other at the same time. The respiratory movements observed in these investigations were coordinated and simultaneously involved the muscles of the thorax and the diaphragm. They therefore cannot be the result of a stimulation of efferent and afferent fiber tracts.

BEATON and MAGOUN (1941) found two circumscribed regions in the respiratory center of apes; on electrical stimulation of one region an inspiratory apnea and on stimulation of the other an expiratory apnea occurred with regularity.

According to PITTS (1946), the ventral portion of the formatio reticularis contains the center of inspiration and the dorsal portion the center of expiration. Inhibitory connections between the two centers cause alternating contractions of the inspiratory and expiratory muscles. When the lungs are expanded during inspiration, impulses from the pulmonary receptors passing through the afferent fibers of the vagus stimulate the expiration center; this mechanism inhibits the inspiration center so that an expiration occurs and, vice versa, the excitation of the expiration center inhibits the inspiration center. Since the neurons of this center are the most sensitive, the inspiration is the predominant phase. During quiet and regular breathing the excitation of the exhalation center chiefly effects a repression of the inspiration while the expiration essentially occurs in a passive manner. But if the breathing is more excited, the expiration muscles participate actively.

Two theories attempt to explain the rhythm of the respiratory center. One theory ascribes automatism to the center; excitations are supposed to be formed in rhythmic repetition even if the afferent excitations are suspended. On the other hand, PITTS assumes on the basis of his investigations that regular breathing is caused by the fact that the tonically functioning inspiration center is continuously inhibited by afferent stimuli of the vagus.

MARCKWALD (1887) and LEHMANN (1888) produced periodic CHEYNE-STOKES respiration in a rabbit by serially sectioning the medulla oblongata. Periodic breathing occurred as soon as the incision reached the externally visible point of the alae cinereae. Aquatic mammals (just as the real amphibia) breathe periodically, a long-known fact. SWINDLE (1926) was able to demonstrate in polar bears and seals that the respiratory rhythm does not essentially change when they are decapitated. He therefore assumed that the spinal cord centers of these animals are still comparatively independent of the centers in the brain stem and are only regulated by them.

According to LUMSDEN, periodic breathing does not suffice in terrestrial animals; rather, they require a higher center that assumes responsibility for uninterrupted respiration. This is the task of the pneumotaxic center, which regularly limits the apnea and thus achieves the breathing pattern of the terrestrial animals.

In suffocating animals the parts of the respiratory center fail in the same sequence from top to bottom as is the case in the serial section experiments (LUMSDEN). The pneumotaxic center vanishes first, prolonged inspirations occur, then the apnea center fails, and finally gasping respiration appears as the last pattern.

It may be debatable whether the individual partial centers of the respiratory center are actually distributed in the brain stem in this simple manner, as LUMSDEN assumes; his conclusions have been repeatedly challenged (ANTHONY, 1939). But my investigations in the infant indicate that a periodic and a gasping center may function as lower partial centers of the respiratory center.

SPIEGEL and ENGHOFF (1928), who experimented with dogs and cats, did not succeed in permanently changing the respiration by operation on the cerebral hemispheres and the mesencephalon. On the other hand, periodic breathing occurred in animals in which only the rhombencephalon and at most the lowest portion of the mesencephalon were preserved. Neither investigator was able to find a special center for KUSSMAUL's respiration. They therefore place the point of attack of the lesions causing this form of respiration in the area in which periodic respiration originates. The two patterns of respiration are produced by the differences in the eliciting causes and not those in the location.

SCHOEN (1928) confirmed in rabbits what LUMSDEN had found in cats, i.e., that respiration does not essentially change as long as the posterior part of the pons is not removed. Expiration predominated just as soon as the cerebral hemispheres and the striatum were removed. Periodic breathing (produced by decreasing the oxygen supply or by intensifying the damage through simultaneous carbon dioxide absorption) was easier to elicit when the cerebral hemispheres were present than when they were absent. In a later publication SCHOEN states that periodic breathing occurs after section at the pons and that deep regular respiration with less ventilation sets in after a lower section.

SCHALTENBRAND and GIRNDT (1925) found periodic motor discharges accompanied by periodic respiratory changes immediately after the operation in thalamus preparations of cats; during movements the breathing was deepened and accelerated, the eyes were opened, and the tonus of the limbs was increased. During the resting intervals the opposite occurred. SCHOEN also described the combination of periodic motor discharges with periodically changed respiration in decerebrated animals.

Another approach to the investigation of the respiratory center is to poison the animals with paralyzing or stimulating substances; the point of attack of the poison can again be ascertained by the gradual removal of brain sections. In this manner certain respiratory patterns, such as periodic and panting respiration, could likewise be produced (SCHOEN, 1928; THIEMANN, 1928). Suffocation leads to the same results. BAGLIONI demonstrated that low oxygen content of the water could elicit periodic respiration of the CHEYNE-STOKES type in fish. Other experiments are described subsequently (see p. 334). In these suffocation experiments swallowing and laryngeal movements, which must be regarded as the lowest form of respiration, occur at the end, as is shown later.

The existing findings from animal experiments, of course, leave many questions still open. However, we can at least conclude from these experiments that the respiratory centers of the higher mammal have phylogenetically been constructed from different superimposed components. Filling the needs of a changed mode of life, new superimposed centers and their corresponding respiratory patterns were formed when the aquatic animals became terrestrial. The oldest centers have not disappeared, but they are usually inhibited in their functions by the superimposed centers. If for some reason the latter are eliminated, the lower centers substitute and maintain the respiration through the patterns characteristic for them. They somehow constitute a last resort for the body to maintain vital respiration and thus life.

In another connection, FULTON (1952) calls the process on which this is based "the law of the development of functional tracts." According to this law, rostral segments of the brain control the caudal portions; after removal of the higher portions many functions of the lower segments are "liberated" or "disinhibited"; thus, it becomes easier to examine these functions. Investigation of these functions shows that each level is established to guide distinct functions but not to supervise anatomic units; in other words, the activity is the main thing and to a lesser degree muscle function. As a consequence of the gradually progressive "encephalization," the cerebral cortex of man has taken over a greater number of tasks than that of the other mammals.

While an abundance of experimental investigations on the structure of the respiratory center in animals is available to us, corresponding observations in man are scarce. As far as we know, the human respiratory center has the same structure as that of animals.

An experiment by KEHRER (1891) showed that the human and the animal respiratory centers are located in the same area. KEHRER sectioned the medulla oblongata of a newborn infant with severe congenital brain damage who was, however, still breathing. The breathing stopped when the point of the calamus scriptorius was reached. The most important part of the respiratory center in the higher mammals is at the same spot.

Periodic respiration, especially that of the CHEYNE-STOKES type, has long been the object of special attention. Little is known of this phenomenon, however, since experiments on human beings are out of the question. However, in young premature infants there is a rather frequent clinical condition that I regard as a "disintegration of the respiratory center" into its phylogenetic and ontogenetic components. If a component other than the main center takes control, lower forms of respiration that are easily recognizable clinically result. *Respiratory disturbances of the premature infant therefore offer an opportunity not present later in life to study the human respiratory center.*

Many findings in the infant which in themselves would be difficult to understand correspond to similar processes in higher and lower animals and particularly in their embryos. It is self-evident, however, that it cannot be said that we are dealing with exactly the same central arrangements. The anatomic structure and the pattern of living of a human embryo and of a fish embryo, for example, are too different from each other to allow any such statement. But we can recognize that the basic structure of the human respiratory center is similar to that of the animal. *Thus, the respiratory disorders of premature infants gain an importance that transcends the field of pediatrics.* Furthermore, respiratory disorders are important because — contrary to an opinion formerly widely held — they play a significant role in infant mortality.

RESPIRATORY DISORDERS OF INFANTS

(26, 27, 33, 34, 39, 42, 45, 46, 49, 50, 84, 86, 123)

In prematurely born infants, ventilation of the lungs is already impeded by the incomplete maturation of the lung tissues. According to ROSSIER (1959), alveoles do not exist until after twenty-four to twenty-six weeks of gestation; there are only canals, covered with cubic cells, that carry the capillaries from the respiratory surface. The canals themselves are still filled with a not very well differentiated mesenchyma.

Aspiration of amniotic fluid, atelectasis, hyaline membranes, hemorrhages, and later aspiration pneumonia, reduce the respiratory surface still more and easily lead to hypoxia, to which the immature respiratory center is especially sensitive. The resulting respiratory disturbances in turn reduce the aeration of the lungs. Thus, a vicious circle which progressively increases the hypoxia but which can be interrupted, at least temporarily, by a supply of oxygen occurs.

Since the respiratory disorders of premature infants repeatedly show the same clinical pattern, they must be caused by the functioning of preformed centers. An external crude force which affects the region of the respiratory center during birth cannot cause a systematic disintegration of the respiratory center, nor produce such a delicately connected vital interplay between respiratory and sucking center as is represented by pendulum induction (see p. 435). However, the respiratory disturbances of the prematurely born were explained particularly on the basis of perinatal intracranial hemorrhages by YLPPÖ (1919), LUMSDEN (1924), SCHWARTZ (1927), and DOLLINGER (1927).

"These hemorrhages in the region of the cerebellum and medulla oblongata have special significance in relation to the frequent attacks of asphyxia in prematurely born infants... as to the etiology of these attacks the first thought should be directed to cerebral injuries" (YLPPÖ, 1923, 1931).

CREUTZFELDT and PEIPER [1932 *(42)*] examined the brain stems histologically in serial sections of premature infants who had evidently died from such disturbances and found hemorrhages only in exceptional cases (see p. 553). We see in our findings a refutation of the statements of YLPPÖ, SCHWARTZ, and others that the letal respiratory disturbances of premature infants must be a consequence of hemorrhages in the region of the respiratory center, but we do not deny their importance (CREUTZFELDT and PEIPER, 1932, p. 35):

"Any damage that impairs the general state of health attacks the respiratory center and diminishes its excitability. It therefore happens that from injuries, such as infections, chills, and perhaps also extended paranatal cerebral hemorrhages, severe potentially fatal respiratory disturbances may develop."

HIRVENSALO (1949) of YLPPÖ's clinic has rejected the validity of our investigations, but arrives at the same results with the same procedures based on comparative clinical and pathological-anatomic examinations of 54 premature infants and 20 born at term: Minor hemorrhages, which can be demonstrated only histologically, in the medulla oblongata and the pons are quite frequent in prematures. They are of no significance as a cause of respiratory disorders, but they are an expression of the damage to the tissue in the region of the respiratory centers. On ESSBACH's suggestion, RÜHL (1952) examined the entire formatio reticularis for hemorrhages by continuous serial sections; he did this in 50 mature and premature newborns. Erythrodiapedetic minor hemorrhages in the cortex (circular hemorrhages) were found in 31 of the 50 newborns; they were distinctly more frequent and severe in the premature infants than in those born at term. However, the hemorrhages were all so insignificant that the functioning of the respiratory center appears not to have been threatened.

RESPIRATORY PATTERNS

It is characteristic of the respiration of premature infants that the pattern and rhythm of each individual breath changes easily. The resultant breathing, which in itself presents a completely regular appearance, is maintained for some time, until a new type of breathing — mostly the previous one — occurs. I term these respiratory phenomena, which remain in themselves unchanged and return regularly, "respiratory patterns." They let us infer fluctuations of central excitability and can be produced in a similar manner in animals by operations on the respiratory center or by toxic substances. Smooth transitions between the individual patterns of respiration usually do not occur; the point at which one pattern is replaced by another is often clearly indicated in the graph. The respiratory center acts, so to speak, like a piano on which different keys are hit successively. Just as two keys can be hit at the same time, respiration may be regulated by two centers at the same time.

Mature (Regular) Pattern

Regular respiration like that seen in the adult, in which each breath is equal to the preceding one, can frequently be found in mature or premature infants even at the earliest time of life; it is most distinct during sleep in a completely undisturbed environment.

If the respiration of quietly sleeping premature infants is traced, the height of each individual breath can fluctuate frequently, apart from the change of the various respiratory patterns, which is described later. The base line of the respiratory tracings is frequently not straight but shows slow waxing and waning which may last through several respirations (see first half of Fig. 107). These waves, which can also be found somewhat less frequently in newborns born at term, were first described by VALLOIS and FLEIG (1905), who attributed them to central influences. ECKSTEIN and ROMINGER (1921, 1922), who also described them, see in them a direct sign of the diaphragmatic function. I am also inclined to interpret them as centrally conditioned changes in the tonus of the diaphragm, where the act of breathing originates. Even when these breaths cease completely for some length of time, as is the case in gasping respiration or in agony, slow or quick diaphragmatic movements may still occur (Figs. 118, 120, and 121).

Periodic Respiration

(26, 34, 39, 45, 46, 50, 51)

Periodic respiration, which includes the form described by CHEYNE and STOKES, has been known the longest; this pattern consists of periods of slowly increasing and decreasing respiratory excursions with the separate periods divided by pauses (Figs. 107 and 108). In premature infants I was able to find only traces of BIOT's respiratory pattern, which consists of periods of respiration of equal depth separated from each other by pauses, and I did not see this pattern in its pure form. However, the periodically accelerated form is frequently found in premature infants. In this form periods of accelerated breathing alternate with periods of decelerated respiration (Fig. 109). The various forms of periodic respiration are all interrelated, show all kinds of gradations of intermediate types and occur under the same conditions, viz., the prematurely born infant, the meningitic patient, and the adult suffering from heart disease. They evidently present only different stages of the same condition

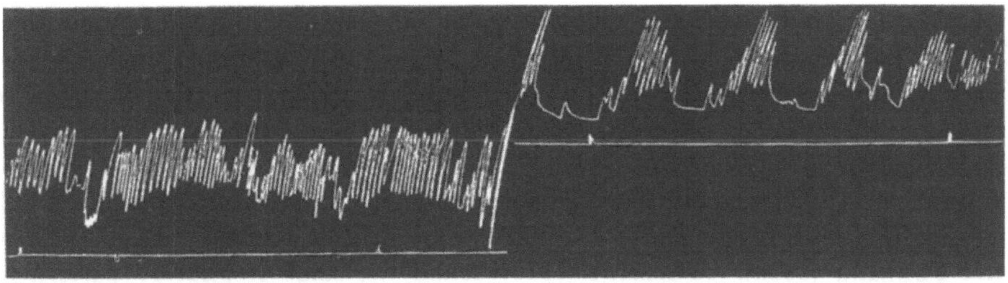

Before feeding After feeding

Fig. 107. Respiration during bottle feeding (time in minutes).

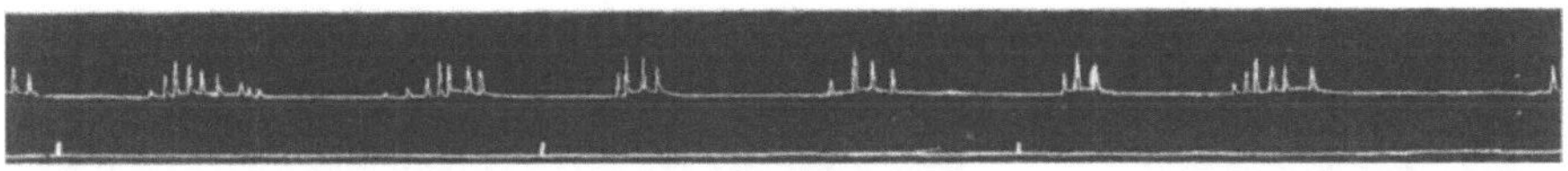

Fig. 108. Periodic respiration of a dying premature infant, 4 hours before death (time in minutes – lower tracing).

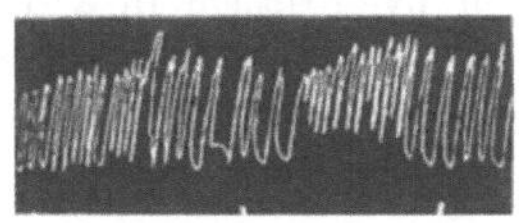

Fig. 109. Periodic alternation of accelerated and decelerated respiration in a 7-day-old premature infant weighing 2180 gm (time in 15-second intervals).

and therefore are closely related, as MARCKWALD (1887), DEDEK (1913), ECKSTEIN and ROMINGER (1921, 1922), and S. WASSERMANN (1922, 1923) have already stressed. I term the center causing this periodic respiration the "periodic center."

BABAK (1911, 1921) and DEDEK (1913) regarded this periodic center as an ontogenetically and phylogenetically lower subdivision of the respiratory center. BABAK speaks of a primary periodicity in the function of the respiratory center. DEDEK considers the regular and uninterrupted respiratory rhythm to be a secondary, later, and higher form of the functioning of the respiratory center, whereas the original rhythmic functioning peculiar to the respiratory center is periodic. Periodic respiration is characteristic of many healthy animals, such as aquatic mammals (whales, seals, polar bears), of bats, marmots, and mice during hibernation, and of amphibians and reptiles, such as frogs, lizards, and salamanders (MARCKWALD; SIEFERT, 1896; BABAK, 1913; SWINDLE, 1926; and others).

I devoted great effort to eliciting periodic respiration in rabbits. Finally I succeeded when I brought the intact animal into a completely quiet environment and, keeping all sensory stimuli away, examined it there (Fig. 110). The tendency to periodic

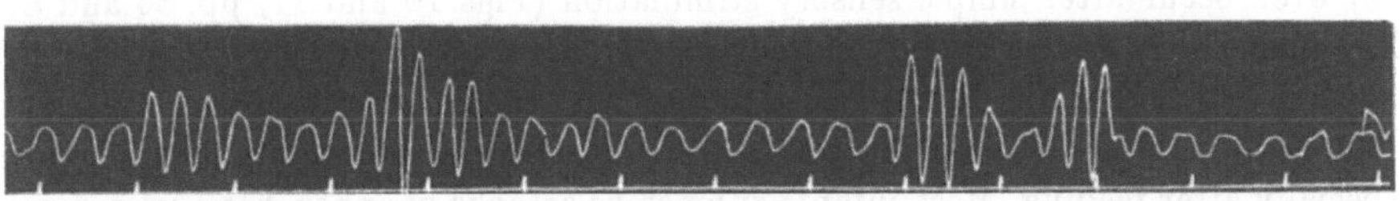

Fig. 110. Periodic respiration of a healthy rabbit at complete rest (time in seconds).

breathing is quite variable and fluctuates in the same animal at different times, too. In frogs and salamanders, BABAK observed periodic respiratory movements of the larynx in the absence of external stimuli and during complete inner rest. According to SIEFERT, suffocation is the most important cause of periodic respiration in batrachians and mammals.

WINSLOW (1799) already observed intrauterine periodic respirations of animal fetuses. This observation was later confirmed by BONAR, BARNETT, BLUMENFELD, and FENNING (1938). CAVAZZANI (1910) found the same respiratory pattern in newborn puppies and SALMI (1930) in premature rabbits obtained by Cesarean section and in mature newborn rabbits. Usually this pattern disappears during the

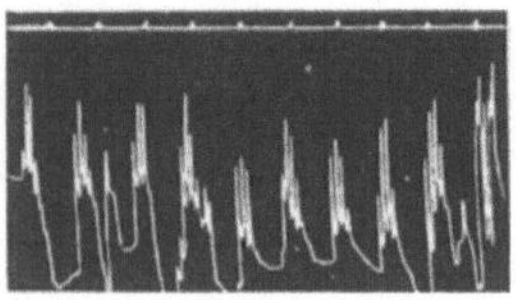

Fig. 111. Periodic breathing in pylorospasm in a 4-week-old infant weighing 2840 gm (time in 20-second intervals).

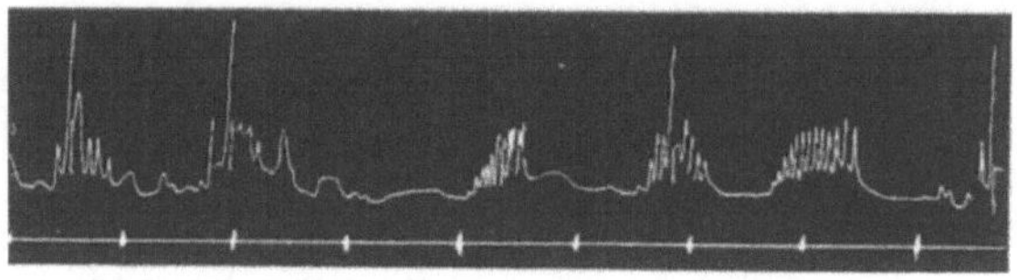

Fig. 112. Singultus during periodic respiration in a 17-day-old premature infant weighing 1900 gm (time in 15-second intervals).

first days of life. SALMI attributes this type of breathing not to a central process but to fatigue of the respiratory muscles. According to WINTERSTEIN (1921), periodic respiration is probably widespread among newborn animals.

Periodic respiration is not always pathological in man. MOSSO (1878, 1885) found it quite often in sleeping adults; I myself repeatedly observed it in healthy sleeping infants. AHLFELD's (1888, 1891, 1905) observations show that intrauterine respiratory movements may become periodic too. He even reported a "periodically accelerated intrauterine respiration."

The occurrence of periodic breathing in premature infants has long been known. The more underweight the infant and the poorer the general state of health, the more likely it is to occur. Hunger and weight loss favor this pattern in premature rabbits (SALMI, 1932). In infants who have a tendency to this type of breathing, certain causes may change mature (regular) to periodic respiration. This may be caused by warm baths and elevated temperatures (MENDELSSOHN, 1913; ECKSTEIN and PAFFRATH, 1928; and SALMI and VUORI, 1930). That feeding alone can cause periodic respiration is shown in Fig. 107. I saw the change from regular to periodic respiration occurring after belching or vomiting. Periodic respiratory movements may even occur after simple sensory stimulation (Figs. 10 and 11, pp. 86 and 87). The intrinsically reduced excitability of the brain stem is so depleted by this neural exention that the functional ability of the respiratory center sinks to a lower level. This explains the occurrence of gasping respiration under similar conditions, especially after feeding. Most infants subject to attacks of apnea have in general a tendency to periodic breathing. This respiratory pattern can be endured for quite some length of time, since it permits adequate gas exchange. The infant may survive even if the respiration is predominantly periodic for weeks.

Breton and Gérard-Lefêvre (1958) also trace periodic breathing to the immaturity of the respiratory center. Its appearance is facilitated by lack of sleep or oxygen. This type of respiration disappears, as Alison (1959) has also found, after oxygen is supplied. Hyperthermia, shock, birth trauma, and cerebral hemorrhages may easily cause attacks of apnea.

Graham and Morris (1929) described the occurrence of periodic respiration in pylorospasm, an observation which I am able to confirm (Fig. 111). Here the respiratory center may have a tendency to disintegrate because of severe hunger.

Periodic respiration during tuberculous meningitis has been known for a long time. Eckstein and Rominger (1921, 1922) postulate, as I do, an incomplete paralysis of the respiratory center and associate the periodic respiration of the premature with the particular qualities of the respiratory center at this stage of development.

Periodic respiration is so frequent in tuberculous meningitis that Biot's respiratory pattern has been erroneously called the "meningitic" pattern. In meningitis occurring after infancy other reflexes can also be released as an expression of a breakdown of cerebral function. These reflexes are usually present in young healthy infants but are later inhibited by higher centers that have matured in the meantime (see p. 299).

In the toxic form of diphtheria, periodic respiration sometimes occurs in the last days before death; this respiration is accompanied by a clouded sensorium. It can be best observed when the child is at complete rest; it disappears as soon as he is aroused from his stupor and can temporarily yield again without visible cause to the regular breathing pattern. However, death occurs at the latest some days after the first appearance of this respiratory pattern.

Periodic respiration, whether it occurs in premature infants, meningitic patients, or adults suffering from heart disease, shows an additional characteristic. Aside from the respiration, other nervous processes may also become unbalanced.

Thiemich (1898) described distinct periodic breathing in two children suffering from tuberculous meningitis; in these cases the size of the pupils fluctuated simultaneously with the periods of respiration. During respiratory arrest the pupils were medium-sized or somewhat constricted, and when respiration set in they dilated considerably. If respiration stopped, they contracted again.

S. Wassermann (1922, 1923) summarized these phenomena, on which Mosso had already compiled the observations reported in the literature, under the title "Cheyne-Stokes symptom complex." Accompanying the periodic respiration of the adult, fluctuations may occur in the heart action, the width of the pupils, the movements of the eyeballs, coughing, sighing and moaning, urinary urgency, reflex irritability, and consciousness; motility particularly shows periodicity.

I was able to show some of these periodic fluctuations in premature infants with periodic respiration. Often groaning and coughing, less frequently singultus (Fig. 112), and sometimes also the accompanying movements of the lower jaw fluctuate together with the respiratory periods. The block in the respiratory center, in which the periodic center frequently participates, is described on page 339.

GOOD and PEIPER [1934 *(50)*] made a similar observation in regard to the heart action when they compared the respiration and the heart action of an eight-week-old hydrocephalic infant weighing 2200 gm. The infant, showing only minimal reaction, presented distinct CHEYNE-STOKES respiration with protracted intervals of apnea. Because of the length of the electrocardiographic tracing the results are summarized in Fig. 113. The number of heart beats increased regularly when the respiration stopped; the frequency was 150 per minute during respiratory arrest and 100 per minute during breathing, i.e., the heart beat more rapidly when breathing stopped. However, we observed another premature infant whose heart action remained regular

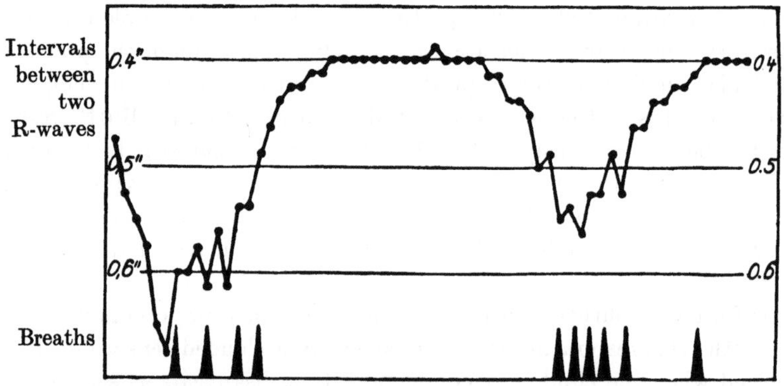

Fig. 113. Periodic respiration and periodic heart action in a 2-month-old hydrocephalic child of 2200 gm. The ordinate shows the time intervals between single heart beats in 1/10-second intervals; on the abscissa the heart beats are represented simply according to their numbers. On the lower line the single respirations are sketched in their time relation to the heart beats: respiratory arrest during accelerated heart action and breathing during decelerated heart action.

despite distinct periodic respiration. The heart beat of the adult may fluctuate in the same manner during CHEYNE-STOKES respiration (WENCKEBACH, 1927).

In the adult these fluctuations are regarded as centrally conditioned and are placed in the same category as those likewise centrally conditioned phenomena that have been described. WENCKEBACH and S. WASSERMANN (1922, 1923) assume that periodic stimulations flow to the heart from the vagus center. They base their opinion on VERWORN's (1903) findings, i.e., that the respiratory center is capable of transmitting intracentral periodic stimulations to the neighboring vagus center and can thus force its own periodicity upon it. According to VERWORN, excitability and the excitation of the vagus center increase during suffocation; the stimuli radiating from the respiratory center to the vagus center are increased.

In our observations we also assume a periodic stimulation of the vagus center; we were even able by another observation to demonstrate clearly the influence of the vagus on the heart action in the hydrocephalic infant mentioned: Following pressure on the anterior fontanel there appeared, after an initial acceleration, a distinct deceleration which lasted longer than the pressure.

Mere observation of premature infants and meningitic children reveals a striking periodic motility disturbance accompanying the periodic respiration. The movements consist of clonic, always identical spasms of the face, arms, or legs. Their intensity fluctuates in different children between minor twitches and seizure-like fits. They generally appear at the end of a respiratory arrest, just when breathing starts again, but can occasionally precede this phase (Figs. 114 and 115) and may even occur in the middle of the respiratory period. In periodically accelerated respiration they

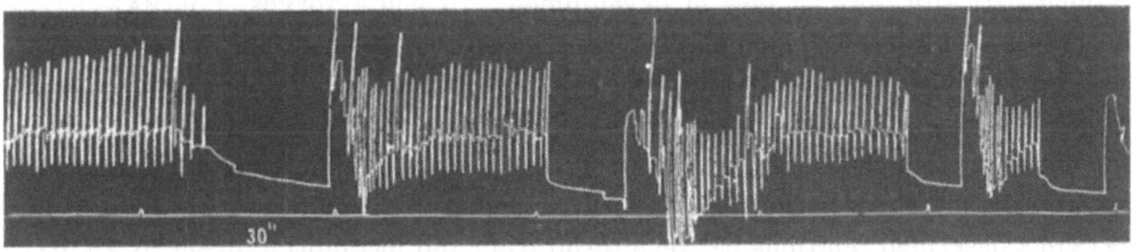

Fig. 114. Periodic panting respiration of a dying premature infant, 3 hours before death (time in 30-second intervals). After each respiratory arrest, generalized movements occur.

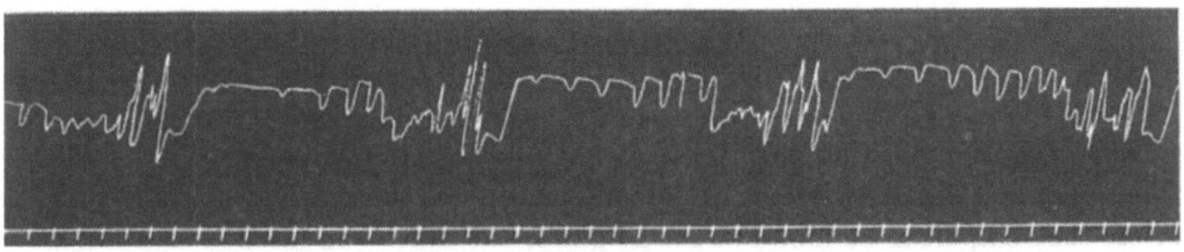

Fig. 115. Periodic respiratory and movement disturbance in a 5-month-old infant with pneumococcic meningitis (time in seconds). Periodically accelerated respiration; generalized seizures that change the respiratory pattern occur each time at the beginning of the acceleration.

appear at the beginning of the acceleration (Fig. 127). This periodic movement disorder was also described by Saint-Anne Dargassies in 1955. Salmi (1930) observed a corresponding periodically appearing motor restlessness of several seconds' duration in prematurely born rabbits. The respiration became periodic at the same time or immediately afterward. Breathing and motility showed the same relationship observed in decerebrate animals by Schaltenbrandt and Girndt (1925) and Schoen (1927, 1928).

Thus, the function of the respiratory center, the vagus center, and the motility centers may rhythmically fluctuate at the same time in the brain stem of the young infant; the fluctuations of the respiratory center are the prerequisite for those of the other centers, but they are not necessarily accompanied by these fluctuations. The different centers seized by the excitation are not always excited at the same time; rather, the irritation may at times spread like waves and then disappear in the same manner. It remains to be seen how far the mutual adjustment of the various rhythms amounts to a relative coordination in the sense of VON HOLST's hypothesis (see p. 422).

All these periodic fluctuations of excitation are distinctly centrally conditioned. SALMI (1932) and YLPPÖ (1954), who see the easy exhaustion of the respiratory muscles as the cause of periodic respiration, did not explain the CHEYNE-STOKES symptom complex and pendulum induction between respiratory center and sucking center (see p. 435); both phenomena prove that we are dealing with central processes. SALMI was unsuccessful in finding support for his hypothesis in experiments on animals.

Generalized movements simultaneous with respiration are not infrequent in phylogenetically lower stages of development. In fish embryos, POLIMANTI (1912) describes respiratory movements accompanied by general body movements. It seemed that the entire central nervous system had acquired this rhythmic ability. Later this association disappeared. BABAK (1911) sees an "auxiliary breathing device" in similar generalized movements that he observed in fish embryos even prior to the appearance of their actual respiratory movements. BARCROFT (1942) and W. F. WINDLE described relations between respiration and locomotion in sheep. In decerebrate dogs, P. D. OLEFIRENKO observed periodic extensor movements of the extremities that corresponded to the respiratory rhythm.

Wheezing Respiration

(39)

This type of breathing consists of short inhalations and prolonged groaning or wheezing exhalations. MAYR (1862) already mentioned their occurrence in pneumonia, where they can often be found at the height of the illness. In infancy and childhood, wheezing respiration is generally accompanied by respiratory movements of the alae nasi (see p. 370). This is not so marked in prematurely born infants.

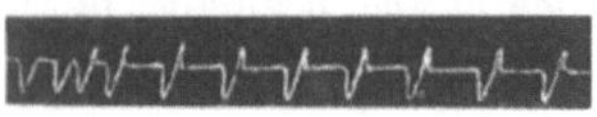

Fig. 116. Wheezing respiration of a premature infant.

The tracings of wheezing respiration are unmistakable (Fig. 116): Each breath follows the other without any interval. Inspirations are rapid and short and are immediately followed by expirations. The expiration is prolonged and has three phases: At first a rapid expiration occurs but proceeds only approximately halfway, then there is an arrest of respiration so that the graph becomes horizontal, and finally the last phase of the expiration occurs just as rapidly as the first. The child groans particularly during the middle phase of the expiration.

The wheezing respirations may either follow each other without interruption or may combine into periodic respiration (Fig. 114). ECKSTEIN and ROMINGER (1921, 1922) made tracings of wheezing respiration in pneumonia of infants, and HOFBAUER (1925) recorded corresponding graphs of the adult. This respiratory pattern was also observed in animals. For example, SCHOEN (1928) reported (Fig. 5 of his book) "a typical agonal wheezing respiration" of the CHEYNE-STOKES type that he produced in a rabbit after making an incision through the middle of the pons. SALMI (1932) (Fig. 4 of his paper) reproduced the graph of wheezing respiration of a dying, prematurely born rabbit. The animal had been taken by Cesarean section from the uterus of its already dead mother.

There is no question that severe pneumonia in infancy is usually accompanied by wheezing respiration. The audible wheezing of the expirium often permits a

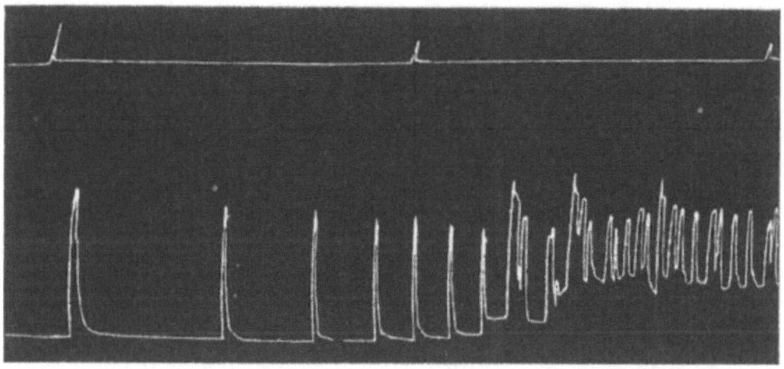

Fig. 117. Abating attack of suffocation in a premature infant (time in minutes). At first, gasping breaths with irregular periods of respiratory arrest; then transition to regular respiration. The respiratory tracing is interrupted by some generalized movements.

positive diagnosis of the disease. In the premature infant the appearance of this respiratory pattern is always a bad sign. Immediately after birth a sharp, purely expiratory sound is heard, the "Knorksen,"* which is a sign of a high degree of aspiration of amniotic fluid with atelectasis or hyaline membranes; newborns with these phenomena are in great danger.

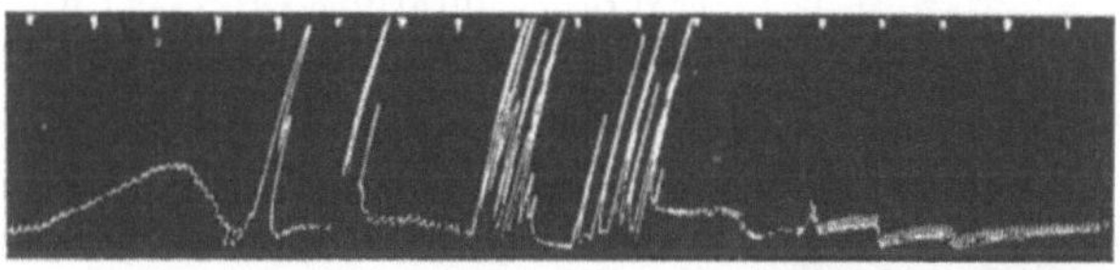

Fig. 118. Respiration and heart action recorded from the abdomen of a premature infant (time in 10-second intervals). The heart action is considerably accelerated after repeated gasping respiration. Diaphragmatic movements sometimes occur between the individual gasping breaths.

Whether wheezing respiration constitutes a breakdown of the respiratory center is undecided. Its appearance in pneumonia suggests the possibility that peripheral stimuli may play a role in the occurrence of this respiratory pattern.

Gasping Respiration

(27, 33, 34, 40, 42, 45, 46, 49, 50, 67, 79, 88, 114, 122)

For a long time the extremely important gasping respiration did not receive any attention in clinical studies at all, and very little notice was taken of it in animal experimentation. I mean by "gasping respiration" a respiratory pattern

* See footnote, p. 312.

composed of individual gasping breaths; they are separated from each other by periods of respiratory arrest varying in duration. The gasping breath is considerably shorter and deeper than that of normal respiration, as can be seen on the recorded tracing (Fig. 117). When the gasping respiration is at its height, the lower jaw drops down and the mouth is opened with each inspiration. Sometimes a smacking sound can be heard; the head is jerked backward, and the tongue is retracted. The closing of the mouth and the sliding back of the head to the original position during expiration seem to be caused by a decrease in muscle tonus (Fig. 119). Thus, gasping respiration fully corresponds to the popular expression "the last gasp." The same form of respiration can be observed in the normal adult after a voluntary or involuntary respiratory arrest of prolonged duration, for example, after diving.

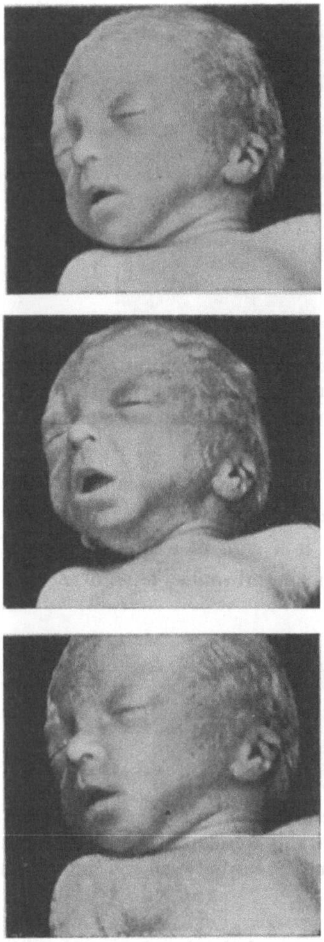

a b c

Fig. 119. Apnoic attack in a 1-day-old premature infant weighing 1650 gm (frames of a film). (a) At rest; (b) gasping breath with opening of the mouth and raising of the head; (c) at rest. [According to PEIPER, HEMPEL, and THOMAS, 1951 *(114)*]

The gasping breath is always preceded by a prolonged respiratory arrest in the premature infant too (Figs. 117–121). The interval may last for some minutes. The longest pause I measured was 2 minutes, but no doubt longer respiratory arrests take place. These pauses are finally interrupted by one respiration and less frequently by several respirations (Fig. 118) which succeed each other in short intervals and are again followed by respiratory arrests. Sometimes the diaphragm is completely at rest during the intervals (Fig. 117). In other graphs slight abrupt movements occurring at short intervals can be seen (Fig. 118) and between them bigger round waves are occasionally apparent as an expression of slower and stronger diaphragmatic innervations (Figs. 118 and 120). Infrequently, the diaphragmatic tonus is changed to such an extent that each gasping breath originates from a different base line (Figs. 120 and 121). In contrast to gasping respiration, all these intervening movements of the diaphragm supply the lungs with only inadequate amounts of oxygen. These movements bring to mind the results of RECHNITZER's (1927, 1929) and CONDORELLI's (1927) experiments; these authors observed involuntary diaphragmatic movements in healthy adults after a long-lasting voluntary respiratory arrest. These movements constantly increased in size and frequency, but were ineffective since the respiratory arrest was continued.

In Fig. 121 a short attack of gasping respiration is recorded from its onset to the end. It can clearly be seen that the single breaths follow each other, as always, at irregular intervals.

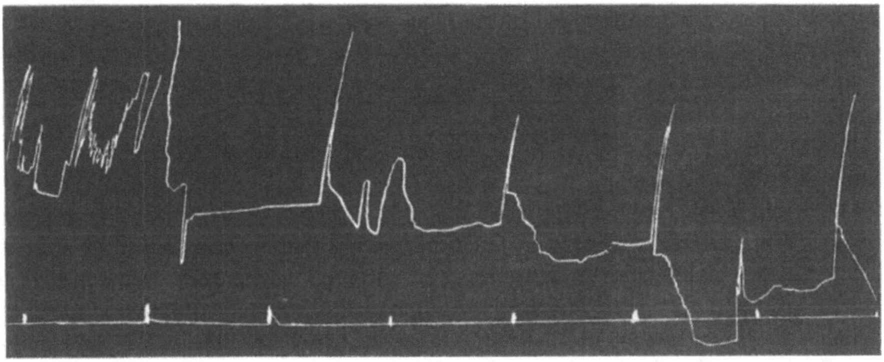

Fig. 120. Attack of suffocation following an attack of whooping cough in a premature infant (time in minutes). Five gasping breaths with respiratory arrests lasting some minutes.

During birth the child experiences a stage of increasing suffocation because of progressive embarrassment of the placental circulation. Therefore, extrauterine respiration always begins as gasping respiration, as my co-worker L. SCHMIDT [1950 *(16a)*] has noted. This pattern usually progresses after a short time to the normal and regular form of respiration. The gasping respiration of the newborn has been recorded in tracings by L. SCHMIDT (1950) (Fig. 142, p. 361) and on films (Fig. 143, p. 363) by PEIPER, HEMPEL, and THOMAS [1951 *(114)*].

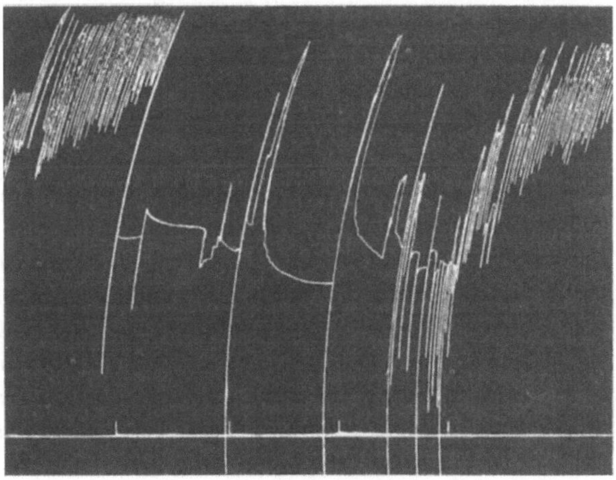

Fig. 121. Attack of total suffocation in a premature infant (time in minutes). Minor diaphragmatic movements during the pauses in the gasping respiration.

Young animals are considerably more resistant to suffocation than older ones. REISS and HAUROWITZ (1929) reported that newborn mice can remain in a room filled with pure illuminating gas for approximately 1 hour without suffocating, whereas adult mice die after only one minute. Similar differences existed when injections of lethal doses of strychnine were given or the head and spinal cord were clamped off or when the animals were poisoned with prussic acid. However, no difference was found in the effect of chloroform. The animals were considered to be dead when they stopped breathing and were no longer excitable by mechanical stimuli. When the thorax was opened, the heart was usually still beating in young animals, whereas it had stopped in the older ones.

FAZEKAS, ALEXANDER, and HIMWICH (1940) saw fetuses of suffocated mothers breathe and stay alive. The respiration of rat sucklings that had been held under water for 40 minutes was not damaged after this experience. Newborn puppies survived suffocation when their arterial blood was practically exhausted of oxygen in the first minutes after suffocation.

OPITZ and K. M. SCHNEIDER (1950) also found that newborn animals can tolerate oxygen deprivation or suffocation much longer than adult animals are able to do. These authors summarize the statements of various investigators; the adults of all animal species investigated survived only 3 minutes in a pure nitrogen atmosphere, whereas newborn rats lived 50 minutes, cats 25, dogs 23, rabbits 17, and guinea pigs 7 minutes. Gasping respiration lasted 485 seconds in decapitated newborn rats, but only 40 seconds in adult rats. Human embryos can also be revived after a long period of time has elapsed. Viable fetuses have been delivered by Cesarean section after sudden accidents as long as 15 to 20 minutes after the mother's death. The situation is similar in neonates born at term. WULF (1958) described an overdue infant who took only a few gasping breaths in the first 4 minutes after a spontaneous delivery, then breathed regularly, and first cried after 8 minutes without benefit of artificial respiration. Immediately after birth the oxygen tension was 2.4 mm Hg in the vena umbilicalis and 1.6 mm Hg in the arteria umbilicalis. The adult loses consciousness at an arterial oxygen pressure below 30 mm Hg.

An observation of MARBY (1959) shows that occasionally a long hypoxemic state after birth may be tolerated without detrimental aftereffects: A premature infant weighing 1490 gm was born without pulsations in the umbilical cord. Since he did not breathe, he was treated with endotracheal artificial respiration for 15 minutes and then was believed to be dead. To demonstrate the attempt at resuscitation endotracheal respiration was again started 25 minutes after birth. Five minutes later the umbilical cord started to pulsate, and 40 minutes after birth the first breath occurred; 20 minutes later the respiration was regular. At the age of 20 months the child was physically and mentally well developed.

OPITZ (1950) and WULF (1958) see, in the ability of the embryo and newborn to survive a long-lasting interruption of blood supply without detriment, an adaptive device that protects the child from the danger of suffocation during birth. The cerebral cortex is not yet functioning and therefore is not so sensitive as later in life. The brain of the newborn is not indifferent to prolonged oxygen deprivation. The encephalodystrophic foci in the newborn's brain are the lasting results of transitory circulatory disturbances during birth.

The appearance of gasping respiration depends on the co-effect of the glomus caroticum, as TSCHIRGI and GERHARD (1947) showed in their animal experimentation. They filled the glomus caroticum of a rat with the blood of another rat and changed the type of gas inhaled by each of these rats so that the centers of the medulla oblongata and the glomus caroticum of the same animal were exposed to different conditions. A simultaneous lack of oxygen in the brain and in the glomus caroticum is a prerequisite for the appearance of gasping respiration.

Gasping respiration frequently occurs in premature infants during the first period of life. One speaks of "asphyctic" seizures, but this term is misleading. Literally translated, "asphyctic" means "pulseless." But the pulse cannot be judged even in healthy young infants. The heart action is quite adequate at the beginning of the seizure — as we shall show later — and deteriorates only after a long duration of the attack. Since the tracings reported prove that we are dealing with respiratory arrests, the term "suffocation" (apnoic) attacks would be more appropriate for these conditions.

The obstetrician uses the term "asphyxia" of the newborn erroneously since in reality we are dealing with paralysis of the respiratory center. The wrong term may cause a wrong therapeutic approach: Respiration must be stimulated; circulatory stimulants are unnecessary. Their application means a waste of the time which is so valuable in this perilous situation.

Gasping respiration is accompanied by very severe life-threatening phenomena and often ends in death. At the onset the skin shows a bluish discoloration and finally becomes deadly pale if the attack lasts very long. The blue and white "asphyxia" of the newborn corresponds to this condition. The pupils are markedly dilated and no longer react to light. The body muscles become completely flaccid and the spontaneous movements, with the exception of the diaphragmatic movement previously described, disappear. The tendon reflexes and all other reflexes that can be elicited in infants are extinguished; only the respiratory center can at times still be reflex stimulated. Often a strong stimulus may still evoke a new gasping breath. The most favorable moment for this occurrence is after a prolonged respiratory arrest, whereas the identical stimulus may not have any effect immediately after a breath. Evidently the respiratory center is nonexcitable for some time after a breath, but otherwise we are dealing with the same process that also changes the rhythm of normal respiration (see p. 24). The stimuli are still transmitted to the gasping center, which, however, is more resistant to stimulation than other centers.

If gasping respiration changes again into a higher respiratory pattern, a motor restlessness often occurs at the beginning of the new pattern (Fig.88). This restlessness consists of an irregular and usually slight twitching of the face, trunk, and extremities.

The discoloration of the skin and the long arrest of respiration and of spontaneous movements, as well as the flaccidity of the muscles, can lead laymen and even physicians to believe that death has already occurred. The most important differential sign, the continuation of the heart contractions, cannot be easily recognized because of the decrease or disappearance of the heart sounds. However, if the child is observed continuously for some length of time, gasping respiration may be noted from time to time. It is therefore better to wait in doubtful cases than to pronounce the child dead prematurely. It is, of course, important to be acquainted with this pattern of dying, which is frequently found in the prematurely born.

Physicians who pronounce such an infant dead and therefore discontinue treatment may get into serious difficulties. I saw an infant die 3 hours and another one, weighing only 670 gm, even 42 hours after the signing of a death certificate. Such incidents could be avoided if all doctors and midwives were familiar with suffocation attacks.

Gasping respiration may also occur in dying mature infants who had not previously suffered from suffocation attacks but from other diseases. A breakdown of the respiratory center may also occasionally occur in dying older children.

Gasping respiration frequently is soon followed by death, but it also may last for 15 minutes and finally again be replaced by normal breathing. But if it continues for several hours — I once observed it to last 12 hours — all attempts at treatment will eventually prove unsuccessful. Lobeline, which may be effective in the beginning, does not show any results and death is inevitable.

Gasping respiration, like periodic respiration, may be caused by the intake of food, particularly when the infant himself is sucking, but occasionally also when tube feeding is used. The aspiration of vomited stomach contents may play a role (see p. 447). Quite often loud screaming, severe coughing spells, or "breath holding" (see p. 371) may have an unfavorable effect. Severe attacks of suffocation may easily follow whooping cough spells in normal infants but even more so in premature infants (Fig. 120, see p. 329), as described in detail elsewhere (see p. 374). The danger that a prematurely born infant may have a spell of gasping respiration and die of it is increased in proportion to the degree of his immaturity and his birth weight. Gasping respiration of long duration is, even in the prematurely born, very rare after the first four to six weeks of life. Up to that time all conditions that cause a deterioration of the general state of health are accompanied by the danger of an attack of suffocation. In this group belong alimentary disturbances; undernourishment, which can easily be caused by an untreated sucking deficiency; icterus neonatorum; and infections of various kinds, particularly pneumonia, which so often causes death. Premature infants who have been moribund for a long time always show gasping respiration before death. I cannot confirm YLPPÖ's assertion (1931) of a connection between congenital cardiac anomalies and attacks of suffocation.

Gasping respiration is facilitated by the immaturity of the respiratory centers; because of this immaturity the higher centers function to a lesser degree and are very sensitive. They therefore break down as soon as they are impaired by general disorders. Gasping respiration becomes a substitute mechanism and maintains life for some time until the higher center, which usually inhibits the gasping center, has recovered or until death has occurred. This process corresponds to the "law of the greater vulnerability of the later genetic acquisition" (VON MONAKOW and MOURGE, 1930).

HUMPHREY (1953) also reports that fetuses and newborns are less sensitive to oxygen deprivation than adults and that they therefore can survive longer when they are deprived of oxygen. In other respects the brain function of fetuses and of adults is identical; when oxygen deprivation progresses, cerebral activity is equally increased in both for a short time and finally is extinguished. In this process it is noted that the cerebral cortex, the cerebellum, and the higher motor centers are in need of more oxygen than the lower centers and therefore are the first to break down. This explains the fact that with progressive lack of oxygen brain function regresses to a stage characteristic of a young fetus. Correspondingly, plantar reflexes of the human fetus disappear prior to the palmar reflexes, which are established later in the course of development (HOOKER, 1958).

MINKOWSKI's findings (1928) on human fetuses obtained by Cesarean section are further proof of the fact that the gasping respiration of the premature is a lower form of breathing. The first respiratory movements were observed in fetuses (18 cm) at four months of gestation; their respiratory pattern differed absolutely from the higher form. Their breaths consisted of deep, sometimes spastic inspiratory movements of the thorax; at the same time the mouth was opened and occasionally the head was jerked backward; the thorax collapsed during expiration and the mouth was closed. Immediately after separation from the umbilical cord the child still breathed regularly; later only sporadic respirations separated by pauses of up to several minutes occurred. This description by MINKOWSKI shows clearly that very immature human fetuses have as their lowest and first pattern of respiration the same gasping respiration to which premature infants later regress so easily. HOOKER (1952) observed gasping respiration in dying human fetuses (13 weeks of gestation).

A kind of gasping for air can be seen even in fish under similar conditions. If the oxygen content of the water becomes inadequate their respiratory pattern changes completely. They swim to the surface in order to gasp for air. They maintain life in this way; they would die if they were prevented by a fence from coming near the surface. WINTERSTEIN (1921) therefore calls this breathing pattern of the fish "emergency respiration." Phylogenetically, air respiration developed from this kind of breathing (see p. 568). The same gasping for air can be observed in live fish on dry land: From time to time the mouth is opened with a smacking sound.

According to REMANE (1954), air respiration was phylogenetically acquired by swallowing air: If the oxygen content of the water becomes inadequate, the fish employ the described gasping for air. The inhaled air bubbles collect in the pocket of the oral and upper intestinal regions. In the ancestors of the Teleostei, a group of fish having lungs and air bladders, the posterior gill pockets contain such air bubbles. From these organs the lungs probably developed in fresh water regions having a temporarily inadequate supply of oxygen.

When frogs and lizards are exposed to hydrogen respiration, a wide opening of the mouth and backward jerking of the head occur with each new inspiration; the pauses become progressively longer (SIEFERT, 1896).

Considering the patterns of the tracings and the conditions under which the gasping center starts to function, it can be placed on the same level with the "gasping center" which in LUMSDEN's (1923, 1924) experiments on cats still maintained the respiration after sections through the brain stem. In both situations we are dealing with short deep breaths succeeding each other at irregular intervals. SCHOEN (1927, 1928) describes a respiratory form in decerebrated cats that conforms to the gasping respiration of the premature infant but which he calls "panting respiration."

CAMMANN and PEIPER [1932 *(46)*] tried to develop a nonsurgical procedure to isolate the gasping center from its association with the other centers. Since it is more resistant than the higher part-centers, the experimental approach of poisoning the animals was considered. This poisoning was intended to achieve a level

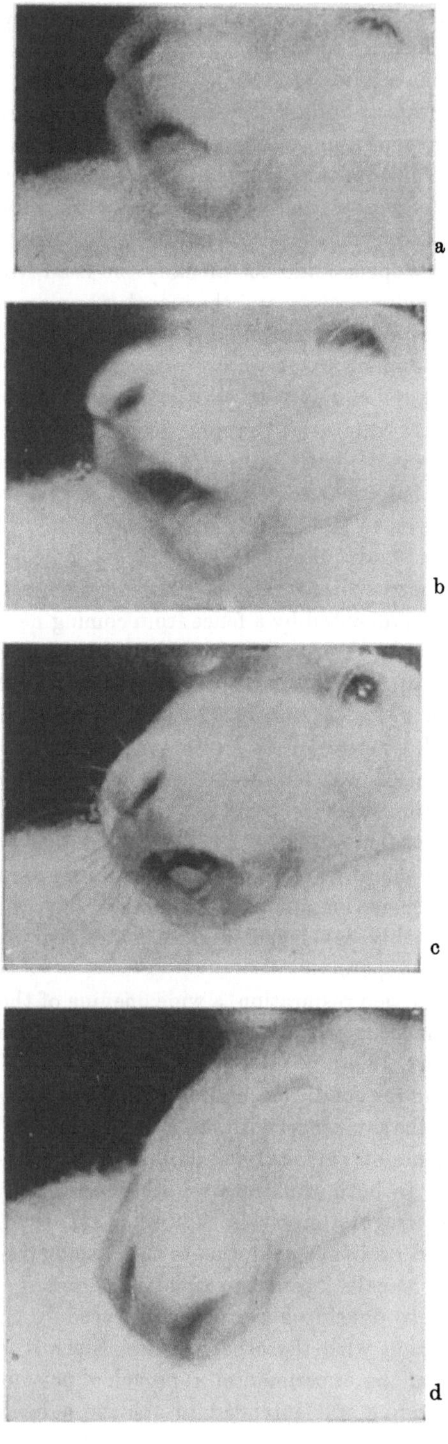

sufficient to cause a deficiency of the more sensitive higher centers without hindering the functioning of the gasping center. It was therefore important that the poisoning be adjusted to exactly graded stages and that it be interrupted at any time as soon as the desired stage was attained. The procedure employed by us corresponds in principle to that used by BARCROFT (1942) and VON HOLST (1939), who investigated by means of graduated anesthesia or suffocation of the central nervous system the hierarchic sequence of centers which are superimposed one on another.

We made pigs and rabbits breathe carbon dioxide in an air-tight chamber or through a funnel, from which procedure we anticipated a special effect on the respiratory center. We succeeded in causing a distinct attack of gasping respiration in every case.

At first the animals reacted to being covered with the carbon dioxide funnel by showing respiratory arrest (KRATSCHMER's reflex, see p. 51). At the same time the diaphragmatic tonus was decreased, as could be seen in the downward trend in the graph. The respiratory arrest was followed by markedly decelerated breaths, which at first were shallow and later became deeper. The deepening of the respiration was evidently inhibited by the carbon dioxide, for if the funnel was temporarily removed so that ordi-

Fig. 122. Gasping respiration of a suffocating rabbit (frames from a film taken through an air-tight bell glass). (a) At rest; (b) beginning of the gasping breath: mouth and eyes half open, head slightly raised; (c) height of gasping respiration: eyes wide open, pupils dilated, head raised; (d) at rest after the gasping breath: mouth and eyes closed, head lowered. [After PEIPER, HEMPEL, and THOMAS, 1951 *(114)*] nary air could be inhaled, the respiration deepened immediately. During deepened respiration the auxiliary respiratory muscles became engaged: The alae nasi, lips, thorax, and flanks were vigorously active. A strong motor restlessness occurred at irregular intervals.

After a duration of some minutes the respiratory pattern changed completely: The animal resorted to gasping respiration, the individual breaths became deeper and were of shorter duration than in normal respiration, and the breaths were separated from each other by longer periods of respiratory arrest of varying duration (Fig. 123).

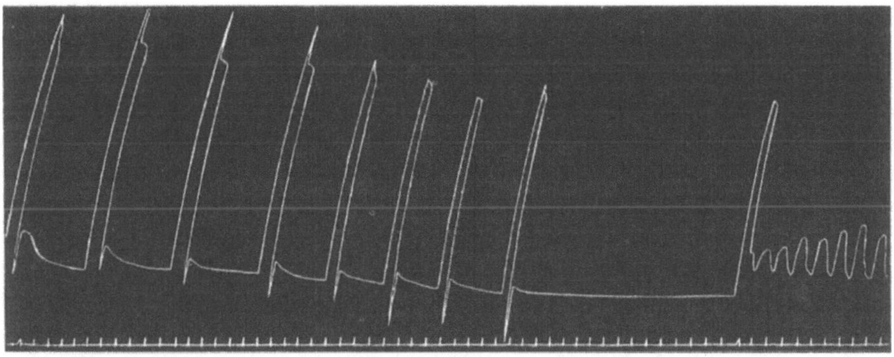

Fig. 123. Rabbit after inhalation of pure nitrogen. Transition from gasping respiration to the highest (normal) respiratory pattern (time in seconds).

During each inspiration the animal opened its mouth wide, frequently with an audible gasp (Fig. 122).

During the periods of respiratory arrest the diaphragm either remained motionless or showed evidence of irregular major and minor fluctuations in tonus. The mucous membranes became pale, and the pupils were dilated. At the same time the muscle tonus and the reflex excitability disappeared. At first only the respiratory center could still be excited by intense stimuli.

If the carbon dioxide supply was interrupted at this time, the animal could frequently still be kept alive. The gasping breaths became more frequent and finally followed each other without pause. Temporarily they alternated with regular breaths (block in the respiratory center, see p. 339), and they were finally replaced by the mature respiratory pattern. The transition could also occur directly without a block (Fig. 123). When mature respiration started again, the mucous membranes regained their color, the pupils contracted, and muscle tonus and reflex excitability returned. Artificial respiration was repeatedly necessary to overcome respiratory arrests of long duration; in some cases this was not successful. Animals that had already been used several times for these experiments were especially endangered.

During the gasping respiration these animals behaved exactly as the premature infant does during a choking spell. This similarity extends to the clinical picture, the tracings of the gasping respiration (sporadic deep and rapid breaths with opening of the mouth and irregular intervals between breaths), the dilatation of the pupils, the disappearance of muscle tonus, the reflex excitability, and to the pallor of the mucous membranes. In premature infants the skin also turns pale ("asphyxia pallida"); the blood content of animal skin cannot be judged.

This gasping, which is caused by the paralysis or exhaustion of the higher part-centers, represents a transitory emergency respiration when life is in danger. Death occurs rapidly if this phase is not overcome either spontaneously or by artificial means.

Gasping respiration occurred in the experiments described either because of carbon dioxide poisoning or because of a lack of oxygen. To differentiate between these two possibilities we made the animals breathe pure nitrogen under identical experimental conditions. In this manner the same gasping respiration was produced as in the carbon dioxide experiments; only the initial reflex respiratory arrest, which we always observed during carbon dioxide inhalation, was absent. Thus, gasping respiration occurs in animals because of a lack of oxygen in the air. It is noteworthy that periodic respiration also occurs because of lack of oxygen in the blood, according to UHLENBRUCK (1928) and SCHOEN (1928).

Our findings do not justify the view that suffocation attacks in premature infants are due exclusively to anoxia, for the air these children breathe always contains sufficient oxygen. But experience shows that a supply of oxygen has a very favorable effect on the spells of suffocation of premature infants. Evidently the higher centers are more sensitive to the lack of oxygen than the lower gasping center and therefore break down at blood oxygen levels at which the gasping center can still function. Choking spells may also occur directly because of aspiration of blood or mucus from the upper respiratory tract or of vomited stomach contents. Therefore, oxygen inhalation and suction of the respiratory tracts and of the stomach are the most important preventive measures (see p. 448).

Death frequently occurs in premature infants shortly after a suffocation attack. The central nervous system stops functioning during the attack. The answers given in the literature to the question of how heart function is affected by the attacks are contradictory. VON REUSS (1927) and VON JASCHKE (1917) state that the heart beat slows down during the attack, but YLPPÖ (1923) and ROHMER (1934) maintain that the pulse at first is still good; it is accelerated at the beginning but at the same time becomes increasingly weak. However, it is so difficult to feel the pulse in infants, especially in those prematurely born, that heart action cannot be judged by the quality of the pulse. Clinically one has to depend on judging the heart sounds.

The heart sounds slow down during the attack and become duller and weaker, but they are still demonstrable for a rather long time. If respiration stops completely, the heart sounds finally become inaudible even though the heart action may not be completely extinguished. After the final respiratory arrest and the disappearance of the heart sounds, heart movements may at times still be recognized in the apex region when the child is placed on his left side. In autopsy of newborns done only a few hours after the cessation of visible vital signs, hearts that were still beating have repeatedly been found (NEUGEBAUER, 1898, 1899; MARCHAND, 1899; OPITZ, 1899; PEISER, 1899). KULJABKO (1930) demonstrated, by removing and rinsing the heart after death, that the ability for heart movement persists in the infant, as well as in the animal, far beyond death. Occasionally he was able to make the heart beat even 24 hours later. Cardiac action potentials have been demonstrated in adults half an hour and even longer after respiratory arrest (VON HÖSSLIN, 1931).

Even in the air-conducted respiratory graphs reproduced (e. g., in Fig. 118, p. 327), heart action is recognizable as small waves during the respiratory arrests. They become appreciably larger immediately after gasping breaths. GOOD and PEIPER [1934 *(50)*] and PEIPER [1953 *(122)*] examined heart action in man and animals electrocardiographically during the suffocation attack itself. We always found the heart action considerably slower during the gasping respiration of the premature infant.

While the heart rate of the healthy premature is approximately 130 beats per minute, we have observed a frequency decrease to values as low as 36; the infant temporarily recovered from this serious condition, the respiration became normal, and the heart rate increased to 171 beats per minute. The electrocardiogram shows that we are always dealing with a sinus bradycardia. We have not found ventricular fibrillations. There is always a danger that the muscle action potentials appearing in convulsions may be confused with these cardiac phenomena.

The heart rate was never accelerated during gasping respiration, but we observed in a premature infant (weight 2000 gm, age 7 days) a heart rate of 240 per minute 50 minutes prior to his death. At the same time an excitatory state existed. The heart action outlasted the respiration in every dying premature infant whom we examined electrocardiographically. The time difference fluctuated considerably; often it amounted to 15 minutes and once even to one and a half hours.

A definite rule for the cessation of heart action cannot be established. The EKG usually remains normal after the last breath, and only the heart rate decreases. Occasionally minor transitory accelerations were observed. We often found a high T wave during gasping respiration and even after the last breath. Before the final cardiac arrest the potentials of the EKG usually became smaller. Atypical ventricular complexes occurred. Several times we observed nodal rhythm and occasionally a complete heart block. We did not observe a breakdown of the stimulus conduction system during gasping respiration. Corresponding investigations were made with the same results by AKEREN (1949) in newborns and by KÖTTGEN and FEYERABEND (1952) in premature infants.

The pulse and respiration rates during an attack of suffocation (pulse 36, respiration once in several minutes) are among the lowest values ever seen in man. All reflexes, all movements, and muscle tonus disappear simultaneously. The body temperature decreases to room temperature if artificial warmth is not supplied. The EKG indicates apparent death (suspended animation). Restoration is still possible, but it is less likely to occur and will last a shorter time if the vital signs have almost disappeared and the respiratory arrest has lasted for a long time. I doubt whether this condition can be compared to the "life without respiration" described by PARROT (1877), since actually a definite, though very slight, respiration still exists. Although PARROT's description is often mentioned, we do not have any more recent observations.

WENCKEBACH (1927) sees as the cause of the pathological sinus bradycardia either the direct decrease of the automatic stimulation or an indirect inhibition caused by the vagus, while in his opinion a third possibility, the reduction of the accelerans tonus, is out of the question. He gives a differential diagnostic criterion for the first two possibilities: Heart action becomes arrhythmic in a vagus bradycardia, whereas it remains regular in a deceleration in the stimulus formation in the sinus node. In our cases the stimulus formation always remained regular during gasping respiration as long as it was not influenced by the respiration itself. Nevertheless, we would like to leave open the question of whether the bradycardia was conditioned by the vagus. Possibly the vagus of the young infant functions in a manner different from that of later life; perhaps its influence on the heart rate develops only gradually. SOLTMANN (1877) attributes the increased heart rate during the first period of life to a lower vagus tonus. We do not have any more recent investigations in this area.

In animal experiments (rabbit), GOOD and PEIPER [1934 *(50)*] also saw pronounced bradycardia during gasping respiration caused by carbon dioxide inhalation. The heart rate, which was 200 to 260 beats per minute before the experiment, sank to 54 during a distinct gasping respiration. This deceleration was occasionally interrupted by a transitory slight acceleration. The T wave often became considerably higher during gasping respiration. We also frequently saw it increase after a respiratory arrest.

The interval between two R waves was markedly reduced by each gasping respiration. The much more rapid inhalation before the experiment did not influence the cardiac rhythm. VERWORN (1903) reported similar observations, only he did not yet use the term "gasping breaths." In our experiments the respiratory arrhythmia that occurs during gasping respiration could not be demonstrated in animals on which a vagotomy had been performed. Therefore, this arrhythmia must depend on the vagus. But we found a decrease of the respiratory rate also in vagotomized animals that had been poisoned by carbon dioxide; therefore, it does not depend, or perhaps we should say does not completely depend, on an increase in the vagus tonus during gasping respiration. We were able to demonstrate bradycardia in premature infants during gasping respiration, as has been mentioned, but never a corresponding respiratory arrhythmia.

After the final cessation of respiration in experimental animals, we repeatedly observed a long-lasting cardiac arrest that was temporarily followed by renewed and much slower heart movements. Ventricular and auricular fibrillations or heart block were not infrequently observed after the last breath. An incomplete block still occurred in a vagotomized rabbit during gasping respiration. We did not observe this in prematurely born infants.

STENGER (1952) thinks that the apnoic attacks of both premature and newborn infants are provoked by reflexes similar to KRATSCHMER's reflex which inhibit respiration through the mediation of the physiological apnea reflex arc. He assumes that the receptors are located in the region of the sensory vagus portion and the trigeminus. However, in my opinion, for which I gave my reasons above, gasping respiration is a lower respiratory pattern. As clinical observation and the graphs reproduced show, individual gasps not only appear at the end of the attack but they are the only form of respiration that maintains life during the attacks. An attack may last only a short time, but it may also last minutes, half an hour, or even several hours. Is it possible that during all this time apnea reflexes inhibit the respiration?

During the early periods of life, attacks of cyanosis frequently occur in very immature or brain-damaged premature infants. Sometimes a bluish discoloration of the face, already visible in the beginning, increases; sometimes infants who had a rosy color show a sudden change in color. The other body parts usually also become slightly cyanotic, or at least the blood flow to the skin is decreased. This disturbance in the blood flow often passes by itself so that the infant after some time again has a rosy appearance. It is advisable to manipulate an infant with such a disturbance a little and thus to improve his general condition, for the cyanosis may be a forerunner of an apnoic attack that can be prevented by suppression of the cyanosis. During cyanotic attacks the respiratory movements give very little information; the respiratory rhythm may be regular or irregular. Sometimes periodic respiration with long pauses may occur. Graphic recordings of the respiration and the EKGs during attacks of cyanosis are not available as yet.

Block in the Respiratory Center

(34, 35, 39, 49)

Further proof that the human respiratory center is composed of different components lies in the fact that at times some of these components are released, so to speak, from this aggregate and direct the respiration simultaneously and independently of each other. Such a situation can arise only after cessation or interruption of the inhibitions which are otherwise imposed by the higher centers on the lower components. I call this condition a "block in the respiratory center" to express its relationship to heart block, which comes about in a similar manner. The principal difference is that in the respiratory block the inhibitions originating in the higher centers disappear, while in the heart block the normal contraction stimulus emanating from the sinus is interrupted. Just as in the conduction system of the heart, we can recognize a complete and an incomplete block in the respiratory center.

The existence of a block in the respiratory center was proved even in early animal experimentation, although this term was not used. After performing operations in the brain stem area of animals, SWINDLE (1926) found superimposed respiration, which he ascribed to simultaneous functioning of several centers. SCHOEN observed, also in experimental animals, that in respiration which was irregular and severely affected by lobeline several breaths were periodically interspersed in the slow respiratory pattern. SCHOEN had the impression that it was as if a second rhythm had been switched on and brought into the leading rhythm by an excitation of other areas. When perusing the older literature we occasionally find a block in the respiratory center clearly evident in the graphs even when the author does not really spell out such an interpretation [for example, SIEFERT (1896), Figs. 16 and 17, in which the respiration of lizards under the influence of ether is illustrated].

Premature infants show a block in the respiratory center when the center is beginning to break down. But this block represents only a level to which the function of the respiratory center is momentarily lowered. I have found it to last 15 minutes at the most and often for a considerably shorter period.

In the three types of blocks that I have observed in the respiratory tracings of the young infant, the gasping center was always involved in some way.

Block Between the Periodic and Gasping Centers

Relatively often I have observed a block between the periodic and gasping centers in premature infants who were moribund for a long time. When these two centers influence respiration simultaneously, a graph is not needed to recognize the condition. The outward appearance of the infant is sufficient. Occasional gasping breaths can be observed during periodic respiration. This respiratory pattern can be easily distinguished from periodic respiration by the opening of the mouth and the backward jerking of the head. If the movements of the lower jaw are recorded separately from those of the abdomen, the graph of the jaw movements gives a pure representation of the activity of the gasping center; in the graphs of the abdominal movements the effects of the gasping and periodic centers are simultaneously recorded. This makes it easy to judge the time relation between the effects of the two centers. An incomplete block occurs if the gasping center, still dependent on the periodic center to a certain degree (i.e., always after a respiratory arrest), interferes. In Fig. 124 only one, or at the most two, gasping breaths can be seen. In Fig. 125 the gasping breaths appear in regular groups at the height of the respiratory period, corresponding to the CHEYNE-STOKES symptom complex (see p. 323). In a complete block both centers regulate respiration independently of each other (Fig. 126).

The occurrence of such a block gives some clue as to how the lower part-centers are inhibited by the higher centers. The simplest explanation would be that it is more difficult to stimulate the lower centers than the higher by oxygen deprivation or increased carbon dioxide levels in the blood. Thus, the activity of the higher centers, which adequately aerate the blood, would prevent the lower centers from reaching their excitatory threshold. However, this hypothesis does not explain all observations, for during a complete block both centers function simultaneously, i.e., at the same oxygen and carbon dioxide content of the blood. Therefore, I postulate nervous inhibitions that originate in the higher centers but are interrupted during the block.

Block Within the Gasping Center

Accompanying movements of the lower jaw and the tongue are absent in the highest (normal) respiratory pattern of man. But they appear in yawning, coughing, sneezing, vomiting, and choking. The movements of the diaphragm, the tongue, and the lower jaw also show close connections during the feeding of the infant. In frogs and lizards there is respiratory activity of the floor of the mouth, which has a rhythm different from that of the highest respiratory pattern (see p. 348).

There is a close phylogenetic connection between the centers for diaphragmatic movement and those for jaw movement. This causes them to act together during gasping respiration. In experiments on animals SCHOEN succeeded in separating the functions of these two centers. The respiratory movements of the lower jaw and the tongue were thereby distinctly shown to be twice as fast as the basic diaphragmatic rhythm in bilaterally vagotomized thalamus preparations of cats. SCHOEN further observed an increased number of jaw movements without correspond-

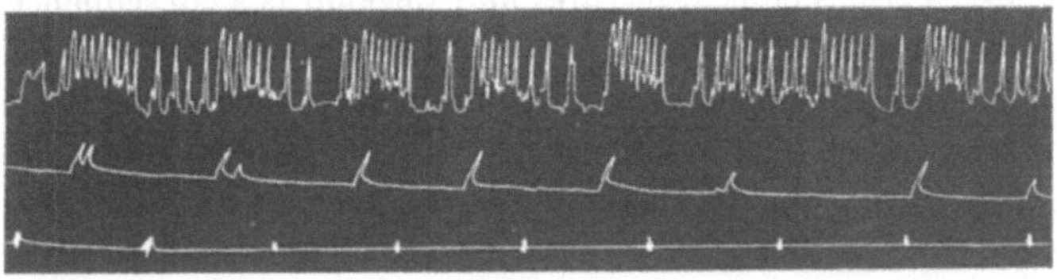

Fig. 124. Diaphragmatic (upper tracing) and lower jaw (lower tracing) movements (time in minutes). Block between periodic center and the center of lower jaw movements.

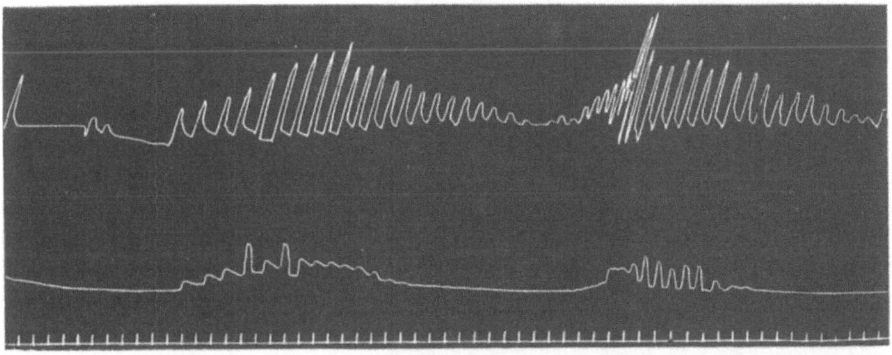

Fig. 125. Diaphragmatic (upper tracing) and lower jaw (lower tracing) movements. Periodic diaphragmatic movements (time in seconds). Lower jaw movements appear only at the height of the respiratory period.

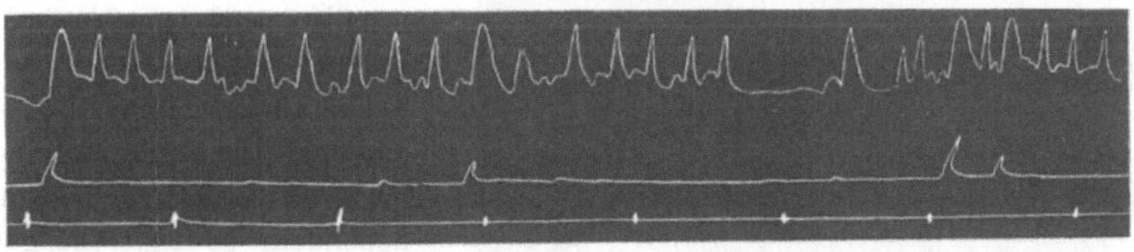

Fig. 126. Diaphragmatic (upper tracing) and lower jaw (lower tracing) movements. Incomplete block of the respiratory center (time in 15-second intervals). In the upper tracing are recorded some gasps characterized by their height and width, in addition to the usual breaths. Only the gasping breaths are accompanied by movements of the lower jaw.

ing diaphragmatic movements in decerebrate animals which had a tendency to "asphyctic" seizures during gasping respiration (termed by him wheezing respiration). These jaw movements were caused by a clonic spasm of the musculus digastricus. The increasing suffocation loosened the close coupling of the centers for the diaphragm and the lower jaw by causing spasm of the auxiliary respiratory muscles earlier than that of the diaphragm.

The same disintegration between the respiratory movements of the diaphragm and those of the lower jaw can be clinically observed in dying premature infants. With each gasping breath the mouth is usually opened and the tongue is retracted, as has been described. This connection between the respiratory movements of the diaphragm and the accompanying movements of the lower jaw may be completely or partially lost. Various clinical pictures may develop as a consequence of a separation of the two centers.

In a complete block the mouth and diaphragm move independently of each other; gasping movements of the mouth occur without corresponding diaphragmatic movements. At first glance a distinct gasping respiration seems to be present: From time to time the infant opens his mouth at irregular intervals with the tongue retracted each time. However, if the diaphragmatic movements are simultaneously recorded, it is found that the gasping movements of the mouth are not accompanied by the diaphragmatic movements that usually characterize gasping respiration. Rather, the diaphragmatic movements show a completely irregular pattern, as seen in Fig. 127. The infant endured this exceedingly incomplete respiration for half a day and then died. The two components that usually work together during gasping respiration were separated in their activity. The center on which the accompanying oral movements depend became independent so that the gasping movements of the mouth occurred without the corresponding diaphragmatic movements.

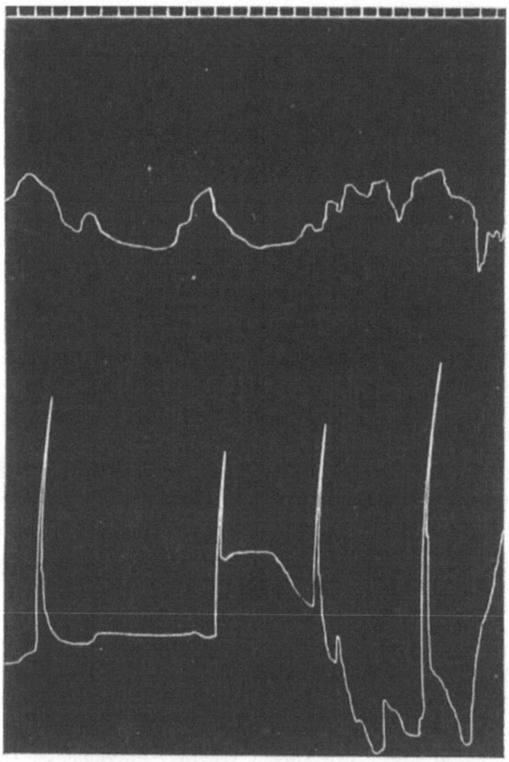

Fig. 127. Diaphragmatic (upper tracing) and lower jaw (lower tracing) movements independent of each other (time in seconds). Complete block in the respiratory center.

However, if the diaphragmatic movements of gasping respiration occur without the corresponding movements of the lower jaw, a familiar condition which has not yet been explained occurs: an attack of singultus.

In an incomplete breakdown, diaphragmatic and respiratory movements coincide only temporarily. I observed this phenomenon in fully developed and incipient gasping respiration.

The individual gasping breaths are separated from each other by distinct periods of respiratory arrest; a gasping opening of the mouth corresponds to each diaphragmatic movement. But there is a certain trend to double or triple these gasping respirations in an irregular sequence (Fig. 128) so that groups of gasping breaths are thus separated from each other by periods of respiratory arrest. If the two components of the gasping center work in their usual synchronized way, the movements of the diaphragm and of the lower jaw double or triple at the same rhythm. This connection can be temporarily lost so that the two movements no longer correspond with each other. At the end of the graph in Fig. 128 it is indicated that the mouth was opened three times while the diaphragm contracted only twice. The same occurrence is indicated in the middle of the same tracing. On the other hand, the diaphragmatic movements indicated in Fig. 129 were often repeated while the mouth was opened only once, but the opening was wider than usual. We are dealing here with a transitional form of gasping respiration in which the periods of respiratory arrest between the individual breaths are still absent.

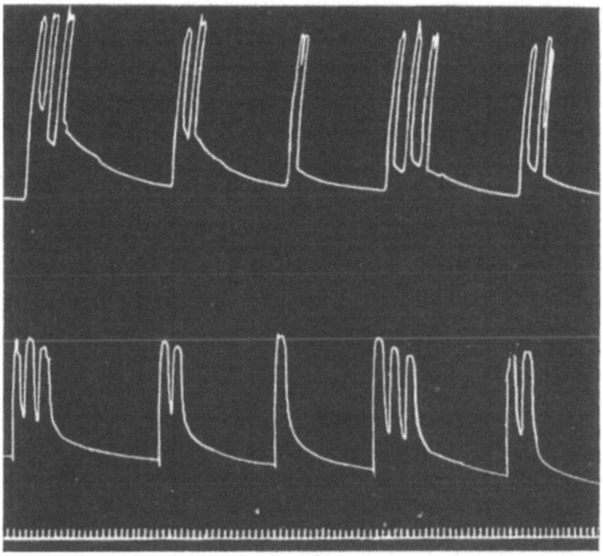

Fig. 128. Upper tracing: lower jaw movements. Lower tracing: diaphragmatic movements. Gasping respiration (time in seconds). The breaths, separated by periods of respiratory arrest, tend to double or triple. At the end of the tracing there is an incomplete block: three mouth openings (top) correspond to two diaphragmatic movements (bottom). Indication of a similar occurrence in the middle of the graph.

Singultus (Hiccup)

(34, 35, 46, 3a)

> Two things cannot be kept secret:
> A hiccup and love.
>
> H. Mercurialis (1583)

The singultus (hiccup) is formed by a group of involuntary rapid and vigorous spasms of the diaphragm that follow each other at short and always irregular intervals. The sudden penetration of air into the airways causes the well-known noise in the rima glottidis.

Without question singultus occurs most frequently in man during the first months of life. Some anxious mothers bring their babies to the physician just because of hiccups. Hiccups may frequently be found even in the prematurely born. The immature nervous system favors the appearance of respiratory disturbances and of singultus alike. An anencephalus whom I observed occasionally experienced a distinct attack of singultus without any special external stimulus during his first weeks of life. SALMI (1932) observed singultus attacks in six prematurely born rabbits.

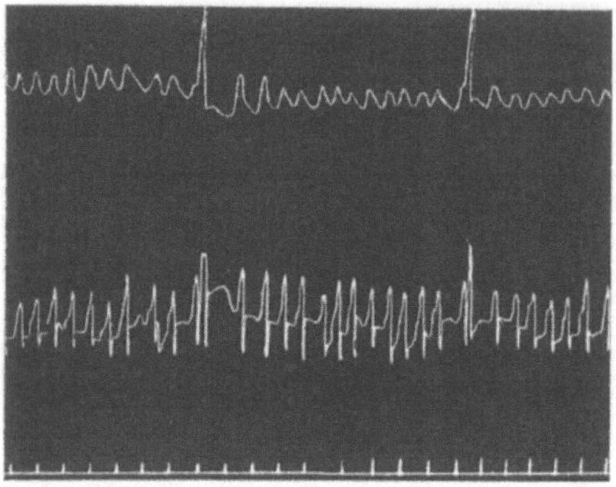

Fig. 129. Incomplete block between movements of the lower jaw (upper tracing) and diaphragm (lower tracing) (time in seconds). The diaphragmatic movements are constantly accompanied by minor movements of the lower jaw. Occasionally there is a double diaphragmatic movement for only a single opening of the mouth.

The tendency to singultus fluctuates in infants. Usually genuine attacks lasting approximately 10 minutes occur; they appear when the child is in good health and disturb the infant only to a minor extent. The attack is most likely to occur when the infant has been awake for some time, for example, when he has been picked up or his diapers have been changed. Feeding is an especially frequent cause.

I was able to observe the onset of attack several times. At first the infant experienced only individual singultus jolts separated by intervals of several minutes. The real attack developed only gradually in approximately 10 minutes. Now the individual spasms occurred at irregular intervals of a few seconds. Sometimes their intensity remained almost unchanged, but in other cases irregular or periodic fluctuations occurred.

The extensive literature on singultus has been concerned almost exclusively with its external cause, its relation to encephalitis, and its treatment. The question as to the real nature of this phenomenon is usually left untouched. Only DUMPERT (1929) tried to explain the phenomenon; he sees it as a remnant of a phylogenetic organization indicating the original function of the diaphragm, i.e., blood circulation in its purest form. In lower vertebrates the diaphragm supports the venous circulation and fills the heart with blood. Only after the lungs have been pushed out of the abdominal cavity through the diaphragm in the course of the embryonic development of the mammals can the diaphragm play its role as a respiratory muscle (KEITH, 1905).

According to DUMPERT, singultus assists blood circulation. HOFBAUER (1925) regards singultus as an "abnormal" respiratory movement. It appears to me that mere observation shows that singultus is actually a respiratory movement; nothing is known about its beneficial effect on the circulation. Furthermore, it does not occur when the circulation is embarrassed. In animals it can be elicited by poisoning of the respiratory center, as will be shown (see p. 347).

That singultus is really a respiratory movement can be seen by the fact that a special respiratory pattern, singultus respiration, exists. It consists of only singultus spasms, while the usual respiratory movements cease completely. RECHNITZER (1927, 1929) observed an adult patient whose breathing had stopped and whose oxygen

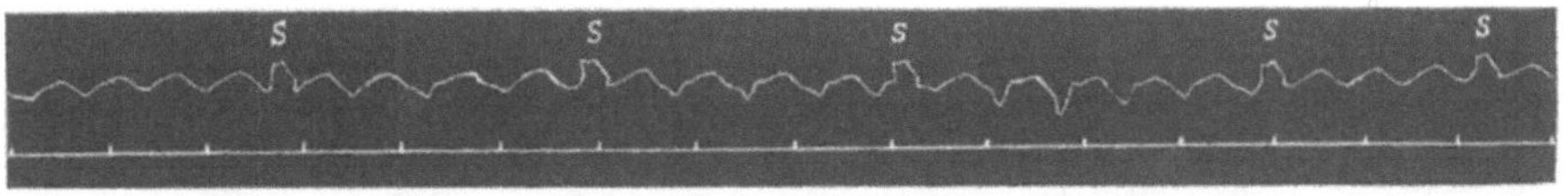

Fig. 130. Singultus (S = singultus) attack of a premature infant (time in seconds). The singultus wave always appears between two respirations. The number of respirations between the singultus waves fluctuates. The new breath starts only at the end of the hiccup.

requirements were satisfied for many hours by irregular singultus so that he survived. According to RECHNITZER, the singultus center substituted, so to speak, for the ordinary respiratory center. RECHNITZER considers singultus to be an emergency respiration, as described in the previously mentioned respiratory patterns. HINK (1895) observed a newborn whose respiration did not start at first but for some time was replaced by singultus. Elsewhere I shall discuss the intrauterine singultus spasms of the fetus.

These findings indicate the close relationship between the singultus center and the respiratory center. RECHNITZER examined in detail the mutual influence of the centers in adults but — except for the cases mentioned — could find no evidence that singultus depends on or influences respiration. SPÜHLER (1954) reproduced the respiratory graph of an adult in which the singultus spasms regularly coincided with the expiration. In the infant I was frequently able to demonstrate a relationship between singultus and respiration.

The singultus wave may occur in any segment of the normal respiratory pattern, i.e., during inspiration, expiration, or — most frequently — respiratory arrest. It clearly depends on the respiratory movements. In Fig. 130 the singultus movement is shown to occur regularly in the pause after a varying number of complete breaths. On the other hand, it influences the respiratory rhythm, since the new breath can start only after the termination of the singultus wave. In this case the singultus wave is interposed between the breaths of normal respiration. VON HOLST (1939, p. 297) acknowledged that we are dealing with a form of relative coordination (see p. 422) which he first demonstrated in the rhythm of the fins of decerebrated bone fish and later frequently also in other animals and man. "This pattern is completely analogous to that of two fin rhythms between which relative coordination exists, while one of them may only occasionally assert itself.

The relationship between singultus and the normal respiratory pattern shows fluctuations even during the same attack, as can be seen distinctly in Fig. 131. There the breath succeeding the singultus wave at times coincides and at times is delayed. This delay can frequently be recognized by mere observation without the recording of the movements. The individual waves succeed each other at irregular intervals; the intensity of the diaphragmatic movement may also fluctuate (Fig. 131).

I observed, as did RECHNITZER, complete disappearance of true respiratory movements during a singultus attack. Of course, this condition was rarely seen later in life as compared to the so frequently occurring singultus in infancy. In the graphs shown in Figs. 132—134, made in a six-week-old infant suffering from pylorospasm, pure "singultus respiration" can be clearly recognized. The child had

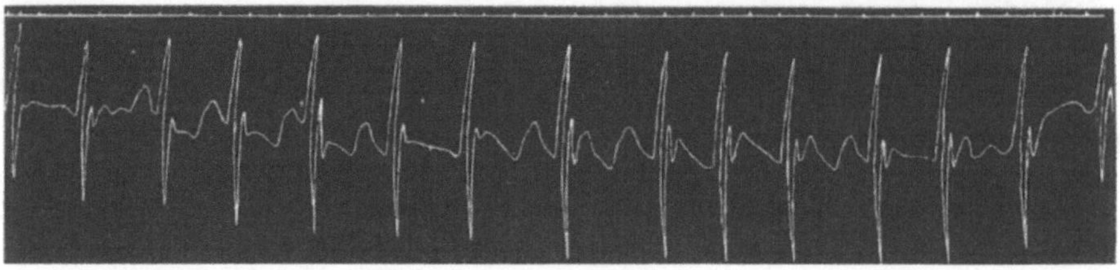

Fig. 131. Singultus and normal respiration mutually influencing each other in a 6-week-old infant (weighing 2300 gm) with pylorospasm (time in seconds). The singultus jolt generally occurs during respiratory arrest. In several instances the respiration fails to appear after a singultus wave (for example, between the two first waves).

been suffering for a long time from frequent spells of singultus, which always occurred at the same time of the day. On the fourth day of observation a singultus attack began at the usual time; it started in the same manner as the previous ones, but it led to a complete arrest of real respiration and lasted for about 15 minutes. The child did not become cyanotic and retained his entire reflex excitability, which is contrary to the pattern in suffocation attacks. Singultus respiration was still observed several times during the following weeks. I found this pattern also in premature infants, but only in short spells. My co-worker K. DIETEL [1954, 1955 *(3a)*] observed it as a transitory intrauterine respiration and made a graphic recording of it (see p. 355).

Since singultus respiration can completely replace the normal respiratory pattern, the same faculty must be attributed to the singultus center as to the other part-centers of respiration: It can maintain respiration and thus life when the other centers fail in their function. It is inhibited by the highest center, just as they are. If this inhibition is broken down, an incomplete block occurs (respiration and singultus become independent of each other). I have found an incomplete block can frequently be observed during infancy. RECHNITZER's observations have led to the assumption that a complete block can occur in adults. The frequency of singultus in infancy can be easily reconciled with the breakdown of the respiratory center so characteristic for this developmental stage. The origin of singultus in adults in the course of a number of surgical diseases, such as peritonitis, should be sought in an irritation of the singultus center from the periphery.

MACKENZIE (1907) and RECHNITZER (1927, 1929) described and recorded the coincidence of CHEYNE-STOKES respiration and singultus in adults. I found very similar singultus processes at the height of the respiratory periods in a premature infant with CHEYNE-STOKES respiration. In the successive respiratory periods they were at times present and at times absent (Fig. 112, p. 322), i.e., indicating an incomplete block between the periodic and singultus centers.

FRUGONI (1908, 1910) observed singultus which was at times rhythmic and at times arrhythmic in a dying adult suffering from epidemic meningitis. In speaking of a "dissociation of respiration" and a "bulbus agony," he approaches my views.

Later experiences have finally brought me to the conclusion that *in the last analysis singultus is only a special form of gasping respiration.* Especially in the young

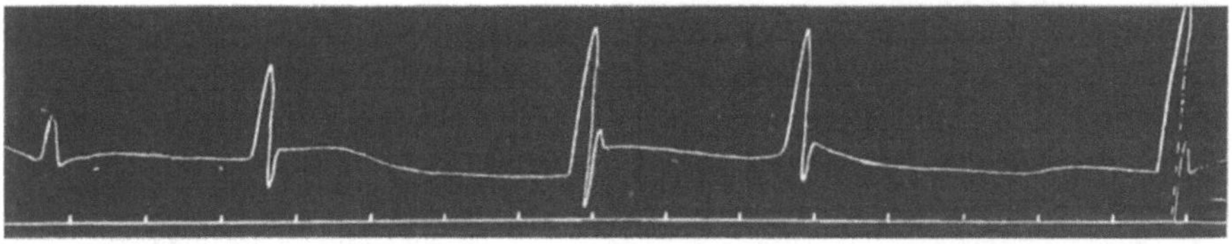

Fig. 132. Pure singultus respiration on rapidly rotating writing drum (time in seconds).

infant the close connection of both respiratory patterns can be recognized. At this age, in contrast to later in life, each singultus spasm is accompanied by a slight drop of the lower jaw and a minimal retraction of the tongue, just as is more clearly seen in gasping respiration. The diaphragmatic movements themselves — short strong jolts at irregular intervals — are identical in both respiratory patterns. Finally, CAMMANN and PEIPER [1932 *(46)*] observed in experimental animals (rabbits) that a singultus attack preceded the onset of gasping respiration on slow withdrawal of oxygen and simultaneous slow administration of carbon dioxide. Under these circumstances the transition from the highest (normal) form of respiration to gasping respiration took considerably longer than in the experiments previously described. Often sporadic hiccups (Fig. 135) could be recognized even during the normal respiratory pattern. In the beginning they were not accompanied by opening of the mouth. These singultus spasms at first recurred at irregular intervals approximately every fortieth to sixtieth breath. The diaphragmatic spasm substituted for a breath. Afterward, the first successive respirations were mostly somewhat diminished (Fig. 135). Gradually the intervals between the diaphragmatic spasms decreased, and they in turn became gasping breaths since they were accompanied by opening of the mouth. In most cases a state of fairly long duration developed in which individual gasping breaths of unequal size alternated irregularly with breaths of the normal respiration pattern (block in the respiratory center, Fig. 136).

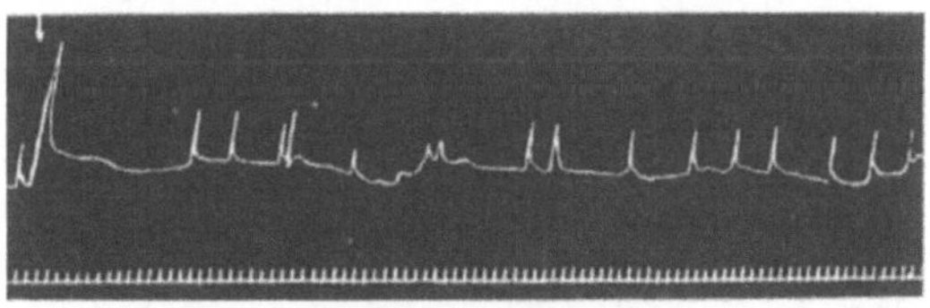

Fig. 133. The same as in Fig. 132 in slow rotation (time in seconds).

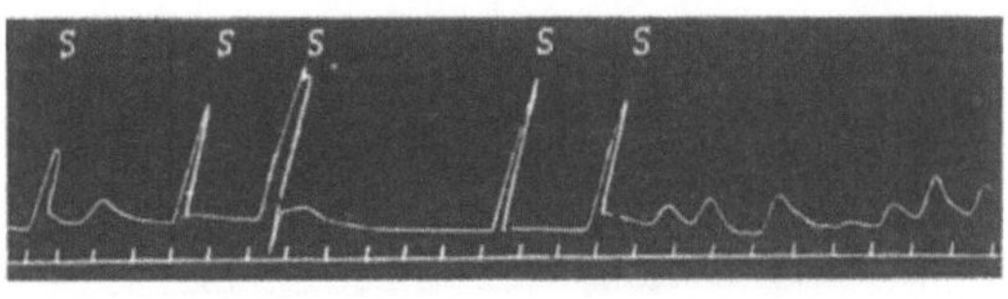

Fig. 134. Transition from singultus (S = singultus) respiration to normal respiratory pattern (time in seconds).

From this state the picture of pure gasping respiration finally developed. SCHOEN (1928) described a similar transition from normal breathing to gasping respiration in decerebrated cats during their agony.

Thus, the singultus center is a subdivision of the gasping center. It directs the movements of the diaphragm which are characteristic of gasping respiration. The movements of the tongue and of the lower jaw depend on another part-center which is only incompletely active during the singultus attack of the infant and is completely inactive later in life.

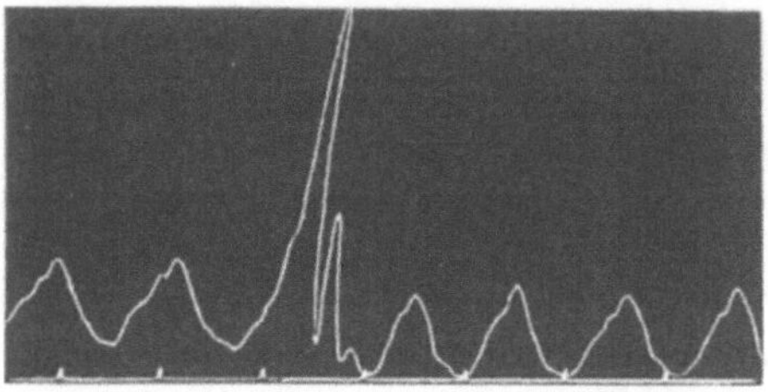

Fig. 135. Rabbit during slow withdrawal of oxygen (time in seconds). Sporadic singultus jolts recurring at fairly lengthy intervals without mouth opening, and preceding the onset of gasping respiration.

Swallowing Respiration (Frog Breathing)

(33, 39)

A close relationship exists between respiration and swallowing. In the primitive vertebrates the gills developed from pockets of the anterior intestinal segments and the lungs were later formed from the posterior gills.

According to BETHE (1925), the frog breathes in the following manner: Rhythmic movements of the floor of the mouth draw the air into the oral cavity through the nostrils, which are provided with valves. From time to time the nostrils close, and simultaneously the entrance to the larynx is opened; with the aid of contractions of the abdominal muscles air is pressed from the lungs into the oral cavity (expiration). Contraction of the muscles lining the oral cavity and the larynx immediately follows, forcing this air mixture back into the lungs (inspiration). The entrance to the larynx closes again, and the whole process is repeated, i.e., swallowing of air (Fig. 138, p. 351).

Similar respiratory movements of the floor of the mouth are also found in reptiles, which already have rib (thoracic) respiration. Swallowing respiration starts to disappear in reptiles but occurs distinctly during suffocation (SIEFERT, see p. 364). According to BABAK and KÜHNOVA (1909), who observed mature salamanders during oxygen deprivation, the salamander's pulmonary respiratory center relinquishes its function earlier and recovers later than its laryngeal respiratory center. According to SIEFERT, lizards do not breath at all during hibernation. If he warmed them, movements of the floor of the mouth, gradually increasing in strength, occurred at

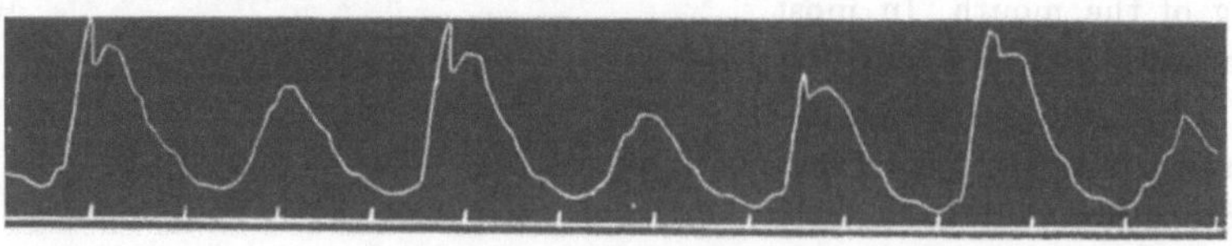

Fig. 136. Rabbit in air with increased carbon dioxide content under a bell glass (time in seconds). Onset of gasping respiration: gasping breaths alternating irregularly with regular breaths (block in the respiratory center).

first and then the first weak thoracic respiratory movements appeared. SIEFERT therefore believes that the center for laryngeal movement is more excitable than the respiratory center proper.

Phylogenetically the respiration of mammals developed in the following stages: regression of amphibian swallowing respiration, leading to its final disappearance; this has already occurred in the reptile stage, in which truncal respiration develops. In the beginning this respiratory pattern was based predominantly on the diaphragmatic sucking function and was later supported by rib movements.

According to WINTERSTEIN (1921), swallowing respiration serves mostly as an emergency respiration in reptiles but is the most important form of breathing for batrachians of various types. In mammals, swallowing respiratory movements seem to have a special place because of their peculiar rhythm, their occurrence in respiratory embarrassment, and their continuance during respiratory movements or in the intervals between respiratory periods. One is reminded of the swallowing respiration of reptiles, which also appears during respiratory embarrassment.

Of course, one cannot expect to find as a regressive pattern in higher mammals respiratory forms that conform completely to frog breathing. But some vestiges of it may even occur in man in various forms, such as the respiration with the floor of the mouth (see p. 365) and the swallowing respiration which occur during suffocation and dying, as well as during the infant's feeding (here respiration and swallowing are coupled) (see p. 438); the previously described block in the gasping center (see p. 340) which causes the lower jaw and the diaphragm to move independently of each other; and finally yawning, which always ends with a swallowing movement (see p. 386). According to GARSCHE (1950), the gasping-swallowing reflex appears in premature infants only at a stage of development corresponding phylogenetically to the animal groups at the level of swallowing respiration.

The functions of the swallowing and respiratory centers are still definitely connected with each other in higher mammals. GARLAND (1886) and DUCCESHI (1905) state that during suffocation in dogs there occur esophageal contractions which accompany each breath until death. To explain this phenomenon both investigators, independently of each other, point to the close connection between swallowing and breathing in the lower vertebrates. KAHN (1905) also observed complete swallowing movements in a nonanesthetized suffocating animal; however, he did not trace these movements to a central origin but regarded them as the result of the increased salivation accompanying the violent movements of the suffocating animals. In apes that were poisoned with illuminating gas, JELLINEK (1929) saw swallowing movements that began during terminal respiration and lasted beyond respiration proper. STEINER (1886, 1887) finally showed in experiments on animals that the respiratory center can be stimulated from the swallowing center. When he stimulated the superior laryngeal nerve, respiratory arrest occurred at the height of expiration; however, this was interrupted by slight, periodically recurring swallowing movements that corresponded to the missing respiratory movements. DUCCESHI (1905) assumed that the stimulus can spread from the excited respiratory center to the swallowing center.

In the infant, the swallowing and respiratory centers are also connected in their functions (see p. 438). Even in the adult the swallowing center can come to the aid of the endangered respiratory center. SCHWARZ (1905), JELLINEK (1929), and EBBECKE (1952) reported that in voluntary respiratory arrest, such as occurs in endurance underwater contests, an absolutely irresistible impulse to swallow that has to be satisfied occurs when respiratory embarrassment becomes pronounced. During swallowing the almost intolerable impulse to breathe stops completely so that respiratory arrest can still be endured for a while. Among 100 patients who suffered from respiratory paralysis due to poliomyelitis and used swallowing respiration (frog breathing) when they were outside the iron lung, one patient who had known this method since childhood was found. He had used this method to be able to stay under water longer when diving (DAIL, AFFELDT, COLLIER, and HONDO, 1955).

When the respiratory center of man breaks down, laryngeal movements may appear, just as they do in animals. RAUTENBERG (1919) described a woman believed

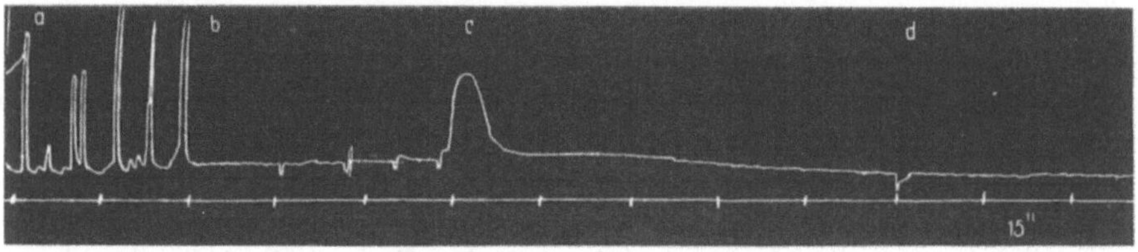

Fig. 137. Terminal respiration during death from pyodermia in a 4-week-old premature infant weighing 1470 gm. At first gasping respiration (a), then sporadic laryngeal movements (b and d), and finally diaphragmatic movement (c).

to be dead in whom laryngeal movements were observed 14 hours after she was placed in the coffin (see p. 363). Then resuscitation attempts were made and were successful. L. R. MÜLLER (1930) also observed occasional swallowing movements after respiratory arrest.

According to JELLINEK (1929), swallowing movements may occur in dying newborns. I could not find any other such claims in the literature known to me. But when I paid attention to this phenomenon, I was able at times, though rarely, to observe laryngeal movements persisting beyond the last breath (Fig. 137). I did not investigate whether they were really accompanied by swallowing movements, but I consider it to be possible.

Later experience proved that a distinct swallowing respiration may develop in the premature infant. However, it is less frequent than gasping respiration and represents an even lower stage of cerebral function. Such an observation is reported here:

The girl (p. 1710/55) is admitted at the age of 1 hour 27 minutes; weight 1200 gm, temperature 33°C. She is very flaccid and shows moderate signs of aspiration of amniotic fluid and defective pulmonary respiration. At first the respiratory rate fluctuates between 48 and 60 per minute. On the third day of life several attacks of cyanosis and finally an attack of apnea occur. Following this attack the infant screams shrilly several times, makes chewing and also less pronounced licking movements, and finally presents the clinical picture of swallowing respiration: Distinct swallowing movements characterized by visible and perceptible upward laryngeal motions appear at somewhat irregular intervals, approximately 6 to 10 times per minute; breaths cannot be noted between these swallowing movements. The mouth has a kind of snout-like appearance, but is wide and not pointed. The baby opens her eyes wide at almost regular intervals; twitching of her whole body is noted but is less pronounced than during a seizure and more marked than in "Stäupchen."*

On preceding days some blood was occasionally obtained when suction was applied. Now suddenly great amounts of light red foamy blood come from the nose; even prior to this event a great amount of crepitation and metallic fine rales were heard over both lungs. Shortly afterward the child dies at the age of 68 hours.

Autopsy shows extensive intraalveolar hemorrhages in both lungs and minor leptomeningeal hemorrhages.

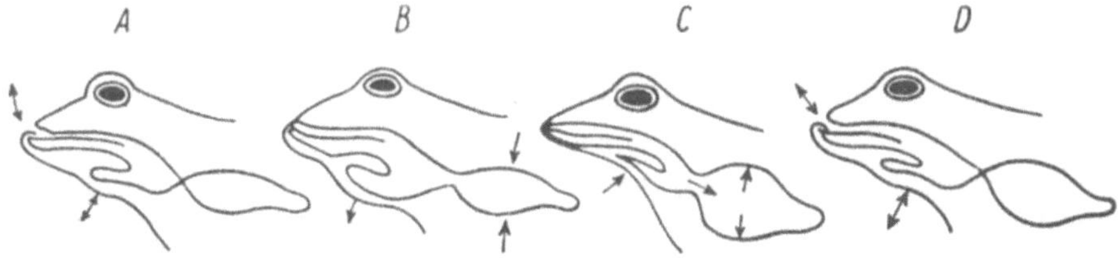

Fig. 138. Diagram of frog breathing (according to HERTER, 1947). (A) Renewal of oral air by laryngeal oscillations while the nose is opened. (B) After closing of the nose and opening of the larynx the used air is pressed from the lungs by intestinal pressure: expiration. (C) By raising of the floor of the mouth the air mixture is swallowed into the lungs: inspiration. (D) Closing of the larynx, opening of the nose, renewed onset of laryngeal oscillations.

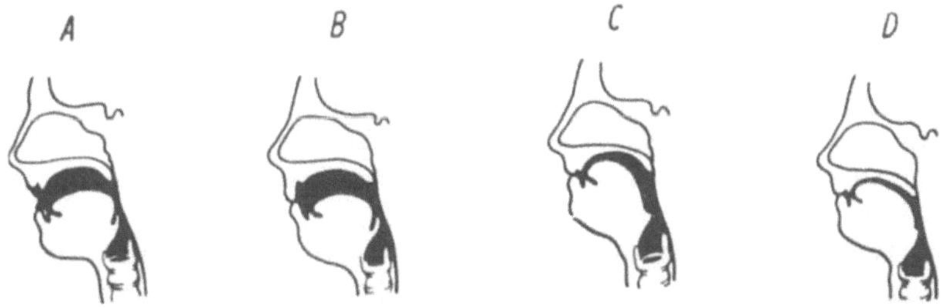

Fig. 139. Diagram of glossopharyngeal ("frog") breathing in weaning from the iron lung (according to PFIZER SPECTRUM, 1954). (A) A small amount of air is taken into the mouth and pharynx. (B) The lips are pressed together in order to imprison the air. (C) The air is pressed through the larynx with the aid of the tongue, the pharynx, and occasionally also the cheeks. (D) Closing of the larynx retains the air in the lungs while mouth and pharynx are again filled with air.

Swallowing respiration is the last safeguard before death due to respiratory arrest occurs. As experience has shown, this respiratory pattern occasionally suffices to maintain life for a while.

During weaning from the iron lung a respiratory pattern exactly corresponding to swallowing respiration has repeatedly been observed in adults. Since this pattern alone can maintain the necessary air exchange for some time, it is also of clinical importance. According to a note in the PFIZER SPECTRUM (1954), many

* See footnote, p. 249

patients have learned frog breathing. Frogs swallow the air from the mouth into the lungs, so to speak (Fig. 138). Similarly, frog breathing (Fig. 139) pumps air into the lungs by the use of the mouth, the cheeks, the tongue, the pharynx, and the larynx. At first a small amount of air is admitted to the mouth and larynx; then the lips are pressed together and the soft palate is lifted to let the air pass. The air is then pressed through the pharynx with the aid of the tongue, the pharynx itself, and occasionally also the cheeks. The closing of the larynx now keeps the air in the lungs, while mouth and pharynx are again filled with air which will be pressed into the lungs in the same manner. This process is repeated approximately a dozen times until the lungs are adequately filled with air. Although the iron lung can aerate better, it has been found that frog breathing is quite sufficient to provide air for many hours and facilitates weaning from the iron lung.

Therefore, frog breathing (tongue-pharynx breathing, glossopharyngeal respiration) is taught to poliomyelitis patients who have respiratory paralysis (Affeldt, Collier, Crane, and Farr, 1956). Dail, Zumwalt, and Adkins (1956) prepared a special guide for this method. The procedure makes use of the mouth and pharynx as a pumping mechanism to press the air down into the lungs. However, a prerequisite is that the muscles of the tongue, the soft palate, the pharynx, and the larynx are still functioning. This respiratory method is particularly indicated if the faculty of independent breathing is impaired by paralysis of the respiratory muscles, if it is desirable to free the patients from respiratory devices for some length of time so that they can practice correct breathing with their own muscles, and if the abdominal muscles are so weakened that the patient is unable to cough. With this procedure the adult can swallow approximately 70 ml of air into his lungs so that the median vital capacity of 11% is raised to 50% of the normal value (according to the measurements of Affeldt and co-workers). This substitute breathing cannot be performed during sleep.

Dail, Affeldt, Collier, and Hondo (1955) reported on more than 100 patients who had respiratory paralysis due to poliomyelitis and who used frog breathing. The advantages of this procedure are that it allows respiration without using the respiratory muscles proper and that no technical devices are needed. It is an emergency respiration which gives a certain security to the patient, psychologically and physically. It prolongs the period the patient can spend without technical devices and facilitates his sitting, walking, speaking, and coughing. The youngest child in whom frog breathing was seen was six years old. Most probably he had already learned this type of respiration two years earlier.

According to Kirschsieper (1954), swallowing respiration of this type developed in a 12-year-old girl who had spent nine months without interruption in an iron lung because of poliomyelitic respiratory paralysis. During her practicing of independent breathing she started to make chewing movements with her lower jaw when trying to take several breaths in succession. These movements were similar to those used in gasping respiration, but the jerking back of the head was absent. The breathing became deeper after some practice and finally the girl was able to stay out of the iron lung for several hours. Fluoroscopy showed that a sequence of chewing movements closely succeeding one another made the diaphragm move downward by little jerks. During expiration it jerked upward again with one big jolt. During the chewing movements the child swallowed air into the trachea and the lungs against the abdominal pressure several times in succession. KIRSCHSIEPER assumes that the path for the preformed but usually latent swallowing respiration was centrally paved by the gasping respiration which was frequently observed prior to the beginning of this pattern.

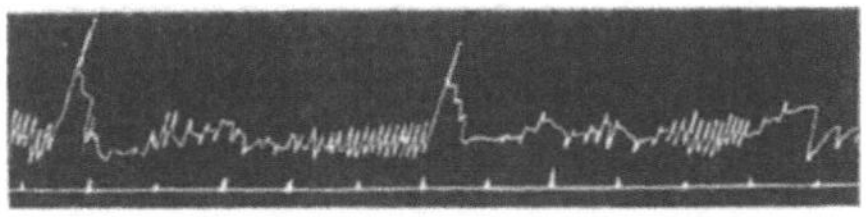

Fig. 140. Constant respiratory irregularity in a 22-day-old premature infant (time in 10-second intervals). In succession: highest (normal) respiratory pattern, sigh with a compensatory pause, wheezing respiration, normal respiratory pattern, etc.

BOSMA, LIND, and GENTZ (1959) took rapid successive x-ray films of the newborn's first breath in order to record the pharyngeal movements which accompany the first penetration of air into the lungs. In the pharynx they found an air accumulation which was temporarily closed off from the oral cavity by the pressing of the tongue's posterior part against the palate. This closing prevents the air from leaving the lungs again and can facilitate further air distribution by increasing the intrathoracic pressure before expiration. BOSMA, LIND, and GENTZ see in this process relations to the frog breathing of poliomyelitis patients with respiratory paralysis.

Constant Irregularity

(39)

Irregular respiratory graphs occur easily, because the infant's general movements disturb the recording. But even if this source of error is eliminated and the infant is examined when completely immobile during sleep, graphs which do not fit into the classifications given here are occasionally obtained. In the respiratory patterns previously described we can predict at each point of the graph with a certain degree of probability what the next breath will be like; this is not possible when dealing with constant irregularity. But even in these respiratory graphs certain patterns can be distinguished.

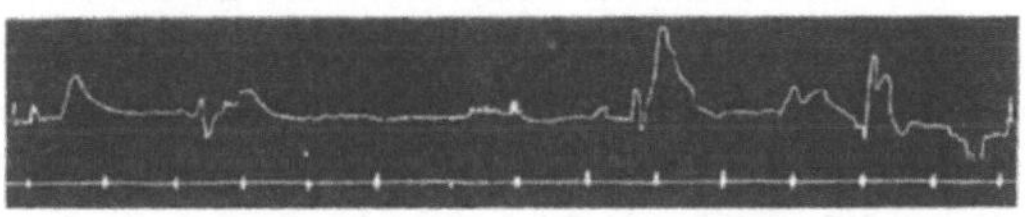

Fig. 141. Constant respiratory irregularity with long respiratory arrests in a 21-day-old premature infant (time in 10-second intervals).

There is a form of constant irregularity in which several respiratory patterns recur again and again at intervals of several seconds, each pattern being readily identifiable by its characteristic graph. Thus, Fig. 140 illustrates a wide variety of respiratory patterns, i.e., the highest (normal) respiratory pattern, periodic respiration, panting respiration, and sighing, in addition to long respiratory arrests. I was not able to find any gasping respiration or singultus in this tracing.

In a second tracing of constant irregularity which also showed long respiratory arrests, the individual respiratory movements had so few characteristics that I was not able to classify them among the patterns described (Fig. 141). It appears premature to establish new respiratory patterns based on only a few graphs without the support of experiments on animals.

I found the patterns of constant irregularity described only infrequently, and only in prematures who were already in poor condition and who expired soon afterward. An irregular respiratory pattern which defies description was also observed in a premature infant by ASILON (1959). A different pattern of constant irregularity occurs during epileptic seizures (see p. 382).

Intrauterine Respiratory Movements

(3a)

The cause of the first breath after birth has been variously considered to be lack of oxygen, carbon dioxide poisoning, or the new sensory stimuli that affect the infant. However, the infant is actually already breathing a long time before he is born. The first statements on this question were made by WINSLOW (1799), who opened the abdominal cavity of pregnant dogs and cats. The fetuses, which were attached to the intact umbilical cord and floated in the amniotic fluid, seemed to inhale this fluid by dilating and contracting their nostrils and thorax. A respiratory arrest usually occurred after two, three, or four breaths. Thus, this appeared to be a respiratory pattern which today we call periodic breathing. According to SCHEEL (1799), who reported on these experiments, the newborn child has already wanted to exercise his respiratory muscles earlier so that they may not fail him later, since after birth his well-being and life depend on their proper functioning. In 1813, LECLARD (1815) observed after premature opening of the uterus that the fetus was breathing and that each inspiration was accompanied by opening of the mouth, dilatation of the nostrils, and raising of the thoracic walls. PREYER (1885) also saw some distinct respiratory movements under similar conditions in fetuses of cats and rabbits whose placental circulation was unimpaired.

Intrauterine respiration was demonstrated by a long-forgotten experiment performed by GEYL which has been repeated in more recent times in a basically identical manner. In 1880 GEYL injected aniline blue into the embryonic membranes after opening the abdominal cavity of a pregnant rabbit. The dye was later found in the lungs of three dead fetuses and one baby rabbit born alive.

The intrauterine respiratory movements of man were first described by the obstetrician AHLFELD (1888, 1891, 1905). These are rhythmic movements of the abdominal walls in pregnant women, occurring mostly in the umbilical region. According to AHLFELD the rate is 61 and according to K. REIFFERSCHEID (1911) 48.4 movements per minute; frequently they show periodicity and can be recorded, as AHLFELD demonstrated. It can be shown that they are independent of the mother's pulse and respiration. Whoever has observed for himself these easily visible movements, which occur in the majority of pregnant women, finds it difficult to believe that their existence and nature were doubted for so long. They were generally accepted only when K. REIFFERSCHEID (1911) confirmed AHLFELD's statements. K. DIETEL (see p. 311) reported on the frequency of the intrauterine respiratory movements.

According to POTTER (1953), amniotic fluid enters the fetal lungs under normal conditions after the fifteenth week of gestation.

Singultus can also occur as a special respiratory pattern before birth (see p. 345) (MERMANN, 1880, 1887; IKEDA, 1893; AHLFELD, since 1888; HINK, 1895; ROTHSCHILD, 1911; and my own observations). We note strong jerks of equal strength at irregular intervals. ROTHSCHILD placed his hand in the uterus and was able to show that the movements originate directly from the fetus's thorax. NORMAN (1942) found fetal singultus in 33 of 48 pregnancies; the earliest appearance was on the one hundred and twenty-fifth day before birth. The duration varied from 1 to 29 minutes, the average being 8 minutes. The greatest frequency per minute was 45; before its complete disappearance the frequency of the singultus decreased to approximately 10 per minute in the fetuses, just as it does in newborn infants.

According to K. and V. DIETEL (1959), intrauterine singultus has essentially the same symptomatology as does singultus of the infant (see p. 346). They observed pure singultus respiration, alternation between singultus periods and periodic respiration, and the simultaneous occurrence of both patterns. They see a relative coordination in the sense of VON HOLST's hypothesis (see p. 422) in these relations. The intrauterine singultus attacks they observed lasted up to 10 minutes. A pregnant woman counted 600 jolts which corresponded to a duration of approximately 50 minutes. Attacks could be provoked by tapping the abdominal wall of the pregnant mother; this was first successful in the seventh month of pregnancy.

KLEMPERER (1933) conducted similar experiments without having knowledge of GEYL's data. He opened the abdominal cavity and the embryonic membranes in pregnant guinea pigs immersed in water at body temperature to which starch had been added, and let the escaping fetuses swim in the fluid with their mouths under water. When he later tested the respiratory tracts with iodine, he was able to detect starch in the air passages down to the lungs in a number of cases, even if he had seen no respiratory movements prior to this examination. In another experimental series he replaced a part of the amniotic fluid with starch by puncturing the embryonic membranes without opening them and then closing the abdominal incision. After birth, starch could also often be demonstrated in the air passages and lungs as an indication that the amniotic fluid had been aspirated into the lungs.

SNYDER and ROSENFELD (1937) conducted extensive animal experimentation on this question. They rendered the abdominal and perineal regions of pregnant rabbits insensitive by transecting the lumbar spine and immersed the bodies of the animals in warm Ringer's solution. Then the abdominal cavity was opened so that the fetuses could be observed through the intact transparent uterus wall. Under local anesthesia a tube was attached to the trachea of the mother animal so that various gas mixtures could be introduced.

In the intact uterus the fetuses showed respiratory movements of the thorax and abdomen which in general conformed to those seen after birth except that they were somewhat weaker. Two respiratory patterns were seen. One was characterized by rapid shallow respirations (approximately 60 per minute) predominantly by thoracic movements, and the other, rarer pattern by sudden deep deflections occurring at a rate of only approximately 15 per minute. (This, in my opinion, was gasping respira tion.) There were great differences among the fetuses. Some kept quiet, while others took up to 60 breaths per minute. The respiration increased in proportion to the age of the fetuses. It was not stimulated by the experiment, but could also be demonstrated in intact animals if they were carefully observed.

In experiments on animals, BONAR, BARNETT, BLUMENFELD, and FENNING (1938) similarly observed three forms of intrauterine respiration that correspond exactly to the respiratory patterns of premature infants as I have described them, i.e., regular rhythmic respiration of approximately equal intensity, then periodically increasing respiration followed by pauses, and finally gasping with the mouth wide open occurring at irregular intervals and often accompanied by generalized movements. If several fetuses were in the uterus, individual fetuses showed different respiratory patterns, as noted in the experiments of SNYDER and ROSENFELD (1937, 1938).

Sedatives given to the mother animal which did not change her respiration impaired or inhibited intrauterine respiration (SNYDER and ROSENFELD). BONAR, BARNETT, BLUMENFELD, and FENNING made the identical observation with morphine, ether, chloral hydrate, and other medicaments. This explains the fact that the fetal respiratory movements were discovered as early as the end of the eighteenth century, but later could no longer be found when the animals were anesthetized during operations.

Much was written in the early literature (compiled by PREYER, 1882, 1885) on the cause of the first breath after birth. We have two opposing opinions: According to one view, the blood becomes venous, i. e., lacks oxygen and accumulates carbon dioxide because of the interruption of the placental circulation, and thus stimulates the respiratory center. Others think that mechanical or cold stimuli to the skin are responsible for the first breath.

SNYDER and ROSENFELD investigated the question of whether the first breath after birth depended on the gas content of the blood by letting the mother animal breathe various gas mixtures and observing the effects on the fetus's respiration. Oxygen deprivation of the mother caused by inhalation of a mixture of 4% oxygen and 5% carbon dioxide in nitrogen immediately slowed down or inhibited the fetus's respiration although it considerably stimulated the mother's own respiration. If the oxygen-deficient mixture was again replaced by air, the normal respiratory pattern returned. But oxygen deficiency of long duration or of excessive degree constantly decreased the respiration. Stimulation or onset of intrauterine respiration because of oxygen deficiency could not be demonstrated.

The addition of carbon dioxide to the air inhaled by the mother did not influence the fetus's respiration to any considerable degree, although occasionally it accelerated the breathing. Breathing was not started in fetuses which had not breathed before. However, the respiration of the mother animal was greatly stimulated.

PREYER saw the first breath after birth as a reflex elicited by the considerable cooling, the skin stimulation, etc. AHLFELD (1891) refuted this opinion by arranging for birth to take place in a warm bath. Likewise, oxygen deficiency or carbon dioxide intoxication cannot be the cause, for the respiratory center of the fetus is always inhibited by lack of oxygen and is not stimulated by an excess of carbon dioxide, although the respiration of the mother animal is intensified in both cases.

HASELHORST and STROMBERGER (1932) and EASTMAN (1932, 1933, 1936) determined the gas content of the blood of the umbilical cord at the moment of the first breath and thus directly tested the correctness of the theory on the newborn human infant. It was found that the onset of respiration was completely independent of whether the oxygen or carbon dioxide content of the blood was high or low at this particular moment. In addition, the gas content of the blood of the umbilical cord before the onset of breathing scarcely differed from that at the time of the first breath.

EASTMAN reports an observation which causes all chemical theories on the onset of the first breath after birth to collapse. In deep cervical Cesarean section under local anesthesia, it is possible to take the child's face out through a small incision in the uterus without disturbing the placental circulation and to let him cry strongly while in the uterus. Incidentally, PREYER (1882) made a similar observation in experiments on animals. If the head of the fetus in a pregnant guinea pig is freed, an inspiration occurs immediately. During this intrauterine air respiration the placental circulation is also maintained, so that oxygen hunger cannot be the only cause of the respiratory movements. If, as previously reported, some fetuses in the uterus breathe and at the same time others do not breathe, the gas content of the umbilical blood cannot be decisive for their reactions.

WULF (1959) made multiparae inhale pure oxygen 10 to 15 minutes prior to the expected birth. All the newborn infants were born with a rosy skin color; gasping respiration started immediately after birth and the first cry occurred 15 seconds after birth at the latest, followed by rhythmic respiratory movements. The analysis of the gas content of the blood of the vena umbilicalis showed high oxygen and low carbon dioxide tension and low pH values. Thus, gasping respiration and the first cry are not elicited by progressive suffocation in eupnoic infants.

In general, the respiratory center occupies a special position at this developmental stage which makes it inadmissible to presuppose the same excitability at this stage as at the adult level. Compared to the mature infant, the human fetus is in a state of considerable oxygen deprivation, as HASELHORST and STROMBERGER (1932) have demonstrated.

The whole question as to the cause of the first breath is posed in an incorrect manner, for respiration is already temporarily functional long before birth. Breathing is inhibited during birth and starts again immediately afterward.

HENDERSON (1929) recommended inhalation of a mixture of carbon dioxide and oxygen as a treatment for threatening suffocation after birth, and even demanded legal provisions ordering that every newborn infant, whether well or sick, should inhale carbon dioxide daily in order to be protected against atelectasis and pneumonia. EASTMAN (1936) warned against this procedure, since he had found excessive amounts of carbon dioxide in the blood of suffocating newborn infants. In addition, infants are said to have an extremely low blood pH in this condition so that carbon dioxide intake would increase the already existing acidosis to a dangerous extent. According to KANE and KREISELMAN (1933), EASTMAN, and others, oxygen is just as effective alone as in combination with carbon dioxide; it may even be more effective. The newborn infant needs only oxygen; carbon dioxide is unnecessary and may even be dangerous. CAMMANN and PEIPER [1932 *(45)*] found administration of carbon dioxide ineffective in suffocation attacks of premature infants, which present a clinical picture identical to that of suffocation in newborn infants.

SNYDER and ROSENFELD showed that the fetus inhales amniotic fluid into its lungs by its breathing. After exposing the uterus of rabbits by an abdominal incision, they injected India ink into the embryonic membranes of the breathing fetuses in a manner similar to GEYL's method. Histological examination showed that this ink had already penetrated into the pulmonary alveoli after several minutes. On the other hand, control animals in which the respiration had been arrested by anesthetics showed the lungs to be free of India ink. Inhalation of amniotic fluid into the lungs is thus not an accidental occurrence during birth, but the ordinary result of intrauterine respiration, which lets the amniotic fluid flow into the lungs and thus prepares the dilatation of the future air passages. Experiments on human and animal fetuses which had not yet breathed showed that the pulmonary alveoli are not empty but are filled with amniotic fluid, since they always contain cells or detritus, marks of amniotic fluid. W. REIFFERSCHEID (1941) repeated ROSENFELD and SNYDER's experiments on animals; he injected India ink or dye into the amniotic membranes of pregnant guinea pigs and obtained the same results.

Similar tests in man were made by EHRHARDT (1939) and W. REIFFERSCHEID and SCHMIEMANN (1939, 1940). Prior to a necessary interruption of pregnancy, EHRHARDT injected thorotrast through the abdominal wall into the uterus of the mother. When on the following day the fetus was delivered by Cesarean section and taken from the amniotic membranes, the thorium was demonstrable by x-rays in the fetus's lungs at least in traces, and at times even in considerable amounts. W. REIFFERSCHEID and SCHMIEMANN were able to demonstrate the same findings in identical experiments. There is no doubt that the thorotrast penetrated into the lungs by means of the intrauterine respiratory movements and was there increased in amount after the resorption of the amniotic fluid. By injecting umbrathor into the amniotic cavity of guinea pigs, W. REIFFERSCHEID succeeded in increasing the content of this contrast medium in the fetal lungs to an extent such that the intrauterine respiratory movements could be directly observed by fluoroscopy.

Respiration of Prematurely Born Infants

(39, 4a)

Although the respiratory center of man is functioning and active long before birth, it does not reach maturity until after birth. Therefore, immature respiratory patterns are not infrequent in the neonate born at term and are especially common in premature infants.

The lowest respiratory pattern, which can be found in premature human fetuses after Cesarean section as early as in the fourth month of gestation, is gasping respiration, as can be concluded from MINKOWSKI's and HOOKER's findings (see p. 543). Some premature infants never pass this stage, but die because of the immaturity of their respiratory centers.

DEDEK (1913) was probably the first to attribute the respiratory disturbances of premature infants, which had already been well known for a long time, to the immaturity of their respiratory centers. To prove this point he cited the comparative physiology of the respiratory movements. His work is forgotten today. BABAK (1911, 1921) and DEDEK (1913) jointly followed the development of respiratory movements through their successive developmental stages in embryos of fish and batrachians. At first they found isolated irregular openings of the mouth, movements which later were grouped together in more or less regular periods, until finally a regular respiration without pauses ensued. They regarded the regular uninterrupted respiratory pattern as a later, higher function of the respiratory center, which originally functioned periodically. Starting from this idea DEDEK examined the respiration of premature infants, and in them he also found the various forms of periodic respiration. These findings, which agreed very well with the experiments on animals described previously, were ascribed by DEDEK to the immaturity of the respiratory center.

ECKSTEIN and ROMINGER (1921) recorded the periodic breathing of premature infants. They stress that fluctuations may be completely absent even during observations lasting several hours. Thus, the respiration is not much different from that of an older infant. They occasionally also found periodic fluctuations in mature infants during their first three months of life. SALMI and VUORI (1930) re-examined the respiration of premature infants. They considered CHEYNE-STOKES respiration, which they saw appear in these infants alternately with normal breathing, to be physiological.

The irregular respiration of young infants was already frequently mentioned in the early pediatric literature. ECKERLEIN (1890) was rarely able to observe regular respiratory rhythm that lasted more than 15 to 30 minutes. However, in his investigations the infant was uncovered, which influenced the respiratory rhythm.

My own observations led me to conclude that no premature infant is able to maintain the mature respiratory pattern continuously. For this purpose I regard an infant as premature if he weighs less than 1500 gm. However, a single observation is sometimes not sufficient. If I examined the infants several times, especially after feeding (Fig. 107, p. 321), even otherwise healthy babies showed distinct, and at times quite considerable, respiratory disturbances, i.e., respiratory arrests in particular, as well as periodic breathing, sighing, or other irregularities. I attribute the lability of the respiratory center at this level of development to its immaturity.

My co-worker V. DIETEL [1952 *(4a)*] made respiratory graphs of 50 premature infants in order to determine at what developmental stage immature periodic respiration changes into mature (regular) respiration as seen in the newborn infant born at term and in sleeping adults. Eighty percent of the infants started to breathe regularly at the predicted time of birth. It may therefore be assumed that the respiratory center is fully matured at this time, i.e., at birth.

SALMI's (1932) investigations on the respiration of premature rabbits permit a comparison with the human premature. According to his investigations the lowest form is gasping respiration; this is followed by wheezing respiration. The most frequent form is periodic respiration, which occurs chiefly and most distinctly in weak animals during or after attacks of restlessness or during agony. This pattern usually disappears during the first days of life. Periodic respiration may also occur in rabbits born at term. The extent depends on the birth weight: The lower the birth weight, the more distinct the periodic respiration. Singultus and periodic motor restlessness frequently occur in prematurely born animals. We therefore find in the premature rabbit all the respiratory patterns that produce the clinical picture of central respiratory disturbances in the premature human infant.

"First" and Last Breaths

(33, 39, 46, 114, 4a, 16a)

> "When an animal is born and begins its struggle for independent existence, its first efforts consist in breathing. Breathing is life. The onset of breathing is the beginning of life."
>
> D. H. Barron

The contrast between birth and death is reflected in the function of the respiratory center, for in both cases it passes through the same stages of excitability, only in an opposite direction.

To reach conclusions which could be pertinent to man, Barcroft (1942) examined the onset of respiration in sheep embryos by exposing — under spinal anesthesia — the amnion sac of pregnant sheep which were immersed in saline solution at body temperature. He was thus able to observe the fetus through the transparent membranes. He distinguishes four stages of the onset of respiration between the thirty-fifth and sixty-fifth day of gestation:

First stage (spasm): When the face of the fetus is touched on the thirty-fifth day, the head jerks, moves slightly backward, and turns toward the stimulus. The excitation spreads to the region of the trigeminus and on the thirty-eighth day also involves the diaphragm.

Second stage (rhythm): Approximately on the fortieth day the same stimulus evokes a rhythm of successive spasms which gradually increase in number from 40 to 60. During this time the intercostal muscles also start to contract.

Third stage (separation): After the forty-fifth day, stimulation of the face causes an extended reaction of the head, trunk, extremities, diaphragm, and thorax; the diaphragmatic and thoracic movements last longer than the other body movements. Finally a pause occurs between the end of the generalized movements and the beginning of the rhythm. This interval increases with progressing pregnancy.

Fourth stage (inhibition): After the sixtieth day the fetus becomes torpid. Even strong stimuli cause only minor local responses. But as soon as the umbilical cord is clamped off the fetus loses its torpidity and gradually returns successively to the third, second, and first stages, provided the clamping off lasts long enough. Thus, the disinhibition occurs in a sequence opposite to that in which the centers developed.

During birth two processes occur: Inhibition is abolished by asphyxia; this arouses in the fetus a tendency to make respiratory movements. At the same time sensory stimuli, which were missing in the uterus, increase the general sensitivity. Adequate results can be achieved only if the brain is sufficiently supplied with oxygen.

According to BARCROFT the same four stages can also be seen in man during birth: A human newborn delivered by Cesarean section under general chloroform anesthesia presents the same picture as a sheep does under similar conditions. He starts life with a series of sporadic gasps. An infant who starts breathing as soon as his head appears represents the fourth stage.

We are indebted to AHLFELD (1891) for a few graphs in which he traced the first breaths of the newborn infant immediately after birth; he did this while investigating another problem. Certainly some of these tracings, which are distorted by generalized movements, can hardly be interpreted today, but undoubtedly we can

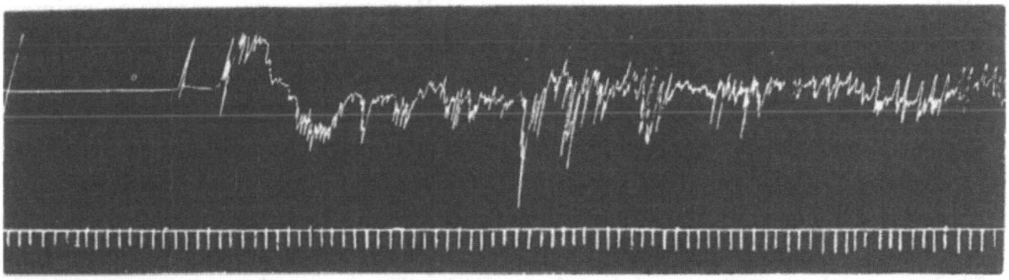

Fig. 142. Eighteen-year-old primipara. Spontaneous birth of a baby girl, length 48 cm and weighing 2960 gm. Duration of delivery 3¼ hours. Expulsion time 30 minutes. Respiratory graph, inspiration upward. Time in seconds. The graph shows — after an initial apnea — a distinct gasp which after another respiratory arrest is suceeded by a second gasp. Afterward a long cry, followed by a number of shallow, nongasping breaths. Then the picture of crying respiration appears. [According to L. SCHMIDT]

even now recognize a block in the respiratory center in two of his graphs (Table 1, Graph 3 and Table 2, Graph 3 of his paper). In both cases we are dealing with regular but superficial respiration which is interrupted at intervals by deeper breaths. With the aid of the graph alone we can no longer decide today whether these breaths were gasping respirations or only hiccups. A similar block in the respiratory center apparently is not infrequent in newborn infants, as is seen in the following description by VON JASCHKE (1917). "The first cry is followed by several breaths; the respiration then usually soon becomes quieter but still shows such irregularities that deep gasping inspirations are followed by completely superficial breaths and even by prolonged apnea."

According to the clinical and kymographic investigations of my co-worker L. SCHMIDT [1950 *(16a)*], respiration after birth always starts as a gasping respiration (Fig. 142). This pattern lasted between 13 and 45 seconds in infants born at term and in those who were approximately four weeks premature, up to 6½ minutes in infants who were still more premature, and considerably longer after a difficult delivery and if ether was used.

The dangerous condition of the newborn infant whose respiration does not start immediately after birth has long been called "asphyxia" ("pulselessness") in obstetrics. Actually it is apnea, i.e., a respiratory arrest; the circulation fails only after the respiration. Obstetricians have therefore repeatedly objected to the term asphyxia (BAISCH, 1924; PHILLIPP, 1956; PSCHYREMBEL, 1956; WULF, 1959). The primary task of starting respiration is easily sidetracked in a wrong direction because of this false terminology.

WULF speaks of eupnea, dyspnea, and apnea in newborn infants. He terms a newborn infant fit and eupnoic if the first gasping breath occurs within 30 seconds after birth and the first loud cry 1½ minutes at the latest after birth, if rhythmic respiration follows the first cry without external influence, if the pallor or the cyanosis of the trunk is replaced by a rosy color in 4 minutes at the latest, and if the pulse and turgor of the umbilical cord are still satisfactory 30 seconds after birth.

GIBBERD (1953) came to similar conclusions when trying to interpret "asphyxia" of the newborn human infant with the aid of BARCROFT's experiments on animals: Anoxia decreases the brain function. It has been stated that anoxia might stimulate the medulla. This is a misconception; it only can abolish the inhibitory influence of higher centers. If during progressive anoxia the medulla does not receive enough oxygen, respiration is decreased; this further decreases the normal oxygen supply of the blood. If the anoxia is severe and of prolonged duration the medulla is permanently damaged and the infant will not recover. In "asphyxia" the medulla is the first to die. Gasping respiration is an indication of a severe depression of the medulla. If the gasps become less and less frequent in the course of 5 minutes and if the intervals between breaths increase to 30 seconds or more, the damage to the medulla is probably irreparable.

KRUKENBERG's (1920) photograph of a neonate who is just taking his first breath shows that the first breath is actually a "gasp": The head arched backward and the wide open mouth unmistakably indicate gasping respiration. L. SCHMIDT *(16a)* reproduces a similar illustration. PEIPER, HEMPEL, and THOMAS [1951 *(114)*] filmed the birth of a child and thus demonstrated that the first breath of the newborn is a gasp (Fig. 143).

The philosophical interpretation of the newborn's first cry was discussed elsewhere (see p. 8). According to RUNGE (1895), the newborn infant cries only after having taken several deep breaths. GUTZMANN (1924), however, stated that the first cry occurs simultaneously with the first breath. All 20 newborns who were especially observed by L. SCHMIDT *(16a)* for this purpose cried only *after* the first extrauterine breath. The same could be noted on the films taken by PEIPER, HEMPEL, and THOMAS (Fig. 143). GESELL (1945) assumes the first breath to be an extensor-startle which causes an expiration; the inspiration supposedly occurs only afterward. Our observations and L. SCHMIDT's respiratory graphs contradict these views.

BOSMA and LIND (1960) examined the first breath of the newborn in the lateral position with cinematographic x-rays; they also found the gasp and, in addition, a swallowing respiration which they termed "glossopharyngeal respiration." Air is transported from the pharynx to the esophagus by this mechanism, just as a morsel is. It corresponds to voluntary frog breathing in poliomyelitis (see p. 352). The same respiratory movements occurred in a 39-day-old infant during focal brain stem seizures lasting 1 to 1½ minutes (BOSMA, 1960).

Fox (1929) observed the birth of a chimpanzee which at first did not start to breathe: "It appeared to those present as though the mother breathed into the mouth of her baby."

During the process of dying there occurs, instead of construction, a disintegration of the respiratory center, this phenomenon being colloquially termed the "last gasp." Frequently the function of the respiratory center decreases gradually in the dying person so that the individual respiratory patterns follow one another: For example, in Fig. 144, a respiratory graph of a dying premature infant, one can see normal and gasping respiration in succession. In Fig. 137 (see p. 350), which also traces the last breaths of a dying premature infant, gasping respirations toward the end and finally only laryngeal movements can be seen. There is no question that some respiratory patterns may often be skipped to some extent during the process of dying.

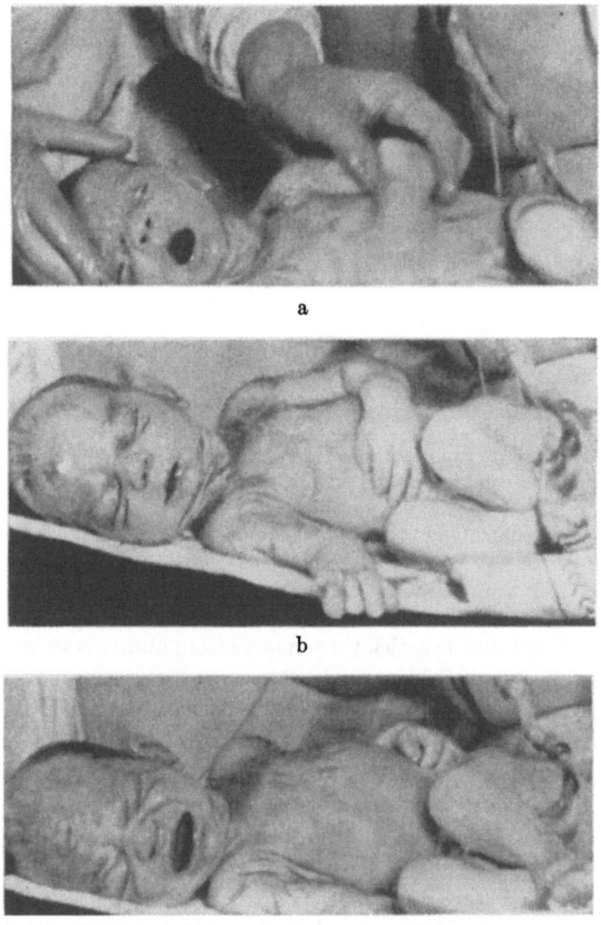

a

b

c

Fig. 143. "First" distinct gasping breath of a newborn infant whose umbilical cord has not yet been cut, immediately after forceps delivery (frame of a film). (a) Gasp; (b) respiratory arrest; (c) first cry.

A similar disintegration also occurs in the older child and in the adult. RAUTENBERG (1919) described a woman believed to be dead in whom slight laryngeal movements were observed 14 hours after she had been placed in her coffin. Resuscitation attempts led to short, jerk-like inspirations occurring every 2 to 3 minutes, i.e., gasping respiration. Subsequently, irregular respiration and finally the normal respiratory pattern appeared. The respiratory center passed successively through all different stages of excitability.

During dying we see not so much a successive failing of the individual part-centers of respiration, but rather a general disintegration of cerebral function during which all reflexes gradually fade away.

A suffocating animal shows identical disintegration of the respiratory center. For example, SIEFERT (1896) describes the behavior of lizards immersed in pure hydrogen (periodic respiration is physiological in lizards). After an initial excitatory phase the respiratory embarrassment manifested itself in a manner similar to that seen in frogs, i.e., in wide opening of the mouth and backward jerking of the head; the pauses in the respiration became longer. Increasing respiratory embarrassment then led to

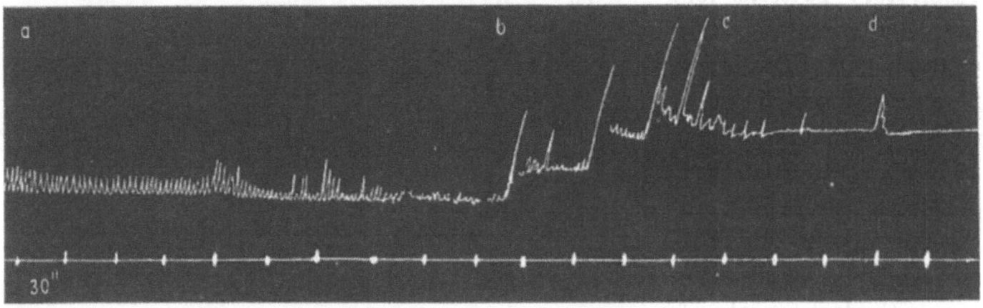

Fig. 144. Terminal respiration while dying in a 21-day-old infant weighing 2000 gm (time in 30-second intervals). (a–b) Irregular normal respiration; (b–c) gasping respiration; (d) last breath. The continuation of heart function beyond the last breath is just barely recognizable in the graph. The respiration was traced from the left side of the chest.

intense swallowing breaths and to laryngeal movements, at first once or twice between two breaths, then in increasing numbers so that 5 to 10 laryngeal movements were always interspersed between two thoracic movements. BABAK and KÜHNOVA (1909) made similar observations on salamanders: In oxygen deprivation the pulmonary respiratory center interrupted its activity earlier than the laryngeal center, and when oxygen was given its recovery from paralysis definitely occurred later. From these observations it can be concluded that the blueprints of the respiratory centers of man and reptiles are identical, despite other differences. CAMMANN and PEIPER described the onset of gasping respiration in suffocating pigs and rabbits shortly before their death [*(46)*; see p. 334].

GARSCHE (1954; see p. 458) clinically demonstrated the excitability of the swallowing center, the oldest bulbar center, with the aid of graded air stimuli. In the mature newborn infant, the air stimulus elicits an excitation in the swallowing center which interrupts the respiration only during swallowing. It is a sign of maturity if immediately afterward the infant again breathes rhythmically. In premature infants, however, the respiratory center is inhibited for a much longer time by the excitation of the swallowing center. The younger the premature infant is, the more prolonged the respiratory arrest seems to be. After the respiratory arrest the respiration—especially in young premature infants—does not become rhythmic, but is resumed as periodic breathing or even as gasping respiration. While swallowing occurs briefly and rapidly as a response to the stimulus in otherwise healthy premature infants, feeble infants show a markedly retarded worm-like swallowing movement if the swallowing center is insufficiently supplied with oxygen, because of the breakdown of the respiratory center. The slow execution of the swallowing reflex in response to an air blast represents a terminal sign. This reflex is usually the last that can be elicited after all others have become extinguished.

RESPIRATORY DISORDERS OF LATER CHILDHOOD

Although the mature respiratory center is more firmly composed and cannot be disturbed in its structure as easily as the immature center, it still may break down after infancy.

Floor-of-the-Mouth Respiration, Inspiratory Retractions

(67, 114, 8a, 10a)

Not infrequently, a premonitory stage of gasping respiration occurs in certain diseases which threaten to be fatal, particularly in diphtheria and pneumonia before death. In diphtheria this preliminary stage may develop because of laryngeal stenosis. If the latter starts gradually, each inspiration will at first be accompanied by barely visible movements of the lower jaw. At the same time the sensorium becomes clouded. Gradually the accompanying movements of the lower jaw become more distinct until finally floor-of-the-mouth respiration develops if no surgical intervention has occurred. This respiration may change to gasping respiration and is a consequence of oxygen deprivation, for it immediately disappears when the stenosis is relieved by surgical intervention, which the infant's condition urgently requires. The administration of oxygen may also have a certain degree of success.

Besides these movements of the lower jaw, which may also be termed mandibular respiration, the muscles of the floor of the mouth are often independently raised and lowered. This can be most distinctly noted at the angle between chin and neck [Peiper, Hempel, and Thomas, 1951 *(14)*]. Therefore, these movements of the soft tissues (Figs. 145 and 146) can in the strict sense be called floor-of-the-mouth respiration. In Fig. 146, the impressive respiratory movements of the muscles of the floor of the mouth in a 10-month-old infant with severe pertussis pneumonia can be recognized. The sketch, which was made by superimposing tracings of two frames of a film, shows the raising and lowering of the floor of the mouth in its end positions as well as the jugular retraction and the thoracic movements.

When the mouth is opened in floor-of-the-mouth breathing, the tongue will also be raised and lowered. This indicates the relations which exist between floor-of-the-mouth respiration as a preliminary stage of gasping respiration and gasping respiration itself in the pure form.

The appearance of floor-of-the-mouth respiration, for example during pneumonia, is always a poor prognostic sign; it has been our experience that older children whose cerebral function is lowered to this stage may be lost more easily than younger children.

Floor-of-the-mouth respiration and mandibular respiration may occur separately, as well as simultaneously. Not infrequently under these conditions, especially in pneumonia, there occur epigastric, intercostal, and jugular retractions accompanying

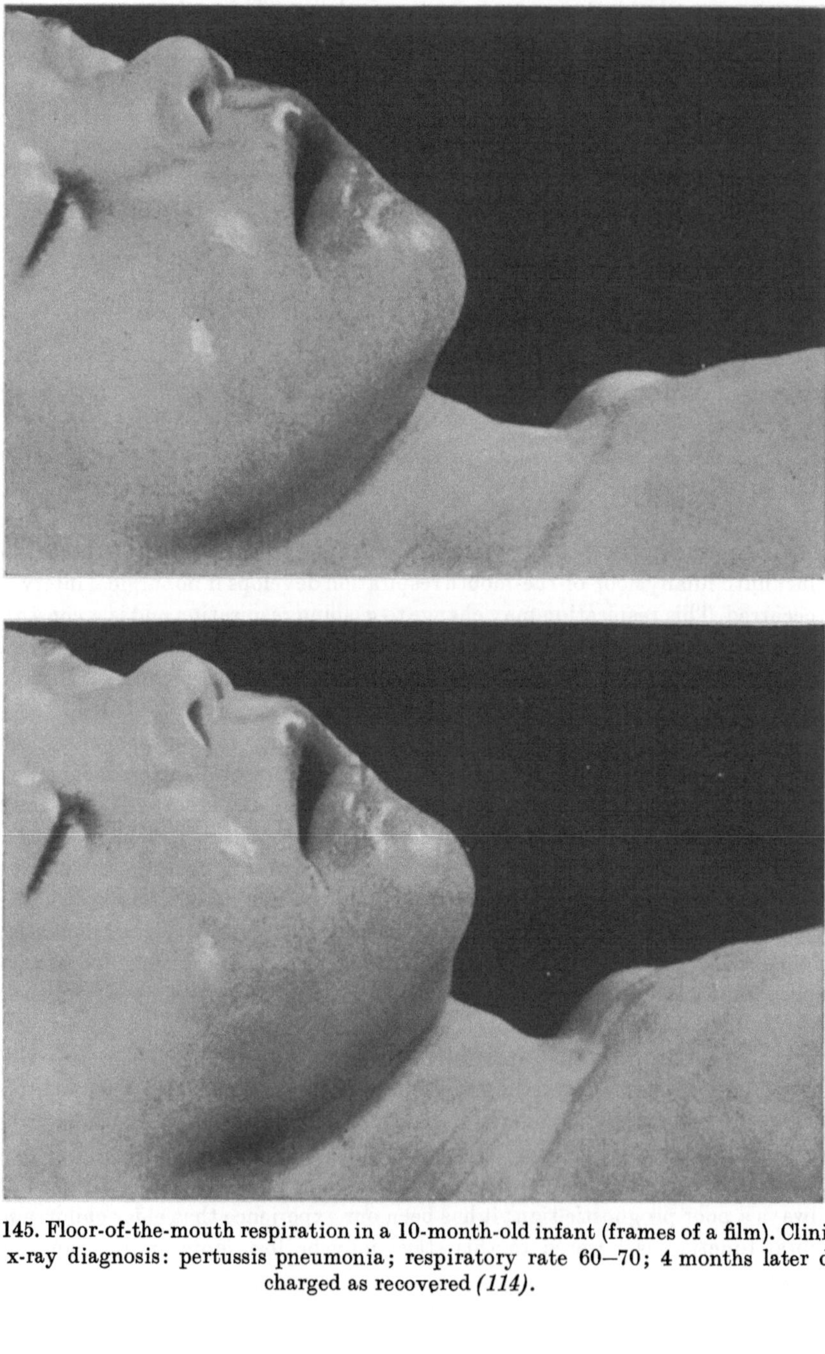

Fig. 145. Floor-of-the-mouth respiration in a 10-month-old infant (frames of a film). Clinical and x-ray diagnosis: pertussis pneumonia; respiratory rate 60–70; 4 months later discharged as recovered *(114)*.

every breath [HEMPEL and THOMAS, 1951 *(8a)*]. These phenomena – as well as the respiratory rate – can be evaluated as signs of pneumonia, but only in the quiet child. If a child is so rapidly and carelessly uncovered for the examination that he becomes restless and starts to cry, the same retraction will occur even without pulmonary involvement. Therefore, one should always wait until the child has calmed down again. Epigastric, intercostal, and jugular retractions are shown in frames of a film in Figs. 147 and 148.

Older children show inspiratory retractions when they suffer from croup, pseudocroup, aspiration of a foreign body, or obstruction of the upper air passages from another cause. The inspiratory retraction becomes more distinct, the more the obstruction, which is usually extrathoracic, impedes or even makes inspiration

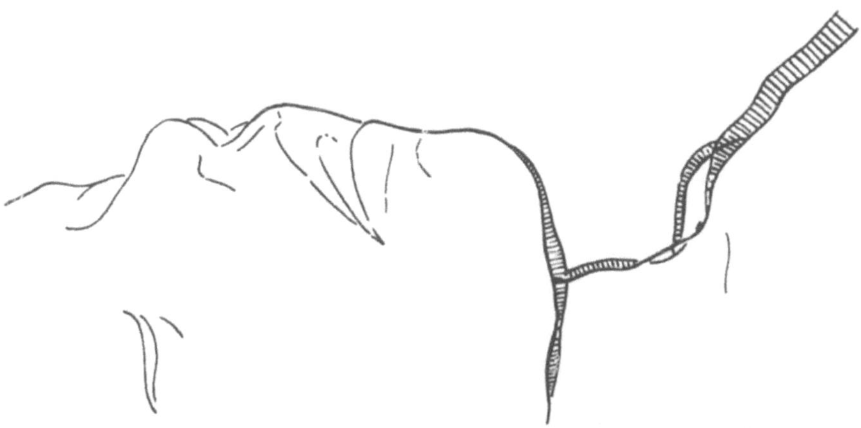

Fig. 146. Sketch of floor-of-the-mouth respiration based on superimposed tracings of the frames of the film in Fig. 145. The movements of the floor-of-the-mouth muscles and the jugular retractions, both marked by shaded lines, can be recognized easily *(114)*.

impossible. On the other hand, in pneumonia at this age inspiratory retraction is not so distinct as it is in young infants, whose chests bulge considerably because of the severe pulmonary involvement. In the absence of pneumonia it may also occur when the abdomen is distended, though less distinctly. Distinct though less conspicuous inspiratory retraction may be observed, despite a considerable decrease of the respiratory surface, in interstitial pneumonia of seriously ill premature infants whose respiration has become accelerated and very shallow.

KRUMPEL [1952 *(10a)*] observed gasping respiration caused by pharyngeal stenosis in a sleeping 6-year-old girl. The child had snored only occasionally during her first 3 years of life; at the end of her third year of life she developed a threatening pharyngeal stenosis following whooping cough. Thereafter she was under continuous medical treatment. After the third adenotomy, which was done under general anesthesia, a physician observed the respiratory disorder for the first time and sent her to a hospital.

In the waking state the only remarkable sign in the girl was distinct oral breathing. If she was asleep, irregular gargling and snoring sounds could be heard even from a great distance. The child, struggling for air in her sleep, appeared as though she were about to suffocate any moment. At times she reared up. Painful stimuli could not wake her. The corneal and patellar reflexes were decreased; the pupils were dilated and did not react; BABINSKI's sign was present. Four to six gasping breaths occurred at intervals of 4 to 5 seconds, accompanied by dropping of the lower jaw, slight backward sinking of the tongue, jerk-like backward movements of the head, and increasingly deepening jugular, intercostal, and epigastric retractions (Fig. 149); general cyanosis was noted until a deep inspiration accompanied by loud noises occurred. This was followed by visible relaxation of all muscles, which lasted 10 to 15 seconds. The same process recurred again and again during sleep.

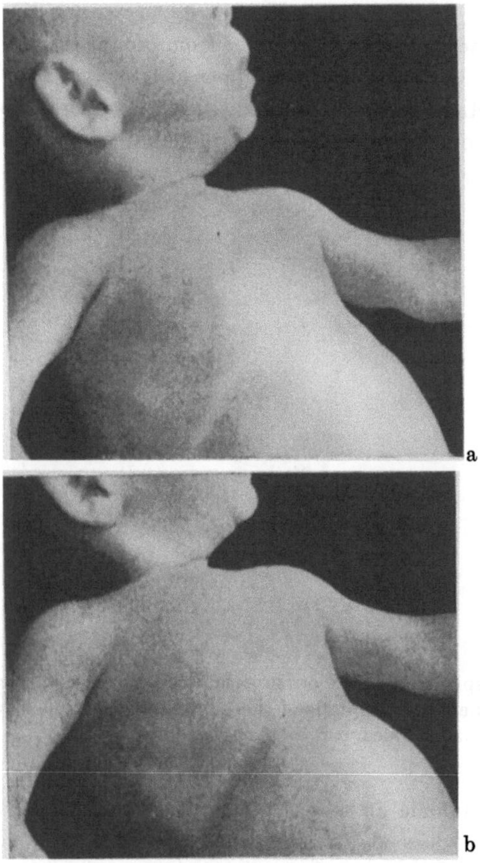

Fig. 147. Inspiratory retractions in a premature infant with interstitial pneumonia. Respiratory rate 62. Frames of a film [according to HEMPEL and THOMAS *(8a)*]. (a) Expiration; (b) inspiration with intercostal and epigastric retractions.

When the oral cavity of the sleeping child was palpated with the finger it was found that the pharynx was tightly closed in the back by medium-sized tonsils, the soft palate, and the tongue. If the soft palate was cautiously lifted without waking the child, she could breathe better. Therefore, a thin catheter, introduced through the nose and brought out through the mouth, was fixed to the face so that the soft palate was lifted from the posterior pharyngeal wall and the tongue. The noise of the stenosis became weaker and the retraction more shallow. A tonsillectomy was performed with the result that the snoring diminished and the retraction became less pronounced. The gasping respiration disappeared. Four months later scar formation was noted, so that the soft palate gave the appearance of having ruffles. The child had clinically recovered.

TAKAGI and BOSMA (1960) describe a 16-week-old infant who — despite normal shape and size of the lower jaw and tongue — could not clasp the nipple with his mouth and suck milk through it. The infant could be fed in the erect position.

I have occasionally observed a glossoptosis such as was described by ULLRICH (1935) and BARTELHEIMER (1936), in which because of relaxation of the musculature or micrognathia the tongue falls back as in general anesthesia and thus obstructs the hypopharynx. In this situation the first signs of gasping respiration may also appear. As is done in general anesthesia, grasping and pulling the tongue forward will relieve the stridor.

Gasping respiration occurred in a four-year-old epileptic boy under similar conditions. During each seizure the child aspirated the tongue to the back of the mouth so that it closed the laryngeal entrance. Threatening suffocation again and again caused extremely severe gasping respiration, during which the child appeared to be dying. The condition could be immediately relieved each time by pressing down the lower jaw. It was not necessary to pull the tongue forward. After a few free breaths this dangerous condition subsided. At first the seizures occurred repeatedly so that the child had to be constantly watched. They subsided under phenobarbital medication.

The floor-of-the-mouth respiration described may also occur in toxic diphtheria and severe pneumonia after infancy, but distinct gasping respiration usually does not develop. If floor-of-the-mouth respiration has developed, the prognosis for survival is poor. These children nearly always die, though they may sometimes be kept alive for several days.

a

b

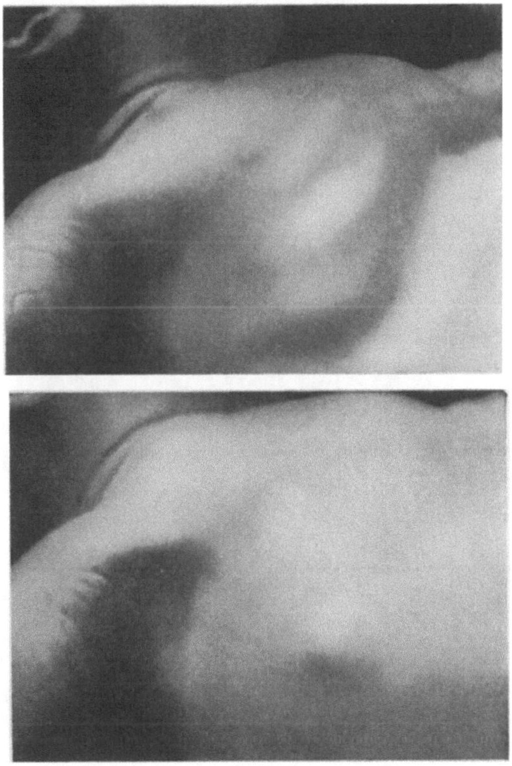

Fig. 148. Intercostal and epigastric retractions in a 10-month-old infant with whooping cough and pneumonia [frames of a film according to HEMPEL and THOMAS *(8a)*]. Respiratory rate 60 to 70. (a) Inspiration with intercostal and epigastric retractions; (b) expiration.

This preliminary stage of gasping respiration differs distinctly in one aspect from the suffocation attack of the premature infant; in the latter the individual gasping breaths are separated by prolonged respiratory arrests. In the older child, however, they follow one another without pause. The respiratory rate is often increased, while it is always markedly lowered during the suffocation attack.

Floor-of-the-mouth respiration is peculiar to fish, amphibians (frogs), and reptiles (lizards) (see Fig. 138, p. 351). Its appearance in a sick child can therefore be regarded as a regression to that developmental stage.

Respiration by Movements of the Alae Nasi

The younger a child suffering from pneumonia is, the more frequently does respiration by movements of the alae nasi occur. It is an important pathological sign in infancy. Even mere observation shows that the alae nasi move in the same rhythm as does the diaphragm.

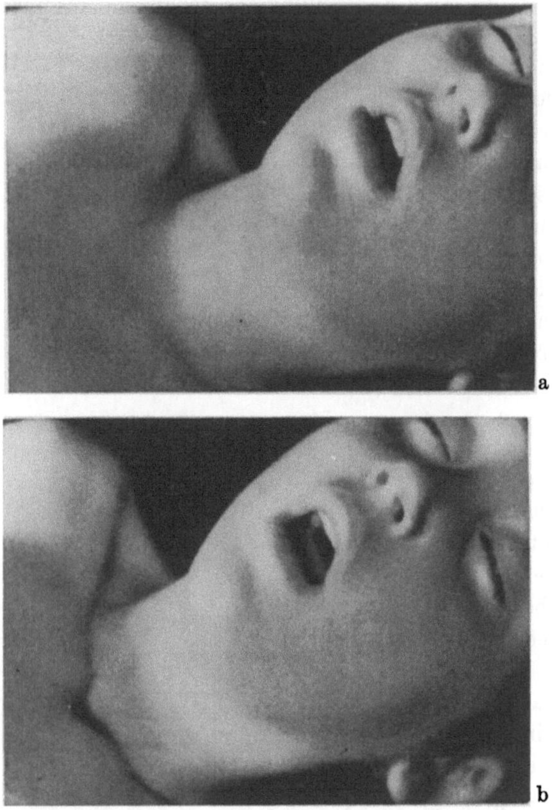

Fig. 149. Gasping respiration in a 6-year-old girl with pharyngeal stenosis [frames of a film according to KRUMPEL *(10a)*]. (a) Asleep with distinct oral respiration. (b) Asleep with gasping respiration: lower jaw dropped, mouth wide open, tongue slightly raised and somewhat dropped to the back, deep jugular retractions. Intercostal and epigastric retractions not visible in this illustration occurred simultaneously.

SECKEL (1931) recorded the child's respiration by movements of the alae nasi. He distinguished two forms: an active pre-inspiratory and a passive expiratory form.

In the pre-inspiratory form, the flaring of the alae nasi precedes thoracic expansion by approximately one sixth of the duration of a breath. This form occurs as a pathological reflex in pneumonia, miliary tuberculosis, and paralysis of the respiratory muscles. The receptors of this reflex are located in the region of the pulmonary automatism. Alae nasi breathing is usually thought to be a "nasal auxiliary respiration." The enlargement of the entrance is said to facilitate the passage of air. This view, which was expounded, for example, by ENGEL (1924), is rejected by SECKEL, primarily because the nasal passages are already free in the pre-inspiratory form of alae nasi respiration and because only the antra of the nose, in themselves spacious, are enlarged, and not the narrow nasal passages.

SECKEL considers alae nasi breathing to be an early respiratory form which is disinhibited by pulmonary disease. WINSLOW (see p. 354) described the working of the alae nasi during the intrauterine respiratory movements of cats and dogs, which after birth no longer exhibit alae nasi respiration. According to BAGLIONI (1911), the opening of the vocal cords in the breathing of frogs (described on p. 351) is coupled with the closing of the outer nares through the closing of the mouth. Because of this the pulmonary space is never directly connected with the outer air. SECKEL further stresses that the alae nasi of amphibians, some reptiles, many aquatic mammals, and the majority of the rodents move rhythmically even during regular breathing. According to some authors, among them BREHM (1883), the nostrils of the seal open during each inspiration; they are then immediately closed and remain tightly shut until the next air exchange. This occurs even when the animal is on land. The same is true of other aquatic mammals, the sirenians and the walrus. This opening and closing of the nostrils, originally of vital importance for amphibians during their sojourn in water, is usually inhibited in animals who breathe on land. Based on this consideration, SECKEL regards pre-inspiratory alae nasi breathing as an incomplete opening and closing of the nares which corresponds to the alae nasi breathing of amphibians. In the young child respiratory embarrassment disinhibits the phylogenetically very old alae nasi breathing.

In expiratory breathing through movements of the alae nasi, the alae nasi are simply inflated by the air current during each expiration. This occurs during loud screaming, coughing, and other violent respiratory movements, and as simple participation of the alae nasi during increased respiratory activity.

Breath-Holding

(79, 107)

Attacks of breath-holding or, as IBRAHIM (1911, 1938) called them, "respiratory affect spasms" usually occur in the second or third year of life; however, IBRAHIM observed these attacks even in infants. The youngest child whom I observed [1939 *(79)*] was 3 weeks old. Although these fits occur rather frequently, only in exceptional cases are children brought to the physician just because of such attacks; experienced mothers know how to cope with them without a physician. But if an only child of an inexperienced mother has severe attacks, he may be sent to the hospital because of this condition, which is regarded by the mother as highly dangerous.

The precipitating stress for the attack is usually trivial; anything which can make a child cry, such as a slap or a scolding, the denial of a wish, a fall, a fright, or a medical examination, may precipitate this phenomenon. In some children these attacks are rare and may occur only at intervals of several weeks, but in others they are considerably more frequent, especially when the environment responds inappropriately. The more frightened one is and the more one gives in to the wishes of the child in order to avoid the attack, which is thought to be dangerous, the more often it will occur. If the people around him remain calm, or if the child is even scolded when he approaches a breath-holding phase during screaming, the attacks become less frequent and soon disappear.

I cannot confirm the statement that only neuropathic children are subject to these attacks. Rather, it is a harmless habit of otherwise healthy children who later develop quite normally. The physician, however, sees mainly children who come from a nervous environment.

The attack always develops from the child's screaming. This fact is important, since one has to depend on the statements of the parents for the diagnosis; this history distinguishes breath-holding from an epileptic seizure.

At first the child screams with mounting excitement. Suddenly breathing stops, the head falls backward, and the body is extended. If the child has been sitting or standing prior to the attack, he will fall to the floor. He will no longer react. Because of the respiratory arrest his face and especially his lips turn bluish-red. Frequently the child does not remain completely motionless during the attack, but makes twitching seizure-like movements with the trunk or extremities. It is easily understandable that a frightened mother thinks her child in mortal danger or even dead. But finally the respiration returns, the child cries and ventilates his lungs. Instantly the bluish-red discoloration of the face disappears and is replaced by a pronounced pallor. After the attack the child is flaccid, tired, and weak until he finally recuperates. Some children have bowel incontinence during the attack. A physician who observes the attack from its beginning will rarely mistake this for an epileptic seizure. But this differentiation is difficult to make from the description given by an inexperienced mother.

Rousseau (1762) may have been the first to describe an attack of breath-holding:

"I shall never forget how I once observed one of these obnoxious screamers whom his nurse had slapped. He suddenly became quiet; I thought he was intimidated. I said to myself: One day this will be a servile mind which will be influenced only by sternness. But I was wrong; the unhappy child was suffocating from anger and had lost his breath; I saw him turn purple. A moment later a piercing scream erupted. All the signs of indignation and desperation of this age could be heard in his screaming; I was afraid he would succumb in this excitement... I am sure a firebrand accidentally falling on the child's hand could not have excited him more than the rather light slap which had been delivered with the apparent intention of chastising him."

Osiander (1796) also recognized breath-holding:

"Some children have the habit of holding their breath while crying and thus frightening their parents. The best remedy for this is a strong slap with the rod on the back or the buttocks. As soon as it starts you can also splash cold water on the child's face or set him down quickly on a cold floor."

The descriptions in textbooks sometimes differ from this account. According to Hamburger (1939), children turn blue or pale during the attack; according to Glanzmann (1939), they first turn pale and then blue; according to Ibrahim, with whom I agree, they turn bluish-red during the respiratory arrest and pale after they start to breathe again.

According to Biedert (1902) and Gött (1931), the respiratory arrest follows the expiration, but according to Neumann (1905), Stier (1918), Finkelstein (1938), and Glanzmann (1939) it follows the inspiration. Ibrahim was an exponent of the first view at the beginning and of the second view later. These contradictions arise from the fact that it is difficult for the physician to observe an attack. Stier, who reported on 29 children exhibiting breath-holding, had himself never observed an attack, but had to rely on the statements of the parents.

To clarify these contradictions I made graphic recordings of the respiration of an eight-month-old infant before and during an attack. The attacks occurred when he was fed mashed bananas, which he did not like, against his will. It was thus possible to elicit an attack after the arrangements for the recording of the respiration were made.

Breathing always stopped at the end of an expiration (Fig. 150); the respiratory arrest was always terminated with an inspiration. However, the diaphragm was only once immobilized as in the recording reproduced here. In the other instances the

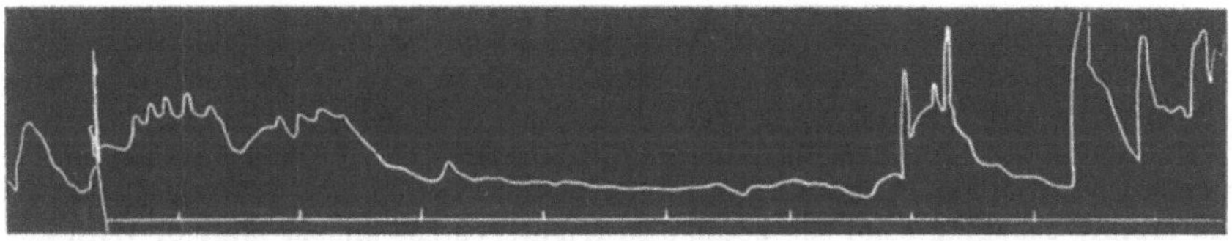

Fig. 150. Attack of breath-holding in an eight-month-old infant (time in 15-second intervals). First, quiet breathing; then the recording was interrupted until the beginning of the attack. At the beginning of the continuous tracing the child is still crying. Then follows an attack of breath-holding with complete respiratory arrest; diaphragm in the expiratory position. At the end of the respiratory arrest gasping respiration; the breaths are shorter and deeper than in normal respiration.

respiratory arrest, which lasted 25 seconds at the most, was interrupted by considerable movement.

The first breath of an infant after a respiratory arrest due to breath-holding corresponds absolutely in the pattern of its tracing to the gasping breaths of a suffocation attack. One has only to compare the short and deep first breath after the attack with a breath in the same tracing which was recorded during the calm before the onset of the attack. This breath lasted much longer and was distinctly less deep than the gasping breath. Evidently the state of excitation of the respiratory center sinks to the level of gasping respiration because of the lack of oxygen during breath-holding.

Usually the respiratory arrest is terminated by a gasping breath. The child continues to cry, but gasping breaths no longer occur. However, in a nine-month-old infant a true suffocation attack followed the breath-holding, and gasping respiration continued for several minutes during this attack, causing the same clinical picture as seen in the suffocation attack of the premature. The child had just recovered from severe whooping cough; transition of the pertussis attacks to periods of gasping respiration of long duration had previously also occurred in this child.

Breath-holding is very seldom followed by an attack of suffocation. We must suspect the parents' statement that the child's breath-holding lasted for several minutes, for the usual respiratory arrest does not last so long.

In their electroencephalographic findings, BAMBERGER and MATTHES (1959) do not see any indication that breath-holding is related to epileptic patterns. Only in 3 of 28 children examined were they able to find an abnormal EEG in the free intervals. Two of these children suffered, in addition, from grand mal epilepsy independent of breath-holding; the third had damage due to birth trauma. In a three-and-a-half-year-old boy breath-holding lasting 19 seconds occurred during the taking of the EEG. The first half of the EEG could not be evaluated because of the child's restlessness. High slow waves were seen in the second half of the tracings, but not seizure patterns. Low, E. L. Gibbs and F. R. A. Gibbs (1955) reported identical findings.

Breath-holding is harmless despite its dangerous appearance. Since according to Ibrahim no death has yet been observed, I cite the following article from the "Rheinische Landeszeitung" (1940):

Aachen, February 13th: A mother wanted to bathe her two children, Marianne, four years old, and Hedwig, one and a half. Marianne had suffered for some time from crying spells during which she held her breath and became blue. While the mother was with her two children in the bathroom, Marianne suddenly started to cry and again held her breath. The mother placed the child on her back on the floor to help her. When she observed the child rapidly changing color she became so frightened she almost fainted. With her last strength she took Hedwig, who was already in the bath, out of the tub and then fell unconscious to the floor. In doing this she dragged Hedwig along and fell on top of Marianne, who was lying on the floor. In the meantime the water overflowed and seeped through the floor. The attention of the people who lived on the floor below was attracted by the water seeping through, and when there was no reply to their knocking they finally opened the door of the apartment. They found the woman unconscious, holding Hedwig in her arms and lying on top of Marianne. Fortunately attempts to resuscitate her and also the one-and-a-half-year-old Hedwig were successful, but Marianne had suffocated beneath her mother.

In this case death had evidently occurred only by a chain of unfortunate circumstances. One can still maintain the view that a breath-holding attack which is not complicated by external events is not dangerous, frightening as it may appear. However, this description clearly shows how much an inexperienced and anxious mother can be frightened by such an attack.

Bridge, Livingstone, and Tietze (1934) reported that among a group of 83 children admitted to a hospital because of breath-holding attacks, there were 3 who prior to these episodes had suffered once or several times from convulsions with no relation to breath-holding, and 2 who later suffered from epilepsy. In exceptional cases, a severe progressive cerebral disease may be the cause of breath-holding (Wiedemann and Tolksdorf, 1955).

Whooping Cough Attack

(88)

Coughing protects the body by removing mucus or foreign bodies from the air passages. Like sneezing, it is an intensified expiration. True coughing is usually preceded by a deep inspiration, so that the air passages are abundantly filled with air. The attack itself is composed of a number of coughing jolts and occasional interspersed inspirations. At each individual coughing jolt the rima glottidis is at first closed for a moment while the diaphragm is suddenly pushed against the thoracic cavity by a prolonged contraction of the abdominal muscles. The pressure in the thoracic cavity is increased because of this jolt as long as the rima glottidis remains closed. Then the closed rima is suddenly opened and the abdominal muscles, which are still active, press a strong air current accompanied by the well-known coughing sound toward the mouth. This mechanism is more effective in cleaning the air passages than normal expiration, since in the latter the air is not under increased pressure. The complicated and exactly timed interplay of the different muscle groups for each coughing jolt suggests the existence of a regulatory mechanism in a coughing center.

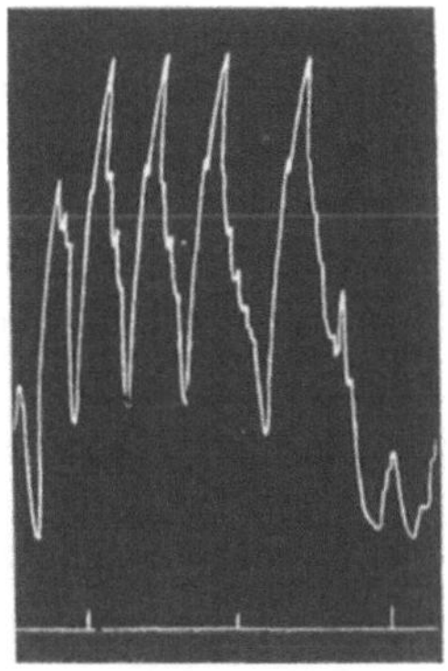

Fig. 151
Whooping cough attack (time in 5-second intervals): Crowing inspirations; several coughing jolts are combined into an expiration. Most of the inspirations are interrupted briefly by coughing spasms.

Fully developed whooping cough differs so distinctly from ordinary coughing that one will always be able to recognize it after having heard it once. Pertussis is described in every pediatric textbook, but we must recognize that the individual descriptions differ in essential points. The "reprise," an unfortunately chosen foreign expression, is thought to be of special significance. In Germany some people interpret it as the noisy inspiration, others as a repetition of the entire attack. Each of these opinions can be traced to French authorities: RILLIET and BARTHEZ (1853) speak of reprises in the sense of repetitions; HUTINEL and DARRÉ (1909) and NOBÉCOURT and BOULANGER (1934), in their contributions to handbooks, denote the noisy inspirations with this term.

When the respiratory movements of a whooping cough attack are recorded graphically, the tracings obtained are just as multiform as is the clinical picture in general. Some children suffer greatly from these attacks, their lives even being endangered, while others are only temporarily distressed. The attacks are more severe at the height of the disease than at the beginning or end. The disease is usually most severe in infancy, although the "whoop" (crowing inspiration) is rarely heard at that age.

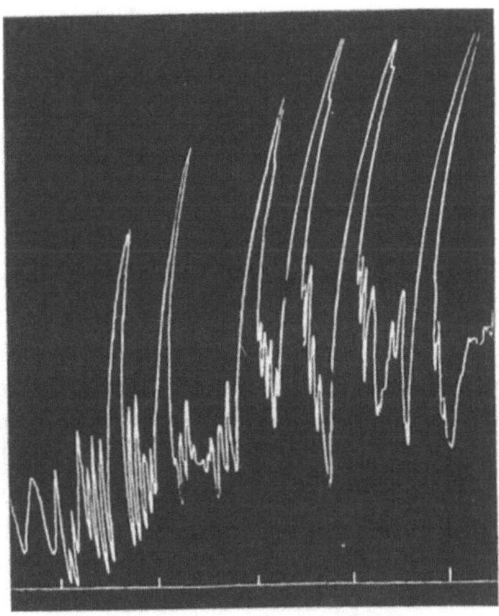

Fig. 152. Beginning of a whooping cough attack (time in 5-second intervals). The condition of Fig. 151 is reached only in the second half of the graph. The position of the diaphragm is lower after the attack; the graph is higher than in normal respiration.

The effect on the child depends on his general state of health, the condition of his nervous system (CZERNY, 1939), his age, the number of coughs that follow one another without intervening inspirations, the duration of the entire attack including the repetitions, and the frequency of vomiting.

Figure 151 illustrates an attack of a moderate degree of severity which corresponds approximately to the usual description. The deep inspirations, mostly connected with "whoops," the short coughing jolts which are combined into a prolonged expiration, and the multiple repetitions of the same process, one immediately after another, are quite characteristic. It should be stressed, however, that a regular tracing such as is seen in Fig. 151 cannot always be obtained; rather, the individual coughing jolts are often separated from each other by more or less complete respiratory movements. The individual coughing jolts often do not combine into one expiration. Figure 152 shows the beginning of an attack. The characteristics of the whooping cough attack develop gradually and only the second part of the graph conforms in principle to Fig. 151.

The more coughing jolts follow one another without pause, the greater are the intervals between the crowing inspirations. According to HEUBNER (1911), whose

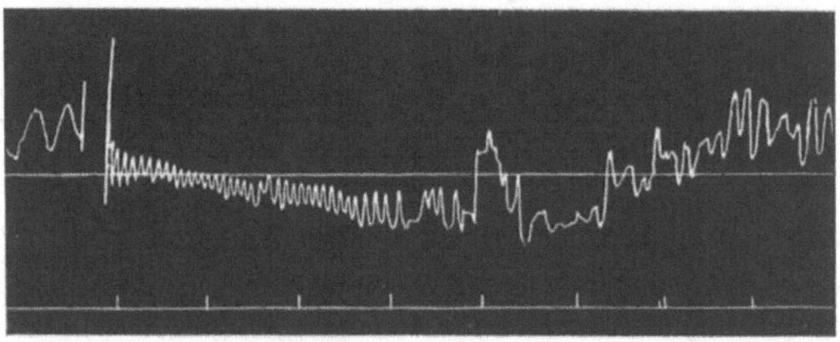

Fig. 153. Very severe, prolonged whooping cough attack; the beginning was not recorded (time in 5-second intervals). At the end of the attack the diaphragm is higher (the graph is lower) than in normal respiration at the beginning of the graph.

excellent description of whooping cough should be especially mentioned, we may have 4, 6, or even more coughing jolts before a new inspiration occurs. These figures may apply to attacks of mild or moderate degree (Figs. 151 and 152), but in very severe attacks they may even be considerably higher. Thus, I found 40 uninterrupted successive coughing jolts in a period of 23 seconds (Fig. 153) in the most prolonged uninterrupted coughing sequence that I was able to record. Since the onset of this attack was not recorded, the total number must actually have been even higher.

If inspiration fails to appear for such a length of time, there must be a marked oxygen decrease in the blood. The child is also harassed and exhausted by the recurring great effort connected with the spasms of the abdominal muscles. The position of the diaphragm immediately prior to the relieving inspiration is of importance for the effect on the child. To make a comparison I first recorded the resting respiration as well as possible and then, without changing the leads, during the attack itself. Considerable differences between the various attacks could be noted. In mild or moderately severe attacks the position of the diaphragm is not changed much. Occasionally it is even somewhat lower at the end of the attack than before (Fig. 152). Thus, the lungs may still be well filled with air at the end of the coughing spell. But when the inspiration comes very late the continuously repeated coughing jolts gradually raise the diaphragm (corresponding to a lowering of the graph in Fig. 153), the lungs contain increasingly less air, and the blood is pressed from the thorax into the exceedingly full veins of the head and neck.

Knöpfelmacher (1931) states:

"Apnea may occur during an attack of violent coughing at the height of the expiration, probably caused by a spastic closure of the rima glottidis."

According to this description, which is similar to those found in many textbooks, the respiratory center continues to function, but successful inspiration is prevented by the spastic closure of the rima glottidis. It may be just an accident that such a respiratory arrest did not occur in my recordings; perhaps the number of the most severe attacks I was able to record was too small. In any case, a respiratory arrest which occurs in this manner does not belong to the picture of a fully developed whooping cough attack.

Of course, very severe and possibly fatal respiratory arrests may occur following whooping cough attacks in infancy, but they are very rare. They are likewise attributed to a prolonged closure of the rima glottidis. In this connection Feer (1938) speaks directly of a rima glottidis spasm. However, we are actually dealing with a completely different clinical entity. A whooping cough attack causes a breakdown of the respiratory center, so that an attack of gasping respiration occurs.

Stenger (1950), on the other hand, attributes the "expiratory apnea" at the end of a whooping cough attack to a strong stimulation of the sensory vagus. I cannot concur; neither can I agree with his interpretation of the apnea attacks (see p. 338). We are not dealing with the occurrence of a single gasp, but with the regression to a lower form of respiration, i.e., gasping respiration, as the graphs reproduced here show.

In a two-month-old mature infant I was able to show graphically the progressive disintegration of the respiratory center caused by whooping cough attacks (Fig. 154a—c). The baby suffered from extremely severe whooping cough spells which always led to gasping respiration. The attacks usually occurred in connection with feeding, so that they could be anticipated with a certain degree of probability. The infant was able only to drink slowly from the bottle; the graphs reproduced were recorded during the intervals between feedings. Immediately prior to drinking the respiration was accelerated (72 per minute compared to 35—45 per minute in the healthy infant), but otherwise regular. But during drinking, distinct periodic respiration was soon noted (Fig. 154a). After the child had taken the bottle for a while a severe whooping cough attack occurred, during which (as can be seen in Fig. 154) breathing stopped completely while the diaphragm remained in the expiratory position except for a transitory diaphragmatic movement after approximately 8 seconds. At the same time the movements of the infant stopped, the skin became bluish-pale, and his pupils were dilated. It was impossible to continue the tracings of the respiratory movements at this time because artificial respiration and oxygen had to be given. After this treatment, which was combined with injections of lobeline and nikethamide, breathing gradually returned, at first as gasping respiration. Figure 154c shows how this respiratory pattern, characterized by its short deep breaths at irregular intervals, gradually changed over to normal breathing. During this transition and shortly afterward the original skin color returned and the child recovered visibly. To avoid the onset of gasping respiration if possible, the child was now tube fed. All went well at first, but then the breathing stopped completely without even a coughing spell after a tube was inserted. All attempts at resuscitation were futile and the baby died.

The graphic recordings during the child's feeding showed the following stages of respiration: normal breathing, periodic respiration (Fig. 154a), whooping cough

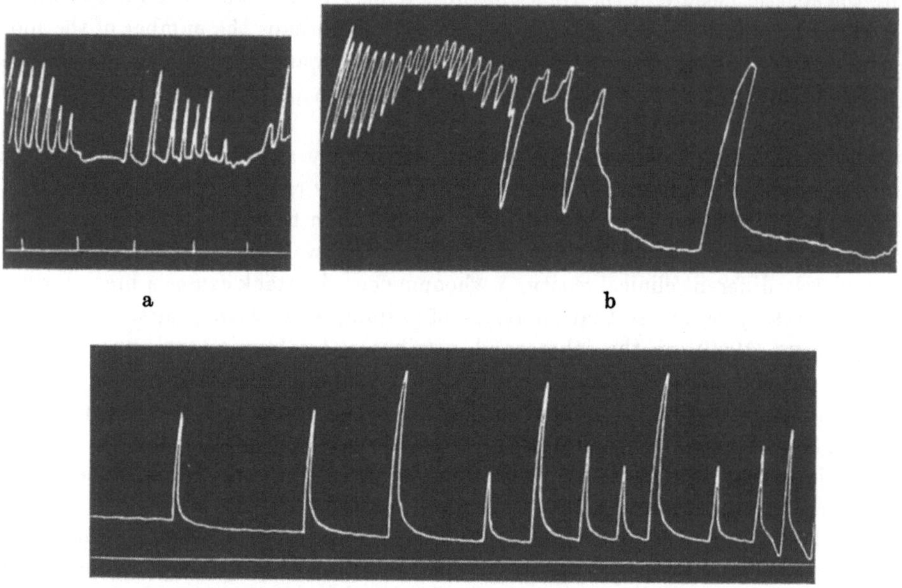

a b

c

Fig. 154. (a) Periodic respiration occurring after feeding (time in 10-second intervals). (b) Attack of whooping cough with following respiratory arrest, temporarily interrupted by a diaphragmatic movement (time as in 154a). (c) Transition from gasping respiration to normal breathing (time as in 154a).

attack (Fig. 154b), respiratory arrest (Fig. 154b), gasping respiration (Fig. 154c), and again normal breathing (Fig. 154c). The progressive disintegration of the respiratory center was at first initiated by the effort at food intake and was aggravated by the whooping cough attack, which completely used up the reserve of nervous energy. Figure 120 (see p. 329) shows a similar attack of suffocation following a whooping cough attack in a premature infant.

A nine-month-old infant born at term also suffered from severe *suffocation attacks* caused by whooping cough. In this child the coughing spells were only temporarily interrupted by gasping respiration; as soon as he was partially able to breathe again, the interrupted whooping cough attack started anew and in its further course again caused gasping respiration. This pattern was repeated several times until the entire attack subsided. During the gasping respiration, gross jerking of the arms and legs and twitching of the trunk muscles appeared. In spite of the exceedingly threatening suffocation attacks, which repeatedly made artificial respiration, oxygen, and respiratory center stimulants necessary, the infant was quite alert during the intervals. There was no spasmophilia.

A suffocation attack in this infant took the same course as in a premature infant: He turned pale, no longer responded, his muscle tonus became poor and his movements subsided, occasionally some muscles twitched with involvement of the diaphragm, the reflexes disappeared, the pupils were dilated, and marked gasping respiration appeared. Since in general the respiratory center has the same composition, the same ontogenetically and phylogenetically lower respiratory patterns, which are disinhibited by its disintegration, are always released. The external precipitating cause is unimportant; what matters is that the respiratory center loses a sufficient amount of nervous energy.

When the nine-month-old infant described above was showing some signs of improvement, attacks of breath-holding independent of the still existing whooping cough occurred. As is common in such cases, they occurred following disagreeable experiences. At the end of these attacks the infant also showed all the characteristics of gasping respiration, but he overcame these states more easily so that artificial respiration became unnecessary. I have previously indicated (see p. 373) that after prolonged breath-holding the first breath is gasping. However, this was the only child in whom I observed a marked prolonged attack of gasping respiration after a breath-holding attack.

Both immaturity and whooping cough favor the onset of gasping respiration. Therefore, it is understandable that premature infants are in grave danger when taken ill with whooping cough. Repeated suffocation attacks, which follow whooping cough spells and to which the infants sooner or later succumb, dominate the clinical picture. The younger the infants are, the more they are endangered; however, there are considerable differences among individual infants. One infant may suffer severe whooping cough attacks without lapsing into gasping respiration, while the latter may occur in another, perhaps even older, child after nearly every coughing spell. The reason for this difference in reactions is unknown. The nine-month-old infant described above, whose whooping cough attacks turned into gasping respiration, had a twin brother who suffered from severe whooping cough but had no suffocation attacks. In my experience, there is no more danger in this regard after infancy.

The tremendous effort which each whooping cough attack requires from the infant is to blame for the fact that even in mature infants it causes gasping respiration, a respiratory form which usually occurs only in premature babies. It deprives the respiratory center of so much nervous energy that the center is reduced to the level found in the premature.

Textbooks have mentioned that an ominous suffocation attack may follow a whooping cough spell in young infants; only the connection between these two phenomena has been variously interpreted. HEUBNER (1911) wrote:

"As in some cases of laryngospasm, the whooping cough spell stops suddenly; the infant gasps several times for air and dies (in the presence of the physician, as I have once experienced myself). Immediate application of fifteen minutes of artificial respiration, in combination with various kinds of stimulants, does not succeed in bringing him back to life. The last-mentioned case is that of an eight-week-old infant who had had whooping cough for only a week (the attacks accompanied by vomiting made the diagnosis of whooping cough certain)."

FINKELSTEIN (1938, p. 598) observed extremely severe suffocation attacks after whooping cough in a four-month-old infant. However, it seems doubtful to me that death occurred, as he thought, because of a "cardiac syncope" during the last attack, for the children die of paralysis of the respiratory center during gasping respiration. The cardiac action is affected only subsequently (see p. 337).

Artificial respiration always proved to be the most effective treatment for suffocation attacks in the infant. This would have been impossible if the attacks had been caused by a spasm of the rima glottidis. In whooping cough, as in prematurity, gasping respiration has nothing at all to do with a spasm of the laryngeal muscles, but is caused by disintegration of the respiratory center. But another phenomenon that occurs during a whooping cough attack could be related to a narrowing of the rima glottidis – the so-called "whoop," the crowing inspiration. This is said to be so characteristic of whooping cough that it alone can confirm the diagnosis. But according to my experience this is not absolutely correct, for during severe coughing spells I have occasionally heard isolated but distinct "whoops" which had nothing at all to do with whooping cough.

How does this inspiratory noise originate? It is usually assumed that the mechanism is a spastic closing of the rima glottidis. Since this has not yet been proved, I considered the possibility that the noise might be caused by transitory paralysis of the recurrens nerve. Meanwhile, at the suggestion of STENGER (1950), KIRSTEIN conducted a laryngoscopic examination of a four-year-old child during a whooping cough attack: The movements of the glottis were normal before the attack; during the attack the rima glottidis opened immediately before each coughing jolt. The extent amounted to approximately one third of the glottis opening during normal expiration. During the crowing inspiration the glottis remained tightly closed; only in its posterior quarter was there a narrow gap which was completely closed for a short time at the end of the inspiration. These findings could be regularly observed as long as the inspiration was crowing; they absolutely excluded the possibility of flaccid paralysis.

Another criterion for the fully developed whooping cough attack after infancy is said to be the tendency of the attack to recur after a short pause that sometimes lasts only a few minutes. This sign alone, however, is not sufficient to allow the diagnosis.

If the attack starts without external cause, a tickling sensation in the region of the larynx will precede the attack. However, it is often elicited by an external cause, such as feeding or the excitement due to the visit of the physician. CZERNY (1915) states that nervous excitability is generally increased during whooping cough. One can elicit attacks in patients suffering from whooping cough by merely touching the posterior pharyngeal wall with a tongue depresser, i.e., from a region where stimulation is usually ineffectual. The reflexogenic zone of coughing – usually restricted to the fossa interarytaenoidea and the bifurcation – is enlarged during the illness. We are dealing here with a central process, the phenomenon of dominance (see p. 531): The excited center attracts stimulations which ordinarily follow other pathways through the central nervous system and thus increases its own excitatory state. In the same manner the hyperexcitable coughing center responds during whooping cough to stimuli to which it would not otherwise respond.

The coughing center may be deprived of its nervous energy by a similar process so that it becomes inactive: When the curtain falls in the theater, there starts among the spectators a general clearing of the throat and coughing that did not occur during the performance. Evidently the intense attention diverts the nervous energy from the coughing center.

During severe whooping cough attacks the excitability of the coughing center is markedly increased, possibly because of peripheral stimuli. Since an inspiration is impossible during the coughing jolt, the respiratory and the coughing center alternate so that one center is inactive when the other is active. The center of inspiration directs the inspirations rhythmically, but the coughing center is usually active only during expiration; as the graphs show, the expiration, when compared with the respiration at rest, can be augmented (Fig. 153), reduced (Figs. 151 and 152), or prolonged by the coughing jolts. But if the intense state of excitation of the coughing center prevents inhalation for some length of time, the chemical stimulus of the blood on the respiratory center becomes so strong that it finally overcomes the inhibition coming from the coughing center and causes the relieving inspiration. The intense state of excitement, however, is by no means abolished; rather, immediately after the inspiration renewed coughing jolts occur. A discharge of the coughing center is only accomplished after the same process has been repeated several times.

The whooping cough attack therefore represents a struggle between the greatly overstimulated coughing center and the center of inspiration for the regulation of respiration. The excitatory states of the two struggling centers alternate in opposing directions. While the coughing center is discharging, the excitatory state in the respiratory center increases; during inspiration, which lasts only a limited time and ends automatically, the intensely irritated coughing center strives for a discharge. Sometimes the state of excitement becomes so strong that some coughing jolts interrupt the inspiration (Fig. 151). Stenger (1950) interprets these interferences as a consequence of a spasm of the laryngeal muscles. One could conceive that the greatly irritated coughing center seizes the supply of nervous energy — which is limited in young infants anyway — and thus tears it away from the respiratory center. This center disintegrates and gasping respiration develops. In the case described above, the respiration was alternately dominated by the gasping and the coughing center. The points of reversal at which one center took over the lead from the other can be seen quite clearly in the graphs.

When choking and vomiting occur at the end of the attack, the excitation has leaped from the coughing center to the vomiting center. The close relation between these two centers — one of which empties the lungs and the trachea, and the other the stomach, toward the mouth — is of ontogenetic origin: The lungs are formed from the gastrointestinal system (see p. 348).

In Fig. 155 we see vomiting in the form of short, violent diaphragmatic contractions, which are characteristic of vomiting (see p. 470), at the end of several coughing jolts.

Whooping cough attacks are occasionally combined with sneezing attacks, as has been described in textbooks. Sneezing differs from coughing mainly in that the contraction of the abdominal muscles has to burst open the lock of the roof of the pharynx instead of that of the rima glottidis. Sneezing movements of the abdominal muscles and coughing jolts cannot be distinguished from each other in the graphs.

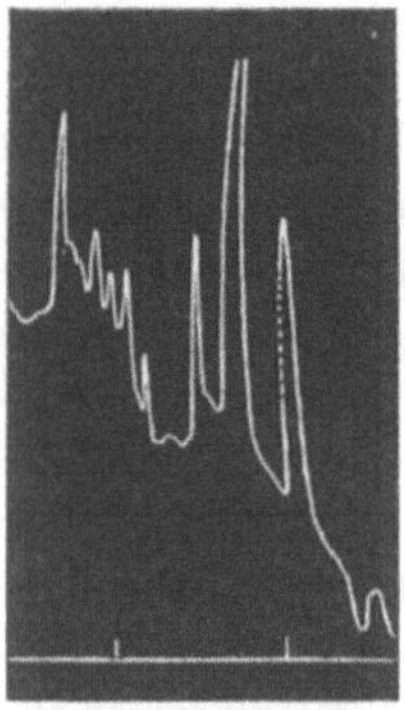

Fig. 155. Whooping cough jolts, followed by three vomiting movements of the abdominal wall (time in 5-second intervals).

Sighing Respiration

(39)

An individual sigh produces a characteristic graph. It rises considerably above the other peaks; the descent, which is frequently followed by a compensatory respiratory arrest, occurs gradually. During inspiration the familiar sighing sound can be heard. Sighs generally occur during normal breathing, but they may also accompany CHEYNE-STOKES respiration, as MOND and S. WASSERMANN (1927) have shown.

Tuberculous meningitis not infrequently begins with repeated deep sighs, according to MAYR (1862) and HENOCH (1881). ECKSTEIN and ROMINGER (1921, 1922) graphically recorded the sighing respiration occurring in this disease. I was occasionally able to demonstrate this type of breathing also in simple febrile infections of the respiratory tracts.

In premature infants, whose nervous reactivity has a certain similarity to that of meningitic children, sighs occur occasionally, though infrequently, whereas they are not infrequently found in healthy newborn infants born at term. I saw sighs recurring more or less periodically in the constant respiratory irregularity of premature infants (Fig. 140, p. 353). In the adult they usually occur sporadically but occasionally in groups of two or three (according to MOND and WASSERMANN). I also observed double sighs in premature infants. MOND and WASSERMANN, who described sighing respiration in adult cardiac patients, see in it an indication of respiratory failure due to the insufficient aeration of the blood.

The condition of uninterrupted sobbing so frequently seen in small children also belongs in this group. It has previously been described (see p. 117).

Respiration in Mental Deficiency and Epilepsy

(37)

The brain damage which causes mental deficiency frequently also disturbs the central function in other ways, by causing paralyses, pareses or spasms, motor disturbances or seizures, tremors, disorders of the muscle tonus and of the unconditioned or conditioned reflexes.

During infancy mental deficiency is frequently accompanied by a disorder of the brain stem which expresses itself in delayed onset or complete absence of static development. Even the lower rhythmic centers of respiration and sucking are not infrequently disturbed.

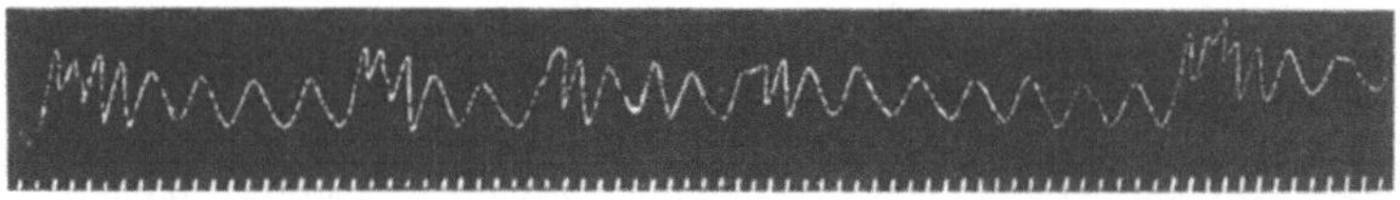

Fig. 156. Periodic respiratory and motor disturbance during sleep in untreated myxedema (time in seconds). With every respiratory acceleration the infant twitches and shows movements similar to fright reactions.

In the history of patients suffering from Little's disease, which is often accompanied by feeble-mindedness, an "asphyxia" (more correctly: respiratory arrest) is frequently stated to have occurred in the newborn or premature infant, or after a difficult delivery. It is not self-evident, as is so frequently assumed, that the suffocation of the newborn infant is the cause of cerebral hemorrhages and has thus produced the brain damage causing the disease; rather, there is the possibility that the original brain damage has another cause and produces the suffocation attack of the newborn infant only secondarily.

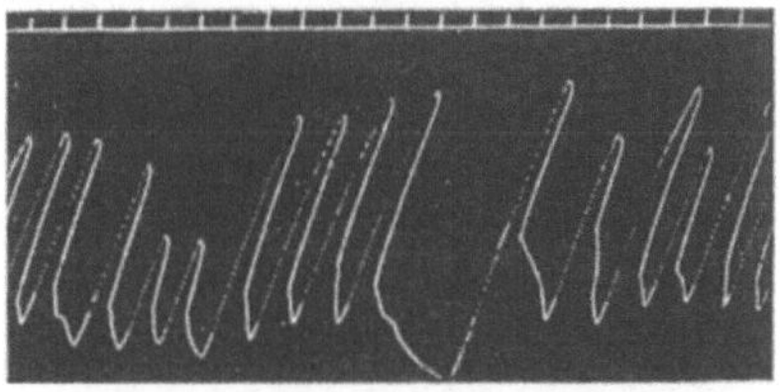

Fig. 157. Constant irregularity during immobile sleep (time in seconds). Epilepsy, age 4 weeks.

In feeble-minded children suffering from seizures, the respiratory rhythm is not infrequently disturbed during sleep when all movements have ceased. In a child nearly three years old who suffered from (untreated) myxedema, a distinct periodic motor disturbance similar to a fright reaction occurred during deep sleep; during this motor disturbance the respiratory rhythm was accelerated each time (Fig. 156).

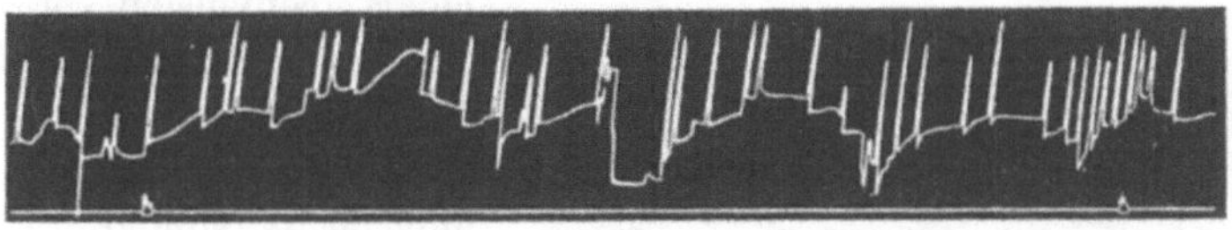

Fig. 158. Respiration during an attack of unconsciousness and atonia without motion disturbance in an eight-month-old epileptic infant (time 1 minute).

Sometimes these children show constant irregularity (Fig. 157). The rhythmic disturbances may be accompanied by sporadic muscle twitches. The breaths, which immediately follow one another, are of varied depth and are separated by irregular intervals and sometimes by pauses of long duration. The base line from which the individual breaths start often has the appearance of an irregular wavy line.

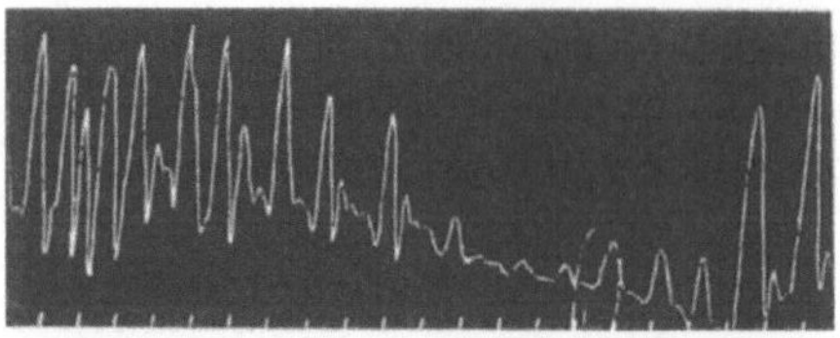

Fig. 159. Constant irregularity with respiratory arrests in a five-month-old epileptic infant during seizure (time in seconds).

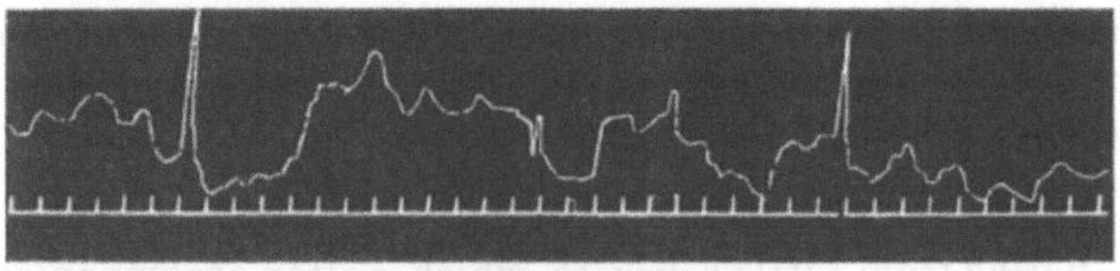

Fig. 160. Constant irregularity of respiration during epileptic seizure in a two-and-a-half-year-old feeble-minded child (time in seconds).

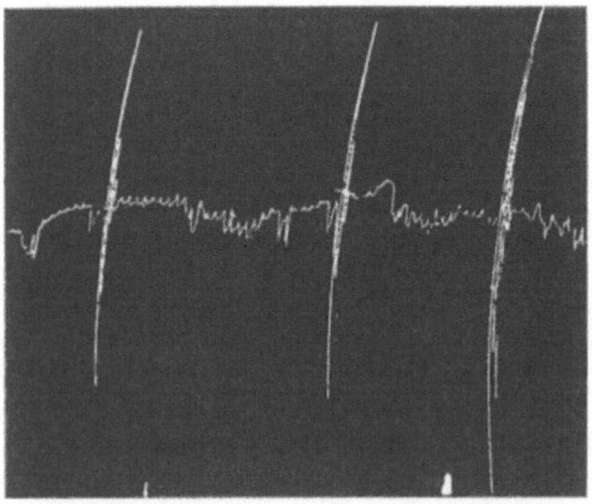

Fig. 161. Short, violent diaphragmatic contraction, interrupted by long respiratory arrests with visible cardiac movements in the same infant as in Fig. 158 during epileptic seizures (time in 15-second intervals).

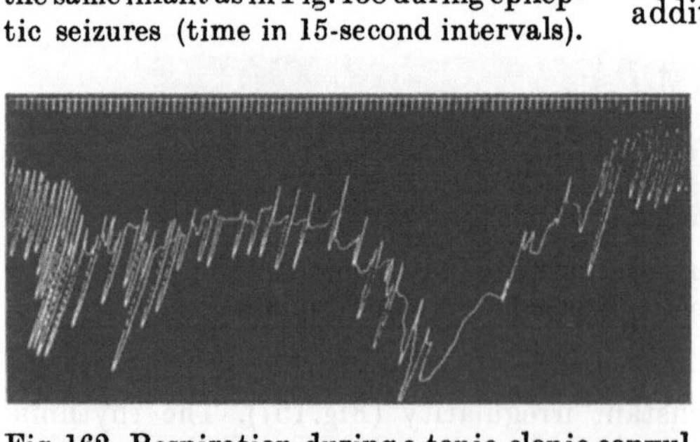

Fig. 162. Respiration during a tonic-clonic convulsion in the same infant as in Fig. 157 (time in seconds). The motor disturbance continues after the respiratory disturbance. Occasional wheezing respiration.

Constant irregularity is most marked during a true epileptic seizure (Figs. 158–160). Henoch (1881) already knew of this condition. He speaks of a "contraction of the respiratory muscles with frightening pauses in the respiration, alternating with very rapid respiratory movements." If these arrests last for some length of time (Figs. 161 and 162), the face becomes red and a moderate degree of cyanosis may occur. However, I never observed an apnea as dangerous as that which occurs in the suffocation attacks of the premature infant.

During seizures the usual uninterrupted sequence of inspirations and expirations was in general maintained. Occasionally a longer lasting inspiratory position of the thorax occurred (Figs. 163 and 164). In these cases the respiratory act consisted of expiration and inspiration followed by a short respiratory arrest.

The seizure may cause a kind of tonic contraction of the diaphragm (Fig. 162). But while the diaphragm gradually contracts and then again expands, some sporadic superimposed movements occur in addition to this big movement. A kind of clonic convulsion of the diaphragm may also be seen (Fig. 165). In addition to the real respirations, some rather violent diaphragmatic movements similar to singultus attacks (see p. 346) occur periodically. But clinically this is not singultus. As in the singultus attack, here too one had the impression that respiration was being simultaneously regulated by two centers.

Occasionally respiration changes

so much during an epileptic seizure that the graph does not have the appearance of a respiratory recording (Figs. 160 and 161). Long respiratory arrests, during

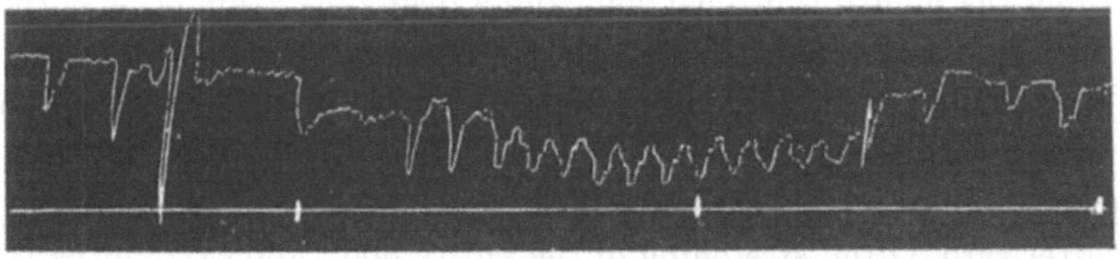

Fig. 163. Respiration during a tonic-clonic seizure in the same infant as in Fig. 158 (time in 15-second intervals). At the start and at the end of the graph the inspiratory position of the thorax persists for some time.

which only minimal or no diaphragmatic movements can be recognized, are interrupted by short, violent diaphragmatic thrusts. Surprisingly enough these irregular respiratory movements are sufficient to satisfy the body's need for oxygen. As far as we can judge clinically these jolts are apparently not restricted to the diaphragm; they sometimes involve the entire musculature. Respiratory and motor movements have no exact time relation during the seizure. In some cases the respiratory disturbance ceases prior to the motor disturbance, but in other cases the respiratory rhythm has already changed prior to the motor discharges.

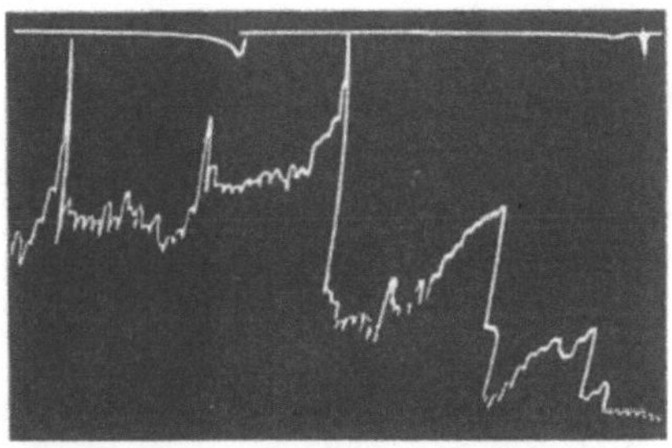

Fig. 164. Respiration during epileptic seizure in the same infant as in Fig. 158 (time in 15-second intervals). Long lasting respiratory arrest with visible cardiac movements; violent movements consisting of expiration and inspiration in the intervals.

Disintegration of the respiratory center is very rare during an epileptic attack; rather, the breathing becomes irregular and grossly distorted. The respiratory center is usually influenced by the motor centers. During a seizure the excitation of the motor centers penetrates into the respiratory center, which thus comes under the influence of violent irregular stimuli. Therefore, release of the lower partial centers of respiration cannot be expected. The encephalitic respiratory disorders, which have a similar appearance, may have a similar origin (Pierre-Marie and Lévy, 1922; Turner and Critchley, 1925; Suckow, 1924; et al.).

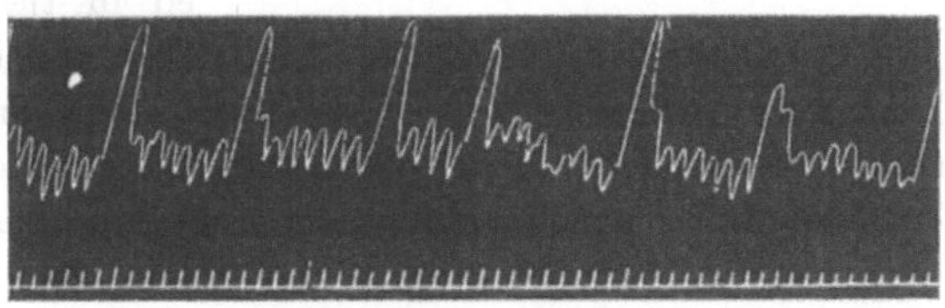

Fig. 165. Periodic diaphragmatic spasms between normal breaths (no singultus) in a one-year-old epileptic child (time in seconds). The child is awake, motionless, and follows with his eyes.

Yawning

(43)

In yawning the mouth is opened wide and a deep inhalation occurs; the thorax is arrested for some seconds in the inspiratory position; several interspersed shallow breaths may follow; then exhalation occurs and the mouth is closed. At times a brief respiratory arrest may follow. Most descriptions overlook the fact that a swallowing movement still occurs at the end. The eyes are closed during yawning and begin to water afterward. Often stretching of the entire body and extremities is connected with yawning. As HEUSNER (1946) has demonstrated, yawning is accompanied by acceleration of the heart action and by vasomotor constriction.

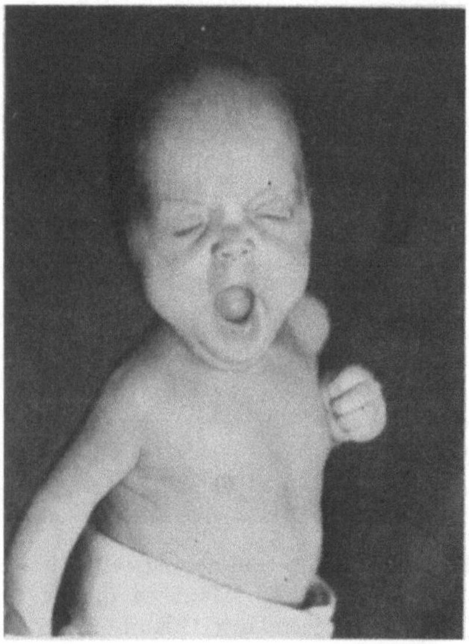

Fig. 166. Yawning infant, birth weight 870 gm. At the time the picture was taken the infant was 142 days old and weighed 2780 gm.

The adult yawns before and after sleeping, especially when sleep is prematurely interrupted, or when he is tired or bored. He is able to yawn voluntarily. LEWY (1921) and ENGELHARDT (1941) mention the yawning which occurs during hunger. That it is contagious also indicates that it is influenced by conditioned reflexes as well. Sometimes yawning may occur even when it is only mentioned, or if one reads or thinks about it. Men and animals yawn during fatal hemorrhage.

Yawning can frequently be observed in infants (Fig. 166), but is also distinctly seen in premature infants weighing as little as 600 gm; thus, we are dealing with an ontogenetically lower nervous process. LEWY assumed the yawning center to be located in the corpus striatum. According to DUMPERT (1929), it is in the not too deeply located subcortical centers of the brain stem, since it ceases functioning during sleep in contrast to the respiratory and vascular centers. He assumes that we are dealing with a primary automatism of the kind postulated by VOGT which is at the service of the brain center most highly developed at any given time.

Today, however, we have better information on the location of the yawning center. We gained this knowledge by observing human monsters born without cerebral hemispheres (GAMPER, 1926; CATEL and KRAUSPE, 1930). The infant described by CATEL and KRAUSPE was able to yawn and stretch himself, although his central nervous system was developed only to the level of the medulla oblongata. Therefore, the yawning center cannot be in a region which is higher than the medulla. It is probably located near the respiratory center, with which its function is closely connected.

GAMPER, whose human "mesencephalic" monster was also able to yawn, likewise rejects the predominant importance of the cerebral hemispheres for yawning. According to his hypothesis the yawn-stretch reflex is subcortical and is very closely connected with the other subcortical mechanisms, which alert the function of the entire brain during the process of waking and in the awakened state. But this hypothesis does not dispute the possibility that yawning can also be elicited by the cerebral hemispheres.

The wide distribution of yawning among animals also indicates that it is a nervous process of a lower order. According to HEMPELMANN (1926), wide opening of the mouth connected with deep breathing and stretching of the whole body may frequently be observed in mammals, especially in apes, beasts of prey (dogs!), and ungulates, and less frequently in birds. Among the reptiles, the crocodiles quite frequently open their jaws wide and keep them open for some time. While similar movements are not known to occur in amphibians, they may be found more frequently in fish, which open their mouths wide and at the same time push the mobile part of their jaw bones forward. It is doubtful, of course, whether these movements have the same meaning in lower animals as they have in man and mammals. In invertebrates, which have a completely different body structure, comparable movements cannot be studied. HEINROTH (1930) also observed in snakes an apparent yawning with wide opening of the mouth, but without inspiration.

The purpose of yawning is still doubtful. Increased oxygen need is not satisfied by yawning, as HAUPTMANN (1920) had already realized. The respiratory arrest which follows deep respiration prevents increased oxygen inhalation. DUMPERT (1929) gives another explanation: Yawning and stretching represent a direct vascular reaction caused by insufficient cerebral circulation. This reaction squeezes the blood from the veins into the arteries and thus effects a better circulation in the brain. According to this view we yawn and stretch each time the cerebral blood flow needs to be improved. With the aid of the plethysmogram, ENGELHARDT (1941) demonstrated that the arm shows better circulation during yawning than afterward.

It is difficult to reconcile DUMPERT's opinion with the findings of CZERNY (1896), BRODMANN (1902), and BERGER (1904—1907). According to these authors the cerebral circulation is better during sleep. If yawning has the same effect it should favor sleep. It is thus only logical that LEWY disputes the abundance of cerebral blood flow during sleep. DUMPERT tries to overcome this difficulty with the rather unsatisfactory hypothesis that the cerebral vessels relax during sleep, while they automatically expand during mental effort. Therefore, the nutritional conditions for the brain differ in these two cases. According to GAMPER, who regards DUMPERT's opinion as being too narrow, yawning and stretching have the effect of bellows which fan the glimmering glow.

W. R. HESS (1931) and WALDVOGEL (1945) were able to elicit yawning in cats repeatedly by electrical stimulation of the diencephalon. The region adjacent to the third ventricle between the anterior commissure and the infundibulum proved to be most sensitive. According to WALDVOGEL, muscle tonus and lung expansion (and with it the readiness for action) decrease during somnolence. Therefore, they must be restored from time to time. Yawning and stretching effect tonus increase and deepening of respiration and thus are best suited to do this. According to C. and H. SELBACH (1953), the stretch syndrome (a tonic motion process with yawning and inspiratory arrest) should be similarly regarded as a consequence of a regulatory circuit between the respiratory center and the extrapyramidal motor system. Therefore, it occurs only if the excitability of the respiratory center is decreased and if it becomes necessary to restore it to a higher state of functioning with the aid of the center-activating carbon dioxide, which originates in the muscle metabolism.

Anyone who has investigated yawning scientifically considers it self-evident that its purpose is to fend off fatigue. But sleep, which is initiated and caused by fatigue, is of vital importance. It would be strange indeed if man possessed a special nervous mechanism to protect himself from sleep. In addition the aim would not be achieved in this manner. There is a distinct sensation of fatigue which causes yawning. This sensation is not decreased by yawning and one cannot remain awake by yawning. The statement that one yawns in order not to be tired is contradicted by everyday experience; yawning is rather the consequence of being tired.

During fatigue, which leads to yawning, the excitatory state of the brain is reduced. The yawning which occurs during fatal hemorrhage can also be traced to a lowering of the central excitability. Thus, yawning is nothing else but a respiratory movement released by the decrease of the nervous excitability in the respiratory center; respiration has regressed to a lower stage.

Consequently, the yawning center is a component of the respiratory center which is usually inhibited. Again an ontogenetically and phylogenetically lower partial center is disinhibited by the disintegration of the respiratory center. The swallowing movement which follows the respiratory movement brings yawning in close relation to the phylogenetically very old "swallowing respiration" (see p. 348). Lower animals breathe by swallowing. It is therefore no coincidence that the rooting reflex of the young infant, which is intimately connected with his food intake, has frequently been compared with yawning on the basis of its external appearance and the way it is evoked (POPPER, 1920; GAMPER and UNTERSTEINER, 1924). POPPER observed — and I agree with him — that in infancy a stroking near the corner of the mouth (a stimulus which usually elicits the rooting reflex) sometimes produces yawning. This reaction is absent in adults.

CHRISTOFFEL (1951) also sees in yawning an "original form of respiration" which can be demonstrated after the beginning of the fourth fetal month. He assumes the existence of some unexplained relations to the drinking of amniotic fluid, although his detailed description of yawning and stretching does not mention the final swallowing movement which is always present. Neither do the theories which relate yawning to the improvement of the blood circulation in the brain take this swallowing movement into account.

Yawning is caused by disintegration of the respiratory center. The inhibitions which usually are imposed on the subordinate yawning center by the highest component of the respiratory center relax when the nervous excitatory state is lessened, so that the yawning center can temporarily regulate the respiratory center. As soon as the yawning center is discharged, its function ceases and the previous respiratory pattern is again released. The adult is able to yawn once voluntarily, but he cannot do it twice in succession. The yawning center can become excitable again only after an interval of short duration.

My interpretation makes the question as to the purpose of yawning superfluous. It cannot be satisfactorily answered as long as one sees in yawning a defense against fatigue. I see in yawning only the consequence or the expression of a state of fatigue which, however, is not influenced by it. Thus, yawning is on a similar level with singultus, which I also attribute to disintegration of the respiratory center (see p. 343). The frequency of yawning in premature infants can be explained by the fact that immaturity favors the breakdown of the respiratory center.

MUMFORD (1897) gave a remarkable explanation for the stretching which accompanies yawning in adults and which can also not infrequently be seen in premature infants. He starts from the "paddle movements" (see p. 218) which occur easily in young infants when they are held horizontally, head and abdomen supported, so that their hands and feet can touch a base. This movement is combined all through life with the yawning movement into a total movement which can be regarded as one of the oldest patterns; according to MUMFORD, stretching is thus a real "paddle movement."

SUMMARY

Formation and Disintegration of the Human Respiratory Center

When the vertebrates left the water for dry land in the course of evolution, the respiratory center adjusted to the changed conditions of life by forming new superimposed partial centers which integrated newer, more suitable patterns of respiration. Thus, the respiratory center of the higher mammals was gradually built up from different partial centers in the course of phylogenesis. The ontogenetic development followed the same pattern. The older centers have not disappeared; they are simply inhibited by the function of the newly formed higher centers. As soon as these higher centers are inactivated for some reason, their inhibiting force terminates also. Now the lower, released centers again become manifest and maintain an emergency respiration with its own characteristic patterns until the newer higher centers have recovered or until death has occurred.

The respiratory center of man is thus a physiological and not an anatomic entity. It is composed of several partial centers which can be localized physiologically and which are formed ontogenetically in the same sequence as they appeared phylogenetically. This development is not complete in the premature infant, so that immature patterns of respiration corresponding to lower phylogenetic forms occur. The last-formed section of the immature respiratory center is always the most sensitive. Damage of different kinds may inactivate the highest partial centers. The more immature the infant is, the more frequently does "disintegration of the respiratory center" occur.

During this disintegration periodic respiration, gasping respiration, and swallowing movements appear successively. If two partial centers are involved in directing the respiration, a block in the respiratory center occurs. Singultus, which can often be found in the young infant because of the immaturity of his respiratory center but which also frequently occurs in adults, belongs in this group. The singultus center is a division of the gasping center, as can be observed by comparing both respiratory patterns. It regulates the movements of the diaphragm, which are characteristic of gasping respiration, whereas the movements of the tongue and the lower jaw depend on another partial center. This center functions only incompletely in the young infant and not at all later in life. The only known methods of producing singultus attacks in experiments on animals consist of reducing the oxygen supply or poisoning with carbon dioxide. In these experiments, singultus precedes gasping respiration.

The same lower respiratory forms are seen in intrauterine breathing, in the respiration of fetuses and premature infants, and in both initial and terminal breathing.

Floor-of-the-mouth respiration and alae nasi breathing, respiratory patterns which also may appear after the neonatal period during severe illness, especially during pneumonia, represent further forms of regression to ontogenetically and phylogenetically lower stages.

Breath-holding and whooping cough attacks may deprive the respiratory center of its nervous excitability to such a degree that it reverts to the stage of gasping respiration. Withdrawal of oxygen and poisoning with carbon dioxide regularly produce the same results in experiments on animals. The same neurological phenomena appear in man and animals during gasping respiration (disappearance of reflexes, dilatation of the pupils, disappearance of muscle tone, changes in cardiac action, etc.).

If the excitement reaches the respiratory center during an epileptic seizure, a kind of distorted respiration appears which is not related to the disintegration of the respiratory center.

In addition, the mammals are still capable of the phylogenetically and ontogenetically lowest form of respiration: swallowing respiration. It is the physiological breathing pattern in the frog and may occur as a last emergency respiration in suffocating mammals. When poliomyelitic paralysis has affected the respiratory muscles but has spared the pharyngeal muscles, patients may supply themselves periodically, sometimes for hours, with sufficient oxygen by "frog breathing," which can be learned.

Yawning represents a phylogenetically old respiratory form related to swallowing respiration and is ordinarily inhibited by the higher components of the respiratory center. If these inhibitions diminish because of fatigue or fatal hemorrhage, the released yawning center temporarily takes the lead over the respiratory center. Therefore, yawning does not have the aim of avoiding fatigue, but is rather a consequence of fatigue.

REFERENCES

AFFELDT, J. E., R. COLLIER, M. G. CRANE, and A. F. FARR: Referat Zentralblatt Kinderhk. *55*, 62 (1956).

AHLFELD, F.: Verh. Dtsch. Ges. Gynäk. *2*, 203 (1888); Festschrift für Carl Ludwig (Memorial Volume in Honor of Carl Ludwig). p. 1. Marburg, 1891; Mschr. Geburtsh. *21*, 143 (1905).

AKEREN, V.: Acta paediatr. *38*, 669 (1949).
ANTHONY, A. I.: Dtsch. med. Wschr. *1939*, 1572.
BABAK, E.: Wintersteins Handbuch der vergleichenden Physiologie (in Winterstein's Handbook of Comparative Physiology). Vol. 2, p. 2. Jena, 1921.
–, and M. KÜHNOVA: Pflügers Arch. *130*, 444 (1909).
BAGLIONI, S.: Arch. Anat. u. Physiol. *1900*, 33.
–, Zschr. Physiol. *7*, 177 (1907); Erg. Physiol. *11*, 539 (1911).
BAISCH, K.: In Döderlein, A., Handbuch der Geburtshilfe (in Döderlein's Handbook of Obstetrics). 2nd ed. Vol. 2, p. 946. Munich, 1924.
BAMBERGER, P., and A. MATTHES: Anfälle im Kindesalter (Seizures in Childhood). p. 488. Basel, New York, 1959.
BARCROFT, J.: Lancet *243*, 117 (Aug. 1, 1942).
BARRON, D. H.: according to Barcroft.
BARTELHEIMER, H.: Kinderärztl. Prax. *7*, 53 (1936).
BEATON, L. E., and H. W. MAGOUN: Amer. J. Physiol. *134*, 177 (1941).
BERGER, H.: Körperliche Äußerung psychischer Zustände (Physical Manifestation of Psychic Conditions). 2 Vols. Jena, 1904–07.
BETHE, A.: Handbuch der normalen und pathologischen Physiologie (Handbook of Normal and Pathological Physiology). Vol. 2. Berlin, 1925.
BIEDERT, P.: Lehrbuch der Kinderkrankheiten (Textbook of Pediatrics). 12th ed., p. 370. Stuttgart, 1902.
BONAR, E. BARNETT, C. M. BLUMENFELD, and C. FENNING: Amer. J. Dis. Childr. *55*,1(1938).
BOSMA, J. F.: Acta paediatr. *49*, Suppl. 123, 56 (1960).
–, and J. LIND: Acta paediatr. *49*, Suppl. 123, 18 (1960).
–, J. LIND, and N. GENTZ: Acta paediatr. *48*, Suppl. 117, 117 (1959).
BREHM, A. E.: Tierleben. Die Säugetiere (Life of Animals, The Mammals). Vol. 3. Leipzig, 1883.
BRETON, A., and GÉRARD-LEVÊVRE: Pédiatrie *13*, 767 (1958).
BRIDGE, E. M., S. LIVINGSTON, and C. TIETZE: J. Pediatr. *23*, 539 (1934).
BRODMANN, K.: J. Psychol. Leipzig *1*, 10 (1902).
BROWN-SÉQUARD, E.: J. Physiol. Paris *1*, 217 (1858); *3*, 151 (1860).
CAMERER, J.: Dtsch. Zschr. ger. Med. *29*, 333 (1937/38).
CATEL, W., and C. A. KRAUSPE: Jb. Kinderhk. *129*, 1 (1930).
CAVAZZANI, E.: Arch. Fisiol. Firenze *8*, 313 (1910).
CHRISTOFFEL, H.: Schweiz. med. Wschr. *1951*, 514.
CONDORELLI, L., and E. RECHNITZER: Wien. Arch. inn. Med. *14*, 49 (1927).
CZERNY, A.: Jb. Kinderhk. *41*, 337 (1896); *81*, 473 (1915); Die Pädiatrie meiner Zeit (Pediatrics in My Time). p. 107. Berlin, 1939.
DAIL, C. W., J. E. AFFELDT, C. R. COLLIER, and C. HONDO: J. Amer. Med. Assoc. *158*, I 445 (1955).
–, M. ZUMWALDT, and H. ATKINS: Anweisung für die Unterrichtung in der glossopharyngealen Atmung (Froschatmung). Hsg. Deutsche Vereinigung zur Bekämpfung der Kinderlähmung [Instructions for the Teaching of Glossopharyngeal Respiration (Frog Breathing). Published by the German Association to Combat Poliomyelitis]. Düsseldorf, 1956.
DEBRÉ, R., J. MARIE, M. LAMY, and DE FONT-RÉAULX: Ref. Zbl. Kinderhk. *35*, 505 (1939).
DEDEK: Fol. neurobiol. (Leipzig) *7*, 539 (1913).
DEMMING, J., and J. P. HANNER: Amer. J. Dis. Childr. *49*, 109 (1935); *51*, 823 (1936).
DIETEL, K. and V.: Kinderärztl. Prax. *27*, 453 (1959).
DOLLINGER, A.: Erg. inn. Med. *31*, 373 (1927).
DUCCESHI, V.: Zbl. Physiol. *19*, 889 (1905).
DUMPERT, V.: Dtsch. Zschr. Nervenhk. *110*, 106 (1929).
EASTMAN, N. J.: Bull. John Hopkins Hosp. *50*, 39 (1932); *53*, 246 (1933); Amer. J. Obstetr. *31*, 563 (1936); Internat. Clin. Philadelphia *2*, 274 (1936).
EBBECKE, U.: Naturwiss. *39*, 222 (1952).
ECKERLEIN: Zschr. Geburtsh. *19*, 121 (1890).

Eckstein, A., and H. Paffrath: Zschr. Kinderhk. *46*, 307 (1928).
—, and E. Rominger: Zschr. Kinderhk. *28*, 1 (1921); Arch. Kinderhk. *70*, 258 (1922).
Ehrhardt, K.: Anat. Anz. *87*; Erg. H. 420 (1939); Münch. med. Wschr. *1939*, 915.
Engel, S.: Pfaundler-Schloßmanns Handbuch der Kinderheilkunde (Pfaundler-Schlossmann's Handbook of Pediatrics). 3rd ed., Vol. 3, p. 500. Leipzig, 1924.
Engelhardt, A.: Zschr. Biol. *100*, 596 (1941).
Fazekas, J. F., F. A. D. Alexander, and H. E. Himwich: Am. J. Physiol. *134*, 281 (1941).
Feer, E.: Lehrbuch der Kinderheilkunde (Textbook of Pediatrics). 12th ed., p. 626. Jena, 1938.
Finkelstein, H.: Lehrbuch der Säuglingskrankheiten (Textbook of Diseases of Infancy). 4th ed., pp. 531 and 598. Amsterdam, 1938.
Fox: in R. M. and A. W. Yerkes, The Great Apes. p. 269. New Haven, 1929.
Frugoni, C.: Neurol. Zbl. *27*, 202 (1908); Berl. klin. Wschr. *1910*, 1005.
Fulton, J. F.: Physiologie des Nervensystems (Physiology of the Nervous System). pp. VII and 217. Stuttgart, 1952.
Gamper, E.: Zschr. Neurol. *102*, 154; *104*, 49 (1926).
—, and Untersteiner: Arch. Psychiatr. *71*, 282 (1924).
Garland, after S. H. and S. P. Gage: Amer. Naturalist *20*, 233 (1886).
Garsche, R.: Mschr. Kinderhk. *98*, 197 (1950).
Gesell, A.: The Embryology of Behavior. p. 84. New York, London, 1945.
Geyl: Arch. Gynäk. *15*, 388 (1880).
Gibberd, G. F., in: Anoxia of the Newborn Infant. Symposium. Published by J. F. Delafresnaye and T. E. Oppé. Oxford, 1953.
Glanzmann, E.: Einführung in die Kinderheilkunde (Introduction to Pediatrics). Vol. 1, p. 481. Berlin, 1939.
Gött, T.: Pfaundler-Schloßmanns Handbuch der Kinderheilkunde (Pfaundler-Schlossmann's Handbook of Pediatrics). 4th ed., Vol. 4, p. 451. Berlin, 1931.
Graham, S., and N. Morris: Arch. Dis. Child. *4*, 335 (1929).
Gregor, K.: Arch. Kinderhk. *35*, 272 (1903).
Gutzmann, H.: Sprachheilkunde (Therapy of Language Disorders). 3rd ed., p. 92. Berlin, 1924.
Hamburger, F.: Die Neurosen des Kindesalters (The Neuroses of Childhood). p. 234. Stuttgart, 1939.
Haselhorst, F., and K. Stromberger: Zschr. Geburtsh. *102*, 16 (1932).
Hasse, C.: Arch. Anat. u. Physiol. *1893*, 293.
Hauptmann: Neurol. Zbl. *1920*, No. 23.
Heinroth, O.: S. ber. Ges. Naturfreunde. Berlin, Feb., 1930.
Henderson, V. E., and T. A. Sweet: Amer. J. Physiol. *91*, 94 (1929).
Henderson, Y.: J. Amer. Med. Assoc. *74*, 783 (1920); *90*, 583 (1928); *96*, 495 (1931); Atmung (Respiration). Leipzig, 1941.
Henoch, E.: Vorlesungen über Kinderkrankheiten (Lectures on Pediatrics). Berlin, 1881.
Herter, K.: Vergleichende Physiologie der Tiere (Comparative Physiology of the Animals). *1*, 89. Berlin, 1947. Göschen Collection.
Hess, W. R.: Die Regulierung der Atmung (The Regulation of Respiration). Leipzig, 1931.
Heubner, O.: Lehrbuch der Kinderheilkunde (Textbook of Pediatrics). 3rd ed., 2, 608. Leipzig, 1911.
Heusner, A. P.: Physiol. Reviews *26*, 156 (1946).
Hink, A.: Zbl. Gynäk. *1895*, 121.
Hirvensalo, M.: Acta paediatr. *37*, Suppl. I (Suppl. 73) (1949).
Hofbauer, L.: Pathologische Physiologie der Atmung. Handbuch der normalen und pathologischen Physiologie (Pathological Physiology of Respiration, Handbook of Normal and Pathological Physiology). Vol. 2, p. 350. Berlin, 1925.
Hondo, C.: J. Amer. Med. Assoc. *158*, I 445 (1955).

HOOKER, D.: The Prenatal Origin of Behavior. Porter Lectures, Series 18. pp. 74 and 80. Kansas, 1952; Evidence of Prenatal Function of the Central Nervous System in Man. p. 32. New York, 1958.
HUMPHREY, T.: J. Psychology *35*, 3 (1953).
VON HÖSSLIN, H.: Erg. inn. Med. *39* (1931).
HUTINEL and DARRÉ: In Marfan, Maladies des enfants (in Marfan's Diseases of Children). Vol. 1, p. 463. Paris, 1909.
IBRAHIM, J.: Zschr. Neurol. *5*, 388 (1911); in Feers Lehrbuch der Kinderheilkunde (in Feer's Textbook of Pediatrics). 1st ed., p. 484 (1911) and 12th ed., p. 522. Jena, 1938.
IKEDA: Zbl. Gynäk. *1893*, 1089.
VON JASCHKE, R. T.: W. Stöckels Lehrbuch der Geburtshilfe (in W. Stöckel's Textbook of Obstetrics). 3rd ed., p. 849. Leipzig, 1930.
–, Physiologie, Pflege und Ernährung der Neugeborenen (Physiology, Care, and Nutrition of the Newborn). Wiesbaden, 1917.
JELLINEK, S.: Med. Welt *1929*, 528.
KAHN, R. H.: Zbl. Physiol. *19*, 990 (1905).
KANE, F. H., and I. KREISELMAN: Amer. J. Obstetr. *20*, 826 (1933).
KEHRER, F. A.: Zschr. Biol. *28*, 450 (1891).
KEITH, A.: J. Anat. Physiol. *39*, 243 (1905).
KIRCHHOFF, H. W.: Arch. Kinderhk. *158*, 238 (1958).
KIRSCHSIEPER, H. M.: Arch. Kinderhk. *148*, 146 (1954).
KLEMPERER, H. H.: Arch. Gynäk. *154*, 108 (1933).
KNÖFPELMACHER, W.: In Pfaundler-Schloßmanns Handbuch der Kinderheilkunde (in Pfaundler-Schlossmann's Handbook of Pediatrics). 4th ed., Vol. 2, p. 396. Leipzig, 1931.
KÖTTGEN, H. U., and H. FEYERABEND: Mschr. Kinderhk. *100*, 244 (1952).
KRUKENBERG, H.: Der Gesichtsausdruck der Menschen (The Facial Expression of Man) 2nd ed., p. 130. Stuttgart, 1920.
KULJABKO: Pflügers Arch. Physiol. *97*, 539 (1903).
LANGENDORFF, O.: Arch. Physiol. *1880*, 518; in Nagels Handbuch der Physiologie (in Nagel's Handbook of Physiology). Vol. 4, p. 337 (1909).
LECLARD: Dtsch. Arch. Physiol. *1*, 154 (1815).
LEHMANN, C.: Pflügers Arch. Physiol. *42*, 284 (1888).
LEWY, E.: Zschr. Neurol. *72*, (1921).
LOEB, J.: Einleitung in die vgl. Gehirnphysiologie und vgl. Psychologie d. wirbellosen Tiere (Introduction to the Comparative Cerebral Physiology and Comparative Psychology of the Invertebrates). p. 83. Leipzig, 1899.
–, Pflügers Arch. Physiol. *96*, 536 (1903).
LOW, N. L., E. L. GIBBS, and F. R. A. GIBBS: Pediatrics *15*, 595 (1955).
LUMSDEN, T.: J. Physiol. (Brit.) *57*, 153 and 354 (1923); *58*, 81 (1924); Proc. Soc. Med., London 17 II, No. 8, 15 (1924).
MARBY, C. C.: J. Pediatr. *55*, 211 (1959).
MACKENZIE, J.: J. Physiol. (Brit.) *36*, XIII (1907).
MARCHAND, F.: Zbl. Gynäk. *1899*, 65.
MARCKWALD, F.: Zschr. Biol. *23*, 220 (1887).
MARX, H.: Medizinische *1959*, 1472.
MAYER, C.: Zschr. Biol. *73*, 101 (1921).
MAYR, F.: Jb. Kinderhk. (Vienna) *5*, 119 (1862).
MENDELSSOHN, A.: Zschr. Kinderhk. *5*, 283 (1913).
MERCURIALIS, H.: De morbis puerorum tractatus. 2. Buch, Kap. 18 (1538). Deutsche Übersetzung in dem Arztneybuch von P. Uffenbach [Tractate on the Diseases of Children, 2nd Book, Chapter 18 (1538), German Translation in the Medicine Book of P. Uffenbach]. p. 123. Frankfurt am Main, 1605.
MERMANN, A.: Zbl. Gynäk. *1880*, 377; *1887*, 622.

VON MONAKOW, C., and R. MOURGUE: Biologische Einführung in das Studium der Neurologie und Psychopathologie (Biological Introduction to the Study of Neurology and Psychopathology). p. 190. Stuttgart and Leipzig, 1930.
MOND, H., and S. WASSERMANN: Wien. Arch. inn. Med. *14*, 335 (1927).
MOSSO, A.: Arch. Physiol. *1878*, 441; *1885*, 39.
MÜLLER, L. R.: Münch. med. Wschr. *1930*, 1999.
MUMFORD, A. A.: Brain *20*, 290 (1897).
NEUGEBAUER, F.: Zbl. Gynäk. *1898*, 1281; *1899*, 465.
NEUMANN, H.: Arch. Kinderhk. *42*, 99 (1905).
NOBÉCOURT, P., and G. BOULANGER: In Traité des maladies des enfants (in Textbook of Pediatrics). Vol. 3, p. 280. Paris, 1934.
NORMAN, H. N.: J. Comp. Psychol. *34*, 63 (1942).
OBERHOLZER, R. J. H., and W. O. TOFANI: Handbook of Physiology. Sect. 1. Neurophysiology, Vol. II, p. 1112. Washington, 1960.
OPITZ, E.: Zbl. Gynäk. *1899*, 6.
—, and M. SCHNEIDER: Erg. Physiol. *46*, 126 (1950).
OSIANDER, F. B.: Lehrbuch der Hebammenkunst (Textbook of Midwifery). p. 664. Göttingen, 1796.
PARROT, G.: Clinique des Nouveau-nés. Paris 1877.
PEISER, E.: Zbl. Gynäk. *1899*, 1033.
PFIZER SPECTRUM: J. Amer. Med. Assoc. 30/1/1954. 154, No. 5. ADV. Anzeigenteil, p. 26.
PIERRE, MARIE, and G. LÉVY: Rev. Neurol. *38*, 1233 (1922).
PITTS, R. F.: Amer. J. Physiol. *134*, 192 (1941); Physiological Reviews *26*, 609 (1945).
—, H. W. MAGOUN, and S. W. RANSON: Amer. J. Physiol. *126*, 672 and 689 (1939).
POLIMANTI, O.: Zschr. Biol. *57*, 237 (1912).
POPPER, E.: Adv. Psychiatr. *63*, 231 (1920).
POTTER, E. L.: Adv. Pediatr. *6*, 157 (1953), after Zbl. Kinderhk. *48*, 322 (1954).
—, Pathology of the Fetus and the Newborn. p. 239. Chicago, 1953.
—, Fetal and Neonatal Death. p. 17. Chicago, 1950.
PREYER, W.: Zschr. Geburtsh. *7*, 241 (1882); Spezielle Physiologie des Embryo (Special Physiology of the Embryo). Leipzig, 1885.
RAUTENBERG, E.: Dtsch. med. Wschr. *1919*, 1277.
RECHNITZER, E.: Wien. Arch. inn. Med. *14*, 353 (1927); Klin. Wschr. *1929*, II, 2361.
REIFFERSCHEID, K.: Pflügers Arch. Physiol. *140* (1911); Dtsch. med. Wschr. *1911*, No. 19.
REIFFERSCHEID, W.: Zschr. Geburtsh., Stuttgart *122*, 316 (1941).
—, and R. SCHMIEMANN: Zbl. Gynäk. *1939*, No. 3; *1940*, No. 2.
REISS, M., and F. HAUROWITZ: Klin. Wschr. *1929*, I, 743.
REMANE, A., in G. HEBERER: Die Evolution der Organismen (The Evolution of Organisms). 2nd ed. Stuttgart, 1954–59.
VON REUSS, A.: in Halban and Seitz, Biologie und Pathologie des Weibes (Biology and Pathology of the Female). Vol. 8, 2, p. 703. Berlin and Vienna, 1927.
RILLIET, R., and E. BARTHEZ: Traité clinique et practique des maladies des enfants (Textbook of the Clinic and Practice of Pediatrics). 2nd ed., Vol. 2, p. 621. Paris, 1853.
ROHMER, P.: in Nobécourt and Babonneix, Traité de médicine des enfants (Textbook of Pediatrics). Vol. 1. Paris, 1934.
ROKITANSKY: Österr. med. Jb. *1874*, 30.
ROTHSCHILD, J.: Zbl. Gynäk. *1911*, 100.
ROUSSEAU, J. J.: Emil. 2nd ed., Vol. 1, p. 54 of German translation. Langensalza, 1882.
RÜHL, FRITZ: Untersuchungen zur Frage der geburtstraumatischen Blutungen der Medulla oblongata (Investigations on the Question of Hemorrhages in the Medulla Oblongata caused by Birth Trauma). Thesis. Leipzig, 1951.
RUNGE, M.: Berl. klin. Wschr. *1895*, 93.
SAINT-ANNE DARGASSIES: Revue neurol. *93*, 331 (1955).
SALMI, T.: Duodecim Ser., Vol. 18. 1, Helsinki, 1932.
—, and E. E. VUORI: Acta paediatr., Stockholm *9*, 432 (1930).

SCHALTENBRAND, G., and O. GIRNDT: Pflügers Arch. Physiol. *209*, 333 (1925).
SCHEEL: See Winslow.
SCHOEN, R.: Verh. Dtsch. Ges. inn. Med. *39*, 75, Kongr. Wiesbaden, 1927; Aussprache Dtsch. med. Wschr. *1928*, 1744; Arch. exper. Path. *135*, H. 3/4; *138*, 339 (1928).
SCHWARTZ, P.: Erg. inn. Med. *31*, 165 (1927).
SCHWARZ, G.: Zbl. Phyiol. *19*, 995 (1905).
SECKEL, H.: Jb. Kinderhk. *133*, 163 (1931).
SELBACH, C. and H.: Mschr. Psychiatr. *125*, 671 (1953).
SHAW, D. A., and F. R. HOPKINS: Amer. J. Dis. Childr. *42*, 335 (1931).
SIEFERT, E.: Pflügers Arch. Physiol. *64*, 321 (1896).
SNYDER, F. F., and M. ROSENFELD: Amer. J. Physiol. *119*, 153 (1937); J. Amer. Med. Assoc. *108*, 1946 (1937); Proc. Soc. Exper. Biol. a. Med. (Am.) *36*, 45 (1937); Amer. J. Obstetr. *36*, 363 (1938).
SOLTMANN, O.: Jb. Kinderhk. *11*, 101 (1877).
SPIEGEL, E. A., and H. ENGHOFF: Zschr. exper. Med. *47*, 193 (1928).
SPÜHLER, O.: Die Erkrankung des Zwerchfells. Handbuch der inn. Medizin (Diseases of the Diaphragm, Handbook of Internal Medicine). 4th ed., 4, 4, p. 655. Berlin-Göttingen-Heidelberg, 1954.
STEINER, I.: Biol. Zbl. *6*, 678 (1886/87).
STENGER, K.: Mschr. Kinderheilk. *98*, 462 (1950); *100*, 435 (1952); Kinderärztl. Prax., p. 47. Sonderheft (1950).
STIER, E.: Abh. Neuropath. Kindesalter *1*, No. 6 (1918).
SUCKOW: Mschr. Psychiatr. *56*, 317 (1924).
SWINDLE, P. F.: Amer. J. Physiol. *79*, 188 (1926).
TEREGULOW, A. G.: Pflügers Arch. Physiol. *221*, 486 (1929).
THIEMANN, F.: Arch. exper. Path. *135*, 213 (1928).
THIEMICH, M.: Jahrbuch Kinderheilk. *47*, 455 (1898); Pfaundler-Schloßmanns Handbuch der Kinderheilkunde (Pfaundler-Schlossmann's Handbook of Pediatrics). II, 2, p. 846. Leipzig, 1906.
TRAPPE, H. B.: Über das Gähnen (Yawning). Thesis. Würzburg, 1929.
TRENDELENBURG, W.: Pflügers Arch. Physiol. *135*, 469 (1910).
TSCHIRGI, R. D., and R. W. GERHARD: Amer. J. Physiol. *150*, 358 (1947).
TURNER and CRITCHLEY: Brain *48*, 72 (1925).
UHLENBRUCK: Zschr. exper. Med. *59*, 656 (1928).
ULLRICH, G.: Dtsch. med. Wschr. *1935*, 1033.
VALLOIS, L., and C. FLEIG: Compt. rend. Acad. Sci. Paris *140*, 1422 (1905).
VERWORN, M.: Arch. Anat. u. Physiol. *1903*, 65.
VORMITTAG, S.: Mschr. Kinderhk. *58*, 249 (1933).
WALDVOGEL, W.: Helvet. physiol. acta *3*, 329 (1945).
WASSERMANN, S.: Wien. Arch. inn. Med. *4*, 415 (1922); *5*, 221 and 283 (1923).
WENCKEBACH, K. F., and H. WINTERBERG: Die unregelmäßige Herztätigkeit (Irregular Heart Action). Leipzig, 1927.
WERTHEIMER, E.: J. Anat. et Physiol. *22*, 458 (1886); *23*, 567 (1887).
WIEDEMANN, H. R., and M. TOLKSDORF: Kinderärztl. Prax. *23*, 548 (1955).
WINSLOW, in D. PAUL SCHEEL: Über Beschaffenheit und Nutzen des Fruchtwassers in der Luftröhre der menschlichen Früchte (On Quality and Purpose of the Amniotic Fluid in the Trachea of Human Fetuses) (Latin). Hafniae, 1799.
WINTERSTEIN, H.: Wintersteins Handbuch der vergleichenden Physiologie (Handbook of Comparative Physiology). Vol. 1, 2. Jena, 1921.
WULF, H.: Klin. Wschr. *1958*, 234; Zschr. Geb.hilfe u. Gyn. *152*, 98 (1959).
YLPPÖ, A.: Zschr. Kinderhk. *20*, 212; *24*, 1 (1919); Pfaundler-Schloßmanns Handbuch d. Kinderheilkunde (Pfaundler-Schlossmann's Handbook of Pediatrics). Vol. 1, 3rd ed. Berlin, 1923, 4th ed. Berlin, 1931; Fanconi-Wallgren, Lehrbuch der Pädiatrie (Fanconi-Wallgren, Textbook of Pediatrics). p. 225. Basel, 1950; Arch. Kinderheilk. *149*, 110 (1954).

9. THE NEUROLOGY OF FOOD INTAKE

(134)

The first reaction which can be elicited in the human fetus by an external stimulus is the opening of the mouth when the trigeminal area is touched, a phenomenon which emphasizes the importance of food intake. This reaction can even be demonstrated in a nine-and-a-half-week-old fetus, calculated from the last menstruation (KINGSBURY, 1924) (also see p. 409).

THE SUCKING ACT

(62, 82)

I use the term "sucking act" to designate collectively all nervous and mechanical processes which participate in the food intake of the infant. The special pattern of food intake is so characteristic for the early period of life that the term "suckling" was derived from this function. Mammals* also derive their name from this term. It matters little that the sucking act consists only partially of real sucking. Since the expression is commonly used, it would be useless to search for a new expression which would characterize the act better. But this should not lead to the frequent error of seeing the physical process of sucking as the only essential part of the infant's food intake.

The sucking act is not merely a reflex that is easily initiated by a number of external stimuli. Ontogenetically and phylogenetically it is very close to another rhythmic function – respiration. Despite the crossing of the respiratory and nutritional pathways in the pharynx, the infant is able to breathe and suck simultaneously only because the two processes are closely coupled.

Role of the Mother

(62)

The ductus lactiferi converge at the nipple and join into increasingly bigger branches until approximately 15 to 20 ductus lactiferi terminate in the papilla of the nipple. Before they enter there, they enlarge into the sinus lactiferi. The nipple serves as a closure and as a mouthpiece for drinking (BASCH, 1893).

* "Suckling" ("Säugling") is the German word for "infant"; "Säugetiere" ("sucking animals") is the German term for "mammals" – The translators.

The nipple is closed by the subareolaris muscle, which cannot be clearly separated from the muscles of the nipple. According to SEITZ (1925) and VON JASCHKE, a double network of circular fibers can be recognized. The external circle encloses the whole papilla and thus the milk ducts; the inner circle surrounds groups of two or three ducts or even occasionally a single duct. The milk ducts are interwoven into the subareolaris muscle in such a manner that some of the milk sinuses are still located in its distribution. According to SEITZ, the end ducts are not simply closed by a circular sphincter; since the sphincter fibers actually cross each other, this suggests an additional constricting effect. It is unknown whether the muscles have a common innervation or if some muscle bundles have a special task and accordingly are innervated individually. VON PFAUNDLER indicated the second possibility.

The sphincter tonus suffices to prevent milk flow from the nipple unless galactorrhea exists. The sphincter tonus relaxes during nursing; this can be inferred from the fact that during feeding many women show milk flow from the other breast, sometimes in considerable quantity. A donkey can be milked only when the foal is simultaneously nursed (CZERNY and KELLER, 1925, I, 156).

That the nipple is erectile on psychic or mechanical stimulation makes it still more useful as a mouthpiece. According to VON HERFF (1889), the erection is caused by reflex hyperemia of the nipple. The sphincter simultaneously relaxes in some parts and contracts in others, thus giving a certain support to the nipple. According to BASCH's (1893, 1894, 1903, 1912) investigations, which were later confirmed by SCHUMACHER (1923), the erection is caused only by the contraction of the subareolaris muscles and those of the nipples while the vessels play a subordinate role. The erection makes the nipple longer, thinner, and harder. SCHUMACHER found that its average height was 0.5 to 1.3 cm in the erect state and 0.0 to 1.0 cm in the resting state. The width of the mammary areola is reduced by the contraction of the muscles from 3.0–7.9 to 2.8–7.0 cm. VON HERFF stated that erection usually occurs in both nipples when one is stimulated. In general BASCH and SCHUMACHER were not able to demonstrate such a consensual reaction. In any case, the erection of a low nipple facilitates the grasping of the infant. The nipple erects and relaxes again with each complete sucking movement of the infant (BASCH), a process which aids the emptying of the breast.

The nipple is not only independently erectile, but is also lengthened by the infant's sucking. Disregarding the independent erection of the nipple, GUNTHER (1945) measured the degree to which it could be extended by suction at a negative pressure of 60 mm Hg. In 150 women the values fluctuated between 2.0 and 3.75 cm; only a few of the mothers with low elasticity were able to nurse their infants.

Since the mammary gland is locked by the nipple, human milk secretion cannot be observed over a period of time in the same manner as the flow of saliva. But the relations between milk production and milking times have been studied in dairies for economic reasons. As a result two opposing theories have arisen. The older theory, the "*drawing off*" theory, is still recognized today by scientists and experts: When all the cavities of the udder are filled with milk the secretion of milk stops, and only when the secreted milk is drawn from the udder to the alveoli by sucking or milking can new milk be produced. The milk obtained during milking is therefore said to be the result of milk production which takes place uniformly during the intervals between milkings and slows up somewhat because of the increasing internal pressure. Milking constitutes only a mechanical procedure which presses the secreted milk from the milk ducts into the cisterna and from there to the outside. The advocates of the "drawing off" theory do not deny that nervous stimuli occur during sucking and milking, but these stimuli are said to serve the sole purpose of collecting the milk from the various parts of the udder into the cisterna.

In contrast, FLEISCHMANN (1876) stresses that even the full udder of the best cow cannot contain more than 3 liters of milk. Moreover, during the first half of their lactation period all good cows produce much more milk on the average if milked only once a day. Therefore only part — perhaps half — of the milk obtained could have been present in the udder, while the other half must be secreted during milking. It is suggested that, in addition to secreting milk continuously, the mammary glands become more effective between milking times in order to function more intensely for a short time during milking. FLEISCHMANN's theory is called the "*two-phase*" theory: During the first phase the mammary gland, after recovering from the preceding milking, gradually increases its milk production, but ceases to function when the cavities of the udder are fairly well filled. The second phase occurs during milking time, when a brief but turbulent interval of milk production caused by the mechanical stimulation of the udder occurs. The effect of these stimuli gradually decreases during milking because the glands are exhausted. The mammary gland has the ability to accumulate during the interval between two milkings a certain amount of "tension," which is rapidly exhausted during milking. Of the two phases, which produce about the same amount of milk, one lasts 10 to 15 minutes and the other 8 to 12 hours.

The validity of FLEISCHMANN's theory was doubted at first (F. HOFFMANN, 1881), but NÜESCH (1904) supplied the evidence to make it feasible. NÜESCH found that on the average a cow produced 10 liters of milk daily; of this amount, 5.5 liters was obtained by milking in the morning. If a cow was slaughtered in the morning before milking time, only 2770 gm of milk could be obtained from the udder, that is, approximately half the amount. The other half must have been produced during milking.

However, other investigators have not found such distinct differences. GAINES and SANMAN (1927) determined the amount of milk obtained by milking 3 cows and then slaughtering them. The proportions between the amounts obtained by milking and those found in the udder after slaughtering were 100:100, 100:96, and 100:108. SWELL (1927) and FILIPOVIC (1928) obtained similar results. WEIGMANN (1932) does not consider it very plausible that the conversion of milk-producing substances from the blood into milk (frequently 5 to 7 liters or more of milk are produced) can happen during the few minutes of milking. It therefore appears questionable whether these still very vague views on milk production can be applied to man.

In the mother, nursing develops conditioned reflexes which facilitate milk flow and perhaps also milk production in the same manner as PAVLOV has demonstrated for salivary flow. According to VON PFAUNDLER (1909), the breasts may start to drip while the mother is only preparing to nurse her baby; even thinking of the baby or of nursing may lead to this phenomenon. While the baby feeds at one breast, the other frequently begins to flow also. I saw milk spurt out of one mother's breast in a jet before she started nursing, as she was just thinking of her baby. Conditioned reflexes also occur in milking: The flow of milk is reflex evoked when the mother animal sees her baby as it approaches to suck, or when the animal hears the clatter of the milk pail or the splattering of milk into the pail as the cow in the next stall is milked (GRIMMER, 1910).

According to VON STETTEN-AYSTÄTTEN (1924), milk flow during milking can be improved in the cow and in the goat by music. It was increased by an average of 6%. Corresponding experiments with women were unsuccessful (NEBERT and KOCH, 1926). The sudden drying up or decrease of milk flow caused by nervous excitement (for example, sudden fright) can be explained by the hypothesis that conditioned reflexes play a role in milk production (the outflow). According to PAVLOV, similar inhibitions frequently occur in conditioned reflexes.

WALLER (1943) characterizes the reflex flow of milk as the draught reflex: This reflex greatly facilitates the infant's effort to obtain milk by sucking. However, if the breast contains too much milk, the draught reflex may be inhibited and the draining of the breasts by the infant's sucking may be made impossible. According to RICHARDSON (1948), the réflexe de jaillissement ("spurting-out" reflex) is elicited by stimulation of the sensitive nerve ending in the nipple. If this reflex is absent, the infant is unable to drain the breast completely even if the nipple is well formed. If the milk pressure in the acini is too weak the reflex cannot occur, and if the pressure is too strong the reflex is inhibited.

ROEHRIG (1876) thought he could accelerate the milk production of a goat by electrical stimulation; VON HERFF (1889) tried the same experiment with a woman. Today we have discarded their statements. ECKHARDT (1858) observed that the amounts of milk of a lactating goat before and after the nervous sympathicus had been cut were unchanged. GOLTZ (1874) transected the lumbar cord of a pregnant bitch and resected the medulla from the third thoracic vertebra down in another. Both animals gave birth to healthy puppies and were able to nurse and raise them. BASCH (1910) ascribed little importance to the influence of the peripheral and sympathetic nerves on milk secretion. During sucking and milking, the role of the sympathicus may consist of transmitting reflexes which are required for uniform and continuous function of the mammary gland. The mammary gland is otherwise independent of the nervous system; it is the hormonal system, rather, which influences milk secretion. When one of the Blazek Siamese twins — who were attached in the sacral and coccygeal region — became pregnant, milk flow also occurred in the nonpregnant twin.

INGELBRECHT (1935) transected the spinal cords of lactating white rats between the lowest thoracic and the highest lumbar vertebrae, rendering the six lowest mammary glands insensitive. If the sensitive upper mammary glands were covered with adhesive, the litter was forced to starve. But if he left two sensitive glands accessible, all other glands continued to secrete milk.

In the opinion of MEITES and TURNER (1948), lactation is a very complicated process which depends on hormonal, nervous, and environmental influences. The growth of the mammary glands and the onset of milk secretion depend on the endocrine system, whereas the continuation of lactation which already exists depends on a neuroendocrine system.

That milk flow occurs in many women in the intervals between nursing proves that the breast can be emptied without the action of the infant (KEHRER, 1875; VON HERFF, 1889; LEUBE, 1892). Although a certain flaccidity of the subareolaris muscle may play a role in this phenomenon, the main force is the excreting pressure (SEITZ, 1925), which may be as high as 30 cm H_2O in animals according to ROEHRIG (1876) and NÜESCH (1904), or even considerably higher according to ZSCHOKKE. This pressure has not as yet been measured in women. If the nursing infant is suddenly removed from the breast, milk may still flow for some time in drops or in a jet. VON PFAUNDLER considers the transmitted blood pressure, the filling pressure of the elastic wall, and the contraction of the fiber cells of the compressor and expressor muscles to be the driving forces in milk excretion. The net of elastic fibers which surrounds all outer ducts of the mammary glands is stretched in the filled breast; as soon as the lock is loosened, the stretched fibers contract and press the milk forward. It is questionable whether similarly functioning muscle fibers exist in the gland tissue (SEITZ, 1925).

VON PFAUNDLER regards milk production as a continuous process. However, if the two-phase theory of FLEISCHMANN is applied to man, then the pressure of the milk secreted during nursing should push the milk already produced forward. It has been mentioned previously that conditioned reflexes also participate in milk production and excretion.

It would be difficult to demonstrate in individual cases how the different forces combine into this pattern. It certainly is not correct to consider only the easily demonstrable activity of the infant and to ignore the function of the mother's mammary gland. The value obtained in determining the mechanical force needed to empty the mammary gland with a pump cannot automatically be said to be the same as that actually exerted by the infant in nursing.

The mammary gland continues to function only when it is emptied regularly. Usually the infant takes care of this; however, for various reasons mother's milk is often given in bottles to infants in institutions and it therefore becomes extremely important to drain the mother's breast artificially. There is no longer any doubt that the milk production of the breast can be maintained in this manner. Just as in dairies, two procedures are used: expression (milking) and pumping. In dairies the advantages and disadvantages of the two procedures are determined only by economic considerations, which cannot be applied to man. In man, the decision depends only on the appropriateness of the procedure.

In children's hospitals and nursing homes, the expression or milking procedure is in general use today. One wet nurse shows the procedure to the next, and a certain tradition develops. Therefore, the institutions actually do not need milk pumps. Difficulties develop only with very sensitive women, who are not particularly suited to becoming wet nurses anyhow. In these cases milk pumps can be successfully employed. Their action depends on rhythmic sucking alone or on combined rhythmic sucking and expression, so that the milk cup compresses the mammary gland as it is applied. Milk pumps were designed by IBRAHIM (1904), KAUPE (1907, 1909), VON JASCHKE (1909), SCHERBACK (1910), SCHEER (1926, 1935), and THOENES (1929).

The danger of contamination of the milk is less with the expression procedure than with the use of the pump, since the rim of the glass cup is contaminated by contact with the skin and therefore transmits bacteria to the milk when the milk is poured over the rim. It is therefore better for mother's-milk banks not to use milk pumps.

Role of the Infant

(62)

Anatomy of the Sucking Act

The oral cavity of the infant shows anatomic characteristics relevant to the sucking act.

The corpus adiposum buccae, which was first described by BICHAT (1801), is said to have a connection with the sucking act (GEHEWE; HENKE, 1877). This corpus adiposum is a fatty mass the size of a hazelnut which is separated from the surrounding adipose tissue by a capsule and is connected by a stem with the adipose tissue of the subtemporal fossa (HEIDERICH, 1934). On both sides the corpora adiposa are located on the buccinator and masseter muscles below the fascia parotidomasseterica. In the newborn infant they are the same size as the adult's or even bigger. Even if up to 97% of the adipose tissue in other body parts in a very immature infant has been resorbed the adiposa will still be preserved (RANKE, 1884), but may finally disappear. LEHNDORFF (1907) found the corpus adiposum buccae to be poorer in the fatty acids which melt easily, and richer in those which are more difficult to melt, than the subcutaneous adipose tissue. In atrophy the subcutaneous fat, which melts easily, is affected earlier than the fat of the corpus adiposum buccae, which is more difficult to melt. According to RANKE and LITINSKI (1902), the corpus plays a vital role in infancy: It is said to serve as a kind of prop for the buccinator muscle so that the cheek retains its hardness during sucking and is not sucked in between the jaws. HENKE calls the corpus adiposum buccae a lid which acts like a valve over the gaps between the posterior parts of the alveolar rims when the nipple comes between them. Because of its anatomic location, FORSTER (1904) rejected the view that the corpus adiposum is associated with the sucking act and stated that it could represent only a fat repository. Contemporary views of the sucking act support FORSTER's opinion. LORENZ (1943) rejects the idea that the corpus adiposum has any mechanical importance, since it is present far beyond infancy and cannot be found in any other mammal. He thinks that the corpus adiposum buccae serves the purpose of enticing us to pinch the baby's cheeks.

It is more justifiable to assume a connection between the sucking pads of the newborn infant's lips and the sucking act. LUSCHKA (1863) was the first to describe these formations. He distinguished external and internal zones on the surface of the newborn infant's lips. The external zones, which are 2 mm in height at the most, appear uniform and even (pars glabra) to the naked eye. The internal zones are larger, and when the mouth is moderately closed have a height of 4 mm in the upper lip and 3 mm in the lower lip. They are soft, irregularly padded, and finely fissured (pars villosa). They are sharply demarcated from the outer zones but they change over with no distinct demarcations to the inner surface of the lips forming, so to speak, their outward portions. W. KRAUSE (1879) and NEUSTÄTTER (1895) had already connected the pars villosa with the sucking act. But only after VON PFAUNDLER (1924) referred to the pars villosa did it become generally known. He described it in the following manner: On each lip it consists of a longitudinal line which is fairly sharply demarcated against the anterior red part of the lip. It is divided by radiating fissures into several symmetrical flat pads which laterally become more narrow. The mucosa over these pads appears lusterless and macerated. VON PFAUNDLER and VON JASCHKE illustrated this really quite striking phenomenon in their textbook (1931, p. 307). According to VON PFAUNDLER, these lip pads disappear early in bottle-fed children, while they persist months after infancy in breast-fed children. VON JASCHKE stresses that the sucking pads are most prominent immediately after sucking. He regards them as an indication of incomplete development of the lips; therefore, they are more distinct in premature infants. According to VON PFAUNDLER, they assist in enclosing the nipple with the jaws.

The sucking pads are pressed to the nipple with a special "sucking muscle of the lips," the musculus labii proprius (according to W. KRAUSE). In the lower lip the fiber bundles take an oblique ascending backward course from the skin. The muscle is well developed in the newborn and is probably later arrested in its growth. It is said to serve sucking, but not speech. Its designation as a sucking muscle has been discontinued and rightly so. So many muscles share in the sucking act that it is not possible to choose one and designate it as the sucking muscle. STIRNIMANN (1936) found sucking pads, i.e., swellings of the red parts of the lips, which could be most distinctly noted in the lower lip, in 30 of 100 newborn infants during their first food intake as a response to gustatory, olfactory, and tactile stimuli. LUSCHKA's lip sucking pad is said to develop from these swellings after the first day of life while a desquamation of the external epithelial layer takes place during the first two weeks of life.

An importance similar to that of the lip sucking pad is attributed to the membrana gingivalis of the newborn infant. This formation is likewise credited with assisting in sealing off the oral cavity around the nipple. Therefore, it was named the "labium tertium" (a pair of assisting third lips). ROBIN and MAGITOT (1860) describe the membrana gingivalis as a mucosa fold on the free edge of the gums in the region of the eyetooth buds of both jaws. This membrane is most distinct in the lower jaw and is sometimes not visible at all in the upper jaw. It disappears in the third or fourth month of life. If it is well developed, the right projection of the lower jaw is usually connected with the left projection by a membranous rim 1 to 3 mm in height. Since this formation contains many vessels it can be slightly erected and moved forward or backward. ROBIN and MAGITOT were the first to relate it to the sucking act. VON SIEBOLD (1861), M. HERZ (1865), and later VON PFAUNDLER and VON JASCHKE, confirmed these statements. According to VON JASCHKE, in whose book an illustration of this formation can be found, the membrana gingivalis can be best observed when the sucking infant is suddenly taken away from the breast.

It seems to me that the position of the jaws in relation to each other is also importtant for the sucking act. In BASCH's and VON JASCHKE's illustrations the infant grasps the nipple with his lips and not with his jaws. In reality the nipple is grasped by lips and jaws and the pressure during sucking is produced mainly by the jaws. One can easily convince oneself of that fact by letting an infant suck on one's finger. The two jaws do not grasp the nipple at the same level like pincers; the upper jaw protrudes over the lower jaw. The difference amounts to about 2 mm, according to measurements which SCHMELTER (1931) made on the living infant. Thus, the nipple is attacked on three different planes: from the lower jaw, from the upper jaw, and from the two lips. To this is added the pressure of the tip of the tongue. This is evidently a better way of squeezing the nipple than if it were only grasped pincer-like.

According to HENKE (1877), ESCHERICH (1888), HASSE (1905), and SELLHEIM (1918), the jaws do not rest on top of one another when the mouth is closed; rather, the tip of the tongue protrudes between them. Laterally they are separated by the very well developed cheek pads. The sucking space described by AUERBACH (1888) is located in the middle of the upper surface of the tongue between the upward arched wall of the palate and the somewhat deepened fold of the tongue. Since the tongue protrudes to the mouth opening, the sucking space during infancy expands to the free edges of the jaws, i.e., farther than later on after the teeth have appeared. The nipple thus has only a short distance from the oral orifice to the sucking space. Therefore, a comparatively short nipple can be used (HASSE).

The available anatomic investigations (HEIDERICH, 1934; O. SCHNEIDER, 1939) demonstrated that the location of the entrance to the larynx is considerably higher in relation to the vertebral column in the newborn infant than later in life. But where it is actually located is not clear from the statements of the anatomists. In x-ray films with the head in the median position, HOFMANN and PEIPER [1937 *(65)*] found the location of the larynx entrance (the base line of the epiglottis) to be at approximately the level of the second cervical vertebra during swallowing (which makes the larynx move somewhat higher), whereas in a child of school age it is located at the level of the third cervical vertebra and in adults between the lower border of the third and the upper border of the fourth cervical vertebra.

Neurology of the Sucking Act

Sucking Center

According to BASCH's (1893, 1894) experiments on animals, the sucking center is located on both sides of the medulla oblongata at the internal border of the corpus restiforme and the brachium conjunctivum and includes the nuclei of the trigeminus, facialis, and hypoglossus. In young rabbits the sucking movement could be eliminated by a lesion in the area between the sensory and motor nuclei of the trigeminus. The centripetal portion of the reflex arc is located in the sensory branch of the trigeminus, the centrifugal portion in the motor part of this nerve and also in the facialis and hypoglossus. It can be assumed that the location of the swallowing center is in the same region.

By its location alone, the sucking center is brought in close relation to the respiratory center, with which its rhythmic function is closely connected, especially since the alimentary and respiratory pathways cross in the pharynx. Phylogenetically, the two centers may have formed one entity.

SOLTMANN's (1876) findings agree with BASCH's investigations. He removed the cerebral hemispheres, including the corpus striatum, in newborn puppies. The sucking movements evoked by a finger placed in the mouth of the puppy were preserved in spite of this procedure. Sucking movements were also observed in human monsters without cerebral hemispheres. The monster described by CATEL and KRAUSPE (1930), whose central nervous system was developed only up to the level of the medulla oblongata, was able to suck and swallow. MINKOWSKI (1928) observed sucking movements in human embryos in the second and third month of gestation. This proves that sucking movements represent one of the lowest nervous functions.

Rooting Reflex

(62, 104, 124, 137)

After infancy man takes his food to his mouth with his hands. The infant is not yet able to do this. But vital food intake is nevertheless well assured for him; he possesses the phylogenetically lower faculty of searching and grasping with his mouth so that he can take his mother's bountiful breast into his mouth without using his hands. This process is assured by a number of reflexes. The ability to respond varies in the same infant with age, fatigue and sleep, state of satiation, and health.

GARSCHE (1950) raised the objection that the evocation of the searching (rooting) reflexes does not always succeed even with the aid of gross stimuli (touching with the fingers, a tap with a hammer). Therefore, the searching for the breast (rooting) scarcely does justice to its appointed task. However, in the young infant the rooting reflexes are no longer so vital and are in the process of regression; they are more distinct in animal offspring. Similarly, tonic palmar and to an even greater extent plantar reflexes (see p. 158) are far more distinct and constant in the young ape than in the human baby. Nonetheless, we cannot fail to recognize in the baby that both reflexes were originally of great importance.

The head movements of the rooting reflex are directed from the place on the lips which has been touched. They must therefore be related to the coordinating system characteristic for the head. The three axes around which the head can turn and which are at 90° angles to each other are the vertical axis (vertex to sole), the sagittal axis (from front to back), and the transverse axis (from right to left).

If the right (left) corner of an infant's mouth is stimulated his head turns, goal-directed, around its vertical axis to the right (left) to the point of stimulation, so that he can grasp the touching object (Fig. 167a and b). If the upper and lower lips are touched in the middle, the head is rotated around its transverse axis, so that the infant's mouth moves away from or approaches his own chest (Fig. 167c and d). If the lower lip is touched, the mouth opens to grasp the touching object. Appropriate movements of the lips cause the nipple to glide into the mouth.

But the possibilities are not exhausted with these movements. Rotation around the sagittal axis of the head cannot usually be elicited by a tactile stimulus, since the mouth is not brought closer to the touching object by this movement. But on appropriate stimuli the head is shifted forward in the direction of its sagittal axis. To evoke this movement it is best to avoid the dorsal position of the newborn infant,

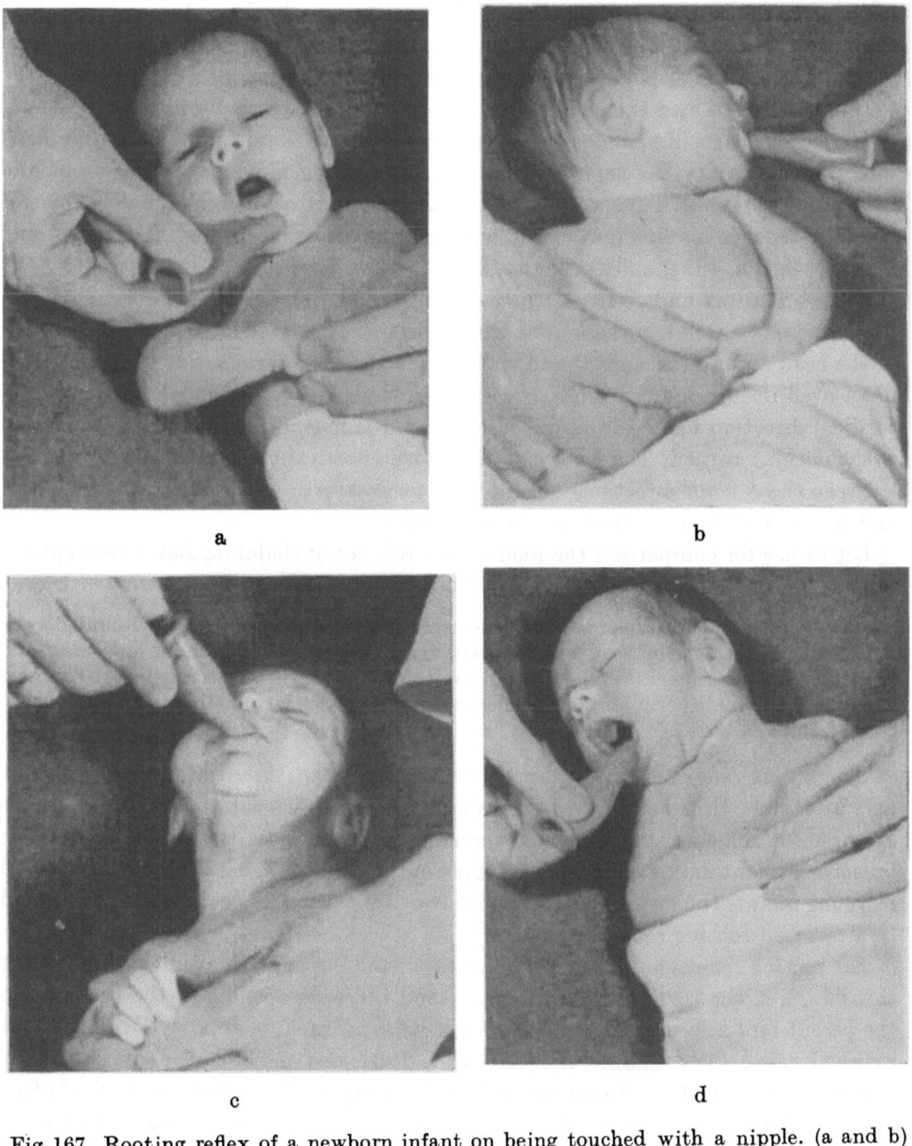

Fig. 167. Rooting reflex of a newborn infant on being touched with a nipple. (a and b) Turning of the head around its vertical axis, to the right and left side. (c and d) Turning of the head around its transverse axis, upward and downward. (a and d) Opening of the mouth to grasp the nipple.

since the head will have to overcome gravity to be raised in this position. The best way is to place the nipple in lateral position in the infant's mouth. After he has begun to suck vigorously, the nipple is cautiously moved out of the oral cavity in the direction of the sagittal axis so that it just barely touches his lips. The infant will follow the receding nipple with his head along the sagittal axis to regain a good hold on it with his mouth. If this experiment is repeated several times, quite distinct forward movements of the head can be elicited. Theoretically, a backward movement along the sagittal axis should also be possible. But this cannot be elicited, since the tactile stimulus cannot be applied to the posterior portion of the mouth.

It is possible to lead the head to and fro in any desired way by repeatedly stimulating various parts of the mouth, as long as the nipple is not allowed to slide back into the oral cavity. As soon as this happens, sucking movements will occur and the rooting reflex will disappear. Thus, the infant is able to grasp the nipple no matter where it touches his lips (automatic oral righting reflex, according to GAMPER and UNTERSTEINER, 1924, and GAMPER, 1926).

ANDRÉ-THOMAS and SAINT-ANNE DARGASSIES (1952) describe the searching for the breast as "a reflex of the four cardinal points." They evidently mean the turning of the head around its sagittal and transverse axes. This term seems to me unfortunate, even if only four cardinal directions existed. But at least one other, a fifth, cardinal direction has been described; the head is moved forward in this direction without being turned. The head can also move effortlessly to any desired direction between these main directions when this is necessary to grasp the nipple. GENTRY and ALDRICH (1948) use the term "rooting reflex."

Let us use for comparison the food intake reflexes of thalamic and mesencephalic preparations of rabbits as described by MAGNUS (1924):

"As soon as the lower lip of an easily excitable animal is touched from the front with any object (finger, carrot), the head will be tossed vigorously forward by a neck movement and the teeth thus pressed against the touching object. If a carrot is pressed against the incisors with the hand, there occur strong chewing movements which last until everything has been chewed and swallowed."

The rooting reflex usually disappears after the infant is satiated and reappears only after some time has elapsed. If the nipple is given to a satiated infant he makes at the most some weak sucking movements and soon loses the nipple again. The breast-fed infant may even push the nipple out with his tongue. I have not observed in young infants the goal-directed turning away of the head which would correspond to a reversed rooting reflex.

The rooting reflex shows some similarities to the "opening movements" of young thrushes, i.e., the tossing up of the heads and the wide opening of the beaks when the parent bird appears on the edge of the nest as a "vacuum activity," i.e., without external cause (TINBERGEN and KUENEN, 1939; see p. 595). The rooting reflex can also be observed as a "vacuum activity": Hungry infants waiting for food often turn their heads excitedly in all directions without an external cause and even without a tactile stimulus to their mouths. Of course, these movements disappear as soon as the nipple has been grasped. Sucking movements also occur quite often as a "vacuum activity" during infancy. They continue far beyond this age during light sleep. The gritting of the teeth during sleep corresponds to sucking movements of the dentate jaws as a "vacuum activity."

The rooting reflex is elicited mostly by tactile stimuli. Radiating heat can also be effective, as SIKORSKI (1908), STIRNIMANN (1939; see p. 42), and SCHENDEL [1944 *(15a)*] have demonstrated. DARWIN (1880) stated that at the age of 32 days his child perceived the mother's nipple at a distance of 75 to 100 mm; in his opinion the infant was guided by olfactory or caloric stimuli. CANESTRINI's (1913) and RIPIN and HETZER's (1930) experiments speak against olfactory stimuli having such an effect.

The rooting reflex, which occurs after tactile stimuli to the oral region, has been rediscovered and described many times. The first to mention this reflex was probably PEPYS (1667).* The obstetrician DE BOER (1802) knew that the healthy newborn infant searches for his mother's breast immediately after birth. VIRCHOW (1871; see p. 265) made the same observation. KUSSMAUL (1859) gave a detailed description:

"A newborn infant five hours old became very restless, moving his head searchingly to and fro with much crying. Just as he stopped crying, but was awake and restless, I softly stroked his left cheek with my index finger without touching his lips. He turned his head rapidly to this side, grasped my finger, and began to suck. I took my finger out and stroked his right cheek; just as rapidly he turned to this side, grasped the finger again, etc... The experiment succeeded best when completely mature infants who were very lively by nature were fully awakened and excited by strong feelings of hunger and thirst."

PREYER (1895) regularly observed in his child a strong turning of the head to both sides during the first weeks of life when the infant was placed at the mother's breast. On the eighth day, when for the first time the child was able to grasp the breast unassisted, the sideways movements of the head gave the appearance that the child was searching for something. According to STIRNIMANN, the righting reactions of the newborn infant vary greatly during the first two days of life. On the tenth to fourteenth day a uniform, exactly directed, and immediately appearing righting reflex occurs.

In some healthy newborns, a reaction which belongs to the rooting reflex occurs on tapping of the facialis, i.e., turning of the head to the stimulated side (CATEL, own observations).

Repeated tapping of the lips causes sucking movements, especially in hungry infants. The position of the lips may change during these movements. ESCHERICH (1888) describes a "mouth phenomenon" which occurs in normal infants during the first two months of life: If the corner of the mouth or the upper lip of the quietly sleeping infant is tapped, the mouth is pointed and arched forward for a short time as if ready to grasp the nipple. Sometimes sucking movements occur. THIEMICH's (1900) lip phenomenon corresponds to ESCHERICH's mouth phenomenon: Tapping of any point of the upper lip, best done several times at short intervals, causes a

* The interesting statement is: "May 3 ... by coach to little Michell's and there saw my poor Betty and her little child, which slept so soundly we could hardly wake it in an hour's time without hurting it, and they tell me what I did not know, that a child (as this do) will hunt and hunt up and down with its mouth if you touch the cheek of it with your finger's end for a nipple, and fit its mouth for sucking, but this hath not sucked yet, she having no nipples" — Ed.

sudden contraction of the musculus orbicularis, so that the mouth is briefly pushed forward like a snout or more like the mouth of a carp. THIEMICH never found the lip phenomenon, which he himself relates to the mouth phenomenon, in healthy children but only in tetanic infants when they were awake. BESSAU (1944) described this reflex as the "Schnuten" (pout) phenomenon, FREUDENBERG (1931) as the "Schnauz" (snout) phenomenon, and BALIASSNIKOWA and MODEL (1932) as the "Rüssel" (snout or trunk) reflex. POPPER (1921) designates the formation of a "snout" as the most, perhaps the only, constant partial phenomenon of the sucking movements. TOULOUSE and VURPAS (1903) and LAMBRANZANI and PIANETTA (1906) termed the same lip movements a "buccal reflex," and regard it as a part of the sucking movement. They observed the reflex in normal infants and in adults with severe cerebral disorders.

Especially striking is the moving of the lips on tactile stimuli which THOMPSON (1903) described as a lip reflex, and GAMPER and UNTERSTEINER (1924) as "oral Einstellautomatismus" (oral righting automatism or automatic oral righting reflex): When the middle of the upper lip of a young infant is touched, the upper lip arches upward like a snout; simultaneously, the mouth is opened and the head is overextended backward. If the corresponding point on the lower lip is touched, the mouth is opened, the lower lip is lowered, and the head is bent forward. On lateral touching, the corner of the mouth is drawn to the stimulated side and the head is turned sideward. This searching with the mouth can be temporarily found in every normal infant; the opening of the previously closed mouth is very striking. POPPER therefore speaks of a "snapping reflex."

INGRAM (1960) has described the movements of the lips: The newborn infant opens his mouth on stimulation; his lips seem to direct themselves toward the stimulus, so that on lateral stimulation they open wider at the stimulated side, and the whole mouth is drawn in this direction. Usually the head is also moved in the same direction; the tongue is extended at least toward the lips, becomes hollow, and forms a trough. When the mouth is turned toward the stimulus as far as possible, the tongue is retracted and the lips are closed. Now the infant makes sucking and swallowing movements. In most infants the reflex changes between the sixth and twelfth weeks; at that time they have learned to breathe through the mouth. Now the tongue is no longer moved forward when the mouth opens; when the mouth closes the movement of the lower jaw becomes more distinct. The succeeding swallowing and sucking movements no longer appear so regularly.

According to ANDRÉ-THOMAS and his co-workers (1954), any painful stimulation in the oral region causes turning of the face to the opposite side, while stroking with the fingers causes turning toward the stimulus. If both stimulations are applied simultaneously on the same side, for example, the right, the head is turned to the left, while the lips and tongue move to the right to grasp the finger. The disagreeable stimulus thus affects only the head movement.

Generally the rooting reflex is more lively in the hungry than in the satiated infant (see p. 472). It has frequently been stated that it can be more easily observed in the sleeping infant. HÄGGSTRÖM (1920) observed the reflex in all infants up to the third or fourth month of life, and in a sleeping child even up to the seventh year.

Occasionally GAMPER and UNTERSTEINER were able to elicit this reflex perfectly in a completely satiated infant who had just been nursed. The reflex was nearly always absent when the infant was occupied, for example, during urination, defecation, vomiting, or crying.

Lower animals grasp only with their mouths; the ability to grasp with the arms appears at a phylogenetically later stage. The rooting reflex therefore occurs particularly in lower developmental stages in which the nervous system is not yet fully developed. MINKOWSKI (1928) observed the "oral reflex," which often introduces the sucking movements, in human embryos of the second and third month of gestation. He classes it with the earliest and most constant fetal reflexes. The reaction could sometimes be elicited in dying infants at a time when all other reflexes had already ceased. The reflexogenic zone occasionally extended beyond the oral and lingual mucosa, so that the reflex could be elicited from the chin, the cheeks, the eyelids, and the forehead. Touch was the eliciting stimulus. BARCROFT (1942) saw a sheep embryo, which was swimming in physiological salt solution while still attached to the umbilical cord and covered with amniotic membranes, turn toward the stimulus when the face was touched. In a newborn opossum, touching of the mouth causes movement of its head toward the stimulated side, i.e., a rooting and sucking movement (according to LANGWORTHY, 1933).

According to HOOKER (1954—1958), the reaction to the stimulation of the oral region is the first and only response to external stimuli in the human newborn infant at the embryonal age of seven and a half to nine and a half weeks: If the oral region is touched with a fine hair the opposite side (more rarely the same side) of the neck is flexed. With progressing fetal age the arms are extended and the trunk is simultaneously turned. At the age of eleven to twelve weeks this reaction increasingly replaces the flexing. At thirteen to thirteen and a half weeks the latter reactions prevail; they will be gradually replaced by local reactions. At the age of nine and a half weeks a rapid, incomplete opening of the mouth occurs when the lower lip is stimulated. At twenty weeks both lips arch forward after stimulation, and at twenty-two weeks they are pursed. Later, opening of the mouth, tongue movements, swallowing, lip and mouth closing, and sucking movements combine to form the reflexes of food intake.

PRECHTL (1958) observed repeated turning of the head to and fro with or without stimulation in premature infants at twenty-eight weeks of gestation and weighing 1200 to 1500 gm. This behavior did not change until the infants had reached an age of three months and a weight of 3000 gm.

The rooting reflex is thus a function of the nervous system of a lower order. Accordingly, it can be found even in severe congenital brain malformations, for example, in the human mesencephalic monster described by GAMPER (1926). In the normal infant the reflex will be inhibited by the functions of the higher centers in the course of further development, but it may be released again in the older child and the adult (OPPENHEIM, 1903, 1913; TOULOUSE and VURPAS, 1903; FÜRNROHR, 1904; POPPER, 1921; GAMPER and UNTERSTEINER, 1924; WIESER, 1957).

The sucking, chewing, and swallowing movements which occur on stroking of the lips of comatose adults and which were described by OPPENHEIM (1903, 1913) as the "eating" reflex have long been known. STRANSKY (1903) observed that the mouth may even open at the approach of an object. These movements have since been frequently described in adults with brain damage. A slight pushing forward of the lips, which grasp and seize the object, occurs when the lips of the patient are touched with a spoon or tongue depressor (according to SCHELLER, 1952). If one attempts to remove the tongue depressor, the lips frequently try to hold it still tighter. Sometimes sucking or chewing movements occur. The extent to which the adult with brain damage can regress to the infantile stage was shown by WIESER (1957). In adults with organic brain damage (brain tumors, atherosclerotic dementia, brain atrophies, and severe retardation), he examined the "orienting reactions," which serve spatial orientation. On light and tactile stimuli he found optically elicited righting movements of the eyes and the head, reaching out for light and tactile stimuli, "magnetic" reactions, wide opening of the mouth, and sucking.

In a 25-year-old adult female suffering from a "contusion" psychosis (post-traumatic psychosis), the reactions were exhausted after a certain time. If the stimulating object was held immobile, she remained standing expectantly for 1 to 3 seconds with mouth open and tongue slightly protruding. Then she closed her mouth and no longer paid any attention to the object. But a response could again be elicited by moving the object to and fro, removing it from the field of vision and then returning it, or moving it close to the patient. The closer the object was brought to her, the more lively was her response. Tactile stimuli were decidedly more effective than optic stimuli. The opening of the mouth and the reaching for the object became more pronounced, and the reactions still continued when a refractory phase for the same reaction to optic stimuli had occurred.

The rooting reflex forms the basis for the development of certain human expressive movements, such as head shaking and nodding and perhaps also smiling (see p. 133).

If the baby is at a distance from his mother's breast he must bring his head nearer to her, for only then can he take the appropriate part of the breast, i.e., the nipple, into his mouth. The reflex described above assists him in this task. The effort of the human baby is supported by his mother, who takes him to her breast. Her assistance, which today seems so self-evident to us, developed very late phylogenetically. Even among the chimpanzees, which among the anthropoids are most similar to man, the baby gets little help from his mother. VON ALLESCH (1921) observed a captive chimpanzee mother with her baby and reported that for a long time finding the breast presented a difficult task for the baby:

"The baby searches for the nipple with its mouth open, first only through head movements, while the rest of the body is more or less 'fastened' onto the mother. Even after several weeks these head movements are not yet goal-directed; rather, the head is turned to all sides without plan and with mouth open. Even when the mouth is at a distance of only 1 cm from the nipple, the baby is not directed to it... it sometimes struggles until it is tired and sinks back to its old place, exhausted by the fruitless effort."

The mother sometimes brings the breast near the baby and then withdraws it again accidentally, since she usually is not concerned with whether the baby is searching for the breast. Corresponding observations by YERKES and TOMILIN (1935) are reported on page 589. CATEL (1950) was able to demonstrate rooting reflexes in a newborn chimpanzee.

In mammals which are phylogenetically lower than the apes, the mother is still less helpful to her baby when it searches for the breast. Therefore, the baby has to make a still greater effort. Since food intake is absolutely vital, it is insured by nature in several ways. Even the human baby carries memories with him, although his mother now helps him. To understand his behavior it is useful to observe the behavior of the young animal. What ways does the animal baby have of moving forward independently and thus approaching its mother's breast?

According to HARTMAN (1920), a ten-day-old opossum embryo – more a worm than a mammal – can crawl 3 inches over difficult terrain immediately after it is freed from its liquid environment. When it has reached the pouch, it is able to find the teat in the midst of a thicket of hairs. According to LANGWORTHY (1933; see p. 239), young opossums paddle forward on a piece of gauze with strong movements of their front legs. GOERLING (1920) conducted similar experiments with young kangaroos; they also found their mother's pouch independently. DATHE (1934) cites an observation made by SCHAAF: The newborn kangaroo baby moved with a peculiar crawl and, searching with its head to and fro along the abdomen of its mother, seemingly followed a damp trail which the mother had licked. After 5 minutes it had reached the opening of the pouch.

Finding the mother's nipple is facilitated in young mammals by a special movement pattern, a pendular movement of the head. PREYER (1895) had already observed that many animal babies move their heads vigorously to and fro at the start of sucking until the nipple is grasped. TILNEY and CASAMAJOR (1924) described similar reflexes in kittens. These reflexes were investigated in greater detail by PRECHTL and SCHLEIDT (1950, 1951). They found (Fig. 169, p. 413) that without external cause young cats, mice, and rats make horizontal pendular movements with their heads, young hamsters and rabbits make midline movements of this type, and puppies do this in both directions to and fro. PRECHTL and SCHLEIDT, as well as VON HOLST, term this phenomenon the "search automatism." It is closely related to the instinctive actions described by LORENZ (1943). The search automatism occurs also as "vacuum activity," i.e., without external cause.

Referring to PEIPER and STIRNIMANN, PRECHTL and SCHLEIDT (1951, 1952) ascribe vacuum activity rooting reflexes to the human baby also, since hungry infants move their heads to and fro if their cheeks, lips, or chins are touched. But there is an essential difference when we compare this phenomenon with that of newborn animals. A hungry infant in the dorsal position turns his head around the spinal axis as a vacuum activity, i.e., without external stimulus. This causes a picture totally different from that in the newborn mammal which, while in the abdominal position, moves its head and neck horizontally sideways or backward and forward without rotation of the spine. The pivot is a fixed point on the cervical spine.

PREYER (1895, p. 203) regularly observed pronounced sideways to and fro turning, almost a wobbling of the head, during the first weeks in his child when he was placed at the breast. When the baby took the breast on the eighth day for the first time without assistance, these sideways movements of the head looked very much as if the child was searching for something. But the same movements were seen when the mouth of the bottle was placed in his mouth on the twenty-seventh day: "Therefore, we are not dealing here with an acquired head movement but with an instinct which causes head movements on sucking on a finger or on the nipple of the breast." According to PRECHTL (1952) these rhythmic searching movements which the infant executes with his head at first appear without external stimulus. However, in the course of development they became increasingly dependent on tactile stimuli in the region of the cheeks and lips, and finally changed into an orienting reflex of the mouth to the source of the stimulus. If elicited repeatedly, the movement became increasingly more rhythmic and more independent of the stimulating cue.

This pendular head movement occurs less frequently in infants in their usual dorsal position. In the abdominal position, infants can frequently raise their heads from the base for short moments with the aid of the labyrinthine righting reflex

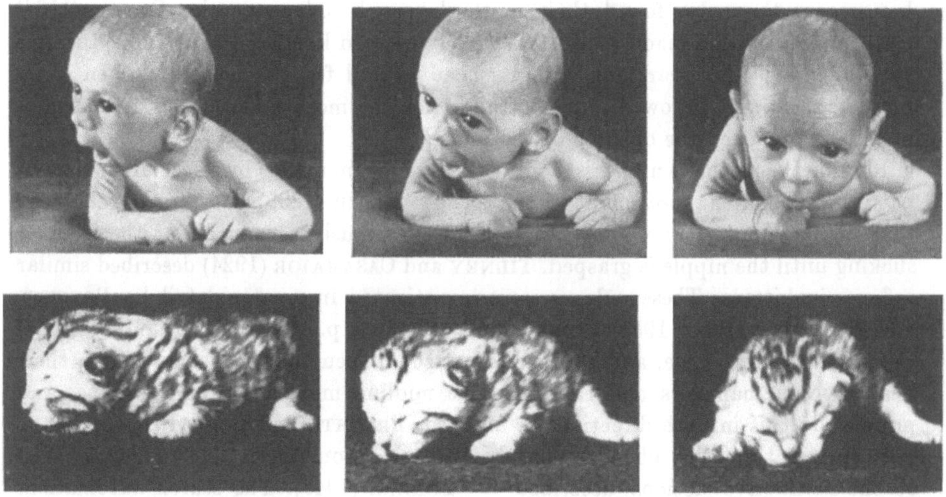

(see p. 175). Pendular head movement is at most only suggested. The mouth usually rubs along the base, which is touched by the opened lips or the tongue; the sucking and licking movements thus evoked prevent further searching for the breast. When older infants in the abdominal position lift their heads (or later their bodies) from the base for a somewhat longer span of time, the pendular movements are usually absent.

GAMPER and UNTERSTEINER (1924), on the other hand, observed pendular head movements in the adult, the case being that of a 26-year-old woman suffering from postencephalitic hyperkinesis; this occurred when the woman was standing and without external stimulus and corresponded to those movements in young mammals. Both investigators recognized the connection with the rooting reflex of the infant. PILLERI (1960) also found a rooting reflex as a vacuum activity in a 67-year-old patient suffering from Pick's cerebral cortical atrophy, in the terminal phase of the disease.

PEIPER [1951 *(115)*] and PEIPER and THOMAS [1953 *(124)*] observed horizontal pendular head movements in two infants suffering from severe congenital disorders of cerebral function. This movement, which corresponded essentially to the pattern described for young mammals, appeared when the infants were in the abdominal position. In the dorsal position both infants exhibited a tendency to open their mouths wide, to distort their lips, to turn their heads excitedly to and fro with mouths wide open — frequently overshooting the stimulating object by this movement — and to snap with their mouths. In addition, searching movements occurred without external stimuli in this position before feeding. They were elicited and intensified by the preparations for feeding, such as the lowering of the bed rails, fastening of the bib, or bringing of the bottle. Essentially the same behavior, though not so distinct, may be found as a conditioned reflex in the normal infant. When we turned these two infants onto their abdomens, pendular searching movements such as are seen in newborn mammals appeared, in addition to the rotating head movements. The head, neck, and upper part of the body rotated around an axis perpendicular to the mid thoracic spine. From this parallel with the searching movements of newborn mammals, we may conclude that the rooting reflex was originally connected with the phylogenetically earlier abdominal position. The observation that it was clearly elicited when the infants were placed in the abdominal position supports this hypothesis.

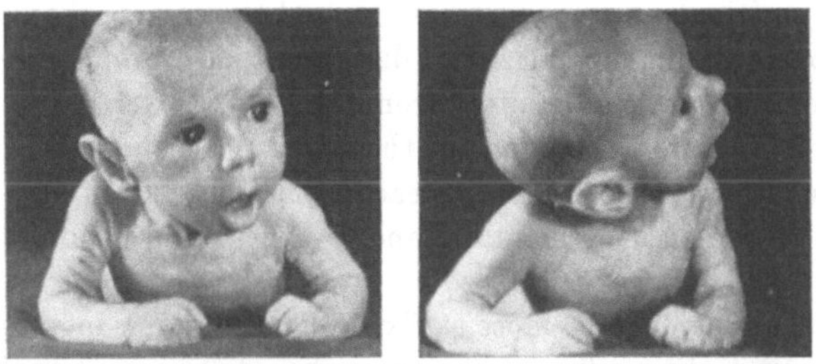

Fig. 168. Rooting reflex as vacuum activity in a three-month-old infant with brain damage (frames of a film). Pendular head movement in horizontal direction with rotation of the head around the vertebral axis and opening of the mouth.

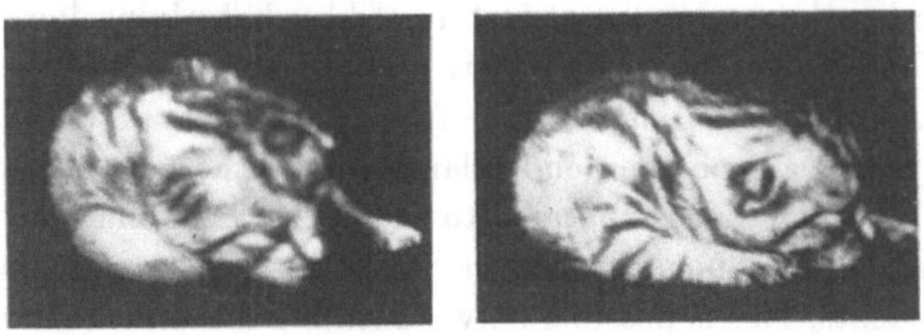

Fig. 169. Rooting reflex of a four-day-old kitten (frames of a film). Pendular head movement in horizontal direction with displacement of the front legs, but without head rotation.

If the infant was in the dorsal position, searching movements could be elicited by tactile stimuli. When brought into the abdominal position (Fig. 168), the infant showed a distinct inclination to turn his head upward and to the left as far as possible; turning to the right occurred less frequently and to a minor degree. The eyes were often moved in the same direction, preceding or following the head movements. Opening of the mouth similar to that seen in searching movements was combined with these head rotations, especially when the infant was hungry. In addition, the lips were frequently pulled in the same direction, preceding the head movement.

The sequence of movements started with the lifting of the head, which had been resting on the arms. If the face was directed toward the left at the beginning, the movement occurred only toward the left and upward. The final position attained was maintained for some time; then the head returned to its starting position. But if the face was resting on the base, a slight turning to the right preceding the upward movement to the left was possible. This almost always occurred if the face was turned to the right before the lifting of the head; however, this pattern was less frequently observed. This sequence of movements was accompanied by licking and sucking movements and touching of the hands with the lips.

Frames of a motion picture film showing the vacuum activity rooting reflex of this infant and that of a four-day-old kitten (Fig. 169) demonstrate the identical instinctive actions in the human infant and in the animal. The infant with brain damage performs the same searching movements in the dorsal position as does the normal infant by rotating his head around the vertebral axis. When he is brought into the abdominal position rotary movements around the vertebral axis as well as pendular head movements in the horizontal direction occur. Simultaneously with the sideward movement, the head is lifted from the base.

In cats, the mother animal is in the lateral position and the kitten in the abdominal; the body axes of the two animals are perpendicular to each other. The kitten finds the nipples of its mother, which are on a horizontal plane, by horizontal pendular movements of the head.

The corresponding conditions in man can be more easily recognized in our infants with brain damage, because their rooting reflex is more pronounced than in normal babies. It is unimportant whether the mother has at first nursed her baby lying down or sitting up. The human infant lies in the arms of his mother, so that only the relative position of mother and child is important. The front of the baby is turned toward his mother. The body axes are not perpendicular to each other, as is the case in the cat; rather, they are more or less parallel to each other. The anatomic structure of his mother's breast does not permit the infant to find the nipple solely by pendular head movements. This is possible only if rotating movements are added.

Cats extend the radius of their searching movements by moving the front legs a little sideways in the direction of the head movement (Fig. 169); for example, in pendular head movements to the right the right front leg is at first placed to the right and if the head movement is very extensive the left front leg follows. If then a pendular head movement occurs to the left, first the right and then the left front leg is brought back to its starting position; sometimes the right front leg is moved sideways across the left leg. Then a corresponding movement sequence to the other side occurs. We are dealing here with limping reactions, which were described previously (see p. 203).

If the kitten does not reach the desired goal, it will not infrequently crawl, searching in a narrow circle until it has returned to its starting position (described by PRECHTL).

The position in which the human infant and the animal baby suck may also play a role in the difference in the sequences of searching movements. For example, rabbits, which lie on their backs under the body of their mother and thus have the source of food above them, make pendular midline movements, as PRECHTL and SCHLEIDT (1950, 1951) have observed.

In addition to the pendular movements illustrated (Fig. 169), newborn kittens exhibit licking movements, which are elicited by the tactile stimulus of the base, and mouth opening. This can lead to a second pattern of searching movements which PRECHTL describes as digging movements: A kitten which has reached the abdominal nipple region of its mother by horizontal pendular head movements begins to execute pushing and digging movements against the breast with its mouth open. The deflections of the simultaneous pendular movements are less than in free pendular movements.

These searching movements also occur in the normal infant when he is in the abdominal position or when he is placed at his mother's breast [PEIPER and THOMAS, 1953 *(124)*; PRECHTL, 1953]. The opened mouth touches the base of the breast lightly and is lifted up again. This procedure, associated with slight pendular movements, is repeated until the lips touch the nipple. Then the mouth is righted in the direction of the tactile stimulus so that he can grasp the nipple. In contrast to those seen in the kitten, these searching movements may occur in the normal infant independently of the nearness of the maternal source of nutrition. However, they increase in strength when the infant is brought in the feeding or a similar position nearer to his mother. The bringing of the infant into the feeding position seems to play an important role; caloric and olfactory stimuli may well have a supporting effect. The touching of the base with the tongue may inhibit real searching movements when the licking and sucking are continued in one area for some length of time. The pattern of searching movements in vacuum activity evoked by the stimulus of the base is found especially in the young infant, who is not yet able to raise his head sufficiently for pendular and rotating movements. As his static capabilities gradually allow the infant to raise his head and to keep it raised, the searching for the breast in vacuum activity disappears. Therefore, this phenomenon can be better observed and demonstrated in infants with brain damage, in whom the more primitive cerebral functions are more pronounced and persist for a longer time.

In an extensive series of investigations, PRECHTL (1958) distinguishes rhythmic nondirected rooting from the infant's arrhythmic directed searching for the breast: In rhythmic rooting in the dorsal position the head is turned to and fro around its vertical axis after the oral region has been touched; in the abdominal position it is simultaneously raised from the base. These movements may also appear in hungry newborn infants without stimulation. Two to three weeks after birth this rhythmic nondirected movement changes to a spatially directed motion. This second movement may at first still have a rhythmic component, but this soon disappears (directed pendulating) and is replaced by a directed turning of the head toward the stimulus, i.e., a directed rooting reflex. Rhythmic rooting persists much longer in the premature infant. It can be similarly elicited by touching any sensitive spot of the oral region.

However, the direction of the rooting reflex depends on the skin area which has been stimulated. If the same area is touched repeatedly and at short intervals, the directed rooting reflex finally disappears (central adaptation). The number of reactions depends on the strength, the frequency, and the exact area of the stimulation. If the stimulation is continued after the onset of central adaptation, the head may turn away from the stimulus (central reorientation). If the right corner and, immediately after the onset of the reaction, the left corner are stimulated, the reaction is not influenced in its sequence pattern. However, if the other side is not stimulated until immediately after the reaction is terminated, an especially intense reaction occurs (rebound mechanism).

As yet, no movements are known which would enable the infant to move forward along his mother's body. The climbing movements described by PRECHTL (see p. 220) do not seem to me to be sufficient for this explicit purpose. Rather, one could suppose that the reflex crawling movements (see p. 217), especially the supporting reactions of the legs (see p. 203), might serve this purpose. With their help the infant could push himself from the inguinal region of his mother in the direction of her breast with his soles. Unfortunately, we do not yet have sufficiently detailed observations of the behavior of the baby monkey on his mother's body.

DARWIN (1872) describes additional reflexes in animal babies which are absent in the human infant: Young cats, dogs, pigs, and probably many other animal babies, push against their mother's mammary glands with their front legs to stimulate more abundant milk secretion or to initiate the milk flow. These tapping movements of the front paws also occur in kittens when they rest on a soft base; their toes are spread and the claws are slightly extended as if sucking at their mother's breast.

Mouth-Hand Coordination

Fish, amphibians and reptiles, and many mammals and birds seize their food directly with their mouths; many beasts of prey catch their food exclusively with the mouth. In contrast to this lower form of food intake, we see another pattern in which the food is at first grasped with the limbs, even reduced to small pieces, and only then seized with the mouth. This higher form of food intake, which is common in mice, squirrels, and monkeys, is facilitated by a more or less erect body posture; the front limbs no longer need to support the body and are thus liberated for other functions. In many birds the hind limbs have taken over this task, since the fore limbs have been transformed into wings. Thus, many birds of prey grasp their flying prey with their talons; chickens scrape the compost heap with their feet in order to pick up worms with their beaks.

Adult man prepares his food and takes it to the mouth with his hands, but the young infant still seizes it exclusively with the aid of the rooting reflex. At a later stage he is inclined to put everything he can reach with his hands, edible or inedible, into his mouth first and to taste it with that organ.

A reflex mouth-hand coordination can be observed quite early in the infant. The palmar-mental reflex described by MARINESCO and RADOVICI (1926) is elicited by scratching the thenar or hypothenar eminence with a rough object: The reflex consists of simultaneous contraction of the chin muscles which lift the chin up. LIPPMANN [1958*(22a)*] was able to elicit this reflex in all newborn infants. MAGNUSSON and WERNSTEDT (1935) found it in 73% of infants during the early weeks of life and even in the last months of the first year. The statements on its frequency after the first year vary greatly. SCHACHTER-NANCY (1937) found the reflex at the age of 6 to 12 years in 28.25% of the children examined.

BABKIN (1956) described a similar reflex in newborn infants: If both palms of an infant lying in the dorsal or lateral position are pressed, the mouth opens, the eyes are closed, and the head is bowed forward (Fig. 14, p. 116). Unilateral pressure evokes only a weak reaction, the head being turned to the side of the stimulus. The reaction to the stimulus was demonstrated in 278 of 311 newborn infants by LIPPMANN. According to BABKIN's descriptions, it weakens during the first month and in most infants disappears in the third month. It can no longer be demonstrated after the fourth month.

However, hands and mouth continue their close relations later in life. I have previously shown that the reflex opening or closing of one sensory organ may spread to another sensory organ as a spreading reaction (see p. 112). Thus, the voluntary opening or closing of the mouth or the eyes often spreads during childhood to the hands and, to a lesser degree, to the feet, which in this way become, so to speak, sensory organs. On the other hand, voluntary closing of the hands may spread to the eyes (see p. 115).

DARWIN (1872) described how some people move their lower jaws in time with the blades when cutting with scissors. He also states that children may occasionally move their tongues in a pattern similar to that of their fingers. WEINERT (1940) put paper and pencil in the hands of a four-year-old chimpanzee and scribbled something for her: "The chimpanzee let her tongue protrude from the corner of her mouth in a manner similar to that of a school child who writes his first letters on a blackboard."

I found a similar accompanying movement in children who were taking their first faltering steps. In order to maintain their threatened equilibrium, they paddle wildly with their arms in the air. At the same time they move their tongues between their lips as if they could make countermovements in this manner.

"Sound gestures" of the tongue occur in baby language. According to PORZIG (1957), "ta" or "ta ta," the infant's responses to impressions which attract their attention or stimulate their desires, have developed from pointing movements of the tongue. Occasionally accompanying movements of the tongue can still be seen in adults when they point to an object.

If the development of cerebral function is inhibited by illness, the original mouth-hand coordination may become pronounced again. FRÜHMANN (1959) describes stereotyped movements corresponding somewhat to BABKIN's reflex in 3 children who suffered from severe cerebral diseases during the first 6 months of life. When the dermatomes C3 to D3 were stimulated, the mouth opened. Since the patients extended their arms and opened their mouths even without external cause, i. e., as vacuum activity, FRÜHMANN assumes that this could be an instinctive movement or at least a remnant of such a motion. He reports that, according to O. KÖNIG, young shrews and kittens open their mouths when they stretch their fore limbs.

A regression of cerebral function may release mouth-hand coordination even in the adult. EBBECKE (1958) observed a patient who awoke from a brief hexobarbital anesthesia. When he had already responded and appeared to have a clear sensorium, he looked without comprehension at a pocket watch shown to him, seized it, and tried to put it in his mouth. RENNERT (1956) saw a snap-claw reflex in adults with progressive paralysis and other lesions of the frontal lobes. According to SCHELLER (1952), seizing with the hands is often connected with grasping with the mouth in adult patients. BENTE and WIESER (1952) and WIESER (1957) reported on similar cases (see p. 576).

Elicitation of Sucking Movements

The ease with which sucking movements can be elicited by certain stimuli has led to the use of the term "sucking reflex." But the onset of these movements without external cause, their rhythm, the relative coordination between sucking, breathing, and swallowing (see p. 420), the pendular induction between respiratory and sucking centers (see p. 434), and the periodicity of the sucking movements (see p. 419) point without any doubt to the fact that we are dealing with an automatism in the sense of the term used by VON HOLST (1939), PRECHTL and SCHLEIDT (1950).

The same tactile stimuli which cause the rooting reflex when applied to the lips and oral region may also evoke rhythmic sucking movements of more or less long duration. They are more easily elicited if the same stimulus is repeated several times. Sucking movements can also be elicited by touching the oral cavity, especially the tongue, the mucosa of the cheeks, the jaws, and the hard palate. If an appropriate oblong object is introduced into the mouth, sucking movements occur. A liquid, even if tasteless, is equally effective.

That gustatory stimuli also cause strong sucking movements has long been known (KUSSMAUL, 1859; GENZMER, 1873; PREYER, 1895). Sweet substances are especially effective, whereas bitter, salty, or sour substances will cause defense reflexes. There are some exceptions. GENZMER and STIRNIMANN (1936) occasionally observed sucking movements in response to weak quinine or saline solutions. But it has not yet been determined whether there are newborn infants who respond only with sucking movements to repeated stimulation with bad tasting substances in a sufficiently high concentration. In ECKSTEIN's (1927) experiments, the majority of infants, including the premature, showed increased sucking movements to a sweet taste and a certain number reacted equally to salty and bitter tastes. But the bitter taste almost always inhibited the sucking movements immediately. With the aid of sucking movements, JENSEN (1932) determined the upper and lower temperature thresholds and the threshold values for the concentration of taste substances in newborn infants. Great fluctuations were seen from child to child; the threshold values of the same child did not change much during the examination. The child drinks amniotic fluid even before birth (see p. 549). DE SNOO (1937) thought that he could improve the drinking by injecting a saccharine solution into the amnion.

Sucking movements, like the rooting reflex, depend on the general condition of the infant, especially the satiation. Whereas they occur frequently and continuously at the beginning of the feeding, they become less frequent later and finally disappear even if more food is offered, i.e., if the stimulus which started the food intake continues without change. Various answers have been given to the question of whether the sucking movements are inhibited by fatigue or by satiation. CRAMER indicated that fatigue was the main cause; PETERS (1902) stressed that breast-fed children fall asleep from fatigue before they are satiated if difficulty in draining the breast exists; SÜSSWEIN (1905) attributes the arrest of sucking movements equally to fatigue and satiation.

To decide this question, KRAWITZKAJA (1950) performed esophagotomies on nursing puppies 1 to 30 days old. These animals showed, 12 hours after the last feeding, sucking movements which lasted 2 hours or more without inhibiting the sucking center. But if milk was introduced into the stomach through the lower esophagus portion, the sucking center was reflex inhibited within 5 to 8 minutes. Cocainization of the stomach wall prevented this inhibition, so that the sucking movements continued in the same manner as before the introduction of the milk.

In the newborn infant, sucking movements can also be caused by stimuli other than oral stimulation, i.e., by pinpricks on the toes, pulling of the hair, and sudden support withdrawal, according to JENSEN (1932); and when the threshold is lowered by touching of the legs, arms, or back, clapping with the hands, slamming a door, etc., according to PRECHTL and SCHLEIDT (1950).

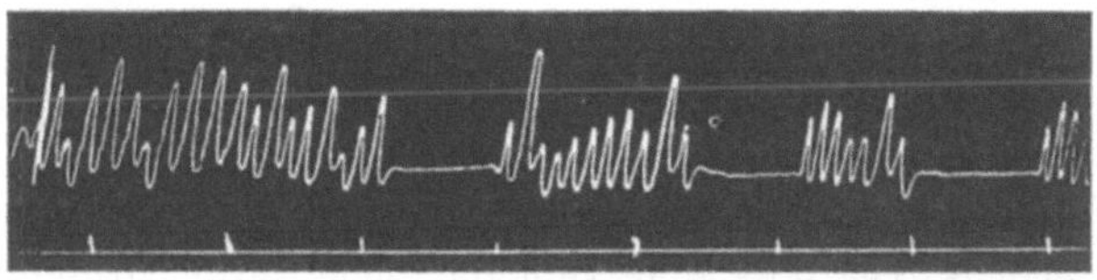

Fig. 170. Transition of regular into periodic sucking movements (time in 5-second intervals).

Two stages of excitability, closely related to those of the respiratory center, can be differentiated in the sucking center of the drinking infant. As SÜSSWEINS (1905) has already described it, the rhythm of the sucking movements change near the end of the feeding so that groups of sucking movements are more or less regularly separated from each other by sucking pauses. The number of sucking movements in a group — approximately 3 to 7 — fluctuates somewhat (Fig. 170). The appearance of the graphs of periodic sucking movements shows great similarity to that of periodic respiration. But still more important is the fact that we are dealing in both cases with a lowering of the excitability in the respective centers. I demonstrated this phenomenon for periodic respiration previously (see p. 320). The same is true for periodic sucking movements. In the normal, healthy infant they occur only at the end of the feeding, when the sucking center has been active for some time. The nervous excitability of premature infants is in itself low; therefore, a priori they have a tendency to periodic respiratory and sucking movements. The available observations absolutely contradict ANDRÉ-THOMAS's (1952, p. 107) statement that the sucking continues rhythmically without pause until the need is satisfied and the infant is satiated.

At the beginning of feeding, healthy infants suck fairly rhythmically and unhurriedly, 40 to 90 times in each of the first 2 or 3 minutes (according to SÜSSWEIN); every or almost every sucking movement is followed by a swallowing movement. If the amount of milk is sufficiently great, a fine rippling in the larynx can be felt with the finger; otherwise, only the upward movement of the larynx can be felt. During the third and fourth minutes of sucking only 1 swallowing movement occurs after every

2 or 3 sucking movements (according to SÜSSWEIN). The sucking is more frequently interrupted by the periodic intervals described above, so that the number of movements per minute is less than it was before. In the following minutes the pauses grow longer and the swallowing movements become more infrequent. Accordingly, milk intake decreases so that half the amount of food is taken during the first 5 minutes. If a normal, healthy infant sucks on a breast which is difficult to drain, the sucking movements are usually slowed down but are sometimes accelerated. Pauses occur more frequently and earlier. The sucking movements are interrupted more often in newborn infants than in older infants (KASAHARA, 1916). According to SÜSSWEIN, weak infants suck more slowly and have a tendency to periodic sucking, as I have stated previously. They suck and swallow in a ratio of 1:1 or 2:1; later swallowing is still more infrequent. Other infants make frequent gasping sucking movements, swallow rarely, and in the intervals rest at the breast as though asleep. Bottle-fed infants drink more regularly and with fewer pauses than breast-fed infants (according to SÜSSWEIN). Inhibition may occur later in this case because they have to exert less effort. PRECHTL and SCHLEIDT (1950) observed periodic sucking movements in young guinea pigs: The sucking occurred in very brief periods consisting of 3 to 7 sucking movements, and was interrupted by pauses of 1 to 2 seconds. The jaws were kept immobile during drinking, while the tongue and the floor of the mouth seemed to move, and swallowing could be recognized.

According to PRECHTL (1956), premature infants and most newborn infants born at term suck with their lips tightly closed around the nipple and each sucking movement causes a negative pressure in the oral cavity. After some weeks this sucking pattern changes to another form: Now the corners of the mouth remain open, while the tongue takes over the whole sucking function and empties the nipple by licking and stroking or milking movements toward the roof of the oral cavity. These two sucking patterns are differentiated as lick-sucking and pump-sucking. The phenomenon of dominance can be demonstrated with the aid of the sucking movements (see p. 530).

Sucking movements occurring during sleep are not infrequent in children of school age, especially after subthreshold awakening stimuli. The gritting of the teeth which not infrequently occurs during sleep in adults, and especially in children suffering from tuberculous meningitis, may also be regarded as a sucking movement of the dentate lower jaws. The inhibition during sleep causes cerebral function to regress to a lower level, a phenomenon that is also evident in yawning (see p. 386).

Concerted Action of Sucking, Breathing, and Swallowing

(38, 56, 62, 65, 74, 75, 80, 83, 93, 112, 134)

During the infant's food intake three centrally directed rhythmic movements arise: sucking, breathing, and swallowing. Since the alimentary and respiratory pathways cross in the pharynx, these three functions must be exactly adjusted to each other. Widely varying opinions have been expressed as to their time relations. To clarify the contradictions I recorded the three rhythms graphically. It is easiest to describe their relations to each other if these three phenomena are compared with each other two at a time.

The hungry infant makes 60 to 80 sucking movements per minute. Each sucking movement immediately follows the preceding one for the first few minutes; only near the end of the feeding do the individual movements combine into smaller or larger groups separated from each other by pauses. This has been described previously.

HALVERSON (1944) observed healthy infants from birth to the age of 18 weeks and found a median value of 72 to 74 sucking movements per minute; the lowest value was 42 and the highest 126. The values fluctuated from infant to infant. Some completely healthy infants sucked fairly slowly, but steadily. In the course of development the number of sucking movements did not change noticeably.

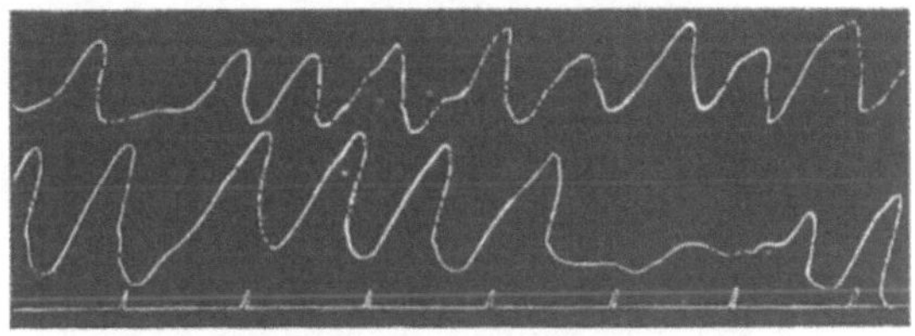

Fig. 171. One sucking movement (lower tracing) corresponds to each breath (upper tracing) (time in seconds). The bottle was temporarily removed during the second half of the graph; this halted the sucking movements for the same length of time.

According to BALINT (1948), most infants try to suck with approximately the same frequency; the "basic frequency" was 48 to 80 per minute. This number fluctuated very little in the same child, sometimes only by 1 or 2, mostly by not more than 4 or 5, sucking movements per 10 seconds. In another higher rhythm the sucking movements always started after an interval with a higher frequency which gradually decreased to the basic frequency. In some infants the higher frequency persisted when they became restless or moody. In addition, half the infants showed a very rapid rhythm of deflections which were smaller than sucking movements and amounted to tongue tremors. This rhythm either appeared when the infant was not sucking or it was superimposed on the sucking movements.

Sucking and Breathing

(38, 74, 75, 80)

If respiratory and sucking movements* are recorded simultaneously, their relations to each other can be best recognized if both patterns are completely regular, an infrequent occurrence. Respiration in particular easily becomes so irregular during sucking that a specific law cannot automatically be recognized. However, if all these irregular graphs are excluded enough remains to permit a comparison.

If the infant has breathed and sucked regularly at all, a definite rule can always be deduced: One or two complete sucking movements correspond to each respiration (Figs. 171 and 172). Either a sucking movement lasts as long as the whole respiratory

* To record respiratory movements a rubber bag is fixed to the abdomen; to record sucking movements another is fastened below the chin.

movement or one sucking movement occurs with inhalation and one with exhalation. Thus, it is no coincidence that the average number of sucking movements is just twice the number of breaths. As long as the respiratory and sucking centers function regularly, their activity is closely coupled. Since both centers are situated near each other in the brain stem, it can be easily understood that the two rhythms adjust to each other.

Colley and Creamer's (1958) statement that respiration and sucking are independent of each other is disproved by the graphs reproduced here.

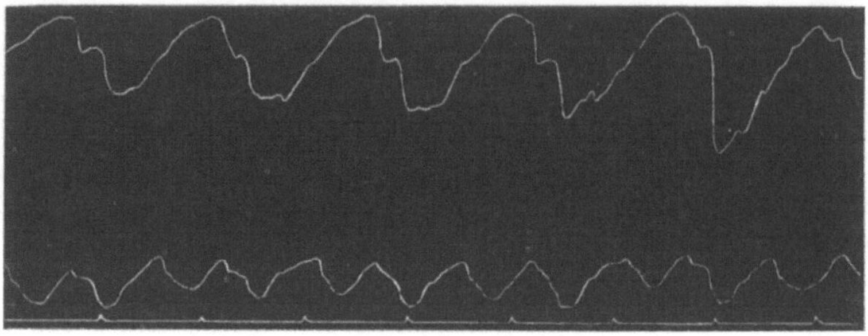

Fig. 172. Two sucking movements (lower tracing) correspond to each breath (upper tracing) (time in seconds).

Halverson (1944) obtained the same findings as I did: In 86 feedings the respiratory rate was lower than the sucking rate in 50 cases and higher in 4, but the respiratory movements tended to keep pace with the sucking movements. In general a low sucking rate was matched by a low respiratory rate. A high respiratory rate corresponded to a high sucking rate, or two sucking movements accompanied one breath. Six infants tended to suck once during one breath; two showed a preference for two sucking movements during one breath. The ratio of sucking movements to respiratory movements was 1:1 in 52 feedings and 2:1 in 23 feedings; mixed forms of one or two (sometimes three) sucking movements to one breath were found in 11 cases. The coordination was equally good for the 1:1 and 2:1 ratios; it was generally best in low and worst in high sucking and respiratory rates.

von Holst (1939) investigated the interplay of the rhythmic movement centers in detail by observing how rhythmic excitations in the spinal cord influence each other. He worked with decerebrate bonefish, recording their rhythmic fin movements through levers. With his method he recognized among the various rhythms important constantly recurring relations which he termed "relative coordination." It was shown that the stronger rhythm could evoke periods or undulations (superposition) in the weaker rhythm or could "capture" it, i.e., induce its complete conformity (magnet effect). But the other rhythm immediately returns to its characteristic pattern as soon as the action of the stronger rhythm ceases. According to von Holst, it may happen "that one automatism splits into several small units." I have similarly described the disintegration of the respiratory center.

The question to be investigated now was whether a corresponding relation between the sucking and the respiratory centers could be demonstrated in the infant and, if this should be the case, which of the two is the stronger and which is the weaker center.

Critical examination of the recordings revealed a real difficulty: In the fish, VON HOLST had severed the spinal cord from the brain by an incision to avoid all interfering side movements and thus preserve pure and absolutely rhythmic fin movements. But to begin with waking infants always have a strong urge to move; this even increases at the moment when they receive the bottle or when it is taken away from them. However, the periods when the sucking movements appear or disappear are precisely those in which the relations of the rhythm to each other are most distinct. In order to obtain satisfactory graphs we therefore had to watch for periods of complete rest. If the infants are too hungry, they cry; if they are sleeping too deeply they are often frightened by the touch of the nipple. Often an experiment which at first begins under favorable conditions is disturbed by external or internal influences during its further course and thus cannot be used. The only way to obtain recordings without disturbances was to make a great number of observations. I started only with recordings in which the respiratory center had functioned as rhythmically as possible (until the beginning of the sucking movements) and in which the sucking center had then done likewise. In the graphs which were chosen in this manner, the same law could always be recognized, an indication that the choice was correct.

VON HOLST has submitted proof that the automatic-rhythmic processes do not, as had been assumed, originate in the motor cells which send the excitations to the muscles but in the "automatic" cells which are interposed in front of the motor cells and which supply stimuli to them. The motor cells can also be stimulated from other areas. If these conceptions, which are primarily valid only for the spinal cord of decerebrate fish, are applied to the medulla oblongata of normal infants, we arrive at a simple principle which may assist in explaining the observed phenomena. In the following descriptions I therefore distinguish an automatic-rhythmic portion and a motor portion in both the sucking and the respiratory center. Both automatic centers emit rhythmic excitations. However, their contact with the muscles is not made directly, but only by way of the motor centers. The latter, on the other hand, are directly connected to the corresponding muscles; they themselves cannot produce stimuli but receive their excitations from some other center, which could be either their own corresponding center, or some other rhythmic partial center, or else from the periphery whence they can be stimulated if the rhythmic partial centers are bypassed.

Figure 173 illustrates an essential part of the phenomena which occur at the onset of sucking movements. In the beginning the stimulation takes the following pattern:

1. *Sucking suspended:*	Rhythmic sucking center		Motor sucking center
	Rhythmic respiratory center	⟶	Motor respiratory center

As soon as the bottle is given, sucking movements which are more frequent than the previous respiratory movements begin. The duration of two respiratory movements corresponds to approximately three sucking movements. This sucking rhythm remains unchanged for the duration of the recording. Its appearance at once disturbs the respiration. The breaths at first become irregular and very shallow, but then increase in size until they definitely attain a frequency and phase identical to those of the sucking movements. A control measurement shows the same for the smaller respiratory movements. The respiratory center is unmistakably led by the sucking center,

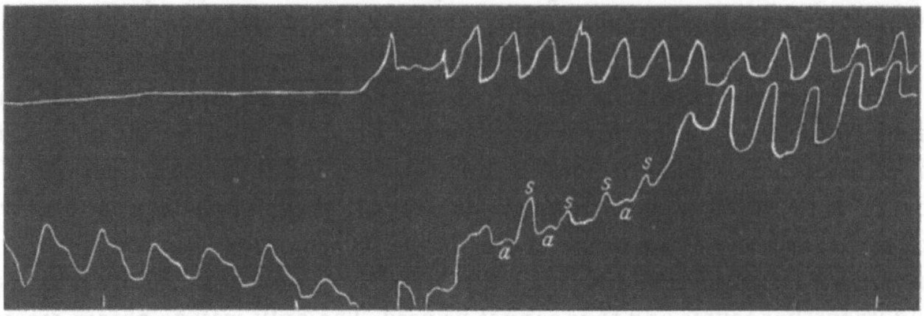

Fig. 173. Upper tracing: Sucking movements; lower tracing: respiratory movements (time in 5-second intervals). Successively: Sucking arrest; disrupted period with superposition, during which the s-waves depend on the sucking center and the a-waves on the respiratory center (which is controlled by the sucking center); finally, dominance of the sucking center.

and this in a rhythm which is more rapid than its original rhythm. The rise of the respiratory tracing after the onset of sucking movements (Fig. 173) is caused by the lowering of the diaphragm, a process which may be noted in some graphs and which is absent in others. The graph reproduced here has been chosen because the relations between sucking and respiratory centers are especially well demonstrated by the rise of the respiratory tracing.

The onset of the sucking movements considerably upsets the respiration in almost all tracings. In most cases a complete respiratory arrest (Figs. 174 and 175) which may last for 5 seconds or more occurs. Incidentally, this arrest also appears in infants after irregular respiration or crying. Diagramatically, the respiratory arrest can be illustrated thus:

2. *Respiratory arrest:* Rhythmic sucking center ——→ Motor sucking center
Rhythmic respiratory center   Motor respiratory center

If the sucking center has taken the lead, the evident influence of the respiratory center on the respiratory movements ceases completely (Figs. 172, 173, and 176). There are two possibilities for the mutual relations between which no final decision can be made:

3. *Direction of respiration by the sucking center*

a. Rhythmic sucking center ——→ Motor sucking center
↓
Rhythmic respiratory center —→ Motor respiratory center

b. Rhythmic sucking center ——→ Motor sucking center (and ↘ Motor respiratory center)
Rhythmic respiratory center   Motor respiratory center

In this condition one or, more rarely, two full sucking movements correspond to one full breath, as was previously stated. Three sucking movements to two breaths, repeated several times in succession, were found as an exception during one session.

von Holst described the "magnetic effect" as the important coordinating principle in the spinal cord. This has to do with an attraction effect which originates in one automatism and influences the frequency of the other; depending on the relation of the phases between the two automatisms at a given moment, the effect will be

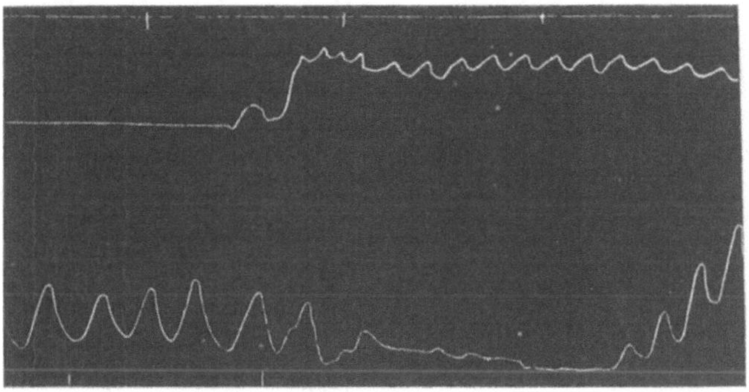

Fig. 174. Upper tracing: Sucking movements; lower tracing: respiratory movements (time in 5-second intervals). Successively: Sucking arrest; respiratory arrest; and control of the respiration by the sucking center.

either decelerating (negative attraction) or accelerating (positive attraction). The attracting automatism tries to transform the attracted in such a way that the automatic cells of the two rhythms function homologously. I was not able to confirm such a magnetic effect in my graphs; on the contrary, the transition from the orig-

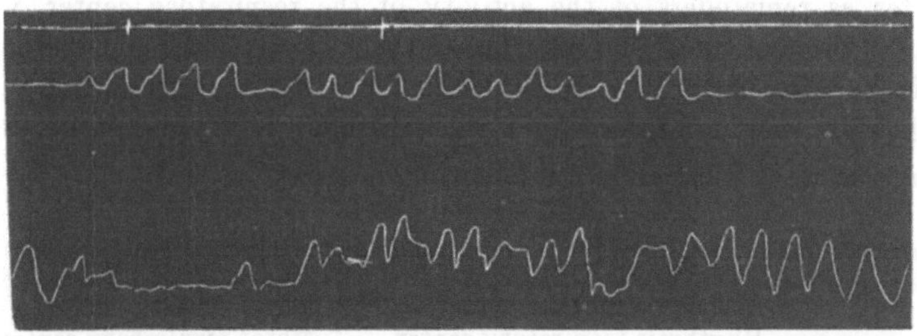

Fig. 175. Upper tracing: Sucking movements; lower tracing: respiratory movements (time in 5-second intervals). Successively: Sucking arrest, respiratory arrest, period of interference with an occasional suggestion of the rhythmic sucking movements, after effect, and sucking rest.

inal respiratory rhythm to the rhythm of the sucking center was never gradual, as Figs. 173 and 174 show. The ratio of respiratory movements to sucking movements is always established suddenly. Occasionally fluctuations between the 1:1 and 1:2 ratios occurred several times in rapid succession in the same tracing.

I call the time from the onset of the sucking movements to the domination of respiration by the sucking center the "disrupted period." The duration and appearance of this stage may vary. In Figs. 173 and 174 this time is comparatively short. But not infrequently it lasted so long that I was able to record only fragments of it with the instruments I had at my disposal. There is a possibility that in some cases this disrupted period may not be superseded at all, so that a fixed relation between the sucking and respiratory centers is not achieved during the entire feeding. The respiration then remains more or less irregular even though the infant sucks regularly.

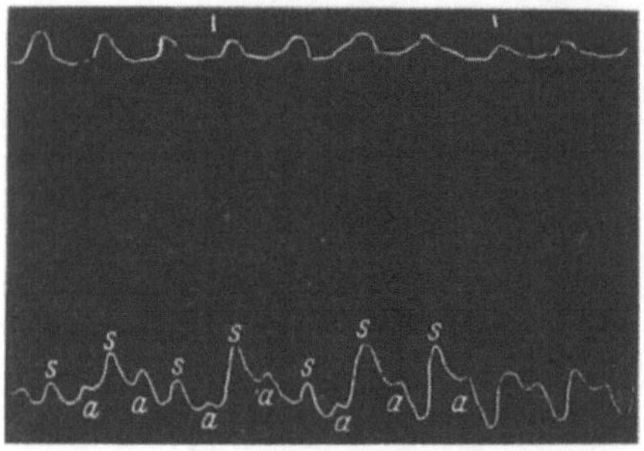

Fig. 176. Upper tracing: Sucking movements; lower tracing: respiratory movements (time in 5-second invervals). Superposition, during which the s-waves depend on the sucking center and the a-waves on the respiratory center, which is directed by the sucking center; later, direction of the respiratory by the sucking center, during which a bigeminy has developed.

During the disrupted period the sucking and respiratory centers struggle for control of the respiratory rhythm. In Fig. 174 the sucking center won immediately at the end of the respiratory arrest. But if the struggle lasts longer, as is usual, an intermediary stage which I term a "superposition" may occur (Figs. 173 and 176). In the respiratory tracing of Fig. 173, one can not only see the gradually intensifying movements which correspond to the rhythm and the phases of the sucking movements, but one can also distinguish initial intervening smaller waves which likewise show the rhythm of the sucking movements (but not of the respiratory movements) and later disappear. I interpret these latter as remainders of the activity of the respiratory center, which has already been brought under the influence of the sucking center and therefore functions in its rhythm, though not in its phases. Figure 176 shows essentially the same, except that the sucking center is not yet victorious.

The superposition is based on the following diagram:

4. *Superposition of the directed respiratory center by the directing sucking center*

Rhythmic sucking center → Motor sucking center
Rhythmic respiratory center → Motor respiratory center

von Holst's "superposition" should not be confused with this superposition. In his superposition, one (the dominant) of the two rhythms of different frequency influences the other (the dependent), so that the dependent rhythm shows periods or undulations. I did not succeed in demonstrating a superposition such as described by von Holst in the infant.

The respiratory rhythm frequently shows still other changes during the disrupted period (for example, Figs. 175 and 177). If no rhythmic movements have appeared I prefer not to hazard interpretation of the irregularity, for the experimental conditions are far more difficult to survey in intact infants than in decerebrate fish. It is possible that the responsiveness of the motor respiratory center varies. In any case the graphs which we obtained have a great similarity to pulse recordings; for example, in the second half of Fig. 176 there is a clearly visible bigeminy.

The next task was to investigate what happened to the respiratory movements after the sucking movements ceased. Figure 175 shows the influence which the onset and the disappearance of the sucking movements exert on the respiratory rhythm. During the sucking arrest the respiration is comparatively slow at the beginning. The onset of the faster sucking movements causes a respiratory arrest at first and then a disrupted period during which the sucking rhythm can already be faintly noted. After the sucking movements have ceased, the respiration becomes regular although initially it still maintains the faster rhythm of the extinguished sucking movements. Finally, the original slower rhythm of respiration which had existed at the beginning of the recording before the onset of the sucking movements again predominates.

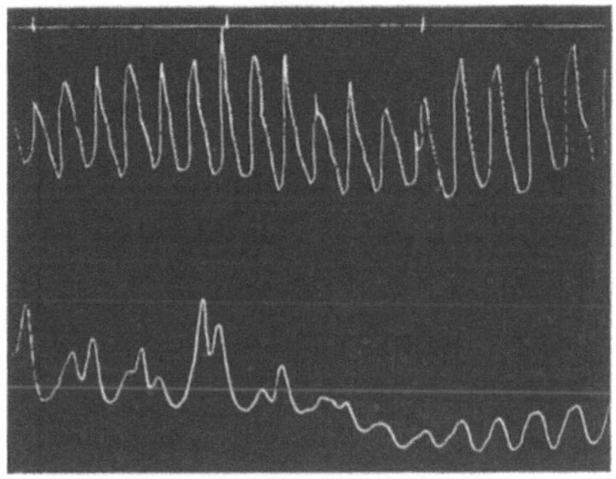

Fig. 177. Upper tracing: Sucking movements; lower tracing: respiratory movements (time in 5-second intervals). Disrupted period then direction of respiration by the sucking center.

I term this intracentral process, which I have frequently observed, the "aftereffect." There are two possible ways of illustrating it diagramatically:

5. *Aftereffect of the sucking center on the respiratory center*

a. Rhythmic sucking center   Motor sucking center
↓
Rhythmic respiratory center → Motor respiratory center

b. Rhythmic sucking center ↘ Motor sucking center
Rhythmic respiratory center   Motor respiratory center

The aftereffect may last even longer than two or three breaths and thus be misleading; it is wrong to expect the original respiratory rhythm immediately after the cessation of the sucking movements while in reality the respiration still continues in the sucking rhythm. The original rhythm of the respiratory center can thus be better demonstrated before the onset of sucking movements than after their disappearance. Not infrequently I made the striking observation that the respiratory graph reflected the sucking rhythm only indistinctly during the disrupted period but distinctly after the disappearance of the sucking movements.

The basic rhythm of the respiration was slower than the rhythm of the sucking movements (Figs. 173–177); the time ratio was approximately 2:3 or 3:4. This ratio was obtained only when the infant was at complete rest. It was absent when the infant, as was often the case, was just beginning to cry and was therefore breathing more rapidly than usual.

According to VON HOLST, two rhythms may influence each other. However, I never succeeded in demonstrating an influence of the respiratory center on the sucking center, however distinct the opposite phenomenon was. Frequently the sucking rhythm in itself became more or less irregular; this finding is not shown on the graphs reproduced here since these, as stated before, were chosen for particular reasons. In any case, the irregularities could not be related with certainty to the respiration.

I speak of a "block" when the sucking and respiratory centers function simultaneously but independently of each other. The more completely the two centers are independent of each other, the more complete is the block.

A complete block, in which the rhythmic sucking center directs only sucking movements and — independently — the rhythmic respiratory center directs only respiratory movements, can be illustrated diagramatically in the following manner:

6. *Complete block*

Rhythmic sucking center ——➤ Motor sucking center
Rhythmic respiratory center —➤ Motor respiratory center

I was never able to demonstrate this complete block during the food intake of a normal infant. In sickness, however, the relation of the two rhythms to one another is easily disturbed. There is very often a motor restlessness which absolutely prevents the appearance of a recognizable rhythm or allows its appearance only for a short time. Rhythms which can be related to each other are observed rarely and only for short periods. The mutual relations frequently change during one and the same feeding, but if enough graphs are reviewed the same law can be recognized again and again. The findings in normal infants and in infants suffering from pneumonia or a nutritional disturbance supplement each other. As can be seen in the graphs reproduced, the respiratory rate was always considerably increased in pneumonia.

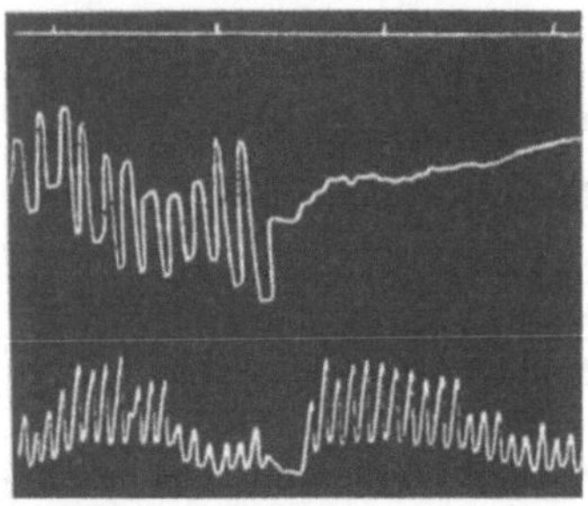

Fig. 178. Block between sucking center (upper tracing) and respiratory center (lower tracing) in pneumonia (time in 5-second intervals). Sucking and respiratory movements are independent of each other.

Figure 178 shows the greatly accelerated respiratory movements independent of the slower sucking movements. Their rhythm remains unchanged even after the sucking movements stop. A complete block exists.

Whereas the sucking and respiratory rates in the healthy infant are in a ratio of 1:1 or 2:1, and perhaps occasionally 3:1, the latter two ratios are decidedly more frequent in the sick infant. If other ratios occur (such as 3:2 or 5:3), the question of whether an interdependence exists at all must remain unanswered. During sickness the duration of the individual respiratory and sucking movements often varies so greatly that exact measurements and comparisons are not possible. Even if they are independent of each other, a definite ratio could be erroneously assumed because of the unavoidable uncertainty of measurements. Therefore, I regard all ratios in which a dominance cannot be definitely interpreted as blocks without attempting to decide whether dependency may still occasionally exist. I did not find a block of the kind described in normal infants, but only in sick children.

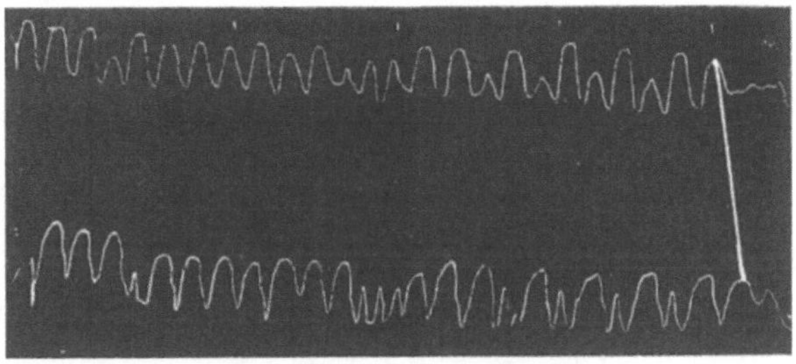

Fig. 179. Dominant sucking movements (upper tracing) imposing their rhythm on the respiration (lower tracing) in pneumonia. The smaller sucking movements are not able to do this, so that the original faster respiratory rhythm again emerges. Direction of respiration by the sucking center alternating with block.

In order to determine which center was dominant, I used the period of onset of the sucking movements in normal infants. In sick children this movement cannot be recorded often enough without disturbances because they are too restless. In lieu of this another phenomenon — which also occurs, though decidedly less frequently, in the normal infant — must be used: If the infant sucks without pause the individual successive sucking movements vary in height and width, sometimes in regular and sometimes in irregular alternation (Fig. 179). In other words: A sucking disturbance in which the sucking center still emits stimuli in a fairly rhythmic succession but of different intensity and duration occurs. Dependent on this intensity the influence on the respiratory center also changes, so that the influence of different excitations can be observed in rapid succession in the same graph. Figure 126 (see p. 341) shows a corresponding incomplete block in the respiratory center: The usual breaths are not connected with the movements of the lower jaw, but the interspersed gasping movements are.

Thus, in Fig. 179 (pneumonia) longer lasting high and shorter lasting low waves can be distinguished in the sucking graph. One respiratory movement corresponds to each high wave. In this instance there is physiological control of the respiratory center by the sucking center. In the lower waves of the sucking graph the excitation evidently does not suffice to influence the respiration; rather, the faster original rhythm of the respiratory movements can always be perceived. In this rhythm approximately 108 respiratory movements correspond to 72 sucking movements per minute. The sucking center thus directs the respiratory center even in a strong excitatory state, although it emits its stimuli in a slower rhythm than does the respiratory center.

This proves that a slower rhythm may direct a faster one and disproves BETHE's view (1938), according to which the rhythm of a nervous system is indubitably determined by its fastest component, provided that the excitation is sufficiently strong. If the dependent part is not able to follow, it is supposed to respond with a partial rhythm and — during the transition to this pattern — with the development of periodicity. Accordingly, it should be inconceivable that a slower central rhythm should prevail over a faster rhythm. But the excitability may vary greatly and the location of the fastest rhythm may change and with it a shifting of the whole system may take place.

It remains uncertain, of course, how far knowledge which we acquire from the freely beating heart or from the medusa can be applied to the medulla oblongata of the young

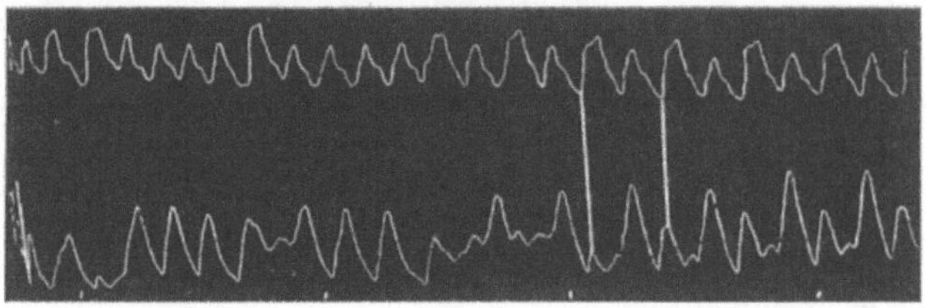

Fig. 180. Alimentary disorder. Alternating block and inhibitory sucking movements (upper tracing) which are characterized by their height and width. The inhibition develops at the beginning of the inhibitory sucking movement and deflects the ascending part of the respiratory movement (lower tracing) on every peak. After the termination of the inhibition a second smaller wave occurs in the respiration as an expression of the remaining excitation in the automatic-rhythmic partial center of the respiratory center (delayed deviation).

infant. But when in a normal, quietly drinking infant the faster functioning sucking center imposes its rhythm on the slower acting respiratory center, this cannot be attributed only to the superior speed of the sucking center. When the respiratory rhythm has become faster than the sucking rhythm because of pneumonia, the dominant role of the sucking center is not disturbed by this change.

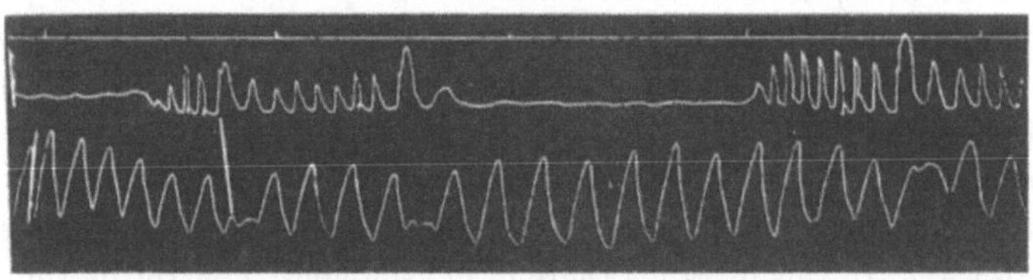

Fig. 181. Pylorospasm prior to operation. Rhythm of the respiratory movement (lower tracing) not visibly influenced by the onset of sucking movement (upper tracing). Respiration is inhibited three times by larger sucking movements.

According to Stirnimann (1947), it is more probable that the respiratory center determines the rhythm because it is biologically more important and functions continuously, while the sucking and swallowing center is active only for a shorter period of time. Since any inhibition of respiration causes a feeling of depression and a certain anxiety, the infant is said to try to avoid this condition. These considerations are clearly disproved by the observations reported here. In this connection we must also remember the "law of the readier responsiveness of the later genetic acquisition" (von Monakow and Mourge, 1930; see p. 332).

In general, the ascendancy of the sucking center can be easily fitted into a broader concept: The sucking center is younger than the respiratory center. Phylogenetically the mammals appear as the last among the vertebrates. This holds for ontogenesis, too, as observation of very premature infants demonstrates. At a birth weight below 1000 gm the sucking center is usually so weak that the infants must be tube fed. But the respiratory centers in these children function at least to a degree that respiration can be somehow maintained. Therefore, the phylogenetically and ontogenetically younger sucking center is functionally dominant over the respiratory center, corresponding to the level by level structuring of human cerebral function.

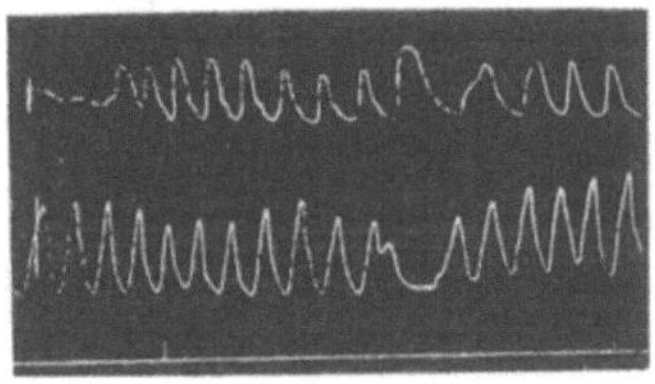

Fig. 182. Pneumonia. Sucking rhythm (upper tracing) and respiratory rhythm (lower tracing) are independent of each other: block between sucking and respiratory centers. Only the strongest and most protracted sucking movement inhibits the respiration.

In a weak excitatory state the sucking center cannot direct the respiratory center; rather, both centers function independently of each other. In this case, a temporary block has occurred. One can deduce from this that the respiratory center always tries to be independent and is prevented from this only by a sufficiently strong excitatory state of the sucking center.

Figure 182 (pneumonia) and Figs. 180, 181, and 183 (alimentary disorders) present a completely different picture; waves of different height and duration appear in the sucking graph. But in the respiratory tracing, inhibitions dependent on the form of the sucking movement appear.

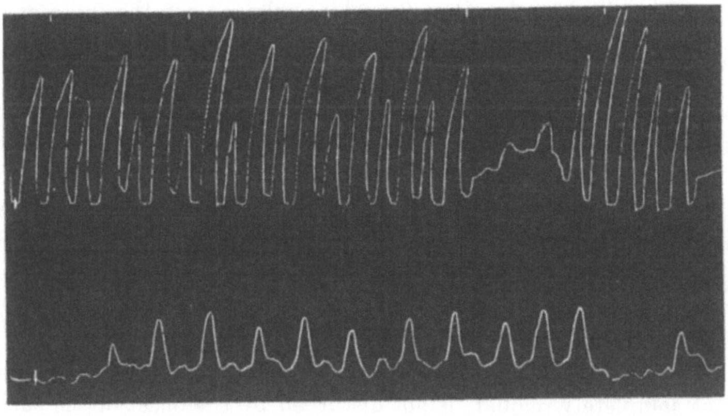

Fig. 183. Alimentary disorder. The bigger and more protracted sucking movements (upper tracing) inhibit respiration (lower tracing); the smaller direct it. In the middle of the graph big and small sucking movements alternate regularly; accordingly, every second respiration is omitted. At the end pendular induction is indicated.

To avoid misunderstandings, it is necessary to explain in more detail the concept of inhibition as used here. I agree with Pavlov that inhibition is not only the absence of an excitation but an active process which is opposed to excitation in the same manner as negative electricity is to positive electricity. Inhibition can be recognized through the extinction (sometimes only the decrease) of an existing excitation. It is unknown what chemical-physical process lies at the basis of inhibition and how far the different inhibitory patterns which we observe clinically represent the identical physiological phenomena.

At what moment does the inhibition which radiates from the sucking center to the respiratory center become effective? The answer can be seen in Figs. 180 and 181: In Fig. 180 we see breaths of approximately equal height separated from each other by irregular intervals. These respiratory movements do not have simple relations to the sucking movements; therefore, a block exists. Between these movements there appear low breaths of various heights, whose rise is suddenly interrupted at some point. These deflections always correspond to inhibitory sucking movements, which are characterized by their height and width. Thus, we are really dealing with inhibitions which suddenly extinguish the available energy. If we now try to determine the segment of the sucking movement at which the respiratory wave has been deflected, we always find the definitely recognizable moment when the inhibitory sucking movement begins. The same can be seen in Fig. 181, and in many other graphs not reproduced here.

The inhibition evidently lasts as long as the descending part of the inhibited breath. According to an estimate — available graphs are recorded too rapidly for exact measurements — the respiration is inhibited between onset and the maximum height of the inhibitory sucking movement, i.e., as long as the lower jaw is moved downward.

Which part of the respiratory center, the automatic-rhythmic or the motor partial center, is seized by the inhibition which originates in the automatic-rhythmic sucking center? Figure 180 gives the answer. As has been shown, the inhibition starts at the onset of the excitation of the sucking center and lasts as long as the ascending part of the sucking wave. When the inhibition of the respiratory movement is terminated, a second shorter wave occurs in the second half of the respiratory graph. This is because the automatic-rhythmic part of the respiratory center, which had just been inhibited, discharges the balance of its excitation after the disinhibition; this occurs somewhat belatedly. The process, which I call "delayed deviation," is not always so clearly visible as in Fig. 180 and may occasionally be absent. The distinctness may vary in the same tracing.

Despite all external differences, Fig. 181 (pylorospasm) presents essentially the same picture as does Fig. 180: The rhythm of the respiratory movement, which in itself is not completely regular, is not disturbed by the onset of the sucking movements, which occurs twice. A block exists. Sucking movements which surpass the others in height and duration appear three times. Each time the respiratory movements are inhibited; the ascending part of the sucking movement again corresponds to the inhibition of the respiratory center. The delayed deviation is only hinted at in this graph.

The automatic-rhythmic part of the sucking center is thus able to send simultaneously an excitation to its own motor partial center and an inhibition to the automatic-rhythmic portion of the respiratory center. The inhibition which reaches the respiratory center extinguishes the excitation originating in that center except for a slight surplus which still appears as delayed deviation after the disinhibition.

In another form of inhibition of the respiratory center by the sucking center, the inhibition of the respiratory center and the excitation of the sucking center do not coincide exactly. Thus, the block in Fig. 182 (pneumonia) is interrupted by an inhibition which is associated with a bigger and longer lasting sucking movement. The beginning of this inhibition does not coincide with the onset of the excitation in the respiratory center as it did in Figs. 180 and 181. This kind of inhibition, like the form previously described, may occur in a multiple pattern.

The ratio between sucking and respiratory movements may also register 2:1 or 3:1, because of the alternation of inhibitory and directing sucking movements in the sucking graph. In this case arrests occur in the respiratory graph, as can be seen in Fig. 183. I did not observe the combination of inhibition and direction in pneumonia, which may be just coincidence, but I saw it in alimentary disorders. The inhibitory sucking movements are bigger and last longer than the directing movements (Fig. 183).

In summarizing these findings, we note that during a sucking movement the sucking center has three possible ways of relating to respiration: It can (1) leave it uninfluenced, (2) direct it, or (3) inhibit it. I assume that we are dealing here with three different excitatory stages of the sucking center. It has always been stressed that the individual sucking movements, when exerting their influence on the respiration in different ways, also differ from each other in height and width according to the intensity and duration of the excitation. However, I never found three different excitatory states in the same tracing, but at the most two which followed one another at times regularly, at times irregularly. For a comparison of all three excitatory conditions, three different curve forms, each with two conditions of excitation, are needed.

I. If inhibition and direction alternate in the same tracing, the inhibitory sucking movements are bigger and broader than the directing movements (Fig. 183). In some exceptional cases no significant differences between them could be recognized. Perhaps the recording rubber bag was improperly located. The opposite relation was never observed.

II. Direction and block alternate in Fig. 179. The directing sucking movements are always bigger and broader than those without influence.

III. If the sucking movements cause inhibition and block in the respiratory graph, the inhibitory sucking movements are always bigger and broader than those without influence (Figs. 180–182).

The different excitatory stages of the sucking center supplement each other to establish a sequence: The inhibitory sucking movements are bigger and broader than the directing movements (I) and these are again bigger and broader than those without influence (II). If this order of size is correct, we should expect that the inhibitory sucking movements will also be bigger and broader than those without influence. This is in fact the case (III).

The excitation of the sucking center is thus strongest when the sucking movement inhibits the respiration, is of medium strength when it directs the respiration, and is lowest when it does not change the respiration. The excitations can thus be compared with each other even if they have not been measured according to an absolute scale.

Of two excitations of varying intensity, the stronger will be of longer duration than the weaker (Figs. 179–181). One could be tempted to use the duration as an absolute criterion, but the available instruments did not allow accurate measurment. Therefore, the question of whether the duration of the stronger excitation lasts longer only when compared with the weaker or does so absolutely cannot be answered.

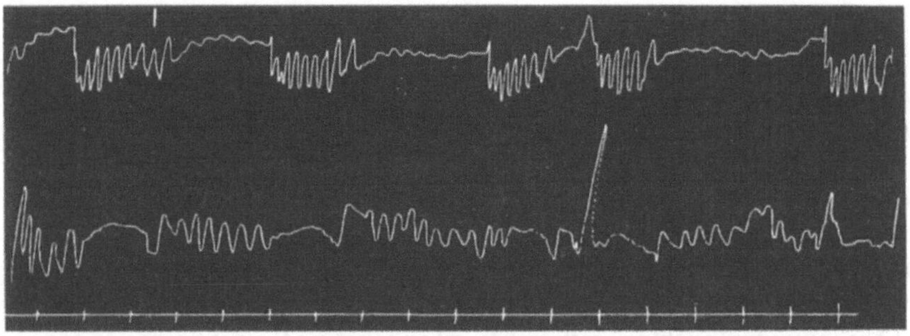

Fig. 184. Pendular induction between respiratory and sucking centers (time in 5-second intervals). During sucking movements (upper tracing) the respiratory movements (lower tracing) stop, and vice versa. The low waves of the sucking graph during the arrest of the sucking movements are respiratory movements which were transmitted from the thorax to the rubber bag below the chin.

In any case the stronger excitation can last considerably longer than the weaker; in Figs. 181 and 182 it lasts about twice as long.

Since the sucking and respiratory centers border on each other in the brain stem of a young infant, and since the existing nervous energy is easily exhausted in the conditions of prematurity and illness, the two rhythms will easily establish relations to each other. In slight excitation, the sucking center will not influence the respiratory center; both rhythms remain independent of each other. But in more intense excitation the sucking center will overpower the respiratory center and force its own rhythm upon it. This direction of the respiratory center by the sucking center can be found especially in healthy infants and represents, if the crossing of the respiratory and alimentary passages in the pharynx is considered, the best solution for concerted action. In the most intense excitation the sucking center will seize all the nervous energy in the brain stem for itself: This energy is especially lowered in the brain stem of the sick infant; thus, there is no nervous energy left for the function of the respiratory center and it will be inhibited. Since we do not have absolute but only relative values for the excitatory strength of the sucking center, this "strength" of the sucking center may consist of a weakness of the respiratory center.

If for the sake of brevity we use the term "sucking disturbance" for those relations which develop between the two centers other than the direction of the respiratory center by the sucking center, the question arises to what degree this represents a pathological process. It is more frequently found in sick than in healthy infants, but can also occur in strong, apparently well nursing infants. We are dealing in part with the processes described here and in part with others which are difficult to interpret. Thus, it would be wrong to relate the dominance of the sucking rhythm only to the healthy infant and the sucking disturbance only to the sick child.

Another sucking disturbance *is pendular induction between the sucking and respiratory centers* [1938 *(75)*]. I never found it, although I looked for it, in healthy

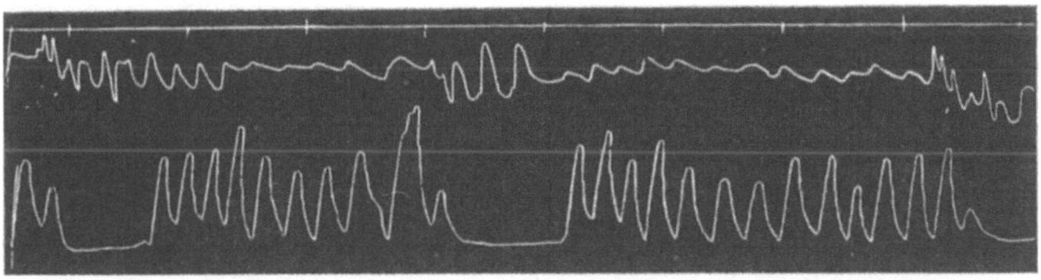

Fig. 185. Pendular induction between respiratory and sucking centers (time in 5-second intervals). During the sucking movements (upper tracing) the respiratory movements (lower tracing) stop, and vice versa; overlapping occurs easily. The same interpretation as in Fig. 184 applies to the small waves in the sucking graph during arrest of the sucking movements.

infants born at term, not even at the end of their feeding when they sucked periodically, but I found it only in weak premature infants or in sick infants. As long as these infants are able to drink regularly, one or two full sucking movements will correspond to one complete respiratory excursion just as in mature healthy infants. After a shorter or longer time — in some children right at the start — periodic sucking movements (as previously described; see p. 320) will appear because of the low excitatory state of the sucking center. The respiration may be so regular during this period that no definite connection with the sucking movements can be recognized. But a very distinct relation may become evident: Respiration becomes just as periodic as sucking, i.e., sucking stops during breathing and vice versa. Thus, the sucking graph in Fig. 184 approximately reproduces the timing of the respiratory tracing. In Fig. 185 a certain superposition can be seen: The beginning and the end of the sucking period correspond more or less to respirations, while its middle phase coincides with a respiratory arrest. In general the periods of respiration last longer than periods of sucking.

The usual relation between sucking and respiration has been replaced by another remarkable relation: States of activity and inactivity alternate between the sucking and respiratory centers. If one of the centers is stimulated the other is inactive, and vice versa. I call this interplay between respiratory center and sucking center "pendular induction."

Pendular induction compensates for the lack of nervous energy which is characteristic of the lowest cerebral activity. The two tasks which the central nervous system must still fulfill on its lowest barely viable level in order to maintain life are respiration and food intake. If the supply of nervous energy is not sufficient to effect uninterrupted respiratory movements, periodic respiration, in which nervous energy is saved during each respiratory arrest, appears. This form of respiration suffices to maintain life. Food intake is just as important as respiration; when lack of nervous energy exists, this also becomes periodic for the same reason. At the lowest stage of viability the nervous energy is not sufficient to supply the respiratory and sucking centers simultaneously, as it can later in life. With the aid of pendular

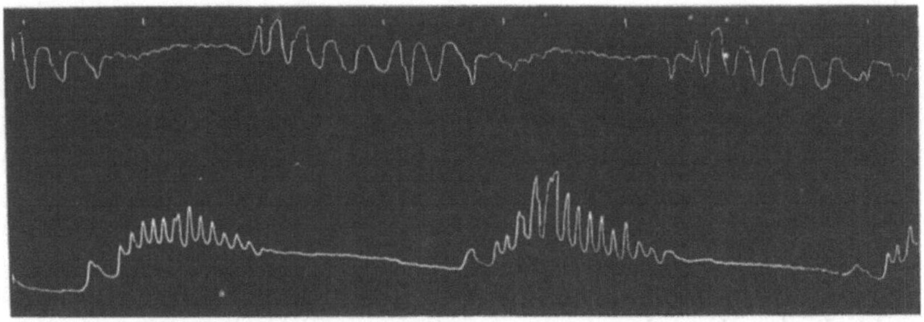

Fig. 186. Pendular induction between sucking and respiratory centers in pneumonia (time in 5-second intervals). In 5 seconds, 4 sucking movements (upper tracing) and 10 respiratory movements (lower tracing). The respiratory period always starts with one slower breath occurring just within the period of the sucking movements.

induction both vital functions can be maintained: In a rhythmic change nervous energy is saved in the sucking center during breathing and in the respiratory center during sucking.

When the respiratory rate is increased in mature infants suffering from pneumonia, pendular induction can also frequently be noted (Figs. 186 and 187). Evidently the nervous energy of these children has been lowered to the level of the premature infant, so that pendular induction must occur to maintain life. In pneumonia it can be recognized merely by clinical observation without graphic recordings: The infants suck, take some noisy breaths in between, and then suck again. The pendular induction seen in Figs. 186 and 187 differs from that in Figs. 184 and 185 in that the respiratory movements are faster than the sucking movements.

In Fig. 187 a transition to pendular induction makes the onset of this state more distinct. The formation of periods in the sucking center occurs in the usual way; however, the breaths are not inhibited, but merely decelerated. It could not be said that the respiratory rhythm was quite the same as that seen in the sucking movements, although a certain similarity may exist.

The deceleration of the respiratory movements under the influence of the sucking center represents a transition to inhibition proper, in which the respiratory movements completely disappear. This intermediate stage is usually skipped. It can still be recognized in Fig. 186 under somewhat different conditions: The period of the respiratory movements always starts with a slower breath which still coincides with the last sucking movements. Thus there is a transitory stage between excitation and inhibition during which the respiration is slowed down. The usual occurrence in pendular induction is inhibition; deceleration is the exception.

At first I regarded pendular induction as an alternation of the excitatory and inhibitory states between the sucking center and the respiratory center, which is entitled to the same rights. But actually the to and fro pendular movements occur only within the sucking center. The periodic excitation of the sucking center inhibits the respiratory center each time and the periodic decrease of excitation in the sucking

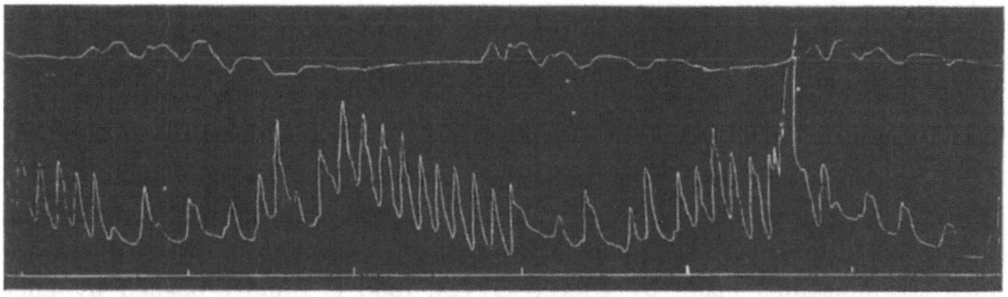

Fig. 187. Pendular induction between sucking center (upper tracing) and respiratory center (lower tracing) in pneumonia (time in 5-second intervals). Complete respiratory arrest does not occur, but only slower breaths are noted. The sucking center does not completely inhibit the respiratory center, but only decelerates its activity.

center releases the respiratory center each time. There are no plausible findings which would indicate that the respiratory center is able to inhibit the sucking center, even though the reverse relation could be recognized.

One would think that sick infants, who show a greatly increased respiratory rate even at rest, would not experience a long respiratory arrest. But just the opposite happens, as Fig. 186 (which by no means illustrates the longest respiratory arrest) proves. Thus, the "turning blue" which occurs during feeding in infants suffering from pneumonia can be explained.

Is the faculty of the sucking center for directing the respiratory center innate or acquired after birth? To answer this question I examined newborn infants during their first food intake. In accordance with the composition of my patient population, mainly premature infants or infants who were admitted because of other diseases were in this group. If the concerted action of the functions was observed during the first food intake (from the bottle), regular rhythmic sucking movements did not always occur during the short observation period (the infants drank only a few grams). In general it was difficult to induce the newborn infants to suck; often they were very restless. Thus, during this short feeding frequently only a "disturbance period" with irregular sucking and respiratory movements occurred; this period did not allow us to recognize any relations. But in other cases, among them even in a premature 9-hour-old newborn infant weighing 1440 gm, sucking and respiratory movements occurred at least transitorily in the same rhythm. Because of the restlessness of the newborn infants and the short observation period, the leading role of the sucking center could not be established with certainty; but because of the findings in older infants this is a probability.

HALVERSON (1944) was also able to demonstrate a transitory coordination between respiratory and sucking movements during the first meal of a newborn infant 8 hours old; in other newborn infants he found it only later. According to a later publication (1946), he observed the occurrence of food intake and respiration without mutual disturbance in two premature infants during their first feeding, an indication that good coordination may be possible even prior to the pregnancy's completion at term.

Breathing and Swallowing

(21, 38, 93, 11a, 24a)

The transportation of food from the oral cavity to the stomach is called "swallowing." The physiology and pathology of this process were described in detail by CATEL (1938); it occurs in the adult in two phases. According to the cinematoradiographic examinations made by KRAUS (1912), food is rapidly pushed by buccopharyngeal swallowing from the oral cavity through the pharynx and a short distance into the esophagus. During the transport of the food the ostium esophageum, a tonic muscle in the region of the seventh cervical and first dorsal vertebrae, opens; usually it closes the entrance to the esophagus. The second phase of swallowing, esophageal swallowing, consists of peristaltic waves which are triggered by the introduction of the food and which last for a time. The ostium esophageum is closed to prevent the regurgitation of food.

The adult is unable to swallow and breathe simultaneously; he must always interrupt his breathing while he swallows, because his respiratory and alimentary passages cross in his pharynx. CLARK's (1921) investigations indicate that the duration of the respiratory arrest during swallowing is approximately 1.5 seconds and that it occurs more often during exhalation than during inhalation. After the respiratory arrest the respiratory movement is usually continued from the point at which it was interrupted.

On the assumption that the anatomic structure in the infant was identical to that of the adult, it was thought that the infant was also unable to swallow and breathe simultaneously (K. VIERORDT, 1877). BASCH (1893) tried to prove this hypothesis by simultaneously recording sucking and breathing. When the infant sucked and breathed regularly, BASCH found a respiratory arrest at certain intervals at the height of expiration coinciding with a low wave. He regarded this wave as an indication of a swallowing movement which necessarily interrupted respiration.

According to ANDRÉ-THOMAS (1952, p. 125), the swallowing movement of the sucking infant temporarily represses the respiration; according to HASSE (1905), VALLOIS and FLEIG (1905), CZERNY (in his lectures), FREUDENBERG (1931), and O. SCHNEIDER (1939), however, infants are able to swallow and breathe simultaneously. HASSE (1905) states that the infant's breathing is not interrupted by swallowing because the food does not slide over the laryngeal entrance, as it does in the adult, but flows on both sides of the elevated larynx into the esophagus.

If an infant is fluoroscoped from the side while he drinks from the bottle, it not infrequently appears as if the fluid is squirted in an arc from the oral cavity into the esophagus. However, LASSRICH (1959) was able to show by cinematoradiographic examinations that the sucking act was completely developed in all its details despite its short duration; the ostium esophageum is opened during swallowing. LASSRICH (p. 265) reports, however, that not every sucking action is followed by a swallowing action; this is contrary to my findings and also to his own statements elsewhere. Only at the start of food intake does a swallowing movement follow each sucking movement; later two or more sucking movements occur for every swallowing movement.

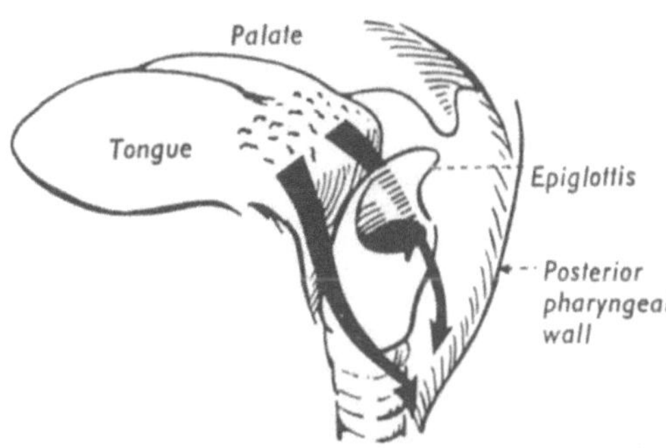

Fig. 188. Swallowing path crossing the respiratory tract. Oblique posterior view. [According to DITTRICH *(24a)*]

During buccopharyngeal swallowing the food to be swallowed has to cross the respiratory passages (Fig. 188). The infant apparently does not have to interrupt his breathing during this event, as does the adult, in whom swallowing is connected with a short expiratory arrest (CLARK, 1921). This characteristic of the infant is related to the higher location of his larynx, which during swallowing is also raised by contraction of the muscles so that the entrance of the larynx lies above the inferior posterior border of the velum palatinum. According to cinematoradiographic examinations (DITTRICH, GRUNER, and HERBST), the epiglottis is placed like a breakwater in front of the larynx during buccopharyngeal swallowing and parts the stream of the material to be swallowed, which passes it on the sides (Fig. 189). Only very big food masses can spill over the epiglottis [DITTRICH, 1957 *(24a)*]. ARDRAN and KEMP (1959) obtained the same findings with the same method.

The baby marsupial, whose larynx reaches like a tube up to the nasopharyngeal region and the choanae, swallows in the same manner. The milk also flows at each side of the laryngeal tube and arrives in the pharynx without disturbing the respiration. The crossing of trachea and esophagus is so perfect that aspiration cannot occur (WUNDER, 1937; HEDINGER, 1953).

When the infant drinks quietly and continuously from the bottle he is, as simple observation shows, able to breathe without interruption during drinking and therefore also just as well during swallowing. He can swallow without having to interrupt his breathing visibly each time as the adult does.

HALVERSON (1944) also stated that sucking and respiratory movements can be executed simultaneously. The sucking movement occurs, as it does in the adult, during inspiration or expiration. He also finds the swallowing movement in the infant to be of short duration. The milk rushes through the pharynx so rapidly that the entrance of the larynx is closed, and the respiration is thus interrupted, for only a fraction of a second. This means that a respiratory arrest is necessary for swallowing. This pause is easily overlooked because of its shortness when the swallowing occurs during one of the normal respiratory arrests, as is the case in the strongly sucking infant. If the swallowing occurs during inspiration or expiration, the respiratory arrest is recognizable as a short pause in the respiratory graph.

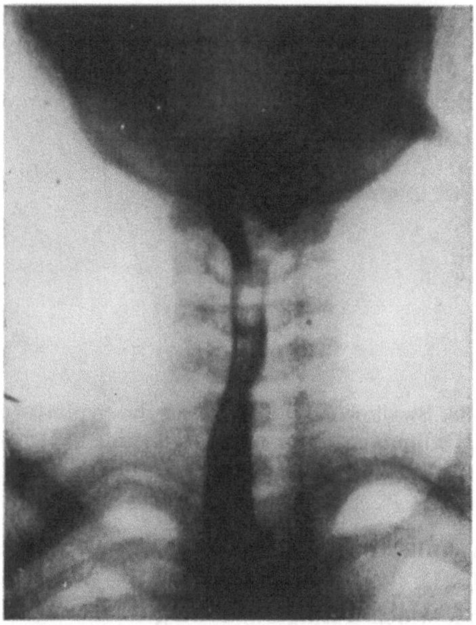

Fig. 189. Swallowing act (frame of a film). Fork formation of the contrast filling because of its flowing around the laryngeal entrance. Anterior view. [According to DITTRICH *(24a)*]

With the aid of high speed cinematoradiography, LASSRICH (1959) found the respiratory passages of bottle-fed infants to be closed by the swallowing act for 8/24–10/24 of a second. This short interruption will escape simple observation. On the basis of their cinematoradiographic examinations, ARDRAN, KEMP, and LIND (1958) dispute the assumption that the infant can drink at the breast without interrupting his breathing. However, there is little probability that the breast-fed infant behaves differently from the bottle-fed.

Premature and sick infants are frequently incapable of rhythmic food intake. Aside from the sucking and breathing disorders described already, a swallowing disturbance may also occur. In order to record the swallowing movements simultaneously with the respiratory and sucking graphs (they cannot be obtained with the aid of rubber bags), I had an investigator feel them with his hand at the infant's larynx and transmit them to the writing drum by movements of a lever. It is in the nature of a recording produced in a manner which is not purely mechanical such as this that the movements, when compared with the others, will always be recorded with some delay. HALVERSON (1946) determined the moment of swallowing by auscultation and recorded it graphically.

I found that if swallowing is weak, respiration can be inhibited by the swallowing act. In Fig. 190, respiratory, sucking, and swallowing movements are simultaneously recorded. Respiration was accelerated (84 breaths per minute) during rest before the experiment in this infant, who suffered from pneumonia. As can be clearly seen in the graph, respiration is interrupted during most of the swallowing movements. But there are exceptions; e.g., the fourth and seventh swallowing movements have no influence at all on respiration. This tracing has been chosen to show both possibilities at the same time. In the same graph the respiration is usually either not at all inhibited or totally inhibited by swallowing.

In Fig. 190 the respiration is interrupted for about 1.5 seconds. This is the same value which CLARK (1921) found in adults.

While the swallowing movements in Fig. 190 are separated by fairly equal intervals, their effect on respiration varies. The sucking movements in this graph do not

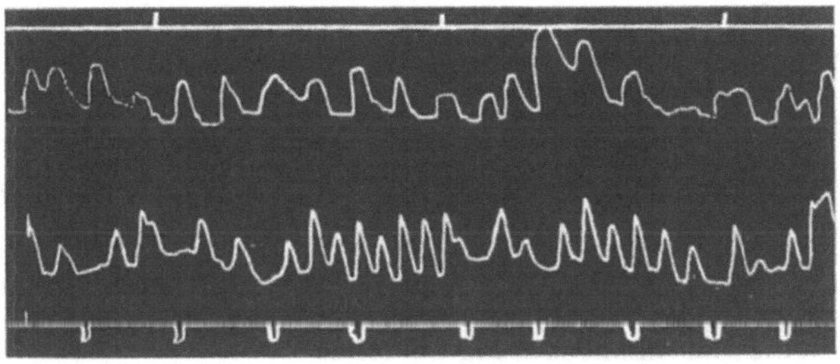

Fig. 190. Full-term infant with pneumonia; 86 breaths per minute at rest. Upper tracing: sucking movements; center tracing: respiratory movements; lower tracing: swallowing movements (time in 5-second intervals). The swallowing movements (with the exception of the fourth and seventh) inhibit respiration.

seem to have any relation to the respiratory and swallowing movements. A control has not been established; a block exists between respiratory and sucking centers (see p. 428).

In Fig. 191 (premature infant) there is periodic breathing. The respiratory arrest after each swallowing movement lasts 5 to 7 seconds. It is an old experience that

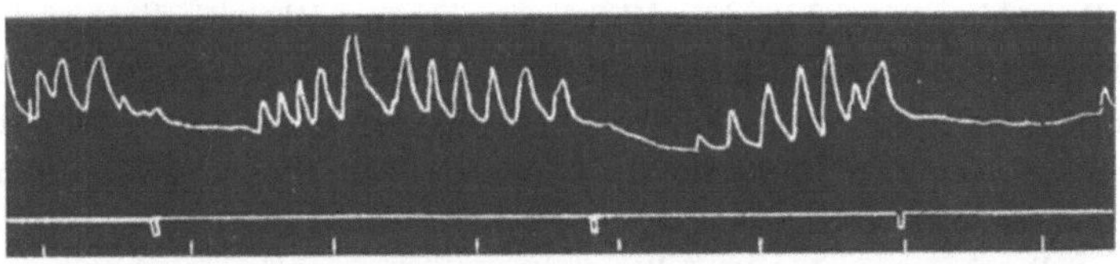

Fig. 191. Respiration (upper tracing) and swallowing (lower tracing) in a premature infant (time in 5-second intervals). Periodic breathing. Extended respiratory arrest following each swallowing movement.

the respiration of premature infants may easily be inhibited after some nervous exertion such as vomiting, coughing spells, or fright reactions. Through the increased activity in another area the immature respiratory center loses so much nervous energy that it can no longer function. This inability to restore spent nervous energy immediately is a particular characteristic of immature cerebral function.

In Fig. 192, which was obtained in the same infant on the same day, the respiratory arrest after the swallowing movement lasts considerably longer, i.e., about 12 seconds. The end of this respiratory arrest coincides with another swallowing movement, which, however, cannot produce a further inhibition. Evidently the excitability of the respiratory center has been so much increased by the long respiratory arrest and the oxygen decrease in the blood thus produced that a renewed inhibition originating in the swallowing center cannot be effective.

According to HALVERSON (1944), the relations between respiration, sucking, and swallowing are not the same in all infants and can change during the same feeding. They depend on the effectiveness of the sucking movements and the amount of effort that is required to draw the milk from the breast. If a swallowing movement occurs after each sucking movement during slow food intake, then this sucking movement has sufficient strength to draw enough milk; the swallowing movement transports the milk over the larynx entrance at a favorable moment. If this is the case, we can expect a constant time relation between the sucking-swallowing sequence and respiration. Good coordination exists when the infant is relaxed and drinks quietly. Motor restlessness makes food intake more difficult.

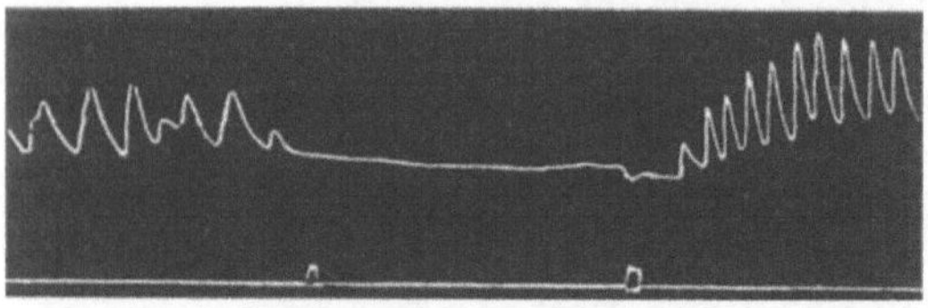

Fig. 192. Respiration (upper tracing) and swallowing (lower tracing) in the same infant as in Fig. 191 (time in 5-second intervals). After the first swallowing movement, long-lasting respiratory arrest which is not prolonged by the second swallowing movement.

WILLI and LÜTHY (1952) described a mature newborn infant who exhibited long periods of respiratory arrest after swallowing. At the age of four days attacks of apnea occurred after swallowing. During drinking respiration ceased with the first swallowing movement; the respiratory center resumed its function only after a long respiratory arrest associated with unconsciousness, seizures, twitches, and cyanosis. Breathing started again with gasping, then with periodic, and finally with regular respiration. The inhibitions of the respiratory center originating in the swallowing center were unusually strong. At autopsy it was found that the respiratory center was severely damaged by a hemorrhage which was probably caused by birth trauma.

While swallowing movements follow every or every other sucking movement in healthy infants who drink well — I have not examined infants who drink arrhythmically — premature infants suffering from a weakness of the swallowing process behave differently. Shortly after the mouth is filled, isolated swallowing movements appear. They occur at increasingly longer intervals and then occasionally in greater frequency again. I have not seen any premature infants who showed no initial swallowing movements.

Signs of a swallowing disorder which usually occur after a sucking disturbance but under the same conditions are: prolonged delay of swallowing movements after the stimulation of a filled mouth; increased duration of mouth filling; increasingly longer intervals between the individual swallowing movements until they finally fade away; flowing of food from the mouth and sometimes also from the nose.

The prognosis in a fully developed swallowing disorder is decidedly poor and the danger of aspiration is great. Therefore, premature infants suffering from swallowing weakness should only be tube fed. Instillation of nourishment into the nose of the weakly sucking infant is absolutely contraindicated. This would considerably increase the danger of aspiration, since the nose represents the only natural respiratory passage of the infant (see p. 447).

GARSCHE (1950) judged the excitability of the swallowing center with the aid of an air stimulus, which affects the oral region and evokes the snap-swallow reflex

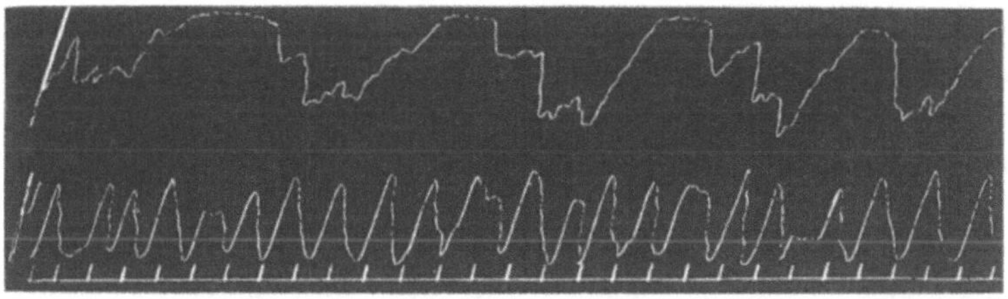

Fig. 193. Respiration (lower tracing) and peristalsis of the esophagus (upper tracing) are independent of each other (time in seconds).

(jaw-tongue reflex and swallowing movement). With this method the relation between the swallowing and respiratory centers can be determined as follows: In the mature newborn infant the air stimulus interrupts the respiration for the duration of the swallowing act. It is a sign of maturity if the infant resumes breathing afterward; this usually occurs in a rhythmic form. There are two possibilities in the premature infant: Either the respiration is inhibited for some length of time by the excitation of the swallowing center, or the respiration which occurs afterward is not regular, but periodic or gasping.

In healthy premature infants GARSCHE found short swallowing movements which occurred immediately after the stimulation; in weak premature infants the movements were delayed and had a slow and worm-like course. The slow course of the swallowing reflex was a preagonal sign which at times appeared even several days before death. This reflex can usually still be evoked when all others have ceased to exist.

LOTH [1952 *(11a)*] examined the gag reflex on touching the posterior wall of the pharynx in more than 100 infants, mostly prematures, and found it absent in only one newborn. The reflex was lively or increased even in severe illness until shortly before death and could still be observed in floor-of-the-mouth or incipient gasping respiration. It was temporarily absent in 3 severely damaged young premature infants.

The association of a strong gag reflex with the inability to suck and to swallow was so impressive that LOTH was inclined to dispute the position of the swallowing reflex as the last reaction that can be evoked before death.

Peristalsis of the esophagus follows the buccopharyngeal swallowing movement in another rhythm. ISBERT and PEIPER [1928 *(20)*] have shown that it can be easily recorded by introducing a small rubber bag into the esophagus; this bag is hermetically attached to a thin rubber tube and picks up the pressure variations. It is difficult to induce regular breathing in infants when a rubber bag is in the esophagus, for as soon as they become restless their breathing is disturbed. However, enough recordings (Fig. 193) were obtained to prove the mutual independence of respiration and esophageal peristalsis.

Sucking and Swallowing

(38, 56, 83, 9a)

The sucking center usually begins to function only if a nipple-like object is placed into the infant's mouth, if nourishment is instilled drop by drop, or if the oral region is mechanically stimulated. In some infants these sucking movements continue long after the eliciting stimulus has ceased. Sucking movements may even occur, especially in hunger, when an external stimulus cannot be noted.

The situation is different in the swallowing center. If the infant is given a non-perforated nipple so that he cannot obtain food, some sporadic swallowing movements may appear from time to time — probably the secreted saliva is swallowed — but a regular swallowing rhythm does not occur. However, as soon as the infant obtains food through a perforated nipple, swallowing movements occur in regular sequence.

The excitatory state of the swallowing center thus depends on the appropriate external stimulus, whereas the respiratory center functions automatically, and the sucking center represents an intermediate stage.

If an infant drinking from the bottle is observed in the lateral position in front of an x-ray screen, it can be recognized that the nourishment always enters the oral cavity from the nipple while the jaws are closed. The findings which Basch (1894) and von Pfaundler (1924) obtained with the aid of a "sucking mirror" are confirmed by these observations. But as far as can be judged from fluoroscopy, food does not seem to enter the oral cavity during the lowering of the mandible, i.e., during real sucking. This view is confirmed by cinematoradiography: According to Ardran and Kemp (1959), it is only at the end of the phase during which the nipple is refilled that some milk flows into the oral cavity; Lassrich (1959) stated that the fluid leaves while the nipple is compressed.

Since the fluid passes the upper portion of the esophagus rapidly, I am inclined to think there is a sphincter-like esophageal ostium in the infant too. If food flows into the infant's oral cavity every time the chin is raised — as frequently happens — this nourishment can be transported into the esophagus in the same rhythm so that each complete sucking movement corresponds to a buccopharyngeal swallowing movement. But frequently the food is retained in the mouth until the second or third swallowing movement, is collected there, and only then squirted into the esophagus. Buccopharyngeal swallowing thus depends in its rhythm on sucking movements and really represents only their continuation.

When sucking movements and peristalsis of the esophagus are recorded simultaneously, it can be recognized that a whole group of sucking movements corresponds to a peristaltic wave of the esophagus (Fig. 194). Since in addition the sucking movements give an idea of the rhythm of buccopharyngeal swallowing, it may be further concluded that the two phases of the swallowing act do not coincide in time, as can also be noted in radiokymographic findings. Peristalsis of the esophagus is started by buccopharyngeal swallowing, but is independent of it in its rhythm. At first the individual sucking movements follow each other without pause (Fig. 194).

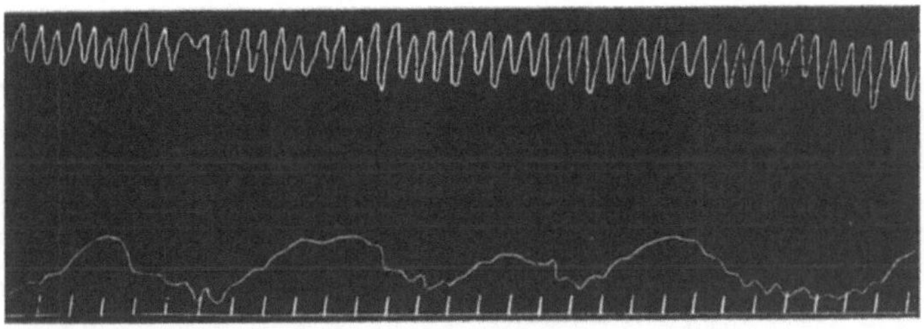

Fig. 194. Sucking movements (upper tracing) and peristalsis of the esophagus (lower tracing) are independent of each other (time in seconds).

It is thus not necessary for the sucking movements to be interrupted during swallowing.

In perfectly rhythmic sucking, swallowing, and respiration, a condition which does not always exist even in strong infants who are drinking well, the ratio of the sucking, swallowing, and respiratory movements is 1:1:1 or 2:2:1. This indicates that the centers of sucking, swallowing, and respiration work in the same rhythm. We have already demonstrated that the respiratory center is directed by the sucking center; we have still to examine how the swallowing center participates in this interplay.

For this purpose it is advisable to compare two groups of sucking movements with each other: those in which swallowing movements are absent and others that are accompanied by swallowing. These two groups must follow each other immediately, a sequence achieved in Fig. 195 by letting the infant suck first on a nonperforated nipple. As soon as the first sucking movements occur, the respiratory rhythm adapts itself to the faster sucking rhythm. The respiratory center is again directed by the sucking center. As soon as food is introduced and the first swallowing movements occur, the sucking movements abruptly become slower and stronger. The swallowing movements accept this new rhythm. I originally thought that the sucking center was directed by the swallowing center, but now I consider the possibility that we are dealing with another process. It may be that the sucking center changes its rhythm if gustatory stimuli are added, but otherwise maintains its leading position. In this case the swallowing rhythm would also depend on it.

In Fig. 195 the adjusted respiratory rhythm changes again under the influence of the decelerated sucking rhythm: After an inhibition which is sometimes incomplete and lasts altogether approximately 7 seconds, new respiratory movements which correspond to the changed sucking rhythm appear. Since the respiration is initially inhibited, this proves beyond doubt that the new rhythm is not induced by the respiratory center.

I never found a deviation from this pattern, but the following should be stressed: When infants are restless during drinking — and they often are — the relations described cannot be demonstrated. They are seen only if regular rhythms of the three centers have existed for some time.

KÖDDING [1940 *(9a)*] showed that the center for eye blinking can also be directed by the sucking center. Eye blinking and sucking movements usually occur independently of each other. But in 3 premature infants an interplay of the two centers

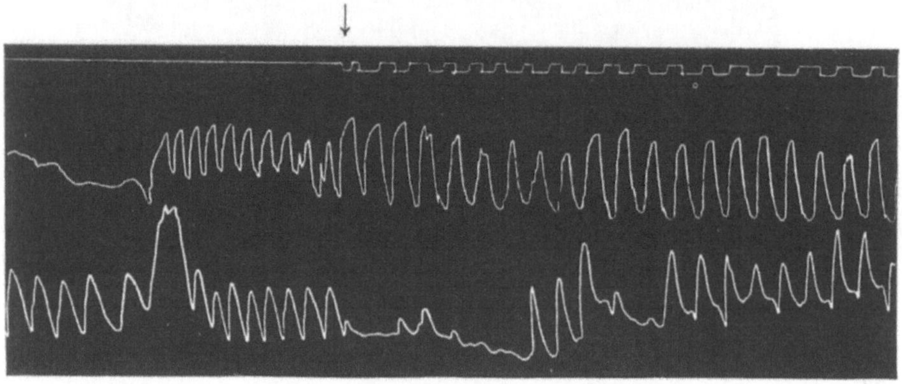

Fig. 195. Upper tracing: swallowing; center tracing: sucking; lower tracing: respiration (time in 5-second intervals). The ↓ sign indicates replacement of the nonperforated nipple with a perforated one. The initial sucking movements impose their rhythm on the respiration. When swallowing movements begin, the sucking movements become slower and stronger. The respiration changes again and follows the new sucking rhythm.

was observed: Each sucking movement was accompanied by one blink. Moreover, the number of blinks may be decreased in premature infants during sucking movements, i.e., inhibition of the blinking center may occur.

PRECHTL and SCHLEIDT (1950) found a relation between sucking movements and blinking in 7 premature infants: The eyes were closed and opened in concurrence with the sucking movements; in 2 of the premature infants the same phenomenon also appeared in the hands. PRECHTL (1953) states that the tonic palmar grasp reflex is coupled to the sucking movements in infants up to the age of four months. The evoking stimulus is the placing of the examiner's index finger across the infant's open hand. Premature infants weighing 1200 to 2500 gm grasp only during simultaneous sucking movements. If sucking movements are absent in a premature infant, the palmar grasp reflex is also absent.

Mature newborn infants weighing 3500 to 4000 gm grasp only weakly before drinking, but strongly during drinking. Toward the end of the feeding the grasping again becomes weaker; it disappears when the infant falls asleep. Some weeks later the palmar grasp reflex can be elicited without simultaneous sucking movements, but it loses its tonic qualities during sleep and also during restlessness; instead, a peculiar, repeated rhythmic opening and closing of the hand appears. During sucking movements the palmar grasp reflex is always tonic. The process is thought to be an intracentral coupling of the excitation between the sucking center and that of the palmar grasp reflex (PRECHTL, 1953).

When the sucking movements become periodic at the end of the feeding of a healthy infant, the swallowing movements become less frequent. Frequently only one swallowing movement occurs during a sucking period composed of several sucking movements. The sucking movements associated with swallowing movement may be slowed down or deepened. In pendular induction (see p. 435), the sucking center naturally cooperates with the swallowing center, and not with the respiratory center.

HALVERSON's (1944) findings are in essential agreement with these observations; in good coordination he found sucking and swallowing movements combined into a rhythmic entity to which the respiration adjusted itself more readily than to one of the individual movements. The rhythmic entity usually consisted of one sucking and one swallowing movement or of two sucking and one swallowing movement. If the sucking and swallowing movements had formed this entity, the strength and duration of the sucking movements remained uniform and one breath corresponded to each unit.

Aspiration

(62, 116, 121, 123, 127, 129, 5a)

Aspiration is the inhalation of more or less liquid material from the upper alimentary passages into the deeper respiratory tracts or the lungs. While the adult and the older child can free their air passages by strong coughing, the premature or seriously ill infant is unable to do this. Thus, he is exposed to the danger of suffocation or of aspiration pneumonia.

Aspiration is facilitated by the dorsal position, which is physiological in infancy. How much the dorsal position increases the danger of aspiration even in adults is proved by observations made at the scene of traffic accidents by CARSTENSEN and EWERWAHN (1957). They examined four severely injured victims with basal skull fractures. Each of the four victims had aspirated a great deal of blood, although immediately after arriving at the accident scenes the two physicians had turned their patients on their abdomens or on their sides.

"The decisive factor was that the victims had been flung on their backs in the accidents and had remained in this position for several minutes."

Aspiration of amniotic fluid by the newborn infant and aspiration of stomach contents by the sick infant represent real dangers. Sterile or infected amniotic fluid with or without meconium and blood can be aspirated into the lungs during birth in considerable quantity, in excess of the physiological aspiration of amniotic fluid during the fetal period (see p. 358).

The clinical signs of nonphysiological, excessive aspiration of amniotic fluid and the associated atelectasis of the lungs [PEIPER and THOMAS, 1951 *(116)*] persist for several days after birth and then slowly disappear. If extensive portions of the lungs are eliminated because of aspiration of amniotic fluid, the newborn infant usually suffocates on the first day of his life, often after only a few hours. Aspiration renders the newborn infant limp, weak, and cyanotic. Inspiratory retraction occurs simultaneously in the epigastrium parallel to the lower thoracic border, in the intercostal spaces, and in the jugulum. The respiratory rate is usually increased. During each expiration a deep sound reminiscent of the expiratory wheeze of severe pneumonia ("Knorksen"*) can be heard. Expiration is prolonged. Preinspiratory dilatation of the nostrils occurs with the "Knorksen." Sometimes a greenish-yellow fluid is vomited or foamy fluid is brought up; this is a sure sign of amniotic fluid aspiration.

Generally death of the premature infant is not caused by aspiration of amniotic fluid alone; rather, several disorders which are clinically or pathologically-anatomically demonstrable and which depend on each other are combined. Hypothermia, centrally induced respiratory, sucking, and swallowing disorders, and intracranial hemorrhages caused by birth trauma belong in this group [ESSBACH and PEIPER, 1953 *(123)*].

As autopsies repeatedly show, aspiration of vomited stomach contents represents a serious danger for premature infants and for infants with severe alimentary disorders. To avoid this danger in premature infants, we do not begin their feedings until 48 hours after birth if their birth weight is under 1000 gm, 36 hours after if it is between 1000 and 1500 gm, and 24 hours after if it is over 1500 gm. We could not decide to let the infants starve still longer, as GAISFORD and SCHOFIELD (1950) have proposed. At first the number of feedings in the premature infant is 8 to 12 and is later reduced to 5 or 6. With this procedure apnoic attacks are curtailed.

THOMAS [1956 *(20a)*] described in detail the importance of suction — best done with an electric suction apparatus — of the air passages and the stomach of imperiled premature infants. Newborn infants suffering from congenital esophageal atresia aspirate the saliva which overflows from the blind upper pouch of the esophagus and die of suffocation rather than starvation if the clinical picture is not recognized early and the esophageal atresia is not repaired [THOMAS, 1954 *(21a)*].

In infancy aspiration of stomach contents occurs mainly in the terminal phases of acute nutritional disturbances, but also in severe infections. Vomited stomach contents are aspirated during swallowing, but food is not. Aspiration can cause sudden death; the infant who is already in a bad state of health suddenly vomits from his mouth and nose and dies immediately afterward. The respiration ceases suddenly or after some gasping breaths. Sometimes the stomach contents are not vomited, but are just brought up and then immediately aspirated. If the infant does not suffocate immediately, the following symptoms and signs appear: strong inspiratory retraction, increased respiratory rate, coarse rales, and (in fluoroscopy) clearly increased density in the lung portions near the hilum.

Aspiration may frequently be demonstrated in autopsies of infants, if one pays attention to this possibility. In 1953, ESSBACH and PEIPER *(123)* found aspiration of amniotic fluid or stomach contents and its complications in 62 of 91 carefully examined premature infants. In 1955, BOLCK and PEIPER *(127)* demonstrated aspiration of stomach contents on postmortem examination in 60 of 70 infants who

---

* See footnote, p. 312.

had died of alimentary disorders. Of the 10 infants who had not aspirated, half were in the sixth to twelfth month of life, significantly older than the average of the infants investigated. Considerable aspiration, which must have directly imperiled life, was found in 24 infants. Another series of observations concerned 42 infants who died with the clinical picture of intestinal paralysis after toxicosis. All showed considerable aspiration [DITTRICH, 1952 *(5a)*]. Continuous gastric drip, if used in time, may be the most adequate means of preventing death from aspiration in infants suffering from severe nutritional disturbances [PEIPER, 1956 *(129)*].

That even healthy newborn infants born at term may easily aspirate food into the deeper respiratory tracts was clearly shown by the Lübeck epidemics of infantile tuberculosis. The primary tubercular pulmonary lesions found in 15 of the 72 infants who died were regarded as "pneumonia due to faulty swallowing" (SCHÜRMANN and KLEINSCHMIDT, 1935). The newborn infants had received the virulent vaccine, not by sucking, but by spoon feeding. Some infants vomited and may have aspirated the vomited material. In 6 of the 15 infants, the noses were pressed together so that the vaccine would glide into the esophagus more easily. SCHÜRMANN and KLEINSCHMIDT stress, quite rightly, that this maneuver actually facilitates aspiration into the trachea. The mothers of 5 infants could not remember whether the children had vomited or whether their noses had been pressed together. Thus, their aspiration of vaccine into the lungs had occurred unnoticed.

Inept management can sometimes cause aspiration. Frequently premature infants suffering from central sucking and swallowing disorders are fed with a spoon or through nasal gavage. Instilling nourishment drop by drop into the nose, a procedure that is still recommended in sucking debility, must be rejected as dangerous under any circumstances. The infant breathes through his nose when he rests and even when he sucks. Therefore, drip instillation of nourishment into the infant's respiratory passages must increase the danger of aspiration considerably.

Especially dangerous for weak infants is the instillation of the oily or fatty nose drops used for treating head colds. Pneumonia caused by aspiration of such medication has been described in the foreign literature (GOODWIN, 1934; IKEDA, 1935; COMBY, 1936; BROMER and WOLMAN, 1939; GRAEF, 1939; and LAUTMANN, 1939) and less frequently in Germany (DONTENWILL, 1953). "Lipoid pneumonia" represents a clinically and pathologically-anatomically well-defined disease entity, the origin of which can be verified by demonstrating cod liver oil, paraffin, or vaseline (petrolatum) histologically. The same disease entity can be produced in experiments on animals. An infant who had received cod liver oil for some length of time acquired chronic lipoid pneumonia from repeated vomiting and aspiration of the oil. If these substances are avoided as a matter of course, lipoid pneumonia can be avoided with certainty too.

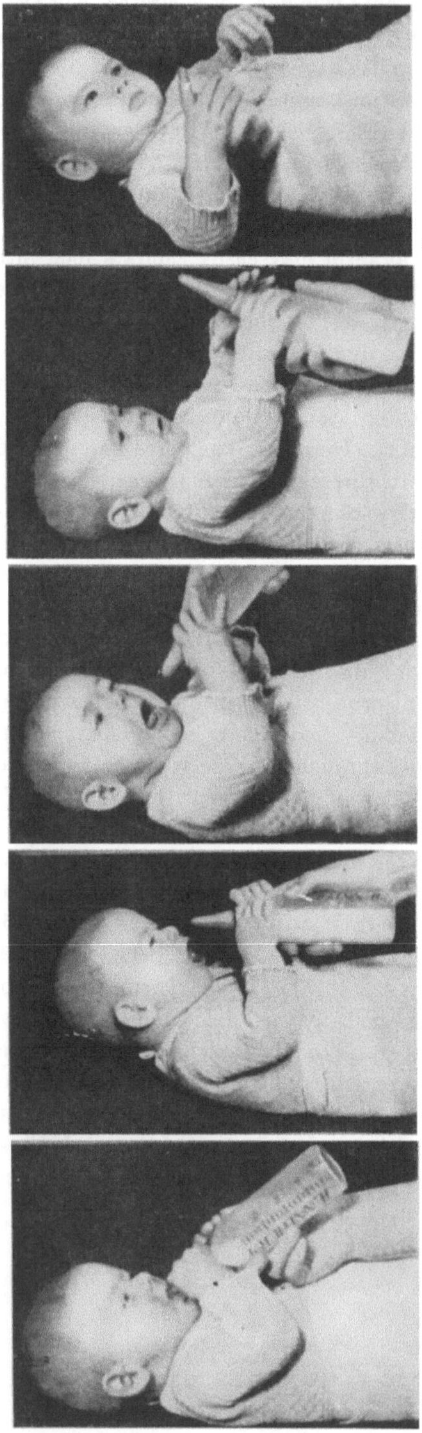

a b c d e

*Conditioned Reflexes of Food Intake**
(53, 66, 137)

VON HELMHOLTZ (1896) stated:

"One can observe that an infant when brought into the sucking position becomes lively and turns his head excitedly to and fro to find the breast; but during the first days he turns away from the breast just as often as toward it, although he can see it in front of him. Evidently at this age he does not know how to interpret his field of vision or the direction of his movements."

The rooting reflex (see p. 404) is an innate unconditioned reflex in the PAVLOVian sense. The touching of the lips is the unconditioned eliciting stimulus. The appearance of the bottle in the infant's visual field at first has no effect. But in the course of the first months the infant who is formula fed sees the bottle before each feeding, before it touches his lips, and before food intake begins. This constitutes the prerequisite for the formation of conditioned reflexes: Gradually the bottle becomes a conditioned stimulator and by its appearance alone, i.e., without tactile stimulation, it may cause a whole series of conditioned anticipatory reactions, among which is the rooting reflex.

* For conditioned reflexes in general see p. 511; for unconditioned reflexes of food intake, especially the rooting reflex, see p. 404.

Fig. 196. Conditioned reactions of a five-month-old bottle-fed infant on seeing the bottle, before it touches his mouth. (a) Infant at rest. (b) He looks at the approaching bottle, his hands reach out, and his mouth is opened. (c) Wide opening of the mouth, protrusion of the tongue, efforts to bring the bottle to his mouth with his hands. (d) The same, the head is raised with mouth wide open and approaches the bottle in the direction of the sagittal axis. (e) Starting of food intake.

The infant's restlessness increases with the approach of the hour of feeding. He cries easily and can be calmed only temporarily. His excitement increases if he becomes aware of the preparations for the meal known to him from previous experience, i.e., when the mother or nurse enters the room, approaches the bed, talks

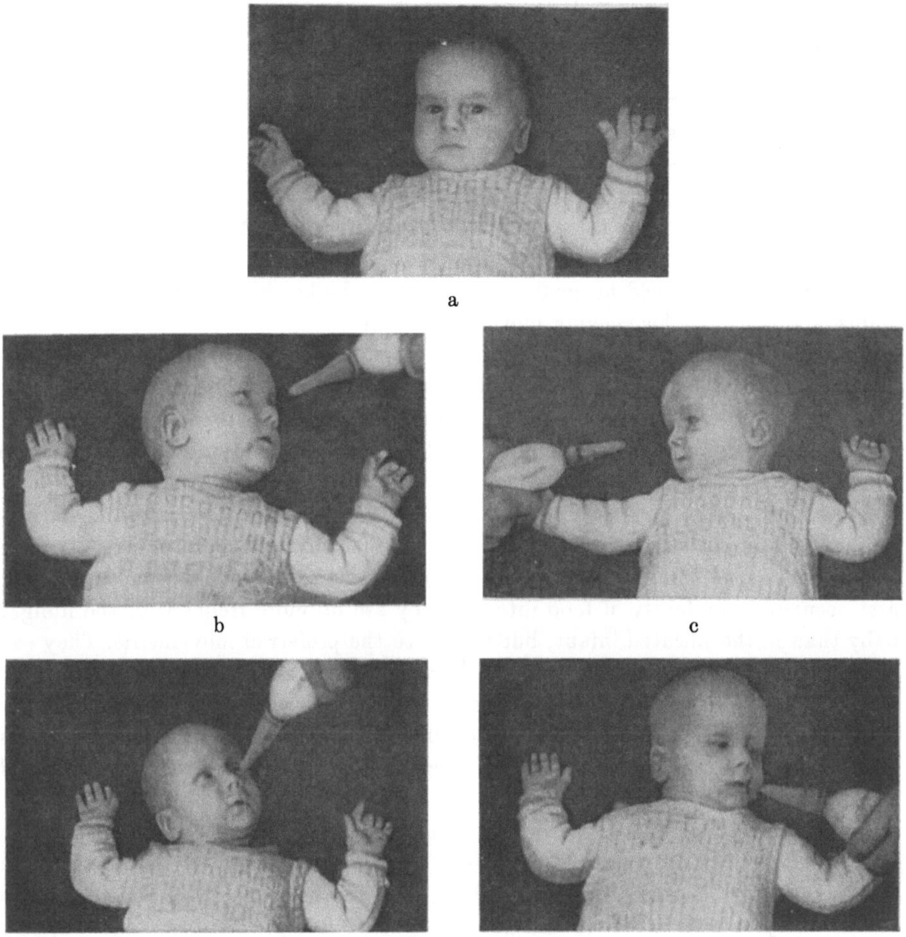

a b c d e

Fig. 197. Conditioned reflexes of a six-month-old bottle-fed infant on seeing a bottle that has not touched the mouth. (b and c) Conditioned turning of the head around its vertical axis to the left and right. (d and e) Conditioned turning of the head around its horizontal axis (mouth directed away from or toward the chest).

to him, or fastens a white bib around his neck. His movements become more stormy when the bottle appears in his field of vision (Fig. 196a–e). He looks at the bottle, stops crying, turns to it, lifts his head, opens his mouth, reaches for it with both hands in order to bring it to his mouth and tries to seize it with his lips, but spoils his efforts because in his excitement he turns his head to and fro several times. Sucking and swallowing movements may already appear. Since all these reactions, which are not innate, occur before the nipple touches his lips, they must be interpreted as conditioned reflexes.

If the bottle is moved to and fro within the visual field of the older infant, he follows it with his head (Fig. 197a–e). Thus, he can easily be induced to make the same movements as are evoked in the rooting reflex (Fig. 167, p. 405). He follows the bottle with his head in a goal-directed manner from right to left and toward and away from his body. Since raising of the head against gravity no longer presents any difficulty, when he is in the dorsal position he can raise his head halfway up to meet the bottle as soon as it is moved toward him along the sagittal axis. Identical movements also occur in the lateral position.

The hungry infant tries to reach his goal rapidly with all the means at his disposal. When unsuccessful he becomes very restless, screams or cries, but quiets down as soon as he has reached his goal, i.e., as soon as he has gotten the nipple into his mouth and can start his sucking movements on it.

All conditioned reflexes of food intake, which also occur in the corresponding manner in breast feeding, can be observed only as long as the infant is hungry. After an adequate feeding the baby is indifferent to the bottle for some length of time. The conditioned reflexes connected with food intake are similarly extinguished in the satiated animal.

According to ECKSTEIN (1927), the young infant's sucking as vacuum activity (without obtaining food) is characterized by regular rhythm, uniform strength, and only slight fatigue. RIPIN and HETZER (1930) state that these sucking movements at first occur independently of food intake. They are no more frequent in the hungry baby than in the satiated infant, but belong to the preferred movements. They can be evoked by various sensory stimuli, such as touch, position change, or loud sound stimuli. Gradually the infant is more and more prepared for food intake. Stimuli which have nothing to do with feeding lose their faculty to evoke sucking movements or other feeding reactions, whereas the latter appear even during the preparation of the meal.

The appearance of the bottle in the visual field of the hungry infant is such an effective stimulus that it can be used to test the infant's ability to differentiate. RÄHLMANN (1903), KRASNOGORSKI (1913), and KASSATKIN (1953) therefore used colored baby bottles to test the infant's color sense (see p. 70).

RUBINOW and FRÄNKL (1934), following LOEBENSTEIN, brought more or less bottle-like objects into the infant's visual field and observed his "bottle-recognition reactions," i.e., focusing, opening of the mouth, movement of the mouth, movement of the arms, raising of the head, etc. The "bottle-recognition reactions" occurred successively in 5 stages, in which the following stimuli became effective:

1. Sudden approach of an object.
2. Approach of a pointed object.
3. Showing of a nipple attached to a bottle or to a bottle-like object.
4. Showing of a bottle filled with white fluid and provided with a nipple.
5. Showing of a bottle filled with white fluid.

Thus, the essential criteria of food intake become more and more effective and the nonessential signs more ineffective. According to BÜHLER, this gradually occurring self-improvement is not acquired from the testing procedure, but appears spontaneously during maturation. The structure of given objects is said to change for the infant, not as a result of external influences, but because his own drives change. But there can be no doubt that in the experiments described conditioned reflexes are formed which appear inevitably in the food intake of every infant. The laws which have proved to be the basis for the formation of PAVLOV's conditioned reflex (see p. 520) can be recognized distinctly in this instance: The young infant is initially in a state of generalization in which all stimuli are effective. After food intake has repeatedly coincided with certain stimuli, these stimuli become the conditioned stimuli, while other stimuli cause an inhibition. This progressive differentiation, which can be demonstrated in the human infant by the feeding experiments described, is not an indication of maturation, as C. BÜHLER thinks, but regularly and regardless of age characterizes the formation of conditioned reflexes.

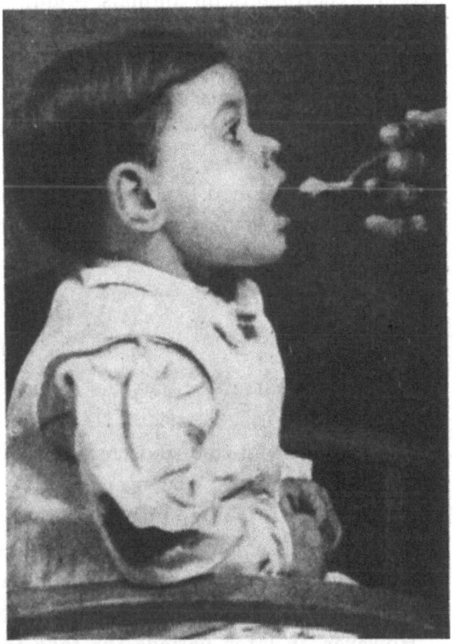

Fig. 198. Conditioned opening of the mouth in a three-year-old infant on approach of a spoon filled with pap.

MARQUIS (1931) reported on conditioned reflexes of food intake during the first 10 days of life. She associated the sucking act with sound stimuli as conditioned eliciting agents. The interpretation of her experiments has been disputed (see p. 516). As has been previously mentioned, sucking movements are at first elicited by stimuli which later no longer have anything to do with food intake.

KROH (1926) demonstrated "acquisitions" even in a fourteen-day-old infant. He observed the behavior of his child, who was alternately breast-fed and bottle-fed. If a nipple with a wide opening was used, the infant refused the next breast feeding, which previously he had gladly accepted; he had also previously refused a nipple with a small opening. He cried and interrupted his sucking repeatedly. A transition period was needed until he adapted himself again. This disturbance did not occur when the breast feeding was preceded by a feeding with a nipple which had a small opening. According to this experiment, each food intake is said to leave a certain disposition which determines the manner of sucking at the next feeding, 4 hours later.

The infant who is already accustomed to strained baby food opens his mouth when the spoon approaches. This too is a conditioned reflex which in this instance occurs at the appearance of the spoon in his visual field. However, if he does not like the contents of the spoon because they are not sweet pap, but salty vegetables, the sight of the substance with the strange taste will inhibit the conditioned reflex.

The conditioned opening of the mouth on the approach of the spoon filled with pap can still be observed very distinctly in the older child who is fed pap by his mother (Fig. 198).

Every mother who observes her child even a little, naturally knows all these striking phenomena. But she pays them little attention simply because they are so common and self-evident. These conditioned reflexes appear—in contrast to PAVLOV's careful experiments on animals — unintentionally and inevitably.

Food intake is a welcome event for everybody from infancy throughout life. During the early period of life it is the most important event that ties the infant to his environment. Since the mother is the mediator of the meal, either by her breast or by the bottle, after a short time her appearance alone becomes the signal for the anticipated pleasure; thus it is not lóng until she herself is joyously greeted. The joy originally conditioned by the feeding ties the infant still more closely to his mother. Of course, we can never demonstrate at what time he becomes consciously aware of this experience.

Breast Feeding

(62)

PETIT (1735) stressed that the infant does not obtain his food only by sucking at the breast. In 1795, ERASMUS DARWIN described the sucking act in a manner which today is still recognized as correct in all its aspects: In his first sucking attempts the infant does not press the nipple softly between his lips and does not suck as an adult would do when sipping milk; rather, he takes the whole nipple into his mouth, presses it together between his gums, and by repeatedly chewing the nipple, presses the milk out in exactly the same manner in which a milkmaid draws the milk from the udder of the cow with her fingers, so to speak.

However, in later periods the food intake of the infant was regarded only as a sucking process and was compared to the function of a suction pump in which a relative vacuum is created by the movement of the piston (ALLIX, 1867; COLLEY and CREAMER, 1958). A backward movement of the tongue was said to correspond to the movement of the piston. BIEDERT (1876) and K. VIERORDT (1877) recognized, however, that the oral cavity is enlarged, not by a drawing back of the tongue, but by a lowering of the mandible. In this "mouth sucking," the first phase of the sucking act, the oral cavity is closed off in back; on the other hand, inspiratory sucking, in which the connection between oral cavity and trachea remains open and the whole thorax participates, may occur in adults, though not in infants. AUERBACH (1888) was the first to observe the difference between these two sucking patterns. He emphasized that when the lips are closed only a small air space exists in the oral cavity because the back of the tongue sticks almost completely to the hard and to the soft palate. Since the infant does not have any teeth, the air space is especially small. The lowering of the mandible can be observed in every sucking infant. According to AUERBACH, it can still be found in children two or three years old when they drink from a feeding bottle or an open cup. The adult, on the other hand, tries to avoid lower jaw movement as much as possible when he is drinking.

Repeated attempts have been made to measure with a manometer the pressure created by the infant's sucking. The values obtained are scarcely comparable with each other because different procedures were used and the concept of the maximum value obtainable was established in various ways. Each lowering of the mandible causes a negative pressure of 4 to 6 cm H_2O (HERZ, 1865; BASCH, 1893; VON PFAUNDLER, 1899). Either the initial pressure is restored immediately after each sucking movement or the pressure sinks step by step during several sucking movements, as can be seen especially well in HARNAPP's (1933) graphs. VON PFAUNDLER prevented the loss of pressure during the relaxation period with a valve. In this procedure the infant attained a fixed measurable value, the "maximal sucking pressure," which in weak infants is 10 to 30 cm H_2O and in strong infants, up to 70 cm. CRAMER (1900) found higher values, i.e., 58 to 140 cm H_2O; BARTH's (1914) values were as high as 82 to 204 cm H_2O.

According to VON PFAUNDLER, the maximal sucking pressure decreases during the meal; it increases again during the interval between meals and seems to correspond to the general state of strength. The sucking energy can be increased by prolonged practice. This often happened spontaneously in breast-fed infants; however, bottle-fed infants forgot how to suck, since food was available to them without the exertion of any special effort.

In these experiments the children usually had the end of a manometer placed in their mouths in some way or other. The children examined by BARTH sucked on a little mouthpiece attached to a manometer and did not suck directly at the breast. We can therefore learn only what sucking strength they developed under the given experimental conditions, not what strength they actually used at the breast.

Polyethylene catheters with a diameter of 1 mm and a lumen of ½ mm are so thin that they can be introduced into the sucking infant's mouth without disturbing him. In this manner, ARDRAN and KEMP (1959) were able to measure the pressure in the mouth, pharynx, and esophagus during food intake. The pressure in the oral cavity in front of the nipple fluctuates with each sucking movement, sometimes regularly, sometimes irregularly. The negative pressure ranged from 150 mm Hg to a maximum of 250 mm Hg.

As recently as 1954, HOTZ said of the infant: "He does not suck, he milks the breast." This statement contradicts every experience and has now also been disproved by measurement of the negative pressure.

It is not known whether milk enters from the breast into the oral cavity during the lowering of the mandible, or whether this always happens or only in some cases. The negative pressure which is necessary to empty the breast has been repeatedly determined, with varying results. BASCH placed a glass sucking bell on the breast, lowered the internal pressure, and determined the exact threshold at which milk came out of the breast. He found a pressure of 54 cm H_2O necessary for a well-functioning breast and from 68 to 81 cm for a breast that was hard to drain. Since he had determined that the infant's sucking strength was only 4 to 14 cm, he concluded that the infant could not overcome the normal tonus of the areola muscles merely by sucking. CRAMER obtained opposite results with the same procedure. According to CRAMER's measurements, the infant is able to develop the negative pressure necessary to suck milk from the breast. LITINSKI (1902) also attributes the main function to the sucking on the basis of his own results. According to BARTH, expression is not absolutely necessary for the sucking act, but it will support food intake. SCHEER (1926) reports that, according to the statements of mothers, the sensations which develop from the infant's sucking seem to correspond to negative pressures of 136 to 204 cm H_2O.

With such contradictory statements, it remains doubtful whether the difficult or easy emptying of the breast can be expressed in figures at all, especially since some breasts are more easily pressed out and others are more easily pumped out (FINKELSTEIN, 1938, p. 57; VON PFAUNDLER, 1924, p. 182). According to BASCH, breasts which yield easily and those which yield only with difficulty differ from each other in the degree of pressure which is necessary to overcome the tonus of the nipple muscles. According to CRAMER and BARTH, the breast which is hard to drain needs a greater sucking strength. CRAMER states that this feature of the breast which yields only with difficulty depends on its milk content, an opinion which has not remained without contradiction. Moreover, the ability of the breast to empty itself frustrates every comparison. This complicated process (see p. 398) can proceed much more easily when the infant is placed at the breast than when a sucking bell is attached. Thus, the mother's conditioned reflexes, on which not only the tonus of the contracting muscles but probably also the formation of new milk during drinking depend, are more easily inhibited by artificial evacuation. The result cannot be clear and definite, since in the last analysis unknown values are compared. VON PFAUNDLER made similar statements: The reasons for easier or more difficult yielding of the breast are not fully known. The coordination of the entire activity of the participating muscles and the mobilization of all reflex paths are decisive for the mode of evacuation of the breast.

VON PFAUNDLER stressed that only the infant is able to drain the breast by sucking. REIMARUS (1773) reports that as an adult he was unsuccessful in sucking at the breast. Perhaps he was in fact only sucking and was therefore unable to obtain anything. However, if children are not weaned at the proper time they continue this practice until school age and obtain milk from the mother's breast without effort.

During the second phase of the sucking act the nipple is pressed out by an upward movement of the lower jaw (BASCH, 1893, 1894). The pressure can be distinctly felt if a finger is introduced into the infant's oral cavity. According to BASCH, the pressure exerted by the jaws of the newborn infant is 200 to 300 gm and during the first two weeks increases to 700 to 800 gm. DREYFUSS obtained similar values, i.e., 500 to 600 gm, by another method. The pressure reaches the lactiferous lobules and ducts of the mother's breast, which have been filled with milk during the preceding sucking. If one lets the infant suck on one's finger, one can feel that the pressure of the lips is only minimal compared with that of the jaws.

WALLER (1943) also states that the infant empties the lactiferous lobules by rhythmically pressing his lips and jaws together. These squeeze the base of the nipple, the areola, and the elongated sucking cone between the retracting tongue and the palate. The infant thus not only sucks from the tip of the nipple, but his lips and jaws enclose the mammary gland near the areola. The nipple is extensible, so that it can enter deep into the oral cavity.

During the raising of the lower jaw, the breast is not only pressed out, but is also milked by movements executed in opposing directions by the infant's tongue and the mother's nipple.

The tongue movement during the sucking act has been variously interpreted. Today it is known that the tongue does not move backward like the piston of a suction pump during sucking. During the downward movement of the lower jaw, the tip of the tongue neither moves downward as BARTH stated nor remains at the edge of the jaw as BARTH assumes, but is extended forward and encloses the nipple from below (HERZ; SÜSSWEIN, 1905; VON PFAUNDLER). This movement can be distinctly felt if the infant is allowed to suck on the finger. It can be seen if the finger is suddenly pulled out during sucking and the mouth is held slightly open. Usually still further sucking movements in which the tongue is distinctly extended forward each time occur. A spoon-like concave cupping of the tongue's upper aspect, into which the lower part of the "sucking cone" (see p. 459) fits, becomes visible. This concave curvature of the tongue is not a result of the air pressure, as AUERBACH and BARTH assume, but is caused by active innervation.

The rising lower jaw presses the tongue like a pillow (SELLHEIM, 1918) against the lower aspect of the sucking cone. The tongue is simultaneously retracted into the mouth and thus milks the nipple by stroking movements. This tongue movement can also be made visible as vacuum activity. The milking (stroking) movement is strengthened still further by a slight sliding back of the nipple which occurs with each raising of the lower jaw. According to BASCH, the nipple erects each time the mandible is lowered and becomes limp again when it is pressed out. If this view is correct, the nipple moves in the opposite direction to the tongue. It has been previously explained that the relative position of the jaws facilitates the milking and expression (see p. 403).

During the second phase of the sucking act milk enters from the nipple into the oral cavity of the infant, as BASCH and VON PFAUNDLER determined with the help of the sucking mirror. The milk is transported to the back by a wave which runs along the upper aspect of the tongue (BASCH).

These findings, which were obtained purely by clinical observation [1936 *(62)*], are confirmed in all essentials by the cinematoradiographic examinations of ARDRAN, KEMP, and LIND (1959): When the lower jaw is raised, the sucking cone is pressed out between the upper gums and the tip of the tongue, which is resting on the lower gum. The tongue presses the sucking cone from below against the hard palate. During this process the cone is reduced to about half its width. When the tongue is moved toward the posterior edge of the sucking cone, the cone becomes shorter and thicker. The larynx is closed each time the upward movement of the tongue toward the sucking cone and the hard palate is completed. Probably the contents of the milk ducts or the cisternae of the sucking cone are pressed out. It has not yet been established what influence sucking has on milk flow from the nipple.

ARDRAN and KEMP (1959) measured the pressure in the oral cavity next to the nipple in breast feeding with a thin catheter and that in front of the nipple in bottle feeding and obtained fairly similar values.

There is a reflex which enables us to obtain a good picture of the sucking act. The sucking act usually consists of rhythmic movements. But occasionally only one isolated movement which represents a kind of fragment of the sucking act is obtained from a single stimulus. The eliciting stimulus is a light tap with the percussion hammer on the cheek near the corner of the mouth; the response is a movement of the tongue and jaws. I have termed this phenomenon the "tongue-jaw reflex" [1936 *(62)*]. A single tap causes a single protrusion and retraction of the tongue with elevation of the lower jaw. It appears that the closing of the jaws occurs before the retraction of the tongue; it is difficult, however, to evaluate such a rapid event exactly. Successive tapping at short intervals makes the tongue protrude more and more until it finally appears between the lips. A transition to rhythmic sucking movements frequently occurs. The movements of the lower jaw and the tongue are not so distinctly visible when the lips are closed, but they can be seen when the infant's mouth is opened.

The tongue-jaw reflex of the infant during the first days of life is at times more and at times less developed and persists slightly for months. But it is necessary to examine a number of newborn infants repeatedly in order to find it present once in particular strength and even then it will rapidly fade again.

GARSCHE (1950) has shown that the tongue-jaw reflex can be regarded as a partial reflex of a functionally significant reflex chain. To study the structure of this total reflex, GARSCHE stimulated the oral region with minimal exactly graded air blasts from an electric hair dryer. On a minimal air stimulus, the mouth was opened and the upper and lower lips were arched outward. When the stimulus was increased, the mouth opening was enlarged by a snapping downward movement of the lower jaw; simultaneously the tongue was moved forward within the oral cavity. At this point a licking reflex corresponding to the tongue-jaw reflex described above occurred. This lower jaw-tongue movement can be evoked much more often and more distinctly by the air stimulus than by tapping. The snapping-licking reflex is accompanied by a swallowing reflex.

According to GARSCHE, the coupled reflex of the oral region and of the larynx can be elicited almost exclusively in premature infants weighing less than 1500 gm. The youngest of these children weighed between 850 and 930 gm. After the eighth month of gestation the reflex is inhibited by the brain stem; thus, only some remnants can be evoked in the newborn infant. Mostly the tongue-jaw movements, and to a lesser degree the lip reflexes, are inhibited. The swallowing reflex is the only reflex which can be demonstrated from the time of its appearance to adulthood (the "immersion reflex," EBBECKE, 1943). Whereas the reflexes described are more and more inhibited in the newborn infant, the rhythmic sucking movements to air stimulation become more distinct. In diseases of the brain, such as chronic encephalitis or tuberculous meningitis (after streptomycin treatment), the inhibitions may cease again, so that the old reflex appears as in the premature infant.

Sucking at the mother's breast was represented in diagrams by BASCH (reproduced also by VON PFAUNDLER, 1924), VON JASCHKE (1931), and HOTZ (Fig. 199a;

1954, reproduced by FANCONI, 1956). These diagrams are wrong in essential aspects. During sucking not only the nipple but the whole "sucking cone" (WETZEL, 1938), a supple formation which the infant's mouth forms from the nipple and areola, is in the baby's mouth. The sucking cone is so deep in the mouth that it is pressed out by the closure of the jaws and not only by the lips. The drinking infant presses his face against the breast of his mother so that no free space exists between his chin and her breast. It is therefore recommended that the mother pull the breast away from the infant's nose with her fingers so that he can breathe unhampered during feeding. The constantly recurring concept that the infant sucks only at his mother's nipple can be disproved by observing any sucking infant (Fig. 199).

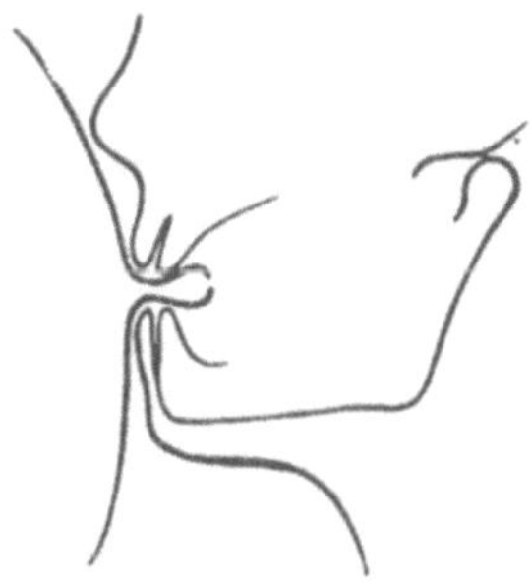

Fig. 199a. Erroneous diagram of the sucking act according to HOTZ (1954). The infant has only the nipple in his mouth.

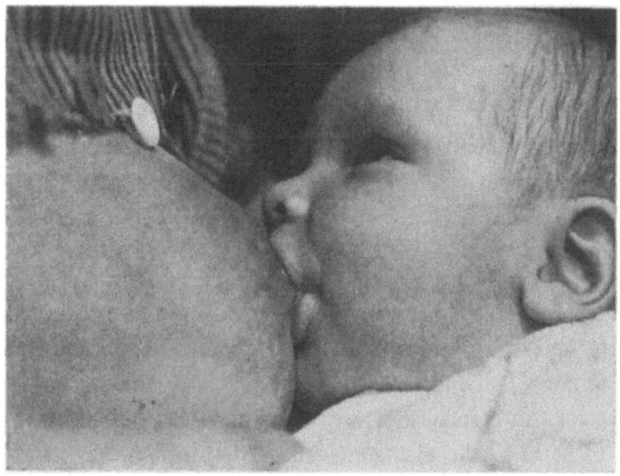

Fig. 199b. Infant sucking at his mother's breast (photograph). The infant has not only the nipple but the whole sucking cone in his mouth.

The sucking diagram presented by ARDRAN, KEMP, and LIND (1958) shows the sucking cone deep in the oral cavity of the infant. In the diagram the oral cavity contains much air, but according to the x-ray films reproduced in the same article it is practically airless.

Bottle Feeding

(62, 110, 134, 143)

When mother's milk was replaced by animal milk the feeding method had to be changed. Although it would be possible to place the infant directly at the teat of the milk producing animal, this procedure would be very cumbersome and the results would be unsatisfactory (BRÜNING, 1908). Thus, milk containers with special drinking attachments which enable the infant to drink have been used for thousands of years. As early as the Bronze Age, i.e., about 1000 to 800 BC, vessels made of clay with a special drinking attachment for infants existed (VON BOKAY, 1936, 1937). BRÜNING reported on Roman clay sucking vessels with a cone-like spout preserved in the museum in Wiesbaden, and VON BOKAY on similar old Roman sucking vessels found in

Hungary. SORANOS of EPHESOS recommended artificial nipples for feeding bottles, but we do not know what form they took.

It took a long time for feeding bottles and nipples to reach their present customary forms. In the Middle Ages, bottles made of glass, clay, horn, or wood in the greatest variety of shapes were used (BRÜNING). Even during the First World War, inflexible attachments made of glass, bone, horn, plastic, or other materials which had been popular before the introduction of rubber nipples were frequently used because of the rubber shortage (VON PFAUNDLER, 1916).

Repeated attempts have been made to improve the feeding bottles and smooth rubber nipples now in use. Many of the proposed changes show complete ignorance of the mechanics of bottle feeding as clarified by VON PFAUNDLER and HARNAPP (1933).

It is easily overlooked that both drinking from the bottle and drinking from the breast are only partially real sucking. VON PFAUNDLER (1926, p. 637) explained that it is nevertheless possible to obtain considerable amounts of milk from the breast by uninterrupted sucking, since the walls of the breast are elastic. But if the bottle has an inflexible attachment, the adult can gain only a little fluid by merely sucking. Only when air bubbles have entered through the opening of the attachment and joined the air space at the bottom of the bottle by rising through the fluid is it again possible to obtain a few cubic centimeters of fluid (VON PFAUNDLER's Draining Type I). No doubt infants can be made to empty the bottle in this manner. I observed infants emptying their bottles with the aid of inflexible nipples almost as fast as usual, although they were not accustomed to them. But the procedure can be considerably simplified by the use of an elastic nipple (VON PFAUNDLER's Draining Type II): The elastic nipple is compressed by the closing of the lips and jaws with the cooperation of the tongue. While the jaws are opened the rubber, freed from the pressure, tends to expand again and thus sucks air from the oral cavity into the bottle. The rubber cover is thus a real "sucker," while the sucking movement which is performed on it plays only a minor role (VON PFAUNDLER).

According to HARNAPP, many infants drain the bottles without sucking by "pressing," i.e., they compress the nipples with their jaws. Most infants obtain the fluid by sucking or a combination of sucking and pressing. Milking (stroking out) plays only a minor role. According to the graphs of the fluctuations of pressure within the bottle which HARNAPP made, the infant who drinks well, especially the older one, obtains his nourishment mostly by pure sucking. This is achieved either by means of short sucking acts, after which the pressure returns each time to its initial values, or by a series of sucking movements, during which the pressure decreases by steps and is then restored to its initial level by the in flow of air.

ARDRAN and KEMP (1959; see p. 457) also determined the pressure in the nipple, the oral cavity, the pharynx, and the esophagus in bottle drinking with the aid of thin polyethylene catheters. When the neck of the nipple was compressed and the milk thus overflowed into the oral cavity, the pressure in the bulb of the nipple increased to approximately 50 mm Hg; when the neck of the nipple was not obstructed, the pressure in the nipple increased only minimally. The pressure in the oral cavity near the tip of the nipple decreased to minus 250 mm Hg at the time the pressure in the nipple reached its highest value. During this phase the tongue was lowered behind the end of the nipple, causing milk to flow into the oral cavity. While the nipple was filled up again, the pressure in the oral cavity at the tip of the nipple increased to its original value. At the end of this phase, when the nipple was again filled, milk again entered the oral cavity. A marked pressure differential between the bulb of the nipple and the oral cavity was not necessary for a good flow of milk. If the hole in the nipple was sufficiently large, a pressure differential of no more than 20 mm Hg was adequate.

The corresponding measurements made by COLLEY and CREAMER (1958) can scarcely be compared with these values, since the infants drank from bottles which were open at both ends.

The cinematoradiographic examinations conducted by ARDRAN, KEMP, and LIND (1959) show that the neck of the nipple is closed by the approaching jaws and that its contents are pressed out by raising the tongue to the soft palate. While the front part of the tongue is raised and compresses the nipple in the anterior portion of the oral cavity, the part of the tongue behind the nipple is lowered, so that milk is sucked in. The contents glide past both sides of the upper laryngeal opening, and when they are pressed from the pharynx the larynx is closed (also see DITTRICH, p. 439).

Cinematoradiography with a picture amplifier permitted a lowering of the radiation intensity for the infant by 1000 times. LASSRICH (1959) was thus able to show the act of sucking from the bottle in all its details. So far only lateral pictures have been made. Essentially the same findings were obtained.

Infants are able to adapt themselves eventually to particular types of drainage. An infant with a cleft palate is able to empty the breast or the bottle although true sucking is impossible. He obtains the milk from the bottle by "pressing" and to a minor degree by milking. On the other hand, the infant may drain the breast under conditions which exclude any pressure (sucking mouthpieces over glass attachments, large sucking bells), as VON PFAUNDLER emphasizes (1924, p. 12). The same is true for the emptying of bottles with inflexible attachments.

The infant has various possible resoures for emptying the breast or bottle, so that the sucking act can be regarded as plastic, as BETHE defines the term. By the psychologists' "method of trial and error," the successful method is "chosen" and maintained. The successful method is the one which requires the least effort to reach its goal. In any case, drinking from the bottle is different in its mechanics from drinking at the breast, even if the function of the mammary gland is disregarded. One should always keep this in mind when a new nipple which is supposed to be a "true imitation of nature" is introduced. So far nobody has been able to prove that the thriving of the infant depends on the mechanics of food intake — with the exception of gross abuses, of course. It is immaterial whether the nipple is as nearly mammiform as possible, for the "sucking cone" on which the infant actually sucks and chews looks different from any artificial nipple. The child will never drink at the breast in the same manner as from the bottle.

I have discussed in detail elsewhere (see p. 607) the infant's preference for the rubber nipple over the breast nipple that arises because in comparison to the breast nipple the rubber nipple is a "supranormal object" in the sense of the term as used by TINBERGEN and LORENZ (1951).

A nipple worthy of recommendation should have the following features: It should be supple, smooth, easy to clean, sterilizable by boiling, and as durable as possible. Since it is an item of daily use, it should also be inexpensive. The smooth rubber nipple cannot be equaled, let alone surpassed, by any other type. Other kinds are certainly more expensive. Smooth nipples are preferable, since they are cheaper and easier to clean.

New types of nipples which are supposed to represent great progress appear on the market continually (Fig. 200), but I doubt that the nipple has been improved in any way whatever in the past several decades.

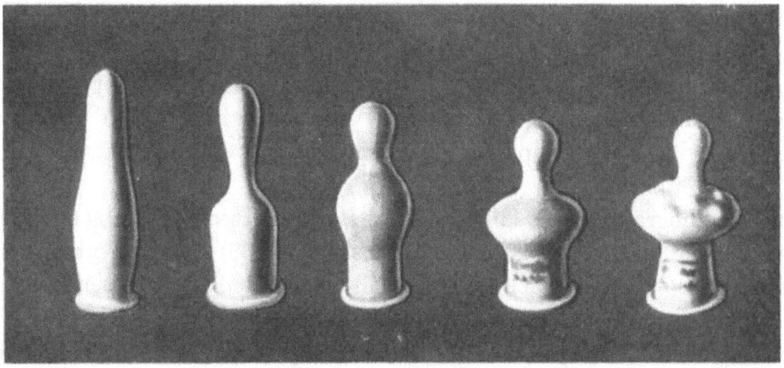

Fig. 200. Popular rubber nipples. A narrow neck makes turning the nipple inside out and cleaning it after each meal difficult.

There is no scientific justification for using special types of nipples recommended today. To increase the sale of the merchandise, their merits are praised and their disadvantages overlooked. For instance, attempts have been made to make drinking more difficult for bottle-fed infants, so that they drink as slowly as breast-fed children do. This can be most easily achieved by a nipple hole of appropriate size. THOMÄ (1942) invented a valve for the feeding bottle which renders food intake more difficult and slows it down. To adjust to the sucking power of the infant he uses valves of different strength, thus taxing the mother's powers of observation considerably. A simple valve which was supposed to regulate the air supply to the bottle was described earlier by A. SCHMIDT (1901, 1904). Valves are more difficult to clean than rubber nipples. FERVERS' (1932) "Natura" nipple also makes cleaning difficult because of its shape. It is certainly possible to keep irregularly shaped nipples clean in a nursery with trained personnel. In daily life, however, the nipple should not create any difficulties for a mother who has little knowledge.

According to DREYFUSS (1928), the popular nipples are disastrous to facial configuration and may even cause lasting damage. During sucking the palate is said to endure regularly a pressure as high as 500 or 600 gm. As a consequence, the space of the nasal chambers may be decreased and the nasal septum deviated. Mouth breathing and a chest constriction may thus be induced, especially in infants suffering from rickets. Therefore, DREYFUSS (1932) recommends the "Pupon" nipple he designed, the tip of which is more resistant. But cannot the same objection be raised against drinking from the breast?

If such serious disadvantages are attributed to the nipple generally used, more than mere theoretical justification should be demanded. It ought not to be too difficult to prove that the malformations described occur from the use of the nipple in infants whose general good health has been demonstrated before the use of the nipple, and that these malformations can be avoided if the recommended nipples are used. Such proof has not been submitted as yet.

According to DREYFUSS (1928), VAHL (1934), and ORAVECZ (1934), the stomach is filled with air to a lesser degree with the use of Pupon and Natura than with the use of ordinary nipples. Even if this statement is correct, it does not prove anything against the common nipple. The important thing is not the size of the air bubble in the stomach, but the thriving of the infant.

The attempts made during the First World War to manufacture supple nonrubber nipples of the usual shape had little success (FINKELSTEIN, 1916; VON PFAUNDLER, 1916). The inflexible attachments made of different materials described above stood the test better. From the survey of VON PFAUNDLER, who illustrated 14 types, we can see how different the forms and shapes of usable nipples can be. GRIMM (1916), SILBERSTEIN (1916), and LANGSTEIN (1917) gave additional descriptions. We found the simple glass nipple designed by OPPENHEIMER (1916) useful: A glass tube has at its free end a circular enlargement 1.5 cm in diameter which is flatly compressed from two sides and has a slit-like opening for the infant's mouth. A piece of cork connects this tube with the bottle. Although one could manage in this way, the substitute nipples rapidly disappeared when rubber was again available

In any case, the mechanical difference between breast and bottle drinking cannot be compensated for by the shape of the nipple. During the effortless drinking from the bottle, too much nourishment may easily be withdrawn (see p. 475), but drinking from the bottle may be rendered more difficult by a smaller opening in the nipple. Milk flows rather uniformly from the bottle into the mouth. When drinking at the breast, the infant receives the greatest amount of food during the first minutes. The infant's efficiency in bottle drinking can be measured to a certain degree, but no appropriate basis has been developed for a comparison with the as yet unmeasured work done by the breast-fed infant. A bigger air bubble develops in the stomach of the bottle-fed baby than in the breast-fed infant (W. R. HESS, 1919).

Phylogenetic Memories

(2)

The mammary glands were formerly regarded as transformed sweat glands. In the duckbill a mammary bag is formed and the glands terminate at its base. The secretion, known as "fatty sweat," is licked up by the young (GRIMMER, 1910). According to this observation, the sucking movements originally developed from a licking movement. PREYER and STIRNIMANN described the newborn infant as already able to lick.

However, on the basis of his investigations BRESSLAU (1931) arrived at another result: He states that the mammary glands did not develop from sweat glands; rather, the glandular regions ontogenetically appearing first are phylogenetically extremely old formations comparable to the "breeding spots" of the birds. Probably they originally served in breeding as a heat source, for which they are especially suited because of the rich development of their blood vessels. These latter facilitated increased growth of the cutaneous glands originating in the epidermis, which thus represent the beginning and not the result of a development.

The interplay of respiratory and swallowing centers is reminiscent of the swallowing respiration of lower species of pulmonary breathers (see p. 348). The lower jaw also drops during inspiration in yawning (see p. 386), in the singultus attack of young infants (see p. 343), in gasping respiration (see p. 327), and in the floor-of-the-mouth respiration of suffocating infants (see p. 365).

Grasping with the mouth, which the young infant is enabled to do by the rooting reflex (see p. 404), is phylogenetically older than grasping with the hands.

Accompanying Neurological Phenomena

(20, 62, 9a)

The sucking act is accompanied by phenomena of the nervous system which may appear in varying intensity but are never completely absent. The disappearance of motor restlessness, which is so characteristic for the awake infant, is especially striking. The infant who just a moment ago became restless when preparations for his feeding were made, who turned his head from one side to the other, opened his mouth, kicked with his arms and legs, breathed more rapidly and perhaps even cried, suddenly lies motionless with limp muscles when food intake has really begun. People speak correctly of "calming"* the infant. Strong and even painful stimuli leave the infant undisturbed to a certain degree. WOLOWIK (1927) found that the stimulation threshold to galvanic current was increased during the feeding (see p. 33). The nervous energy, of which the infant has a comparatively low supply, is seized by the centers of food intake and thus is diverted from the motility centers, evidently a very appropriate phenomenon. KÖDDING [1940 *(9a)*] repeatedly found the number of blinks reduced during sucking. In two premature infants and in two other infants suffering from acute nutritional disturbances, the average number of blinks was 2 to 4 per minute when they were awake; it decreased during sucking movements to 0 to 1 and afterward increased again. These observations were made on several successive days. For the same reasons, the cerebral function of the prematurely born infant may regress during food intake to periodic respiration (see p. 321) or even to the lower gasping respiration. The infant usually falls asleep after his meal.

---

* The German word for "nursing" is "stillen," which literally translated means "to calm" – The translators.

BALIASSNIKOWA and MODEL (1932) described additional neurological phenomena during the sucking act: The pulse rate increases by 16 to 30 per minute, the face is flushed, irregular respiration becomes regular,* and sometimes breathing is interrupted by deep sighs. Defecation occurs immediately after sucking. Erections occur at the beginning of sucking. The luster of the eyes increases during the meal and the body temperature decreases afterward.

Little is known of the influence of the sucking act on salivary flow. It is evidently difficult to obtain saliva during sucking. IBRAHIM (1904) let a newborn infant with a congenital esophagotracheal fistula suck strongly on a nonperforated nipple and obtained a generous portion of saliva when he emptied the blind pouch. The danger of aspiration in congenital esophagotracheal fistula demands early operation. This observation showed that the sucking act produces an abundant salivary flow even in the first days of life. ISBERT and PEIPER [1928 *(21)*] occasionally noted irregular swallowing movements when examining the movements of the esophagus of an awake or half-awake infant. These movements became rather frequent when the infant sucked on a nipple, a sign that saliva was secreted during sucking.

After PAVLOV had demonstrated in experiments on animals that feeding has a stimulating reflex effect on the flow of gastric juice, the next step was to examine a newborn infant to decide whether this reflex is an innate unconditioned or a conditioned reflex acquired later. COHNHEIM and SOETBEER (1902) performed esophagotomies on newborn puppies and obtained – even without feeding them – "psychic" gastric juice. They therefore regard the reflex as innate. Prior to these experiments, VON PFAUNDLER (1899) had demonstrated in an infant that the sucking act stimulates gastric secretion, be it by way of increased salivary flow or through special reflex processes. NOTHMANN (1909) made newborn infants who had not yet drunk suck at empty bottles and in this way was able to obtain gastric juice which contained hydrochloric acid, pepsin, and renin. He therefore also concluded that gastric secretion is an innate reflex which is elicited either from the mouth simultaneously with the sucking movements or by these movements alone, without the necessity of food intake.

TAYLOR (1917) and ANNI SCHMIDT (1927) disputed the conclusiveness of these experiments. ANNI SCHMIDT succeeded in demonstrating flow of gastric juice as a "hunger secretion" even before the beginning of sucking. TAYLOR effected a mock feeding by sucking the food out of the esophagus again; with this procedure he could not find any secretion of gastric juice although the milk had entered the mouth and had been swallowed. These experiments would tend to show that gastric secretion is independent of the sucking act.

It is doubtful whether the sucking act stimulates gastric peristalsis. VON PFAUNDLER and ANNI SCHMIDT compared the emptying times of the stomach in infants after drinking from the bottle and after tube feeding. VON PFAUNDLER obtained the stomach contents and found that after active meals they were almost always less than after tube feeding. However, ANNI SCHMIDT, who based her statements on

* According to my findings (see p. 421), this is not always the case.

x-ray examinations, was not able to find a distinct difference. ISBERT and PEIPER have shown that the peristalsis of the esophagus does not spread to the stomach.

It is the custom in institutions today to breast-feed infants only at their mother's breasts and to feed them with drawn-off milk when the milk comes from a milk donor. It makes a difference for the success of the breast feeding whether you use this method or breast feeding, since infants thrive better when they are breast-fed (CZERNY; W. R. HESS, 1919; FISCHL, 1923). CATEL (1935) similarly found in nutritional experiments on goats that they thrived better when they sucked on the udder than when they were fed with milked milk. The reasons for these differences are not known.

The question of whether sucking as vacuum activity may be harmful because the digestive juices are secreted in vain has been repeatedly discussed. No such damage has been demonstrated clinically.

Finger Sucking

(62)

Thumb sucking exists even in monkey babies (BRANDES, 1939). The ability to put a finger in the mouth and to suck on it is present in man at birth. BLANTON observed a newborn infant putting his finger in his mouth and sucking on it 20 minutes after birth. During the purposeless movements of the newborn infant, his hands come near his mouth. When the lips are touched, sucking movements are apt to occur (PREYER, 1895).

Later the infant is very much inclined to suck on his fingers. Frequently certain patterns develop; for example, he sucks only on two certain fingers of a certain hand, so that calluses develop on the preferred fingers. A nearby object such as a corner of his cover may be preferred, though less frequently. A crying infant calms down when he puts a finger in his mouth and sucks on it; on the other hand, infants become excited easily if their fingers are removed from their mouths. Some infants cannot fall asleep unless they have their fingers in their mouths. When their sleep is disturbed by any stimulus they frequently, without completely awakening, execute some sucking movements during which the fingers are sometimes put into their mouths.

It has long been known that infants can be calmed by letting them suck; nipples of various kinds are used for this purpose. Sucking on the finger or a nipple thus develops into a conditioned reflex which is sometimes quite difficult to give up. It may persist until school age and longer; it has even been described in adults. The older child frequently masturbates during his finger sucking (LINDNER, 1879).

The psychoanalysts regard the child's finger sucking as an expression of his sexual drive and therefore ascribe great importance to it. HUG-HELLMUTH (1921) states:

"When he started to suck the facial expression of my nephew distinctly showed signs of sexual excitement and later of comfortable relaxation as he gradually fell asleep."

With such an attitude there is an understandable concern with finger sucking, which according to W. SCHMIDT (1926) is said to favor mental development during the first months of life. He suggests that finger sucking should be supervised by a pedagogically correct (i.e., a psychoanalytically oriented) environment so that the infant will develop normally in all respects. However, the observation of a single child does not adequately justify such an extreme demand.

A child may put his finger in his mouth when embarrassed or afraid; thus, finger sucking becomes a kind of expressive movement.

Dentists (FISCHER, 1913; KANTOROWICZ, 1930; ORAVECZ, 1934) are of the opinion that finger sucking or the continual use of a nipple as a pacifier endangers the shape of the jaws. The pictures reproduced by these authors indicate considerable changes.

Pathology of the Sucking Act
(62, 75, 80, 93)

As previously described, in respiratory, sucking, and swallowing disorders we find a disturbance of both the rhythm of the required movements and of their coordination. Respiratory, sucking, and swallowing centers are frequently limited in their functional capacity from birth by immaturity and so sensitive that they easily break down in response to any damaging agent. Whether the basic damage is nutritional or an infection (particularly pneumonia), a cerebral hemorrhage due to birth trauma, a cold, or some other disturbance, the clinical picture is always the same. Of course, certain differences occur, since in pneumonia the basic respiratory rhythm is accelerated at the height of the disease, whereas it is decelerated in pylorospasm or hunger from other causes.

The more immature the infant is, the more easily central rhythm disorders occur. Usually respiratory, sucking, and swallowing weaknesses occur simultaneously, either immediately after birth or — as sequelae of various diseases which develop in the newborn infant — after only days or weeks.

Despite different causes, the clinical picture is always the same. Frequently the infants refuse to drink at the beginning of a feeding. If they are roused by shaking or moving of the breast or the nipple, they will make some weak periodic sucking movements and then stop again. If they are roused again, they again suck only for a short time. Even prolonged efforts are ineffectual. They are even damaging, since the infant becomes cold while at the breast. The first step is to increase the number of feedings. But in serious central sucking weaknesses this helps only temporarily. Food intake is not adequate and the infants will starve if no intervention occurs. Frequently a swallowing weakness is associated with the sucking weakness: The food which is introduced into the mouth is not swallowed, but remains there and finally flows out of the mouth or through the nose. Sucking and swallowing weakness can be overcome by tube feeding, which sometimes has to be used for weeks. If there are no other disturbances, particularly no serious respiratory disorders, the nervous irritability will sooner or later improve and the children will learn again to take and to swallow food without artificial feeding. However, HOFMEIER and KÖRNIG (1938) recommended the use of bottle and pipet and discouraged the use of the tube because they were afraid of its shock effect. But clinical experience does not agree with their view. Premature infants must frequently be tube-fed for months but no damage results.

In organic brain damage the sucking act may be difficult to evoke or may even be completely absent. In this condition, in contrast to the situation in central sucking weakness of premature infants, the general health of the child is not impaired and the general nervous irritability is not diminished, but there exist certain more or less distinct deficits of cerebral function.

JOGICHESS (1932) collected data on the early sucking ability in older children suffering from congenital disorders of the nervous system. This function was disturbed in bilateral athetosis, Little's disease, congenital hemiplegia and diplegia, epilepsy, psychopathy, and dysgenesis of undetermined type. In some cases the children had been unable to suck for weeks and in a few cases for the whole period of nursing. If cerebral hemorrhages due to birth trauma were considered to be the cause of these diseases, JOGICHESS did not assume a direct lesion of the sucking center but pictured remote effects from distant foci; a trauma to the sucking center could easily involve the neighboring respiratory center and thus make the infants non-viable. According to KASAHARA (1916), the sucking graphs of feeble-minded infants always show irregularity.

Severe congenital damage caused either by birth trauma or other factors may evidently involve the sucking and respiratory centers. If a sucking weakness persists for long in an otherwise strong infant, organic brain damage should be suspected, but this is by no means proof of a cerebral hemorrhage caused by birth trauma; according to my own experience, Little's disease is especially likely. FINKELSTEIN (1938, p. 73) describes an infant who could not suck until his sixth month but whose intelligence nevertheless later developed very well.

Other sucking disturbances which will be described here by referring to VON PFAUNDLER (1924) are less serious. The infant who is "clumsy in sucking" shows hunger and grasps eagerly for the nipple, but does not get very far with his efforts; therefore, he becomes restless, cross, loses the nipple or pushes it from his mouth, searches for it again, renews his attack with no better success, etc. He is not lacking in hunger or muscle strength. This condition is usually overcome by his learning to suck. The infant who "shys away from the breast" is one who becomes more and more restless at the breast, turns his head, even throws his whole body around obstinately, rears himself, and cannot be induced to take sufficient quantities of nourishment. However, bottle feeding is almost always successful, but letting the infant go hungry is not. VON PFAUNDLER observed this "shying away from the breast" particularly in neuropathic families. The feeding of these infants is very difficult, but the outlook is fairly favorable. The infants who display "voluntary hunger at the breast" make insufficient use of the ample food available, although they have no sucking weakness; they are not clumsy nor do they shy away from the breast. They do not seem to be hungry; the necessary amount of food is not correctly estimated, so to speak.

Sucking may be disturbed by painful diseases of the mucous membranes of the mouth, by a thick coating of the tongue which covers the taste papillae, or by a congested nose or a cold. Short-winded infants usually drink poorly. In general, any severe disease affects food intake.

A harelip or a cleft palate can make the sucking act more difficult, but are not absolute obstacles to sucking at the breast. PETIT (1735) states that an infant born without a palate is able to suck, for he can squeeze the nipple sufficiently with his lips and suck with his tongue. Teeth present at birth may injure the nipple and are better removed.

According to CZERNY and KELLER (1925, I, p. 32), a lower jaw which is too short presents a severe impediment to sucking. These authors stress that this condition is a congenital malformation, and are therefore in opposition to many dentists who attribute the malformation of the jaws to sucking.

Pharmacology of the Sucking Act

(2a)

Since the respiratory center may be stimulated by drugs such as lobeline, icoral, and nikethamide, it seems to be obvious that we should examine whether the sucking center, which is so closely related to the respiratory center, can also be stimulated in this manner. GRÜNDLER (1942), REINER, and BELLET reported on the good effect of nikethamide on the drinking torpor of newborn infants; NEVINNY (1936) and WALLISCHEK (1937) treated sleepy premature infants with pentylenetetrazol-glucose, which made them more lively and made them drink better. Others have used pentylenetetrazol and Veriazol for the same purpose.

However, the effect of nikethamide on the sucking ability of weakly drinking infants has never been exactly measured. Therefore, BRUNN [1944 *(2a)*] determined the average number of sucking movements in sleepy, slow drinking premature infants weighing 1600 to 2500 gm before and after the administration of nikethamide. He determined the average number of sucking movements per minute during 10 meals over a period of 3 days and compared these figures with the corresponding values during the next 3 days after the infant had first received 3 drops — and when this dose proved to be ineffective, 10 drops — of nikethamide. The sucking rate without nikethamide was then studied once more in a few later meals.

On the average the infants made 31.6 sucking movements per minute before taking nikethamide; under the effect of the medication this figure increased to 39.4 and later again decreased to 33.5. Although the stimulating effect of small nikethamide doses was quite distinct in children who had severe sucking weakness and did not drink at all, only large and very irritating doses could elicit sucking movements. The use of this medication proved to be very involved and time consuming, since it had to be given in food or tea some time before feeding. It cannot be used without food or tea because of its bad taste. Since its effect on the sucking center is not very great and since its use is cumbersome, BRUNN questions the advisability of recommending it. But these trials at least show that nikethamide can increase the activity of the sucking center.

OTTO (1955) obtained improvement with Pervitin in 8 of 41 infants who sucked weakly, but he recommends using this medication only in selected cases.

Belching and Vomiting

(40)

By belching, the infant relieves himself of the air which entered his stomach during food intake. Belching is facilitated when the infant is in the erect position, when the air rises to the cardia. KLEE (1919) states that belching is due to a reflex which originates in the cardia as soon as the stomach has reached a certain size. But an additional force is needed to push the air from the stomach through the esophagus. As in vomiting, this force can be exerted only by the stomach or diaphragm.

The diaphragmatic movements during belching can be recorded without difficulty by placing a rubber bag on the anterior abdominal wall and by connecting the bag with a writing drum. This method, however, cannot differentiate to what extent the recorded movements originate in the diaphragm and to what extent in the total abdominal musculature. The major part ought to be ascribed to the diaphragm, since during belching the anterior abdominal wall is arched forward and this can be caused only by lowering of the diaphragm but cannot be explained by contraction of the abdominal muscles.

The graphs (Fig. 201) show that during belching the diaphragm is as a rule briefly and strongly contracted. This is usually followed by a respiratory arrest of several seconds' duration until normal respiration is restored; but since the abdominal circumference has become smaller, the breaths now usually start from a lower base line. In weak infants and particularly in premature infants, a previously regular respiration may easily be periodically accelerated after belching. This occurrence is similar to that which can be observed in weak infants after other nervous events (see p. 322).

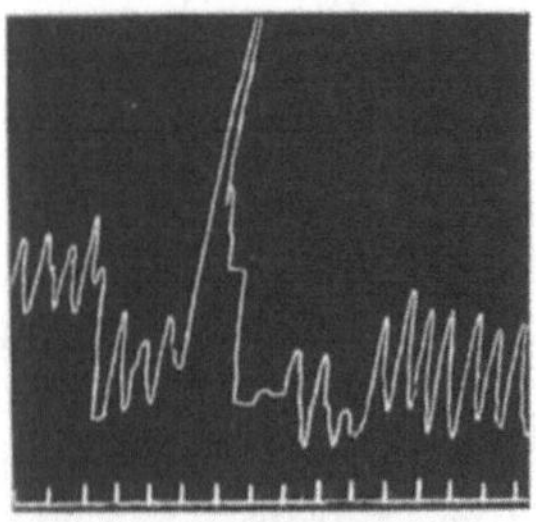

Fig. 201. Two episodes of belching, one shortly after the other, as indicated by the respiratory movements of the anterior abdominal wall (time in seconds). During the first belch the diaphragm was only slightly contracted; during the second considerable contraction occurred.

A sudden contraction of the diaphragm cannot be observed in all graphs. Rather, it may happen that the graph shows a more or less abrupt drop (Fig. 202). This proves that the stomach alone is able to expel the air without the assistance of the diaphragm. It is possible that in belching the stomach always contracts simultaneously with the diaphragm. The assistance of the diaphragm may vary greatly in the same infant (Fig. 201).

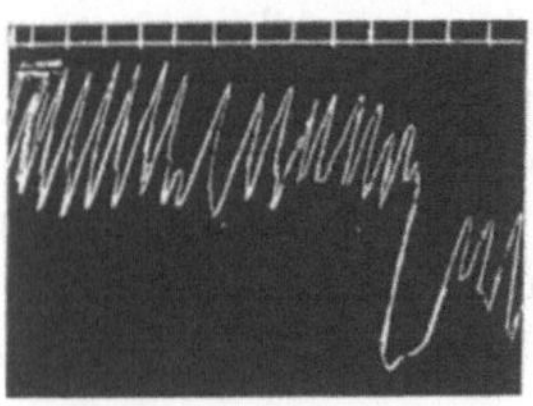

Fig. 202. Belching without a strong diaphragmatic movement, followed by a period of respiratory arrest (time in seconds).

In vomiting, the fluid or solid stomach contents are suddenly pushed back into the esophagus, which no doubt requires considerable strength. It has long been controversial whether this strength comes from the stomach or the diaphragm. Fluoroscopy and recordings of the stomach movements have both been employed in trying to settle this question. Vomiting was usually evoked by apomorphine; in addition, electrical stimulation was used in experiments on animals. Only LEVY-DORN and MÜHLFELDER (1910) reported the case of an adult who could vomit at will and who did not need artificial stimulants for this purpose. The question of the extent to which artificially evoked vomiting corresponds to the usual vomiting act is not discussed anywhere, although the identity of the two events is not self-evident.

As early as 1813, MAGENDIE found that during vomiting the stomach is emptied by the diaphragm and the abdominal muscles. He was still able to bring about vomiting in a dog after he had replaced the stomach with a pigskin bag. The assistance of the abdominal muscles has been disputed at times, but is for the most part acknowledged today. The extent to which the stomach may empty itself independently is uncertain (VON OPENCHOWSKI, 1889; CANNON, 1898; LEVY-DORN and MÜHLFELDER, 1910; HESSE, 1913; KLEE, 1919, 1927; GUINARD, 1898; MILLER, 1912; WEITZ and VOLLERS, 1927).

As yet, there seem to be no investigations on the mechanics of the vomiting act in infancy. HELMREICH (1931) asserts that the abdominal musculature does not play a role in vomiting, but submits no proof for this assertion. It is known conclusively that vomiting is not accompanied by general nausea in infancy.

If immediately after an abundant meal the infant is placed in an oblique downward position, the head lower than the pelvis, vomiting occurs easily. I used this fact to trace the diaphragmatic movements with the aid of a rubber bag placed on the anterior abdominal wall and to record them graphically.

All graphs have an appearance identical to that of Fig. 203; during each vomiting episode the diaphragm contracts briefly and strongly and relaxes immediately afterward. This movement must have a thrusting effect on the stomach contents and therefore this alone can evacuate the stomach, provided the cardia is relaxed. Some participation of the stomach may be necessary, for at intervals real vomiting is sometimes preceded or followed by the aforementioned diaphragmatic movements, which are not accompanied by vomiting.

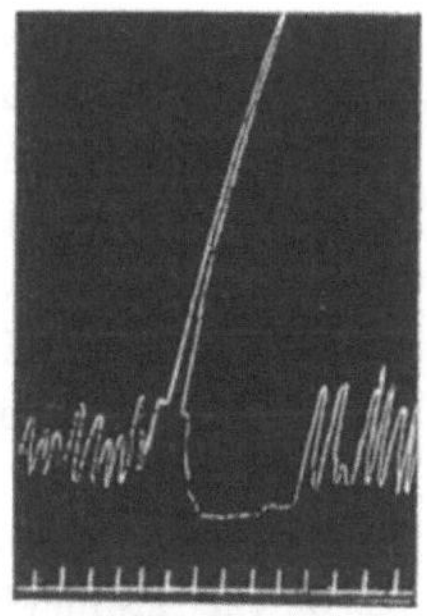

Fig. 203. Strong diaphragmatic movement during vomiting, followed by a prolonged respiratory arrest (time in seconds).

Often the respiration is involved in this process. Although vomiting may suddenly interrupt regular breathing, infants may frequently become very restless prior to vomiting; they start to cough or to cry and vomit when the excitement has reached its height. If the phenomena are not so strong, the respiration becomes irregular prior to vomiting; often a respiratory arrest precedes it. Even more frequently the respiration is interrupted for several breaths after vomiting, which may give some protection against aspiration.

As with belching, the respiration of weak infants tends to become periodic after vomiting. According to GRAHAM and MORRIS (1929) (Fig. 111, p. 322), periodic breathing also frequently occurs in pylorospasm, in addition to real vomiting.

When we compare the activity of the diaphragm during belching and vomiting, no essential differences can be found. As far as the diaphragm is concerned, both processes depend on the same center, the vomiting center. No certain answer can be obtained from the available investigations about the participation of the stomach in belching and vomiting. The observation described above supports the view that belching can occur without the assistance of the diaphragm. The activity of the stomach may consist only of vigorous peristalsis. An antiperistalsis is out of the question, since the peristaltic waves in the stomach are always directed toward the intestines, as both fluoroscopy and observations of the anterior abdominal wall in pylorospasm have shown.

Hunger and Satiation

(20, 104)

A satiated beast is horribly dumb.
GOETHE

Hunger and satiation are vital checking devices that complement each other, thus preserving the individual and dividing his life into more or less regular periods. Hunger forces us to search for food. As the meal progresses, satiation and a steadily increasing aversion to further food intake develop. The game starts anew when hunger occurs again after some hours. When internal processes thus cause the individual to act rationally, he is protected from starvation and excessive food intake.

Despite the differences in the search for food and in food intake in individual animals, basically the same process always takes place. When a certain task is given to an animal in an experiment, for example, when a goal has to be reached by an approach made difficult, it is very useful to utilize hunger as an incentive. The animal must be hungry prior to the test, and the goal must be a desirable kind of food. If the animal is satiated, it will fail at tasks it had previously performed while hungry; it may even completely lose its learning ability [among many other experiments, the observations of rats made by SZYMANSKI (1921) and of chimpanzees by W. KÖHLER (1921) are noteworthy].

Man is no exception. In him too hunger is a potent, I would say the most potent, original impulse, which spares nobody and forces everybody to act. It is this way from the earliest time of life. But since the infant is physically and mentally completely different from the adult, hunger and satiation express themselves in babies in a manner different from that of later life.

The hungry infant is quite excited, perhaps in connection with the hunger peristalsis of the stomach [CARLSON and GINSBURG, 1915; TAYLOR, 1917; ISBERT and PEIPER, 1928 *(20)*]. The infant moves his limbs, throws himself restlessly from one side to the other, and tends to cry loudly with or without external cause. He announces his readiness for the meal so that his environment is always kept aware of his presence and he is so prepared to suck that all kinds of stimuli, even if they have nothing to do with food intake, cause sucking movements. Many infants acquire the habit of sucking at a finger; in this way they become calm. They may also show this habit pattern when they are satiated, but their sucking movements occur more frequently during hunger. When milk — or sometimes only water — is instilled into the hungry infant's mouth, he starts to suck and to swallow animatedly.

According to GINSBURG, TUMPOWSKY, and CARLSON (1915), weak tonic contractions of the fundus ventriculi start about an hour after the feeding. These waves gradually increase in frequency and size until after 2½ to 3 hours they have changed into strong hunger contractions.

RICHARDS (1936) also observed that the longer the interval since the last food intake, the stronger was the peristalsis of the stomach in newborn infants. The body movements do not depend on gastric peristalsis; they are inhibited by a full stomach even if it is filled only with air.

Conditioned reflexes appear in connection with food intake at the age of a few months (see p. 450). Familiar events, such as the bringing of the bottle or the fastening of a white bib, are sufficient to cause mouth opening or sucking and swallowing movements.

The rooting reflex (see p. 404) is characteristic for the young infant: When the oral region is touched with a rubber nipple the infant moves his lips correspondingly and turns his head in a goal-directed manner to the nipple until he actually has it in his mouth and can suck on it. When hungry infants who are ready to suck are very eager, they may shake their heads to and fro several times; these infants may even "search" with their heads as vacuum activity without even being touched. However, the corresponding lip movements never appear unless they are touched.

The behavior of the satiated infant varies greatly from that of the hungry baby. The restlessness has disappeared and his mood is peaceful. He is now more inclined to laugh and has less tendency to cry, and when left alone soon falls asleep. Even adult persons become irritated or angry more easily when hungry than after meals.

The rooting reflex usually disappears completely in thoroughly satiated infants. All stimuli are without effect. Sucking movements may sometimes occur without external cause or when the rubber nipple is put into the infant's mouth, but when compared with those in the state of hunger they are much weaker. The infant is weary of sucking. If some drops of milk are instilled into his mouth, they elicit only weak sucking movements or none at all; sometimes the infant does not swallow and the milk flows out again. KRASNOGORSKI (1931) states that both conditioned and unconditioned salivary flow are distinctly reduced after food intake, compared with the flow before feeding. In similar studies, KANTROW (1937) showed that satiation is connected with complete loss of the conditioned sucking movements. The mere temporal succession of conditioned and unconditioned stimuli does not suffice to cause conditioned sucking movements; it is therefore mandatory to conduct these tests in hungry infants.

As WOLOWIK (1927) has proved, the more time that has elapsed since the last meal, the more the infant's pain sensitivity increases (see p. 33). TISCHLER (1957) has reported that crying depends on the time of food intake (see p. 120).

If an infant is examined at regular intervals after a meal, the sucking movements will gradually become more pronounced. Prerequisites are that the infant be awake prior to the examination, since responses are usually absent during sleep, and that he be really satiated. In formula feeding this often requires considerably greater amounts of food than the babies need for their normal growth. In incomplete satiation, the disappearance of the sucking movements cannot be distinctly observed.

Two observations of this kind in the same infant are presented in Table XI. Rooting was eliminated in both instances by providing considerable amounts of food. Sucking movements evoked by introduction of a milkless nipple into the mouth disappeared for $2\frac{1}{2}$ hours in the first test and for only $^{3}/_{4}$ hour in the second

Table XI: Disappearance of the Food Intake Reflexes after Meals in Hannelore W., Age 3 Months, Weight 4950 gm

Time of Day (PM)	Rooting Reflex	Sucking Movements on	
		Milkless Nipple	Instillation of Milk
1. 5:00 hungry	+	+	+
5:00–5:24	drinks 230½ milk		
6:15	–	–	–
7:00	–	–	–
8:00	–	+	+
9:00	+	+	+
2. 4:45 hungry	(+)	+	+
4:45–4:55	drinks 290½ milk		
4:55	–	+	+
5:15	–	+	+
6:00	–	+	+
6:00–6:17	drinks 150½ milk		
6:17	–	–	(+)
7:00	–	(+)	–
8:00	–	(+)	–
9:00	–	(+)	–
10:00	+	+	+

trial, although the second time the infant had received considerably more milk since he had been given a second feeding. Sucking movements on instilling milk by gavage were absent for 2½ hours in the first trial. In the second trial they did not disappear at all after the first feeding; after the second feeding – 1 hour later – they were at first still just noticeable but later disappeared for several hours. A relation between the amount of food given and the disappearance of sucking movements is thus evident, as can also be noted in additional observations which are not reported here. These mutual relations are not always so distinct as in the example cited; the excitability of the infant is evidently quite often dependent on less clearly recognizable conditions.

In the experiments cited, the relation between food intake and the disappearance of the rooting reflex is still more striking; here also certain fluctuations from feeding to feeding could be observed in the same infant. Sucking movements may be quite weak in a good eater. This fact explains the contradictory statements made by Kussmaul (1859) and Häggström (1920), who found a dependency on hunger, and Popper (1921), who disputed this relation.

The vital orientation to the breast is doubly assured. Sometimes it can be evoked without touch just by bringing a warm object near the mouth [Sikorski, 1908; Stirnimann, 1919; Schendel, 1944 *(15a)*]. The temperature sense thus aids the tactile sense in finding the nipple. Schendel found a distinct orienting movement on heat stimulation alone in 5 of 50 hungry infants; after feeding this could be demonstrated in only one infant. Satiation diminishes the sensitivity to heat stimuli as well as to tactile stimuli.

The examples reported show that the individual reactions do not completely agree. Periodic sucking movements (see p. 419) represent an external sign of increasing satiation. The infant, who up to this point has sucked regularly like a machine, now pauses after every few sucking movements. The increasing satiation has caused a lower excitability of the sucking center, which corresponds to the occurrence of periodic respiration under similar conditions (see p. 321). The periodic sucking movements are frequently accompanied by motor restlessness, sometimes only during the intervals and sometimes also during sucking: The infant moves his trunk, makes fists, opens his hands, and especially moves his head, turning it this way and that and occasionally bowing or raising it.

Through these movements, the nipple may glide out of the mouth unless someone holds the bottle and follows the head movements. If this is done, sucking movements may still occur. In addition to this general motor restlessness, the infant is able to push the nipple away with his tongue. Of course, this does not exclude the possibility that he may continue to drink if an opportunity is offered to him. The breast-fed infant loses the nipple of the breast in the same way during his restlessness after satiation, and is able to eject the nipple with his tongue when he is satiated. Afterward he will lick his lips. The turning toward the source of food, which finds its expression so distinctly in the breast orientation and in the sucking movements during hunger, thus stands in contrast to a rarer turning away at the end of the meal. The development of the gesture of negation from this pattern has been described previously (see p. 133).

On the whole, the sucking act is fitted only to the needs of the breast-fed infant. The bottle-fed baby drinks from the bottle throughout the entire feeding with the same effortless ease; the breast-fed infant has to exert considerable effort. Afterward he is sometimes really exhausted and covered with perspiration. There are certainly variations, depending on whether the mother's breast is easy or hard to drain and whether the infant sucks well or poorly. But compared with bottle feeding, drinking at the breast is much more tiring and exhausting, especially toward the end of the meal. On the average, breast-fed babies therefore need more time for food intake than do bottle-fed infants. In addition to satiation, the increasing fatigue, which is of minor importance in the bottle-fed infant, gradually inhibits further sucking movements.

The breast-fed infant "determines the size of his meal not according to available tables, but by instinct, remains free from disturbances, and shows us the picture of normal development" (Czerny and Keller). When the breast glides from the mouth at the end of the feeding, a search for the breast does not occur. If the mother is not overwhelmed by unnecessary anxious ideas, she considers the nursed and calmed ("gestillte") infant as really satiated without worrying about how much he has drunk.

There can be no doubt that in earlier times when the infant shared his bed with his mother, he was nursed as often as he wanted. The monkey baby receives as a "breast baby" much more frequent breast feedings than today's human "bed infant" does (see p. 590). Scientific opinions have always varied widely on the questión of how often the child should be nursed. According to Guillot (1852), a good wet nurse does not feed her child fewer than 30 times in 24 hours. Besser (1853), however, recommends only 4 feedings per day. Czerny (1893) observed that healthy breast-fed infants who obtained as many and as large feedings as they wanted were satisfied with 5 meals in 24 hours. Therefore, he stipulated that infants should have only 5 meals in 24 hours. Actually, it is easy in most cases to satisfy babies with 5 meals given during the day at intervals of 4 hours and at night after an interval of 8 hours. By regular feedings, a conditioned time distribution in the sense of PAVLOV's theory is achieved. Under CZERNY's influence, this feeding schedule, which was soon extended to bottle-fed infants, has been widely used in Germany.

However, gradually this very strict rule of 5 daily meals was abandoned at times. Weak infants, particularly those born prematurely, are often not capable of absorbing such large meals (they have to be large, since they are fewer in number). In pylorospasm, vomiting can be decreased and often stopped by frequent small feedings (as many as 24 in 24 hours). In toxicosis of the infant, vomiting and thus the danger of aspiration are distinctly diminished by continuous gastric instillation (see p. 449).

ALDRICH and his co-workers (1939, 1945, 1946, see p. 119) have demonstrated that the better the infant's care is adapted to his needs, the more his crying diminishes. GESELL (1945, 1954), who fed infants according to their own wishes and observed them closely, made detailed investigations about their behavior during feedings. The number of meals diminished gradually. A three-week-old infant demanded 7 meals, but later prolonged the duration of his sleep so that at the age of eight weeks he arrived at a schedule of 5 meals. It was surprising how punctually some infants announced their feeding demands.

Referring to these investigations, under the slogan "Self-Demand Feeding," and on the basis of the democratic rights of the individual (TRAINHAM and MONTGOMERY, 1946), it was demanded that the baby himself should determine the time, number, and duration of his feedings. According to ROWAN-LEGG (1949), the infant has his own hunger and sleep rhythm, which varies from child to child and in the same infant from day to day. When the infant's demands as to feedings are obeyed, he cries only rarely; this satisfies mother and child. Most infants adjust to 4-hour intervals. But some may be damaged by the forced schedule, for as soon as nourishment is given by force when not demanded or withheld when desired, mother and child are dissatisfied.

According to TRAINHAM and MONTGOMERY, the unnatural spatial separation of mother and child in maternity hospitals (also see p. 623) favors formula feeding of the newborn infant. They also found that the number of meals decreases when feeding is practiced. During his first day of life the infant drinks 10 to 14 times in 24 hours, but after milk flow has been initiated, much more rarely.

On the other hand, it is absolutely necessary that the infant adjust to the requirements of his environment. Very few mothers today are able to devote all of their time to the infant, even if they are not working. If the mother has to keep house, or if she has several other children, from the outset she will not be able to fulfill all the demands of her youngest child. Even if she were able to do this, it is doubtful whether such child raising would be desirable. Moreover, it makes a great deal of difference whether we are dealing with breast-fed infants — the only kind TRAINHAM and MONTGOMERY mention — or with formula-fed babies, who make considerably greater demands on their mother's time because of the necessary preparation of food. ROWAN-LEGG (1949) also considers "demand feeding" for bottle-fed infants. As GLASER (1948) stresses, this kind of feeding schedule presupposes not only a willing mother with sufficient time and independence, but also a considerable amount of intelligence on the part of the parents and the child's relatives; originally, however, the mother was supposed to follow only her instincts.

Soon loud warnings were heard, e.g., BAKWIN (1946):

"Whereas children were formerly treated too strictly, we now have too little discipline. The reins formerly held too tight are now loosened so far that children have no direction at all. It is just as bad – perhaps even worse – to allow children always to follow their own way than not to grant them enough freedom."

Nor was LAPIN (1954) satisfied with the now customary advice to mothers – "Feed the newborn infant when he is hungry and as much or as little as he likes" – for parents are not able to distinguish between the hunger of the infant and aches, spasms, air swallowing, or other conditions that may lead to crying. They feed the infant each time he cries. Thus develops a new disease, "acute demanditis." An example of this is the ten-day-old newborn infant who receives 23 meals in 24 hours and becomes sick with diarrhea and vomiting. Furthermore, such a method must necessarily harm education, which is ultimately indispensable. LAPIN therefore pleads for "demand feeding within reason." In general, views on infant nutrition have changed considerably in the US since 1950. Formula feeding on demand has been abandoned. Mothers should keep to certain hours for feeding and should prefer breast feeding (HANNAPPEL, 1960).

This new point of view substantially approaches the procedure which I recommend from my own experience. CZERNY's demand for 4-hour intervals between meals in natural and formula feeding seems to answer basically the needs of most infants today. It is also satisfactory for the mother who has more work today and cannot devote all her time to her child. But this schedule should not be adhered to too strictly in specific cases; particularly in the beginning, the number of meals should be increased in weakly sucking infants and especially in premature infants. This number will in time be reduced automatically by demand feeding, a tendency which should be supported.

It has long been known that the breast-fed infant has to exert more effort than the bottle-fed. CZERNY and KELLER stress that an infant nourished by both methods easily becomes spoiled by the ease of bottle feeding, and that he will therefore not suck strongly at the breast and will give up as soon as the breast is hard to drain or yields too little milk. The difference in effort between sucking at the breast and bottle drinking is so great that most infants prefer the bottle. Therefore, bottle drinking should be made more difficult by supplying a nipple with small holes. Since the bottle-fed infant does not tire so early, the rooting reflex disappears only after the intake of great, sometimes excessive, amounts of nourishment. But even this does not prevent further feeding. The infant has been fed too long if turning away from the food is either not recognized or not observed. This is only possible because the intake of the necessary amount of food does not tire the infant sufficiently. Thus, quantities of milk far beyond the amount needed are easily given, particularly in formula feeding. In the early era of a scientific approach to pediatrics, bottle-fed infants received, to their detriment, far greater quantities of nourishment and fluids than are today thought to be appropriate. Adequate amounts can easily be given. Thus, an infant who was thriving on 800 gm per day drank 1200 gm in 5 feedings without being urged; if he had been urged or if the number of feedings had been increased, he would have taken even more. It is true that vomiting occurs easily after intake of excessive quantities of food; this gives some protection. It is only in nursing that overfeeding is not a danger. Breast-fed infants generally adjust to a moderate amount of food, which makes them thrive and makes restriction of the quantity unnecessary.

The automatic protection of satiation, which makes the infant too tired to suck, generally occurs only in the breast-fed infant, who really does not need this protection so much. The bottle-fed infant, who receives a far more dangerous nourishment, is much less protected. This shows another disadvantage of formula feeding, as compared to natural nutrition.

It is therefore necessary to prescribe the quantity of nourishment in formula feeding and not to leave this to the discretion of the infant or the mother. This has the disadvantage that the bottle-fed child receives exactly the same previously determined amount of food at each meal, while the quantities received from the breast vary from day to day and meal to meal according to the needs of the infant. This simplification, which cannot practically be avoided in the formula method, is unfavorable.

The body acts in self-protection when acute infections of any kind reduce food intake. But if sucking and swallowing weakness occurs in a young weak premature infant and makes food intake difficult or impossible, one has to resort to tube feeding, since the infant would otherwise starve.

Worried parents frequently bring children of all ages to the physician because they have no appetite. This loss of appetite is (provided the parents are not making exaggerated demands on their children's abilities) a sign of neuropathy, which should be treated as an entity and not with worthless "appetite-increasing" medicines alone.

Nursing

(62, 94)

In nursing, mother and child fit together like the key and the keyhole — individually they make no sense, combined they serve a definite purpose. As previously described, during sucking the mouth of the infant forms a "sucking cone" from the pliable breast. Mother and child simultaneously adjust to each other in their nervous activity to guarantee the transfer of milk from the breast to the baby's oral cavity. During feeding and the following pause, exactly the right amount of milk is formed in the breast to satisfy the infant's hunger and with it his nutritional demand. When the infant needs more milk during his growth, the amount of milk will increase. For twins, the breast adjusts without any difficulty to the double output. The infant demands his needs from the breast, and the breast conforms to his demand.

Mother and child adjust to each other in a very simple way without either being conscious of the process: If the infant does not obtain enough food, he remains hungry and sucks more energetically and for a longer time. Nature invented the very useful arrangement whereby the more thoroughly the mother's breast is drained, the more milk it gives. Thus, the greater demand automatically creates a greater supply. On the other hand, if there is too much milk the breast will be drained only inadequately by the quickly satiated infant. A damming up of milk thus occurs which causes a decrease in the quantity of milk secreted. Just as a baker adjusts his production of rolls to the demands of his customers, the infant's demands regulate and control the amounts of milk supplied.

Even SORANOS OF EPHESOS knew that an insufficient milk flow can be improved by placing a strongly sucking infant at the breast. For some time it was uncertain whether the natural stimulus of sucking or the mere draining of the breast was the decisive factor. Today, when mothers empty their breasts artificially by milking or pumping in many mother's-milk banks and even initiate milk flow from the breasts in this manner, we know that the decisive factor is the thorough emptying of the breast, while the manner in which it is emptied is of less importance. HELBICH (1911) was the first to prove this fact. Incidentally, the same conditions also exist in the cow. Women learn in institutions to drain their breasts artificially, but at home many cannot manage it as well as they do nursing. Considering that milk drawn off artificially is not so good as milk drunk directly at the breast (see p. 466), the child should be placed at the breast if at all possible.

The interplay between the baby and his mother's breast can be disturbed by a child's illness. The younger the child, the more easily the excitability of his sucking center is decreased by disease, especially by nutritional disorders and infections of any kind. This will cause sucking weakness (see p. 467). Infants who suck weakly are not able to empty a full productive breast that is rich in milk. They do not obtain enough nourishment and become still weaker in their sucking because of malnutrition. The milk is dammed up in the breast, a dangerous condition which in turn reduces the milk secretion. When it was still customary to hire wet nurses for sick infants, the same child occasionally exhausted several wet nurses, who lost their milk flow because of the sucking weakness of the nursed infant. But this can be easily remedied by artificial draining of the breast. It is self-evident that premature infants with sucking weakness are not able to keep their mother's breasts productive.

When the mother cannot satisfy her child by giving him one breast, the danger arises that the other breast, if given then, will not be sufficiently drained. Therefore, it is mandatory that the breast offered last should be given first at the next feeding.

If we disregard diseases, which nature could not anticipate, we have to admire the simple and effective way in which the demand regulates the supply. But a purposeful interplay such as nursing does not take its course according to a prescribed pattern; rather sometimes more and sometimes less time, effort, and patience are needed until mother and child are adapted to each other. They should be allowed this time even if the gain in weight which is usually the sole concern of the environment does not come up to expectations. Frequently, the patience to wait and see how mother and child fare alone is missing. The mother animal that nurses its baby has only to follow its instincts to find the correct way. Some mothers peruse big volumes on infant care, become unsure of themselves, and then do just the wrong thing. To give up the expectation of fast weight gain is a demand which the environment often cannot accept since the importance of weight and of the quantities of nourishment is overrated, or the urge exists to see the weight loss of the newborn infant made up within a certain time. If no baby scales existed — though they certainly should not be abandoned — these difficulties would not exist; nursing would adjust itself automatically, simply because nobody was supervising it. Instead, somebody is there who feels responsible, watches anxiously, and much prefers to take a — seemingly — better route.

Thus, nature's work is frustrated and the bottle is offered as soon as the calculated amounts of milk are not taken by the baby. The child does not get a chance to announce and claim his demands at his mother's breast and stimulate it to greater efforts. On the contrary, since drinking from the bottle is less strenuous, he is inclined to drink more and more from the bottle and less and less from the breast. That the breast then gradually dries up is regarded as a sign of unproductiveness, while the mistake was only that the breast's milk flow was not started correctly or was not maintained.

As yet, there are no effective means of inducing milk flow, which would certainly facilitate our procedures. We therefore have to go the way nature has prescribed to increase the abundance of milk in the mother's breast, and leave the infants on short rations for a time to stimulate them to stronger sucking.

It is, of course, frequently against our feelings to let such a small and weak creature starve, although we immediately reduce an infant's food intake when he shows a nutritional disturbance. Since only the mother and not the infant can be informed about the advantages of breast feeding, we have to get in touch with him in another way. For this purpose we use his hunger. This is a voice, indeed the only voice, which he understands from the beginning. Nature herself gave him this understanding because his life depends on it. How well the hungry infant knows this voice can be learned from the difference in behavior of the hungry and the satiated infant (see p. 472).

Thus, hunger is for the young infant an important and indispensable motivator that governs his entire waking life. The close relations between the infant and his mother that have been described developed only out of necessity. It is not so long ago that man owned no milk animals. Many primitive races are still at this stage today. When America was discovered there was not one milk cow on the whole continent, and yet men grew up under those conditions. Today the ease of procuring substitutes for mother's milk endangers nursing, and with it the infant. Certainly we cannot reverse this development. But we should accept formula feeding only if feeding with mother's milk has proven to be insufficient despite all serious efforts.

Not every mother can nurse her infant completely, but at least she should make every effort if no known contraindications prohibit it. At the onset we may occasionally have to put up with some temporary malnutrition of the baby, since the milk flows abundantly only sometime after birth. However, it would be inappropriate and even dangerous to let premature infants suffering from the described sucking weakness starve in order to induce them to drink better. In this case tube feeding should be started immediately to provide them with the necessary quantities of food.

Whether a mother is able to nurse her child can be determined only by trying; nobody can say so by just looking. Unfortunately, it frequently happens that because her ability to nurse is doubted she is filled with distrust of herself and will worry about her child. No wonder then that the prediction proves to be correct!

Although milk production occurs involuntarily, it is subject to the influence of the nervous system, as is shown by the occurrence of conditioned reflexes in milk production described previously (see p. 398). Nursing is not a simple emptying process. The mother, as well as the infant, has to fulfill a nervous function, since a portion of the milk is produced during nursing. That restlessness, worry, and anxiety have no favorable effect is evident, even if we distrust the numerous "nursery tales" which for obvious reasons were formerly told by wet nurses. A mother needs — if I may say so — a certain devotion to nursing. But how is she to achieve this when she is torn by a thousand anxieties about her nursing? Therefore, she must be encouraged and cheered up rather than discouraged by distrust in her nursing ability.

Frequently, infants who become ill with nutritional disorders after premature weaning are brought to the physician. He has first of all to resume nursing immediately, since this banishes the danger. However, if the infant is admitted to a hospital because of his alimentary disorder, mother and child are separated at a time when their being together is most needed, for the attempt to restart an already dried-up breast by artificial draining always fails despite the best intentions. Although it is frequently not sufficient for the mother to come to the hospital for the baby's nursing, this measure is at least worth a trial. The only natural thing to do is to admit the mother to the hospital with her child, so that she can learn how to nurse or keep the milk flow going under favorable conditions, free from her worries and her household tasks. The additional expense is more than compensated for by the possibility of an earlier discharge. But as yet most children's hospitals do not have the facilities to accept the mothers at the same time. These facilities have been given up because most mothers severely burden orderly hospital management without knowing or comprehending that they do. Nevertheless, Reuss (1939) demanded — and rightly so — that nursing mothers be admitted.

Premastication

(109, 117)

Among the anthropoid apes living in the primeval forest, the mother does not have spoons, knives, and vessels, to say nothing of fire, at her disposal. How does she manage to feed her baby with food supplements during the weaning period under these conditions?

BRANDES (1939) had the opportunity to watch an orangutan mother with her baby in captivity: During the first months she was already giving other food to her baby in addition to mother's milk. She proffered fluids by pressing her lips between those of the baby. Later the same thing occurred regularly with premasticated solid food. The mother did this by instinct, since she did not have the opportunity to learn from the example of others.

In Greek and Roman times the custom of premastication was widespread. The physicians MNESITHEOS OF ATHENS and SORANOS OF EPHESOS rejected premastication. GALEN, however, writes:

"At the onset the infant must be fed with milk alone. But as soon as his front teeth appear, he must become accustomed in some way to accepting more solid food; this is done by the women who, based on their experience, chew the food and then transfer it into the mouths of the children. The food consists of small amounts of bread, later of pudding, meat, and other things."

Premastication was also formerly widespread in our country. This can be seen from the fact that the German language has two expressions for it: "Ätzen" ("to feed") and "Vorkauen" ("to premasticate"). Both expressions were also applied to man; for example, WALDIS (1565) wrote in a poem:

"Es hett ein Weib ein kleines Kind,
wie man derselben noch wohl findt,
des kundts mit etzen and mit seugen
von seinem weinen nimmer schweigen."

["A woman had a baby fair,
The kind you can see anywhere.
With breast and cud she fed him well,
But he would never cease to yell" — Ed.]

(GRIMM's Dictionary)

According to SENNERT (1632), solid food was not to be given before the teeth appeared; then the child received bread, cooked vegetables, and meat, after the food had been premasticated. STAHL (1718), THEBESIUS (1759), SCHÄFFER (1792), GIRTANNER (1794), FAUST (1795), STRUVE (1797), and GOELIS (1810) were opposed to this bad custom; BUFFON (1773), however, claimed that nature herself had introduced premastication in the most different and widely separated countries (Italy, Turkey, nearly all the provinces of Asia, America, the Antilles).

An old woman told me that before 1900 as a maid in the country she had to pre-chew the solid food (bread, meat, potatoes) for the infants. She spat the food into a spoon and fed the infant thus.

Premastication has certainly become less frequent, but it has not completely disappeared. The physician does not become aware of it because he does not expect it and because the mother is afraid to tell him. As late as 1947, a mother premasticated the food for her baby in the waiting room of the pediatric clinic of the University of Greifswald.

The widespread occurrence of premastication can be explained by its prehistoric origin. The kiss, which is essentially a sucking movement, is connected with it. When one mouth touches another mouth, this symbolizes the feeding of the baby by his mother, in the last analysis. Following a similar train of thought, BILZ (1943) termed the human kiss a "vacuum activity performance of archaic premastication." In this connection he cites the following observation: A young mother who gave additional food to her child, and did this with much devotion and tenderness, always involuntarily opened her own mouth, which she had pointed like a snout, when she brought a spoon near her child's mouth.

SUMMARY

Mother and child are in perfect harmony during nursing to ensure vital food intake for the infant. The functions of the mother – aside from the self-evident external care – are the secretion of the milk, which occurs partly during nursing, and the more or less spontaneous emptying of the breast under the influence of conditioned reflexes.

The infant, aided by the rooting reflex, grasps the "sucking cone" which he forms in his mouth from the nipple and its areola. This cone is sucked more deeply into the mouth with each lowering of the mandible and pushed out again with each upward lifting of the jaw. It enters so deeply into the infant's mouth that it can be pressed out by a closing of the jaws. The oral cavity is sealed against the outside by the gums and the lips.

The sucking act occurs in two stages:

The main characteristic of the first stage is the lowering of the mandible while the mouth is kept closed. Thus, a negative pressure in the oral cavity develops which draws the sucking cone deeper into the oral cavity. The tip of the tongue is moved forward. It is not certain whether milk enters from the mammary gland into the infant's oral cavity at this point.

In the second stage, the lower jaw makes the opposite movement: It rises and presses the protruded tip of the tongue against the sucking cone and the cone against the upper jaw. The jaws do not have a pincer-like grip; rather, the lower jaw is retracted in relation to the upper jaw. The tip of the tongue, which is pressed against the sucking cone by the lower jaw, withdraws during this process and thus exerts a milking or stroking movement from below; this is facilitated by the fact that the sucking cone now glides out of the oral cavity more and the previous erection of the nipples subsides. The lips assist in pressing the sucking cone out, but their pressure is only slight. Milk passes from the nipple into the mouth at this stage of sucking. A peristaltic wave runs across the surface of the tongue and transports the milk further.

During the first stage of sucking the infant sucks by lowering the mandible, and during the second stage he presses the sucking cone toward the upper jaw by lifting the lower jaw and at the same time strokes it out with his tongue.

When the sucking and respiratory centers function rhythmically, each full respiratory movement corresponds to one or two full sucking movements. The rhythm of the younger sucking center is forced upon the ontogenetically and phylogenetically older respiratory center, no matter whether the sucking rhythm is faster or – as in pneumonia – slower than the respiratory rhythm. The respiratory center's persistent striving for independence is usually suppressed during the food intake of healthy infants. Sucking disorders, during which the influence of the sucking center on the respiratory center fluctuates according to the degree of excitation in the sucking center, occur more often in sick than in healthy infants. Three different excitatory phases can be observed; they differ from each other in size and duration of the sucking movements and consequently in their influence on respiration. The strongest excitation of the sucking center causes respiratory inhibition, the moderate dominates the respiration, and the weakest excitation has no effect on respiration (block between the respiratory and sucking centers). The inhibition occurs right at the start of the sucking movement.

When the excitability of the sucking center is reduced, the sucking movements will become just as periodic as respiratory movements do under the corresponding conditions. Pendular induction combines both forms. Periods of respiratory and of sucking movements alternate; the respiratory movements are either inhibited or only slowed down during the excitatory phase of the sucking center. Pendular induction is found in weakly sucking premature infants and in mature infants suffering from pneumonia.

In healthy, regularly sucking infants, swallowing does not noticeably interrupt respiration, as is the case in the adult, since the occlusion of the larynx entrance lasts only for a fraction of a second. In addition to respiratory and sucking disorders, swallowing may also be disturbed in premature infants, respiration being inhibited by the swallowing. In a distinct impairment of swallowing, the swallowing movements occur only sporadically and may finally cease completely.

As postmortem findings show, infants suffering from severe nutritional disturbances and young premature infants frequently die from aspiration of vomited gastric contents. This danger is considerably reduced if continuous gastric instillation is used.

During breast feeding, conditioned reflexes which support this process are formed in mother and child.

The mechanism of breast feeding differs essentially from that of bottle feeding; this cannot be changed by nipples of special form.

The sucking center responds to nikethamide and Pervitin.

The diaphragm plays an important role in belching and vomiting. However, the investigations available do not indicate to what extent the stomach also takes an active part in these phenomena.

The reflexes of food intake can be easily elicited and are very distinct in the hungry infant, but they are reduced or totally absent in a satiated condition. Too much food can be distinctly rejected by means of certain movements. Since drinking from the breast is usually more strenuous than drinking from the bottle, there is danger that in bottle feeding greater quantities of the milk formula, which is itself more dangerous, will be taken than in breast feeding.

Provided that the mother's breast has an abundant milk supply, the amount of milk secreted corresponds to the infant's needs and therefore increases with age. This agreement is attained by the insufficient emptying of the breast when it has too much milk and the ensuing retention of milk, which causes a reduction in milk secretion. However, if the milk supply is not adequate the infant will remain hungry and suck more vigorously. The infant's sucking stimulates milk secretion. Untimely addition of milk formulas disturbs the balance between milk secretion and milk consumption. When a sucking weakness exists, the breast has to be emptied artificially to keep it functioning.

The healthy infant receives 5 meals a day by either natural or formula feeding. It is frequently advisable to increase this number of feedings during the first weeks of life.

REFERENCES

von Allesch, G. I.: Naturw. *1921*, 774; S. ber. preuß. Akad. Wiss., Phil. hist. Kl. (July 14, 1921).

Allix, E.: Étude sur la physiologie de la première enfance (Essay on the Physiology of Early Childhood). Paris (1867).

André-Thomas and Saint-Anne Dargassies: Études neurologiques sur le nouveau-né et le jeune nourrisson (Neurological Studies on the Newborn and Young Infant). p. 32. Paris (1952); Examen neurologique du nourrisson (Neurological Examination of the Infant). p. 8. Paris (1955).

–, et. al: Presse méd. *1954*, p. 146.

Ardran, G. M., and F. H. Kemp: Acta paediatr. *48*, 261 (1959); Brit. J. Radiol. *31*, 11 and 156 (1958).

–, F. H. Kemp, and J. Lind: Brit. J. Radiol. *31*, 11 (1958); *31*, 156 (1958); Acta paediatr. *48*, 203 (1959).

Auerbach, L.: Arch. Physiol. *1888*, 59.

Babkin, P. S.: Paediatrie (German edition) Berlin *5*, 64 (1954); Zschr. Neuropath. i psichiat. *56*, 22 (1956) (Russian).

Bakwin, H.: Amer. J. Dis. Childr. *71*, 113 (1946).

Baliassnikowa, N., and M. Model: Zschr. Kinderforsch. *39*, 1 (1932).

Balint, M.: J. Genet. Psychol., Worcester *73*, 57 and 81 (1948).

Barth, H.: Zschr. Kinderhk. *10*, 129 (1914).

Basch, K.: Arch. Gynäk. *44*, 15 (1893); Jb. Kinderhk. *38*, 68 (1894); Erg. Physiol. *2*, No. 1, 326 (1903); Prakt. Erg. Geburtsh. *4*, 293 (1912).

–, Jb. Kinderhk. *64*, 795 (1906); Dtsch. med. Wschr. *1910*, p. 987.

Bässler, R.: Frankfurter Zschr. Pathol. *69*, 37 (1958).

Bellet: According to Brunn.

Bessau, G.: In Feers Lehrbuch der Kinderheilkunde (in Feer's Textbook of Pediatrics). 15th ed., p. 39. Jena, 1944.

Besser, L.: (1853) nach Czerny-Keller, Des Kindes Ernährung usw. (according to Czerny and Keller's The Child's Nutrition, etc.). 2nd ed., *1*, 364. Leipzig and Vienna, 1924.

Bethe, A.: Pflügers Arch. Physiol. *239*, 1 (1938).

Biedert, P.: Dtsch. Arch. klin. Med. *17*, 166 (1876); Berl. klin. Wschr. *1880*, I, 245.

Bilz, R.: Lebensgesetze der Liebe (Vital Laws of Love). pp. 14 and 23. Leipzig (1943).

Blanton, M. G.: Psychol. Rev. (Am.) *24*, 456 (1917).

de Boer, L. J.: Abhandlungen und Versuche geburtshilflichen Inhaltes (Treatises and Investigations of Obstetric Content). Vol. 2, Pt. I, p. 85. Vienna, 1802.

von Bokay, J.: Jb. Kinderhk. *148*, 226 (1937); Wien. klin. Wschr. *1936*, No. 27/28.

Brandes, G.: Buschi. Leipzig, 1939.

Bresslau: Nach H. Eggelin, im Handbuch der vergleichenden Anatomie, herausgegeben von Bolk usw. (according to H. Eggelin, in Handbook of Comparative Anatomy, edited by Bolk et al.). Vol. 1, p. 675. Berlin and Vienna, 1931.

Bromer, R., and J. Wolman: Radiology *32*, 1 (1939).

Brüning, H.: Geschichte der Methodik der künstlichen Säuglingsernährung (History of the Methods of Artificial Nutrition of Infants). Stuttgart, 1908.

BRUNN, E.: Die Wirkung des Coramins auf das Saugvermögen trinkschwacher Frühgeburten (The effect of nikethamide on the sucking ability of weakly drinking premature infants). Kinderärztl. Prax. *15*, 131 (1944) and thesis, Greifswald, 1945.

CANNON: Amer. J. Physiol. *1*, 358 (1898).

CARLSON, A. S., and H. GINSBURG: Amer. J. Physiol. *38*, 29 (1915).

CARSTENSEN, E., and W. J. EWERWAHN: Dtsch. med. Wschr. *1957*, 1338.

CATEL, W.: Normale und pathologische Physiologie und Pathologie der Bewegungsvorgänge im gesamten Verdauungskanal (Normal and Pathological Physiology and Pathology of the Movements in the Entire Intestinal Tract). Pt. 2. Leipzig, 1937.

–, Die Prophylaxe der Ernährung im Säuglingsalter (The prophylaxis of nutrition in infancy). Intern. Verband f. vorbeug. Pädiatrie, Bulletin *10*, p. 4 (1935); Zoologischer Garten, N. F. *16*, 244 (1950); *18*, 19 (1951).

–, and C. A. KRAUSPE: Jb. Kinderhk. *129*, 1 (1930).

CLARK, G. A.: J. Physiol. *54* (1921).

COHNHEIM, O., and F. SOETBEER: Zschr. physik. Chemie *37*, 467 (1902/03).

COLLEY, J. R. T., and B. CREAMER: Brit. Med. J. *1958* II, p. 422.

COMBY, J.: Arch. méd. Enfants *39*, 537 (1936).

CRAMER, H.: Slg. klin. Vortr. *1900*, No. 263, H. 23; Dtsch med. Wschr. *1900* I, 32.

CZERNY, A., and A. KELLER: Des Kindes Ernährung, Ernährungsstörungen und Ernährungstherapie (The Child's Nutrition, Nutritional Disorders, and Nutritional Therapy). 2nd ed., Vol. 1. Leipzig and Vienna, 1925.

DABELOW, A.: Die Milchdrüse. W. Möllendorffs Handb. d. mikrosk. Anatomie des Menschen (The Mammary Gland. In W. Möllendorff's Handbook of the Microscopic Anatomy of Man). *3*, 3, p. 277. Berlin-Göttingen-Heidelberg, 1957.

DARWIN, C.: Der Ausdruck der Gemütsbewegungen bei Menschen und Tieren (The Expression of Emotions in Man and Animals). German translation. p. 47. Stuttgart, 1872; Biographische Skizze eines kleinen Kindes (Biographical Sketch of a Small Child). Minor Works, edited by E. Krause. Vol. 2, p. 134. Leipzig, 1886.

DARWIN, E.: Zoonomie oder Gesetze des organischen Lebens (Zoonomia or Laws of Organic Life). German translation by I. D. Brandis. Vol. 1, p. 256. Hannover, 1795.

DATHE, H.: Zoolog. Garten N. F. *7*, 223 (1934).

DONTENWILL, W.: Mschr. Kinderhk. *101*, 35 (1953).

DREYFUSS, S.: Arch. Méd. Enf. *31*, 261 (1928); Med. Welt *1928*, 1358, 1392.

EBBECKE, U.: Pflügers Arch. Physiol. *246*, 675 (1943); *247*, 222 and 234 (1943).

–, Physiologie des Bewußtseins in entwicklungsgeschichtlicher Betrachtung (Physiology of Consciousness from an Ontogenetic Viewpoint). p. 120. Stuttgart (1959).

ECKHARDT, C.: Beiträge zur Anatomie und Physiologie. 1. Bd. Die Nerven der weiblichen Brustdrüse und ihr Einfluß auf die Milchdrüse (Contributions to Anatomy and Physiology. Vol. 1. The Nerves of the Female Mammary Gland and Their Influence on the Milk Lobules). Gießen, 1858.

ECKSTEIN, A.: Zschr. Kinderhk. *45*, 1 (1927).

ESCHERICH, T.: Münch. med. Wschr. *1888*, I, 687; Die Tetanie der Kinder (The Tetany of Children). p. 57. Vienna and Leipzig, 1909.

FANCONI, G., and A. WALLGREN: Lehrbuch der Pädiatrie (Textbook of Pediatrics). 4th ed., p. 528. Basel, 1956.

FERVERS, C.: Zbl. Gynäk. *1932*, 2424.

FILIPOVIC, S.: Milchwirtsch. Forsch. Orig. *6*, 4 (1928).

FINKELSTEIN, H.: Münch. med. Wschr. *1916*, 1315; Lehrbuch der Säuglingskrankheiten (Textbook of the Diseases of Infants). 4th ed., p. 73. Amsterdam (1938).

FISCHER, G.: In Brüning-Schwalbe, Handbuch der allgemeinen Pathologie und pathologischen Anatomie des Kindes (in Brüning and Schwalbe's Handbook of General Pathology and Pathological Anatomy of the Child). Vol. 1, 2. Wiesbaden (1913).

FISCHL, R.: Mschr. Kinderhk. *24*, 453 (1923).

FLEISCHMANN, W.: Das Molkereiwesen (The Dairy Industry). Braunschweig, 1876.

FORSTER, A.: Arch. Anat. *1904*, 197.

FREUDENBERG, E.: In Pfaundler-Schloßmanns Handbuch der Kinderheilkunde (in Pfaundler-Schloßmann's Handbook of Pediatrics). 4th ed., Vol. 1, p. 785. Berlin, 1931.
FRÜHMANN, E.: N. österreich. Zschr. Kinderhk. *4*, 15 (1959).
FÜRNROHR: Zschr. Neurol. *27*, 375 (1904).
GAINES, W. L., and F. P. SANMAN: Amer. J. Physiol. *80*, 691 (1927).
GAISFORD, W., and S. SCHOFIELD: Brit. Med. J. *1950* I, p. 1404.
GALEN: Nach Oribasios, Oeuvres, Herausg. Bussemaker u. Daremberg (according to Oribasios, Oeuvres, edited by Bussemaker and Daremberg). *3*, 138. Paris, 1856–1876.
GAMPER, E.: Z. Neur. *102*, 154; *104*, 49 (1926).
–, and T. R. UNTERSTEINER: Arch. Psychiatr. *71*, 282 (1924).
GARSCHE, R.: Zschr. Kinderhk. *68*, 587 (1950); Mschr. Kinderhk. *98*, 197 (1950); Pflügers Arch. Physiol. *253*, 72 (1950).
GEHEWE: According to Forster.
GENTRY, E. F., and C. A. ALDRICH: Amer. J. Dis. Childr. *75*, 528 (1948).
GESELL, A.: The Embryology of Behavior. New York, London (1945).
–, The Ontogenesis of Infant Behavior, in: Carmichael, Manual of Child Psychology. 2nd ed. New York, London, 1954.
GINSBURG, H., TUMPOWSKY, and A. J. CARLSON: J. Amer. Med. Assoc. *64*, 1822 (1915 I).
GOERLING: Anat. Record *19*, 259 (1920).
GOETHE, W. v.: Faust. Ges. Ausg. Leipzig o. J. Inselverlag. Paralipomena. No. 61, p. 512.
GOLTZ, F., and J. R. EWALD: Pflügers Arch. Physiol. *63*, 386 (1896).
GOODWIN, T. C.: Amer. J. Dis. Childr. *48*, 309 (1934).
GRAEF, J.: Arch. path. Anat. *28*, 613 (1939).
GRAHAM, S., and N. MORRIS: Arch. Dis. Childh. *4*, 335 (1929).
GRIMM, K.: Münch. med. Wschr. *1916* II, 1315.
GRIMMER, W.: Chemie und Physiologie der Milch (Chemistry and Physiology of Milk). Berlin, 1910.
GRÜNDLER, P.: Hippokrates *1942*, 876.
GUILLOT, N.: J. Kinderkrkh. *19*, 113 (1852).
GUINARD, L.: Étude expér. de Pharmakodynamie comparée sur la Morphine et l'Apomorphine (Experimental study of the comparative pharmacodynamics of morphine and apomorphine). Lyon thesis, No. 107 (1898).
GUNTHER, M.: Lancet *249*, 590 (1945); March 15, 1955, p. 575, No. 6864.
HÄGGSTRÖM, P. F.: Ref. Zbl. Neurol. *19*, 417 (1920).
HALVERSON, H. M.: J. Genet. Psychol., Worcester *64*, 185 (1944); *68*, 205 (1946).
HANNAPPEL, C.: Jb. Jugendpsychiatr. *2*, 231 (1960).
HARNAPP, G. O.: Jb. Kinderhk. *140*, 31 (1933).
HARTMAN, C. G.: Anat. Rec. (Philadelphia) *19*, 251 (1920).
HASSE, C.: Arch. Anat. *1905*, 321.
HEDIGER, H.: Ciba Zschr., No. 59, p. 1968 (1953).
HEIDERICH, F.: Peter-Wetzel-Heiderichs Handbuch der Anatomie des Kindes (Peter, Wetzel, and Heiderich's Handbook of Child Anatomy). Vol. 1, 3rd printing. Munich, 1934.
HELBICH, H.: Mschr. Kinderhk. *10*, 391 (1911); Beih. Arch. Kinderhk. 1939, 19.
VON HELMHOLTZ, H.: Handbuch der physiologischen Optik (Handbook of Physiological Optics). 2nd ed., p. 599. Hamburg and Leipzig, 1896.
HELMREICH, E.: Physiologie des Kindesalters (Physiology of Infancy). Vol. 1, p. 271. Berlin (1931).
HENKE, W.: Gerhardts Handbuch der Kinderkrankheiten (Gerhardt's Handbook of Childhood Diseases). Vol. 1, p. 299. Tübingen, 1877.
VON HERFF, O.: Beiträge zur Lehre von der Galaktorrhoe (Contributions to the Theory of Galactorrhea). Habilitat thesis. Berlin, 1889.
HERZ, M.: Jb. Kinderhk. *7*, 46 (1865 II).
HESS, R.: Zschr. Kinderhk. *19*, 1 (1919).
HESSE, O.: Pflügers Arch. Physiol. *152*, 1 (1913).

HOFMANN, F.: Die angebliche Neubildung von Milch während des Melkens (The Alleged Formation of New Milk During Milking). Thesis. Leipzig, 1881.
HOFMEIER, K., and K. KÖRNIG: Mschr. Kinderhk. *73*, 397 (1938).
VON HOLST, E.: Erg. Physiol. *42*, 228 (1939).
HOOKER, D.: Proc. Assoc. Res. Nerv. Ment. Dis. *33*, 98 (1954).
HOTZ, R.: Orthodontie in der täglichen Praxis (Orthodontia in Daily Practice). p. 26. Bern and Stuttgart, 1954.
HUG-HELLMUTH, H.: Aus dem Seelenleben des Kindes (From the Psychological Life of the Child). 2nd ed. Leipzig and Vienna, 1921.
IBRAHIM, H.: Münch. med. Wschr. *1904* II, 1056.
IKEDA, K.: Amer. J. Dis. Childr. *49*, 985 (1935).
INGELBRECHT, T.: Comptes rend. Soc. Biol. *120*, 1309 (1935 II).
INGRAM, T. T. S.: Little Club Summer Meeting, Groningen, 1960 (unpublished).
VON JASCHKE, R. T.: Zbl. Gynäk. *1909*, 556; Physiologie, Pflege und Ernährung des Neugeborenen (Physiology, Care, and Nutrition of the Newborn Infant). Wiesbaden, 1917; Halban-Seitz, Biologie des Weibes (Halban-Seitz: Biology of the Female). Vol. 5, Pt. 2. Berlin and Vienna, 1926; Pfaundler-Schloßmanns Handbuch der Kinderheilkunde (Pfaundler-Schloßmann's Handbook of Pediatrics). 4th ed., Vol. 1. Berlin, 1931.
JENSEN, K.: Genet. Psychol. Monogr., Worcester *12*, No. 5/6 (1932).
JOGICHESS, M.: Mschr. Kinderhk. *54*, 381 (1932).
KANTOROWICZ, A.: Erg. inn. Med. *38*, 467 (1930).
KANTROW, R. W.: Studies in Infant Behavior III. University of Iowa Studies. Studies in Child Welfare *13*, No. 3 (Iowa, 1937).
KASAHARA, M.: Amer. J. Dis. Childr. *12*, 73 (1916).
KAUPE, W.: Münch. med. Wschr. *1907* I, 126; *1909* I, 343.
KEHRER, F. A.: Beitr. Geburtsh. *1875*, No. 4.
KINGSBURY, B. F.: Anat. Rec. *27*, 305 (1924).
KLEE, P.: Arch. klin. Med. *128*, 204 (1919); Handbuch der normalen und pathologischen Physiologie (Handbook of Normal and Pathological Physiology). Vol. 3. Berlin, 1927.
KÖDDING, I.: Mschr. Kinderhk. *84*, 212 (1940).
KÖHLER, W.: Abderhaldens Handbuch der biologischen Arbeitsmethoden (in Abderhalden's Handbook of Biological Methods). Sect. VI, Pt. D, No. 1, 49th printing, p. 39. Berlin and Vienna, 1921.
KRASNOGORSKI, N. I.: Erg. inn. Med. *39*, 613 (1931).
KRAUS, F.: Dtsch. med. Wschr. *1912*, 393; Zschr. exper. Path. u. Ther. *10*, 363 (1912).
KRAUSE, C. F. T.: Handbuch der menschlichen Anatomie (Handbook of Human Anatomy). 3rd ed. revised by W. Krause. p. 787. Hannover, 1879.
KRAWITZKAJA, R. S.: Physiol. Zschr. der UdSSR (Russian) *37*, 47 (about 1950).
KROH, O.: Zschr. Physiol., Leipzig *100*, 325 (1926).
KUSSMAUL, A.: Über das Seelenleben des neugeborenen Menschen (On the Psychology of the Newborn Infant). Leipzig, 1859; 3rd ed., 1895.
LAMBRANZANI, R., and C. PIANETTA: Rev. Psychiatr. et Psychol. expér. *10*, 148 (1906).
LANGSTEIN, L.: Therap. Mhefte *31*, 363 (1917).
LANGWORTHY, O. R.: Contributions to Embryology *24*, No. 139 (Carnegie Institute of Washington, 1933).
LAPIN, J. H.: J. Pediatr. *45*, 583 (1954).
LASSRICH, M. A.: In Fr. Linneweh, die physiologische Entwicklung des Kindes (in Fr. Linneweh's The Physiological Development of the Child). p. 257. Berlin-Göttingen-Heidelberg, 1959.
LAUTMANN, F.: Rev. frç. Puéricult. *6*, 63 (1939).
LEHNDORFF, H.: Jb. Kinderhk. *66*, 286 (1907).
LELONG, M.: Semaine des Hopitaux *1955* I, 1771.
LEUBE, W.: Arch. Gynäk. *43*, 10 (1892).
LEVY-DORN and S. MÜHLFELDER: Berl. klin. Wschr. *1910*, 388.
LINDNER, S.: Jb. Kinderhk. *14*, 68 (1879).

LITINSKI, O.: Über den Saugakt bei Säuglingen (On the Sucking Act of Infants). Thesis, St. Petersburg. Ref. Jber. Physiol. *11*, 86 (1902).
LOEBENSTEIN, F.: Formunterscheidung im ersten Lebensjahr (Form Differentiation During the First Year of Life). Thesis, Leipzig.
LORENZ, K.: Zschr. Tierpsychol. *5*, 235 (1943).
LUSCHKA, H.: Zschr. rat. Med., Series *3*, *18*, 188 (1863).
MAGENDIE, P.: Lehrbuch der Physiologie (Textbook of Physiology). German translation. Vol. 2, p. 110. Tübingen (1826).
MAGNUS, R.: Körperstellung (Body Posture). p. 204. Berlin, 1924.
MAGNUSSON, J., and W. WERNSTEDT: Acta paediatr. 17, Suppl. *1*, 241 (1935).
MARINESCO, G., and A. RADOVICI: Revue neurol. *27*, 237 (1920).
MARQUIS, D. P.: According to Pratt.
MEITES and TURNER: According to Dabelow.
MILLER, F. R.: Pflügers Arch. Physiol. *143*, 1 (1912).
MORO, E.: Wien. klin. Wschr. *1906* I, 637.
NEBERT and E. W. KOCH: Mschr. Kinderhk. *33*, 385 (1926).
NEUSTÄTTER: Jena. Zschr. Naturw. *29*, 345 (1895).
NEVINNY, H.: Münch. med. Wschr. *1936*, 799.
NOTHMANN, H.: Arch. Kinderhk. *51*, 123 (1909).
NÜESCH, A.: Über das sog. Aufziehen der Milch bei der Kuh (On the So-Called Drawing-Up of Milk in the Cow). Thesis. Zürich, 1904.
VON OPENCHOWSKI, T.: Zbl. Physiol. *3*, 1 (1889).
OPPENHEIM, H.: Mschr. Psychiatr. *14*, 241 (1903); Lehrbuch der Nervenkrankheiten (Textbook of Nervous Diseases). 6th ed., p. 877. Berlin (1913).
OPPENHEIMER, K.: Münch. med. Wschr. *1916* I, 995.
ORAVECZ, P.: Orvosképzés (Hungarian) *24*, Hainiss-Sonderh. 197 (1934).
OTTO, F. M. G.: Med. Klin. *1955*, 1748.
PEPYS, S.: According to Pratt.
PETERS, H.: Arch. Kinderhk. *33*, 295 (1902).
PETIT: Der kgl. Akademie der Wissenschaften in Paris; Anatomische, Chemische u. Botanische Abhandlung (The Royal Academy of Sciences in Paris: Anatomic, Chemical, and Botanical Essay). Pt. 9. (1735–37). German translation. Breslau (1760). Reprint: Das Nest (The Nest). p. 2. Lindau am Bodensee (April, 1948).
VON PFAUNDLER, M.: Verh. Ges. Kinderhk. *16*. Munich, 1899; Wien. klin. Wschr. *1899*, 1012; Sommerfelds Handbuch der Milchkunde (Sommerfeld's Handbook of Milk Science). p. 108. Wiesbaden, 1909; Mon. Bl. Säuglingsfürs. *8*, 33 (1916); Physiologie, Pflege und Ernährung des Neugeborenen (Physiology, Care, and Nutrition of the Newborn). 2nd ed. Munich, 1924; Handbuch der normalen und pathologischen Physiologie (Handbook of Normal and Pathological Physiology). Vol. 14, 1. Berlin, 1926.
PILLERI, G.: Arch. Psychiatrie *200*, 603 (1960).
POPPER, W.: Arch. Psychiatr. *63*, 231 (1921).
PORZIG, W.: Das Wunder der Sprache (The Miracle of Language). 2nd ed., p. 24. Bern, 1957.
PRATT, K. C.: Murchison's Handbook of Psychology. 2nd ed. Worcester, Massachusetts. Clark Univ. Press, 1933.
PRECHTL, H. F. R.: Experientia (Basel) *8*, 6, 220 (1952); Naturwissensch. *39*, 140 (1952); *40*, 347 (1953); Mschr. Kinderhk. *101*, 519 (1953); Klin. Wschr. *1956*, 281.
–, Behaviour *13*, 212 (1958).
–, and W. M. SCHLEIDT: Zschr. vgl. Physiol. *32*, 257 (1950); *33*, 53 (1951).
PREYER, W.: Die Seele des Kindes (The Soul of the Child). 4th ed., p. 203. Leipzig (1895).
RANKE, H.: Virchows Arch. path. Anat. *97*, 527 (1884).
REIMARUS, H. S.: Allgemeine Betrachtungen über die Triebe der Tiere, hauptsächlich über ihre Kunsttriebe (General Contemplations on the Instincts of Animals, Especially Their Artistic Instincts). 3rd ed., p. 343. Hamburg (1773).
REINER: According to Brunn.

RENNERT, H.: Nervenarzt *27*, 467 (1956).
REUSS, A.: Dtsch. med. Wschr. *1939*, 1410; Wien. med. Wschr. *1939*, 596.
RICHARDS, T. W.: Human Biol. *8*, 368 (1936).
RICHARDSON: According to Lelong.
RIPIN, R., and H. HETZER: Zschr. Psychol., Leipzig *118*, 82 (1930).
ROBIN, C., and E. MAGITOT: Gaz. méd. France *1860*, 251.
ROEHRIG, A.: Virchows Arch. path. Anat. *79*, 119 (1876).
ROWAN-LEGG, C. K.: Canad. Med. Assoc. J. *60*, 388 (1949).
RUBINOW, O., and L. FRÄNKL: Zschr. Psychol., Leipzig *133*, 1 (1934).
SCHACHTER-NANCY, M.: Rev. franç. Péd. *13*, 180 (1937).
SCHEER, K.: Mschr. Kinderhk. *33*, 433 (1926); Kinderkrankheiten und Ernährungslehre (Children's Diseases and Nutritional Science). p. 36. Dresden and Leipzig, 1935.
SCHELLER, H.: In R. Wartenberg, Die Untersuchung der Reflexe (in R. Wartenberg's The Examination of Reflexes). p. 184. Stuttgart, 1952.
SCHERBACK, L.: Zbl. Gynäk. *1910*, II.
SCHMELTER, E.: Über den Saugakt mit besonderer Berücksichtigung der Kieferstellung beim Neugeborenen (On the Sucking Act with Special Reference to the Location of the Jaws in Newborn Infants). Thesis, Berlin, 1931.
SCHMIDT, A.: Münch. med. Wschr. *1901*, I, 22; *1904*, II, 2141.
SCHMIDT, ANNI: Zschr. Kinderhk. *45*, 19 (1927).
SCHMIDT, W.: Imago (Vienna) *12*, 377 (1926).
SCHNEIDER, O.: Zschr. Anat. Entw.gesch. *109*, 230 (1939).
SCHUMACHER, P.: Zbl. Gynäk. *1923*, 465.
SCHÜRMANN, P., and H. KLEINSCHMIDT: Pathologie und Klinik der Lübecker Säuglingstuberkuloseerkrankungen (Pathology and Clinic of the Tuberculosis Epidemics of Infants in Lübeck). Arbeiten Reichs. ges. amt *69*, Berlin, 1935.
SEITZ, A.: Arch. Gynäk. *123*, 46 (1925).
SELLHEIM, H.: Arch. Kinderhk. *66*, 312 (1918).
VON SIEBOLD, E.: Mschr. Geburtsh. *18*, 234 (1861).
SIKORSKI, I. A.: Die seelische Entwicklung des Kindes (The Psychological Development of the Infant). 2nd ed. Leipzig, 1908.
SILBERSTEIN, Münch. med. Wschr. *1916*, II, 1411.
DE SNOO, K.: Mschr. Geburtsh. *105*, 88 (1937).
SOLTMANN, O.: Jb. Kinderhk. *9*, 106 (1876).
SORANOS OF EPHESOS: Die Gynäkologie (Gynecology). German translation by Lüneburg. p. 73. Munich, 1894.
STEINBACHER, G.: Zschr. Tierpsychol. *4*, 40 (1940/41).
VON STETTEN-AYSTÄTTEN, B.: Einwirkung von Schallreizen auf die Milchsekretion von Kuh und Ziege (The Influence of Sound Stimuli on the Milk Secretion of Cows and Goats). Thesis, Halle, 1924.
STIRNIMANN, F.: Jb. Kinderhk. *148*, 211 (1936); *149*, 326 (1937); Kinderärztl. Prax. *7*, 210 (1936).
—, Das Kind und seine früheste Umwelt (The Child and His Earliest Environment). p. 16. Basel, 1947.
STRANSKY, E.: Mschr. Psychiatr. *13*, 470 (1903).
SÜSSWEIN, J.: Arch. Kinderhk. *40*, 68 (1905).
SWELL, W. W.: J. Dairy Science *10*, 1 (1927).
SZYMANSKI, I. S.: Zschr. Physiol. *149*, 111 (1920); Abderhaldens Handbuch der biologischen Arbeitsmethoden (in Abderhalden's Handbook of Biological Methods). Sect. VI, Pt. D, No. 1, Issue 49, p. 39. Berlin and Vienna, 1921.
TAHAGI, Y., and J. F. BOSMA: Acta paediatr. *49*, Suppl. 113, 62 (1960).
TAYLOR, T.: Amer. J. Dis. Childr. *14*, 258 (1917).
THIEMICH, M.: Jb. Kinderhk. *51*, 225 (1900).
THOENES, F.: Münch. med. Wschr. *1929*, II, 1464.
THOMÄ, F.: Zbl. Gynäk. *1942*, 235.

THOMPSON, J.: According to Pratt.
TILNEY, F., and L. CASAMAJOR: According to Pratt.
TOULOUSE, E., and VURPAS: Compt. rend. Soc. Biol. Paris *55*, 952 (1903).
TRAINHAM, G., and J. C. MONTGOMERY: Am. J. Nurs. *46*, 767 (1946).
VAHL, V.: Orvosképzés (Hungarian) *24*, Hainiss-Sonderheft 197 (1934).
VALLOIS, K., and C. FLEIG: Compt. rend. Acad. Sc. Paris *140*, 1422 (1905).
VIERORDT, K.: Gerhardts Handbuch der Kinderkrankheiten (in Gerhardt's Handbook of Childhood Diseases). 2nd ed., Vol. 1. Tübingen, 1877.
WALLER, H. K.: Lancet *244*, 69 (1943).
WALLISCHEK, C.: Med. Klin. *1937*, 1440.
WEIGMANN, H.: In W. Fleischmanns Lehrbuch der Milchwirtschaft (in W. Fleischmann's Textbook of Dairy Management). 7th ed., p. 156. Berlin, 1932.
WEINERT, H.: Der geistige Aufstieg der Menschheit (The Mental Rise of Mankind). Stuttgart, 1940.
WEITZ, W., and W. VOLLERS: Zschr. exper. Med. *54*, 152 (1927).
WETZEL, G.: In Peter-Heiderich-Wetzels Handbuch der Anatomie des Kindes (in Peter, Heiderich, and Wetzel's Handbook of Child Anatomy). Vol. 1, p. 635. Munich, 1938.
WIESER, S.: Arch. Psychiatr. *195*, 373 (1957); Der Abbau der oralen Leistungen (The Disintegration of Oral Performance). Ann. Univ. Saraviensis *5*, 211 (1957).
WILLI, H., and F. LÜTHY: Schweiz. med. Wschr. *1952*, 397.
WOLOWIK, A. B.: Jb. Kinderhk. *115*, 185 (1927).
WUNDER, W.: Erg. Biol. *14*, 317 (1937).
ZSCHOKKE: Cited by Grimmer.

10. SLEEP OR SLEEP-LIKE STATES

SLEEP

(1, 3, 7, 8, 9, 51, 76, 78, 7a)

Need for Sleep

The younger an infant is, the longer he usually sleeps. HELLBRÜGGE, LANGE, and RUTENFRANZ (1959) estimate that the newborn infant sleeps approximately 15 to 16.5 hours per day. However, statements on the sleeping time vary considerably in individual cases; it evidently differs greatly from child to child. VON PFAUNDLER (1906) estimates the sleeping time to be 20 to 16 hours for the newborn infant, 16 to 12 hours for the young infant, and 11 to 9 hours for the older child. According to VON ECONOMO (1926), night sleep is approximately 10 hours at the age of six to eighteen and 7 to 8 hours in the adult. Night sleep decreases starting at the age of fifty to fifty-five and lasts only 3 to 4 hours at the age of sixty-five to seventy-five. However, considerable fluctuations of these values are found from person to person.

According to KRASNOGORSKI (1948), children need the following amounts of sleep at different ages:

Age (years)	Hours per Day
6 to 10	11½
11	11
12	10½
13	10
14 to 15	9½

BOGATSCHENKO (1952) gives the following minimum hours of sleep needed by school children:

Age (years)	Hours
7	11 to 12
8	10.5 to 11
9	10.5 to 11
10	10 to 10.5
11 to 12	9 to 10
13 to 14	9 to 9.5
15	9.5 to 8.5
16 to 17	8.5 to 8

In MOORE and UCKO's (1957) investigations, 70% of the infants stopped waking up during the night at the age of three months, another 13% stopped at six months, and 10% never completely overcame their nightly restlessness during the first year of life. During the first three months, diseases seemed to have no lasting effect on night sleep; very severe disturbances affected it several successive nights at the most, sleep then being restored again. Hunger favored the restlessness. However, seasonal differences in light or temperature or a noisy environment had little influence.

HELLBRÜGGE's survey (1959) gives lower values for the necessary sleep at different ages:

Age (years)	Range hrs/mins		hrs/mins
2	13/30		
5 and 6	11/30		
7	9/44	to	11/38
8	9/50	to	11
9	9/25	to	10/52
10	9	to	10/37
11	8/42	to	10/35
12	8/36	to	10/30
13	8/15	to	10/10
14	7/30	to	9/54
15	7/42	to	9/14
16	7/30	to	8/54
17	7/18	to	8/30
18 and 19	7/1	to	8/46

C. BÜHLER and HETZER (1927) made more exact measurements of sleep in infancy (see Table XII). They obtained their values by direct and continuous observations. PETZOLDT (1930) gives the following average minimum values for the duration of sleep: 19 hours for the first three months, 17 for the next three months, 15 for the sixth to ninth month, and 13 hours for the ninth to twelfth month. REYNOLDS (1937) observed 59 young children for six weeks. Each was sleeping in a separate room. The average sleeping time was 12½ hours for the two- to three-year-old, 11½ for the three- to four-year-old, and 11 hours for the four- to five-year-old.

Table XII. Sleep Duration in Percentage of the Day (24 hours)

Age (months)	0	1–3	4–6	7–9	10–12
According to BÜHLER and HETZER	80	60	55	53	49
According to PETZOLDT	—	79	70	63	54

GESELL (1945; see p. 476) investigated the sleeping and waking periods in infants who chose the time of their feedings themselves. Great differences were noted from child to child; one child is punctual and another not. The number of meals drops slowly. The nightly interruptions of sleep disappear. But sleeping during the late morning and in the afternoon is usually kept up during the first year of life.

The duration of sleep depends on the state of health. Markedly cachectic infants in particular need much sleep and neuropathic infants very little (PETZOLDT). The environment influences sleep. According to BAYER (1927), fresh air increases the duration and depth of sleep. According to PETZOLDT (1930), infants in hospitals get measurably less sleep than those on the outside because they are disturbed often. BOGATSCHENKO (1952) found this to be true in older children also. However, KLEITMAN and ENGELMANN (according to HELLBRÜGGE, 1959) found the amount of sleep just as low in infants who were at home. According to these authors, the sleeping time for younger infants is about 14 to 15 hours and that of older infants about 12 to 13 hours per day. During the second year of life it is about 13½ hours; by the sixth year of life it gradually drops to about 11½ hours. From approximately 10 to 11 hours in the seventh year, it drops to 9 to 10 hours at the age of eleven and to 8 to 9 hours at the age of sixteen. This shows that adolescents also need more sleep than adults (HELLBRÜGGE, LANGE, and RUTENFRANZ, 1959).

The young child's great need for sleep is explained in different ways. According to PREYER, it is caused by a lack of oxygen in the brain because of the consumption of great amounts of milk, a circumstance which would also tire an adult. CANESTRINI (1913) also attributes the infant's great need for sleep to the one-sided nutrition with milk. However, an infant fed only farina sleeps just as much. PREYER, K. BÜHLER, and KOFFKA made use of SOLTMANN's experiments on animals for an explanation. SOLTMANN proved that the graphs of muscle contractions of the newborn animal are comparable to those of a tired adult animal. However, the tests on the excitability of an excised animal muscle cannot be applied to the cerebral function of the newborn human infant.

In my opinion the young infant's great need for sleep is to be attributed to the immaturity of his brain. The youngest, only newly activated centers are not yet able to function without interruption; they are easily exhausted by their own activity or by the consumption of nervous energy in another cerebral area, and they must rest until they have recovered again. The inability to function caused by immaturity and the inactivity during sleep are closely related to each other. It is thus understandable that the newly activated centers in the infant's immature brain tend to regress to the state they were in prior to their active participation in cerebral events. Since immaturity depends on age, the need for sleep is especially great in young infants.

According to STIRNIMANN (1947, p. 21), the excitability of the newborn infant is very low for the first 10 to 20 minutes after birth, but increases in 1 to 2 hours. But then the infant falls into a deep sleep which even strong stimuli cannot interrupt. During this time newborn infants practically never cry but cling to the deep sleep which they need so much after birth. The first cause for awakening is usually the evacuation of meconium. Gradually the sensitivity to external stimuli increases and also the readiness to react with defense movements and expressions of emotions, at first only by low whimpering, later by loud crying. It is only then that the newborn infant grasps the nipple tightly and that the discomfort of hunger leads him to sucking.

Premature infants are hard to rouse from sleep and immediately fall asleep again when left alone. This by no means excludes their waking by themselves, moving strongly, and crying. The faculty of being awake thus exists long before the birth of the mature child. Even the unborn child alternates between sleeping and waking if we start from the assumption that he too, like the mature or premature newborn infant, is capable of strong movements only when he is awake. When the abdomen of a pregnant woman lying motionless on her back is observed, violent movements of the fetus not infrequently appear with no recognizable external cause. They may be so strong that they wake the mother from her sleep. Moreover, the longer lasting motor restlessness of the fetus can also be caused by strong sound stimuli (see p. 90). Thus, the infant can be awakened from his sleep before and after birth.

Parents do not always correctly understand their children's great need for sleep. Often children are seen playing in the street when they should have been in bed long ago. They cannot possibly be fresh the next day in school. Parents cannot always avoid taking their children on a trip late in the evening. But the sight of a sleepy child staggering around on a train is surely not pleasing. In most cases it should be possible for the parents to come home earlier by cutting their excursion or visit short. They should make this sacrifice for their child, for the child is more entitled to his sleep than the adult to his pleasure. Under no circumstances should adults be condoned in taking their children to restaurants or places of amusement late in the evening.

Parents must see to it that their child has sufficient sleep even if he himself is not yet ready to go to bed, as so frequently happens. When he is absorbed in a game in the evening, he loses the feeling for his own sleeping needs and begs for permission to stay up longer. When the parents yield, the struggle about going to bed will be repeated every evening, a struggle in which sometimes the child and sometimes the parents are the victors. All this can be avoided if the child is regularly put to bed at a definite hour.

A child who is tired in the evening does not show any physical peculiarities. The case of the overtired child who for some reason has not had enough sleep is different. His face is pale, as those who know him well will especially notice. His eyes are dull and are deep in their sockets. He sulks, is inclined to cry, and "behaves badly." Not infrequently he will not admit that he is tired; the well-meant advice to go to bed is answered with screaming even if he is not able to keep his eyes open. However, if one is not confused by this behavior and forces the child to go to bed against his will, he will again be friendly and happy after he has had enough sleep. His sick appearance and his "naughtiness" have disappeared.

The adult usually changes between waking and sleeping only once, and at the most twice, in 24 hours: He rests during the night and works during the day. SZYMANSKI (1912, 1920) termed this rhythm, which likewise occurs in many animals equipped with vision, "monophasic," and contrasted it with a rhythm he called "polyphasic," in which the change between sleeping and waking occurs several times during the same day. Like so many animals, the human infant also has a polyphasic rhythm, as SZYMANSKI clearly indicated. Since the change from sleeping to waking occurs more frequently in the decorticate dog than in the intact dog (KLEITMAN), the same behavior in the infant may be related to the lack of function of his cerebral hemispheres. In SZYMANSKI's experiments, the infants were not fed regularly and were placed at the breast only when they had cried for at least 15 minutes. Under these conditions, an average of 5 or 6 activity and rest periods distributed rather regularly over 24 hours occurred. In all, the infants had 10 hours of activity and 14 hours of rest. However, it should be considered that in these experiments physiological conditions did not exist, as can be seen from the overextended activity periods. It is just not customary to let infants cry continuously for 15 minutes; instead, they are cared for immediately.

ASERINSKY and KLEITMAN (1955) followed the eye movements of infants one to seven months old by direct observation and the movements of the whole body by actographic tracings. Several movement cycles, which were periodically repeated after 50 to 60 minutes, developed in the course of 24 hours. Each cycle started with a period of increased body and eye movements and was followed by a rest period of approximately 23 minutes. Even with increasing age the rest period remained uninfluenced by stimuli, whereas the movement period depended on the environmental stimuli and on maturation. If meals were given on demand, the intervals between them were whole multiples of the movement cycles. At night the activity and rest phases of a movement cycle and the intervals between meals became longer.

"The infant sleeps during a movement cycle (approximately an hour), then rouses violently, afterward either continuing his sleep for the duration of another movement cycle or staying awake until his needs have been fulfilled. Finally the infant sleeps for one or two movement cycles during the day and for five or more cycles at night."

During the first weeks of life the sleeping and waking periods are distributed rather irregularly throughout the day and night hours. But a certain difference exists even in the youngest infants. In the third week of life, the infants slept 6.4 hours during the day and 8.5 hours at night. Then the ratio changed more and more in favor of night sleep. At the end of six months, the infants slept 3.6 hours during the day and 10.1 hours at night.

HELLBRÜGGE and LANGE (1959) confirmed and supplemented these investigations extensively: Until the third week, the sleeping and waking periods were regularly distributed over the day and night hours. Starting from the fourth week, sleep was more prevalent at night and waking during the day. This shift was still more evident by the eighth month. At the end of the first year the day-night periodicity of waking-sleeping distribution was clearly established: More than 90% of the infants were awake at 7:00, 9:00 and 11:00 AM. About 30% were awake at 1:00 PM, and almost 100% at 3:00 and 5:00 PM; only 10 to 15% of the infants woke up between 7:00 PM and 5:00 AM. At 6:00 AM, 96% of the young children were found awake, but only 43% of the school children and 8% of the adolescents. Thus, sleeping is shifted more and more to the morning hours during development (HELLBRÜGGE, 1959). Compared with that of mature babies of the same age, the development of the day and night sleeping and waking rhythm was distinctly delayed in premature infants (HELLBRÜGGE, LANGE, and RUTENFRANZ, 1959).

It has been found practical to feed infants 5 times at 4-hour intervals and to let them sleep uninterruptedly for 8 hours during the night. Even young infants become accustomed to this conditioned time schedule (see p. 519), despite the longer interval at night.

Phenomena Accompanying Sleep

The most important phenomena which accompany sleep (loss of response, decrease of cardiac rate, lowering of blood pressure, closing of the eyes, slowing down of respiration, and narrowing of the pupils) are present in the adult and child alike. The eyeballs are not infrequently found to be turned inward and upward in the sleeping infant; when the eyelids are passively opened only the white sclera is visible. The same eye position occurs in agony, and sometimes in normal infants when they suck. The common denominator in these conditions is the elimination of higher brain centers (fatigue in sleep, the process of dying in extremis, and withdrawal of nervous energy by the sucking act). According to PLOTKE (1880), the eyeballs frequently move individually and independently of each other during sleep.

The preferred position of the infant is characteristic when awake and especially during sleep (see p. 169). Usually the muscles of the infant relax during sleep just as those of the adult do. But this does not exclude tonic excitatory states; EBBECKE (1926) stresses that lid closing during sleep presupposes not only relaxation of the eyelids but also lasting contraction of the ring muscles.

GÖTT (1955) declares that the pattern of falling asleep, especially the position, sucking habits, or covering of the oral region with the hand, actually constitutes a family characteristic. Among 14 healthy pairs of twins, he always found discordant behavior in the 6 dizygotic twins of the same sex, and always concordant in the 8 identical twins.

GARSCHE (1954) summarized clearly the work on the EEG of children during different stages of sleep. The stage of "falling asleep" is accompanied by changes in the basic activity. The frequency — especially in the older child — may slow down and the amplitude may decrease. In the newborn and in the infant during the first month of life, no clear distinction can be made between waking and sleeping EEG's. At certain stages of the infant's development specific cerebral potentials ("drowsiness waves") frequently occur during somnolence. These are of slow frequency and high amplitude and occur to a great extent in regular succession in groups.

At the end of this falling asleep stage (first transition stage), low amplitude beta waves in short bursts can be seen — but not always — in infants. With increasing depth of sleep, the so-called "spindles" (waves of a constant frequency, mostly 14/sec) appear at increasingly shorter intervals as characteristics of light sleep. The essential sign of the second transition stage, moderately deep sleep, is the simultaneous appearance of spindle activity and slow activity (theta and delta waves). In deep sleep the EEG consists exclusively of delta waves. The individual stages of sleep do not follow each other in succession, but always overlap one another. This overlapping lasts longer in the infant than in the adult. The delta wave stage is usually followed by a stage of low voltage, the so-called "null stage." The depth of sleep in this stage is well known: The delta waves dissolve into intermittent runs which become shorter and shorter and finally disappear. Beta and theta waves occur only infrequently in short stretches. The null stage usually lasts several hours in the adult and 15 to 30 minutes in the infant. During the first year of life, the sleeping hours described generally occur in reverse order until awakening. The moment of awakening cannot be exactly determined in the EEG.

Central Regulation of Sleep

KRASNOGORSKI (1931) demonstrated that sleep occurs in the infant — just as in PAVLOV's experimental animals — because of an inhibition which extends over the cortical and subcortical centers. It is conditioned inhibition which most of all causes sleep when repeated immediately. The inhibitory readiness of the central nervous system gradually increases in the child from morning to evening; accordingly, the inhibitions spread more and more and cause sleep more easily. Sleep inhibition originates in the most fatigued portion of the cortex and spreads from there to the other sections of the cortex and to the deeper centers. As soon as the inhibition in the subcortical centers has become so strong that the influx of stimuli from the outside is interrupted, sleep occurs. Thus, sleep starts in the cortex of the cerebral hemispheres, but is the result of inhibitory events in the subcortical centers.

There is no doubt that this interpretation gives a clear picture of the onset of sleep, provided functioning cerebral hemispheres exist. But certain facts contradict the statement that sleep always originates in the cerebral cortex. Young infants and premature infants, whose cerebral hemispheres are not yet functioning, and some human monsters without cerebral hemispheres (for example, those described by GAMPER) show a change between waking and sleeping. This has also been observed in decorticate dogs and apes. Therefore, the subcortical centers must also have the capacity to evoke sleep inhibition independently and without the influence of the cerebral hemispheres.

VON ECONOMO (1926) demonstrated inflammations in the region of the mesencephalon and diencephalon in encephalitis lethargica and therefore assumed the center of sleep regulation to be in the border region between diencephalon and mesencephalon, the involvement of this center causing the sleep disturbance of this form of encephalitis. Referring to PAVLOV, he assumed that an inhibition of the cerebral hemispheres originating in this center blocks the stimulus conduction and that the numerous sympathetic, parasympathetic, and vegetative centers of the immediately adjacent floor of the diencephalon are thus influenced. Arousal is caused by disinhibition.

In experiments on animals, W. R. HESS (1933, 1949) was able to demonstrate a "sleep center" in the region of the thalamus, i.e., in the wall of the third ventricle: On electrical stimulation from fine electrodes which he inserted in this region, the animal went to its usual sleeping place, assumed its sleeping position, and fell asleep, but could be awakened at any time. As in natural sleep, it remained in contact with its surroundings. The Japanese investigator AKIMOTO and his co-workers (1959) have shown that when the same region is stimulated with other types of currents and currents of greater strength sleep is not elicited, but rather an arousal reaction.

Responsiveness During Sleep, Depth of Sleep
(6, 7, 8, 9, 51)

The famous sleep before midnight.
C. MORGENSTERN

According to ARISTOTLE (On Sleep and Waking), BURDACH (1838), EXNER (1879), ROSENBACH (1880), and CZERNY (1892), the child sleeps especially deeply compared with later years. The proof for this statement is still hard to establish today. However, proof was established for BURDACH's statement: "Sleep is deepest at its beginning, soft and quiet during its course, lightest at the end."

In 1862, KOHLSCHÜTTER devised a method of measuring the depth of sleep by an arousal threshold to graded sound stimuli. MÖNNINGHOF and PIESBERGEN (1883) and many others conducted these experiments in adults.

CZERNY (1892) determined the depth of sleep in the child with electrical arousal stimuli applied to the skin, i.e., with faradic opening induction currents from a variable induction apparatus. He found that during their first weeks of life healthy infants sleep uninterruptedly for three hours; older infants have somewhat longer sleep periods. During this period the depth of sleep rapidly increases at first and gradually becomes lower again during the following hours until awakening. But in older children or in adults the depth of sleep shows two peaks over a period of approximately 8 hours: The first is higher and occurs soon after falling asleep and the second is minor and appears in the early morning hours. CZERNY regarded this as proof that the adult's sleep actually consists of two phases which in infancy are still separated.

The EEG confirmed these findings. According to GARSCHE (1954), two sleeping periods were occasionally observed in the EEG of an infant during an uninterrupted 8-hour sleep: The first period ends approximately 4 hours after falling asleep with the appearance of waking activity, but usually no recognizable awakening occurs. This is followed by a second period in which the individual stages of sleep succeed one another.

According to CZERNY's measurements, the depth of sleep is connected with the body temperature. The increasing depth of sleep 1 or 2 hours after falling asleep is often accompanied by perspiration, which lowers the body temperature. The mother, looking at her child once more before going to bed, finds him covered with perspiration; she is frightened because she has heard about the night sweats of consumptives and the next morning she takes him to the physician. The physician should know that this type of sweating is not pathological.

WÖHLISCH (1957) rejected the measurements of sleeping depth customary until then and established a "depth-of-sleep scale" essentially from the ratio of sleeping threshold to waking threshold. The waking threshold is the stimulus strength which the subject is just able to perceive at a fixed time of day, and the sleeping threshold is the stimulus strength which is felt by the subject at a fixed time during sleep and which is usually sufficient to wake him up.

From the ratio of these two values, WÖHLISCH obtains a physically nondimensional and abstract figure. At his suggestion, JÄHNICHEN (1955) tried to determine this depth-of-sleep figure in 100 children three days to fifteen years old. He worked, as CZERNY had done, with electrical skin stimulation, i.e., square wave impulses generated by an electronic stimulator. In children under three years of age, he was able to ascertain the arousal threshold but not the waking threshold, so that it was impossible to determine the depth-of-sleep figure for this age. The sleeping thresholds of infants and little children did not differ from those of older children. It was possible to determine the arousal threshold and the waking threshold in children between three and fifteen years old. The waking threshold, which had been thought to be unchangeable, fluctuated considerably according to the time of day; it proved to be independent of age. From JÄHNICHEN's observations, it seems that the belief in an especially great depth of sleep in infancy is not justified, at least in relation to electrical skin stimuli.

Especially deep sleep was found in epileptics and in a child suffering from chorea minor. In a thirteen-year-old girl with nocturnal enuresis, the deepest sleep of all those recorded was encountered during the first hours of sleep, i.e., the time of enuresis.

By using a nondimensional figure as a measure of the depth of sleep, it was possible to compare the depths of sleep in relation to various sensory areas. It was found that the senses of pain and of pressure are more deeply asleep than the auditory sense. According to WINTERSTEIN (1957), this is related to the dangers which may arise if the sleeper is not awakened in time: For a creature living in hostile surroundings, a danger which can produce pressure or pain stimuli is much greater than a peril which is perceptible from a distance as an auditory stimulus. The former danger does not allow much increase in the threshold.

According to HEDIGER (1959), the receptors are in general not completely eliminated in the sleeping animal; rather, the animal remains attentive to stimuli which may be caused by its enemies. Harmless and dangerous stimuli can be distinguished even during sleep, and only hostile stimuli are followed by awakening, startle, and perhaps speedy flight. Since the nondomesticated animal is always threatened by enemies, it has to be constantly on guard.

Man also to a certain degree still watches over the stimuli of his surroundings during sleep. "The mother awakens at the slightest movement or the faintest sound of the infant" (BURDACH, 1838, p. 486). How constantly the sleeping infant is aware of the presence of his mother is described on p. 606. Figure 1, p. 24 and Figs. 10 and 11, pp. 86 and 87, show the changes in respiratory rhythm that occur in the sleeping infant under the influence of sound stimuli without his awakening.

The miller who is awakened by the stopping of his mill has become proverbial. A corresponding occurrence is the re-evocation of extinguished sucking movements by the soundless elimination of a light source (phenomenon of dominance, see p. 531).

Even if the sleeping infant does not wake up, stimuli may evoke defense movements. Thus, either local or general reactions, which correspond to the skin protection reflexes (see p. 30), occur after painful stimuli (pinpricks). The simplest reaction to a pinprick is the flight movement. From the same position the leg is sometimes flexed and sometimes extended; the head is turned to the other side. The stimulated area may also be scratched with the ipsilateral hand. PFLÜGER (1853) was already familiar with these movements of the sleeping child and compared them to the reactions of decerebrate frogs to local stimulation.

A general restlessness occurs after stronger pain stimulation. The movement is not restricted to the stimulated body part, but spreads to other areas. Thus, when the face is stimulated the legs may be flexed or the arms extended or the whole body may be rolled over on the other side. The same occurs on stimulation of other senses, for example, on sound stimuli. The response shows that the depth of sleep has considerably decreased and that waking is not far off. Not infrequently the body is rolled over to the other side after some generalized movements, the head is flexed forward, and the arms and legs are drawn up so that the infant finally assumes a new sleeping position. If all additional external stimuli then subside, the former depth of sleep rapidly returns. However, if new stimuli continue, the sleeper finally awakens. Sleep can finally be interrupted by a summation of subthreshold arousal stimuli. This shows that the individual stimulus, even if it does not cause awakening, temporarily reduces the depth of sleep.

During the half-awake state, some constant inner stimuli previously without effect may become effective. The sleeper thus suddenly starts to cough on stimulation of any sensory organ. In the child, such a stimulus sometimes causes peculiar rhythmic movements of the mouth which TRÖMNER (1921) described as "tasting movements." There is no reason why the sleeping child should suddenly taste something. The phenomenon can be explained differently: Some children put their fingers in their mouths during sleep. If such a child is examined, it is found that he starts to suck on his finger on any stimulation, is calmed by this sucking, and falls asleep again. Other children put their fingers into their mouths to suck on them only when they are stimulated, and again others show only the rhythmic sucking movements. Even the rooting reflex can be elicited during sleep (see p. 404). HÄGGSTRÖM (1920) noted it as late as in the seventh year. The mature brain has a tendency to regress to a lower stage of development during sleep.

Similar phenomena may still be found in the adult: The gritting of the teeth during sleep corresponds to sucking movements of the dentate lower jaw.

During normal sleep the (unconditioned) subcortical reflexes are more or less decreased; the (conditioned) reflexes of the cortex cannot be elicited at all. ROSENBACH (1880) noted that the patellar, abdominal, and cremasteric reflexes disappeared during very deep sleep; the defense movements became less distinct and palmar and plantar reflexes were more difficult to elicit. According to BICKEL (1902) and FINK [1945 *(7a)*], BABINSKI's sign can be elicited during very deep sleep.

According to my findings [1924 *(7)*], the galvanic skin reflex (VERAGUTH's psychogalvanic reflex), which passes through the truncus sympathicus and can be demonstrated with a sensitive galvanometer, depends on the depth of sleep. With equal stimulus strength it disappears during deep sleep, but is present in light sleep. Stimuli which reduce the depth of sleep cause the reappearance of the reflex. It cannot be proved that it is coupled to conscious processes. The reaction pattern of this reflex is the same during sleep as that of the dilatation of the pupils to be described later, since both are caused by sympathicus innervation. The sleeper's various bodily responses to stimuli are independent of each other.

The deeper the sleep, the narrower the pupils are. They do continue to respond, however, but the narrower the pupil, the stronger the stimulus must be (SANDER, 1877, 1879; RÄHLMANN and WITKOWSKI, 1877, 1878; PLOTKE, 1880; PIETRUSKY, 1922). This order of events can be best observed when the child is suddenly awakened. The pupils will even dilate then if they are illuminated. Therefore, light is always blinding on first awakening. The usual reaction to light occurs only shortly a little while later.

Another reflex of the autonomic nervous system can be observed in the cerebral pulsation [1924 *(8)*]. I tested two children, a seven-year-old girl in whom a bony defect of the skull had been created because of epilepsy and a one-and-one-half-year-old rachitic child in whom the closing of the anterior fontanel was delayed. When the cerebral plethysmogram of these children was recorded during sleep, a fluctuation of the brain volume on external stimulation occurred with great regularity. The volume increased for a short time and the pulsations simultaneously became smaller; then the volume rapidly diminished below its previous value while the pulsations regained their former size. The decrease of the brain volume lasted approximately three to four times as long as the preceding increase.

Mere observation of the sleeping child shows a change in the respiratory rhythm on external stimulation [1924 *(6)*; 1934 *(51)*]. This phenomenon can serve to demonstrate the newborn infant's excitability to sound stimuli (see p. 24). In incompletely interrupted sleep, stimulation of the respiratory center sometimes causes a temporary acceleration, slowing, or constriction of breathing, but occasionally a deepened expiration. But usually another very significant response occurs: At first a very deep breath followed by a respiratory arrest and less frequently by some shallow breaths can be noted (Figs. 10 and 11, pp. 86 and 87). It can be clearly recognized in Fig. 11 that the deepest breath is followed by the longest respiratory arrest. After this response to the stimulus has subsided, i.e., after approximately 5 to 10 seconds, the previous respiration may be restored. But frequently the following breaths remain irregular in depth and quite often periodic respiration even follows the first reaction (Fig. 10). This lasts for about two to three periods of equal strength and then changes to the previous respiratory pattern.

The most frequent effect of the sensory stimulus is thus a strong excitation of the respiratory center which is followed by a strong inhibition. The periodic respiration which then occurs reflects repeated changes between weaker excitations and inhibitions. All these phenomena occur successively in the respiratory center. They prove that the nervous energy, i.e., the ability to perform nervous activity, is limited.

If the stimulation is repeated a second time shortly after the first response has subsided and the respiration has again calmed down, a second reaction may be completely absent even if the second stimulation is considerably stronger than the first (Fig. 11). Sometimes the respiratory center becomes unresponsive only after the second or third stimulation, provided the child was not awakened by the stimuli. The unresponsiveness lasts only a short time, but is still present 10 to 15 seconds after the first stimulation. It seems obvious to interpret this phenomenon as an inhibition. However, there is no doubt that in this case we are not dealing with an inhibition of the respiratory center itself, since it continues its automatic function undisturbed despite the stimulation. Under corresponding conditions, the phenomenon of a temporary unresponsiveness to the same stimulus after a fright reaction can be demonstrated even in the adult.

Pavlov described a similar occurrence in the waking animal: If a new stimulus suddenly appears in the surroundings, an investigatory reflex occurs. The animal directs its sensory receptors toward the new stimulus, and looks up or listens ("What-is-the-matter?" reflex). If the same stimulus is repeated at short intervals or if it is in effect for some length of time, the investigatory reflex gradually becomes weaker and finally disappears.

I. F. WAGNER (1937) distinguishes seven stages of transition in newborn infants in their reactions to pain, touch, smell, and sound. They range from deepest sleep, during which the infant is completely calm, breathes regularly, and does not move his eyelids or his mouth, to the state of being awake, when generalized movements occur, the eyes are open, and the mouth is moved. The lighter the sleep of the infant, the more widespread the reactions are and the longer they last.

Sleep Disturbances

(76)

As the infant falls asleep, conditioned reflexes connected with the process can be formed easily and unintentionally; if they are disrupted, a state of excitement occurs and the process of falling asleep is halted. One child falls asleep only if his mother is present, another even wants her to hold his hand as he falls asleep, another demands that the light be kept on in the room, etc. If such a demand is granted for only a few successive evenings, the conditioned reflex is formed and later on tears will have to be shed to extinguish this reflex again. I knew a young child who could fall asleep only when he was holding a yellow rag in his hands. He rejected any other color. Some conditioned reflexes of the first year of life may persist until school age, as, for example, the habit of sucking on a rubber nipple or on a certain finger while falling asleep.

Many children sleep through the whole night without interruption. They are put to bed in the evening and regularly wake up in the morning without being heard from in the interim. Others are not so peaceful. They wake up suddenly after a more or less long sleep, call their mothers or scream for some time, and cannot be calmed. This sleep disturbance (see p. 605) may recur many nights in succession. However, undisturbed nights always follow these periods of restlessness. This behavior may indicate the beginning of a contagious disease, but frequently the mother knows of no reason for it and the physician may also not be able to find any pathology. We are not speaking here of the disease entity of the true pavor nocturnus, but of an everyday condition of normal children who show no indication of neuropathy, either now or later. Children of the same family may act differently during this critical time of the first years of life.

The cradle, which was a very good way to calm the crying infant, has disappeared today (see p. 606). However, the nipple, which has the same effect, has not been abolished. It is not as dangerous as it was long thought to be.

Apart from the fact that the cradle and the cradle song are unfortunately vanished ornaments of German folk art, it is regrettable that the mother has been deprived of a harmless and effective calming agent without receiving a substitute for it. However, even today the experienced mother does not forego the calming influence of rocking, even though she has no cradle. She calms her child by rocking him in her arms or in his baby carriage. It is almost always possible to calm a screaming baby quickly in this manner. The mother likes to make use of this calming effect at night: The baby carriage with the baby is next to her bed. As soon as the child stirs, a tired hand is stretched out to move the carriage back and forth. The child calms down, but as soon as the movement stops he starts to cry again. Sometimes this lasts for hours and even through the whole night.

What advice can the physician give when the mother consults him on the nocturnal restlessness of her child? Some physicians recommend placing the baby in a distant room and letting him scream as long as he likes. It has occasionally happened that an infant who was so disposed screamed so long that he suffered a rupture. But apart from this, a mother does not like to ignore her child. It would often be better to separate the parents' bedrooms, but this is not always feasible. If people are nervous, it is not so much the degree and loudness of the crying as the mere fact that the child is crying which is disturbing, even if several soundproof walls are in between. Therefore, a sedative is not infrequently indicated. If the infant receives a sufficient amount, three people can sleep undisturbed through one tablet, not to speak of the neighbors, and family peace is restored again. The omission of the afternoon nap in the young child is on occasion surprisingly effective. Sufficient exercise in the open air (not in the house) brings about the necessary fatigue.

It is impossible to determine at what age children begin to dream. According to HOCHE (1926), they may dream very early, but definitely at the age of two. He himself cannot remember any time in his earliest childhood when he did not dream.

Children have very vivid dreams. Nightmares cause the disease entity of pavor nocturnus, in which the children startle in their sleep, usually at the same time each night, and cry, frightened by their dreams. RICHARD WAGNER suffered from this disorder, as he wrote in his memoirs:

> "Until my later boyhood years, no night passed that I did not awaken, screaming terribly from a ghostly dream. This never ended until a human voice told me to be quiet."

HYPNOSIS

(7)

The child is suggestible to such a degree that physical pain, for example, can be abolished by mere blowing on the painful spot. Suggestion is an important aid in education.

True hypnosis of the child scarcely differs from that of the adult. The statements on the age at which the first hypnoses can be achieved vary greatly according to the criteria demanded for a successful hypnosis. The hypnosis of a six-year-old child usually corresponds to that of an adult. Something like it can be found even at the third year of life.

In my experience, the hypnotized child always shows the galvanic skin reflex [1924 *(7)*]. The reflex does not disappear even if one tries to eliminate by appropriate suggestion the sensory area which is to be stimulated. However, in the hypnotized adult the galvanic reflex was noted to disappear, though not without exception, if the stimulated sense was eliminated by the appropriate suggestion.

Tonic Immobility ("Animal Hypnosis")

(1, 2, 78)

Science became familiar with "animal hypnosis" through the work of SCHWENTER (1636) and FATHER KIRCHER (1646). Both authors described how a chicken would remain motionless for some time if it was pressed to the ground and a line starting from its beak was quickly drawn on the ground with chalk. The animal freezes in this uncomfortable position until it is awakened again by a special stimulus. FATHER KIRCHER's "miraculous experiment of bewitching a chicken" has been repeated frequently. Besides the chicken, various animals of many species can be brought to temporary immobility. Especially suited to this are locusts, spiders, crabs, snakes, frogs, turtles, and crocodiles; still other birds in addition to the chicken, such as swans, owls, eagle owls; and finally guinea pigs, rabbits, dogs, and monkeys. VÖLGYESI (1938) even hypnotized a lion in captivity. The hypnosis was achieved by having the animal focus on an object as it was moved past, by rocking the animal from side to side, by suddenly changing the animal's position, especially by turning it on its back, by Mesmer's magnetic passes, by regular stroking with the hand, or by frightening the animal.

The "hypnotized" animal loses its mobility and freezes in an uncomfortable and unaccustomed position. Thus, the Egyptian snake charmer can change a poisonous snake into a "staff" by a hypnotizing grip, a procedure that makes a great impression on lay people. Lizards, crocodiles, and birds stay in a fixed position on their backs, crabs can be placed upside down in curious postures, etc. Even animals as shy as lizards or chickens lose their flight reflexes in the frozen state. The centers for postural correction or locomotion are suddenly inhibited; sensory activity, reflex excitability, and muscle tone change.

The freezing of the animals may last for hours. In PREYER's (1890) experiments, the hypnotized frog remained immobile more than 24 hours. In many cases the animals even died, although they were protected against dehydration. Sudden stimuli of various kinds, such as sounds, air blasts, electrical stimuli, or sprinkling with acids, immediately interrupt the immobility; the animal awakens and runs away.

In the course of time, "animal hypnosis" has been described under the most varied names, such as animal magnetism, akinesis or immobilization, reaction inhibition, tonic immobility, catatonia, catalepsy, cataplexy, etc., depending on which sign was regarded as characteristic for the condition.

Some forms of "animal hypnosis" have great biological significance: The snake "charms" its victim. Some animals "play dead" when they are in danger. However, in most cases, for example in SCHWENTER's and KIRCHER's experiments, the purpose of the freezing cannot be recognized. According to STEINIGER (1936), to whom we are indebted for the latest major work on "animal hypnosis," a comprehensive explanation of its nature which would do justice to all individual cases cannot yet be given. Phenomena which have no biological relation to each other may even be included in this group.

PAVLOV already recognized that internal inhibition, sleep, and hypnosis are all caused by the same process, i.e., a central inhibition. The only difference is the degree to which this inhibition extends over the brain. During a slow experimentally enforced transition from waking into deep sleep, a number of stages corresponding in important points to human hypnosis can be observed in dogs.

PAVLOV states:

"In this kind of sleep the activity of the cerebral hemispheres persists. All conditioned reflexes remain ineffective and there is no reaction to any new stimuli (orientation reflex) unless they are of excessive strength, but all the time the experimental animal maintains a completely active posture. It stands with eyes open but fixed and with the head in the normal erect position on stiff legs without supporting itself by the straps; it stands immobile for minutes or hours."

PAVLOV drew the obvious conclusions in reference to human hypnosis. VÖLGYESI (1938) developed the theory still further. He considers the nature of hypnosis to be such that different cerebral areas are eliminated layer by layer down to the medulla oblongata starting with the youngest and highest brain portion, the frontal lobes, while at the same time the functioning of those centers still alert is enhanced.

The method of "animal hypnosis" can often be applied to the young child who, as previously mentioned, cannot yet be hypnotized in the same way as the adult.

The well-known entomologist FABRE tells about his school years in southern France, when with youthful mischievousness he and his friends hypnotized whole flocks of chickens by seizing the animals, putting their heads under their plumage, moving them up and down in this position, and then left them as dead or dying on the lawn, to the anger of the peasants. This procedure has proved quite effective with birds. VÖLGYESI was able to immobilize even wild birds by this method. A mother who calms her child proceeds in a similar way (see p. 606).

A further characteristic of "animal hypnosis" is the inhibition of motion in the previously very mobile animal. MANGOLD and ECKSTEIN's hypnosis apparatus consists of a device to turn the animals (rabbits, guinea pigs, dogs, and cats) suddenly on their backs. An infant may frequently be calmed in this manner, although not without exception. For this purpose the crying infant is held so that one hand of the examiner covers the back and the other the chest and abdomen. If the infant is now turned head over heels from the dorsal into the abdominal position, he stops crying; however, he will not become completely immobile. If he is turned back, he starts to cry again. With infants who react well, this experiment can be successfully repeated many times in succession. Since MANGOLD (1914, 1920) and STEINIGER (1936) described movements during "animal hypnosis," complete immobility can no longer be regarded as an absolutely necessary characteristic of this condition.

A manipulation — which is not unobjectionable because of the suffocation danger — is used in infant nurseries. The crying infant is placed on his stomach and this generally calms him down. Another, though less frequently successful, method consists of shouting briefly and loudly at the infant. Many crying infants immediately calm down when their faces are covered with a piece of cloth.

At approximately the second to fourth year of life, a quite striking phenomenon occurs in some children: If one limb of the sitting or lying child is brought into an uncomfortable position, for example, if one arm or one leg is raised, this being done without verbal suggestion, the limb remains in this position for a long time without showing signs of fatigue. The illustrations (Figs. 204 and 205) of a three-year-old child reproduced here show such positions. The raised limbs are not completely immobile, but sway more or less to and fro and finally are slowly lowered. LESAGE (1912) observed a child two years and ten months old who kept his arm raised for 45 minutes; he frequently saw durations of up to ½ hour.

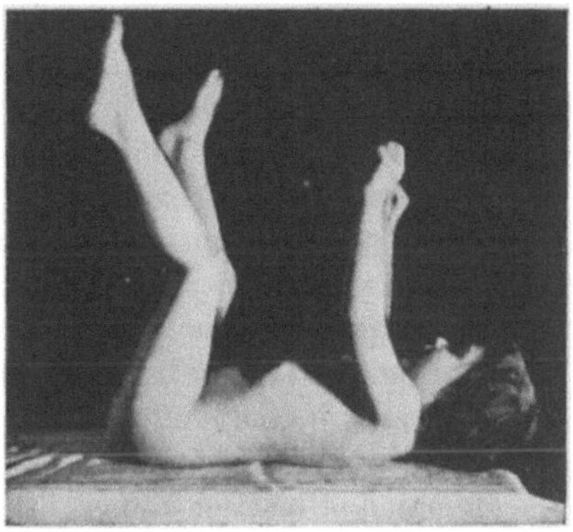

Fig. 204. Tonic immobility in a three-year-old girl in the dorsal position. The legs remained raised for 10 minutes, the arms for 2.

In the literature this phenomenon is evaluated in various ways. The French (LESAGE) regard it as a physiological phenomenon, but EPSTEIN (1897) and THIEMICH (1898) believe it to be pathological. EPSTEIN's conjecture that it may be attributed to rickets could not be proved. Dull, apathetic children are particularly suited for these experiments, but they are not necessarily retarded. Their minimal motility impulse in comparison with other children of the same age — also apart from this experiment — is striking. If a little wooden block is placed on the head of such a child (when he is sitting), it takes a long time until it would falls, while it stay on the head of a healthy child for only a few seconds (CZERNY).

If one explores this phenomenon, the condition described can be found frequently. HERMANN (1919) observed it in 8% of children between the ages of twenty months to six years; LESAGE was able to demonstrate it in all tested children between the ages of one and one half to two years. I found it especially in institutionalized children and in this case attribute it to the lack of intellectual stimulation (hospitalism, see p. 631).

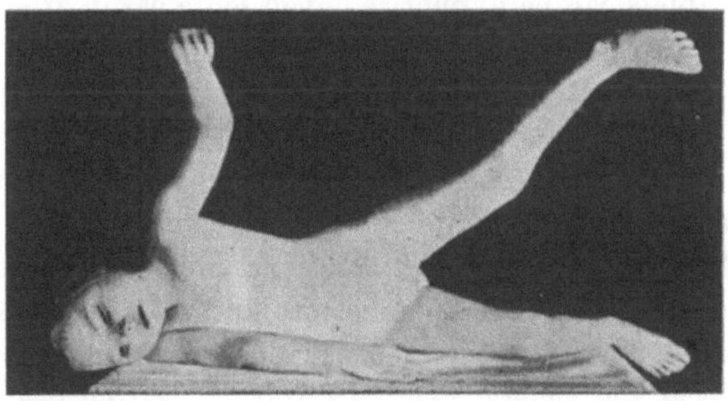

Fig. 205. Tonic immobility in the lateral position. The leg remained raised for 6 minutes, the arm for 1.

This phenomenon did not become well known and was therefore repeatedly rediscovered, for example, by SURANYI (1932), VON MORITZ and J. KÖNIG (1938), and ROGATZ and PICKER (1939). EPSTEIN (1897) was the first to describe it in any detail.

The condition corresponds absolutely to the phenomena of "animal hypnosis." All investigators stress that they were not able to demonstrate any pathological-psychological or neurological signs in the immobile children. It seems to me, however, that there is a certain rapport between examiner and child which corresponds to that of true human hypnosis.

The phenomena of "animal hypnosis" have been described here only as far as they can be externally determined. However, in many cases the content of consciousness of the "hypnotized" animal has been investigated. PREYER thus states that the inhibitory reaction is a cataplexy, a fright-freezing; STEINIGER interprets it similarly as an effect of fear. However, the question of the content of consciousness can be answered just as little for the animal as it can for the infant. I therefore regard the attempt to interpret "animal hypnosis" of the infant psychologically as useless. Perhaps a psychological explanation would be more promising in the young child, since he can talk. Evidently he is not frightened during this condition of immobility.

"Animal hypnosis" is caused by physiological stimuli; in human hypnosis, during which attempts are made to change the content of consciousness by suggestion, there is a characteristic rapport between hypnotizer and hypnotized. However, a rapport is also suggested in the "hypnosis" of higher animals, but not in that of insects (HABERLAND, 1926, STEINIGER, 1936, VÖLGYESI, 1938). STEINIGER observed that the inhibition of reaction always disappeared in young sea gulls when no attention was paid to them. While they remained immobile for half an hour or longer if observed, they immediately became mobile if they were not observed and, cleverly taking advantage of the situation, made goal-directed attempts to escape. When a rapport is also suggested in the hypnosis of the young child as described, this corresponds to similar phenomena in the "hypnosis" of higher animals.

Hypnosis is not obviated by the fact that the cerebral hemispheres of the infant are not yet functioning. Even in vertebrates, the removal of the cerebral hemispheres does not abolish the susceptibility to hypnosis [VERWORN's (1898) experiments on chickens, SZYMANSKI's (1912, 1920) on rabbits]. According to SPIEGEL (1927), rabbits can be hypnotized even more easily after this operation.

SUMMARY

The child's need for sleep decreases with increasing age. Sleeping and waking states alternate more often in the course of 24 hours in the infant than in the adult. The infant has a characteristic preferred posture while awake and particularly during sleep. The sequence of depth of sleep is essentially the same in children and adults. Under the influence of sensory stimuli which do not cause complete awakening, local and general reactions appear. During deep sleep, the pupillary reaction, the galvanic skin reflex, and the tendon reflexes disappear; BABINSKI's sign may at times be elicited. Sleep disturbances are especially frequent in childhood and require an understanding environment. Sleep inhibition in the PAVLOVian sense must originate in the brain stem in the young infant, since his cerebral hemispheres are not yet functioning.

Simple hypnosis succeeds after the sixth year of life. Methods used in "animal hypnosis" can be employed in the infant, such as calming him by rocking or by turning. Tonic immobility not infrequently occurs in the young child. It is favored if the child lives in an institution.

REFERENCES

von Allesch, G. J.: Naturw. *1921*, 774.

Akimoto, H. N.: According to M. Schneider and H. Hirsch: Med. Klin. *1959*, p. 933.

Aristotle: Über Schlaf und Wachen. Kleine Schriften zur Seelenkunde (On Sleep and Waking. Minor Treatises on Psychology). p. 86. Paderborn, 1947.

Aserinsky, E., and N. Kleitman: J. Appl. Physiol. *8*, 11 (1955); according to Hellbrügge and Lange.

Bayer, W.: Jb. Kinderhk. *117*, 125 (1927).

Bickel: Zschr. Nervenhk. *22*, 163 (1902).

Bleckmann, K. H.: Der Schlaf des Kindes (The Sleep of the Child). Göttingen, 1955.

Bogatschenko, L. S.: Pawlow Zschr. höh. Nerventätigk. (German) *2*, 269 (1952).

Buckman, S. S.: The Nineteenth Century *36*, 1894 II, p. 733.

Burdach, H. F.: Die Physiologie als Erfahrungswissenschaft (Physiology as an Empirical Science). 2nd ed., Vol. 3, p. 479. Berlin, 1838.

Bühler, C., and H. Hetzer: Inventar der Verhaltensweisen des ersten Lebensjahres (Inventory of Behavior During the First Year of Life). Jena, 1929.

Czerny, A.: Jb. Kinderhk. *33*, 1 (1892).

Ebbecke, U.: Handbuch der Physiologie (Handbook of Physiology). Vol. 17, p. 563. Berlin, 1926.

von Economo, E.: Ibid., p. 591.

Epstein, A.: Verh. 13. Versammlung Ges. Kinderhk. p. 38. Wiesbaden, 1897.

Exner, S.: Hermanns Handbuch der Physiologie (in Hermann's Handbook of Physiology). II, 2, p. 293. Leipzig, 1879.

Fabre, I. H.: According to Mangold (1914), p. 13.

Gamper, E.: Zschr. Neurol. *102*, 154 (1926); *104*, 49 (1926).

Garsche, R.: Elektroenzephalographie, in Brocks Biologische Daten für den Kinderarzt (Electroencephalography, in Brock's Biological Data for the Pediatrician). 2nd ed., *2*, 856. Berlin-Göttingen-Heidelberg, 1954.

Gött, H.: Homo *5*, 60 (1955).

Haberland, A. F. O.: Beitr. klin. Chir. *135*, 370 (1926).

Häggström, P. F. N.: Ref. Zbl. Neurol. *19*, 417 (1920).

Hediger, H.: Med. Klin. *1959*, 938.

Hellbrügge, T.: Med. Klin. *1959*, 954.

—, and J. Lange: Med. Klin. *1959*, 946.

—, J. Lange, and J. Rutenfranz: Schlafen und Wachen in der kindlichen Entwicklung (Sleeping and Waking in the Child's Development). Beih. z. Arch. f. Kinderheilk., No. 39. Stuttgart, 1959.

Hermann, E.: Mschr. Kinderhk. *15*, 533 (1919).

Hess, W. R.: Klin. Wschr. *1933*, 129; Das Zwischenhirn (The Diencephalon). Basel, 1949.

Hoche, A.: Handbuch der Physiologie (Handbook of Physiology). Vol. 17, p. 622. Berlin, 1926.

Jähnichen, S.: Über Schlaftiefenmessungen. Übersicht und Versuche im Kindesalter (On Measurements of the Depth of Sleep. Review and Experiments in Infancy). Doctoral thesis, Würzburg, 1955.

Kohlschütter: Messungen der Festigkeit des Schlafes (Measurements of the Depth of Sleep). Thesis, Leipzig, 1862.

Krasnogorski, N.: Erg. inn. Med. *39*, 613 (1931).

—, (1948); see Bogatschenko.

Lesage: Krankheiten des Säuglings (Diseases of the Infant). German translation. p. 87. Leipzig, 1912.

Mangold, E.: Hypnose und Katalepsie bei Tieren im Vergleich zur menschlichen Hypnose (Hypnosis and Catalepsy in Animals in Comparison to Human Hypnosis). Jena, 1914; Erg. Physiol. *18*, 79 (1920).

Mönninghoff and Piesbergen: Zschr. Biol. *19*, 114 (1883).

MOORE, T., and L. E. UCKO: Arch. Dis. Childh. *32*, 333 (1957).
VON MORITZ, D., and J. KÖNIG: Wien. klin. Wschr. *1938*, 709.
MORGENSTERN, C.: Palmström. Berlin, 1920.
PETZOLDT, G.: Mschr. Kinderhk. *45*, 193 (1930).
VON PFAUNDLER, M.: Pfaundler-Schloßmanns Handbuch der Kinderheilkunde (Pfaundler-Schlossmann's Handbook of Pediatrics). I, 1, p. 157. Leipzig, 1906.
PFLÜGER, E.: Die sensorischen Funktionen des Rückenmarks der Wirbeltiere (The Sensory Functions of the Spinal Cord of Vertebrates). p. 134. Berlin, 1859.
PIETRUSKY, Ref. Zbl. Physiol. *14*, 397 (1922).
PLOSS, H.: Das kleine Kind vom Tragbett bis zum ersten Schritt (The Infant from the Portable Bed to His First Step). p. 68. Leipzig, 1881.
PLOTKE: Arch. Psychiatr. *10*, 205 (1880).
PREYER, W.: Die tierische Hypnose (Animal Hypnosis) (1890); Die geistige Entwicklung in der ersten Kindheit (Mental Development in Early Infancy). p. 36. Stuttgart, Berlin, Leipzig, 1893.
RÄHLMANN and WITKOWSKI: Arch. Anat. Physiol. *1877*, 455; *1878*, 109.
REYNOLDS, M. M.: Ref. Zbl. Kinderhk. *32*, 9 (1937).
ROGATZ, J. L., and M. PICKER: Arch. Pediatr., New York *56*, 444 (1939).
ROSENBACH, O.: Zschr. klin. Med. *1*, 358 (1880).
SANDER: Arch. Psychiatr. *7*, 652 (1877); *9*, 129 (1879).
SPIEGEL, E. A.: Der Tonus der Skelettmuskulatur, Monograph (The Tonus of the Skeletal Musculature, Monograph). Ges. geb. Neurologie. Vol. 51. Berlin, 1927.
STEINIGER, F.: Erg. Biol. *13*, 348 (1936).
SURANYI, J.: Wien. klin. Wschr. *1932*, 234.
SZYMANSKI, R. S.: Pflügers Arch. Physiol. *149*, 11 (1912); Zschr. Physiol. *18*, 105 (1920).
THIEMICH, M.: According to E. Hermann (1919).
TRÖMNER: Problem des Schlafes, in Löwenfelds Grenzfragen (The Problem of Sleep, in Löwenfeld's Borderline Problems). p. 21. Wiesbaden, 1921.
VERWORN, M.: Die sog. Hypnose der Tiere. Beitrag zur Physiologie des Zentralnervensystems I (So-Called Animal Hypnosis. Contribution to the Physiology of the Central Nervous System I). Jena, 1898.
VÖLGYESI, F.: Menschen- und Tierhypnose mit Berücksichtigung der Stammes- und Einzelentwicklung des Gehirns (Human and Animal Hypnosis in Consideration of the Phylogenesis and Ontogenesis of the Brain). Zürich and Leipzig, 1938.
WAGNER, I. F.: J. Genet. Psychol. 51, 17 (1937).
WÖHLISCH, E.: Klin. Wschr. *1957*, pp. 480 and 705.
WINTERSTEIN, H.: Schlaf und Traum (Sleep and Dream). 2nd ed. Heidelberg, 1957.

11. CONDITIONED REFLEXES

(1, 3, 19, 53, 66, 68, 118, 120, 137)

A burnt child dreads the fire.
PROVERB

PRINCIPLES

DAVID HARTLEY (1705–1757) systematized the psychology of association to a certain extent as far as it was then developed and attempted to provide it with a physiological basis: When two stimuli, A and B, successively affect two different sensory organs, A will activate certain nerve bundles and brain portions, and B others. In addition, the two affected parts of the brain are connected with each other. If stimulus A is repeated, it activates the brain portion previously stimulated by B even if stimulus B should not recur. Although effect B itself does not occur as such, a weaker image of this effect is created, a memory (according to STÖRIG, 1954).

In 1853 LOTZE described the schema of the conditioned reflex still more clearly:

> "If we assume that a glowing body at first affects the retina only with its light, it will evoke no reflex effect other than that of either a turning of the eyeball to bring its image into the area of central (acute) vision, or if the stimulus is very intense, a closing of the eye. But if the same glowing object were to touch the skin of the body, it would cause a strong reflex movement to evade the pain. If this experience has occurred once, the mere light impression will thereafter be connected with the concept of painful glowing heat and will cause this reflex movement. Thus, a property of the stimulus which is not at all physically effective at the moment can nevertheless have a decisive influence on the type of reaction which ensues..."

PAVLOV states (1909, III, 1, p. 83; IV, p. 120):

> "What the genius of VON HELMHOLTZ* denoted with the famous term 'unconscious conclusion' evidently corresponds to the mechanism of the conditioned reflex."

And again (Selected Works, p. 422):

> "Since VON HELMHOLTZ did not know conditioned reflexes, he called them unconscious conclusions."

According to VON HELMHOLTZ (1867, 1896, 1912), the unconscious coupling of sensory perceptions to their causes occurs with the force of necessity. It allows us to

---

* VON HELMHOLTZ speaks of unconscious conclusions as early as in the first edition of his Handbook of Physiological Optics. It appeared in 1856–1867 and not, as the German editor of PAVLOV's Complete Works (IV, 120) states, in 1909–1911.

judge that a certain object of a certain quality exists at a certain place outside of us. Impressions received by our senses which are frequently repeated in the same manner and follow one another in the same manner reinforce each other. Changing sensations which occur only accidentally usually recede and disappear if their impression is not emphazised and deepened by specific affects associated with them.

After the turn of the century, PAVLOV (1849–1936) investigated conditioned reflexes with a whole staff of co-workers in specially established laboratories. In 1926 and 1932, he published his findings in two big reports in German. TRENDELENBURG (1937) published a summarizing review of the vast scientific material contained in the second volume, which was very difficult to survey, and gave in detail the reasons for his own position in this field. Since 1954, PAVLOV's complete works have been translated into German. ISCHLONDSKY (1930) gave a very lucid review of this topic in German, and RAZRAN (1933) in English. According to GESELL, 1500 papers on conditioned reflexes had already been published by 1937. At the First International Congress on Child Psychiatry in Paris (1937), the sessions of the entire first day were devoted to this topic. From these reports, the position of the scientists of the various countries in regard to this theory can be realized. Russian papers on PAVLOV's theory, as far as they refer to pediatrics, were made accessible by VAIL (1953) in a bibliography and by KASSATKIN (1957) in a summary review in German.

PAVLOV repeatedly stressed the importance of his theory for the psychology of man, but he himself never related it to childhood. His co-worker KRASNOGORSKI, however, has investigated the conditioned reflexes in the child since 1907, and summarized his findings in German in 1913, 1931, and 1935, and in Russian in 1954. For some time KRASNOGORSKI was a co-worker of my teacher CZERNY, who was thus kept in touch with these new investigations. CZERNY recognized their importance for the normal and sick child and he supported them after 1910. IBRAHIM (1911) and HAMBURGER (1935) followed CZERNY in this position.

As experimental animals, PAVLOV used mainly the dog and in later years also the higher apes. He distinguishes in principle between unconditioned (innate) and conditioned (acquired through experience) reflexes.

The unconditioned reflex develops in the following manner:

"A certain agent from outside or inside the organism strikes any receptor nerve. This impact on the nervous process is transformed into the phenomenon of nerve excitation. This excitatory process climbs along the nerve fibers, as on ladders, to the central nervous system. From here it is led – thanks to established connections – through other communicating tracts to the functioning organ, where it is now transformed into the specific process of the appropriate organ. In this way each agent is connected with a certain function of the organism by the law of cause and effect."

Unconditioned reflexes are innate, for example, the narrowing of the pupils to light, the withdrawal of a limb at a pinprick, or the flowing of saliva when food enters the mouth. Unconditioned reflexes, which are already present in great numbers in the newborn infant, pass through the brain stem without involving the cerebral hemispheres, and allow a certain rough adjustment to the stimuli of the outside world. They serve the purpose of rendering the threatening stimulus harmless (for example, flight reflex) or of strengthening the welcome stimulus (for example, rooting reflex). According to PAVLOV, who in this instance follows SPENCER's view (1855), the instincts are chains of unconditioned reflexes. The presence and strength of these reflexes vary with the excitability of the reflex centers so that, for example, the reflexes of food intake in hunger are different from those in satiation (see p. 473).

Unconditioned reflexes protect the higher animal against damage within certain limits, but they are not sufficient to keep it alive. Left alone, the animal will soon perish if the higher brain functions are removed and only the unconditioned reflexes are left at its disposal. The newborn human infant, who can respond to the stimuli of his environment only by unconditioned reflexes, is viable only because his lack of functioning cerebral hemispheres is compensated for by the function of the cerebral hemispheres of his mother, in other words, because the mother takes care of her child, who cannot yet take care of himself.

Conditioned reflexes are acquired only after birth. The cerebral cortex divides the incoming stimulations into very minute components and puts them together again into new formations, which become connected with the motor, secretory, vasomotor, and sexual organs. Since conditioned reflexes are formed, continuously supervised, and transformed under the influence of internal and external conditions, they achieve an incomparably more subtle adjustment to the inside and outside world than do the rigid unconditioned reflexes.

The appropriateness of the conditioned reflexes was made clear by PAVLOV in the following example: It would no doubt be very involved and impractical if each participant in a telephone circuit were to be connected with every other party by a separate line. However, the process is greatly simplified if each participant can be connected with any desired party according to requirements and disconnected again by a main switchboard. These temporary and transitory connections correspond to conditioned reflexes.

PAVLOV developed a special method of demonstrating conditioned reflexes in the dog: He transferred the duct of a salivary gland to the outside and collected the saliva so that the secretion could be exactly measured as to time and amount. If food was introduced into the mouth of an animal prepared thus, a salivary flow occurred — as an unconditioned reflex. PAVLOV then connected the unconditioned stimulus with another, the conditioned, stimulus, which in itself had nothing to do with food intake; for example, he rang a bell each time before food intake. It did not take long until the ringing of a bell alone — without feeding — was sufficient to start the salivary flow. A conditioned reflex, i.e., an unstable connection which is not innate but is acquired through the influence of the environment, was thus established. It can be extinguished again if the feeding after the ringing of the bell is omitted several times.

Conditioned reflexes are based on unconditioned reflexes. The conditioned salivary reflexes have been most thoroughly investigated, because the quantity of saliva and the time sequence of the secretion can be determined relatively easily. Apart from these phenomena, conditioned reflex connections of the kidneys, the liver, the spleen, the heart, and the blood vessels, the digestion, the metabolism, etc., have been demonstrated (BYKOW).

GIURGEA (1953) was able to establish conditioned reflexes by direct influence on the cerebral cortex. He attached electrodes to the motor and visual cortical areas in dogs. The electrode in the motor cortical region of the left hemisphere was so placed that the dog lifted the right front leg on weak electrical stimulation. When the visual cortical region was stimulated, the dog showed no reaction or at the most a general orientation reaction. GIURGEA then applied repeated stimulations to both cortical areas so that, just as in the formation of conditioned reflexes, the visual area of the cortex was always stimulated 3 seconds prior to the motor area. After 20 repetitions the dog lifted its leg even on stimulation of the visual field alone.

KRASNOGORSKI (since 1907) applied PAVLOV's method to children, at first by investigating oral and swallowing movements. Later he conducted the saliva to the outside with small collecting vessels fixed by suction to the orifices of the salivary glands, and was thus able to check and measure the salivary flow of all glands (see p. 49).

Even the investigation of unconditioned salivary flow showed noteworthy results. The effect of different foods varies in different children of the same age and weight. The quantity of saliva secreted also depends on the amount of the stimulant and the duration of its effect. The slower the child eats, the more saliva is secreted. If the child chews only on one side, the saliva flow is greater on that side than on the other. Mechanical stimulation of the interior of the cheeks does not have any influence on the salivary flow, whereas corresponding stimulation of the anterior half of the tongue increases the secretion of the submaxillary gland and stimulation of the posterior half of the tongue increases the secretion from the parotid gland. As soon as the food gets between the incisors, saliva from the submaxillary glands is secreted onto it. When the food reaches the posterior portion of the oral cavity and the molars, saliva starts to flow from the parotid glands. The mechanical stimulation is followed by the chemical excitation, as additional investigations have indicated (see p. 49). The temperature of the stimulus also changes the saliva flow.

In children one to fourteen years old, the unconditioned function of the salivary glands increases with age. The quantity of the secreted ferments depends on the amount of saliva and on the chemical properties of the stimulants. Food which begins to undergo chemical digestion in the mouth (zwieback) causes the flow of saliva of greater ferment content than rejected substances, for example, acids. The more active the salivary gland, the more alkaline the saliva is. Unconditioned saliva flow is diminished after satiation and in thirst. In thirst, salty substances cause a greater reduction of saliva flow than do other substances.

Infectious diseases, such as scarlet fever, measles, and especially typhoid fever, influence salivary flow considerably. In mumps, the unconditioned salivary flow of the diseased gland is greatly reduced and this reduction lasts longer than would be assumed from the clinical signs.

Development of Conditioned Reflexes

The age at which conditioned reflexes can be established is preceded by a stage in which the proof of their formation is questionable. Therefore, we cannot exactly answer the question of when conditioned reflexes are formed for the first time. With the progress of research, this time of first occurrence was set at an increasingly earlier age. KRASNOGORSKI stated in 1913 that the first of the conditioned reflexes can be evoked in the second half of the first year; according to a later statement in 1932 (as reported by WENGER), this can be done after the third month. In 1952, he wrote that conditioned reflexes can be obtained during the first four weeks of life, but only in a very restricted way because healthy newborn infants — and premature infants to an even greater degree — are awake only for a short time, tire easily, and quickly fall asleep again. In contrast to the poor excitability of the cerebral cortex, the extracortical excitability and the tendency to a generalized spread of excitation and inhibition are especially high in newborns and infants (according to KRASNOGORSKI, 1952). According to IWANOW-SMOLENSKY (1951), the physiological brain activity of the newborn infant occurs without any visible participation of the cerebral cortex.

In each of his investigations, KASSATKIN (1935, 1937, 1953, 1955) placed the beginning of the conditioned reflexes at an earlier age. He used the nutritional reactions (appearance of sucking movements on touching of the lips with the nipple) as unconditioned reflexes and the palpebral reflex (shutting of the lids on air blast) as a defense reaction. According to his investigations, all sensory organs form conditioned reflexes during the second half of the first month and the first half of the second month of life. When one starts to establish conditioned reflexes simultaneously for the various sensory organs, the reflexes will not appear at the same time but in a specific sequence: 1) Conditioned auditory reflexes; 2) conditioned vestibulary reflexes to rocking in a cradle; 3) the conditioned optic reflex; and 4) the conditioned tactile reflex of the skin. In between the conditioned olfactory and taste reflexes appear.

Should one succeed in establishing a conditioned reflex before the fourth week of life, the reflex will be very inconstant and will not be present at each observation. If attempts to establish the reflex are already started at the tenth day of life, numerous reinforcements are needed to form it. However, if the same trials are started only after the first month of life the conditioned reflex may develop even on the first day of the experiments and with only few reinforcements. Permanence, distinctness, and rapidity of reflex formation varied from child to child at all ages, but the reflex was finally permanently established in all children.

Conditioned auditory reflexes could be established in premature infants between birth and the time gestation should normally have ended. Thus, the first signs of a conditioned auditory reflex could be demonstrated in the fourth week of life in an infant born six weeks premature; three weeks later it was clearly established.

SYKOWA began attempts to form a conditioned reflex to sound in premature and mature infants on their first day of life. In mature newborn infants the conditioned nutritional reflex appeared on their tenth to twelfth day and the conditioned defensive blink reflex on the twentieth to twenty-second day. In premature or damaged infants it took longer to establish the conditioned reflexes; immaturity and the degree of damage played an important part.

According to POLIKANINA and PROBATOWA (1957), a conditioned motor nutritional reflex to light is formed at different times after birth in premature infants of differing maturity: In infants whose gestation time was shortened by one to one and one half months, the conditioned reflex occurred on the twenty-second to thirty-ninth day; in those born two to two and one half months before term, on the thirty-fourth to forty-seventh day; and in those born three months prematurely, on the fifty-ninth to seventy-fifth day.

The time of formation of the conditioned reflex depends not only on the degree of prematurity at birth but also on the intrauterine development, the degree of immaturity, and the general condition of the infant during birth. Among equally premature children, those having the lowest birth weight take the longest to establish conditioned reflexes. In infants of varying degrees of prematurity the conditioned motor nutritional reflex is more easily established over the path of the acoustic analyzer than over the optic analyzer.

DASCHKOWSKAJA (1953) observed the first distinct conditioned reactions in healthy newborn infants at the age of seven to nine days. A 10% sugar solution was given as an unconditioned stimulus and the sound of an electric bell was used as the conditioning factor. The — very inconstant — reaction consisted of sucking and blinking movements. She did not succeed in obtaining conditioned sucking to a red light at this age.

DENISOVA and FIGURIN (1935), who established a defense reflex instead of the usual nutritional reflex to a puff of air in the face, observed the first conditioned reflexes at the age of five to six weeks.

MARQUIS (1931) tried to elicit conditioned reflexes in newborn infants two to nine days old by sounding a buzzer for 5 seconds before each meal. After five days of practice 8 newborn infants started to make sucking movements and to open their mouths solely at the sound of the buzzer. In four experiments on other newborns the stimulation with the buzzer alone did not show any results.

WENGER (1936) used a procedure that permitted exact grading of the stimuli; he also found conditioned reflexes in newborn infants. He conditionally connected the unconditioned lid closing reflex on light stimulation with a vibratory stimulus. In these experiments, made on 3 newborn infants, he found conditioned lid closing reflexes at the most in 77% and on the average in 69% of the trials. Comparative studies indicated that the reaction was conditioned when over 29% of the tests showed a positive result. Conditioned reflexes first appeared on the fifth day after 124 stimulations distributed over three days. In another series of experiments the unconditioned reflex, drawing back on electrical stimulation, was connected with a vibratory stimulus. The experiment produced a conditioned reflex in a newborn infant on the ninth day. The same unconditioned reflex was connected with sound and light stimuli in other newborn infants. Conditioned reflexes could be obtained in this manner, too. However, they were more difficult to establish and to maintain in newborn infants than in older children.

D. and C. WICKENS (1940) tried in numerous experiments to establish conditioned reflexes in newborn infants under the age of 10 days. They worked with three groups of 12 infants each; one was the experimental group and the other two constituted the control groups. For three days the experimental group received 12 paired stimulations daily with a buzzer as the conditioned stimulus and an electric current as the unconditioned stimulus. The responses consisted of movements of the legs, feet, and toes. After these experiments the newborn infants were tested with the buzzer alone. The experimental group seemed to show conditioned reactions, extinction, and return of excitability. But one of the control groups which had received 12 electrical stimulations alone without the buzzer on three successive days behaved in exactly the same manner. These experiments therefore did not allow the conclusion that a conditioned reaction had occurred.

Going one step further back, Ray (1932) and Sontag and Wallace (1934) tried, though in vain, to establish conditioned reflexes even in the fetus. For this purpose Ray used the tap of a hammer as the unconditioned stimulus and a vibrator as the conditioned stimulus. Sontag and Wallace attached to the pregnant mother's abdomen a vibrator which as an unconditioned stimulator caused fetal movements; a tuning fork fastened to a wooden block served as a conditioned, though ineffective, stimulator.

Spelt (1938, 1948) also tried to establish conditioned reflexes in 16 fetuses before birth. Fourteen of the 16 women were already beyond the seventh month of pregnancy. After the position of the head, arms, and legs had been determined by x-rays, recording instruments which transcribed the fetal movements were attached to suitable points. The sound of a rattle served as the unconditioned stimulus and a vibrator attached to the mother's abdomen as the conditioning stimulus. Between 15 and 20 paired stimulations were needed to achieve 3 or 4 responses to the vibrator alone; with increasing practice it was possible to elicit 11 conditioned reflexes in succession. The reflex could be extinguished; it recovered and could still be elicited again after three weeks. These experiments are in distinct contrast to all other attempts to establish conditioned reflexes in newborn infants, for in the infants conditioned by other workers they were only uncertain and very weak. The findings of Spelt seem unconvincing to me and have yet to be confirmed.

According to Pavlov's experiments on dogs, the conditioned reflexes depend on the function of the cortex of the cerebral hemispheres. If conditioned reflexes can be demonstrated during earliest infancy or perhaps even before that, during gestation, i.e., at developmental stages in which the existence of cerebral hemisphere function is still disputed, the presence of similar processes of a less complicated nature in the lower brain regions must be assumed. In this regard the young infant behaves like an animal without a highly developed nervous system. In the ontogenesis of the individual a process which occurred in the phylogenetic development is repeated: The faculty of establishing conditioned reflexes is transferred to the newly formed higher brain regions. In Pavlov's laboratory, Zeliony established conditioned reflexes in dogs even in the absence of cerebral hemispheres. He assumes that lower regions of the central nervous system play a part in the development of these reflexes.

Pavlov worked mostly with dogs, and also with higher apes during his last years. His experiments were extended to many other animals. The question remains of how far down in the animal kingdom the ability to form conditioned reflexes extends. Alverdes (1939) and his co-worker Bramstedt found this faculty in unicellular animals (paramecia), but Grabowski (1939) was not able to confirm their findings.

FRANZISKET (1950) reported that he had succeeded in eliciting conditioned reflexes in spinal cord preparations of frogs, but TRINCKER (1952) doubted his results.

WORONIN (1954) conducted investigations on various higher animal species and determined the plasticity of the nervous processes by changing the significance of the signals of the stimuli. This change was accomplished after the conditioned reflexes had been formed by connecting the offer of food with the prior negative stimulus instead of with the positive. After a series of such trials, the animals adjusted by no longer reacting to the formerly positive stimulus and vice versa. In this way it could be demonstrated that the plasticity of the nervous processes is approximately identical in fish, turtles, birds, rabbits, and dogs. Apes, especially chimpanzees, were exceptions; their nervous reactions were by far the most versatile. Several such reversals of signals produced quite poor results in fish; the result was somewhat better in turtles, even better in birds and rabbits, still better in dogs and baboons, and again by far the best in chimpanzees. From this we can conclude that the more highly developed the animal's nervous system is, the more adaptable are its excitations and inhibitions and the more rapidly do these adjust to new tasks.

When an attempt was made to establish conditioned inhibitions, most of the animals mentioned did not differ from each other. Exceptions were turtles, in which inhibitions were formed very slowly, and chimpanzees, in which they occurred very rapidly.

WORONIN (1951) observed that at the age of two weeks a baby monkey which receives food every two hours awakens and cries, sucks, and makes biting movements exactly at the times of food intake.

According to WORONIN (1953), the saliva method is unsuccessful in forming conditioned reflexes in apes, since they continuously secrete saliva to food residues remaining in their mouths. He therefore used IWANOW-SMOLENSKY's "grasping method." In this procedure the animal learns to press or not to press a bulb or a lever on a certain sign. The conditioned reflex movements are recorded graphically.

Imitation plays a great role in apes and monkeys. Thus, when a positive reflex and its differentiation are established in one ape in the presence of others, the same conditioned reflexes also occur in the other apes without further application. The ape responds to every significant change in his environment with violent excitement and disturbance of the conditioned reflexes. In a rhesus monkey, the conditioned reflex was still present after an interval of 8 years.

TROSCHICHIN (1952) was able to establish conditioned reflexes to olfactory stimuli in newborn dogs (puppies). The mother animal was rubbed with a substance with persistent odor (oil of cloves, camphor, peppermint, menthol, etc.) some hours before and twice daily after giving birth. The puppies were lying under the mother. To determine the formation of a conditioned reflex, they were put in a box, away from the mother, for 1 to 1½ hours. Then they were placed on a table and a piece of fur or muslin drenched with a smelly substance was held at a distance of 2 to 4 cm from their noses. If the odor was the same as that of the mother, the puppy crawled toward the piece of fur; in some cases it even made sucking movements. However, when the odor was different, distinct defense movements occurred: breath holding, remaining on one spot, retreating from the source of the odor, and whimpering.

PANTSCHEKOWA (1956) tried to establish conditioned reflexes to the sound of an electric bell in young white rats. The general movement reaction was recorded. The animals reacted to the sound stimulus starting at the fifth to sixth day of life. A conditioned reaction could be formed starting at the tenth to fifteenth day. The younger the animals were, the more paired stimulations were needed. At the age of seventeen to nineteen days the young rats needed the same number of paired stimulations to form conditioned reflexes as do adult rats.

Physiology of Conditioned Reflexes

According to KRASNOGORSKI, a fundamental difference between the conditioned reflexes of infants and those of animals is their extraordinarily rapid formation in infants. Healthy infants need only a few stimulations to establish a conditioned motor reaction.

In my experience there is a very simple method of establishing a conditioned reflex in an older infant: If one claps his hands several times in succession over the infant's eyes, his eyes will be closed each time (unconditioned reflex). If the hands are then moved without contact, the eyes will also be closed. A conditioned reflex is thus formed between the movement of the hands, which by itself is ineffective, and the closing of the eyes.

The first conditioned associations which can be distinctly observed in infancy are tied to food intake. If the child is fed regularly, there often develops even in the first month a conditioned time schedule that divides his life into certain time periods according to food intake: The child demands his food at regular intervals of 4 hours, but with a nightly interval of 8 hours.

The establishment of a conditioned reflex associated with the time of food intake was investigated in detail in the newborn infant by BYSTROLETOWA (1954). If food was offered every 3, 3½, or 4 hours, the general movements and sucking movements increased some minutes before the feeding, just as in WORONIN's experiment with monkeys (see p. 518). The infants frequently awoke before feeding. When the meals were given regularly, the periods of rest also alternated regularly with periods of motor restlessness. Conditioned reactions could be demonstrated starting on the third to fifth day of life and became more distinct after the seventh to eighth day.

Even the usual preparations for feeding, such as the tying of a white bib around his neck, excites the infant; he turns his head from side to side, opens his mouth, sucks, and swallows. Depending on how he is fed, he becomes excited at the sight of his mother's breast or of the bottle. He strains with his head and the upper part of his body to reach the source of milk (Fig. 196, p. 450). The formerly unconditioned rooting reflex (Fig. 167, p. 405) has become conditioned: The infant turns his head to the source of milk without being touched (Fig. 197, p. 451). Infants and older children (Fig. 198, p. 453) open their mouths when the familiar spoon with pap is brought near.

The main period for conditioned reflexes begins when cerebral hemisphere function develops in the course of maturation. Then the same laws by which conditioned reflexes occur in animals can also be demonstrated in the child with the aid of the conditioned salivary reflexes (KRASNOGORSKI).

The two opposite basic processes which continuously operate in the cortex of the cerebral hemispheres are *excitation* and *inhibition* (PAVLOV). At first the external stimulus causes either an excitation or an inhibition in a certain point of the cortex. After their formation the two processes have the tendency to spread over additional parts of the cerebral cortex *(generalization)* and then to assemble again in one point *(concentration)*.

An experimental series performed on the skin shows this law very distinctly: If mechanical stimulation of a certain area of skin is chosen as the conditioning stimulus, the stimulation of any skin area near or distant from the original point will at first cause the identical salivary flow (generalization). If the experiment is repeated frequently, a progressive concentric narrowing around the original spot of the skin from which the reflex can be elicited occurs and finally only the stimulation of this original spot is still effective, while the reflex can no longer be obtained even from nearby areas (concentration). According to KRASNOGORSKI, the generalization of conditioned reflexes is comparatively pronounced in young children; later, with the development of cortical function, it becomes more and more limited. It is most pronounced in children with brain damage.

Conditioned reflexes to visual, olfactory, acoustic, and other stimuli behave just as do those to skin stimulation. For example, if a tone of a certain pitch is made the conditioning stimulus, at first any tone of any other pitch, indeed any sound at all, causes the same effect. But gradually, if the experiment is continued, the effective range of tone pitch narrows more and more. Other tones lose their effectiveness, especially those most remote from the original tone. Finally only the original tone remains as an effective stimulus. If the limits of constriction are determined, a measure for the differentiating ability for tones is obtained.

During the second half of the first year of life, conditioned reflexes from all sensory organs (eye, ear, nose, skin, etc.) can be developed; however, their formation occurs slowly at first. KRASNOGORSKI was thus able to investigate the infant's differentiating ability in detail. The differentiation of tones was difficult, but the child was able to differentiate forms and movements of objects well.

The earliest attempts to establish conditioned reflexes in infants had no connection with PAVLOV's work. They had already been done in 1903. At that time RÄHLMANN examined the differentiating ability for colors by associating food intake with the showing of certain colors. He wanted to find out if the infant's color sense develops only after birth or is already present in the newborn. To this end he painted the feeding bottles with different colors; for example, of two bottles the red was always filled with milk, while the green bottle was empty. Both bottles were offered at the same time. At first the infants were not sure, sometimes reaching for one, sometimes for the other bottle. Later, however, they turned with certainty to the correct, red bottle. By this approach RÄHLMANN was able to prove that infants, provided they are at all capable of reaching out, can differentiate among all colors offered. The children acquired this faculty at varying times. Some were already capable of differentiating at the beginning of the sixth month. RÄHLMANN was not acquainted with the concept of conditioned reflexes when he conducted his experiments and naturally did not use this term, but in principle he used the right approach in his methodology.

Without knowing of RÄHLMANN's work, KRASNOGORSKI (1913) later tested the infant's color sense by a similar method and obtained the same results. KASSATKIN (1953) broadened these experiments: The conditioned nutritional reflex appeared at first as sucking and later an extension of the hand toward the source of food. Thus, sucking movements at the sight of a blue (full) bottle as a positive reflex and the reflex differentiation at the sight of a red (empty) bottle were established in two infants. A stable differentiation was achieved in the first infant at the age of approximately five months and in the second at six months. A month later, when both infants were able to reach distinctly for all offered objects, they immediately reached for the blue bottle but not for the empty red one. Thus, they immediately transferred the conditioned reflex which was established for the sucking movement to the reaching for the bottle. PAVLOV did not succeed in establishing conditioned reflexes to colors in dogs. Objections to the interpretation of color experiments in children which do not consider the relative brightness values of the colors are discussed on p. 70. A method of examining color differentiation ability in infants which takes this source of error into consideration is described on p. 71.

It has long been known that temperature stimuli cause only weak conditioned reactions. USSOLZWEW and TERECHOWA (1958) were able to establish a conditioned temperature reflex with 16 to 24 combinations in the second month of life. At the age of two and one half to four and one half months only 2 to 8 combinations were needed. The stability of these reflexes increased in different infants with varying speed and reached its height at the age of five to six months. The stability depended mainly on the chronological age and not so much on the number of combinations. Differentiation of temperature stimuli proved to be very unstable.

BOGEN (1907) observed conditioned gastric secretion in a three-and-one-half-year-old child in whom a gastrostomy was performed because of an esophageal atresia; the blowing of a trumpet was the conditioned stimulus for the gastric secretion. NOTHMANN (1909) obtained conditioned flow of gastric secretion on mere presentation of food in an infant at the age of nine months, but not earlier. In contrast to these reports, KITAIGORADSKAJA (1937, 1939) did not succeed in effecting gastric secretion by showing food or by mechanical stimulation of the gastric mucosa in children under the age of two years. She was successful only in older children and in adults.

KRATSCHKOWSKAJA was able to demonstrate a conditioned-reflex nutritional leucocytosis in newborn infants: After the eighth or ninth day of life the number of leucocytes in the peripheral blood smear is increased before and after regular food intake. If the meals were given regularly at other times of the day, the leucocytosis adjusted to the new hours after two to three days.

SAMKOWA and TSCHERNOUSSOWA produced a conditioned-reflex diuresis. In infants four to seven months old, the amount of urine increased after drinking of yogurt, whereas the quantity did not change when this food was given with a spoon. However, the urine increased when the infant drank the same quantity of this food from the bottle.

Watson (1924; see p. 136) established conditioned affects at the end of the first year of life by associating fright stimuli, i.e., the booming thud of an iron hammer on an iron pipe behind the child, with the presentation of a white rat with which the child had played for weeks.

The joy which the infant shows at the sight of his mother is originally conditioned by her bringing food (see p. 454).

Pavlov stresses the great importance of the nutritional center, which like the respiratory center is located in the medulla oblongata, but which works in close association with the cortex and is very sensitive to the composition of the blood. This was demonstrated by first establishing a conditioned reflex to specific food substances in an animal. If the animal was overfed with this substance for a long time, the conditioned reflexes for this food disappeared, but they remained for other types of food. By this method Krasnogorski was able to demonstrate in the child a special sensitivity to carbohydrates, fat, albumin, salt, alkali, acids, and water. Subcortical excitation stimulates the striving for the required food; the high degree of cortical excitability permits the finding of this food and thus the filling of the need in an appropriate manner.

The cerebral processes which occur during differentiation were investigated in detail by Pavlov and his co-workers. The cerebral hemispheres have the ability to analyze the individual components of the complicated influences of the outside world which reach them in the form of stimuli. Only then will certain stimuli be connected with organ functions. Therefore, the presence in the cortex of analyzers for visual, acoustic, gustatory, and similar sensations must be assumed. The differentiating ability for the afferent stimuli depends on the degree of sensitivity of these analyzers. Thus, there is a corresponding area in the cerebral cortex for the quality of each sensory stimulus. Lesions of the analyzers restrict or abolish their functions.

The analyzing ability of different mammals varies greatly. For example, the dog has a very fine differentiating ability for odors and sounds. The dog can distinguish between half tones and can recognize different tones without direct comparison, i.e., he has "absolute" sense of pitch, a faculty which only few men possess. Conditioned reflexes to sounds inaudible to man can be established in the dog. In man the analyzers of the visual sense are very sensitive.

The formation of conditioned reflexes may be represented in the following manner: The unconditioned stimulus evokes a strong excitation in the brain stem, the "focal point of great activity." If any stimulus reaches an analyzer of the cerebral cortex during this stage of excitation, the excitation will force its way from there to the focal point of great activity which is connected with the responding organ. After repeated use this path is so facilitated that the connection will also occur without the unconditioned stimulus.

In the cerebral hemispheres the conditioned stimulus at first excites a certain point of the analyzer and from there is directed to the brain stem. Additionally, the excitement within the analyzer radiates into neighboring fields, so that these are excited as soon as new excitations penetrate after application of similar stimuli. The excitation is also directed from the stimulated area to the focal point of great activity. It thus happens that initially when conditioned reflexes are established, stimuli even other than the original can cause organ activity (generalization). However, the excitation in the cerebral cortex gradually retreats again to the area of the analyzer originally stimulated, so that the stimulation of the other, now inhibited, areas becomes ineffective (concentration). The analyzers may definitely be regarded as being spatially arranged. The migration of the excitement from one place to another can be distinctly followed in its time sequence.

The inhibitory processes obey the same laws as do those of excitation. Pavlov distinguishes different forms among them according to the manner of their formation. They were also demonstrated in childhood by Krasnogorski.

The conditioned reflex is abolished absolutely if the conditioned stimulus is repeatedly applied without the unconditioned stimulus. It is then inhibited, but will return after some time. This "getting accustomed" is particularly representative of the more subtle adjustment to the constantly changing stimuli of the environment.

Trace conditioned reflexes are not connected to the stimuli themselves, but to their traces; this is done by letting the unconditioned stimulus become effective only a certain time after the conditioned stimulus. In this case the reflex appears after an equally long latent period. An inhibition which suppresses the conditioned reflex during the latent period has developed at the corresponding point of the cerebral cortex. Trace reflexes are hard to establish in dogs and they are soon extinguished again. However, trace reflexes are just as easily established as other conditioned reflexes in children older than one year and they are far less easily extinguished than in the dog (Krasnogorski).

The described differentiation of various stimuli is caused by differential inhibition. The "wrong" stimulus, which is not accompanied by an unconditioned stimulus, causes an inhibition in the corresponding parts of the analyzer. Krasnogorski states that the cortex of the infant's cerebral hemispheres is able to differentiate extremely fast.

Conditioned inhibition is accomplished in the following manner: At first a conditioned reflex is established. Another stimulus is then used simultaneously with this conditioned stimulus, but no unconditioned stimulus is added to this combination. After a sufficient number of repetitions this second stimulus acquires the faculty of inhibiting the reaction if it is used in combination with the first stimulus; however, if the first stimulus is used alone its faculty remains unchanged. In another combination the second stimulus may even cause an inhibition.

Kassatkin (1953) investigated in detail what relation age may have to the development of possibilities of forming internal inhibitions in differentiating between varied stimuli: Differentiations of the same degree of difficulty are more speedily achieved in older than in younger children. The subtlety and stability depend on the interplay between the afferent system and the responding organ. In the course of ontogenesis this interplay changes continuously: It remains permanent only in adults at the height of the development. In early childhood the rapid change between excitation and inhibition and the tendency to generalization also affect the conditioned reflexes. There are vast differences between children. Differentiation may be formed at once in one child, whereas in other children of the same age the identical differentiation needs a longer time for its development. The stability of the process also differs from child to child. Whereas in some children the establishment of a rather stable differentiation may be successful, in others it remains fluctuating, and in still others it may not succeed at all.

Conditioned stimuli which coincide several times with chloral hydrate anesthesia gain such strongly inhibitory qualities that they can put the child to sleep even when used without chloral hydrate (KRASNOGORSKI; LEWIN, 1935).

The nature of stimulation and inhibition is not known. PAVLOV asserted that the chemical and physical nature of the different types of inhibition is uniform but that the conditions of their origin vary, a hypothesis which was questioned by TRENDELENBURG. According to BIRJUKOW (1955), stimulation and inhibition develop continuously in their similarity and diversity in the course of phylogenesis and may be differently related at different stages of development. Naturally the conditions are most complicated in the higher animals. We can already speak of independent processes of excitation and inhibition in the lower animals. A certain independence is also maintained in higher animals. "When one speaks of their intimate physical and chemical nature, one has to acknowledge their uniformity." This fact is supposed to insure the possibility of transition from excitation to inhibition and vice versa.

Inhibition is recognized by the lowered excitability of the cells and reflex arcs and is in general less firm and weaker than excitation. Its spreading and contraction occur according to the same laws as excitation. KRASNOGORSKI demonstrated this rule by attaching spaced instruments for mechanical skin stimuli to a dog's leg. In appropriate preliminary tests he conditioned these stimuli for salivary flow. Then he produced a conditioned inhibition for the most distal area. If immediately after an inhibition he now tested the other skin areas from which excitations could be evoked, the stimulation remained ineffective; at first the neighboring and later the more distant skin areas remained nonexcitable. After some time, excitability returned in reverse order. Thus, the inhibition had first spread to the corresponding cortical portions and then had concentrated in the original area (generalization and concentration).

Man, like the animal, falls asleep easily during the formation of these inhibitory processes. PAVLOV investigated this initially coincidental and disturbing phenomenon in more detail and was finally able to produce sleep at will by an appropriate experimental design. He therefore regarded sleep as a form of internal inhibition which spreads over great segments of the cerebral cortex and the brain stem.

According to PAVLOV and KRASNOGORSKI, the cortical cells possess the important faculty of protecting themselves against fatigue or exhaustion. As soon as this danger sets in they reduce their excitability, i.e., during falling asleep those cortical cells which worked most during the day and are therefore tired are the first to be less excitable. Thus, fields reduced excitability occur in the motion, light, sound, and skin analyzers. The environment simultaneously promotes the lowering of excitability in the cerebral cortex during the process of falling asleep. The reclining position arrests the excitatory current which enters from the motor organs, the closing of the eyes keeps light stimuli out, and sound stimuli decrease in the stillness. The analyzers, which are tired from their daily work, thus become gradually less excitable. As soon as a focus of decreased excitability has been formed in the cerebral cortex, this focus enlarges and also encroaches on the subcortical centers. In interrupting the centripetal tracts in that region, it severs the connection between cortex and body, and sleep supervenes.

Sleep disorders and sleeplessness are attributed to foci of increased excitability in the subcortical centers or in the cortex of the cerebral hemispheres. The central regulation of sleep is discussed on p. 498.

PAVLOV obtained conditions of tonic immobility, which also appear in the young child (see p. 505), in dogs by producing inhibitions in them.

Excitation and inhibition influence one another to a great extent. A long lasting excitation of a certain area of the cerebral cortex can thus be followed by an inhibition and vice versa. The excitation which is evoked in one part of the cerebral cortex also leads to simultaneous inhibition in other places. This phenomenon is termed *induction*. We have previously described how the urge-to move decreases and reflex excitability is lowered in the infant during food intake (see pp. 33 and 473). The nervous energy is used up by the centers of food intake so that none is left for the motor centers.

I have previously described additional examples and reproduced graphic tracings of induction in the rhythmic function of the respiratory and sucking centers, i.e., respiratory arrest after a deep breath (see p. 86) and pendular induction between the respiratory and sucking centers (see p. 435).

PAVLOV was able to demonstrate induction even more distinctly in animals. He succeeded in making an electric current the conditioned stimulus for the food reaction so that finally very painful currents which caused burns and would otherwise have evoked violent defense reactions produced a salivary flow with no display of excitement.

According to PAVLOV, a tremendous number of conditioned reflex systems are spread over the cerebral cortex. They are in dynamic equilibrium and, in contrast to unconditioned reflexes, change constantly in response to environmental influences.

KRASNOGORSKI demonstrated the influence of the food intake on the production of conditioned reflexes in infants. The excitability of the cerebral cortex is diminished immediately after the noon meal, but one hour later the cortex functions as usual. Constipation depresses cortical activity. The influence of different medicaments was tested. Alcohol reduces conditioned and unconditioned reflexes; inhibitory reactions may be disinhibited by alcohol. Chloral hydrate and, to a lesser degree, barbital diminish the subcortical reflexes. Bromine also lowers subcortical excitability, but increases that of the cerebral cortex. Stimuli such as caffeine, coffee, and tea increase the excitability of the cortical centers and disinhibit their inhibitory reactions. Carbon dioxide lowers and oxygen increases the activity of the cerebral cortex.

LEWIN (1935) examined the influence of phenobarbital on conditioned reflexes in children and was able to distinguish different stages of excitability: At a dosage of 0.02 to 0.03 gm a neutral stage or slight increase of conditioned reflexes occurs, at 0.03 to 0.05 gm lowering, at 0.05 to 0.1 gm increase by weakening of the inhibitions, and at 0.1 gm and more marked decrease because of beginning sleep. Phenobarbital influenced mainly the conditioned and only in higher doses the unconditioned reflexes.

When PAVLOV interpreted his findings, he did not try to explain the internal world of his animals with the usual psychological explanations. Like the American school of behaviorists, he confined himself to objective investigation of the phenomena under clearly defined experimental conditions; he was satisfied to approach them only from the outside, so to speak, and to interpret them exclusively with the aid of the physiological concepts of excitation and inhibition. It is undeniable that he owes the logical structure of his theory particularly to clear and consistent adherence to this principle.

Nevertheless, many objections have been raised to this exclusively objective approach (STRAUS, 1935; TRENDELENBURG, 1937; SCHRÖDER, 1938). It is undoubtedly necessary to take the objective approach in lower animals. However, since in higher animals, especially in the anthropoid apes, we must look for a connection with the world of experience of man, TRENDELENBURG sees no alternative but that

"when observing higher animals the objective, and when observing lower animals the subjective, viewpoint be kept more and more in the background, and in general that the idea of inseparable existence of the physical and psychological, of objective and subjective, be accepted."

PAVLOV himself made concessions in this direction [Works III, p. 86 (1909)]:

"I do not reject psychology as a science of the inner world of man."

And in Wednesday Colloquium II, 394 (May 30, 1934):

"Formerly I made very crass statements about psychology and avoided all psychological expressions. This was actually a passion. Later I resigned myself a little... It would be foolish to reject subjective psychology. It is, of course, self-evident that it exists. Psychology as a formulation of the phenomena of our subjective world is completely in its right, and it would be stupid to dispute it. We act on this basis; our entire social and personal life is constructed on it. There can be no discussion about it. The real problem is the analysis of this subjective world. Naturally, we must regard psychological analysis as being insufficient... However, psychology as a study of the reflection of reality, as a subjective world which is formulated in a certain way, is, of course, a necessity."

Conditioned reflexes and associations form the basis of the first education which the child receives (CZERNY, since 1910). When the infant is to be toilet trained, this is accompanied by certain sounds while he is held on the potty (see p. 527). Some mothers who try hard succeed in the third or fourth month; however, complete toilet training is not attained in most cases before the second year of life. VON ALLESCH (1921) described (see p. 590) how even the chimpanzee mother "holds out"* the baby away from her body.

Similarly, objects and actions which are forbidden to the child are associated with certain words; the mere use of these expressions is finally sufficient to inhibit the

* German descriptive expression for toilet training – The translators.

forbidden intention. The child thus acquires his first inhibitions through training. For example, I observed a two-year-old child who reached for the forbidden scissors, then said "no, no" and put them down again.

Language (Second Signal System)

Pavlov represents the totality of higher cerebral function in the following manner (1932, Works III, 2, p. 451): In higher animals and man, lying within the hemispheres is the first site for the complicated interrelations between the individual and his environment, the subcortical region, with its highly developed unconditioned reflexes, instincts, tendencies, affects and emotions. These reflexes are evoked by comparatively few unconditioned external stimuli which affect the infant at birth. The second site is supplied by the cerebral hemispheres without the frontal lobes. The conditioned reflexes which pass through the cortex of the cerebral hemispheres make possible more comprehensive orientation in the environment, and with it more subtle adaptation. They form the only signal system of the animal and the first signal system of man. In the frontal lobes of man, a second signal system is added: the signalization of the signals. Pavlov writes (1935, Works III, 2, p. 551):

> "In the ontogenesis of man, extraordinary signals of a second order were created, developed, and perfected, signals of these primary signals in the form of spoken, heard, and visible words. These new signals marked everything that men immediately observed in their environment and in their internal world and they were used, not only for communication with others, but also by themselves. The predominance of these new signals was naturally implied by the enormous importance of words, although words were and remained only the second signals of reality."

Thus does the infant learn his first words toward the end of his first year of life according to the scheme of the conditioned reflex. For example, every mother eagerly tries to teach the meaning of the word "a-a" or "poo-poo" to her child as soon as possible by letting him hear this word each time she sets him on the potty. Since the same word sound always coincides with the maneuver the child finally learns to understand the association and uses this word for bowel movement and feces. Thus, the unconditioned spinal reflex of defecation and urination becomes conditionally connected with toileting. Disturbances of this process lead to enuresis and encopresis and can be overcome with the aid of newly formed conditioned reflexes.

It is a sign of our times that the first word which the child pronounces is "auto." The first "tricks" which he learns consist of conditioned connections. For example, at the end of his first year of life he can point on command to his nose, to his ear, or to his eyes, offer his hand, or show how tall he is by raising his arms. When the word "ticktock" is said, he looks at the clock. At first only certain movements of the child are conditionally connected with certain sounds. Understanding of words always precedes the ability to pronounce them.

These observations are supplemented by Krasnogorski's (1935) experiments: The multiformity of the environment continuously demands conditioned reinforcements, weakening, acceleration, or deceleration of reactions. These changes could also be evoked in the child by replacing the conditioned stimuli with corresponding words, for example, the word "faster." These conditioned reflexes proved to be very labile in children. Constant practice was needed to keep them functioning well. The reflexes could not be formed in feeble-minded children.

Iwanow-Smolensky (1935) tested the formation of conditioned reflexes in children five to fifteen years old by connecting language and thinking functions; for example, he replaced the sound of a bell by the written or spoken word "bell," or by a picture of a bell. An immediate influence of the cerebral cortex on the vegetative nervous system could be proved in this way; in this process the word spoken or thought of acted as a conditioned stimulus.

According to Krasnogorski (1950, 1952), the second signal system synthesizes the activities of the first and directs them; on the other hand, the first system constantly improves the activity of the second when the functioning of the latter does not correspond to the stimuli actually present. For example, when a conditioned reflex to red light causing a conditioned salivary flow is established in a healthy child and when during the presentation of the red light the verbal stimulus contradicts reality, such as the sentence "This is not a red light," the red light will cause the usual salivary reflex. In healthy children the actually effective stimulus always outweighs the verbal stimulus. In sick children or in those who can be easily hypnotized, the verbal stimulus may render the actually existing stimulus ineffective.

Skorunskaja similarly investigated the relations between the two signal systems in children and adolescents eight to nineteen years old and in adults by establishing conditioned reflexes to colors and to the names of these colors. She then connected a color stimulus with a word that was the name for another color; for example, she showed the color yellow and spoke the word "red." In these experiments all subjects were guided at first by the color. However, with increasing chronological age the number of those who were guided by the "wrong" word increased.

Krasnogorski (1956) demonstrated that in the young child who is learning to speak, the words heard from the environment combine as signals in the verbal-acoustic analyzer with the first signals of reality, abstract them from reality, and thus allow a very broad generalization by the formation of word signals. Just as conditioned reflexes are transmitted by language, words and their associations are brought to life from the start of the second year by numerous motor conditioned reflexes. In this way, very complicated words and word groups may be mastered.

The verbal sense depends on the conditioned associations of the first signal system which were formed by the word in question and on the relations of this word to other words. If words coincide with the effects of other words, they form stable connections with them. Finally the conditioned stimulus of the first signal system is replaced by the word and is abstracted from reality which, however, influences its meaning.

Through the formation of conditioned reflexes, it can be determined to what extent the young child is able to differentiate between multitudes of objects of the same kind [Peiper, 1927 *(19)*]: Red discs were pasted on light gray pasteboard squares so that 1, 2, or 3 discs were arranged in various combinations on 5 pasteboards each. In each test the boards were changed or rotated 90°. Behind the "correct" disc was a chocolate. A child two years old had to chose between 1 and 2 discs. After 40 trials he chose 2 discs 20 times in succession without an error. In the corresponding test with 2 and 3 discs he failed. A child two years and one month old made the correct choice as long as he had to chose between identical boards with 1 or 2 discs. But as soon as the boards were changed so that the discs appeared in another arrangement he failed.

KOEHLER (1942, 1953, 1956) used a similar procedure when he demonstrated "undesignated thinking" in children and higher animals. His co-worker HASSMANN (1952) offered 5 bowls with 3, 4, 5, 6, or 7 thumbtacks on the lids to squirrels; in each of these tests a bowl with one of the five numbers of thumbtacks on its lid was the "model." The test animal grasped the bait lying on the model lid; it then opened only the lid which had exactly as many thumbtacks as were on the model lid, choosing from the five different numbers only the one which was the same as the model. The control task, choosing the one with 3 points from four equally numbered with 4, 4, 4, 3, 4 points and then the one with 4 points from 3, 4, 3, 3, 3, etc., was solved by a squirrel within the numbers 0 to 6. Two children apparently discovered the principle of model choosing after a few trials, but before they could verbalize it, they had already lost this ability. This also happened at times in birds. However, once they had verbalized it – one child said "This one is just like that one" and pointed with "this one" to the bowl to be opened and with "that one" to the model lid – they absolutely retained this faculty. BRAUN (1959) made similar investigations in parrots.

According to KOEHLER, our language is not simply learned and is even less a prerequisite for every concept formation. Rather, it has been phylogenetically and ontogenetically preceded by undesignated thinking, which many animals also possess. At this stage man and animal already have the faculty of understanding the meaning of the spoken word. This animal-human undesignated thinking reflects the relations of the environment and psychic inner conditions so that species-preserving planning of future actions, social agreements, and many other phenomena become possible. KOEHLER thus speaks of innate and acquired concepts in animals, but stresses that the question of consciousness must be kept out of the discussion.

There are many animals that acquire a certain understanding of language: They listen when their names are called and can execute certain verbal commands, but they progress no further. The young child also passes through a stage when he already has verbal understanding but cannot speak yet. In him this is a preliminary stage of speaking.

After childhood we can learn a foreign language only through instruction. However, the child learns his mother tongue from his mother playfully and without effort just by their speaking to each other. Children in institutions talk later than those at home because they are not spoken to enough. But language lessons would not be the answer. The number of attendants should be increased to a degree that will give the child his right.

Pathology of Conditioned Reflexes

Conditioned reflexes, invisible at first, develop in the infant imperceptibly and are not planned by the environment. However, as soon as the conditioned stimuli fail to appear for any reason whatever, the cerebral function is disturbed. PAVLOV demonstrated in the dog that even neuroses may arise in this way. The child may, for example, have difficulties when he is accustomed to a certain nurse at home and now has to be put to bed, be fed, or be toilet trained by another person. CZERNY reported on an infant who would take his nourishment only when somebody sang at the same time. A very frequently occurring nervous disturbance in childhood is homesickness (see p. 617); conditioned associations participate in its formation.

At the borderline of pathology is so-called "breath-holding." IBRAHIM stresses that it belongs to the conditioned reflexes (see p. 371).

According to CZERNY, the speed with which conditioned reflexes are formed in a certain child shows the condition of his nervous system. Of course, age also plays an important role. In the first year of life more practice is needed than later. Apart from this, great differences in the acquisition of conditioned reflexes are noted from child to child. Children who need frequently repeated practice are termed difficult to train. In marked mental retardation, conditioned reflexes cannot be established at all. In the healthy child, however, they can be formed just as easily as they can be inhibited again when the conditioned stimulus ceases. If the practice is started anew, they can be rapidly restored. A pathological nervous system may also betray itself by the pattern of extinction of conditioned reflexes. That is to say, there are children in whom acquired associations persist for a conspicuously long time, and then there are neuropaths in whom these phenomena are "fixed" and apparently can no longer be extinguished (CZERNY; IBRAHIM, 1911).

Some nervous disorders consist in the persistence of conditioned reflexes or associations after the unconditioned stimuli have already been absent for a long time. One example is pavor nocturnus. A child under my observation was awakened at night by the dripping of water from a leaking pipe in the ceiling. During the following nights he awoke screaming at the same time. Here it was sufficient for a conditioned stimulus to be operative only once.

According to CZERNY and KELLER (1925), habitual vomiting, rumination, and constipation are expressions of conditioned reflexes. The tendency to vomit, which is more or less physiological in early infancy, persists as a conditioned reflex in nervous children. Even older nervous children are easily conditioned to vomiting: At first the vomiting is caused by enforced food intake, and later mere talk about eating more may become the conditioned stimulus for vomiting.

Certain forms of habit tics also originate from conditioned associations. I observed a girl who had the habit of throwing her hair back with a certain movement of the head in order to free her eyes. The movement persisted when her hair was so arranged that it no longer hindered her seeing. Evidently the persistence of these pathological conditioned reflexes presupposes a special disposition. In neurologically healthy persons, an inhibition rapidly forms when the conditioned stimulus is absent, whereas in the neuropath the conditioned association persists.

In whooping cough, children become gradually accustomed to the special coughing pattern; these attacks can be attributed to the disease only at the onset and later they are more and more nervously conditioned. At this time, strongly suggestive measures such as the painting of the throat with silver nitrate become effective. If shortly after recovery from whooping cough children are taken ill with common upper respiratory infections, the irritation of coughing may again easily release the old conditioned reflex, and whooping cough attacks may recur.

BAYER (1932) reported on the formation of additional conditioned reflexes in children suffering from whooping cough. At first he left the sick child uninfluenced in a private room for two to three days. Then a loud bell was sounded each time a coughing spell occurred. Night and day special nurses had to see that the bell was sounded for a certain length of time at each sign of a beginning coughing spell. After a few days the duration of the coughing attack adjusted itself to the duration of the sound. If the sounding bell was now discontinued, the coughing disappeared in several cases, even when the child was transferred a few days later back to the general whooping cough ward. Although this observation is of scientific value only, this nevertheless shows how effective conditioned reflexes can become in whooping cough.

To a certain extent the situation is similar in bronchial asthma. There is no doubt that the asthma attack may occur as an unconditioned hypersensitivity reaction to allergens; however, in childhood this relation can be convincingly demonstrated only in exceptional cases. The attack may also come about as a conditioned reflex when the environment gives the child the impression that on a certain occasion, for example, on his return from the sanatorium, an attack will occur. The conditions may be similar to those in an observation made by BESSAU (1934): A young lady was so hypersensitive to apples that even a small quantity would cause urticaria. BESSAU told her that a piece of cake which he had offered her contained apples. After a few minutes a severe urticaria appeared, although the cake was free of apples.

Some well-known clinical entities are due to the lack of conditioned reflexes. As CZERNY recognized, bed wetting and fecal incontinence are caused by the fact that appropriate conditioned reflexes were not formed at the right time or were lost again. For example, many newly toilet trained children lose this achievement when they suffer from severe infectious diseases. Feeble-minded children present a special problem. The more severe their mental subnormality is, the more difficult it is to establish conditioned reflexes in them and thus to toilet train them. Older intelligent children suffering from enuresis can be treated by making impressive stimuli, such as faradization, the conditioned inhibitory stimuli of the pathological organ function.

The Phenomenon of Dominance

(68, 72)

The phenomenon of *dominance* is closely related to the conditioned reflexes. It is the process by which an excited cerebral area diverts excitations originating from external sensory stimuli from their pathways within the brain, attracts them, and thus increases its own excitation. Dominance can be demonstrated in the following manner:

If a focus becomes very excited, a stimulus is exerted which is really destined for another center and should cause a corresponding reflex there. However, whenever this reflex more or less fails to develop, the activity of the dominating center is increased.

The phenomenon of dominance was first demonstrated in frogs with the aid of the clasping reflex. According to ALBERTONI, the male frog may be subjected to different stimuli during the clasping reflex without being induced to leave the female. It even seems as though these stimuli strengthen the clasping even more. According to UFLAND (1925, 1929), a natural dominance exists in the pars brachialis of the spinal cord in the male during the mating period: The stimulation of any part of the trunk, head, or limbs increased the embracing reflex although the local reflex was preserved to a certain extent. UCHTOMSKY and WINOGRADOW created an artificial dominance in the spinal cord of the frog by local poisoning with strychnine or phenol. They did this at the right side of the pars lumbalis. If they then stimulated a spot on the frog's left side, a corresponding movement occurred not on the left but on the right side.

UCHTOMSKY conducted the following experiment: Water was poured by force into a cat's pharynx for a certain length of time to enforce strong and frequent swallowing. This caused a strong excitation of the swallowing center. If the cortical area which effects flexion of the hind leg was now stimulated with faradic current, leg movement did not occur; rather, the swallowing movements became more pronounced.

I repeated this experiment in the young infant with appropriate changes [1937 *(68)*]. In order to bring the sucking center into an excitatory state of such degree as to make it suitable for the demonstration of dominance, the bottle was suddenly removed during drinking. This causes an increased invisible state of excitation of the sucking center which occurs immediately after the sucking movements stop and thus makes the center accessible to stimulations of other types. Various infants react differently to the removal of the bottle. Their behavior depends in part on whether they are at the beginning of the meal or already near the end. But apart from this, the nervous characteristics of the individual infant are always distinctly evident even if these experiments are conducted on different days. Many children become very excited after the bottle is removed, so that it is useless to apply other stimuli since their effect cannot be recognized. In other infants the sucking movements continue for some time after the removal of the bottle, so that the influence of new stimuli likewise cannot be demonstrated. In still others the sucking movements continue for a short time more, but then disappear. Not infrequently, however, they reappear after a short time without an external stimulus; periodic sucking movements which make the infants unsuited for the demonstration of dominance may occur. However, if the sucking movements last for a short time after the removal of the bottle and then disappear, we have the moment which is favorable for the experiment. This pattern is probably the most common. Of course, it must be ascertained before the experiment how the children will react to the removal of the bottle.

I stroked the sole of the foot with a pin or rang a table bell as a sensory stimulus. The experiment must be conducted in a quiet room, since any other sensory stimulus would interfere. A second observer must be present to note whether sucking movements actually occur on stimulation; reading the motion recordings, which are made with the aid of a rubber bag attached to the mandible, cannot always indicate with certainty whether the tracings were caused by sucking movements or by general motor restlessness following the stimulus.

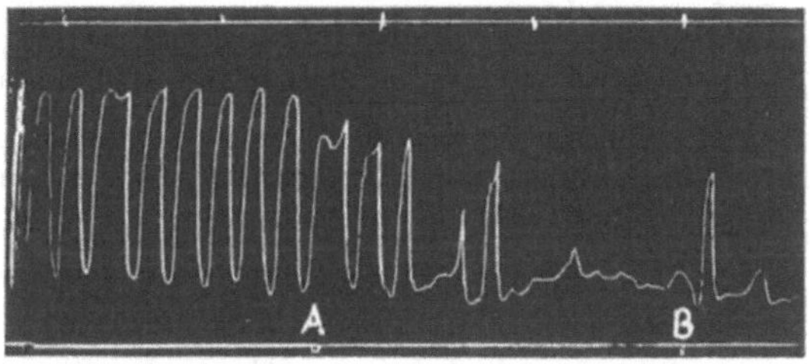

Fig. 206. Phenomenon of dominance (time in 5-second intervals). The sucking movements gradually disappear after the removal of the bottle at A. On the sounding of a table bell at B, there is a distinct sucking movement. The smaller waves are not suggestive of sucking movements.

Figures 206 and 207 show that sucking movements can be evoked by tactile or sound stimuli under these conditions and that the phenomenon of dominance can thus be demonstrated. The sensory stimuli were applied approximately 5 to 15 seconds after the disappearance of the sucking movements. They sometimes caused only one (Fig. 206) and sometimes several sucking movements (Fig. 207). The reaction occurred approximately 1 second after the stimulation.

The infants in whom the phenomenon of dominance could be demonstrated were one and one half to four months old. Babies who showed this phenomenon once usually also showed it on other days, but there were exceptions which could not be easily explained. The phenomenon could nearly always be demonstrated in the same child several times at the same session.

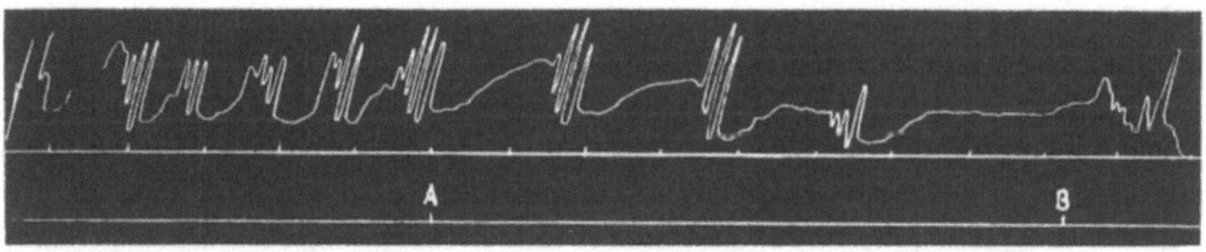

Fig. 207. Phenomenon of dominance (time in 5-second intervals). On removal of the bottle at A the periods become longer; finally the movements disappear. At B a pin is stroked over the sole and this is followed by a resumption of sucking movements.

It is well known that sometimes any type of sensory stimuli, though they have nothing to do with food intake, may evoke sucking movements in young infants. The experiments cited allow deeper insight into the intracentral activity, for they show that a center whose excitatory state is not visible can be brought to visible excitation by a sensory stimulus which in itself is indifferent. The phenomenon of dominance, which occurs only in the lower cerebral regions, is therefore closely related to the conditioned reflexes, which are connected with the brain stem at lower stages and to the cerebral cortex at higher stages. As has previously been described, the conditioned reflex is caused when an excitation which in itself is ineffective enters the cerebral cortex, there encounters an area in a state of great excitation, "the focal point of great activity," and is attracted to it. A pathway on which the previously ineffective stimulus will now regularly travel to the excited focus is thus formed between the reflex center of the ineffective stimulus and the focal point of great activity. ISCHLONDSKY (1930) states:

"The principle remains the same (as in the phenomenon of dominance). In all cases we are dealing with the competition of the nerve centers and with their struggle for dominance at a given moment."

I doubt, however, whether one could agree with ISCHLONDSKY and regard the phenomenon of dominance as a special case of conditioned reflex coupling. In any case, a process which has something in common with the formation of conditioned reflexes can actually be demonstrated in the lower brain regions of the young infant: An excited center attracts excitations which really should follow another path and thus increases it own excitation.

The behavior of the infant under the influence of sensory stimuli appears to the observer to depend on chance, but this is not so. On the contrary, the phenomenon of dominance determines the direction in which excitation spreads within the brain. The excitation ties itself to that reflex activity which is dominant at a given moment (ISCHLONDSKY).

The equilibrium between external stimuli and the excitatory state of the brain can also be disturbed by the sudden extinction of an existing stimulus. We become aware of such a process if during a boat trip the turning of the propeller, which is otherwise completely disregarded, suddenly stops. Conditioned reflexes were established in experiments on animals by the extinction of stimuli, such as the switching off of an electric light or the cessation of a sound.

From this, it remained to be investigated whether the newly extinguished action of the sucking center could be started again by the disappearance of a permanent stimulus in a manner similar to the appearance of a new stimulus. In these experiments [1938 *(72)*], light and sound stimuli were used. In the experiments with light, a 40-candlepower electric lamp was directed at the face of a drinking infant from a distance of approximately 50 cm. If the sucking movements ceased after the removal of the bottle, the light was soundlessly switched off. This did not leave the room in complete darkness; there was still a weak light which permitted direct observation of any sucking movements of the child which might occur and supervision of the recording apparatus.

It was found that sucking movements may actually occur following the turning off of the light; the phenomenon of dominance may thus also be evoked by the extinction of a sensory stimulus. In a series of additional experiments, I used an electric bell as a source of sound in a similar manner. The bell was started at the beginning of the experiment and stopped when sucking movements ceased on removal of the bottle. Dominance could not be demonstrated with certainty with this method.

SUMMARY

Innate unconditioned reflexes are transmitted through the brain stem (subcortex). The next higher stage consists of conditioned reflexes, the first signal system; they are transmitted through the cortex of the cerebral hemispheres with the exception of the frontal lobes. Words represent the signal of signals, and therefore comprise the second signal system; it is coupled to the function of the frontal lobes. Man and animal have unconditioned and conditioned reflexes in common; only man possesses the second signal system.

Every higher stage presupposes the existence of the lower stage, is closely connected with it, and considerably improves the adjustment to the environment. The newborn infant has almost exclusively unconditioned reflexes. The ability to form conditioned reflexes increases during the first year of life, and the second signal system begins to appear at the end of the first year. Thus, the brain function of the young child develops in accordance with phylogenesis in successive stages which PAVLOV has compared to layers or stories of a building.

In the child, conditioned reflexes follow laws which are identical to those known from experiments on animals. The characteristics of the child's brain function are reflected in the particular speed with which conditioned reflexes can be formed and in the increased tendency to generalization. In the newborn infant, whose nervous behavior is still predominantly dependent on unconditioned reflexes, conditioned reflexes are transitory and therefore cannot be demonstrated with certainty.

The conditioned division of time, which is dependent on food intake from early in life, separates the infant's life into periods which vary in length because of the nocturnal interval. When the child is several months old, the conditioned reflexes can be more and more distinctly evoked from all sensory organs; however their full development is attained only during the second year of life. The discriminative ability of the senses can be tested with the aid of the conditioned reflexes. They effect a much more subtle adaptation to the continuously changing influences of the environment than do the rigid unconditioned reflexes. Conditioned reflexes which are unintentionally formed by the environment are: the general restlessness and sucking and swallowing in response to preparations for the meal, and the opening of the mouth and turning of the head on the approach of the nipple, the bottle, or the spoon.

In addition to conditioned reflexes proper, conditioned associations which form the basis for early education are established. In this way the child is trained in the first tricks which he learns, in the first words which he understands or speaks, and in the first inhibitions. The unconditioned spinal cord reflex for the evacuation of bladder and bowel is conditionally associated with toilet training.

Clinical entities in which conditioned reflexes play a more or less important role are: homesickness, vomiting, rumination, constipation, habit tic, pavor nocturnus, whooping cough, enuresis, incontinence of stool, and bronchial asthma.

The phenomenon of dominance can be demonstrated in the sucking center of the infant following the appearance or disappearance of a stimulus.

REFERENCES*

ALBERTONI, P.: According to Ischlondsky.

ALDRICH. C. A.: Amer. J. Dis. Childr. *35*, 36 (1928).

VON ALLESCH, G. I.: Naturw. 1921, 774; Sitzungsber. Preuß. Akad. Wiss. Philosoph. Histor. Klasse, July 14 (1921).

ALVERDES, F.: Zschr. Tierpsychol. *2*, 258 (1939).

BAYER, W.: Med. Welt *1932*, 1668.

BESSAU, G.: Dtsch. med. Wschr. *1934*, 1819.

BIRJUKOW, D. A.: Pawlow-Zschr. f. höhere Nerventätigkeit *5*, 343 (1955).

BOGEN, H.: Jb. Kinderhk. *65*, 733 (1907).

BRAUN, H.: Zschr. Tierpsychol. *9*, 40 (1952).

BYKOW, K. M.: Großhirnrinde und innere Organe (The Cortex of the Cerebral Hemispheres and the Internal Organs). German translation. Berlin.

BYSTROLETOWA, G. N.: Pawlow-Zschr. f. höhere Nerventätigkeit *4*, 361 (1954).

CZERNY, A.: Straßbg. med. Ztg. *7*, 230 (1910); Zschr. Kinderforsch. *20*, 1 (1914/15); Der Arzt als Erzieher des Kindes (The Physician as Educator of the Child). 4th ed. Leipzig and Vienna, 1916; and A. Keller: Des Kindes Ernährung. Handbuch (The Nutrition of the Child. Handbook). 2nd ed. II, p. 280. Leipzig and Vienna, 1925.

DASCHKOWSKAJA, W. S.: Pawlow-Zschr. f. höhere Nerventätigkeit *3*, 299 (1953).

DENISOVA, M. P., and N. L. FIGURIN: Sovet. Pediatr., No. 6, 96 (1935).

FRANZISKET, L.: Naturwiss. *1950*, 93; *1952*, 92; Zschr. vgl. Physiol. *33*, 142 (1951).

GAMPER, E.: Zschr. Neurol. *102*, 154; *104*, 49 (1926).

GESELL, A.: See Premier congrès internat. de Psychiatr. infant.

GIURGEA, C.: (1953) nach L. Pickenhain in Müller-Hegemann, Neuroseprobleme in Klinik und Praxis (According to L. Pickenhain in Müller-Hegemann's Neurosis Problems in the Clinic and in Practice). p. 15. Berlin, 1959.

GRABOWSKI, U.: Zschr. Tierpsychol. *2*, 265 (1939).

HAMBURGER, F.: Arch. Kinderhk. *104*, 92 (1935); Med. Klin. *1935* I, 265.

HARTLEY, D.: According to H. J. Störig: Kleine Weltgeschichte der Wissenschaft (Small World History of Science). p. 717. Stuttgart, 1954.

HASSMANN, M.: Zschr. Tierpsychol. *9*, 294 (1952).

VON HELMHOLTZ, H.: Handbuch der physiologischen Optik (Handbook of Physiological Optics). pp. 430 and 447. Leipzig (1867); 2nd ed., p. 601 (1896); 3rd ed., pp. 3, 5, 21, and 439. Hamburg and Vienna, 1912.

IBRAHIM, J.: Neurol. Zbl. *30*, 710 (1911); Zschr. Neurol. *5*, 388 (1911).

ISCHLONDSKY, N. E.: Neuropsyche und Gehirn (Neuropsyche and Brain). Berlin and Vienna, 1930.

IWANOW-SMOLENSKY, A. G.: Fiziol. Ž. *19*, 133 (1935) (Russian).

—, Dtsch. Gesd.wes. *6*, 1373 (1951); Grundzüge der Pathophysiologie der höheren Nerventätigkeit (German) (Outline of the Pathophysiology of Higher Nerve Function). Berlin, 1954.

KASSATKIN, N. I.: Ref. Zbl. Kinderhk. *32*, 9 (1937); and A. M. Levikowa: J genet. Psychol. *12*, 416 (1935); Frühe bedingte Reflexe beim Kinde (Early Conditioned Reflexes in the Child). Mediz. Schrifttum d. SU. No. 16. Berlin, 1953; Grundriß der Entwicklung der höheren Nerventätigkeit des Kindes im ersten Lebensjahr (Outline of the Development of Higher Nerve Function in the Child in His First Year of Life). Berlin, 1955.

—, Pawlow-Zschr. f. höhere Nerventätigkeit *7*, 491 (1957).

KITAIGORADSKAJA: According to Bykow.

KOEHLER, O.: Z. Tierpsychol. *5*, 575 (1942/43); Vierteljahrsschrift Naturforsch. Gesellsch. Zürich *98*, 242 (1953); Arbeitstag über die zentrale Regulierung der Organismen (Workshop on the Central Regulation of the Organism). Leipzig, December, 1955. Berlin;

* Transliterations of Russian proper names follow German practice — Ed.

Fondation Singer Polignac, l'instinct dans le comportement des animaux et de l'homme (Singer Polignac Foundation: Instinct in Human and Animal Behavior). p. 647. Paris, 1956; F. Linneweh, Die physiologische Entwicklung des Kindes (in F. Linneweh's The Physiological Development of the Child). p. 3. Berlin-Göttingen-Heidelberg, 1959.

KRASNOGORSKI, N. I.: Jb. Kinderhk. *69*, 1 (1909); *78*, 373 (1913); *144*, 255 (1935); Erg. inn. Med. *39*, 613 (1931).

–, Zum 50jährigen Bestehen der Pawlowschen Lehre von der höheren Nerventätigkeit (Russian) (On the Fiftieth Anniversary of Pavlov's Theory of Higher Nerve Function). Pediatrija, Moscow *1952*, 1, 3 and 2, 3.

–, Arbeiten über die Erforschung der höchsten Nerventätigkeit des Menschen und der Tiere (Russian) (Essays on the Study of the Highest Nerve Function in Man and Animals). State Medical Literature Press, Moscow, 1954.

–, Pawlow-Zschr. f. höhere Nerventätigkeit *6*, 289 (1956).

KRATSCHKOWSKAJA: According to Kassatkin (1957).

LEWIN, S. L.: Fiziol. Ž. *19*, 614 and German summary 621 (1935).

LOTZE, H.: Göttingische Gelehrte Anzeigen *3*, 1756 (1853).

MARQUIS, D. P.: According to Wenger.

NOTHMANN, H.: Arch. Kinderhk. *51*, 123 (1909).

ORBELI, L. A.: Pawlow-Zschr. f. höhere Nerventätigkeit *5*, 354 (1955).

PANTSCHEKOWA, E. F.: Pawlow-Zschr. f. höhere Nerventätigkeit *6*, 154 (1956).

PAVLOV, I. P.: Die höchste Nerventätigkeit von Tieren (The Highest Nerve Function of Animals). German translation. Munich (1926).

–, Vorlesungen über die Arbeit der Großhirnhemisphären (Lectures on the Function of the Cerebral Hemispheres). German translation. Leningrad, 1932.

–, Sämtliche Werke (German) (Complete Works). Berlin III (1953), IV (1954). Ausgewählte Werke (German) (Selected Works). Berlin, 1953.

–, Mittwochkolloquien (Wednesday Colloquia). German translation. Vol. 3. Berlin, 1956.

POLIKANINA, P. I., and L. E. PROBATOWA: Pawlow-Zschr. f. höhere Nerventätigkeit *7*, 418 (1957).

Premier congrès internat. de Psychiatr. infant. à Paris. I. Lille (1937) (Wallon, Dublineau, Wintsch, Gesell, Peiper, Marinesco, Jonesco-Sisesti and Kreindler, de Jong, Gemelli, Ischlondsky, de Sanctis, Ibrahim, Meignant).

RÄHLMANN, E.: Ophth. Klin. *7*, 21 (1903).

RAY, W. S.: Child Developm. Baltimore *3*, 175 (1932), according to Wenger.

–, Cited according to Wenger; also see Carmichael, p. 209.

RAZRAN, G. H. S.: Arch. Psychol., New York. No. 148. April (1933).

SCHRÖDER, P.: Münch. med. Wschr. *1938*, 445.

SAMKOWA, M. A., and A. N. TSCHERNOUSSOWA: According to Kassatkin (1957).

SKORUNSKAJA, T. N.: Pawlow-Zschr. f. höhere Nerventätigkeit *8*, 207 (1958).

SONTAG, L. W., and R. F. WALLACE: Amer. J. Dis. Childr. *48*, 1050 (1934).

SPELT, D. K.: Psychol. Bull. *35*, 712 (1938); J. Exper. Psychol. *38*, 338 (1948).

SPENCER, H.: Prinzipien der Psychologie (Principles of Psychology). 2nd ed. German translation. Stuttgart.

STRAUS, E.: Vom Sinn der Sinne (On the Meaning of the Senses). Berlin, 1935. 2nd ed. Berlin-Göttingen-Heidelberg, 1956.

SYKOWA, O. P.: According to Kassatkin (1957).

TRENDELENBURG, W.: I. P. PAWLOWs Lehre von den Verrichtungen des Großhirns (I. P. Pavlov's Theory of the Functions of the Cerebral Hemispheres). Abh. preuß. Akad. Wiss., Berl., Physik.-math. Kl., 1937.

TRINCKER, D.: Naturwiss. *39*, 92 (1952).

TROSCHICHIN, W. A.: Pawlow-Zschr. f. höhere Nerventätigkeit (German) *2*, 743 (1952).

UCHTOMSKY, A.: According to Ischlondsky.

UFLAND, I. M.: Pflügers Arch. Physiol. *208*, 49 (1925); *221*, 605 (1929).

VAIL, V. S.: Pädiatrie (German edition, Berlin) *2*, 88 (1953).

WATSON, J. B.: Psychology from the Standpoint of a Behaviorist. 2nd ed. Philadelphia and London, 1924.
—, Der Behaviorismus (Behaviorism). German translation. Berlin and Leipzig, 1930.
WENGER, M. A.: Studies in Infant Behavior III. University of Iowa Studies *12*, No. 1. Iowa City. University of Iowa Press, 1936.
WICKENS, D., and C. WICKENS: J. exper. Psychol. *26*, 94 (1940); Ref. Mschr. Kinderhk. *91*, 398 (1942).
WINOGRADOW, M. I.: According to Ischlondsky.
WORONIN, L. G.: Pawlow-Zschr. f. höhere Nerventätigkeit *1*, 213 (1951).
—, Pawlow-Zschr. f. höhere Nerventätigkeit (German) *2*, 73 (1953); Dtsch. Gesd.wes. *9*, No. 52, Supplement (1954).
ZELIONY: According to E. von Skramlik, Dtsch. Gesd.wes. *6*, 887 (1951).

12. NEUROLOGICAL CHARACTERISTICS OF CERTAIN DEVELOPMENTAL PHASES

TISSUE MATURATION AND FUNCTIONAL ABILITY OF THE BRAIN

(1, 3)

The newborn infant's brain is comparatively large; it comprises 12% of the newborn infant's body weight, but only 2% of that of the adult. Accordingly, the total weight of the body increases approximately 12 times from the neonatal period to adulthood, whereas the weight of the brain increases only 3.8 times. Because of these ratios of size and because of its outward appearance, the newborn infant's brain seems to be rather complete, suggesting that all its sections are functional just as those of the liver, the lungs, the kidneys, or other organs. But this is not so; the most important and largest portions of the brain, the cerebral hemispheres, have not yet begun to function and even the lower parts of the brain are not yet able to function fully.

The brain of the newborn infant is histologically immature. The cells are not yet completely differentiated so that it is difficult to demarcate the individual cell layers clearly from each other; myelinization has developed only to a minor degree and in some places not at all.

In comparing the anatomic findings and the onset of function in the different brain portions, it is important to consider that the immaturity at birth is not evenly extended over the brain; the ontogenetically and phylogenetically oldest sections of the brain are histologically the most mature. Maturation progresses from the medulla oblongata over the basal ganglia and the mesencephalon toward the cortex of the cerebral hemispheres. The ontogenetically younger tracts also show a retarded development of their older sections. Those brain portions that have to fulfill vital tasks at an early age are the most mature. This applies particularly to the region of the respiratory center in the medulla oblongata.

According to FLECHSIG (1876, 1920), the fibers of a specific tract are simultaneously myelinated; moreover, myelinization always progresses in the direction of the pathway, i.e., toward the brain in sensory tracts and away from the brain in motor tracts. Myelinization usually occurs earlier in the sensory tracts than in the motor tracts.* FLECHSIG attributes great importance to the sequence in which the sensory

* LANGWORTHY (1933), on the other hand, states that in all mammals the ventral roots of the spinal cord are myelinated prior to the dorsal roots and the efferent pathways of the cerebral hemispheres prior to the afferent tracts.

tracts of the cerebral hemispheres become myelinated; the earliest to develop are the central pathways of tactile and olfactory sensation, then the optic, and last the auditory tracts. FLECHSIG did not clarify the maturation of the vestibular pathway. FLECHSIG, W. STERN (1923), and KOFFKA (1925) drew conclusions about the functional ability of the individual senses from their state of myelinization (see p. 94).

According to A. WESTPHAL (1894, 1897), the motor cranial nerves of the newborn infant are already myelinated, whereas the nerves of sensation and those of the sensory organs and the mixed nerves, with the exception of the auditory nerve, are not myelinated. The least developed is the optic nerve within the orbital cavity. The myelinization of the cranial nerves is not completed before the ninth or tenth month of life. In general the cranial nerves mature earlier than the peripheral nerves do; the latter are fully developed only in the second or third year of life.

Myelinization is accelerated when the tracts fulfill their physiological tasks. This was demonstrated by AMBRON and HELD's (1896) and HYDÉN's (1952) experiments on animals, and A. WESTPHAL's observations (1894, 1897) on prematurely born infants (see p. 66).

Myelinization of the pyramidal tracts occurs at different rates in various infants. According to FLECHSIG, myelinated pyramidal tracts have already been found in some newborn infants in the gyrus centralis anterior and in their course through the capsula interna, the mesencephalon, and the medulla oblongata. In the spinal cord, fibers of the anterior pyramidal tracts are already myelinated in contrast to those of the lateral tracts.

According to A. WESTPHAL, the peripheral spinal nerves of the newborn infant are thinner and more difficult to stain than those of the adult. The further development, which consists chiefly of myelinization, progresses steadily from birth. In the second and third week of life myelinization has markedly increased; however, a condition similar to that in the adult is not reached before the second or third year of life. A. WESTPHAL especially stresses the varying rapidity of myelinization in different instances.

Contrary to FLECHSIG's view, the opinion that the nervous tracts are functional only if myelinization has been completed is no longer held. ANGULO Y GONZALES (1929) found movements and certain reflexes in fetuses of rats many days before myelinization. According to MINKOWSKI (1928) and LANGWORTHY (1933), the first myelinization appears in man in the fourth or fifth month of fetal life, i.e., long after the first reactions and movements have occurred. For example, MINKOWSKI observed the palmar grasp reflex even at two and one half months of gestation, closing of the mouth and sucking movements on touching of the lips and the tongue at two to three months, and the spinal reflex beginning with the third month. The earliest movements — those of the right hand — were observed by YANASE (1907) in an embryo 2 cm long (estimated age, six weeks).

MINKOWSKI, ANGULO Y GONZALES, and LANGWORTHY thus concluded that reflex functions are possible long before any tracts are myelinated. This reflex activity consists of weak, often tremulous, movements which spread over large sections of the body. The area which receives the stimulus is large and the motor reaction is widespread. As soon as the central nervous system has developed more fully and has become more mature, more exact responses to the reflexes occur. Although reactions can be demonstrated in an unmyelinated nervous system, the sequence in which myelinization occurs represents an important clue to the stages in which the various brain regions become functional (FLECHSIG).

The aforementioned investigators occupied themselves chiefly with the myelinization of the nervous tracts; the no less important maturation of cells has also been related to the beginning of nervous activity. According to these investigations, the morphological changes of the ganglion cells form the basis for the beginning of cerebral function and allow us to draw conclusions on the developmental processes in childhood. The slow development of a function is caused by the slow maturation of the anatomic area. When these investigators tried to find relations between the anatomic maturation of the cells and the initiation of function, they proceeded from rather arbitrary assumptions:

"Toward the end of the first year of life the first successful attempts at regulating bodily equilibrium can be observed, and during the second year of life the attempts to master the complicated execution of the upright gait."

My detailed description, however, shows that the infant attains the faculty of maintaining his equilibrium very gradually, starting with the labyrinthine righting reflex of the head in the neonatal period and progressing to the first independent step at the age of one year. It is therefore not possible to single out the achievement of a certain developmental stage and to call this the first successful attempt. Conscious meaningful attempts at articulation that can be regarded as verbal expressions cannot be recognized before the end of the second year of life; this is said to be connected with the fact that BROCA's area matures only in the seventeenth to twentieth month. The maturation of WERNICKE's centers is supposed to occur at the same time, which corresponds to the observation that the first intelligent understanding of acoustically or verbally differentiated stimuli falls in this period or even somewhat earlier. Actually no one will ever be able to determine when the first word is consciously uttered.

Thus, the attempt to bring anatomic maturation and the physiological beginning of function into close agreement has failed.

In a detailed review, CARMICHAEL (1954) summarized the numerous writings dealing with the prerequisites for the onset and the first development of behavior in man and animal. The only conclusion he was able to draw from the many theories and experiments is that we are dealing with a very complicated process which is impossible to trace to one single cause. Anatomic maturation, whether it is expressed in myelinization, formation of neurofibers, or the maturation of the ganglion cells, is important, as are chronaxie, internal secretion, and many other phenomena, but none of them is decisive in itself. CARMICHAEL therefore prefers to describe as unequivocally as possible the responses to stimuli of which a fetus is capable at any stage of its development instead of compressing the entire development into one formula. On the basis of careful investigations in living human embryos, HOOKER (1958) also concluded that even with the newest methods and microscopes it cannot be determined when the synapses between the neurons begin functioning.

EMBRYO AND FETUS

(1, 2)

Behavior before birth has long attracted attention. There are numerous statements in the Bible and in the writings of antiquity (ARISTOTLE) on this problem, as can be found in NEEDHAM's (1931) and CARMICHAEL's (1933, 1954) summaries. GOECKEL (1911) described the evaluation given to the life of the unborn child through the ages by jurisprudence and the church.

PREYER (1885) was the first to make fundamental investigations of the behavior of vertebrate embryos. Some isolated reports on the behavior of the human embryo existed even earlier. MINKOWSKI, who summarized the existing literature in 1928, made detailed investigations, as did COGHILL (1929) and LANGWORTHY (1933). In 1933 and 1954, CARMICHAEL reviewed the research on human and animal fetuses in detail and compared the results with his own investigations.

A premature delivery by Caesarean section makes it possible to observe the behavior of nonviable human fetuses for a short time. It is essential that the surgical intervention be performed under local anesthesia. General anesthesia of the mother impairs the nervous behavior of the fetuses most severely and they die so fast that they can only be observed for some minutes in agony, with rapidly reduced excitability. Consequently, the presence of certain functions and reactions is of greater importance than their absence.

According to COGHILL's investigations in the amblystomum, the development of the nervous system in the vertebrates attempts above all to preserve the intactness of the body; in this development, automatisms appear continuously as independent entities which struggle among themselves for power over the individual. On the basis of all available reports on the movements of human embryos and fetuses during the first six months, COGHILL is certain that in man, just as in the amblystomum, nervous behavior develops through the elaboration of a stage which from the beginning is integrated into a whole, and through the development of further stages (reflexes), within the whole. Of course, difficulties arise if the first head-body movement at the earliest stage of the human embryo is compared with the movements of the amblystomum.

The law of the cervicocaudal direction of development (see p. 237), which can be demonstrated so clearly in the infant during the different developmental stages of his locomotion, can also be applied to the origin and spreading of the first movements during the fetal period.

PFLÜGER (1877) observed contractions of the heart even in the third week of a human embryo approximately 4 mm long. YANASE (see p. 540) described the earliest movement in an embryo 2 cm long and approximately six weeks old. In 1837, ERBKAM saw movements in an embryo only 1.6 cm long; these consisted of opening of the mouth, lateral movements of the head, and movements of arms and legs. MINKOWSKI observed slight worm-like movements of the head, trunk, and limbs, which reminded him of athetoid movements, starting at two months of gestation. He observed respiratory movements starting at four months (length of fetus, 18 cm) — occasionally even before the umbilical cord was severed. These were deep inspirations in which the mouth was opened and the head was sometimes overextended backward. Respiratory movements, sometimes coupled together very closely in pairs, appeared at rather long intervals of as much as several minutes. Thus, gasping respiration (see p. 333) appears first, while the higher stages of the respiratory center may occasionally cause periodic or rhythmic respiration, but in general these stages appear later. Gasping respiration is altogether incompatible with life if it lasts for any length of time and causes the quick death of embryos and fetuses. However, the heart, which does not need the brain for its action, usually still remains active in fetuses and immature infants long after respiration has ceased.

Reflexes caused by skin stimulation tend to spread over extensive sections of the body. These reactions become more specific and more limited with progressing age. By destroying the spinal cord, MINKOWSKI was able to prove that the stimulation in far-spreading reflexes, that is to say the spreading from the arms to the legs and vice versa, is transmitted by the spinal cord.

Skin reflexes can be easily elicited in the face. Stroking of one lid, for example, causes a bilateral contraction of the musculus orbicularis oculi. MINKOWSKI was even able to elicit a palmar grasp reflex at two and one half months of gestation by stroking the palm; BOLAFFIO and ARTOM (1924) were not able to demonstrate it before the seventh month. The spinal reflex with participation of the limbs was often present at three months of gestation and the abdominal reflex accompanied by a flexion of both legs and often also by a raising of both arms at five to six months. The varying behavior of BABINSKI's sign was previously described on p. 285.

After the second to third month of gestation, MINKOWSKI saw closing of the mouth and sometimes rhythmic sucking movements on touching the mucosa of the lips or the tongue. The reflex zone occasionally extended to the cheeks, lids, and chin. This reflex, which is one of the earliest, most reliable, and most constant, could sometimes still be elicited when all other reflexes were extinguished. It disappeared after the destruction of the medulla oblongata. On the other hand, BOLAFFIO and ARTOM found a sucking reflex only after the sixth month.

I observed a fetus 26 cm long making distinct sucking and swallowing movements when fluid was brought into his mouth, but noted no facial grimaces when he received quinine. Pinpricks evoked distinct reactions only in the facial muscles, and did not influence the other parts of the body. Hence, reflexes of food intake can be evoked at a strikingly early developmental stage.

A comprehensive, planned study was conducted by HOOKER (1936 until 1958; also see pp. 34 and 409), who observed the behavior of 149 human embryos and fetuses. He carefully used the tip of a fine hair as a tactile stimulus. He found that the first reflex can be elicited between the middle of the seventh and the beginning of the eighth week of gestation in embryos 20 to 23 mm long: On stimulation of the mouth and the alae nasi, the neck is flexed to the other side. At eight weeks the flexion spreads in the caudal direction to the upper part of the trunk; at eight and one half to nine and one half weeks it extends over the whole trunk, while the excitable skin region gradually becomes larger. After eight and one half weeks the pelvis participates in the flexion. Occasionally ipsilateral flexions occur at this time. As soon as the flexion of the trunk involves the shoulders, both arms are extended backward at the shoulder joints. On stimulation of the oral region, extension replacing flexion of the trunk and accompanied by inward rotation of both arms gradually appears at nine and one half to ten and one half weeks and more distinctly at eleven to twelve weeks. The face now moves away from the stimulus instead of being turned to the side. This response persists until the age of thirteen to thirteen and one half weeks.

HOOKER sees in this embryonal and fetal behavior a counterpart to the observations of COGHILL (see p. 542), who observed in the amblystomum a generalized reaction (total pattern) preceding an individual reaction (partial pattern). The arm and leg reflexes also occur in man not individually at first but in association with trunk movements.

Certain local reflexes can be demonstrated only later: At nine and one half weeks the mouth opens on stimulation of the edge of the lower lip. At ten to ten and one half weeks the upper eyelid responds with a kind of squint, a contraction of most of the fibers of the orbicularis, to a slight touch. The same response occurs at twelve and one half to thirteen weeks when the upper, and more rarely the lower, lip is touched. At ten and one half weeks stimulation of both lips sometimes causes swallowing movements. At eleven weeks the sensitive region of the facial skin extends so far that wrinkling of the forehead may occasionally occur on stimulation of the upper eyelid or an area near the eyebrows. At fourteen weeks the tongue is retracted on touch. At twelve and one half weeks the mouth closes when it is touched; if this is repeated several times, swallowing occurs. At thirteen weeks the mouth is closed even on touching the lower lip. Between thirteen and fourteen weeks almost none but local reactions can be obtained when the trigeminal skin area is stimulated. Stimulation of the lateral portion of the lower lip or of the alae nasi causes a contraction of the musculus quadratus labii superior and of the alae nasi in which the upper lip and the alae nasi are raised as if in a scornful smile. Usually the head is turned away from the stimulus. This "avoiding reaction" (corresponding to the skin protection reflexes, see p. 30), which appears, as previously described, as neck bending to the other side at seven and one half weeks, occurs at thirteen to fourteen weeks only as a local reflex without participation of the trunk.

Reflexes which occur on stimulation of the mouth (opening of the mouth, tongue movements, swallowing, lip and mouth closing) gradually combine to form the reflexes of food intake: At seventeen weeks the upper lip protrudes when the mucosa is stimulated; at twenty weeks the lower lip joins in this movement. The earliest sucking movements are observed at twenty-nine weeks.

Arms and legs are at first moved only with the trunk and not independently before ten and one half weeks. At this time the arms are sometimes extended on stimulation of the shoulders. Stimulation of the palm causes rapid, incomplete closing of the fingers, but not of the thumb. At eleven weeks every embryo shows flexion of the fingers on stimulation of the palm. At this age the wrist is usually flexed. The flexion of the fingers is sometimes accompanied by flexion of the elbow, inward rotation of the arm, or pronation of the forearm, or by all three movements simultaneously. At eleven and one half weeks the arm is occasionally abducted and again adducted on returning to the rest position. Abduction and adduction at the shoulder do not appear as independent movements before the twenty-seventh week.

According to HALVERSON (1937; compare p. 255), grasping in the infant consists of closing the fingers and seizing the object. These two components appear successively in embryonal and fetal life: At ten and one half weeks an incomplete and at fourteen weeks a complete closing of the fingers occurs in some fetuses; at eighteen weeks a glass rod or a hair is seized in the closed fist for the first time. The grip becomes stronger, so that at twenty-seven weeks almost the entire body weight can be carried.

HOOKER found the earliest reflexes on stimulation of the soles, i.e., a plantar flexion of all toes, at the age of ten and one half weeks (see p. 286 for further observations by this author). In progressive oxygen deprivation, the palmar reflex is extinguished prior to the reflexes evoked from the trigeminal region, and the plantar reflex prior to the palmar reflex, i.e., the ontogenetically and phylogenetically younger reflexes cease before the older reactions.

According to HOOKER and HUMPHREY (1954), the local reflexes elicited by the stimulation of the trigeminus appear in the same sequence in which food intake developed phylogenetically: mouth opening (nine and one half weeks), swallowing (ten and one half weeks), and preceding closing of the lips (twelve and one half weeks). A low vertebrate living in the water needs only to open its mouth to transport food into its nutritional canal. When the food is in the mouth, swallowing will then aid the process. For the terrestrial animal, mouth closing, if it precedes swallowing, facilitates the swallowing of live nourishment. In some lower vertebrates, as for example the frogs, movements of the tongue facilitate food intake. These movements appear quite early in human embryos. A bag-like contraction of the lips and sucking movements appear only as the last local reaction to trigeminal stimulation in the young mammal.

The trigeminal reflexes are thus ontogenetically and phylogenetically the oldest. They are the last to be extinguished in progressive oxygen deprivation (HOOKER, 1952; HUMPHREY, 1953). This fundamental rule also applies to the partial centers of the respiratory center, as I have previously shown (see p. 389).

According to MERREM (1960), the trigeminus assumes another special position in the course of phylogenesis: In the mammal it receives important sensory impressions from the environment, since it is the nerve nearest to the mouth. The sensitivity of the dog's or horse's mouth is well known. Everybody knows that a pinprick on the face is felt differently from one on the hand. MERREM attributes the susceptibility of the trigeminus to severe neuralgias to this special position.

According to PREYER (1885), the ability to respond with local movements to a certain stimulus applied to the body surface can even be found in guinea pig fetuses. According to CARMICHAEL (1946), who confirms these statements, this is not learned but develops as the result of a phylogenetically determined growth of the nervous system. Localization occasionally takes place without the possibility of previous practice for this faculty. Weak and strong skin stimuli have totally different effects; a weak stimulus usually evokes fewer general movements than a stronger stimulus does.

According to CARMICHAEL (1946), movements of the human embryo occurring without external stimuli are first observed after nine to twelve weeks of gestation.

This ability to act "spontaneously" has led adult man to nearly all the accomplishments which can be regarded as foundations of human freedom. It has been thought that this behavior is not caused by anything but is itself the cause. However, the first "spontaneous" function of the fetus should not be regarded as a phenomenon without a cause. There is certainly no reason to assume that it originates in certain new, nonbiological forces which are called "volition" by unscientific writers. Such a reaction must be regarded as the natural result of the growth of the fetal organs... From a very early stage of behavior, the organism is set in motion not only by external but also by internal forces. It changes its adjustment as a result of internally induced movements. This thereby changes the nature of the external stimulation, although no changes have taken place in the external energy field. Thus, the sensory channels are stimulated anew; a continuous process of adaptation to internal and external stimuli develops.

HOOKER (1954) also states that the area of the maxillary and mandibular trigeminal branches is excitable at an earlier time than the region of the ophthalmic branch. The first sensory field is restricted to the lips and the alae nasi. The first responses to stimulation of the oral region are not localized, but spread to all neuromuscular mechanisms which are capable of functioning. While at first restricted to the neck muscles, the stimulus response spreads in the cervicocaudal direction to the long dorsal muscles and the muscles of the shoulder and pelvic girdles. The responses of the dorsal muscles continue during the fetal period for a rather long time, but gradually decrease with the development of the local reflexes. They appear mostly on the side opposite to the stimulus.

HOOKER (1954) and HUMPHREY and HOOKER (1959) further demonstrated that two body areas can be stimulated simultaneously after it has become possible to elicit finger and toe reflexes by external stimulation of the palms and the soles of the foot. Attempts at simultaneous stimulation of face and hand in human embryos and fetuses at ten and one half to thirteen weeks, and of hand and foot at thirteen to thirteen and one half weeks, showed distinct dominance of the proximal (those situated near the oral region) over the distal areas. Simultaneous stimulation of mouth and hand caused reactions only in the oral region and of hand and foot only in the hand. However, all fields proved to be excitable on individual stimulation. The older reflex arcs were excitable prior to the younger ones, in accordance with the cervicocaudal direction of development (also see p. 237).

MINKOWSKI found no pupillary light reflexes in any fetuses (10 to 34 cm long, corresponding to an age of three to seven months).

Tonic neck reflexes could regularly be demonstrated in older fetuses, but only indistinctly in younger ones. MORO's reflex (see p. 151) was also present. The movements, irregular at first, became more constant in later months of gestation.

The tendon reflexes were hard to judge. Tapping of the tendons was mostly without result.

The athetoid movement pattern was previously described on p. 250. According to BOLAFFIO and ARTOM, the cortex of the cerebral hemispheres and the pyramidal tracts do not respond to electrical stimulation until the sixth month of gestation, whereas the pons, medulla oblongata, and spinal cord are already excitable.

Stimuli coming from the outside to the spinal cord do not meet in circumscribed tracts at first, but radiate in different directions so that various reactions occur. With progressing development, the reactions are restricted more and more to the stimulated body area and follow certain laws of transmission (MINKOWSKI). This development is not yet completed in infancy; the individual movement gradually evolves from the general motor restlessness.

According to MINKOWSKI, the muscles of fetuses can be directly stimulated by tapping even at an early age. After approximately the third month of gestation, they can also be stimulated by galvanic and faradic currents. The muscle response is less lively on stimulation from the nerve. After the fourth month the nerves are able to transmit the electrical excitation despite the lack of myelinization.

PREMATURE INFANTS

(1, 4, 39, 84, 99, 123, 131, 132)

The neurological characteristics of premature infants, i.e., infants with a birth weight up to 2500 gm, have been repeatedly described, for example, by SAINT-ANNE DARGASSIES (1955, 1957), ROSENBAUM (1954, 1958), and WOHLMUTH and PETRORINI (1958). Nevertheless, it has so far hardly been possible to differentiate neurologically among the various developmental stages which would approximately correspond to birth weights of 600 gm, 1000 gm, 2000 gm, and 3000 gm. The available data differ greatly. Moreover, such a comparison is made much more difficult by the fact that diseases may easily lower the excitability in small premature infants, or may even prohibit examination. Many premature infants die soon after birth because of birth injuries; others — though also damaged — recover and some days afterward show a neurological picture completely different from that seen immediately after birth. Hypoxia, which may arise from many causes in premature infants in the first weeks of life, proved to be especially disturbing. Nervous excitability disappears altogether during an apnoic attack. Icterus neonatorum, which appears some days after birth, may again markedly lower the excitability. Thus, we must forego a summarizing description and can only mention some details.

MINKOWSKI (1928) describes the weak cry of a premature infant 28 cm long in the sixth fetal month as the first audible sound of a human voice. Premature infants weighing approximately 650 gm can already cough, sneeze, and yawn, can have an attack of severe singultus, and may cry audibly. The eye-neck reflex can be demonstrated [PEIPER, 1926 *(15)*; LOTH, 1952 *(11a)*]. GARSCHE (1950) found the snapping-swallowing reflex (see p. 443) almost exclusively in premature infants weighing less than 1500 gm. The youngest infants weighed 850 to 930 gm. After the eighth month of gestation, the reflex is inhibited by the brain stem so that only remnants of it can be found in the mature newborn infant.

At a birth weight of less than 1000 gm, the brain is still so immature that respiratory, sucking, and swallowing disorders often occur. Sucking disturbances can be overcome by tube feeding, but respiratory disorders frequently lead to death from attacks of suffocation. I was able to observe pendular induction between the respiratory and sucking centers only in weak premature infants and in sick infants (see p. 552). The question of birth damage is discussed on p. 435 and thereafter.

Immature infants behave differently in many ways from mature babies. They have a greater need for sleep, are more difficult to arouse, and have a longer reaction time. Lower brain functions, such as athetoid movements, slight motor restlessness ("Stäupchen," see p. 249), and asymmetric tonic neck reflexes of the limbs, occur more frequently in them than in mature newborn infants. Inhibitory phenomena can already be observed.

Suranyi (1957) describes a spastic forced position which appears eight to thirty-six hours after birth and lasts for five to six days as a "frog-leg" sign occurring chiefly in prematures. In this position the legs are strongly abducted at the hips and flexed at the knees to an angle of less than 90°. This sign was found in almost one third of all premature infants, most frequently in those weighing 1000 to 1700 gm at birth.

In the first days of life the EEG shows characteristics identical to those of the mature newborn infant, except that the tracings of wave potentials are frequently still slower and more irregular (Mai, Schütz, and H. W. Müller, 1951; Garsche, 1954). In the first days of life an overirritability to mechanical stimulation, either on direct muscle stimulation or by excitation from the nerve, is frequently found (see p. 300).

At the time of his "real" day of birth, the premature infant is superior in many ways to the newborn infant born at term. His movements are more plentiful and more varied. He can move freely, uninhibited by hypertonia. The muscle tonus is of median strength, but the legs are still rather passive. The primary automatisms are distinct, but still show a special pattern in the legs: The crossed wiping reflex is still weak. The supporting reaction is hesitant and impaired by insufficient plantar support. The stepping movements are not yet exactly rhythmic. In the abdominal position the premature infant lifts his nose by turning his head sideways. He can turn on his side independently, turn his head, turn toward a soft light, and hear the human voice (Saint-Anne Dargassies, 1955).

UNBORN INFANTS

(10, 3a)

At the operation of an extrauterine pregnancy in the third month, Krauss (1951) had an opportunity to observe the behavior of a human fetus which was still supplied with placental blood through the umbilical cord. This is a condition which Barcroft (see p. 360), for example, produced artificially in experiments on animals: Through the transparent cupola of the chorion, which had left the tube, one could see how the fetus, swimming in the amnion, executed lively spontaneous movements for several minutes; especially striking was the fact that Moro reflexes could be elicited by shaking, tapping, and touching the amnion. These appeared in the form of nearly symmetrical movements of all four extremities, similar to those seen in the defense and fright reflexes of newborn infants. The stimulus was transmitted by the amniotic fluid. If the amnion was kept completely quiet and free from shocks, the fetus also became quiet, again responding to the next stimulus with the same movements.

W. N. KELLOGG (1941; see p. 237) and WALTHER (1951) recorded the movements of the human fetus through the mother's abdominal wall and thus obtained continuous graphs. According to WALTHER, the fetus has a centrally regulated motion rhythm with the highest number of movements occurring between 5:00 and 12:00 PM; this period is followed by the period of least mobility with complete intervals of rest until 6:00 AM. During the day the fetus remains rather quiet. The fetal and maternal rest periods coincide to a great extent. The two influence each other in their waking and sleeping patterns. KELLOGG found a distinct increase in movements with increasing fetal age.

SONTAG and WALLACE (1935) investigated the influence of the pregnant mother's cigarette smoking on her unborn child's cardiac function. They found an increase in the cardiac rate of 5 ± 0.19. My co-worker DOERFEL [1951 *(6a)*] was able to confirm these statements: The cardiac rate of the fetus increased $1\frac{1}{2}$ minutes after the mother had started to smoke, and reached its peak with an increase of 7 beats per minute in the fifth minute of smoking. The effect of nicotine outlasts the smoking time by 7 to 9 minutes.

The fetus drinks amniotic fluid while in the womb; therefore, such contents of this fluid as lanugo and epidermis scales may be found in the gastrointestinal tract. DE SNOO (1937) tapped the amnion through the abdominal wall of a pregnant woman and replaced part of the amniotic fluid with a saccharine solution to stimulate the fetus to drink, and with a solution of methylene blue to observe the absorption. The injected substances entered the gastrointestinal tract of the fetus together with the amniotic fluid, were absorbed there and taken into the fetal circulation, entered the circulation of the mother through the placenta, and were finally excreted through her kidneys. By continuously examining the mother's urine, it was possible to follow up the drinking of the fetus exactly as to time and amount. It was shown that the fetus drinks only at long intervals. The quantity of amniotic fluid absorbed daily is estimated by DE SNOO as approximately 1 liter. The quantity of amniotic fluid depends entirely on the function of the amniotic epithelium and the amounts which the fetus drinks.

EHRHARDT (1937, 1939) arrived at the same results by radiographic procedures. Prior to a necessary interruption of pregnancy, he injected thorotrast into the amniotic fluid of pregnant women and subsequently checked the location of the contrast agent. In the next two days, most of the thorotrast entered the gastrointestinal tract of the fetus. According to EHRHARDT, the amounts absorbed are rather large, since an amount equal to several times the total amniotic fluid content is consumed in a period of 15 hours. The drinking of amniotic fluid could be demonstrated as early as the fifth month of gestation.

A hydramnion may occur if the fetus is unable to drink the amniotic fluid because it suffers from a congenital esophageal atresia without a tracheoesophageal fistula [H. R. SCHMIDT, 1926; H. MEYER, 1929; FUKAS, 1940; THOMAS, 1954 *(21a)*]. ROSSIER (1955) found a hydramnion in 14 of 21 malformations of the esophagus. SAGER (1957) even observed a hydramnion in congenital duodenal atresia.

The intrauterine respiratory movements (AHLFELD, 1888) were previously described in detail on p. 354.

In general, the responsiveness of fetuses will be no less than that of those of the same fetal age who for some reason were prematurely born: Rather, a higher excitability should be expected before the child has to go through the birth process. Although the child cannot be observed so clearly in his mother's womb, his movements can be recognized through the abdominal wall.

Strong fetal movements occur from time to time even if the mother remains still for some time. This everyday observation alone refutes the frequently stated opinion that children sleep continuously before birth. Some mothers of several children who had also given birth to a hyperactive idiot told me that this child moved more during pregnancy than the other children did. PREYER obtained similar statements from a mother who had given birth to three healthy and three microcephalic children. After the birth of the first microcephalic baby, she was able to predict correctly each time she was going to give birth to another microcephalic child.

Responses of the unborn child to external stimuli can be observed only under special conditions. On a sudden loud sound stimulus (automobile horn), the fetus reacts with distinct movements (see p. 90). If the same stimulus is repeated several times in rapid succession, very soon responses no longer occur; this is in accordance with the fright reaction of the infant after birth. Such movements may become uncomfortable to the mother when she attends indoor concerts.

SONTAG and WALLACE (1934) found that the mobility of the fetus increases when the mother's abdominal wall is stimulated by vibration. For example, 15 seconds after onset of this stimulus, it increased by 97% in one fetus, and in three succeeding 15-second periods by 59%, 17%, and again 17%. If the stimulus was repeated several times, the fetus became unresponsive. These two investigators determined the influence of vibration on the fetal cardiac rate in eight pregnant women. Six fetuses showed an increase and two a decrease. The average increase amounted to 0.3 heart beats four months before birth, 7.8 beats three months before birth, 13.4 beats two months before birth, and 14.3 beats per minute one month before birth. The unborn child thus responds to sound and vibration stimuli. DE SNOO's (1937) statement that the fetus drinks sweetened amniotic fluid better than unsweetened has still to be documented by more accurate evidence. Attempts to form conditioned reflexes in the unborn child (see p. 517) have so far usually failed, as did those in the newborn infant. According to SONTAG (1940), apart from sound stimuli, the mother's fatigue and emotions affect fetal mobility. Moreover, he describes two cases in which the emotions of the mother caused malnutrition of the newborn infant.

INFANTS DURING BIRTH

The birth process is usually more normal in healthy than in sick infants; it is frequently abnormal in premature infants, twins, and infants with torticollis or paralysis caused by spina bifida cystica.

"A dead baby cannot help himself" (JUSTINE SIEGEMUNDIN, 1648–1705, according to SELLHEIM, 1925). Deliveries of dead babies are therefore more difficult than those of living. But how does the healthy infant aid delivery? SELLHEIM (1917) found that newborn infants tend to move in the cephalad direction. If during the night they lay on their backs, lightly covered, in the morning they were found to have clearly moved to the head of the bed, never to the foot. During parturition, the infant probably assists by a more or less strong extension of the head. This can be regularly observed after birth if the head is pushed down, and it can also be expected to occur during the delivery (SELLHEIM, 1924).

The expulsion of the baby may be aided by supporting reactions of the legs (see p. 194), with which the baby is able to push himself strongly away from the uterine wall. In normal birth the neck righting reflex of the body (see p. 166) and in breech presentation the body righting reflex of the head (see p. 167) promote the rotation of the infant's body.

According to LANGREDER (1949, 1950, 1958), crawling and stepping movements, the spinal reflex, and MORO's reflex can be demonstrated even before birth in the uterus. From recordings through the abdominal wall and observations of pregnant women and of women in labor, as well as of newborn and premature infants, it seems probable that the fetus adapts himself to the birth canal and adjusts his head by these reflexes, thus taking an active part in parturition.

In newborn infants, PRECHTL and KNOL (1958) investigated the influence of breech presentation on the leg movements and the plantar reflexes by comparing babies born in breech presentation with newborn infants delivered in vertex presentation. After simple breech position, extension of the legs prevailed during rest and during movements without external cause. Scratching caused an extension, rather than the flight reflex. After simple breech-foot position and simple foot position, the flexion reflex was increased and the extension reflex was diminished or absent. A lesion in the afferent fibers of the spinal apparatus is assumed to be the basis of these phenomena.

MATURE NEWBORN INFANTS

(1, 3, 4, 11, 17, 18, 25, 39, 47)

For the child, birth means a tremendous readjustment in respiration, circulatory system, digestion, temperature regulation, sensory perception, motility, and other important physiological tasks. In addition, the birth process may easily cause more or less distinct pathological changes which cannot always be separated from the physiological phenomena, since we do not know the infant's condition before birth. The observation of infants delivered by Caesarean section is of no help, for these infants have not gone through the normal birth process; they show increased content of amniotic fluid in the lungs, since their chests were not compressed during birth and thus were not freed of a portion of the amniotic fluid which is always present in the air passages and lungs.

At the stage of the newborn infant, the cerebral hemispheres are not yet functional. The movements are not yet regulated by the cerebral cortex and the pyramidal tracts, are not yet inhibited by the corpus striatum, and are therefore uncoordinated and — with the exception of food intake — aimless. The mass movements of the first months correspond to those of a pallidum preparation (see p. 265).

In the first days of life, the muscles are often completely atonic. If one tries to sit the child up, he slumps. However, a muscular hypertonia which is probably connected with the absence of a functional cortex of the cerebral hemispheres (see p. 269) soon develops.

The vital centers of the medulla oblongata, especially the respiratory center (see p. 313) and the centers of food intake (see p. 408), are fully developed in the newborn infant.

Some reflexes of the position and motion sense, such as MORO's reflex, the supporting reactions of the legs, and stepping and crawling movements, can be distinctly elicited. During these movements and especially in MORO's reflex, a coarse arm tremor which has not yet been explained can not infrequently be noted.

All the senses of the newborn infant can be stimulated and all sensory nerves can transmit the excitation to the brain. The reflex arc passes through the thalamus opticus and the pallidum without the participation of the cerebral hemispheres. The newborn infant's nervous behavior toward the outside world is regulated mainly by the senses of the skin, taste, and the position and motion senses. The higher senses, eye and ear, are less prominent at first, although they are functional from birth. Phylogenetic development may have occurred in the same sequence.

The newborn infant and his senses are entirely oriented to his mother, who forms his only natural environment (see p. 600). He already can — within certain limits — even respond purposefully to external stimuli by increasing his perceptive capacity for useful stimuli (for example, at the beginning of food intake) or by decreasing it for unwelcome or damaging stimuli (for example, at the end of food intake or when exposed to painful stimuli). Within certain limits, he can thus protect himself against damage.

Analyzers of a simple type (see p. 522) must be assumed to exist in the brain stem of the young infant, whereas the more delicate analyzers of the cerebral cortex begin functioning only during the first year of life. The responses of the newborn infant consist of innate (unconditioned) reflexes. The ability to establish conditioned reflexes and to acquire a more delicate adaptation to the stimuli of the external world depends on the cerebral hemispheres, which mature later. The newborn infant's brain has this ability only to a minor degree.

Viability, Brain Damage Caused by Birth Trauma

(4, 18, 26, 39, 42, 47, 86, 99, 116, 123, 131, 132, 135, 20a)

The mortality rate in premature infants, i.e., those with a birth weight up to 2500 gm, is greater, the more immature and the nearer to the fetal stage they are. Even at the turn of the century, a newborn infant weighing 1200 gm rarely survived. Since then, the dividing line between nonviable fetuses and already viable premature infants (expressed in birth weight) has been markedly lowered. For obvious reasons mortality among fetuses having the lowest birth weights must remain high, but premature infants with low birth weights are far rarer than those with higher birth weights.

According to a statistical summary compiled by DUNHAM (1955), 10.96% of premature infants with a birth weight of less than 1000 gm survived in the years from 1922 to 1940 (HESS and LUNDEEN), 20.5% from 1948 to 1951 (LEVINÉ and DANN), and 10.4% from 1945 to 1947 (BAIN). Of premature infants with a birth weight from 2100 to 2500 gm, however, more than 90% survived. The lowest birth weights of surviving premature infants which are reliable to any degree were those given by FAKIM (1950) — 454 gm — and FISCHER-BAN (1931) — 600 gm. Such infants would formerly have been called miscarriages (abortions), considered as nonviable, and abandoned. The limit of viability cannot be stated in figures.

Seriously damaging factors such as prolonged labor or forced delivery, general anesthesia of the mother, suffocation and artificial resuscitation, hemorrhages, encephalodystrophies, respiratory disturbances of the lungs such as aspiration of amniotic fluid, atelectases, and hyaline membranes as well as sepsis, malformations, etc., may occur in many combinations. The more immature and sensitive the respiratory center is, the more these factors may injure it and thus cause death.

Breathing is the most important task of the newborn infant immediately after birth. If he does not start breathing a few minutes after birth, his fate is sealed. Thus, life is determined from birth to death by the function of the respiratory center and the related function of the brain.

Of course, most paranatal fatalities are ascribed to "asphyxia" by obstetricians. Asphyxia means pulselessness, but newborn infants die from failure of the respiratory center. This incorrect term threatens to lead treatment in the wrong direction. The respiration, not the circulation, must be started and kept going.

Intrauterine respiration (see p. 354) is visible through the abdominal wall of the mother only in the last months of pregnancy. It is not known when respiration first occurs, but it appears much later than the heart contractions. As previously mentioned, PFLÜGER (1877) was able to demonstrate heart contractions in a human embryo approximately 4 mm long in the third week of gestation. The circulatory system thus fulfills its vital role long before birth, whereas the intrauterine respiration — as far as we know, it is not vital — does not appear until later. Accordingly, the heart often continues to function in dying premature infants long after the last breath, as has been shown by electrocardiographic findings (see p. 337).

Human fetuses obtained alive by operations die within a few minutes with the symptomatology of gasping respiration; their respiratory centers can sustain even this low respiratory pattern only for a very short time.

Death usually occurs in premature infants in the form of an apnoic attack; either the infant cannot survive the stage of gasping respiration, which is also temporarily physiological for the mature newborn infant immediately after birth (L. SCHMIDT, see p. 361), or he regresses to this stage.

The apnoic attacks of premature infants (see p. 327) were attributed by YLPPÖ (1919, 1926, 1931) and SCHWARTZ (1926, 1927) to perinatal hemorrhages in the respiratory center and in nearby areas. CREUTZFELDT and PEIPER [1932 *(42)*] therefore examined by histological serial sections the brains of 7 premature infants who died of severe suffocation attacks. Six cases did not show any hemorrhages in the brain stem; in the seventh, sporadic small blood extravasations with no reaction of the surrounding tissue, i.e., so-called agonal hemorrhages, were found. A moderate hemorrhage from the vena terminalis was also present. According to these findings, the apnoic attacks are not caused by hemorrhages into the region of the respiratory center. Our findings were confirmed by HIRVENSALO (1949, YLPPÖ's clinic) and RÜHL (1952) (see p. 319).

In most cases, CREUTZFELDT and PEIPER found hemorrhages which, like those described above, were minimal and demonstrable only microscopically in other parts of the brain. Since there was no glial reaction, they could only have occurred shortly before death. They were probably connected with suffocation, since all premature infants examined had undergone a protracted period of gasping respiration.

BARTH (1933) suffocated pregnant guinea pigs with carbon dioxide and found extensive hemorrhages in the fetal ventricles and meninges. Thus, suffocation alone can also cause or intensify hemorrhages in premature infants.

In general, the occurrence of hemorrhages is favored by injuries of the most varied types: Immaturity, fragility of the blood vessels, disorders of blood coagulation, mechanical effects of birth, suffocation, and forceful resuscitation attempts combine in different ways and cause hemorrhages.

Intracranial hemorrhages have long been known. They were described by the French investigators DENIS (1826), BILLARD (1828), CRUVEILHIER (1829–1835), as well as by VIRCHOW, WEBER (1851–1854), BEDNAR (1851), and WEST (1853). YLPPÖ (1919, 1926) and SCHWARTZ (1926, 1927) later emphasized their importance. The data on the frequency of their occurrence in premature infants vary considerably. According to a report by FISCHER-WASELS (1924), SCHWARTZ found them in 95% of the cases examined. He declared these hemorrhages to be the cause of death of nearly all newborn infants dying at birth or during the first week of life. HEDRÉN, however, noted intracranial hemorrhages in only 6.5% of 1000 autopsies on newborn infants. YLPPÖ (1919) found more or less extensive hemorrhages in nearly all premature infants weighing less than 1500 gm. The source of these hemorrhages is often a tear in the tentorium (DENIS; BENEKE, 1910; E. KEHRER, 1939). WEBER (1851), LITZMANN (1880), SCHAEFFER (1897), and HAUSBRANDT and MEIER (1935) demonstrated hemorrhages in the spinal canal.

The extent and location of intracranial hemorrhages vary greatly. They may be epidural, intradural, or subdural; they are easily caused by a tear in the falx and especially in the tentorium. They often expand throughout the pia mater. At the base of the skull they are frequently caused by an injury to the vena magna Galeni. In the brain they are not infrequently punctate; extended subependymal hemorrhages frequently originate from a tear of the vena terminalis. Hemorrhages into the ventricles are caused by ruptures of cerebral hemorrhages or by injuries of the choroid plexus. Cerebral hemorrhages caused by lack of oxygen predominate in premature infants and mechanically caused meningeal hemorrhages in mature newborn infants (GRÖNTOFT, 1953). Possibly the frequency and appearance of hemorrhages have changed in recent decades because Schultze's method of artificial respiration and similar forceful resuscitation procedures have been recognized as dangerous and their use therefore discontinued.

Repeated searches have been made for signs of an existing intracranial hemorrhage. The jumping jack phenomenon (lightning-fast lifting of the arms and legs on tapping of the chest) cited by YLPPÖ (1919) could always be demonstrated in sleeping newborn infants by ISBERT and PEIPER [1927 *(17)*]. The swallowing weakness also cited by YLPPÖ is quite often found in hemorrhages.

The negative response to testing of the vestibular function is regarded by BERBERICH and WIECHERS (1924) and VOSS (1927) as a sign of birth injury. According to DOLLINGER (1926), this test is even a perfect clinical procedure for the ready recognition of minor intracranial hemorrhages. However, CATEL (1928) absolutely rejects these conclusions on the basis of his investigations. The vestibularis of immature infants may react differently on different days under the same conditions. A. STERN (1925) ranks spontaneous nystagmus with the configuration of the skull as an extremely sensitive sign of brain injury. DOLLINGER (1926) declared that "Stäupchen," the minimal movements which occur occasionally in all premature infants, are an expression of birth-incurred brain injuries (see p. 249).

However, according to THIELE (1954), the only sure, completely reliable, and very sensitive sign of cerebral damage is corneal clouding (see p. 54). It is hard to understand why the cornea of all things should be a mirror of the processes which take place in the brain. Actually the younger and the more underweight premature infants are, the more frequently a delicate membrana iridopupillaris is temporarily present [SCHMÖGER, 1955 *(17a)*]. SEEFELDER (1938) states:

> "Such a clouding may actually still be so pronounced in older fetuses, for example, in the eighth month of gestation, that even an experienced investigator may be misled in assuming pathology of the cornea. However, anatomic examination cannot show even a trace of corneal cloudirg."

YLPPÖ, SCHWARTZ, and DOLLINGER (1927) related the tonic neck and labyrinthine reflexes, BABINSKI's sign, singultus, yawning, and the peculiarity of movements to hemorrhages in newborn infants. However, all these phenomena are connected with the immaturity of the brain, as I have shown. Furthermore, if squinting, cachectic appearance, and atrophy are related to intracranial hemorrhages, the objection must be raised that the same phenomena also occur in infants dying after the neonatal period.

Actually, there is no reliable neurological sign for the presence of an intracranial hemorrhage. Only rarely can a decision be made whether the hemorrhage is unilateral or bilateral and which side is affected. Bulging of the big fontanel indicates a supratentorial location.

Lumbar puncture is not a reliable procedure for establishing or excluding such a hemorrhage. The spinal tap itself may cause a hemorrhage, whereas an intracerebral or meningeal hemorrhage does not have to extend to the cerebrospinal fluid spaces.

Unlike the adult, the neonate rarely shows focal signs. Many brain regions, for example, the cerebral hemispheres, are not yet functional because of their immaturity and are therefore "silent."

The distinctness and time of appearance of the clinical signs depend on the location and extent of cerebral damage and on the increase in intracranial pressure. Large hemorrhages cause death during birth or immediately thereafter. Continuation of the bleeding and increase in cerebral pressure may not appear as suspicious signs until hours or days postpartum.

Large hemorrhages usually cause severe general disorders immediately after birth: The newborn infants are not red like lobsters, but pale and cyanotic, and are limp and more or less unresponsive. Infants with severe brain damage present the picture of shock. Usually more or less extended pulmonary disorders exist at the same time. The onset of breathing is delayed and the gasping respiration which always occurs immediately after birth persists longer. The healthy newborn infant cries loudly shortly after his first breath; the infant damaged at birth whimpers and moans quietly. His condition soon becomes threatening. If oxygen intake does not improve soon, he dies after a short while. If he survives, his respiration tends to be irregular for days and even weeks, and easily regresses to gasping respiration, especially if other disorders of different kinds occur. Meanwhile, the infant often becomes temporarily excited and moans, cries shrilly, moves violently, and trembles. Individual muscle groups begin to twitch and nystagmus appears; convulsions may occur in more mature infants.

The blood flow in the skin of the face, as well as in other body parts, changes rapidly; the newborn infant sometimes appears pale and sometimes cyanotic. In other infants, the skin color remains reddish-cyanotic for days and then becomes pale.

The temperature of infants injured during birth easily becomes subnormal despite the usual warmth precautions. Occasionally inspiratory stridor occurs. Vomiting can be variously interpreted and is therefore of no use as a diagnostic sign. Temporary pronounced mechanical hyperirritability, as manifested by the facialis, plexus, and peroneus phenomena (see p. 300), indicates the presence of hemorrhages. The ambiguous BABINSKI sign is of no value. As in other brain lesions, these infants envelop the nipple poorly and make more chewing than sucking movements. Swallowing inability indicates a cerebral hemorrhage. Spastic paralyses and epileptiform seizures cannot yet be observed in premature infants during the first days of life; transitory flaccid paralyses are usually caused by lesions of peripheral nerves.

Newborn infants suffering from intracranial hemorrhages are frequently semiconscious; their eyes stay open and show a somewhat fixed gaze. The babies are either hyperexcitable or only slightly responsive to the usual stimuli. Newborn infants suffering from incipient meningitis behave similarly.

In brief, the most important signs which should rouse suspicion of an intracranial hemorrhage are: pronounced limpness initially, later hyperexcitability with motor restlessness, swallowing disorders, changing blood flow in the skin or prolonged reddish-cyanotic discoloration of the skin.

These statements refer to actual newborn infants and particularly to premature infants whose lives were greatly imperiled by the hemorrhages, so that we were frequently able to compare the clinical data with the autopsy findings. In order to also recognize the less serious birth-incurred brain lesions which newborn infants usually survive, PRECHTL and DIJKSTRA (1959) compared the neurological behavior of mature newborn infants who underwent difficult birth with a group of mature newborn infants who experienced easy birth. These authors found deviations from the healthy control group in 58% (reflex differences between right and left, hypokinesia, hyperirritability and trembling, hypo- and hypertonia, absence of MORO's reflex and of the oral reflexes, and epileptiform seizures).

Of the newborn infants who underwent difficult birth, 52% were healthy and 48% were impaired. Both groups were re-examined at the age of two to four years; reflex differences between right and left, pareses, spasticity, ataxia, choreiform syndromes, considerable developmental retardation without known external cause, and epileptiform seizures demonstrated clinically or by EEG were considered to be pathological.

In these re-examinations, 8% of the newborn infants originally thought to be healthy showed neurological disorders, while 68% of those who were already impaired after birth also showed impairment later. This difference is significant. Fifty-two percent of the newborn infants suffering from hypoxia before or during birth were impaired at the time of re-examination, but only 17% of those without visible signs of hypoxia were impaired later. Thirty-eight percent of the newborn infants who had not shown neurological impairment were later disturbed in their behavior. However, 70% of those with neurological disorders also showed impaired behavior later.

There is another frequent but less widespread brain disease of the newborn infant and especially of the premature which is not distinguishable clinically from hemorrhages. It was originally described by VIRCHOW (1867, 1868, 1883) as encephalitis interstitialis and today is usually called encephalodystrophy, after SIEGMUND (1923). To what extent these terms apply to the disorder may be left unanswered (GUILLERY, 1923; SCHWARTZ, 1926, 1927; RYDBERG, 1932). In encephalodystrophy we are dealing with sporadic or numerous myelin foci ranging from the size of a pinhead to a lentil, yellowish-white in color, and located near the ventricles. They are attributed to locally circumscribed circulatory disorders which occur during birth and which disrupt the oxygen supply of the affected area to such a degree that a permanent lesion occurs.

To find the cause of death in premature infants, ESSBACH and PEIPER [1953 *(123)*] collected and compared clinical observations and pathological-anatomic findings, including extensive histological examinations, in the same children. Statistics on the autopsy findings — SCHWARTZ examined only the brains and never the inner organs — are altogether insufficient for understanding the total pathological process. According to our investigations, the most frequent life-threatening phenomena in the first days of life of premature infants consist of lowering of the body temperature, central respiratory, sucking, and swallowing disorders, vomiting, signs of amniotic fluid aspiration and, less frequently, of aspiration of stomach contents or subsequent pneumonia, increased tendency to bleeding, visible hemorrhages or signs which point to an invisible hemorrhage. The pathologist can demonstrate only part of these disorders in the postmortem findings. He finds simple or infected amniotic fluid aspiration, atelectasis of the lungs, hyaline membranes, pneumonias and pulmonary hemorrhages, aspiration of stomach contents, intracranial hemorrhages, and encephalodystrophies. In some cases additional, sometimes decisive, findings may be seen, such as sepsis, listeriosis, etc. Syphilis is no longer important. Multiple lesions are found more frequently than single lesions. This makes diagnosis, already difficult in itself, still more difficult.

Severe malformations such as transposition of the large cardiac vessels without a compensatory connection between the right and left cardiac spaces, congenital absence of the kidneys, of the trachea, or of the intrahepatic bile ducts, and cystic degeneration of the kidneys are not compatible with life after birth. All these malformations are extremely rare. Congenital esophageal atresia, absence of the extrahepatic bile ducts, and prolapse of the gastrointestinal tract or of the bladder can now often be repaired by early surgical intervention, so that the child can survive.

As a rule, death cannot be attributed to a single lesion; usually there is a combination of several clinically or pathologically-anatomically demonstrable disorders which depend on each other. As previously mentioned, many premature infants die of central respiratory disorders caused by the immaturity of the respiratory center. The functional incapacity of the immature nervous centers and tracts can be determined only by observation, but cannot be demonstrated anatomically (see p. 539). Conditions which stand in interrelationship with prematurity are intracranial hemorrhages and dangerous pulmonary diseases such as atelectases, hyaline membranes, aspiration and aspiration pneumonias, focal pneumonias, and pulmonary hemorrhages. These diseases prevent aeration of the lungs, thus causing suffocation and in turn promoting the appearance of central respiratory disturbances and hemorrhages. It is impossible to decide in any detail to what degree death may be attributed to one or the other difficulty.

Severely ill newborn infants urgently need painstaking nursing care and continued supervision. If hemorrhages can be proved or are only suspected, absolute rest is indicated. Bathing must be omitted. Severe illness demands agnosis (Bürger, 1953), i.e., foregoing a diagnosis which can be made only by disturbing the child. Simple harmless observation frequently reveals more than drastic diagnostic procedures. Procedures such as lumbar punctures or x-rays are only dangerous without being helpful and should therefore be omitted. Obtaining the most exact diagnosis possible could destroy the last chances for survival. How great these chances are cannot always be judged with certainty even by an experienced diagnostician. Some apparently lost infants recover surprisingly if they are well cared for, i.e., protected against cooling and given oxygen. Suction of the contents of the upper air passages and of the stomach may sometimes be vital. This must be done repeatedly if necessary, and by an experienced person [Thomas, 1956 *(20a)*]. However, all forced attempts at resuscitation are extremely harmful, since they cause lowering of the infant's temperature and intensify the hemorrhage.

A given newborn infant's viability is thus by no means an invariable factor destined by fate, but in damaged newborn infants depends directly on the physician's insight and knowledge. I refer especially to the book by Abramson (1960), who with 24 co-workers described the resuscitation of the newborn infant in detail. We are dealing here with one of the most important questions of pediatrics and obstetrics.

Little is known with certainty about which of the permanent brain damages that can be observed later are attributable to perinatal cerebral hemorrhages. Schwartz (1926, 1927) regarded such pathological-anatomic disease entities as porencephaly, diffuse lobar cerebral sclerosis, cerebellar atrophy, status marmoratus, and congenital absence of basal ganglia as residuals of cerebral hemorrhages. According to VOSS (1926, 1927), congenital deaf-mutism belongs in part in this group. There is no convincing evidence that these diseases originated in this way, just as there is no proof that diplegia spastica infantilis, which LITTLE (1862) attributed to a premature or difficult delivery and asphyxia, occurs in this manner. The spasticity of the extremities in LITTLE's disease is still absent in the newborn infant and can be demonstrated only after approximately six weeks. THUMS (1939) found, on the basis of extensive examinations on unselected twins, that hereditary influences do not play a major role in this disease.

INFANTS

The most essential neurological differences between newborn infants and adults are gradually compensated for in the course of the first year of life. The age at which individual changes occur varies greatly from child to child and therefore cannot be exactly stated.

The athetoid movement pattern and the lower form of periodic respiration are already disappearing in the newborn infant born at term.

The sensory reactions are innate, but utilization of the sensory perceptions improves continuously, while vision and hearing attain increasingly more influence on the behavior. The facial expression develops from the sensory reactions.

Directed gaze (fixating) begins at the age of three months and pupillary fluctuation at the age of four to five months.

Whereas the newborn is barely able to right his head temporarily in space (labyrinthine righting reflex of head), the child can walk independently at the age of one year. This development occurs through many intermediate stages which are ingeniously interconnected. The innate preliminary stages of locomotion, i.e., reflex stepping, mounting, crawling, and swimming movements, can no longer be elicited after four to five months. With the aid of reflex groups which gradually appear and partially disappear again, corresponding to the sequences in MAGNUS's and RADEMAKER's experiments on animals after excision of brain areas, the infant learns during the first year of life to turn on his stomach, to creep and to crawl, to sit, to sit up and to slide, to stand, to stand up, and finally, toward the end of the first year of life, to walk independently.

At the neonatal stage the infant's highest functional centers are located in the pallidum; in the course of the first year the cerebral hemispheres gradually gain influence on brain function. The mass movements of the first months, which are regulated by the extrapyramidal system, are joined at approximately the fourth or fifth month by goal-directed movements which depend on the cortical centers and the pyramidal tracts, while brakes are applied to the mass movements by the development of the corpus striatum. The first movement of this kind is reaching with the hand for a seen object. During the whole first year, the urge to move is still quite uninhibited and greater than it is at any time later in life.

Of the clinically important reflexes, the eye-neck reflex is extinguished in the first three months. MORO's reflex can be found only rarely after three months and disappears after nine months. Contralateral adductor contraction on eliciting of the patellar reflex usually disappears in the seventh month, the crossed flexor reflex of the legs in the seventh to twelfth month, the palmar and plantar grasp reflexes, which even earlier can barely be distinguished from voluntary grasping, at the age of one year. Overextension of the big toe (BABINSKI's sign) becomes less frequent during the first year of life.

As a newborn the infant sleeps 80% of his time and at the age of one year only 50%. The preferred position of the infant is abandoned between approximately the ninth and twelfth months of life.

At the time of weaning – after the ninth month the infant should no longer be breast fed – the sucking movements disappear, but they persist if the child is bottle fed for a prolonged period of time. Thus, food intake at the age of nine months and later is similar to that of the adult.

Conditioned reflexes, which if present at all are weak and difficult to demonstrate in the newborn infant, appear more and more distinctly after some months and for some time constitute the highest brain function. With their aid the first phase of training is possible, as for example the conditioned division of time during the first months of life. The beginning of toilet training in some infants may be partially successful in approximately the ninth to twelfth month. The first words are learned as conditioned associations in the fourth to sixth month. The highest human faculty, intelligent action, shows its beginnings at the end of the first year of life.

Whereas the mother alone represents the environment for the young infant, his world expands through the progressively better utilization of sensory perceptions and independent locomotion.

CHILDREN AFTER INFANCY

Children at the end of the first year of life differ only little neurologically from the adult.

The motility drive gradually decreases, but is still considerably greater in the small child than it is in the school child; the school child, in turn, moves still more than the adult. The movements themselves become more balanced and more economical in their expenditure of strength, until a new alteration occurs in puberty.

Some reflexes, such as the reaction movements of the big toe (BABINSKI), LERI's forearm sign, and MAYER's reflex of the metacarpophalangeal joints, attain their permanent pattern only after several years.

Independent gait is accomplished at the end of the first year of life; ascending steps and jumping are learned only later. Unaided elevation from the dorsal position reaches its permanent pattern at the age of four or five years.

The development of cerebral function, which starts so dramatically in infancy is consummated not in the neurological but rather in the intellectual sphere. At the end of the first year of life, the child raises himself above the highest animal stage with the aid of this development.

A specific chronological age at which all portions of the child's nervous system have reached their full maturity cannot be given.

SUMMARY

The functional ability of the immature nervous tracts and centers cannot be assessed anatomically.

Human fetuses are nonviable because their immature respiratory centers can maintain gasping respiration only for a short time and soon fail completely. The high mortality rate in premature infants of low birth weight is also caused by the failure of the respiratory center. In the newborn infant, the following factors frequently combine in mutually influencing each other: immaturity; general anesthesia of the mother; central respiratory, sucking, and swallowing disorders; aspiration; atelectases; hyaline membranes; pneumonias; intracranial hemorrhages; encephalodystrophies; and malformations. Usually several disorders occur simultaneously, damaging the sensitive respiratory center and preventing the ventilation of the lungs, and may thus cause death in the form of an apnoic attack.

The significance of traumatic (perinatal) hemorrhages at birth for the newborn infant's viability has frequently been overestimated. No neurological sign exists to demonstrate their presence with certainty. Limpness, followed by overexcitability with motor restlessness, swallowing disturbances, and great variations in the blood flow or reddish-cyanotic discoloration of the skin are suspicious symptoms.

In human fetuses, the reflexes tend to spread over wide areas. During development they become more precise and more limited. The lowest form of breathing is gasping respiration. The oral region is the most sensitive to skin stimulation. The closing of the mouth on touch belongs to the earliest, most stable, and most resistant reflexes.

Special movement types, tremors or "Stäupchen" (see p. 249), and athetoid movements are characteristic for premature infants. Fatal respiratory, sucking, and swallowing disorders occur easily because of the immaturity of their brains.

Fetuses breathe and drink amniotic fluid. They can be stimulated by loud sounds and vibrations.

The behavior of the newborn infant, whose cerebral hemispheres are not yet functional, corresponds to that of a pallidum preparation. The sucking and respiratory centers are functioning adequately. All senses can be stimulated; the reflex arcs pass through the brain stem. Within certain limits, the newborn infant can evade unwelcome and reinforce welcome stimuli. He is completely adjusted to his mother and is dependent on her.

Most of the neurological differences between newborn infants and adults have disappeared at the end of the first year of life. Later, cerebral function progresses more in the intellectual than in the neurological sphere.

REFERENCES

ABRAMSON, H.: Resuscitation of the Newborn Infant. St. Louis, 1960.
ALISON, F.: Méd. infant. *66*, 17 (1959).
AMBRON, H., and H. HELD: Arch. Anat. u. Physiol. *1896*, 203.
ANGULO Y GONZALES, A. W.: J. Comp. Neur. *48*, 459 (1929).

ARISTOTLE: Von der Zeugung und Entwicklung der Tiere (On the Procreation and Development of Animals). German translation. Leipzig, 1860.
BARTH, H.: Hals- usw. Heilk. *35*, 1 (1933).
BEDNAR, A.: Die Krankheiten des Neugeborenen (The Diseases of the Newborn Infant). Part 2. Vienna, 1851.
BENEKE, R.: Münch. med. Wschr. *1910*, 2125.
BERBERICH, I., and A. WIECHERS: Zschr. Kinderhk. *38*, 59 (1924).
BILLARD, C.: Traité des maladies des enfants nouveau-nés (Treatise on the Diseases of the Newborn Infant). Paris, 1828.
BOLAFFIO, M., and G. ARTOM: Ref. Zbl. Neurol. *38*, 418 (1924).
BÜRGER, M.: Klinische Fehldiagnosen (Errors in Clinical Diagnoses). p. 54. Stuttgart, 1953.
CARMICHAEL, L.: In Murchison's Handbook of Child Psychology. 2nd ed. Worcester, Massachusetts, Clark Univ. Press and London, Oxford (1933).
–, In P. L. HARRIMAN: Encyclopedia of Psychology. p. 198. New York, 1946.
–, Manual of Child Psychology. 2nd ed., p. 60. New York, London (1954).
CATEL, W.: Mschr. Kinderhk. *38*, 303 (1928); *52*, 1 (1932); *58*, 89 (1933).
COGHILL, G. E.: Arch. Neurol. Psychiatr. *21*, 989 (1929).
CRUVEILHIER, L.: Anatomie path. du corps humain (Pathological Anatomy of the Human Body). Paris, 1829–1835. Lief. Apoplexie des nouveau-nés (Issue on Apoplexy in Newborn Infants).
DENIS: According to Sunde.
DIJKSTRA, J.: De prognostische Betekenis van neurologische Afwijkingen bij pasgeboren Kinderen (The Prognostic Importance of Neurological Deviations in Premature Children). Thesis, Groningen, 1960.
DOLLINGER, A.: Erg. inn. Med. *31*, 373 (1926); Zschr. Kinderhk. *41*, 20 (1926).
DUNHAM, E. C.: Premature Infants. 2nd ed., p. 70. New York (1955).
EHRHARDT, K.: Münch. med. Wschr. *1937*, 1699; *1939*, 915; Anat. Anz. *87*, Erg.-H. 420 (1939); Zbl. Gynäk. *1939*, 146.
ERBKAM: According to Preyer.
FAKIM, H.: Brit. Med. J. 4676, 445 (1950)
FISCHER-BAN, H.: Klin. Wschr. *1931*, 1354.
FISCHER-WASELS, B.: Schweiz. med. Wschr. *1924*, 905.
FLECHSIG, P.: Die Leitungsbahnen im Gehirn und Rückenmark des Menschen (The Pathways in the Brain and Spinal Cord of Man). Leipzig (1876); Anatomie des menschlichen Gehirns und Rückenmarks (Anatomy of the Human Brain and Spinal Cord). Leipzig, 1920.
FÖDERL, V.: Arch. Gynäk. *143*, 598 (1930).
FUKAS, M.: Zentralbl. Gynäk. *64*, 907 (1940).
GARSCHE, R.: Mschr. Kinderhk. *98*, 197 (1950); in Brocks Biologischen Daten für den Kinderarzt (in Brock's Biological Data for the Pediatrician). 2nd ed., *2*, 870. Berlin-Göttingen-Heidelberg, 1954.
GLASER, K.: Amer. J. Dis. Childr. *75*, 309 (1948).
GOECKEL: Die Wandlungen in der Wertung des ungeborenen Kindes (The Changes in the Evaluation of the Unborn Child). Thesis, Heidelberg, 1911; Arch. Gyn. *94*, 91 (1911).
GRÖNTOFT, O.: Acta obstetr. Scand. (Stockholm) *32*, 308 and 458 (1953).
GUILLERY, H., JR.: Zschr. Neurol. *84*, 205 (1923).
HAUSBRANDT, F., and A. MEIER: Frankf. Zschr. Path. *49*, 21 (1935).
HEDRÉN: According to Rydberg.
HOOKER, D.: Yale J. Biol. *8*, 579 (1936).
–, Proc. Amer. Philosoph. Soc. *79*, 597 (1938).
–, Res. Publ. Ass. Nerv. Mental Dis. *29*, 237 (1939).
–, Psychosomat. Med. *4*, 199 (1942).
–, The Origin of Behavior. Ann Arbor, Univ. of Michigan, 1944.
–, The Prenatal Origin of Behavior. Porter Lectures, Series 18. Univ. of Kansas Press, Lawrence, Kansas, 1952.

HOOKER, D.: Res. Publ. Ass. Nerv. Ment. Dis. *33*, 88 (1954).
—, Evidence of Prenatal Function of the Central Nervous System in Man. J. Arthur Lecture on the Evolution of the Human Brain. American Museum of Natural History, New York, 1958.
—, Genetics and Inheritance of Integrated Neur. and Psychiatr. Patterns. Proc. Ass. Res. Nerv. Ment. Dis. *33*, 98 (1954).
—, and T. HUMPHREY: Gazeta médica Portugesa *7*, 95 (1954).
HUMPHREY, T.: J. Psychol. *35*, 3 (1953).
—, and D. HOOKER: J. Comp. Neur. *112*, 75 (1959).
KEHRER, E.: Die intrakraniellen Blutungen bei Neugeborenen (Intracranial Hemorrhages in the Newborn Infants). Beih. z. Z. Geburtshilfe u. Gyn. Vol. 118. Stuttgart, 1939.
KELLOGG, W. N.: J. Genet. Psychol., Worcester *58*, 307 (1941).
KOFFKA, K.: Die Grundlagen der psychischen Entwicklung (The Foundations of Psychic Development). 2nd ed. Osterwieck, 1925.
KRAUSS, W.: Geburtsh. u. Frauenhk. *11*, 557 (1951).
LANDÉ, E.: Zschr. Biol. *81*, 224 (1927).
LANGREDER, W.: Dtsch. med. Wschr. *1949*, 661; Zschr. Geburtsh., Stuttgart *131*, 236 (1949) Arch. Gynäk. *177*, 273 (1950); Kinderärztl. Prax. *26*, 559 (1958).
LANGWORTHY, O. R.: Contributions to Embryol. *24*, No. 139, Carnegie Institute, Washington, 1933.
LITTLE, W. I.: Trans. Obstetr. Soc. *3*, 293 (1862).
LITZMANN, C.: Arch. Gynäk. *16*, 87 (1880).
MAI, H., E. SCHÜTZ, and H. W. MÜLLER: Zschr. Kinderhk. *69*, 251 (1951).
MERREM, G.: Lehrbuch der Neurochirurgie (Textbook of Neurosurgery). p. 307. Berlin, 1960.
MEYER, H.: Zentralbl. Gyn. *1929*, 1562.
MINKOWSKI, M.: In Abderhaldens Handbuch der biologischen Arbeitsmethoden (in Abderhalden's Handbook of Biological Methods). Issue 253, Sect. V, Pt. 5B. Berlin, 1928.
NEEDHAM, J.: Chemical Embryology. Cambridge 1931.
PAVLOV, I. P.: Sämtliche Werke (Complete Works). German translation. Berlin, 1953 and 1954.
PFLÜGER, E.: Pflügers Arch. Physiol. *14*, 628 (1877).
PRECHTL, H. F. R., and J. DIJKSTRA: Proc. Symposium on Prenatal Care. Editor, B. S. ten Berge. Groningen, 1959.
—, and A. R. KNOL: Arch. Psychiatr. Z. ges. Neur. *196*, 542 (1958).
PREYER, W.: Spezielle Physiologie des Embryo (Special Physiology of the Embryo). p. 437. Leipzig, 1885.
ROSENBAUM, S.: Acta med. orient. *13*, 119 (1954); Dtsch. med. Wschr. *1958*, 1128.
ROSSIER, A.: Arch. franç. Pédiatrie *12*, 190 (1955); Méd. inf. *66*, 7 (1959).
RYDBERG, E.: Acta path. microbiol. Scand. *10*, Suppl. (1932).
SAGER, C. A.: Medizinische *1957*, 108.
SCHAEFFER, O.: Arch. Gynäk. *53*, 278 (1897).
SCHMIDT, H. R.: Mschr. Geburtshilfe *72*, 1 (1926).
SCHOLZ, W.: Mschr. Kinderhk. *75*, 5 (1938).
SCHWARTZ, P.: Mschr. Kinderhk. *34*, 511 (1926); Erg. inn. Med. *31*, 165 (1927).
SEEFELDER, R.: in Peter-Wetzel-Heiderich, Handbuch der Anatomie des Kindes (in Peter, Wetzel, and Heiderich, Handbook of Child Anatomy). *2*, 138. Munich, 1938.
SIEGMUND, H.: Virchows Arch. path. Anat. *241*, 237 (1923).
DE SNOO, K.: Mschr. Geburtsh. *105*, 88 (1937).
SONTAG, L. W.: Amer. J. Dis. Childr. *60*, 621 (1940).
—, and R. F. WALLACE: Amer. J. Dis. Childr. *48*, 1050 (1934); *51*, 583 (1936); Amer. J. Obstetr. Gynec. *29*, 77 (1935).
—, and R. F. WALLACE: According to Kellogg.
SAINT-ANNE DARGASSIES: Revue neurol. *93*, 331 (1955); Etudes néonat. *3*, 101 (1954); *4*, 71 (1955); *6*, 11 (1957).

Stern, A.: Arch. Gynäk. *124*, 639 (1925).

Stern, W.: Psychologie d. frühen Kindheit (The Psychology of Early Childhood). 3rd ed. Leipzig, 1923.

Sunde, A.: Acta obstetr. gynec. Scand. *9*, 477 (1930).

Suranyi, G.: Acta medica Acad. scien. Hungar. (Budapest) *11*, 117 (1957).

Thiele, W.: Klinische Untersuchungen zur Problematik der geburtstraumatischen Schädigung des kindlichen Zentralnervensystems (Clinical Investigations on the Problems of Birth Traumatic Damages of the Central Nervous System in Children). Zschr. Geburtsh. *139*, Beiheft. Stuttgart, 1954.

Thums, K.: Monogr. Ges. Geb. Neurol., No. 66. Berlin (1939).

Virchow, R.: Virchows Arch. path. Anat. *38*, 129 (1867); *44*, 476 (1868); Berl. klin. Wschr. *1883*, No. 46.

Voss, O.: Mschr. Kinderhk. *34*, 568 (1926); Acta oto-laryng., Stockholm *11*, 73 (1927).

Walther, O.: Geburtsh. u. Frauenhk. *11*, 1025 (1951); Münch. med. Wschr. *1951*, 1619.

Weber, F.: Beiträge zur pathologischen Anatomie der Neugeborenen (Essays on the Pathological Anatomy of Newborn Infants). Kiel, 1851–1854.

Wenger, M. A.: Studies in Infant Behavior. III. University of Iowa Studies *12*, No. 1. Iowa City, 1936.

West, C.: Pathologie und Therapie der Kinderkrankheiten (Pathology and Therapy of Childhood Diseases). German translation. Berlin, 1853.

Westphal, A.: Arch. Psychiatr. *26*, 1 (1894); *29*, 474 (1897).

Westphal, C.: Neurol. Zbl. *1886*, 361.

Wohlmuth, G., and R. Petrorini: Mschr. Kinderhk. *106*, 334 (1958).

Yanase, J.: Pflügers Arch. Physiol. *119*, 451 (1907).

Ylppö, A.: Zschr. Kinderhk. *20*, 212 (1919); Mschr. Kinderhk. *34*, 502 (1926); Pfaundler-Schloßmanns Handbuch der Kinderheilkunde (Pfaundler and Schloßmann's Handbook of Pediatrics). 4th ed., Vol. 1. Berlin, 1931; Erg.-Werk. Vol. 1. Berlin, 1942.

13. EVOLUTION AND DISSOLUTION OF CEREBRAL FUNCTION

PHYLOGENESIS AND ONTOGENESIS

(39, 52, 95, 141)

> "There is a history in all men's lives,
> Figuring the nature of the times deceased."
> SHAKESPEARE, King Henry IV: II, 3, 1

> "Youth has always to start anew from the beginning and to pass as an individual through the epochs of world culture." GOETHE

The view that embryonic growth recapitulates phylogenesis has been attributed even to ARISTOTLE. Certainly there are some sentences in the second book of his treatise, "On the Procreation and Development of Animals," which support this theory:

"The animal and the man do not develop at the same time, nor do the animal and the horse, and it is the same with the other animals, for the goal develops last; what is characteristic is the goal of development of each."

Doubtless this indicates correct recognition that general characteristics can be distinguished earlier in the course of embryonic growth than those which are specific for the race and the species.

ARISTOTLE distinguishes three stages in the aquisition of mental functions in the growing fetus, i.e., those of nutrition, sensation, and thinking.

"Originally all such animal embryos seem to lead a kind of vegetable existence. It is only later that one can speak of their having a sensation-soul and a thought-soul."

The attitude of the church in relation to stillbirth and to baptism of the unborn child was connected with this question of the appearance of the soul in the growing fetus. For centuries, ARISTOTLE's view on this question was authoritative for the church (for references see GOECKEL, 1911; NEEDHAM, 1931; and CARMICHAEL, 1933).

It seems to me rather arbitrary for NEEDHAM to conclude from passages like those cited above that ARISTOTLE had already anticipated the theory of recapitulation "in clear words." Actually, he had no knowledge of the evolution of species. HAECKEL did not regard him as one of his predecessors, although he discussed his views on animal development in detail; however, HAECKEL completely overlooked the extensive evolutionary knowledge of VOIGT (1817), his predecessor in Jena.

Clear concepts could be formed only after CHARLES DARWIN (1859) established his theory of evolution. The anatomic development of the embryo corresponds in many ways to the course of phylogenesis, according to FRITZ MÜLLER (1864) and HAECKEL (1866). Of course, no embryo repeats the image of its ancestors in all traits; rather, each organ passes through its own individual development. Since atavistic reconstructions of various types also combine in the same embryo, forms develop which have never before lived in such shape on the earth (WEINERT, 1932, 1944). Later adjustments may even change earlier developmental stages. Thus, even HAECKEL had to modify his "fundamental biogenetic law" with the concept of cenogenesis (modification by early adaptation).

Today's scientists return to the view of KARL ERNST VON BAER (1828): The embryo of a higher animal species never reproduces another species, but only the embryo of that species. We have demonstrated only an embryonic similarity and a repetition of characteristics in the corresponding stages of embryonic development (GOERTTLER, 1950). "No higher form repeats the adult stage of a lower form" (HOOKER, 1958). Nevertheless, "the findings identified with the fundamental biogenetic law, formerly generalized too much, frequently confirm and supplement the morphologically and palaeontologically deduced family trees in a very impressive way" (RENSCH, 1943). Anatomic examples for this statement are the occurrence of gills in amphibian embryos, especially in those that remain in the mother's body (viviparous forms), the appearance of teeth anlagen in vertebrates which are without teeth in the adult stage, the intermediate jaw, etc.

However, other, nonanatomic findings have been used in this connection. BILZ (1944) tried to establish the basis for a palaeopsychology and thus for a psychology of human emotions. He sees — arising from their prehistoric significance — vestiges of primordial function in the kiss, premastication, pappy food, the preferred position of the infant, EBBECKE's immersion reflex, as well as in head scratching and yawning during fatigue. A palaeopsychology, a phylogenesis of the mind, was established by VON EICKSTAEDT (1959). EDINGER (1937, 1949) made the first attempts at establishing a palaeoneurology.

Connections such as those mentioned above can be demonstrated in many ways in the development of the cerebral function of the child.

THE ASCENT TO MAN

(141)

The period in which the older vertebrate species developed is subdivided by RÜGER (1943) as follows:

Fish	from the Lower Silurian Age	(450 million years ago)
Amphibia	from the Upper Devonian Age	(330 million years ago)
Reptiles	from the lower Carboniferous Age	(310 million years ago)

In the course of phylogenesis, the cerebral hemispheres, whose function has made man a human being, appear in rudimentary form in the amphibia; the reptiles are the first to have a cerebral cortex. In the ascending line of the mammals, the cerebral hemispheres grow to a huge formation which finally covers brain stem, mesencephalon, and cerebellum. In man they are considerably larger than in the anthropoid apes. The frontal lobes comprise 29% of the total surface of the brain in man, but only 17% in the chimpanzee, and only 18% in monkeys (SCHWIDETZKY, 1959).

According to VON MONAKOW (1914) and VON MONAKOW and MOURGUE (1930), the development of brain function proceeds in the ascending line of vertebrates to the anterior end of the medulla, i.e., to the only area where in the lower vertebrates space is still available for the new acquisition of nervous formations. In higher mammals, new faculties develop at the expense of old, which become rudimentary; these new abilities take over their task in an improved manner, but bequeath the original possession of spinal reflex arcs with their primitive associated connections to the old faculties.

Phylogenetically, the brain develops from the anterior sections of the ectodermal medullary canal. These sections are first subdivided into three primary cerebral vesicles, located one behind the other: the rhombencephalon (the rhomboidal brain), the mesencephalon (the midbrain), and the prosencephalon (the forebrain). Later the rhombencephalon is subdivided into the myelencephalon (the afterbrain with the medulla oblongata) and the metencephalon (the hindbrain with the pons and cerebellum). The prosencephalon is again divided into the diencephalon (the 'tween brain with the thalami) and the telencephalon (the endbrain with the corpora striata, fornix, the cerebral hemispheres, and the olfactory brain). The cerebral hemispheres finally embrace the brain stem from above by growing extensively in all directions.

The newborn infant has, corresponding to his relatively large cranium, a relatively large brain. From birth to adulthood the weight of the brain increases only four times, whereas the body weight increases 21 times during the same period.

However, the performance of the newborn infant's brain does not correspond to its size. The individual cerebral regions begin to function in the same sequence in which they have developed ontogenetically and phylogenetically. At birth, the functional ability which ascends from the brain stem up into the brain has not yet reached the cerebral hemispheres.

In the newborn infant the corpora striata, which inhibit the mass movements of the pallidum, and the pyramidal tract system, which is a phylogenetically younger acquisition, do not yet have any influence on movement; movements are therefore uninhibited and — with the exception of movements during food intake — aimless. The cerebral hemispheres do not participate in the performance of movements.

The developmental structuring of the brain from different successive phylogenetic and ontogenetic levels is also apparent in other details. It is especially evident in the development of respiration (see p. 313).

The fishes, because of their gills, are still completely restricted to the water. In the ontogenesis of the amphibia, the phylogenetic transition from larval gill respiration to breathing with the lungs, which are formed during the metamorphosis, is repeated. Reptiles breathe only with their lungs.

The respiratory tracts and lungs of the earliest vertebrates develop from a protuberance of the anterior intestinal section. Similarly, the anterior intestinal wall of the human embryo bulges and forms a groove and finally a protuberance which later is transformed into the primary lung vesicles.

"Swallowing respiration," which as the lowest respiratory pattern still connects swallowing with breathing, corresponds to the originally close anatomic relations between the intestines and the lungs. The center of swallowing respiration is the lowest stage in the composition of the respiratory center, which is successively built up from different partial centers.

The development of air breathing in the course of phylogenesis is traced back to this lowest form of breathing: As has previously been described (see p. 333), fish in oxygen-deficient water ascend to the surface to gasp for breath. In this "emergency breathing" (WINTERSTEIN, 1921), air bubbles collect in the pockets of the mouth and anterior intestinal regions. In the ancestors of the Teleostei, a group of fish with lungs and air-bladder organs, the gills contain such air bubbles. According to REMANE (1954), the lungs with their air passages could have developed, in fresh-water areas temporarily deficient in oxygen, from such pockets, which were filled with air from time to time.

Lung breathers of a low order, such as frogs and lizards, have one center for laryngeal movements and another for lung breathing, which in these species consists of swallowing movements (see p. 348). According to WINTERSTEIN (1921), this kind of breathing still serves as an emergency respiration in reptiles. The same is true for the "frog breathing" of poliomyelitic patients whose respiratory muscles proper are paralyzed (see p. 352).

In man, swallowing and breathing still combine during food intake in infancy (see p. 438), during yawning, which is always terminated by a swallowing movement (see p. 386), and during suffocation (see p. 349). According to GARSCHE (1950), the snap-swallow reflex (see p. 443) can be elicited almost exclusively in premature infants weighing less than 1500 gm.

The gasping respiration of premature infants (see p. 327) is closely related to swallowing respiration. PEIPER – with CAMMAN (see p. 333) and GOOD (see p. 337) – was able to elicit gasping respiration in mammals (pigs and rabbits) in all cases if the animal inhaled carbon dioxide under a bell glass or remained without a supply of fresh air. In fish which are taken out of water, a few sporadic gasps still appear as an emergency respiration. MINKOWSKI (1928) found gasping respiration exclusively in the most premature human fetuses who could breathe at all.

Suffocation extinguishes the higher partial centers of the respiratory center, which are more sensitive to oxygen deprivation, and causes the less sensitive lower centers, which can still maintain an emergency respiration temporarily, to take their place. Ontogenetically and phylogenetically lower respiratory patterns appear in the form of floor-of-the-mouth respiration (see p. 365), respiration by movements of the alae nasi (see p. 370), and finally gasping respiration (see p. 327). Floor-of-the-mouth respiration, with its mandibular movements, is already reminiscent of gasping respiration, to which it finally changes. SECKEL (1931) associates alae nasi respiration with the movements of the alae nasi which constitute the highest form of respiration in amphibia and aquatic mammals (seals).

Between the low level pattern of gasping respiration and the highest respiration form, in which the breaths follow one another without pause, stands periodic respiration. The more immature premature infants are, the more frequently it appears, but it may easily be replaced by the dangerous gasping respiration if they do poorly. Periodic breathing is found physiologically in aquatic mammals, in sleeping mammals, especially during hibernation, and also in reptiles and amphibia (see p. 321).

When thoracic breathing is later added to the diaphragmatic respiration of early life in the infant, phylogenesis repeats itself (see p. 310).

The sucking center is phylogenetically and ontogenetically younger than the respiratory center (see p. 431). The (younger) sucking center is able to inhibit the (older) respiratory center or to enforce its own rhythm on it. However, a dependency of the sucking center on the respiratory center has never been demonstrated. This again confirms the general rule that the younger center determines the function of the older center.

Up to the reptile stage, the vertebrates grasp only with their mouths. Many higher mammals have not developed any other form of grasping (see p. 255). The newborn human infant also grasps for an object only with his mouth (see p. 404). In kittens, young mice, rats, rabbits, and hamsters, PRECHTL and SCHLEIDT (1950, 1951) found pendular head movements which serve to help find the mother's nipple. THOMAS and PEIPER [1953 *(101)*] observed identical movements in two infants with severe cerebral malformations (see p. 412). Traces of this behavior may also be found in healthy infants.

The palmar and plantar grasp reflexes of the young infant (see p. 158) constitute special grasping patterns. However, no spatially directed movement as in grasping with the mouth occurs, only a flexion of the fingers and toes. Both reflexes still today help the baby monkey to hold onto its mother's fur. These reflexes have persisted in a weaker form — more distinctly in the hands than in the feet — in the human infant, although man's loss of body hair after the phylogenetic separation from his ape ancestors has made them useless.

The vestibular reaction described by MORO as the clasping reflex does not have the significance of a phylogenetic vestige he attributed to it (see p. 154).

A detailed description has been given of how independent locomotion develops in the child in the same sequence as it did in his phylogenetic ancestors (see p. 239). As the fishes, amphibia, reptiles, and most of the mammals do, the infant first moves forward lying on his belly. Creeping and, later, crawling in crossed coordination (see p. 225) can be demonstrated in the amphibia and reptiles in the same manner. The cervicocaudal developmental direction (see p. 237) is expressed in the fact that at each stage of independent locomotion — from creeping to crawling to the erect gait — the arms always precede the legs by one developmental stage.

According to HINSCHE (1943), the tonic skin reflexes of the vertebral column appeared at the stage of the lower vertebrates with locomotion on all four extremities: In amphibia and reptiles, the dorsal aspect is still more sensitive to environmental stimuli than the frontal aspect, which is turned to the ground. In many lower vertebrates, there is a corresponding very intense sensitivity of the dorsal aspect which gives them information on the environment as a sensory organ, so to speak. At this stage, movements of the spine can be evoked and regulated by weak skin stimuli. Thus, the tonic skin reflexes of the vertebral column develop; they cause rotations as well as lordotic, kyphotic, and scoliotic curvatures, and thus adapt the animal to the external conditions. When our ancestors achieved erect posture through progressve development, the anterior aspect of the body became more accessible to stimuli and the forelimbs left the ground and took over the task of intercepting stimuli, which formerly reached only the back, while the dorsal skin, transposed from the top to the back, became more and more insensitive.

In the young child, HINSCHE still sees in the dorsal postural reaction, the nucleus of which is the spinal reflex, original autonomous movements of the vertebral column; as remnants of the old reflex network, which was tonically regulated from the skin and extended over the entire body, they are elicited by a weak skin stimulation. According to PRECHTL (1953), the spinal reflex (see p. 163), which is present in every infant, constitutes a significant protective movement in the amphibia.

The oldest known mammals — of the size of a mouse or a rat — originated in the Triassic Period, the oldest insect eaters in the Cretaceous Period. HEBERER (1955) assumes that already at that time a group of original insect eaters climbed into the trees by grasping with hands and feet. The way for the half-erect body position characteristic of the higher primates was paved in the monkeys.

The preferred posture of the infant (see Fig. 31, p. 169) is reminiscent of the posture of monkeys, which cling to branches with their hands in the same manner (FOERSTER, 1913).

Phylogenetically, the lower senses developed prior to the higher senses. Although all senses are functional in the newborn infant, the nervous phenomena are accordingly at first predominantly dependent on the lower senses, even though reactions can be evoked by way of the eyes and ears. The most important senses of the newborn infant are the skin and taste senses (see p. 93); their stimulation effects distinct and appropriate reactions during the first months. Thus, sucking movements develop at first chiefly by mechanical stimulation of the mouth and by the entry of food into the mouth. It takes months before conditioned reflexes appear on stimulation of the eye and ear during food intake (see p. 450).

During its early life the baby monkey, which lives as a "breast baby" (see p. 588) on its mother's body, does not need the higher senses. Its main tasks of searching for and sucking on the nipple, of holding on and of monitoring its mother's presence (see p. 604), are solved by the skin and taste senses, which are stimulated by direct contact. Knowledge of a more remote environment, which is transmitted by the higher senses, is needed only when the baby leaves its mother's body and becomes independent.

In civilized races, the breast infant becomes a bed infant (see p. 611) who returns to his mother's body only for feeding — provided he is breast fed at all. Mothers of races of a lower culture even today carry their children around with them on their bodies, as monkey mothers must do in order not to lose their babies. It is an unnatural achievement for the human baby to have to spend his life in bed. He has in no way adjusted to the bed; rather, his wish to be carried around becomes clearly evident again and again. Calming by rocking or pacifiers is reminiscent of the time when mother and child were physically more closely associated (see p. 607).

Rigid reflexes at first tie the young infant to a narrow living space, i.e., to his mother (see p. 600). Since he is attuned only to her in his body and in his behavior, separation from her means peril to life. The ancestors of the anthropoid ape were correspondingly "stenoic," i.e., restricted to a certain limited living space, as the anthropoid apes still are today. A decrease in many innate mechanisms was necessary to make man in a short time, geologically speaking, the most "euroic" of all creatures, one who gets along equally well on the ice of the Antarctic Circle and in the virgin forests of the tropics (LORENZ, 1954).

The child goes through the same development when he leaves his mother. This makes him a creature of curiosity open to the world who collects and utilizes experiences. In animals the behavior of curiosity — if present at all — is of only short duration during youth, but in man it is maintained until old age. Neoteny, an inhibition of development in animals in which characteristics of youth remain until old age (see p. 243), is a prerequisite for becoming a human being. According to LORENZ (1954), in man it involves not only morphological characteristics (BOLK, 1926; SCHINDEWOLF, 1928) but also the behavior of curiosity.

Conditioned reflexes constitute the highest brain function permanently in many mammals, but only temporarily in the infant. Through the mechanism that allows any one reaction to be tied to any sensory stimulus when needed, the individual can adjust to the changing demands of his environment better than he could if each stimulus could only rigidly elicit the same reaction each time. Conditioned reflexes can be observed even in lower animals, but they reach their highest development at the stage of the higher mammals.

According to HEBERER (1955, 1959) and VON KROGH (1957), the last step in becoming a human being, "humanization," started in the middle of the Miocene Epoch with the branching off from the anthropoid ape ancestors who remained in the trees and started to swing. By descending from the trees to earth and acquiring an erect gait, the ancestors of man completed humanization at the beginning of the Pliocene Epoch, after perhaps ten million years of transition from animal to human in which neither animal nor human behavior could be clearly differentiated.

Thus, from ancestors which had lived as quadrupeds on the ground during the reptilean period, tree dwellers developed which climbed trees in an obliquely erect body position, grasping as quadrumanes and quadrupeds. Returned again to the ground, they became completely erect and became bimanous and bipedal human beings.

The attainment of erect posture required a tremendous rearrangement of the body and a corresponding nervous reorganization. The knee joints were extended, while the hip joints were simultaneously overextended backward (see Fig. 208). The center of gravity of the upper half of the body and of the head, which was still located far too much to the front, was transposed backward by the backward bending of the lumbar vertebral column and by the development of the promontorium. The more pronounced lumbar curvature necessitated further compensatory thoracic and cervical curvatures to bring the bulk of the upper part of the body and the head forward over the feet and thus into balance (MOLLIER, 1924).

When the quadruped looks forward in the direction of his locomotion, the direction of his gaze is parallel to his vertebral column. The upright position lifts the head and neck, which had been stretched forward by the quadruped, above the vertebral column. If the eyes were still to look in the direction of the vertebral column, they would now have to look upward since they would be rotated 90°. However, they still look forward because the direction of gaze has now become perpendicular to the vertebral column (see Fig. 208). This is accomplished by a transformation of the skull and forward relocation of the point at which the skull is supported by the vertebral column (see Fig. 208, *3*). The quadruped must lift head and neck against gravity with strong neck muscles to approximately the horizontal plane; in erect man, the head is already nearly in equilibrium on top of the vertebral column (see Fig. 208, *3*). Keeping it in balance with the aid of the labyrinthine righting reflex demands no great muscle strength.

Fig. 208. Phylogenetic development of the erect posture. [According to MOLLIER]

Only man, not the erect quadruped, can attain such a balanced position of the head. A transverse plane (see Fig. 208) passed through the foramen magnum (located forward in man) would divide the cranium into two approximately equal sections, whereas in the erect quadruped the frontal part would be considerably larger. Compared to the quadruped, the human face is flattened because of the disappearance of the projecting mouth and the decrease in size of the muscles of mastication, whereas the posterior half of the cranium has enlarged through the increase of the cerebral hemispheres.

The forelimbs were freed from the task of locomotion by the righting of the body so that the hands became even better grasping tools. The weight transformed the hind limbs, which until then had grasped in the functions of standing, walking, and running.

Fish grasp their prey with their mouths. At the stage of the amphibians and reptiles, the newly developed limbs serve locomotion, but food is still grasped with the mouth. Some mammals, for example, the squirrels and monkeys, can grasp their food with their "hands" and bring it to their mouths. This development is repeated during infancy: At first the infant searches and grasps the nourishing breast of his mother only with his mouth, the rooting reflex assisting him in this task. Only later does he learn to reach for food with his hands and bring it to his mouth.

In the transition period from animal to human, man began first to use tools and later to manufacture them. According to HEBERER (1959), man possessed recognizable tools several hundred thousand years before he used fire. Of course, the preserved tools made of stone or bone keep better through hundreds of millennia than carbonized objects. Peking Man roasted his prey even in the second warm period between the Mindel and the Riss glacial periods, 429,000 to 236,000 years ago. We do not know how and when man made the most important discovery which was granted to him, the maintenance and utilization of fire. Neither do we know how – probably in connection with the manufacture of tools and the use of fire – language and intellect appeared.

In any case, the erect gait with the freeing of the hands, language, and intellect constitute the most important achievements by which man gained his advantage over animals.* The infant arrives at a corresponding transition stage from animal to man at approximately the end of the first year of life: At this age he has mastered the upright gait and is starting to talk and to act intelligently. Any determination of the moment when he has passed this animal-man transition stage and has become a human being would be only an arbitrary one.

At the lowest stage of language development, the infant acquired language understanding through the establishment of conditioned reflexes to certain frequently repeated words (see p. 527). It is not difficult to teach a certain understanding of human language also to higher animals by "training" them, i.e., by establishing certain reflexes to certain words. For example, dogs learn without any difficulty to obey verbal commands. However, it is not possible to teach even the higher species of apes to use given words meaningfully; using words is a stage which the child attains near the end of his first year of life. Here the difference between man and animal becomes clearly evident.

According to W. STERN (1928), it is practical understanding which develops first in the child and then theoretical. The first indications depend on external or internal stimuli which are effective at a given moment. However, foresight and anticipation of future strivings and decisions are absent at first. VON MONAKOW and MOURGUE (1930) state:

> "The child knows no past or future; his desires concern only objects of the immediately present moment."

This practical intellect needs the impulse of strong emotions which must be satisfied (see p. 141). According to KÖHLER (1915–1918), the chimpanzee behaves in the same manner; it learns best when it wants to satisfy its hunger (see p. 472).

---

* Of course, gorillas free in their natural habitat also walk erect (see p. 241).

Theoretical intellect, the aim of which is a result of cogitation, develops in the infant after practical intellect.

In the battle over the theory of evolution, it was declared impossible, even ridiculous, that human intellect could have developed from the animal preliminary stage in the course of phylogenesis. Yet every infant develops from the animal to the human stage before our very eyes.

DISSOLUTION OF CEREBRAL FUNCTION

(39, 43, 52, 55, 67)

VON MONAKOW and MOURGUE (1930) contrasted the law of fragmentary evolution with the law of fragmentary dissolution.

The cerebral function of the infant is constructed in successive layers. The older centers which have regulated cerebral function for some time are not eliminated by the development and maturation of new and younger centers, but are only inhibited. The control of nervous behavior returns to them if for any reason the higher centers are extinguished and the inhibitions which they impose thus disappear. Therefore, the same signs of breakdown may occur long after infancy as a result of the greatest variety of diseases, such as poisoning, suffocation especially, and also hypoglycemic shock, degenerations, hemorrhages, inflammations of the brain and the meninges, etc.

Dissolution of the respiratory center (see p. 363) can occur even in the adult. Periodic respiration was first observed in adults suffering from heart diseases by CHEYNE and STOKES. This respiratory pattern is an expression of the lowering of nervous excitability in the respiratory center. It may be accompanied in adults and in premature infants by periodic singultus, periodic generalized movements, and periodic disturbances of the cardiac rhythm. It is also found in meningitis, where it has been described as "meningitic" (Biot's) respiration, and in toxic diphtheria as an indication that the brain stem is considerably involved in this disease process. MOSSO (1878) found that periodic respiration may occur in healthy adults during deep sleep; in this case the sleep inhibition has also seized the respiratory center and has rendered inactive its highest component, which regulates the regular respiration.

Gasping respiration, which is the only respiratory form in young nonviable fetuses, occurs only under nonphysiological conditions later in life. During the process of dying, especially during the very last breaths, a distinct gasping respiration may occur; the popular expression for this occurrence is the "last gasp."

It is a bad sign if during a pneumonia, accompanying movements of the lower jaw gradually occur with each inhalation. The child rarely survives this condition, although sometimes death may not occur until some days later. The picture of pronounced gasping respiration may not necessarily occur. Death may also be presaged in other diseases by this floor-of-the-mouth respiration, a preliminary stage of true gasping respiration. The younger the child, the more apt it is to occur; it can be endured longer than true gasping respiration, which can rarely be borne more than several hours even by premature infants. This respiratory pattern might be connected with the increasing oxygen deprivation (see p. 365).

The picture of fully developed gasping respiration in a four-year-old epileptic boy was described previously (see p. 369). Gasping respiration probably occurs in every suffocating patient. In experiments on animals, and in adult animals (rabbits) as well, it is distinctly elicited by suffocation. Gasping respiration depends not on carbon dioxide poisoning but on the loss of oxygen, for the inspiration of pure nitrogen also causes it to appear. If it is not too late when the supply of oxygen is re-established in these experiments, gasping respiration disappears and the animals recover.

The individual gasps during the suffocation attack of the premature infant are separated by prolonged periods of respiratory arrest of varying duration; the gasping breaths of older children usually follow one another without interval. The respiratory rate in this instance is often increased, whereas during a suffocation attack it is always considerably decreased. In experiments on animals, floor-of-the-mouth respiration is seen as a transition from fully developed gasping respiration to the normal mature respiratory pattern.

Like gasping respiration, singultus (see p. 343) can also be disinhibited by the failure of higher components of the respiratory center. At no other age does singultus occur so frequently as in the first months of life, when the structure of the respiratory center is not yet compactly integrated. Later, singultus occurs as a sign of the breakdown of the respiratory center, for example, under the influence of alcohol. Peripheral stimuli, for example, from peritonitis or tumors of the phrenic nerve, may also irritate the singultus center so violently that it will assert itself against the main respiratory center.

Yawning, a respiratory movement which is always followed by a swallowing movement, is a regression to the ontogenetically and phylogenetically primordial swallowing respiration. It is still evident in the young infant: He sucks, breathes, and swallows simultaneously without having these functions interfere with each other (see p. 420). The adult yawns when he is tired or is half asleep, i.e., in a condition during which inhibitions spread in the cerebral hemispheres so that brain function regresses to a lower stage.

Many children of school age still make sucking movements when any type of stimulus so disturbs them during sleep that they are half awakened. Some put their fingers into their mouths at the same time. I do not know if any attempts have been made to explain the well-known grinding of the teeth which occurs in many healthy adults and children during deep sleep and frequently in children suffering from meningitis. We are dealing here with sucking movements of the dentate lower jaw. The calming effect of smoking is not attributable solely to the influence of the tobacco but also depends on the elicitation of sucking movements, as BUCKMAN (1894) already recognized.

In 1943, EBBECKE described the immersion reflex in adults: If the face is immersed in cold water, swallowing movements occur. According to GARSCHE (1950), this reflex represents the remainder of a coupled reflex of the oral region and the larynx which can be elicited by an air blast stimulus in all premature infants weighing less than 1500 gm (see p. 458).

According to SCHELLER (1952), in many cases of cerebral disorder in adults, if the lips are touched by some object they are thrust forward like a snout to grasp and hold the object. At the same time, the patient makes sucking and mumbling movements, sucks the object deeper into the oral cavity in this manner, gets it between his teeth, and then executes chewing movements with his jaws.

In adults with brain function impaired by disease or hypoglycemic shock, BENTE and WIESER (1952) and WIESER (1957) observed fixating movements of the eyes on visual and tactile stimuli in the mouth area, reaching and grasping, magnetic reactions, wide opening of the mouth, and sucking. The searching movements these authors observed correspond to the infant's rooting reflex or breast-seeking in vacuum activity; the same patterns of grasping with the mouth, sucking, chewing, and swallowing can also be found in infancy. SCHÖNECKER (1957) observed the appearance of smacking and licking movements in sick adults who had received chlorpromazine. BENTE and WIESER did not regard these phenomena as a disinhibition of phylogenetic layers, but as the expression of a general morbid reduction of brain function with the emergence of preformed survivals of earlier phylogenetic and ontogenetic developmental stages.

The mouth-hand coordination in adults suffering from severe cerebral disorders also belongs to this group (see p. 416).

In slowly progressing degenerative atrophic brain diseases (Pick's disease, Alzheimer's disease, certain types of senile regression), PILLERI (1960) observed the following oral phenomena: wide mouth opening following optic stimulation, optically evoked snapping opening of the mouth on tactile stimuli, sucking (pumping and empty sucking), finger sucking, sucking movements (pump-sucking) in vacuum activity, pendular head movements with rooting reflex, and oral patterns as seen in Klüver-Bucy's sign (tendency to take all kinds of objects into the mouth or to sniff at them).

All these movement patterns assure vital food intake for the infant.

In slow progression of the disorders mentioned, PILLERI found a regular sequence of the movement patterns: They are elicited at first by optic stimuli and may also be evoked in later stages of the illness by tactile stimuli. Finally, the searching automatism with pendular head movements and sucking movements appears in vacuum activity. At the terminal stage, several of the patterns may be present simultaneously in different degrees of distinctness, but with increasing cerebral atrophy they appear successively in the opposite order encountered in the young infant.

In every adult who showed the movement patterns described, pathological partial or total involvement of the temporal lobe was found.

Through the dissolution of the higher brain functions, a lower brain function which is characteristic for the infant is thus released in the older person, a function which had become invisible with progressive maturation of the brain without disappearing completely.

After infancy the mass movements, which are regulated by the pallidum, disappear because they are inhibited by the development of the corpus striatum. Under very intense physical pain these inhibitions are penetrated. New and strong mass movements corresponding absolutely to those of the infant appear.

The well-known preferred position of the infant (see p. 169) depends on the functional inability of the corpus striatum. In mentally retarded children, for example, in feeble-mindedness or in rickets, this position persists far beyond infancy. It reappears in meningitis. It may occur even in adults as a response to deep surprise (see p. 132). Its appearance in adults deprived of the striatum was pointed out by FOERSTER (1921); he interpreted it as a "regression phenomenon." FOERSTER interprets striatal disorders, especially the "climbing movements" characteristic of congenital generalized athetosis, in the same manner. In this connection I refer the reader to my position (see p. 219).

The crossed adductor reflex is always present during the first months of life. WEGENER (1927) found it in 10 to 50% of all healthy persons, but BOLL [1955 *(1a)*] found it only in exceptional cases after the seventh month of life (see p. 284). It may appear unilaterally in spastic paralyses, whereas bilateral occurrence is not considered to be pathological. Its presence later in life indicates either a survival from an earlier developmental stage or a regression to that stage.

A more pronounced tonus of the adductors is characteristic of the first period of life. During the stepping movements of the newborn infant (see p. 212), there is always a tendency to cross the legs so that the heels more or less catch on each other. When the legs are crossed as a result of the adductor spasms in Little's disease, this can be regarded as an arrest at a lower stage of development.

As previously described, the lowering of nervous excitability during meningitis is manifested by the appearance of periodic respiration and the return to the preferred position of the infant. According to GARSCHE (1950; see p. 458), the snap-swallow reflex may be released in chronic encephalopathy following tuberculous meningitis which has been treated with streptomycin for a long time. In such children, WECHSELBERG (1953) elicited sucking movements and sometimes even wide opening of the mouth by touching the oral region or the neighboring area. Mechanical stimulation of other body parts, for example, the arms, occasionally had the same effect. The readiness for this response was so great in two children that they even opened their mouths wide on sound stimuli.

Another early indication of this disease which can be explained in the same way is the homonymous crossed flexion reflex of the legs described by BRUDZINSKI (1908): On passive bending of one leg, the other leg is also flexed almost simultaneously (see p. 294). Although BRUDZINSKI did not find this sign in young healthy children, it can be found regularly in all infants up to the seventh month of life [STENGEL, 1945 *(18a)*] and then it disappears gradually. I was not able to find it in meningitis after the fourth year of life.

The eliciting stimulus in MORO's reflex is a shaking of the head (progressive movement in straight direction), and the reaction consists of uncoordinated movements of both arms and, to a lesser degree, of the legs. MORO used a tap with both hands on both sides of the pillow as a stimulus. If the reflex is elicited in this manner, the reaction disappears after the first month of life and persists only in feeble-minded children. But under special conditions (see p.154) it can still be obtained in children of school age.

The same reflex is also present in the adult. SCHALTENBRAND (1925) quite rightly emphasizes that a typical radius fracture occurs because the person who falls forward extends his arms by reflex action. It is difficult to suppress this reflex in a voluntary fall.

The adult's facial expression is for the most part formed from direct sensory reactions during infancy (see p. 108) and can therefore be traced back to the characteristics of this age. Thus, the expression of bad mood develops from the facial expression of bad taste. The gesture of disgust — to the point of spitting out — and pouting can also be traced to this expression. The sticking out of the tongue is reminiscent of the tongue pushing out the nipple. On the other hand, the kiss represents the sucking movement of the infant (see p. 136). Perhaps smoking became so popular because the necessary sucking movements set in motion extremely old reflexes which lie deeply anchored in every adult.

Affirmative nodding developed from the head movements of the hungry infant during the grasping of the nipple, and negative shaking of the head from the turning away at the end of the meal. The expression of obstinacy with compressed lips and jaws originally represents the rejection of unwanted food (see p. 133).

Theory of Levels

According to Aristotle ("On the Soul"; "Natural History of Animals"), nature progresses so gradually from things without souls to those with souls that the dividing line between them is not noticeable. The kingdom of the soulless objects is followed by that of the plants. Compared to lifeless objects, plants are nearly animated, but compared to animals they are soulless. They are limited to nutrition and propagation and have only the "nourishing" soul. In animals, the feeling soul is added. The lowest type of sensation common to all animals is mediated by the tactile sense. Only some animals possess locomotion, a faculty which combines with sensation in the soul. The highest spiritual function, thinking, occupies a special position as a function of the intellect and can be found only in man. Thus, the different functions of the soul are not present in the same manner in all living beings; rather, they form a sequence and the lowest level is always contained in the higher level.

Although Aristotle recognized the step-by-step structure of the functions of the soul, the idea of a sequence of development in the sense of the theory of evolution was still far from his thinking. He observed only a structured phenomenon, and not a temporal succession (Gompertz, 1941). In his "Physiognomy," Aristotle regarded the heart as the seat of the soul.

Jackson (1884) clearly recognized the evolution of brain function from different superimposed layers: The higher nervous arrangements, which have developed from the lower complexes, keep the latter in check, just as a government which has its origin in the people supervises and guides the people. The "dissolution" is not only a "taking away" of the higher accomplishments, but at the same time a "release" of the lower centers. If the governmental body of this country were to be suddenly dissolved, two causes for complaint would arise: the loss of the service of outstanding persons and the anarchy of the now unsupervised people. The loss of the governmental body corresponds to the breakdown in our patients, and the anarchy to the no longer supervised activity of the next lower stage of reduction.

PAVLOV also distinguished different building "stories" in the brain (III, 1, pp. 109 and 162):

"Depending on the phylogenetic development of the central nervous system, the nervous combination systems of distinct and increasingly more complicated so-called reflex centers were moved more and more closely to the head end of the body; this enabled them to accomplish a continuously improved analysis and synthesis of the stimulating agents in combination with the increasing complexity of the relations of the organism to the environment in progressively widening areas.... (III, 2, p. 393) Thus arises for research the very complicated question of the connection of these different stories."

Elsewhere PAVLOV speaks of the different layers (III, 2, p. 540) or "instances" (III, 2, p. 451) of the brain:

"The cerebral hemispheres, as the highest organ, usually keep under their permanent influence the shifting relation of organism and environment; as constant controllers of the body's executive functions they also dominate the lower brain regions with their instinctive and reflex activity. From this it follows that a more or less chaotic function of the subcortex, deprived of the necessary moderation and necessary harmony with the given circumstances, must be in causative relation to the elimination and weakening of the function of the cerebral hemispheres" (III, 2, p. 444).

JUNG (1953) dealt with these questions in detail: The higher layers cannot function without the lower. The ascending messages are already integrated* by the coordinating function of the lower layers; they reach the higher layers selected, so to speak; there they are again sorted out in order to bring new commands to lower layers. Many events which reach only the centers of the brain stem in lower animals are reported to the cerebral cortex in man. JACKSON's hypothesis allows an explanation of why, after the elimination of higher functions, simple mechanisms such as the increase of reflexes after the elimination of the pyramidal tracts come to the foreground. The higher levels not only influence the lower by inhibitions, but are also able to stimulate them on their own accord.

The higher levels function only when the lower levels are in good working condition. Similarly, the nervous system depends on the nonneural body functions which are themselves directed by the central nervous system. Thus, the arrest of the blood circulation eliminates the function of the nervous system.

Only the lowest levels are able to form a system somewhat sufficient to itself. This independence is restricted to the lowest functions only and appears only temporarily in simple reflexes, which are usually influenced by the higher layers. The short arc of the reflex excitation with its sensory and motor components therefore has numerous side connections and associations with higher centers when the nervous system functions normally.

"The order of the different layers is not only morphologically established but is also of great physiological importance for the central nervous coordination."

---

* In this connection, integration is the combination of different excitatory events into a single process.

These ideas of JUNG's can easily be brought into harmony with the findings which are encountered during the maturation of the brain. Just as the human brain develops anatomically from the brain stem upward to the diencephalon and mesencephalon, to the cerebral hemispheres, and finally to the telencephalon, so also does brain function ascend during individual development from the centers of the brain stem upward. This determines the neurological behavior of the successive developmental stages. One level after the other becomes capable of functioning and assumes direction of brain function with the always indispensable older layers as a foundation.

Only the ontogenetically and phylogenetically older centers of the brain stem have begun to function in the newborn infant, the cerebral hemispheres being still inactive; thus, the total brain is composed of an older, already functioning layer and a younger stratum not yet capable of activity. FOERSTER (1913) named the newborn infant a "pallidum creature" after his highest functioning brain structure.

As even ARISTOTLE recognized, the older layers are included in the younger layers. The younger layers are placed above the older but cannot work without them, whereas the lower layers can function very well without the higher, as embryology and phylogenesis prove. The higher layers are able to inhibit and to guide the lower. Many a function which a lower layer may fulfill is taken over and performed better by a higher layer. In addition, new functions are added. But at each stage the function of the brain remains a uniform whole.

The neurological understanding obtained from the layer structure of brain function corresponds to the psychological theory of levels. For phylogenetic reasons alone, KRAUS (1919, 1926) differentiates the deep, or lower, personality, which is rooted in the brain stem, from the cortical personality. According to H. F. HOFFMANN (1935), ROTHACKER (1941), LERSCH (1938), THIELE (1948), and WELLEK (1950), the levels of the mind developed one after the other. They are superimposed one on another so that the older levels are inhibited and regulated by the younger, but are also able to function independently. The following purely psychologically intended words of H. F. HOFFMANN can be applied without change to the neurological phenomena:

> "Each higher layer is superimposed on and curbs a lower, but also lives on the lower layer's strength. Higher layers take over the functions of the lower layers, which in less developed forms of life, i.e., in primitive organisms, are tasks of the lower layers only."

Following ARISTOTLE, the philosopher NICOLAI HARTMANN (1949) described the four-step structure of the world of reality.

The nucleus of the psychological theory of levels again and again proves to be that the higher levels possess their own laws, though they are not independent of the lower levels but enclose them.

Of course, terms such as "layers" or "steps" cannot be taken literally; they are only images with which we try to make lower and higher brain functions understandable. We will never be able to separate the individual levels as we separate, for example, the earth's strata. In the extremely complicated structure of the human brain — nothing else of its kind is known in animate or inanimate nature — the separate levels are most closely interwoven.

Carmichael (1933, 1954) indicates that important conclusions on certain burning questions of psychology can be drawn from the facts known so far. Spemann (1936) stressed that events occur in the developmental mechanics of the fertilized egg which very much resemble psychic phenomena.

SUMMARY

Just as the human brain is formed in successive stages in the course of phylogenesis and ontogenesis, the individual stages also become successively functional. Brain function thus spreads from the brain stem across the diencephalon to the cerebral hemispheres. The neurological behavior of the different developmental stages depends on this maturation process. The older layers are capable of functioning without the younger, which are not able to become active without the older; the younger layers inhibit, control, and stimulate the older. If the younger layers fail, brain function regresses to lower stages.

Some developmental stages of the child's brain function are phylogenetic vestiges which extend down to the reptiles, amphibia, and fish. These are: the evolution and dissolution of the respiratory center with the appearance of lower respiratory forms (swallowing respiration, gasping respiration, floor-of-the-mouth breathing, singultus, alae nasi respiration, periodic breathing); reflex grasping with the mouth and pendular movements of the head; the tonic palmar and plantar grasp reflexes; the tonic skin reflexes of the vertebral column, among them the spinal reflex; the preferred posture of the infant; locomotion starting from the abdominal position via creeping and crawling to the erect gait; the dependency of brain function first on the lower and later on the higher senses; and the bed infant's striving to reach his mother.

There are two forms of ascent from the animal stage to man: About a million years ago the ancestors of man, having assumed the erect posture, reached the animal-man transition stage, a stage that does not allow us to say whether we are still dealing with animals or already with human beings. The infant, having learned to walk erect, reaches this stage at the end of his first year of life with the beginnings of language and intelligence.

Lower levels of brain function can be disinhibited after infancy and even in the adult by the dissolution of higher levels. This explains the appearance of lower respiratory patterns during severe illness, the yawning in fatigue and fatal hemorrhages, the grinding of the teeth as sucking movements during sleep, sucking movements during sleep, the grasping with the mouth by adults in hypoglycemic shock or in those suffering from organic brain disorders, and finally the appearance of the preferred posture of the infant, the homonymous flexion reflexes of the legs, and periodic respiration as indications of meningitis.

In the facial expressions of the adult we still note the elements of many sensory reactions of infancy.

The theory of levels in psychology corresponds to the neurologically recognizable levels of brain function.

REFERENCES

ARISTOTLE: Über Zeugung und Entwicklung der Tiere (On the Procreation and Development of Animals). Works in Greek and German. Vol. 3, pp. 149 and 736b. Leipzig, 1860; Hauptwerke (Major Works). German translation by W. Nestle. p. 176. On the Soul. Stuttgart, 1953; Physiognomik (Physiognomy). Heidelberg, 1929; Naturgeschichte der Tiere (Natural History of Animals). German translation by Kersch. Vol. 3, 1, p. 119. Stuttgart, 1866.

BABKIN, P. S.: Z. Nevropath. i psichiat. (Russian) *56*, 22 (1956).

VON BAER, K. E.: Über Entwicklungsgeschichte der Tiere (On the Ontogenesis of Animals). *1*, 224. Königsberg, 1828.

BENTE, D., and S. WIESER: Arch. Psychiatr. *188*, 301 (1952).

BILZ, R.: Schweiz. Zschr. Psychol. *3*, 202 (1944).

BOLK, L.: Das Problem der Menschwerdung (The Problem of Humanization). Jena, 1926.

BRUDZINSKI, J.: Wien. klin. Wschr. *1908*, 225.

BUCKMAN, S. S.: The Nineteenth Century *36*, 1894 II, 733.

CARMICHAEL, L.: In Murchison's Handbook of Child Psychology. 2nd ed. London, Oxford, 1933.

—, Manual of Child Psychology. 2nd ed. p. 60. New York and London, 1954.

CHARMA: According to DARWIN (1872).

DARWIN, C.: Die Entstehung der Arten [The Origin of Species (1859)]. Stuttgart; Der Ausdruck der Gemütsbewegungen (The Expression of Emotion). Stuttgart, 1872.

EBBECKE, U.: Pflügers Arch. Physiol. *246*, 675 (1943); *247*, 222 and 234 (1943); Physiologie des Bewußtseins (The Physiology of Consciousness). p. 120. Stuttgart, 1958.

EDINGER, T.: Fortschr. Palaeont. *1*, 235 (1937); Confinia neur. (Basel, New York) *9*, 5 (1949).

VON EICKSTAEDT, E.: In HEBERER (1959), p. 1192.

FOERSTER, O.: Berl. klin. Wschr. *1913*, 1217 and 1255; Z. Neur. *73*, 1 (1921).

FRÜHMANN, E.: Neue Österreich. Zschr. Kinderhk. *4*, 15 (1959).

FÜRNROHR: Z. Neur. *27* (1904).

GIESELER, W.: In HEBERER (1957), p. 951.

GOECKEL, H.: Die Wandlungen in der Beurteilung des ungeborenen Kindes (The Changes in the Evaluation of the Unborn Child). Thesis, Heidelberg (1911); Arch. Gyn. *94*, 91 1911).

GOETHE, J. W. v.: Nach H. Schmidt, Geschichte der Entwicklungslehre (according to H. Schmidt's History of the Theory of Development), p. 408. Leipzig, 1918.

GOMPERTZ, T.: Griechische Denker (Greek Philosophers). 3rd ed. *3*, 126. Berlin and Leipzig, 1931.

HAECKEL, E.: Generelle Morphologie (General Morphology). Berlin, 1866; Anthropogenie (Anthropogeny). 2nd ed. Leipzig, 1874.

HARTMANN, NIKOLAI: Der Aufbau der realen Welt (The Structure of the World of Reality). 2nd ed. Meisenheim am Glan (1949).

HEBERER, G.: Die Evolution der Organismen (The Evolution of Organisms). Jena, 1943; 2nd ed. Stuttgart, 1954—1959; Naturwiss. Rundschau *8*, 373 (1955).

HINSCHE, G.: Psychiatr. Neurol. Wschr. *45*, 281 (1943).

HOFFMANN, H. F.: Die Schichttheorie (The Layer Theory). Stuttgart, 1935.

VON HOLST, E.: Erg. Physiol. *42*, 228 (1939).

HOOKER, D.: Evidence of Prenatal Function of the Central Nervous System in Man. James Arthur Lecture. p. 10. New York, 1958.

JACKSON, H.: Die Croon-Vorlesungen über den Aufbau und Abbau des Nervensystems (1884) (The Croon Lectures on the Evolution and Breakdown of the Nervous System, 1884). German translation by G. Sittig. p. 63. Berlin, 1927.

JUNG, R.: Handb. d. Nervenphysiologie. Handb. der inn. Med. (Handbook of Neurophysiology. Handbook of Internal Medicine). 4th ed. V, 1—3, p. 101. Berlin-Göttingen-Heidelberg, 1953.

KLAATSCH, H.: Der Werdegang der Menschheit (The Evolution of Mankind). 2nd ed. Berlin, 1922.
VON KROGH, C.: In HEBERER (1957), p. 917.
LERSCH, P.: Der Aufbau des Charakters (The Development of Character). Leipzig, 1938.
LORENZ, K.: In HEBERER (1954), p. 131.
MAGNUS, R.: Körperstellung (Body Position). Berlin, 1924.
MAGNUSSON, J. H., and W. WERNSTEDT: Acta paed. *17*, Suppl. 1, 241 (1935).
MARINESCO, G., and A. RADOVICI: Revue neur. *27*, 237 (1920).
MINKOWSKI, M.: In Abderhaldens Handbuch der biolog. Arbeitsmethoden (In Abderhalden's Handbook of Biological Methods). Issue 253, Sect. V, Pt. 5B. Berlin, 1928; Revue neur. *93*, 247 (1955).
MOLLIER, S.: Plastische Anatomie (Plastic Anatomy). p. 123. Munich, 1924.
VON MONAKOW, C.: Die Lokalisation im Großhirn (Localization in the Cerebral Hemispheres). p. 141. Wiesbaden, 1914.
–, and R. MOURGUE: Biologische Einführung in das Studium der Neurologie und Psychopathologie (A Biological Introduction to the Study of Neurology and Psychopathology). pp. 15, 134 and 172. Stuttgart and Leipzig, 1930.
MORO, E.: Münch. med. Wschr. *1918*, 1147; *1920*, 360.
MOSSO, A.: Arch. Physiol. *1878*, 441.
MÜLLER, FRITZ: Für Darwin (On Behalf of Darwin). Leipzig, 1864.
NEEDHAM, J.: Chemical Embryology. p. 64. Cambridge, 1931.
OPPENHEIM, See FÜRNROHR.
PILLERI, G.: Arch. Psychiatr. *200*, 455 and 603 (1960); Schweiz. Arch. Neurol. u. Psychiatr. (in press).
REMANE, A.: In HEBERER (1954), p. 340.
RENNERT, H. P.: Nervenarzt *27*, 467 (1956).
RENSCH, L.: See HEBERER (1943).
RICHTER, C. P.: Arch. Neurol. *26*, 784 (1931).
ROBINSON, L.: The Nineteenth Century *30*, 831 (1891).
ROTHACKER, E.: Die Schichten der Persönlichkeit (The Layers of Personality). 2nd ed. Leipzig, 1941.
RÜGER, L.: See HEBERER (1943).
SCHACHTER-NANCY, M.: Rev. frç. pédiatr. *13*, 180 (1937).
SCHALTENBRAND, G.: Dtsch. Zschr. Nervenhk. *87*, 23 (1925).
SCHELLER, H.: In R. Wartenberg, Die Untersuchung der Reflexe (In R. Wartenberg's The Examination of Reflexes). p. 180. Stuttgart, 1952.
SCHINDEWOLF, H.: Das Problem der Menschwerdung. Ein palaeontologischer Lösungsversuch (The Problem of Humanization. An Attempt at a Palaeontological Solution). 1928.
SCHÖNECKER, M.: Nervenarzt *38*, 35 (1956).
SCHWIDETZKY, I.: Das Menschenbild der Biologie (The Image of Man in Biology). p. 14. Stuttgart, 1959.
SECKEL, H.: Jb. Kinderhk. *133*, 163 (1931).
SPEMANN, H.: Experimentelle Beiträge zu einer Theorie der Entwicklung (Experimental Contributions to a Theory of Development). p. 278. Berlin, 1936.
STERN, W.: Die Intelligenz der Kinder und Jugendlichen (The Intelligence of Children and Adolescents). 4th ed. Leipzig, 1928.
STIRNIMANN, F.: Das Kind und seine frühe Umwelt (The Child and His Early Environment). Basel, 1947.
STRAUSS, H.: J. Psychol. *39*, 111 (1929).
STRÜMPELL, A.: Dtsch. Zschr. Nervenhk. *15*, 254 (1899).
THIELE, R.: Studium generale *1*, 144 (1948).
TINBERGEN, N.: Zschr. Tierpsychol. *4*, 1 (1940–1941).

VOIGT, F. S.: Grundzüge einer Naturgeschichte als Geschichte der Entstehung und weiteren Ausbildung der Naturkörper (An Outline of a Natural History as a History of the Origin and Further Development of Bodies in Nature). Frankfurt am Main, 1817.
WECHSELBERG, K.: Mschr. Kinderheilk. *101*, 224 (1953).
WEGENER, H.: Mschr. Psychiatr. *66*, 342 (1927).
WEINERT, H.: Ursprung der Menschheit (The Origin of Mankind). Stuttgart, 1932. 2nd ed. (1944).
WELLEK, A.: Die Polarität im Aufbau des Charakters (Polarity in the Evolution of Character). Bern, 1950.
WIESER, S.: Ann. Univ. Saraviensis *5*, 211 (1957).
WINTERSTEIN, H.: Handb. der vergleichenden Physiologie (Handbook of Comparative Physiology). 1, 2. Jena, 1921.
VAN WOERKOM, W.: Rev. neurol. *20*, II, 285 (1912).

14. THE CHILD AND HIS ENVIRONMENT

In the laboratory, individual sensory stimuli are applied to test the infant's sensory functions while all other stimuli are excluded as much as possible. Such an experiment usually creates artificial conditions which do not permit application of the results directly to the infant's behavior under natural conditions. Actually, only a limited number of the many stimuli which collectively comprise the external world are effective at a given time. A process of selection which depends to a great extent on age but which may also change with the child's general condition is constantly taking place: If the infant is hungry he tries to reach the breast, and if he is satiated he rejects the breast. There is a difference whether the infant is stimulated during excitement, feeding, rest, or sleep, whether he is well or sick, and whether he is in his accustomed environment or not.

Not only is the infant dependent on his environment, but his environment also depends on him. This extremely important mutual relation cannot be established in the laboratory, but can only be recognized in an environment which corresponds to his actual necessities of life. But what constitutes the infant's natural environment? Certainly not the environment in a children's home or in a laboratory!

The adult is better able to adjust to his environment than the infant or the higher animal; he has made the greater part of the earth's surface habitable and finds his livelihood under the most varied conditions, as far as food, clothing, shelter, and tools are concerned. It is frequently overlooked, to the young child's detriment, that he is far less capable of adjusting himself. The younger he is, the more dependent he is on having the necessities of life provided by someone else; and these are provided under natural conditions. His needs are the same everywhere, no matter how and where his parents may live. In the course of phylogenesis, nature has adapted the child to a certain environment, and this environment is always the same. The infant is adapted in his entirety to his mother, and only to his mother.

If the mutual relations between the child and his environment are disturbed by an unnatural environment which does not completely satisfy his vital needs, certain reactions corresponding to this disturbance develop in the child. I will discuss only the changes in cerebral function, while the reaction to unnatural nutrition, the most important problem of pediatrics, will not be considered.

FROM BABY MONKEY TO HUMAN CHILD

(106, 128)

The newborn of different mammals have completely different vital needs. According to their behavior in reference to their environment, four groups can be distinguished: pouch babies, nesting young (those which remain in the nest), autophagous young (those which at once feed themselves), and breast babies.

The newborn of the marsupials, which are born while still at an early embryonic stage of development, find shelter, protection, and food in their mother's pouch. It has previously been shown (see p. 239) that they have reflexes which aid them to move into their mother's pouch after birth.

Nesting young are also born at a low stage of development, and depend entirely on their mother's care. In many instances they are born in hideaways which are as inaccessible as possible, and are naked, hairless, very much in need of warmth, and often blind at birth. To this group belong the young of rabbits, mice, rats, and — among the beast of prey — the young of lions and cats.

Autophagous young are already hairy at birth and are able to run around immediately after birth. The autophagous young of those ungulates which roam in herds are particularly independent. The mother sheep pauses a moment to give birth and then immediately follows the roaming herd. The lamb, which is fully equipped with its sensory organs, follows the mother.

Although breast babies, which are found among the bats, monkeys, and apes, are born at a rather low stage of development, they are usually born with their eyes open and a complete hair cover. They have special clasp reflexes with which they cling to their mother's body, and they are always carried around by the mother.

Since the structure and the functions of the human body are frequently reminiscent of man's ape-like ancestors, it is worthwhile to compare the data on the infant care of the anthropoids, meager as they are, with our customary baby care, even though the chimpanzee mother should not necessarily be used as a model. It does not matter to what extent the ape can be trained to employ human customs, although the lay person is particularly interested in this aspect. Rather, it is important to know how the ape mother behaves under conditions which are as natural as possible for her, for the methods of care used by the human mother, which exist even in the animal stage, are not actions based on reason, but are much more deeply rooted.

Even though Doflein (1914) includes the human infant as well as the baby monkey among the breast babies, it should be stated that this is only relatively correct; while the baby monkey lives on its mother's body, the human infant spends his life in bed for the most part. I therefore term him a "bed baby." This difference is important because it explains some of his behavioral characteristics.

In starting from the concept of nesting and autophagous babies, Portmann (1944, 1945, 1953) arrives at a completely different view. According to him, the young of reptiles and of lower birds are nesting and those of the higher birds and the lower mammals autophagous, whereas those of the higher mammals are autophagous at birth but during gestation pass a developmental stage corresponding to that of the nestlings: Their sensory organs go through a completely insensate period of complete occlusion which is terminated some time before birth.

According to PORTMANN, the primates pass the nestling stage while still in the mother's womb. As a nestling, man should be born after 5 months of gestation. But prior to birth he matures to the stage of the calf and foal, which he is said to resemble, for example, in the development of his pyramidal tracts. Thus, the human child is born as a "normal premature infant"* although he weighs much more than newborn anthropoids. The nesting stage is already superseded in the mother's womb; the newborn infant is on the road to becoming autophagous, but he is born before this stage is reached. Man owes his wide-awake senses and his abundance of movements to the autophagous stage of the mammals. However, especially because of his prolonged dependency on his parents, he became a kind of new species of nestling, i.e., a secondary nestling. His helplessness does not correspond to the physical immaturity of the nestling, but represents an exceptional condition among the mammals, i.e., the condition of man. The range of development which the human brain has to encompass considerably surpasses that of any other mammal. This factor also increases the dependency and intimate relation between mother and child. Thus, the newborn human infant is said to conform in his physique to the autophagous stage, but to deviate from everything else which we observe otherwise in higher mammals. PORTMANN states:

"A superficial comparison is mostly reminiscent of the nesting stage. Therefore, the biologist must particularly stress the difference of the lower mammals from the nesters. In his mother's womb, man passes the characteristic autophagous † stage of occlusion of the sensory organs. He arrives in the world with senses far advanced in their development, and his motor faculties are also equally great at birth."

"The latent autophagous nature is also evident in that human milk is that of an autophagous animal. *Only thus* ‡ is it possible for it to be relatively easily replaced by cow's milk."

But if this is relatively difficult (pediatricians have extensive experience in this field), then according to this kind of reasoning the human infant is a nestling.

According to PORTMANN, the birth weight of the nestling's brain increases 5 to 10 times by adulthood, and that of the autophagous animal 4 times, and frequently only 2 or 1½ times. PORTMANN regards the brain development of the primates as "the most important and certain sign that permits the differentiation of autophagous and nesting animals." *However, to me it appears mandatory that this question be decided only according to the behavior of the infant and the young animal in reference to the nest.* Certainly one cannot agree with PORTMANN in this regard for he declares newborn primates to be "true autophagi," the same as young antelopes, deer and roe fawns, and foals, except that they do not leave the nest.

* Autophagous babies are not physiological prematures, but pass through the maturation period of the nestlings during their embryonic period (PRECHTL and SCHLEIDT, 1951). Compared to the nesting species, newborn autophagous babies really ought to be termed overdue (PEIPER).

† I assume this should read "nesting" (PEIPER).

‡ Italics mine (PEIPER).

In anthropoids the "breast baby" still lives on its mother's body and is nourished, warmed, protected, defended, and trained by her. Chimpanzees and gorillas living in their natural habitat have not yet attained the stage of a permanent domicile, but construct a nest of branches just for one night even if they return to the same neighborhood; each animal builds a nest for itself alone. Only the mother takes her baby along with her to her own nest; moreover, a baby is never found there alone. Monkey mothers have to carry their babies with them constantly in order not to lose them. One baby is occasionally left in a hurried flight (REICHENOW, 1921; NISSEN, 1931).

According to NISSEN, the newborn chimpanzee baby in its natural habitat is probably just as dependent on parental care as the human infant, although complete dependence no doubt lasts longer in man. The baby monkey evidently possesses the very important faculty of holding onto the adult with its hands and feet (palmar grasp and plantar grasp reflexes, see p. 158) right from the beginning of life. In the forest the baby is never literally carried; the larger animal passively supports the baby's weight while the young animal holds on by its own efforts. During locomotion it is in a horizontal position, back down, while clinging to the adult animal's abdomen. On the ground this kind of locomotion is fairly safe, but high in the treetops it is sometimes dangerous for the baby. Usually a chimpanzee is said to be slow and inhibited when carrying her baby and to warn the baby (how?) before making a violent movement.

We must also assume the same mother-child relationship in man's phylogenetic ancestors. It has only been in historic times that civilization separated the human infant from his mother against his will and placed him in his own bed, from where he still tries with all his strength to get back to his mother. The nestless mother-child relationship of the primates differs absolutely from the relation of an animal baby to the nest.

PORTMANN actually comes to the same conclusion for the anthropoids: Their babies are considered to be autophagous,

"but at the same time it is recognized that – because of the life in the trees, the original modus vivendi of the primates – they remain with the mother animal. Normally they have no nest at all; their mother's body is the scene of their first climbing attempts."

Briefly summarized, the human passes through the following stages, according to PORTMANN: before birth nesting, then development to an autophagous animal. But before this stage has been completely attained, the "normal premature birth" occurs. It makes the latent but "true autophagous animal" (which in reality does not leave its nest) again a secondary nestling. The child thus becomes nesting or autophagous just as needed, a magic formula which could easily explain everything if only the infant had any relation at all to the nest. But precisely this relation is completely absent in primate babies, which are breast babies.

PORTMANN's views can hardly be reconciled with BOLK's (1926) hypothesis of a developmental inhibition of the growing child. According to BOLK, the signs which differentiate man from ape actually appear during the individual development of the primate fetus, but are lost again in the ape because of the progressive differentiation.

What was only a passing stage in the development of the ape becomes a final stage in man. For this reason the fetuses of the monkeys and the fetuses and newborns of the anthropoids have a distinctly human appearance — not because the apes descended from a human looking ancestor, but because man retains this feature of fetal development until the termination of his physical development. In their individual development, however, the other primates still go through a final stage which man no longer experiences. Physically, man is thus a sexually mature primate fetus.

PORTMANN (1956) does not agree with this theory, which has been accepted by many investigators (GEHLEN, 1948; LORENZ, 1943).

On the basis of his experience with the gorilla baby Goma, PORTMANN (1960) comes to the realization "that the concept of an 'autophagous' animal does not do justice to the facts and that we should give another name to this variable of birth behavior of the higher mammals... The babies of the monkeys and anthropoids, of course, cannot literally be considered 'autophagous' if we include their behavior in the definition."

We know about the baby care of the anthropoids from observations made by VON ALLESCH (1921), YERKES and TOMILIN (1935), and STEINBACHER (1940–41) on chimpanzees, BRANDES (1939) on orangutans, and STEINER (1949) on gibbons in captivity. I wish a pediatrician might some time have the opportunity to make similar observations!

The tonic palmar and plantar grasp reflexes and the rooting reflex considerably facilitate the baby's task of holding onto the mother and of finding the nipple with its mouth.

The chimpanzee observed by VON ALLESCH gave birth in the dorsal position quickly and without uttering any cries of pain, removed the baby from between her legs with happy sounds, placed it on her abdomen, and covered it with straw.

The chimpanzee baby observed by STEINBACHER clung to its mother immediately after birth and during the first months was almost continuously pressed against her body with her hand. When the mother lay down in the evening, she crossed her arms and legs over the baby so that only its head remained visible. The young gibbon observed by STEINER clung to its mother's abdomen to the extent that it was warmly and safely bedded in the groove formed by its mother's adducted thighs and abdominal wall.

With the aid of the palmar and plantar grasp reflexes (see p. 158), the monkey baby clings to the fur of its mother's abdomen. It thus accomplishes the task of searching for the breasts, which in this position are not far from its face, and of taking the nipple into its mouth. The food intake of the chimpanzee baby observed by VON ALLESCH seemed to begin on the third day (?) only when the mother raised the restless and faintly whimpering baby slightly and pressed her breast against it. The finding of the breast remained difficult for a long time. Often a chance movement of the mother pushed the baby away from the right spot so that the search had to start anew. Sometimes the baby was assisted by the mother. Goal-directed searching for the mother's breast could not be noticed before the age of three months (see p. 410). According to YERKES and TOMILIN's observations of six chimpanzee babies, the newborn found the breast a few minutes or hours after birth with or without the mother's help and then attempted to suck. The mother's behavior varied from simply watching her baby to actively grasping it and placing it at her breast. STEINBACHER observed a chimpanzee baby vainly trying to find the breast in its mother's hair several hours after birth. The rooting reflex, which is probably just as characteristic of the newborn ape as it is of the newborn human infant, facilitates the finding of the breast.

It has previously been mentioned that reflex crawling movements, which can be assumed to exist in the monkey baby just as they do in the human infant, also facilitate the finding of the breast. These movements are more pronounced in hunger than in satiation. The thighs, the inguinal region, and the feet clinging to the mother's fur probably provide a hold for the soles of the baby's feet.

It is difficult to determine the number of meals which a baby hidden in its mother's fur receives in the course of a day. No exact data exist, but the monkey baby is breast fed considerably more often than the human infant, who in our civilization is usually fed only 5, or at the most 6, times a day (see p. 475). The feedings of BRANDES' (1939) orangutan baby lasted from fractions of a minute to 2 or 3 minutes and occasionally 6 minutes. The orangutan baby is regularly fed premasticated food by the mother; STEINBACHER states that this method is rare in chimpanzees. STEINER's gibbon baby grasped morsels of food from its mother's mouth with its hands.

Even in the first months, the orangutan mother also fed other food to her baby. She gave fluid by pressing her lips between the baby's. The same process was later regularly followed in giving premasticated solid food. Some human mothers still premasticate food for their babies, a method which is not recommended (see p. 481). This was necessary at an earlier stage of historical development, i.e., before the use of spoons, vessels, and fire.

The orangutan mother observed by BRANDES shortened her baby's fingernails by cautiously gnawing them off. According to a widespread superstition, a mother should not cut her baby's fingernails, but should only bite them off. Thumb sucking also exists in monkey babies; BRANDES photographed an orangutan baby engaging in this activity.

VON ALLESCH describes the process of defecation: The mother suddenly raised the baby, which otherwise clung to her, by one leg so that it hung free in mid-air. With the baby, thus "held out,"* the feces could drop to the ground without touching the mother. Afterward the baby was brought back to its former position. Often it was set down again before defecation or during the actual process, and sometimes nothing at all happened. It is easy to believe that in this original form of "holding out" the mother, carrying her baby on her body, recognizes the right moment more easily than can the human mother, whose child is in bed. YERKES and TOMILIN report on similar efforts of the chimpanzee mothers they observed, and STEINER on those of a gibbon mother.

C. F. and M. M. JACOBSEN and YOSHIOKA (1932) compared the locomotor development of a human child and a chimpanzee baby, and obtained the following figures (age in weeks: human/chimpanzee):

* See footnote, p. 526.

1. Crawling developing in the following stages:
   Raising head in abdominal position, chin free 3/3
   Raising head in abdominal position, chest free 9/5
   Pushing with knees or swimming 25/7
   Rolling over 28 / 8 – 10
   Turning to and fro to move forward slightly 37/11
2. Progress to erect posture through the following stages:
   Raising head in dorsal position 15/5
   Sitting unaided for a moment 25/2
   Sitting unaided 31/3
   Standing with support of a railing 42/15
   Pulling up on a railing to standing position 47/15
3. Progress to walking through the following stages:
   Walking with aid 45/17
   Standing unaided 62/20
   Walking unaided 64/25–29

RIESEN and KINDER (1952) made a corresponding but more detailed comparison. On the basis of investigations made by LASHLEY and WATSON, TINKLEPAUGH and HARTMANN, and FOLEY and HINES on *Macacus rhesus*, of their own investigations on chimpanzees, and of GESELL's and THOMPSON's investigations on human children, they arrived at the tabulation presented in Table XIII.

Table XIII

Stage of Locomotor Development	Age (weeks)		
	Macacus	Chimpanzee	Man
Raising head and upper half of body	At birth	2	13
Turning to abdominal position	At birth	11	40
Sitting, head up (continuously)	½	2	16
Supporting on hands	1	11	23
Standing on all fours	1	20	42
Crawling	1	20	45
Sitting without support	9	24	23
Standing erect	10	39	54
Walking erect	–	43	58

In the ascending line from Macacus to chimpanzee to man, the rate of development slows down markedly, just as the first menstruation (also according to RIESEN and KINDER) occurs at an increasingly later age: at the age of three years in *Macacus rhesus*, at the age of eight years and eleven months in the chimpanzee, and at the age of thirteen years in man.

The chimpanzee baby is no doubt more developed at birth than the human baby, and develops faster during the first four months of life. Nevertheless, the two are very similar in their dependency and general behavior at the age of four weeks.

According to YERKES and TOMILIN (1935), the chimpanzee baby learns to walk first as a quadruped and then as a biped. A stage of creeping and crawling may or may not precede walking on all fours. According to YERKES (1954), the young chimpanzee can climb with great skill and run on all fours.

VON ALLESCH and BRANDES observed that the mother puts the baby down from her body gradually, cautiously, and, observing it carefully, first for only a short time, then for longer and longer periods, lets the baby make its attempts to get up and to crawl away. The chimpanzee observed by VON ALLESCH set up a real walking school: She put the young animal in front of her and held it tightly. Walking backward, she forced her baby to follow on three legs. These exercises were performed infrequently at first and then more and more frequently and for distances of 10 m.

According to YERKES and TOMILIN (1935), every chimpanzee mother encourages and supports her baby in its attempts at crawling, standing erect, climbing, walking, and running. The chimpanzee mother observed by STEINBACHER set her baby down in front of her and encouraged it to hold onto her with one hand only. Then the mother retreated, sliding slowly backward and thus forced the still very clumsy baby to follow. Later she grasped both the baby's hands in one hand and made it run to and fro in front of her. After some time she encouraged the baby to walk independently by setting it down in front of her and enticing it by certain sounds. The baby was able to move freely on all fours at the age of nine months.

BRANDES stresses that it would be wrong to interpret actions such as feeding from the mouth, biting off of the fingernails, or making the baby independent of her as carefully considered actions of the apes. The same might well also be applied to the human mother.

According to YERKES (1945), the chimpanzee mother opposes any attempt by a person to occupy himself with her baby and she resists all attempts at forcible separation with extreme force and cunning. Therefore, the greatest difficulties arise as soon as somebody is suspected of planning a separation. Later, when the baby is capable of leaving its mother, the probability of success varies according to the condition of the mother and the baby. The difficulties are probably slight if the mother is inexperienced, stupid, and careless; they are great if she has had repeated bad experiences, and if she is quick-moving, suspicious, and on guard. A chimpanzee mother will permit a person to take her baby only if she trusts that person fully and if the baby stays within her sight.

A human mother may not infrequently be more or less replaced in caring for her child by an energetic other woman who has no children of her own (see p. 604). The same may occur in monkeys. In the rhesus monkeys of India, BRANDES describes the substitution of the mother by the grandmother. The latter had seized the newborn and did not let go of it, so that it starved to death at her milkless breast.

TEMBROCK (1949) reported a case in which a higher ranking female in the chimpanzee cage had taken a baby from its mother; the mother did not dare to take it back. Even the guard was unable to accomplish this either by force, cunning, or friendliness. Nature finally corrected the situation when the baby, attempting in vain to drink milk, bit the false mother's breast rather roughly. Only then did she put the baby aside. After she had been lured away, the baby could be returned to its rightful mother, who immediately reclaimed the baby.

According to YERKES and TOMILIN, five of seven chimpanzee mothers still recognized their babies clearly after a year; the sixth no longer showed any distinct reaction after a week, and the seventh, after six months.

Exactly as in man, the baby's whining is not directed only to females. BRANDES reports how the crying of an ape baby attracted an old male orangutan instead of its absent mother. As soon as he saw the helpless baby, he climbed down, picked it up, and took it back to his sleeping compartment. When the mother tried to take it back, he resisted and showed his teeth. In chimpanzees, the mother usually takes exclusive care of the baby's cleaning, feeding, protection, and supervision (YERKES, 1945).

Phylogenetically, the still incomplete development to the bed infant must have started very gradually. It became possible only after man's ancestors had become used to a steady camping place, which became necessary after they had learned the use of fire.

With the development to a human being proper, the body hair was for the most part lost and the skin became smooth, so that the baby could no longer cling to it. Thus, the infant's close ties to his mother's body loosened more and more, and the appropriate reflexes became weaker without disappearing completely. Despite this development, mothers carried their babies with them for a long time. In many primitive races which form a kind of transition stage, it is customary even today to carry the baby, depending on changing popular customs, on the arm or astride the hips, or to bind him to the back, neck, or side with straps of cloth, or to carry him along in a bag, basket, or similar device, as PLOSS described in detail. However, the mother has always had to use considerably more effort than the ape in order not to lose hold of her baby. At home she usually lays the baby down on a certain resting place, the cradle or the bed. According to PLOSS (1912), however, there are races whose mothers almost never put their babies down, but always carry them around.

THE INFANT'S ENVIRONMENT

(96, 97, 101, 102)

"The infant needs neither stars nor planets; his mother is his planet and his star."

PARACELSUS (1493–1541)

The image of the outside world which every adult creates around himself originates on the basis of sensations and perceptions which follow the stimulations of his senses and are projected outward. It is self-evident to man that this same environment with which he must continuously come to terms also exists, at least some components, for other living beings. VON UEXKÜLL (1920, 1921, 1939, 1940) demonstrated, however, how great a variety of subjective environments exists for various animals. Each possesses its special environmental image which is transmitted through its senses and is characteristic of its species and which surrounds it like a soap bubble from which it can never escape. The environment exists for animals only to the degree that it can stimulate the animal's stimulus receptors (perception-world) and can affect the animal's effective organs (effect-world). Thus, only "earthworm objects" exist in the subjective environment of the earthworm; for example, sun, moon, and stars are absent in this world. In the dog's environment, only "dog objects" exist; among them, the olfactory sphere plays a dominant role. In many instances the animal's image of the outside world will be less complicated than man's; often, however, for example in the dog, it will exceed it. Only man is able to envisage environments alien to his species. Each animal species requires its own special set of questions and the appropriate methods of investigation. Not even all adult humans live in an identical subjective environment. A forest does not exist as an objectively determined environment; there is a different "forest" for the forester, the hunter, the botanist, the hiker, the nature-lover, the woodcutter, the berry-picker, and a fairy tale forest in which Hansel and Gretel lose their way (SOMBART, 1938).

Physically and mentally, the infant is no closer to the adult than many animals are; therefore, he has a completely different subjective environmental image which will change considerably in the course of the years. Grave errors are therefore unavoidable if the adult's subjective environment is assigned to the infant, as is so often done. In lieu of this, let us try to apply VON UEXKÜLL's environmental theory, which is so satisfactory in animal investigations, to the infant and take a look at his environment!

In this connection we will forego any speculation about his content of consciousness; this material is not accessible to research; we cannot even determine exactly when consciousness appears in the growing child (see p. 14).

From the abundance of chemical-physical processes continuously occurring in our environment, our senses are able to filter out only those which, because of their exactly determinable chemical-physical qualities, can stimulate our sensory portals. The sensations caused in this manner lead to perceptions which inform us of the events in our environment. However, we perceive only a part of the many stimuli which constantly reach our sensory organs. We are protected by a second filter which depends on our mood and our attention, which works differently when we are satiated or when we are hungry, which is more restricted in fatigue and during sleep than at the height of waking, and which functions differently in sickness than in health.

Accordingly, the behavior of animals is not determined simply by their sensory organs, but in addition by innate releasing mechanisms (LORENZ; see p. 606) which can be demonstrated clearly and in detail in sham experiments. The animal waits for the eliciting stimulus, which fits into its releasing mechanism as a key fits in the keyhole; yes, it even searches, so to speak, for this eliciting stimulus (appetence behavior).

The young infant is still at an animal stage. His behavior is therefore determined for the most part by his innate releasing mechanisms, which are set for key stimuli — exactly determined and effective only at this age — and which above all serve to regulate the association with his mother in general and with food intake in particular.

The infant's sensory life was previously described in detail (see p. 23). All of the senses are able to receive stimuli at birth. Investigations have shown that all sensory nerves are capable of transmitting excitations to and evoking reactions from the brain. However, while in all such experiments an individual sensory organ is in most cases stimulated by artificially isolated stimuli, the impressions of the outside world usually consist of more complicated influences of a combined stimulating pattern which does not influence one sensory organ, but several simultaneously.

The environment is a summation of individual stimuli for neither the infant nor the adult, different as the image of the outside world may otherwise be in the various age groups. For example, when the hungry young infant searches for his mother's breast he uses all senses appropriate for this purpose, i.e., the tactile, temperature, and taste senses. His movements are appropriate and goal-directed even if he has never sucked before. He strives by all available means for the breast, for the "stimulus-gestalt" of gestalt psychology, which he does not first have to acquire through experience. The newborn who has never sucked behaves as if he had advance knowledge of sucking.

An overwhelming number of similar examples can be found among the animals. Most similar to the process just described is a report by LORENZ (1937) on a young starling which had been raised in a room. The bird had never had occasion to see a fly, much less to catch one. Nevertheless, it suddenly attacked an invisible object, snapped it up in the air, returned with it to its roost, hacked at it with its beak as all starlings do with captured flies, and finally swallowed the invisible object ("vacuum activity").

In the example described, the starling is not motivated to execute this action by an external stimulus; however, the newborn infant usually acts in response to a tactile stimulus. But not infrequently young infants can be observed moving their heads to and fro, "searching" in hungry expectation, without a tactile stimulus.

Even at the level of insects, actions of much greater precision, and of farsighted value, are executed during hunting, in the care of offspring, and on many other occasions; their sense and purpose must be unknown to the acting creature, because it has never learned about their importance until then.

According to VON UEXKÜLL (1940), the spider web is thus a true template of the fly, from the body size of which the spider determines the width of the meshes. The resisting strength of the spun threads corresponds to the vital strength of the fly's flying body. It tightens the axial threads of the web more than the circular threads, so that the fly will be enclosed on impact by the resilient circular threads and will surely be entangled in their sticky droplets. The axial threads are not sticky and serve the spider as a path to its prey. The threads of the cobweb are so fine that the fly, with its coarse differentiating capacity, cannot see them and flies to its destruction without warning. But the spider which weaves the net has never encountered a fly before. It is "fly-like," i.e., it has incorporated components of a fly. Two beings of such different origin as the bumblebee and the snapdragon blossom are so constructed that they fit into each other in all details when the two "gestalt-melodies" melt into each other: as point and counterpoint do, the melody of the snapdragon penetrates the melody of the bumblebee as a theme, and vice versa.

This kind of mutual supplementation to a higher unity also exists between the infant's mouth and his mother's breast, between the sucking act and milk secretion, between the infant's body and the mother's milk, and — on a higher plane — in the mental adjustment of mother and child to each other.

I have previously described in detail (see p. 459) how the mother's breast directly forms a mold for the baby's oral cavity in the shape of the "sucking cone" and thus offers the infant the necessary drinking attachment which alone makes sucking possible. A description has also been given of how milk is produced in the mother's breast during nursing with the aid of unconditioned and conditioned reflexes, and how the amount is determined by the infant's need. Thus, mother and child work in complete harmony during nursing.

The infant stops sucking at the end of the meal; for hours the mother's breast loses its significant qualities, so that the otherwise effective stimuli no longer evoke rooting reflexes or sucking movements (see p. 473). von Uexküll's (1934) theory also recognizes such changes in significance: According to a hermit crab's mood, for example, the same sea anemone may assume a "protecting tone," a "sheltering tone," or an "eating tone." In the first case, the crab may have been deprived of the sea anemone which lived on its shell; the crab therefore replants it back on its shell. In the second case, the shell may have been taken away from the crab; the crab tries, though in vain, to live in the sea anemone. In the third case, when starved it begins to eat the sea anemone (F. Brock, 1939). Similarly, the mother's breast has a "sucking tone" for the hungry infant which it loses after satiation to such a degree that it is pushed out of his mouth (see p. 475).

Food intake most closely links the externally effective sucking act with internal processes such as breathing and swallowing. The sucking act provides the most important occasion for the infant to influence his environment independently. His behavior depends chiefly on his feedings.

The young infant's subjective environment consists of his mother. Since the breast infant from the beginning has the task of clinging to his mother's body, of finding the nipple, and of sucking, he needs first of all the senses of skin and taste, i.e., the lowest senses which transmit the message of the most immediate environment to him because they are stimulated only by direct contact. Visual and auditory senses, on the other hand, are at first of no importance to the breast-fed infant. These higher senses, which bring knowledge of the more distant environment, influence his behavior to an increasing degree only later, when the separation from the mother's body takes place. Thus, the sequence in which the individual senses influence the nervous processes (see p. 92) corresponds absolutely to the child's needs. In the same way as he feels and tastes his mother in the beginning, he later also learns seeing and hearing through her and from her, if by "seeing" and "hearing" is meant recognition by sight and hearing. The mother is physically and spiritually adapted to her child in the same way as the child is to her. With her breast she represents the most important "perceptual figure" and "effective figure" for the young infant. In her person, nature has given to the child unable to help himself a special helper of his own (see p. 600). The infant is adapted to her, his natural environment, in the most minute details of his physique and brain function, though not to the completely different tasks of the adult. What a miracle of adaptation mother's milk and the sucking act alone form! Mother and child supplement each other mutually to a higher vital entity for which language has not yet found a term. According to Fröbel (1826), the first "communal feeling" (in the sense of a feeling of togetherness) is that which connects mother and child. In a similar connection, F. Brock (1939) mentions an "animal-environment monad."

Without his nourishing mother, every infant's life is in danger. The nutritional disorders during the first year of life and the resultant infant mortality after the first days of life derive almost exclusively from this fact. The raising of infants without their mothers in institutions causes the widespread traumatization of institutional life (see p. 623).

According to ALDRICH (1939), the infant's need for food, sleep, fresh air, sunshine, clothing, etc., must be satisfied if he is to thrive. But in addition the child demands motherly warmth, tenderness, freedom from physical restraint after his meal, and sleeping periods which must be adjusted to his own personal rhythm (see p. 476) and not arbitrarily forced on him; he also wants to try out his newly acquired faculties. Only if all his needs are met does he develop according to plan; any lack inhibits this plan. If his care is too rigidly regulated, his physical growth also suffers. Therefore, a child develops better if he is with his mother than if he is in an institution.

If an infant is in a situation where he cannot help himself, he has a very simple means of calling his mother for help: He cries as soon as something does not suit him. His mother, who is always present under natural conditions, then takes care of him. The faculty of crying is therefore of the greatest importance to the infant; yes, it may even save his life, no matter how annoying it may sometimes be to adults.

Considering all these facts, the mother is not to be omitted from our thinking in connection with an infant's environment, so that the development of a child without his mother is inconceivable. However, precisely this has happened again and again in science. The well-known presentations of the psychological development of infants by K. BÜHLER (1929), KOFFKA (1925), W. STERN (1923), and DIX (1939), and the manuals of child psychology by MURCHISON (1933) and CARMICHAEL (1954), say little or nothing of the mother. The physiologist BURDACH (1838) devoted a great amount of space to her in his handbook of physiology. What physiologist thinks of her today?

Pediatrics, however, which again and again is made aware of the relations between mother and child, has long given mother love the place it deserves (IBRAHIM, 1917; MOLL, 1920, 1930; HAMBURGER, 1939). HERDER's beautiful words on the living unity of mother and child are quoted on p. 601.

We are in danger of misjudging young children if we compare their faculties with the final goal to which their development is directed. A child is considered to be more nearly perfect, the nearer he approaches this goal. The earlier stages appear to be imperfect, even inferior, because they are compared with functional ability accomplished later instead of with the demands appropriate for the age. Actually, however, every child born maturely is a perfect being in himself at every stage of his development and is always adjusted admirably to his natural surroundings. Otherwise mankind would have become extinct long ago.

The same error, i.e., of adult man holding himself to be the measure for all living beings, is also found in zoology. VON UEXKÜLL (1921, p. 4) therefore wrote, quite appropriately:

"The question as to a higher or lower degree of perfection in living beings can be posed only if the investigator regards the world around him as a universe containing not only him but all living beings, and as one to which they are all evidently more or less well adjusted.

From this point of view the human world is regarded as the only standard, and therefore the designs of the lower animals seem inferior to those of the higher animals... The individual animal is not adjusted more or less well to its environment; rather, all animals are equally completely adjusted to their environments."

Infants and children were long regarded as small, incomplete adults. Few artists of the past succeeded in portraying a real child, since they rendered them as too old in physique and facial expression (see p. 143). Even today it is often the custom in rural areas to dress little girls in the fashions of adult women. However, children are not small adults, but human beings of a special kind whose needs and subjective environments differ absolutely from those of later years. Even ROUSSEAU (1762) recognized this fact: "Nature wants children to be children before they mature into reasonable beings." ROUSSEAU was not successful in having this knowledge generally accepted.

The more primitive an individual's sphere of imagination, the more likely he is to see himself as the center of the world around which everything revolves. The child thus takes it for granted that he is the center of all events. This standpoint must later be abandoned. Originally, natural science, based on popular belief, regarded the earth as the center of the universe and man as the crown and center of the inhabited world. Whoever speaks today of the "whole world" means – following the old custom – only a part of the earth's surface. This opinion, though, gradually gives place to the increasing insight that the world is quite considerably larger. The adult still has to learn that he lives in another world than youth, and that youth's world must therefore be measured with a different yardstick.

The civilized adult lives in an environment which he himself has created and which has become vital to him. The growing up of a child is the only life process which the modern urbanite can still observe at home in pure form. Since the newborn infant is primarily a creation of nature, he depends completely at first on his nature-given environment, his mother. The many products of civilization which surround the adult are still absent in the newborn infant's subjective environment as long as it remains natural. The younger he is, the more easily his life is endangered if he is forced to come to terms with them.

To the physician falls the task of maintaining the natural environment for the infant, which above all means leaving him with his mother. If outside causes prevent this, we have to do our best without her. We give the infant artificial mother's milk in the form of a cow's milk formula, an artificial breast nipple in the form of a rubber nipple, an artificial mother in the form of a pediatric nurse, and a children's home nursery in lieu of his family. Artificial pap instead of premasticated food has already become indispensable. We combat the danger of temperature loss, which is minimal near the mother's warm body, with clothing, bedding, and even hot water bottles. Who could possibly believe that all these makeshift arrangements can accomplish the same thing nature does? Today it is generally acknowledged, at least in theory, that mother's milk cannot be replaced. It is just as unlikely that mother love can be replaced.

Whereas the adult is under the impression that the infant's environment barely seems to change in the course of the first year of life, the surroundings with which the infant comes in contact actually change considerably. In addition to the lower senses, which chiefly determine his behavior in the beginning, the importance of the higher senses, the eye and the ear, gradually becomes more and more prominent. The visual field develops with the aid of the already existing tactile field (see p. 93).

Gradually the infant learns to recognize graspable objects as such and in this way to assess distances. He has to see them "detached" from his environment, so to speak. This is facilitated if the objects are moved in his visual field, or especially if they are handed to him. The child does not reach out for the unattainable moon, as W. STERN (1923) has so rightly stressed. However, he may reach out with his hands for distant but familiar objects either to ask for them or to point them out, just as the adult does. This may have caused the wrong interpretation that he wishes to take the moon directly into his hands. The proverbial "reaching for the moon" is only a symbolic and poetic expression for the child's so often unrealizable wishes, and has been taken literally in error. Certainly it would be odd if the infant were to desire only the moon and not other distant objects which are far more striking and come into his visual field decidedly more often.

Soon the familiar feeding preparations lead to joyous excitement. The infant now knows as the giver of food not only his mother, but also the bottle, which he can roughly differentiate from differently formed or differently colored objects; he even recognizes the bib which is tied around his neck before each meal. He now follows the bottle with his head on merely seeing it, without being touched by it.

As his development progresses, his food-giving and solicitous mother will be differentiated more and more distinctly from strangers as a result of his growing experience and his increasing ability to utilize this experience. At first he does not see her entire figure, as she appears to the adult, but only her breasts, her face, and her hands. The knowledge that this "mother" has a voice, which usually sounds friendly but at times also unfriendly, comes to the child. But the mother's presence is chiefly connected with agreeable sensory impressions and sensations; disagreeable sensations are removed by her. Close ties thus develop between child and mother in earliest childhood, ties which will be preserved on a higher plane throughout all of life.

From the very beginning, the adult faces those events in his environment with which he must learn to deal with an equable and fairly uniform mood. Special situations are needed to change his mood. The child reacts in an entirely different manner, showing the greatest interest in all events, persons, and objects in his environment. For him the mother is not simply a human being, but an emotionally charged experience. As soon as he sees her, he wants to be taken in her arms, talked to, comforted, praised, fed, or entertained. He laughs when his mother comes and cries when she leaves. His experiences with objects are similar. He not only wants to possess them, but also to grasp them, to touch them, to turn them to and fro, to put them into his mouth, or to make noises with them. If he cannot do this, they do not exist for him at all.

The child later behaves similarly in his first attempts to speak. The first individual words do not simply designate a person, but express a whole sentence. The call "mama!" does not mean just the mother, but may express the wish that the mother should come, should give him something, should play with him, help him, or comfort him. Not until later is this stage of the "one-word sentence" abandoned. That the child turns again to his mother with his first meaningful word shows what his subjective environment consists of.

The child is aided in his development by an intense motility drive and urge for action which do not let him rest while he is awake, but continuously drive him to new experiments (see p. 272). Another characteristic of his behavior is the inability to give his attention to the same object or to the same occupation for any length of time. Only during feeding can he not be distracted. This lack of persistency is still a characteristic of the small child; in later years it is a special sign of feeble-mindedness.

The environment is enormously enlarged by the faculty of independent locomotion.

The small child, too, has his special "perception and action world," which differs greatly from that of the adult. ELSENHANS (1939) writes:

"He sees, hears, and perceives associations corresponding to his sphere of living. The environmental events appear quite different to him. A room consists of table legs, chair legs, and objects to be climbed on. He has a crawling floor or places which may be licked, others on which he hurts himself particularly easily if he falls, and still others which give a fine chance for running or sliding. Lamps represent light points into which one can gaze, but which otherwise cannot be utilized. The people in these rooms consist of long legs or round forms on whose tops mobile faces of expressive importance are enthroned. The young child quite often has to look up to them with head bent backward... Streets consist of cellar openings, water pipes that go into the ground, cellar windows, letter boxes which one would like to reach, and of more or less difficult stairways. Thus, the vertical dimension in the small child's perception world is quite different from that in ours, and therefore he will quite sensibly act differently in it."

THE MOTHER AND HER CHILD

(87, 111, 113, 115, 117, 138)

"The women's love nourishes the child."
(GOETHE, Elpenor, I, 2)

"Even art, divinely born,
Creates nothing higher
Than mother and son."
(SCHILLER, The Bride of Messina, I, 3)

The mother and her child form a living unity. When they are separated they endeavor to reach each other, and they are satisfied only when they are together again. Even PESTALOZZI already knew this:

"The child has needs, the mother has what he needs; she gives him what he needs; she is his world, he sees the world only through her. He hungers; she satisfies his hunger, and now he feels well. The place where he lies is not comfortable; she takes him into her arms and he feels well. To be well and to be with his mother is woven into one and the same concept" (1807). "Mother love is the main force, and affection is the original motive in early development" (1827).

In irreconcilable contrast to this statement, "the complete helplessness of the newborn human infant" is — according to the child psychologists — a self-evident fact which needs no proof (K. BÜHLER, 1922; KOFFKA, 1925; C. BÜHLER and SPIELMANN, 1928; REMPLEIN, 1952; SPITZY, 1958). Also according to the zoologist PORTMANN (1956) and the anthropologist SCHWIDETZKY (1959), the newborn infant is completely helpless.

The anthropologist GEHLEN (1958, p. 35) called man a "creature of want":

"He lacks the hair cover and thus the natural protection against the weather; he also lacks a body formation suited for flight. Man is surpassed by most animals in sensory acuity. His lack of true instincts endangers his life, and during his entire infancy and childhood he is subjected to a disproportionately prolonged need for protection. In other words: If he had lived as a ground creature under natural and primitive conditions among the most agile flight animals and the most dangerous beasts of prey, he would have been exterminated long ago."

The infant belongs in the arms of his mother especially under natural conditions. Genuine and vital instincts tie them to each other.

GEHLEN (p. 89) cited HERDER (1772, p. 26) as a witness:

The newborn infant "expresses neither concepts nor impulses by sounds, such as any other animal of his species does; placed among animals only, he is the most orphaned child of nature. Naked and bare, weak and needy, timid and unarmed; and, what makes the sum of all his misery, deprived of all guides to life... No! Such a contradiction cannot be the nature's way of ordering things."

The explanation for this unnatural contradiction is expressed by HERDER in another statement (pp. 113 and 114), not quoted by GEHLEN:

"The human baby, the infant placed in the world, how much is he a vassal of human assistance and social pity! From a state in which he hung plant-like from his mother's heart, he is thrown onto the earth, the weakest, most helpless being among all animals, *if there were not** motherly breasts to feed him and fatherly knees approaching him to welcome him as a son... Just these pains and these discomforts increase the mother's love. Just these miserable and unlovable qualities of the infant, his weakness, his frailty of temperament, the difficult irksome effort of his upbringing, double the sentiments of his parents."

ROUSSEAU (1761) also knew why nature makes the newborn infant seem so pitiable to us:

"Is there a creature which early in life is weaker, one which is more dependent on its environment and has greater need for pity, love, and protection than the child? Does it not seem that *just for this reason** the first sounds that nature has given to him are crying and lamenting, and that *just for this reason** she has given him such a frail body and such a moving expression that everybody who comes near him takes pity on his weakness and is eager to help him?"

Within certain limits, the infant is able to help himself: The skin protection reflexes (see p. 32) and the body righting reflex of the head and the neck righting reflex of the body (see p. 166) enable him to adapt his body position to certain conditions without outside assistance. The rooting reflex (see p. 404) lets him find the nipple independently. Nature has placed him in his mother's arms. There he has more help at his disposal than man will ever have in his entire life.

* Italics mine (PEIPER).

The infant becomes helpless only if he is taken out of his natural environment and if he is given tasks which may be natural for the adult but are not natural for him. The animal must also perish under unnatural conditions. The fish dies on dry land and the mammal under water. Who is more helpless? Even the adult cannot survive under conditions which are unnatural for him. Thus, in 1847-1848 all the members of Franklin's expedition* starved and froze to death on the icy north coast of America, in an area where even today the aborigines, the Eskimos, know how to survive. Who is more helpless?

Certainly an abandoned infant cannot survive without the aid of others. But he is by no means lost; rather, he becomes a transmitter who by his appearance and his whimpering emits calls for help just as a sinking steamer sends out SOS calls and receives help if humanly possible. And just as the SOS calls are addressed to everyone, the abandoned baby's calls for help find sensitive receivers everywhere. Even HERDER and ROUSSEAU already recognized this.

Similarly, the young bird not yet able to fly, abandoned on a tree, cries and is fed by adult birds even of a different species.

GEHLEN (p. 141) calls the newborn infant unoriented — and overlooks the mother-child relation and the rooting reflex; he calls him stimulus-unprotected — and overlooks the skin protective reflexes; he calls him immobile — and has never seen a newborn infant kick. As a natural being, man is said to be "hopelessly unadapted" because practically he possesses no morphological specialization. What a morphological specialization is the mother's breast!

How much the last step of becoming a human being depended on the higher morphological development of the cerebral hemispheres!

REMPLEIN (1952) characterizes the newborn human infant as poor in instincts, and therefore as rather helpless compared with any newborn animal. But in just this paucity of instincts lies the tremendous advantage which nature gave man for his higher development. He is said to be the only living being whose behavior has not been fitted a priori into certain narrowly circumscribed situations. He possesses a great realm of freedom within which he can use and develop his energies and functions in the most varied directions. This freedom allows his adaptation to the most varied environmental conditions and is said to be a prerequisite for his spiritual conquest of the world.

However, nature adapted the infant to a clearly defined and very narrow environment, i.e., to his mother. Away from her he is immediately in vital danger. Near her, and only near her, his instinctive behavior corresponds completely to his needs. Initially the young child lacks the tremendous advantage which REMPLEIN mentions. Does this also mean he lacks the prerequisite for conquest of the world?

VILLINGER (1956) is convinced of the young child's biological helplessness, imperfection, and defenselessness. He states that the infant is biologically badly equipped and not adjusted to a definite perception and action world. VILLINGER's view is refuted by the mere term "suckling." And as VON UEXKÜLL states (see p. 597), "all animals are equally well adapted to their environments."

* Sir John Franklin's disastrous polar expedition of 1845-47 — Ed.

According to SCHELER (1928), man is "open to the world," i.e., he is not bound, as the animal is, by his instincts and the senses extended in front of him. Based on this statement, GEHLEN (1958) also calls man open to the world because he lacks the animal's adaptation to a fractional environment. LORENZ (1943) sees in the worldwide investigative curiosity of man a continuation of BOLK's "retardation" (see p. 243), i. e., an arrest at a stage at which the young anthropoid remains only temporarily. Though this is no contradiction, it should be emphasized that the young infant, whose environment consists of his mother, has by no means reached the stage of being "open to the world." To look at an infant removed from the framework of his natural environment and confronted with tasks which can be solved only by adults constitutes an idle play of the imagination, a complete misconception of what an infant really is. Even the adult cannot hold his own in an environment to which he is not adapted. From the infant's point of view, it would be equally justifiable to ask how an adult would behave in the baby's situation. The adult might perhaps have to starve to death; at least REIMARUS (1773) reported that as an adult he was unsuccessful in sucking at the breast.

Never are two human beings so intimately dependent on and adjusted to each other as the mother and her child. Motherhood is the only profession nature has meant for woman. The greatest works of art created by the hand of man depict the mother and her child. Playing with dolls already prepares the young girl for her future tasks. The love for her child is an essential, perhaps even the most essential, womanly quality.

In many higher mammals and birds the parents, especially the mothers, care for and protect their offspring. Many an otherwise shy animal will resolutely attack the enemy who threatens its offspring. Despite this behavior of animals, true mother love, according to BUYTENDIJK (1957), is said to exist only in human beings, if by mother love is meant the mother's dedicated – even though caused by circumstances – deep-felt, and solicitous preoccupation with her child. True mother love is said to depend on conditions which never exist in animal life.

GUENTHER (1930) and K. M. SCHNEIDER (1956), on the other hand, devoted whole volumes to the mother love in the animal kingdom. According to SCHNEIDER, the foundations of motherly behavior are deeply rooted in the emotional and instinctive life and are therefore at least similar in man and animal, if not identical. This entitles us to speak of mother love also in certain animals. Members of the same species have inherited firmly established behavioral patterns which are evoked by appropriate stimuli and occur more or less automatically. When the time of confinement approaches, the animal becomes more sensitive and more easily excited about everything which applies to its immediate environment, its nest, and its offspring. It easily suspects persons as enemies who until then were familiars.

Mother love is thus one of the means with which nature provides for the survival of the species. Man has the innate instincts on which it is based in common with many higher animals. In man, reason, absent in the animal, usually improves and facilitates infant care, although to be sure exaggerated mother love may occasionally lead to unreasonable acts. Conversely, reason may influence mothers to act against their instincts, as the history of foundling homes shows in so many unhappy examples

[PEIPER, 1958 *(117)*]. Compared with baby care in the animal, the care of human parents for their children is of considerably longer duration. The child requires years before he becomes physically independent to any degree, decades before he becomes economically independent. All this time he depends on his parents, who see the happiness of their lives in the thriving and prospering of their children.

Women to whom motherhood is denied suffer from a lack of inner satisfaction. Frequently they find a substitute by adopting a child or by making up for it in some other way. Not infrequently two women come to the pediatrician with a sick child. One talks while the other stands mutely by. It is not easy to recognize that the talking woman is certainly not the mother, but a neighbor, an aunt, perhaps occasionally the grandmother,* who has pushed the real mother aside — "She is too young and inexperienced" — and has taken her place. It is not rare for a mother who has raised her own children to adopt another young baby to experience once more the joys of motherhood.

A mother's love for her child has a strong emotional content which is often directly contrary to reason. MOLL (1920, 1930) very strikingly described a "maternity neurosis," which has nothing in common with the real mental disease of delactation psychosis: The mother's anxiety clouds her judgment; she finds no rest day or night because of fear. She does not dare to bathe the baby, to introduce the thermometer, is afraid of colds or malnutrition, believes her thoroughly healthy child to be sick, and runs from one physician to the other. Even a woman physician can fail with her own child.

The close relation of the child to his mother desired by nature is considerably disturbed by the fact that civilization took the child out of his mother's arms and placed him in the baby bed. Serious difficulties arise because the infant does not concur in this procedure and attempts to get back to his mother by crying again and again.

So well has nature equipped the infant for staying on his mother's body — for example, with the tonic grasp reflexes and the rooting reflex — that civilization has yet been unsuccessful to the same degree in adapting him to his new abode, his own bed, which quite recently was forced on him against his will. Phylogenesis can explain that he does not agree with this change. His mother — and only his mother — is still his natural environment; he is adjusted to her with all his senses, just as she always carefully watches him. Nature gave him the desire and the ability to monitor her presence continuously. It was always of the utmost importance for the breast-fed child to announce himself to her immediately to draw her attention to him during a separation which could rapidly become dangerous. Even in the dark of the night, his voice announces to his mother where she can find him.

Day or night, whenever he cries, the mother comes to his aid. An everyday observation shows how much she is adjusted to her child: Let there be a great deal of noise in the street at night, or let a storm with thunder and lightning rage — the mother will not be disturbed in her sleep. But as soon as her child turns, cries, or otherwise moves in the next room, she immediately awakens and looks after him.

---

* The displacement of a rhesus monkey mother by the grandmother is described on p. 592.

Similarly, PAVLOV demonstrated in experiments on animals that sleep can be interrupted by conditioned stimuli, while other stimuli which are stronger in themselves will remain ineffective. For example, a conditioned reflex to a certain sound stimulus has been established in a dog. When the animal was fast asleep, it could not be awakened either by shrill whistling or by loud pounding with the fist on the laboratory door. The familiar tone, however, immediately awakened it and made it alert. PAVLOV (III, 2, p. 632) related this experiment explicitly to the sleeping mother.

The infant's pitiful whimpering constitutes an important protective device for the child who is apparently so helpless. It is this "touching condition" which awakens the motherly sentiments, be it in the real mother or in another female in his environment. In appropriate cases a man's pity may also be aroused. Previously, a description was given (see p. 593) of how this may occur even in the orangutan.

Ancient physicians such as HIPPOCRATES, GALEN, RUFUS, RHAZES, and AVICENNA wrote on the nocturnal restlessness of the child; today it does not attract much interest in the scientific literature, with the exception of pavor nocturnus. But today, no less than in olden times, it still plays an important role in daily life. It is not "naughtiness," but can be explained by the fact that among man's phylogenetic ancestors the mothers carried their babies around with them, as is still the custom today among anthropoids and many primitive races. I observed mothers in India following the local custom of carrying their children astride their left hips while they did heavy physical labor (carrying stones in vessels on their heads). The mutual bond is so deeply rooted in mother and child precisely because it originated so very long ago. When the "breast infant" still lived on his mother's body and was fed, warmed, protected, and trained by her, any separation from her meant a mortal danger. As previously described, the "bed infant" who only returns to his mother for feeding developed out of this stage as a result of civilization.

Many children are restless at night even when they are older. In itself this is not tragic, but it can develop into a tragedy if adults interfere: Father has worked all day and wants to sleep quietly at night, so that the next morning he can return to work refreshed. If he is patient by nature — a virtue not found in everybody — he will perhaps acquiesce for the first or even the second night. But if this behavior continues and the baby disturbs his rest night after night, lack of sleep makes an otherwise peaceful man irritable and angry. The family drama which now develops in the dark of the night always proceeds in a definite sequence: First the child screams, then the father scolds, and finally the mother starts crying because she does not know what to do. This is just the appropriate moment for the neighbors to knock on the wall and ask for quiet. The uproar which thus arises is in no way suited to calm the child. The mother most deserves to be pitied, since everybody blames her inadequacy, and she herself may even have feelings of guilt. In such cases it is a pity that the adult's memory does not reach back into his own early childhood. An especially difficult situation arises in times of a housing shortage when parents or, worse yet, a mother alone, depends on the landlady's good will.

It has previously been stressed that a nightly upheaval may occur even in persons who are not nervous. But what may happen if we are dealing with neuropaths! Late one evening a crying mother appeared in my office with her baby; she had been thrown out of her house in all seriousness by her husband — a physician — because of the infant's screaming.

However, whether the child cries with or without recognizable reason, he will always be immediately calmed as soon as his mother picks him up. True, there is a theory that a crying child should not be picked up but should be left to cry. This approach is understandable in institutions, where it is impossible to carry each crying baby around. However, this has been carried over to the family, although it is against every mother's nature to disregard her child's crying. This crying worries her to the point where she can no longer bear it. Finally, despite all warnings, she picks the baby up and is happy to see how soon he calms down. Less sophisticated women who follow their instincts more than their reason often — even in our country — carry their children around with them.

By crying, the child is obviously trying to call his mother. During his first weeks of life he is not able to recognize her presence by sight or sound, as the older child does, but senses her presence only by the actions which she performs on him or he on her. Therefore, rocking, cradle songs, or nursing have such a calming effect because they convince the baby of his mother's presence. But these measures can be replaced by others which do not require the mother's physical presence. Such deceptive devices have always been used in baby care without their working having been entirely clear.

Even the newborn infant adjusts his behavior to quite definite stimuli. The animal also behaves likewise by responding to certain key stimuli on the basis of its innate endowment. The lock for the key, so to speak, is the innate scheme (term used by LORENZ, 1943, following VON UEXKÜLL), an expression for which TINBERGEN and LORENZ later substituted the term "innate releasing mechanism." According to LORENZ, the sham experiment in the still inexperienced young animal serves as a proof of this mechanism. In such an experiment we draw our conclusions from the fact that a purposely inappropriate stimulus provokes an appropriate course of responses; we gain our knowledge from the animal's wrong actions.

How this can be understood may be shown by the example of the cradle. Probably every nation has learned to calm its children with some sort of cradle. VON ALLESCH (1921) even observed a chimpanzee mother who from the second week on rocked her baby on her thighs or by patting it on its back with her hand; this procedure gradually satisfied the baby more and more. There is no better sedative. It is necessary to rock a healthy young infant in his cradle or in the arms or baby carriage only once when he is on the verge of crying: He immediately quiets down and starts to cry again as soon as the movement stops momentarily. He will surely not cry if it is done right.

How can the calming effect of rocking be explained? According to BUCKMAN (1894), rocking simulates the to and fro motion of branches and is reminiscent of anthropoid ancestors dwelling in the trees. VON PFAUNDLER (1915) explained it more correctly: The cradle is nothing but a replacement for the child's naturally moving bed on his mother's lap, in her arms, or on her back.

Thus, the cradle serves in the last analysis as a sham device, in the LORENZ terminology, which simulates the mother's presence. Since the child cannot yet recognize her with his eyes and ears, he must be convinced of her presence in another way. His position and movement senses serve him well to this end. The success of rocking does not depend on whether the child is rocked under natural conditions on the body of his mother, or in a cradle or otherwise. The young infant does not know the difference; he is calmed by the rocking as though he were in his mother's arms.

Some decades ago science was very successful in banning the cradle from the nursery for reasons which today are no longer deemed valid. PLOSS notes that it was even spoken of as the "cradle of idiocy." Thus, the cradle unfortunately became a museum piece.

The cradle song with which the mother sings her child to sleep is plain and simple. A few frequently repeated calming sounds suffice. Folk cradle songs are also popular. Serious songs also termed cradle songs are heard on the concert stage, where they are sung to a rapt audience by a singer with muted accompaniment, but not at the baby's bed. These cradle songs, whose artistic value naturally is not to be disputed, come from the adult's world and are therefore written and composed only for his world, not for the child's.

Just like the cradle, the still very popular nipple calms the child. Any oblong object with an end somewhat resembling the mother's nipple serves as an imitation of the mother's breast; this may be the well-known nonperforated foodless rubber nipple (pacifier) or a finger. The mere contact of a sham device with the oral region can be effective. To the baby it suggests sucking at his mother's breast and, by this, her presence. The infant remains restless as long as he has nothing in his mouth on which to suck, and calms down again as soon as he sucks on a suitable object; under natural conditions this can be only his mother's breast. It is impossible to cry and to suck at the same time. When the infant awakens and is still half asleep, he makes some sucking movements if the nipple is still in his mouth; then he immediately falls — satisfied, one could say — asleep again. But if the nipple is absent and his mother with it, so to speak, he becomes restless and cries even in this drowsy state. Certainly the nipple affects sensory areas which are quite different from those affected by the cradle, but the result is the same: the simulation of the mother's presence. Food intake, which also has a calming effect, is not simulated, for the hungry infant could be calmed only temporarily by a deceit; even the satiated infant who is accustomed to a nipple remains restless without it.

The young infant has yet another method of calming down: self deception; his own fingers may serve as an imitation of the mother's breast. During the first months of life there is a distinct tendency to grasp with the mouth and to put everything grasped with the hands into the mouth. If his fingers get into his mouth by chance, the baby sucks at them just as he does at the breast or the nipple. Not infrequently he awakens and cries until he has put his finger into his mouth. As soon as this is accomplished, sucking movements which calm him down occur.

LORENZ and TINBERGEN (1951) both referred to artificial key stimuli which trigger the innate releasing mechanism much better than natural stimuli. It is "possible to construct a sham device which surpasses the biologically adequate object in its trigger effect, in some cases even by several fold" (LORENZ). These sham devices are called supranormal objects (KOEHLER and ZAGARUS, 1937; TINBERGEN, 1951).

In the oyster catcher *(Haematopus)*, the key stimuli which originate in the egg and cause breeding behavior in the parent animal consist, among others, of the bluish color of the eggs and the contrast of the gray spots against this background. If such a bird is confronted during a selection experiment with a single bright ultramarine egg with spots as black as tar, ten times as big as its usual egg, the bird immediately chooses this sham device and clings to it, even though his breeding attempts fail because of the egg's size. LORENZ (1953) also finds types of exaggerated trigger stimuli in man, for example, in the "most damnable art of cooking," which attempts to create only supranormal objects, and in sexual life.

For the infant, the rubber nipple is such a supranormal object, a hugely enlarged breast nipple; the bottle's contents are also supranormal. Sweetened with sugar or saccharine, they taste decidedly better than mother's milk. It has long been known that the infant drinks from the bottle better than from the breast. It is as difficult to bring a bottle-fed baby back to the mother's breast as it is easy to change a breast-fed infant over to the bottle.

In using the nipple the infant is not concerned with the milk, but with the reassuring certainty that "everything is in order." The wordless dialogue between mother and child can be expressed in the following words: Half asleep, he calls his mother by his crying: "Mother, are you there?" Thus he does not call, "I am so lonely," or, "I am hungry," but demands his mother. In thereupon picking her child up, carrying him around, rocking him, feeding him, or giving him the nipple, she calms him in a way only he can understand, as if she said: "Here I am, my baby!" By calming down, the child answers: "Well then, everything is all right!" and falls asleep again. Since the mother has the same impression, the two sleep again in the reassuring certainty of being together until the infant again feels the impulse to assure himself of his mother's presence.

If he is in his mother's arms from the beginning, the wordless dialogue is correspondingly shorter: A sound from the baby — a movement or sound from the mother — and the dialogue is ended. The child wants only to reassure himself of his mother's presence. One sucking movement may be sufficient. As a late memory of this desire, sucking movements may occasionally still occur in older children until school age when, without an external cause or under the influence of sensory stimuli, they half awaken out of deep sleep.

For the chimpanzee baby, the relation to its mother is even more vital than for the human child. According to YERKES and TOMILIN (1935), every chimpanzee mother responds immediately as soon as her baby emits a sound. In their family, W. N. and L. A. KELLOGG (1933) simultaneously raised a son from the age of ten months to nineteen months and a chimpanzee baby from the age of seven and one half to sixteen and one half months. During the entire nine-month observation period the chimpanzee baby always tried to remain within seeing or calling distance of a friend and protector. It was much more dependent in this regard than the human child. Whether in the house or outside, it never went far away from people it knew. Locking it up alone or walking faster than it could follow and thus leaving it behind was the most terrible punishment possible. It could not be left alone without grief, whereas the infant often played alone.

According to MITCHELL (1912), the babies of carnivores wish most of all to have company. The loud cries which they emit when left alone call the mother to them; the mother calls her baby with a special sound which is totally different from her usual roaring or growling. Many carnivores, for example, pumas and Syrian cats, use their voices only during breeding time, mostly as calls between mothers and babies. PRECHTL and SCHLEIDT (1951) taped up a baby guinea pig's eyes. The baby remained where it stood and emitted a shrill whistling (voice-emotion sound). On hearing this, the mother animal regularly came to let the baby crawl under her body.

LORENTZ (1951) had an unusual experience with his "goose-child Martina." He was present when it hatched from a wild goose egg which had had artificial incubation at the end. Holding its head obliquely, it looked up at him with big dark eyes. As he moved and spoke a short word, the tiny goose greeted him: With neck outstretched and nape depressed, it rapidly uttered the polysyllabic voice-emotion sound of the gray goose. He now placed the little bird under a sitting domestic goose whose baby geese had just hatched. Immediately a soft sort of questioning whisper sounded from under the goose. The old goose answered reassuringly with the same voice-emotion sound in her intonation. But this was of no avail; the gosling rushed out from under the warming plumage, looked at the foster mother's face, and ran away crying loudly. As soon as LORENZ made a movement, the crying stopped and the gosling approached him with neck outstretched, greeting him joyously. Another attempt to place it unter the domestic goose had the same result. Thus, LORENZ had to take the gosling along with him. In the days following he could not leave it alone for a moment. If left by itself, it would be seized by such a desperate fear and would cry so pitifully that after several attempts he gave in and fashioned a little hanging basket in which he could carry the gosling with him. Especially at night it repeatedly uttered the voice-emotion sound which SELMA LAGERLÖF translated so perfectly into the words: "Here I am; where are you?" It could be calmed only by LORENZ.

A comparison between the "goose child Martina" and the human infant suggests itself. Instinctive knowledge of the infantile impulsive actions of its species is not innate in the baby gray goose; this knowledge is imprinted on it by a single impression, for which it is receptive only at a clearly determined moment. According to LORENZ (1935), this imprint of behavior patterns characteristic for a species has no equal in all of animal psychology.

According to HEDIGER (1953), however, it has been shown that behavioral patterns comparable to the imprints in birds may occur in definite autophagi among the mammals, especially in certain ungulates. The bison hunters of the 1860's on the North American prairie observed that the newborn calves of wounded bison mothers stuck closely to the heels of the hunter's horses, sometimes so closely they were even accidentally kicked. Similar experiences were reported in Africa with buffaloes, antelopes, and zebras. In regard to the behavior of young mouflon sheep, an "imprinted relation to man" is actually spoken of.

The human child is certainly not imprinted in the same manner as the gray gosling. Nevertheless, STIRNIMANN (1947) reported that a breast infant recognizes his mother's voice after two weeks. In his obstetric department the newborn infants did not sleep in their mothers' rooms. When the mothers visited their babies in the nursery at the end of the second week, the babies became restless and began to search with their heads and mouths. This surprised the mothers so much that they repeated the experiment with other infants; they received no response and therefore believed that their voices were recognized by their babies.

STIRNIMANN (1947) reports similar occurrences from CZERNY's old clinic: The mothers slept with their babies in the same room. If one of the breast-fed babies cried, only his own mother awoke; the others slept on undisturbed. ILLINGWORTH (1955) reported similar data. Although REMPLEIN (1952) states that the newborn infant is deaf, he reports elsewhere that plethysmographic examinations of the heads of newborn infants showed that even during the first 24 hours the infants responded with joy to their mother's voices; this response did not occur to the voices of other persons. The only plethysmographic examinations relating to this problem were made by CANESTRINI (1913). However, he reports that newborn infants cannot yet distinguish the voices of their mothers from other voices of equal volume. According to HETZER and TUDOR-HART (1927), also, the mother's voice is not distinguished from other voices by these infants. This question must therefore be left unanswered for the present. According to ANDRÉ-THOMAS (1952, p. 93), the infant does not recognize the voices of his parents before the third month of life.

Anyone who knows the right manipulation can calm a young infant by rocking. Later on, however, the child will scream even more if a stranger tries to come to his aid. By the formation of conditioned reflexes, his nurse's face and manipulations become familiar to him; he adjusts to the regularly recurring periods of the day in which he is cared for, and responds to everything unusual with violent indignation.

He does not have to earn his mother's love, but the mother has the extremely gratifiying task of earning her child's love. Certainly a stranger can completely replace the real mother, but this means that the mother has lost her place. Such a replacement is nevertheless far better for the child than a possibly repeated change of nurses such as unavoidably occurs in institutional care.

The infant quickly adjusts to the calming measures of his environment. He actually demands them by his screaming and remains restless if they are once omitted. The conditioned reflexes which are formed in this manner may become very annoying to the environment.

Of course, it cannot be asserted that the young infant is able to form any distinct conception of his mother's presence. On the contrary, the content of his consciousness is unknown. We can only observe that he acts as if he were calling for his mother.

In any case, the annoying faculty of crying is an important endowment which nature has given to the child for his protection. As long as he is still unable to shout for help, he can only cry or scream. A sick infant cries especially pitifully, differently from the healthy child — an important sign of illness — and by this he provokes in those around him an even stronger impulse to take him up. This crying is later replaced by the conscious call for help. There are still adults who call their mothers when in danger. A report by TEMBROCK (1949) shows what a chimpanzee baby can accomplish by its crying (see p. 127).

Certainly many a child "asks" for his mother more frequently than is necessary and than the adult would prefer. But it must never be forgotten that we are dealing here with a protective device which was once vital in man's past history. That civilization has meanwhile changed the conditions, that the question of the mother's presence has lost its urgency — this the infant has not yet noticed; in his bed he is still attuned to his mother just as much as he was as a breast baby.

As a bed infant the child must call again and again for his mother, since he is happy only in her arms. The former commonly accepted custom of letting the baby sleep in the same bed with his mother cannot be recommended today because of the danger that the infant may be suffocated or crushed to death by his sleeping mother. In the ancient times of the distant past, when beds and sheets did not exist, infants were just as safe with their mothers as the baby monkey is in its mother-nest (see p. 588).

Today it is commonly accepted that the baby should have his own bed. This was different in former days. According to SCHREIBER (1918), interdictions against taking children into one's bed are contained in the resolutions of the Synods of Worms (868) and York (1236) and in the edicts of Fritzlar (1243), Canterbury (1236), Helsingborg (1336), and Upsala (1343–1348). The Synod of Aquileja (1339) stipulated that adults were not allowed to take children less than two years of age into their beds. The aforementioned Synod of Canterbury even requested that the priest inform the women about this edict every Sunday.

As late as in the eighteenth century the child slept with his mother. Old reports tell how often children lost their lives in this manner (ROSEN VON ROSENSTEIN, 1764). The custom was so widespread — especially because it had such ancient origins — and the conditions so bad that legislators many times had to take measures against it. For example, as late as 1817 the general law for the Prussian States decreed, under threat of imprisonment or physical punishment: "Mothers and wet nurses are not allowed to take children under two years of age into their beds at night or to let them sleep with them or others."

Certainly every mother loves to take her child (who has become a bed infant) into her arms at times. But to cover greater distances, today she uses a baby carriage, which keeps the baby at a distance from her body. The baby rides backward, looking into his mother's face. The baby carriage is a recent invention. The ancient Greeks and Romans had none; it is not to be found in pictures of the Madonna. At the beginning of the nineteenth century, the baby was placed in a so-called "Boller carriage" (a wooden chest carriage with four disk wheels). The contemporary type was invented in England in the middle of the last century and rapidly became popular in all civilized countries, although it was originally indignantly rejected by pedestrians, who felt inconvenienced by it [PEIPER, 1951 *(117)*].

The child retains the ideal image of his mother far beyond infancy. The small child likes to call for her at night, for in her arms he is not afraid even in the dark. He becomes homesick for her only (see p. 617); when he is away he longs for her, not for his home, which he leaves unhesitatingly when she leads him by the hand.

The physician must have understanding for a mother's feelings. He must realize how much just the physical examination, necessary and harmless as it may be, sometimes hurts her feelings. That her darling is undressed, percussed, auscultated, and palpated, that the big fontanel ("the lifeline") is touched, and that he is made to sit up or even to stand up, shocks her; these are procedures which many a mother permits only with great misgivings in her firstborn. A widespread popular custom even forbids cutting the nails during the first year of life. The taking of blood samples and other minor procedures should never be done in the mother's presence. A mother who is made to hold her child during such a procedure will easily break down. But a mother who is apprehensive at first becomes more cooperative after she has had several children.

Exaggerated mother love that seeks to satisfy the child's every wish becomes a weakness. As a consequence, it is very difficult for many children to become used to any change in their way of life and they resist changes with all their strength. They do not want to be weaned; they do not want to eat pap or vegetables or to give up the nipple as a pacifier. Other children do not like to be left alone, or they can sleep only with the lights burning, or they do not want to be fed by strangers. This resistance could be overcome, but some mothers are unable to do it even if they are aware they are doing the wrong thing. In former times a mother would not infrequently nurse her child several times a day even when he was in school, or would be unable to make him eat any vegetables, or a child would receive mother's milk exclusively for years until he became severely ill and finally died. Such grave errors no longer happen today, since the principal rules of correct nourishment have become common knowledge. But traces of these errors may often be found even today. That we are dealing only with mistakes in training is proved by the success of institutional treatment.

The older the child is, the more difficult it becomes to satisfy all his wishes, since other conditions and other persons have to be considered to an increasing extent. School has an especially wholesome effect because it forces the individual child to adjust to the group. Classmates are not so considerate as a mother.

The deepest grief a mother can suffer is the loss of her child. The idea that her child could become ill or that her sick child could die worries her to the utmost, whether or not this concern is justified. She would prefer to have her child in her arms so that she can always feel, hear, and see him. Her worries are greatest in the dark of the night. In ancient times mothers' anguish invented terrifying nocturnal ghosts which would come to abduct, to mistreat, or to kill the child [PEIPER, 1951 *(117)*]. Thus, the old Egyptian magic formulas for mother and child (sixteenth century BC) consist mostly of exorcisms of terrible demons which abduct children. In the old Assyrian Labartu Texts (seventh century BC), similar exorcisms of the female demon Labartu, whose terrible appearance has been preserved in stone and bronze, are handed down to us.

For thousands of years, mothers have feared for their children; they still do today. It is not too much to say that the mother of a sick child is just as sick herself. Inner restlessness, loss of sleep and exercise, staying constantly in the sickroom, and insufficient nourishment weaken her body so that it can be noted in her appearance. In such situations the pediatrician always has two patients, and must also take care of the mother who so willingly and unsparingly devotes herself to her child. He cannot permit her to overtax her strength caring for the child and executing the physician's orders. Critically ill children with a hopeless prognosis, all else aside, should be hospitalized just to spare the mother the anguish of having to witness the prolonged process of seeing her child die.

Exaggerated worry about the child easily leads to a lack of confidence in the physician. If results are not immediately forthcoming, a second, a third physician will soon be consulted, and on it goes. The greater the worry, the more rapidly the physicians are changed. Thus an infant only three months of age may already have "used up" half a dozen physicians. That such a change is not good for the child is obvious. It is hard to convince such a mother that her child should be treated in a hospital, although this is often the only chance of helping him. Rather, she thinks her child – just her child – is something special, and would waste away from homesickness among strangers. If she has finally been persuaded with great difficulty to hospitalize her child, she will return in the afternoon or the next day to take him home because the pressing worries disturb her peace of mind. In other instances the child is taken home prematurely because the mother evidently becomes jealous of the nurse, who alienated her child's affections.

It is quite informative to observe how children of such mothers behave in an institution. They adjust very well without any trouble and without making any special demands; during the first days they may show the same temporary signs of bad humor many other children do, i.e., they are homesick. But as soon as the mother appears, the picture changes. The formerly peaceful child becomes restless, obstinate, and demanding. Although he may have readily let the nurse feed him until then, the same attempt now fails if the mother wants to feed him or if he is fed in her presence. Now he resists each spoonful and can be persuaded to take a few bites only with much effort. The more the child resists the feeding, the more excited the mother becomes; she sees her child already starved to death. This makes her more and more incapable of feeding her child. Because of these and similar reasons, many pediatric hospitals do not have any accommodations for the mothers of the children. The only disadvantage is that nursing mothers are also turned away, and that breast feeding is interrupted because of the baby's hospitalization.

It is a rare exception for the mother or the parents not to love their child. Even if he is born out of wedlock or in secret, and by his presence alone constitutes a severe indictment of his mother, mother love remains. Thus, a severe conflict occurs if a mother disclaims her illegitimate child.

The physician encounters exaggerated mother love incomparably more often than indifference or even worse attitudes. In peacetime, severe damage from wrong infant care or neglect occurs most often because the mother is slightly retarded mentally. And it is just those mothers whose mental incompetency is only slight who may greatly endanger their children.

However, even reasonable mothers may deceive themselves about their children's condition from lack of experience and seek the physician's advice much too late. Exaggerated propaganda for nursing may mislead the mother into believing that a breast-fed child is always healthy.

In their concern and their awareness of their own inadequacy, inexperienced mothers are easily influenced by their environment. Their insecurity facilitates the formerly prevalent, unscrupulous propaganda of manufacturers and salesmen of baby supplies.

Friends and relatives are prone to give well-meant advice, with which the pediatrician cannot always agree. Occasionally mothers are even accosted on the street by strangers in regard to their children. For example, sympathy may be expressed by someone seeing the child and exclaiming, "My, how bad your child looks!" It is evident that such remarks are not good for the child. The mother, seeing a reproach in this exclamation, goes to the physician. He increases her alarm even more if he makes one of those favored and usually unfounded diagnoses such as general weakness, weak lungs, cardiac valvular disease (instead of a nonorganic heart murmur), malnutrition (instead of leanness), anemia (instead of paleness), not to mention tuberculosis of the hilum glands (and the tuberculin reaction proves to be negative when checked). In such cases the mother, not the child, is sick. The physician therefore has the not always easy task of reassuring her when he has confirmed the baselessness of her worries.

FATHER AND CHILDREN

(100)

In the extremely intimate relation between a mother and her child, the close connection existing during pregnancy continues for some time. This is similar in all mammals; the mother always takes over most of the care of the baby, even beyond feeding proper. A relationship between a father and his offspring which even approximates this situation practically does not exist in animals. The prerequisite for such a relationship would be monogamy. Monogamy has been shown to exist in some birds, for example, in hawks, cranes, swans, and storks; it is thought possible only in approximately a dozen species of mammals, and is not known in others. To be sure, fatherly instincts occur in isolated cases in some male herd or other social animals. But many male beasts of prey, living alone, would eat their offspring if they were within reach. No male animal seems to have a clear idea of the connection between the satisfaction of its sex drive and its progeny. Therefore, it cannot comprehend what it means to be a father (GERLACH, 1942).

Monogamy and the concomitant bond between the father and his progeny is phylogenetically a very late achievement of man, and has no natural foundation similar to that of mother love. Thus, despite all the love the father might feel for his children, he is tied to them less closely than the mother.

It is therefore no coincidence that poetry praises mother love and mother joy much more frequently than the father's feelings, although most such poems were composed by male poets. Similarly, classic painting, done exclusively by men, has repeatedly portrayed the mother and her child with deep devotion; the father may appear in the picture as another member of the family, but hardly ever alone with his child.

However, the relationship of a father to his grown-up child is among the deepest human relationships, and therefore has frequently been the object of poetic representation. The tragic epic of Hildebrand has been preserved for us from the old German literature; it is the oldest German heroic epic. In many biographies, in dramas (King Lear, Don Carlos), in novels, and in humorous descriptions (W. Busch: Tobias Knopp), this relationship has been treated again and again.

The mother keeps house; the father earns the family's livelihood and represents the family to the outside world. His work usually forces him to be away from home during the day. Thus, the children spend much more time with their mother. From her they receive the first food and from her they learn all their early skills. She is responsible for their first training. It is even more significant that in German we speak of a "mother tongue" and a "fatherland." At least it was so in earlier times. Of course, the picture has changed considerably because of the increasing number of working mothers.

The father, whose name the family carries, usually desires a male child who will carry on the name and the family tradition. The mother, who herself has just recently accepted the new name, does not consider this question so important. The decision of the child's sex, however, is totally independent of the parents' wishes. Thus, many a father is very dissappointed if his hopes are unfulfilled – possibly even several times; yes, it may even happen that he does not want to look at his newborn daughter and that he reproaches the mother, an attitude, incidentally, which usually improves quickly. There is no basis for these reproaches, for a malicious fate causes the sex to depend on the father and not on the mother. Thus, only the father is at fault – if one can speak of being at fault. Most parents, though, are usually unaware of this biological relation. Nothing prevents them from trying their luck again and again. Thus, the wish for a boy has become the father of many a girl.

In China and India, the esteem in which a woman is held depends even today on whether she has borne sons and how many.

Birth is the mother's accomplishment. On this day everything revolves around her, and justly so. As pretentious as the father might be at other times, today he is not the main person. Nobody has time to be concerned about him or to inquire about his wishes. He cannot even complain, for his wife, who must always listen to him and must always have time for him, is not available to him today. But many a woman does not stop worrying about her family even in her most difficult hour.

If the advantages of a delivery at home versus in a hospital are considered, it has to be regarded as an advantage that the father who lives through the delivery at home gets a much deeper understanding of his wife's accomplishment than he can get from visits in a maternity hospital. This knowledge can be nothing but useful in the couple's mutual relations.

Now, finally, the child has arrived. If it is the firstborn, the father has probably never before observed an infant with care and attention; enthusiastic as girls or young women are about babies, the same cannot be said of young men. The sentiments of a father on seeing his child for the first time were excellently described by Heinrich Seidel in "Leberecht Hühnchen," from which I quote an excerpt:

"In regard to the admiration of newborn babies, I am a barbarian, as are most men. This was my son, even my first son, this frog-like reddish creature with the funny folds on his legs, and I loved him and was certainly proud of him. He was also able to cry beautifully and piercingly — he did this with all his soul and body — and he could kick with his little limbs, but he surely was not beautiful. Like all newborn infants he had little similarity to human beings. The eyes of women see differently, and when mother Lore had unwrapped him from his bundles she looked at him with an enthusiastic expression on her face and said with deepest inner conviction: 'A beautiful child, a real angel, and just like his father!' When Leberecht Hühnchen's father objected timidly, he was informed: That women see with different eyes then men, and this you must respect, for it is an expression of the magnificent faculty which God put into woman's soul, of that force which is higher than the mountains and deeper than the sea — it is called mother love."

Many a father who lives through the delivery with fear and apprehension has the feeling afterward that this important event, which has happily come to an end, should now be duly celebrated. Thus, happy as he is, he drinks until he is drunk, with the result that the midwife now has to take care of him, too — or has to fend him off.

In the nursery, the father at first feels quite uncomfortable, superfluous, and clumsy. He does not know what to make of all these things, which are quite familiar to a woman, and he hardly dares to look at his child, not to speak of touching him. If he attempts to take him in his arms, he provokes loud screaming. He does not know the simplest manipulations, and has to be taught that he cannot smoke in the baby's presence. Many a mother takes advantage of this opportunity to let him see clearly that her superiority in this sphere at least is incontestable.

Other fathers overestimate their knowledge in this regard. Unfortunately, the father's anxiety about his children often explodes in reproaches toward the mother. If the child coughs, she has not taken proper care of him and has not protected him against a cold. If he cries at night and keeps his father awake — and perhaps his mother as well — she has not trained him correctly. However, if the child is to be punished, the father objects.

The father who spends all day at work wants to enjoy and play with his children in the evening, a very understandable wish. If it is not too late, an understanding mother will consider this wish and schedule the children's day accordingly. The father, after his not always enjoyable work, wants to see happy faces at least at home and does not enjoy being an avenging angel in the evening for naughtiness which has occurred during the day. This easily causes differences of opinion between the parents.

That the mother is always together with the children has both advantages and disadvantages. If a small child is asked to whom he belongs he always answers, with absolute conviction, "To my mother." She is always preferred to the father. When the parents take a walk with the children, they always crowd on both sides of the mother and, if there are more than two children, fight for the privilege of holding her hand. The father may easily be forgotten if the mother fails to urge the children to do otherwise. Some fathers feel hurt by this neglect, instead of seeing it as a natural expression of the mother's position in regard to her children.

Although the children love their mother most, it is more difficult for her to assert her will and to make them obey. She is much more concerned with all the children's bad habits — the inclination to untidiness, to mention only one — than the father is. If the father is weak, gives in, and permits the children to do what the mother forbids, their esteem for her is decreased to a dangerous extent. They are certainly best off under the strong and sure guidance of both parents, even if some wishes remain unfulfilled.

Quarrels between parents in front of the children are always a heavy psychological burden for the children. Their belief in the highest power which exists for them totters, and there is nothing to replace it. Memories of such experiences will last all their lives. Thus, FRIEDRICH HEBBEL reports in the memoirs of his youth how "scenes of anguish" sometimes occurred between his parents, who otherwise lived in the best of peace: "I can remember no occasion when those scenes, although they never went to the extreme, were not the most terrible thing for me." A scene like this which occurred in the second or third year of the poet's life is one of his earliest memories.

If children hear their father reprimanding their mother or even abusing her, they may easily follow his example. Thus, the mother who has prepared the meal in a way other than the father expected may receive a strong and detailed criticism from him. It is not surprising if the children soon behave in the same manner, and become used to criticizing their mother's cooking! If parents and children start to exchange unkind remarks, the children have usually experienced similar scenes between their parents.

In a divorce, the mother is usually the person who suffers most. The children always suffer, for the disturbance of conjugal peace must endanger their equilibrium, too. It is better if they live only with one parent after a separation, instead of being dragged back and forth still more between conflicting feelings.

The father accompanies the sick child to the physician less frequently than the mother. If he comes alone with the child, he proves to be very clumsy at the ordinary tasks of undressing and dressing the child. Whereas the mother is satisfied if the disease is given a name and the necessary orders are given, the father's thirst for knowledge, especially if he is educated or even half-educated, is less easy to satisfy. He wants above all to know the origin of the disease and often asks more questions than the physician can answer in good conscience.

MOTHERLESS CHILDREN

The younger motherless children are, the more easily they are exposed to dangers to their lives. A child who is completely adjusted to his mother with his whole being will be severely damaged by even a temporary separation. If a separation lasts for any length of time the child will fare best if another woman takes the mother's place. An institution cannot replace a mother.

Homesickness

(92)

Homesickness was first described by JOHANNES HOFER (1678) under the name "nostalgia." In the famous book in which he made physicians acquainted with percussion, AUENBRUGGER (1761) mentions nostalgia as a cause for emaciation in young men impressed into military service. It was not mentioned for a long time after his description. In 1929, SCHWAB and MARBE took it up again. It has not yet been described in the handbooks and textbooks of pediatrics.

Homesickness is a depressive mood caused by a change in the usual environment. It is not alien to the adult, especially in early adulthood, but mainly it is a true children's disease. The younger the child is, the more strongly this depression will manifest itself. Only infancy is an exception, because during the first months of life the infant has not yet established such close relations to his environment that he would be greatly disturbed by a change if only his vital needs are satisfied. However, there are instances when a change in the environment has an effect surprisingly early. I saw a six-month-old infant in an institution who had completely rejected all approaches and laughed for the first time when his mother visited him after a week.

Children are almost always seized by homesickness at the beginning of their stay in a hospital. Of course, there are exceptions. Hyperactive defectives do not show the least shyness toward strangers, even in the doctor's office; neither do they show any signs that the new surroundings displease them in the hospital. However, peculiarly enough, some mentally healthy children also adjust to the new environment without any difficulty or conflict.

Homesick children are negativistic, ill-humored, and whining. At first they are either very restless or their urge to move and desire to talk are completely inhibited. When they are not weeping loudly, they sit mute and motionless in their beds, do not participate in the conversations of the other children, do not respond to any questions, but cry if one talks to them, play neither alone nor with others, and refuse to eat.

The children pass successively through the stages of rebellion, despair, and negativism. The adult can scarcely imagine the intensity of the emotions in a young child who is separated from his mother. His hunger for her presence and love is as great as his physical hunger for food. She is his whole world; without her he is lost. The two-year-old child does not yet know about death, but his behavior is the same whether his mother has died or is only absent (ROBERTSON and BOWLBY, 1952). The younger infant does not feel the loss so distinctly; even a stranger can easily calm him by rocking.

Young children always ask for their mother; after the sixth or seventh year of life they ask to "go home." For phylogenetic reasons, the mother represents home for younger children. The separation occurring because of hospitalization is therefore very painful for them, but they leave their home in their mother's company without resisting.

KLINE (1898) reported on a child who suffered from a homesickness-like disease if he had to stay home while his parents went on a trip. He refused to speak and suffered from vomiting, loss of appetite, and vertigo when he had to cross the room. All efforts to satisfy him were in vain. Thus, the child suffered from homesickness at home because of longing for those who had left home.

If children talk at all after the separation, they repeat — no matter what answer they receive — the question innumerable times: "When is my mother coming?" At first their sadness cannot be successfully dispelled, and later only for brief periods.

It is self-evident that the basic illness, especially if accompanied by fever or pain, makes their disposition still worse.

Unfortunately the child, especially at the beginning of his hospital stay, must undergo many tests which cannot be omitted, although they are not suited to increasing his confidence in the new environment. Tuberculin tests or taking blood samples, not to speak of spinal taps, blood transfusions, and other injections, repeatedly interrupt the child's rest. They make him suspicious and fearful even if the procedures themselves are not particularly painful. On the other hand, the child's behavior makes the physical examination and the various procedures most difficult. An exact neurological examination of small children can be managed only after they have overcome their resistive attitude. Intravenous injections are often unsuccessful because of this general restlessness.

The duration of the homesickness varies. Generally the bad mood is overcome after two to three days, but it may persist for several weeks. In rare instances the children may even maintain their resistive attitude despite a stay of several months' duration and independently of the basic illness. Homesickness is most easily overcome in the company of other children; it lasts longer in children who are in single rooms.

Conversely, according to NITSCHKE (1955), severe reactions of homesickness may occur when infants who did well during a long stay in the hospital return home to their mothers.

Homesickness has its origins in the fact that the new environment differs from the familiar surroundings. Every person's conduct and attitudes from morning to evening — and at night as well — are determined to a great extent by this constant environment. The contacts with other family members, the familiar objects in the room and in the house, the clothing, the food, and the bed cause a regular succession of concepts and actions, the disturbance of which is experienced as a more or less disagreeable sensation. Even in infancy, conditioned reflexes are formed — unintended and unobserved by the mother — in response to regularly recurring events. They are only recognized if reactions of displeasure occur by chance, through omission of the conditioned stimulus. An infant who is regularly fed by his mother will cry and refuse to take the bottle if a stranger gives it to him. In homesickness, many conditioned reflexes are severely disturbed by the change in environment because the conditioned stimuli are missing. In addition to these more or less unconscious processes, the longing for his familiar environment awakens in the child. The younger the child is, the more likely does this environment consist solely of his mother.

Just as in the case of the gray gosling described by LORENZ (see p. 609), the infant receives the imprint of his mother at home and sometimes that of a nurse in the hospital. If the imprinting figure is removed, homesickness occurs.

The change of nurses may deeply affect the child. The pediatric nurse HODEK (1957) writes:

"I remember an especially striking case. Whenever I met this child, a former charge of mine, I always had to disappear rapidly before the child saw me; otherwise, he could not be calmed for days and cried and called continuously for me. I was this child's nurse from his birth to the end of his first year of life; then a young kindergarten teacher took over. In a short time the previously lively creature had changed into a subdued child. He seemed to be petrified."

What homesick children need most is understanding love. Strictness and too much persuasion, although well meant, are not called for. Completely wrong are the unfulfillable promises so often given such as, "Mother will come tomorrow to take you home." It is best to give the impression of overlooking homesickness until the unhappiness disappears by itself. Very appropriate and to be recommended is the company of other children, who distract the child with their conversation and play. Quite often the newcomer forms an attachment to children rather than to adults. But it generally takes a certain length of time until the child is accessible to such influences.

Unwise parents – and where are they not to be found! – interfere with their child's adjustment and sometimes make it totally impossible. Most mothers agree only very reluctantly to a separation, because they anticipate their child's homesickness with fear and themselves suffer from homesickness for their child. Such a mother would therefore prefer to do the nursing in the hospital, in the strong conviction that nobody understands her child's special characteristics as well as his own mother. Particularly when we are dealing with errors in training methods which make a separation of mother and child necessary, the mother cannot be easily convinced of the necessity for hospitalization and of the inappropriateness of her visits. The child will not be able to adjust to the strange environment if he is continuously reminded of his own home. Therefore, the parents' visits, especially at the beginning of hospitalization, are always a heavy strain for the child. At a given moment he might have been content to play in his bed. Then he sees the mother enter, starts to cry, asks to be taken home, and cannot be comforted. After the original fever has subsided, the temperature charts often show marked "visitor peaks" on visiting days as an indication of the excitement family visits can cause. It is always especially painful for both sides to say goodbye. The mother cannot leave her child, either; she starts to cry and thus increases her child's suffering. Therefore, a visit is by no means a pure joy for mother and child.

The mother, who almost always makes her child cry during her visits, concludes from this experience that her child behaves the same the rest of the time. It is therefore good to let her see through a window how happy the child can be when he is not aware of her presence.

How greatly the child's behavior is influenced by his mother's presence may occasionally be seen during office hours when an obstinate young child has to be examined: As long as mama is present he resists the physician desperately with hands and feet, so that it is impossible to examine him. If the mother is sent out of the room, many children immediately calm down and submit to the examination without resistance, though not gladly.

The hospital, of course, cannot forbid parents to visit their children; as long as there are no special reasons, such as the danger of contagion, they must be permitted to come to the hospital. If errors in training bring the child to the hospital, an arrangement should be made with the parents at the time of admission – if they are at all amenable or open to suggestions! – to omit visits, at least at the beginning.

The question of how often parents should visit their children in the hospital is still undecided. It is difficult to reconcile the parents' and child's desire for daily and possibly extended visits with the orderly administration of a hospital.

At camp, the great new experience of being together with a group of children of the same age fills the child with pride and makes it easier for him to overcome his homesickness. It is very useful in this connection that the child is regularly occupied throughout the day, something which unfortunately cannot be accomplished in the same manner in a hospital.

Parents who send their children to a camp often have the understandable wish to visit them there and thus convince themselves personally of their wellbeing. They may even spend their own vacation at the same location. Appropriate warnings seldom succeed. The child, placed in a conflicting situation between his parents and his playmates, always has unfavorable reactions from these visits. The mere thought that the parents are near and can easily be reached greatly disturbs the child's adjustment. At camp he longs for his parents, and when with his parents he wants his friends. Thus, he cannot truly enjoy life and is dragged painfully back and forth, a process which often ends with his departure from the camp. To make visiting more difficult for the parents it is desirable to pick an institution not too near home.

The educational value of homesickness is so great that every child should experience it from time to time. At home he takes his mother's love and unending efforts somewhat for granted, and always demands them. He has known nothing else since birth. This is completely changed when he is away from home. Here he is no longer the center of the world around which everything revolves, but only one of many. Here he must first earn the love and respect of others, so it behooves him to make a good impression. It does not matter whether he recognizes all this consciously or not: Away from home, he realizes its importance and value for the first time in his life — a very wholesome experience!

Under the influence of being away from home, even spoiled only children adjust to the new community without too much difficulty; this is completely unexpected and may even be disappointing to the mother, since she thought she was indispensable. This adjustment makes the children free and more accessible to training than they had been at home, where they were used to getting what they wanted. Here there are no mothers or grandmothers, with whom the children can find refuge when something is not to their liking, who always agree, and who prevent a well-deserved punishment by their pleas. Instead, the ambition to do as others is aroused. For example, at camp there is no opportunity to make special wishes known in the common dining room. Neither is there a need. Thus, the choosy child enthusiastically eats the same dishes with which he found fault at home. Bad habits such as refusing to eat or bed-wetting not infrequently disappear in the hospital without the exertion of effort. At home they resisted all attempts at treatment.

When away from home many children are much more obedient than they are at home, especially at first. Of course, when the charm of the novelty has paled, the child's real nature reasserts itself. Not infrequently it can be noticed from the child's behavior in a hospital that he has already been there too long.

Disagreeable as homesickness might be, as a rule it does not have much importance. In the fortunately rare cases when it does not subside at all, it can endanger recovery no matter what the basic disease may be; a child continues to thrive only in an environment in which he is happy. In such cases, premature discharge from the hospital must be considered.

Pathologically excessive homesickness may cause serious difficulties. It has occasionally happened in a hospital that children, in their exaggerated urge to go home, escaped through doors or windows and ran home in their nightclothes or otherwise insufficient dress. HOMBURGER (1926) described the case of a ten-year-old boy who was barely prevented from committing suicide when he became homesick after being admitted to a hospital. Such extremely unpleasant events, though rare, warn us to be careful. Children who have escaped once should be refused re-admission if their condition permits, since experience shows that their attitude will remain unchanged.

Since puberty frequently coincides with the end of schooling, many children have to leave home to become apprentices or domestics just at the difficult time in which their mental state is already quite unsettled. They suffer particularly from homesickness and have great difficulty in overcoming it. In the belletristic literature (HERMANN KURZ: "The Pale Apollonia") and in the psychiatric literature (JASPERS, 1909), we meet the infantile young housemaid who commits a crime such as killing the child under her care or setting fire to the house just to enforce her return home. The young man away from home and the newly-wed bride are in situations which favor homesickness even in the adult.

RÜMKE (1941) and LINDNER (1940), who hold opposing views, base their opinions on observations on adults. RÜMKE would differentiate real homesickness, the longing for the accustomed environment to which a person belongs, from a number of situations or conditions among which he includes the tie to the mother as a representative of the environment. LINDNER, however, opposes this view; for him homesickness in its true pattern is not connected with a longing for home or homestead, but instead is always tied to a person, above all the mother, if its deepest causes were to be successfully determined. Wife, fiancée, sweetheart, or sister may take the mother's place. The homesick person's most pronounced trait is his tie to his mother, the impossibility of living without her or her image. LINDNER's view is doubtless applicable to the homesick child: Here homesickness is exclusively longing for the mother, for "home" itself is at first an incomprehensible concept.

In 1830, JACOB GRIMM gave his first academic lecture – still in Latin – at Göttingen on homesickness, "desiderium patriae." He stated that the emigrant is constantly drawn back to his homeland by an unconquerable longing. In our native land, we have acquired the most felicitous security for all our life situations, and our abilities and skills in almost all our transactions and enterprises spring in turn from this security. Here our steps and strides are securely grounded; on foreign soil our feet stumble easily. From infancy, our phantasies have been nourished by the sagas and the history of our country. Our indelible memories are connected with home; even the graves encourage us to emulate the virtues of our forefathers. But in nothing is the link of the love for our country more evident than in the common language.

JACOB GRIMM's words remain beautiful and correct even if they cannot yet be applied to the narrow world of thinking of the small child, who frequently suffers from severe homesickness. Since homesickness occurs so early in life it must be more deeply rooted than in rational considerations. "Man can be homesick for the most barren region and the poorest place" (HOMBURGER, 1926). A three-year-old child returned after a vacation of several weeks to his mother's poor garret and exclaimed: "Mother, it is most beautiful where you are!" (KIENE, 1953).

Even mistreated children long to get back to their tormentors (HETZER, 1936). As BOWLBY (1952) writes: "The attachment of children to parents who could be called bad parents by the usual standards is a continual source of astonishment to those who want to help the children." They long for their parents of their own free will, because despite everything they are attached to them. This instinctive feeling of belonging can be influenced only with difficulty or not at all by experience or reasoning. The relationship of a child to his parents (especially his mother) can rarely be destroyed by mistreatment, particularly in early childhood. "Some children suggest a comparison with the beaten dog that returns again and again to the people who chase it away" (HETZER).

McCANN (1941) collected the literature on homesickness. The real literature, though, consists of the innumerable songs with which poets greet their homeland from far away, and of the youthful memoirs and novels in which they depict the image of homesickness. In these literary efforts homeland and youth, glorified by memory, blend together.

INSTITUTIONALIZED CHILDREN

(1, 117)

Under natural conditions the infant's environment consists of his mother; for the young child the expanded environment is his family and his home. In contrast to this natural situation, the removal of young children from their environment and their crowding in institutions is unnatural and may become quite dangerous if no special precautions are taken, since the children do not do so well in such institutions as they do at home. This difference is attributed to so-called "hospitalism." This term comprises damages caused by institutionalization, which to be sure have completely changed in character and effect, but cannot be completely avoided. Younger children are more imperiled than older children.

In times past, damages caused by institutionalization have cost the lives of many infants. Particularly notorious were the foundling homes, which offered mothers an opportunity to get rid of their illegitimate and even their legitimate children through the turnstile box without being identified. But despite all good intentions, the separation of mother and child usually caused the death of the children [PEIPER, 1958 *(117)*].

For a long time the task of raising infants in institutions could not be accomplished at all satisfactorily, since the mortality of the children admitted was alarmingly high. In 1858, the State Foundling Home in Prague thus had a mortality rate of 103.1%. This figure is explained by the fact that not only all 2831 newly admitted children died, but also those surviving from the previous year (STEINERT, 1921). According to the statistics of the Charité-Annals [*17*, 106 (1892)], 174 of 176 children suffering from diseases classified as "atrophia infantum" or a "debilitas vitae" [tabes mesenterica] died in HENOCH's Berlin University Pediatric Clinic in the fiscal year 1890 to 1891. The two remaining children by no means recovered, but were reported as still in the hospital at the end of the fiscal year. FINKELSTEIN (1938) saw 70% of the infants admitted die in the same clinic, then under HEUBNER, in the year 1894—1895. He was able to save the remaining infants from death only by returning them to their families as quickly as possible. In 1898 he was still of the opinion that the mortality in pediatric hospitals under the best conditions could fall below 40%, but this low figure would be exceptional.

In 1909, SCHELBLE reported on an institutional mortality of 90% in originally healthy infants who were institutionalized only for shelter. When these infants were placed individually in homes with ordinary foster mothers, only 10 to 15% of them died. From 1903 to 1906, SCHLOSSMANN (1913, 1920) observed a mortality of 71.5% in 165 infants in a children's home, i.e., in really healthy infants, although the home was kept clean and the nurses had a great sense of responsibility. After a thorough reform in infant care, the mortality fell to 16.8% from 1907 to 1912.

It was often felt urgently necessary to abandon altogether any dangerous attempts to treat infants in hospitals. CZERNY (1939) tells of an announcement which was posted at the entrance of the Emperor Franz Joseph Hospital in Prague at the end of the eighties of the last century: "Infants are admitted only in exceptional cases and only with the special permission of the Director." This permission, however, was never granted. Even as late as 1903, NEUMANN declared the treatment of sick infants in general hospitals to be hopeless. There were even university pediatric departments which did not dare to admit sick infants. The mortality was said to be as high as 49%.

The cause of the high infant mortality can be deduced from a description by KUSSMAUL (1922), who visited Vienna in 1847 when SEMMELWEIS was collecting his experiences on puerperal fever:

> "In the foundling home near BEDNAR, there was ample opportunity to study the most varied and worst forms of septic disease of early infancy. Fortunately, I rarely saw them in children kept at home. As long as the means of fighting sepsis were only insufficiently known, these institutions, founded with the best of intentions, were not much better than murder dens."

Before antisepsis and asepsis existed, before patients with infectious diseases were isolated, children in institutions were the victims of numerous epidemics. The absence of any theory of nutrition and milk sanitation was just as fatal for children in infancy; such a theory was not formulated until this century. Today we have to admire those physicians who did not lose their courage, but continued their efforts until they were finally successful.

The obvious comparison between the effects of breast and bottle feeding on the infant's thriving pointed to the value of proper nutrition. Gradually, formula nutrition was improved. With the expansion of the well-baby clinics, in which the scientific advances of the theory of nutrition were made available to great masses of the population, the mortality during the first year of life fell from 22.6% in Germany in 1900 to 4.1% in the German Democratic Republic in 1959.

The raising of babies in institutions no longer presented fundamental difficulties when buildings suitable for this purpose were erected and staffed with trained physicians and pediatric nurses, strict cleanliness was observed, contagious cases were isolated, and milk kitchens were established.

Despite these advances it was never overlooked that in institutions, too, it is best not to separate mother and child, but to admit the mother so that she can nurse her child. Breast feeding, which can nearly always be continued under even halfway favorable conditions, gives by far the best results. This is of special importance for children coming from poor economic environments and particularly for illegitimate children, whose mortality is always higher than that of legitimate children. Therefore, special institutions to admit unmarried mothers together with their children were established. It frequently happens, however, that the mothers leave these institutions too early or that the infants are admitted without their mothers, which is even more unfavorable for the babies. This shows that even under improved conditions certain drawbacks still exist for institutionalized children, a situation which has always been called "hospitalism." These drawbacks have always existed, but can be decreased by proper measures. They are pronounced in times of economic stress.

There is even a contradiction in the term "suckling home,"* for an infant depends on his mother for sucking, and when the mother is absent the "suckling home" offers only an imperfect and unnatural substitute. Even the great successes of the scientific nutritional theory could not change this fundamental fact. Without his mother, the baby in the infant's home at best obtains milk from a wet nurse temporarily, but sooner or later he will receive cow's milk formulas. Although today the damage is no longer as bad as it was in former times, breast-fed babies will still do better than bottle-fed infants.

Hospitals function in a similar fashion to baby homes; they admit sick infants, but usually do not have facilities for nursing mothers. Thus, mother and child are separated at a time when the sick infant most urgently needs mother's milk. In big cities there is a way out, since the mother usually lives in the same city and can either express her milk or come to the hospital for nursing. In smaller towns this makeshift arrangement is rarely possible, since the mothers often live in the country. The deficiencies are obvious and can be remedied only by making provisions in all pediatric hospitals for the admission of nursing mothers. Even if this is not possible in every single case, it could be done more often than is done today.

In many cases, the facilities for isolating patients still leave much to be desired. Despite these reservations, the results in institutions are incomparably better than they were in former times.

Then came World War II. It increased the dangers for institutionalized children by the air protection measures, which jeopardized regular care and spread contagious diseases. The year 1945, with its tremendous population migration, brought another

* "Säuglingsheim" ("suckling home") is the German term for an institution for infants – The translators.

turn for the worse. The institutions were soon filled far beyond their capacity. Mortality rose again, reminiscent of the past decades of hospitalism. This was caused by the poor condition of the newly admitted babies and young children, who had not been well cared for during the flight and had had insufficient food, sometimes no food at all for several days, because breast feeding had ceased completely. Contagious diseases of all kinds, which spread rapidly under the overcrowded conditions and attacked the nurses as well, also had unfavorable effects; the rapidly dwindling supply of disinfectants, linen, preparations for skin care, and fuel was also a contributing factor. Another disastrous circumstance was the continuing deterioration of the quality of milk, if it was available at all, and the exhaustion of the supplies of powdered buttermilk. The conditions improved only when the stream of refugees abated, the overcrowding of institutions ceased, and a supply of high-quality milk could again be obtained.

Even under the most favorable conditions, children raised in institutions are at a disadvantage compared to children raised at home. According to ERIKSSON (1925), they are retarded in their physical growth and development, they are smaller, thinner, less muscular, more limp, and do not look as well as other children do. Their mortality from measles, diphtheria, pneumonia, and alimentary disorders is increased. ERIKSSON explains this form of institutional damage as arising from the danger of contagion and the uniform and no doubt unnatural mass rearing of the babies. VON PFAUNDLER, from whose institution this publication originated, sees the evil's main point of attack not in the body, but in the child's soul, while the vegetative sphere constitutes the responding organ.

The baby needs his mother's care not only because he must be kept clean and well fed, but also because he has to be surrounded by an environment of friendly affection. BERSOT (1926) writes:

> "The mother's love is the indispensable stimulus which permits the infant to awaken to a life of normal relations. Without his mother, or a person who replaces her, the infant remains sad and indifferent, his muscles develop for walking only slowly, and his face remains expressionless for a long time."

Therefore, the development of cerebral function is impaired in institutionalized children. CZERNY, FINKELSTEIN, FEER, VON PFAUNDLER, BIRK, STEINERT, and others have described hospitalism in detail. At home the mother is always available to the child. She feeds him, satisfies all his needs, rushes to him at every sound, and always has the time and the desire to occupy herself with him, and to play or joke with him. She understands her child's every emotion; rapidly, long before he himself can talk, the child also learns to understand what the mother says to him. Mother and child are so intimately attuned to each other because it is the mother alone who always cares for her child. However, in an institution the nurse always has to care for several children who take up her time one after the other and often are even placed in different rooms. When the institutionalized baby cries, the nurse may be occupied elsewhere, she may even not be present at all, or she may not hear his crying because of the general noise in the room. In addition, several nurses share in the care of the same child in the course of day and night shifts. Nurse and child therefore cannot be attuned to each other as intimately as the mother and her child.

In order to get her work done at all, the pediatric nurse has to schedule her time. She must first take care of her charges' basic physical needs by feeding them, changing their diapers, bathing, and dressing them. This gives her plenty of work, so that she is able to play with the children only occasionally and incidentally. At the most, some "pets" who are to be found in every institution come closest to the position of children at home, insofar as the nurses occupy themselves with them beyond the performance of their basic duties.

The formula-fed infant receives an exactly prescribed amount of food in institutions; thus, his meals are the same size every day, whereas the amounts of food a breast-fed infant receives vary from meal to meal in accordance with his changing needs. Adapting the meals to the individual need is even impossible in institutions. Furthermore, many nurses do not even know that such uniformity is unnatural. PAVLOV demonstrated the degree to which the secretion of saliva and of gastric juice depends on the conditions under which food is served.

As has been previously described (see p. 119, 476), ALDRICH, SUNG, and KNOP (1945) determined that newborn infants in institutions cry much longer and more frequently every day than babies at home. When the baby care was rearranged so that each infant received more attention, the periods of crying diminished considerably. This observation corresponds to one which showed by exact measurements that children sleep less in institutions than at home (see p. 494). All these observations show how difficult it is in an institution to replace the mother, who responds to all her child's needs.

The infant is not a machine which functions the better, the more punctually it is serviced. Rather, he wants to eat more one time, less another. Sometimes he wants to be stimulated, other times to be left alone. Usually he is cheerful and gay, yes, he needs happiness to do well physically; on the other hand, he is grouchy and negativistic at even the slightest indisposition. Certainly the child's every whim should not be satisfied – this would be extreme exaggeration in the other direction – but a certain indulgence of his special wishes is useful.

That even the young child needs stimulation is evident in a fairy tale* reported in the chronicle of SALIMBENE VON PARMA (thirteenth century): Emperor Frederick II wished to find out what language types and patterns boys would have after growing up, if they had never before talked to anybody. He therefore told the wet nurses and nurses to feed, bathe, and clean the children, but otherwise pay them no attention or speak to them. He wished to determine whether they would speak Hebrew, the oldest language, or Greek, Latin, Arabic, or perhaps the language of their parents. His efforts were in vain, since all the boys died, "for they could not live without the clapping of the hands and the cheerful faces of their wet nurses."

Lack of stimulation has an even greater effect in young children after infancy. At home there is always something going on and life is never so dull and monotonous as it is in an institution. Parents, siblings, other relatives, and visitors bring more than enough entertainment and do not let boredom develop. Outings mean new and joyfully welcomed events, natural for the adult, but for the child, experiences

* For a similar fairy tale, see p. 6.

in a never dreamed of colorful world. A dog is really an experience for him; he sees it, strokes it, hears it bark, and becomes frightened and fearful of it. However, the child in an institution knows a dog only from pictures or as a toy. He spends his life in an unchanging environment which, with its regular schedule and continuously repeated daily routine, offers nothing new to him. The furniture in the room may be

Table XIV: Number of Social Contacts in the Course of a Day in Institutions and at Home (according to STURM*)*

Age (in years)	Institution	Home
1	25	137
2	62	261
3	169	243
4	178	279
5	133	263
6	177	226

Table XV: Number of Objects Which a Child Handles in the Course of a Day (according to STURM*)*

Age (in years)	Institution	Home
1	8	24
2	14	71
3	32	55
4	25	72
5	29	57
6	22	51

the most practical, but the room lacks the many objects which a room at home contains, and in its bare plainness it is sometimes more reminiscent of a prison than of a nursery.

There are statistical data for these differences between life in an institution and life at home. STURM (1932) found the number of social contacts to be considerably less in an institution than at home (Table XIV). The disproportion is still more marked when the type of social contact is compared: Short commands predominate in the institution, thorough answers to questions at home. The difference in the number of objects which a child handles in an institution or at home in the course of a day corresponds to these findings, as STURM also demonstrated (Table XV).

According to GRASER (1957), the best development of a child up to his third year is possible only with individual care:

"A child in that age group grows up on his mother's apronstrings and expands his life experiences within her sphere and protection. The learning of independent movements and the miracle of learning to speak as a rule occur undisturbed only if the child is intimately linked to his mother."

Since the lack of stimulation affects all institutionalized children rather equally, their mental backwardness becomes evident only if they are compared with children raised at home. In these comparisons, ERIKSSON (1925) found on the average a deficit in the acquisition of locomotion, speech, and toilet training. Using SCHWAB's tests (1925), H. W. MÜLLER (1925) conducted comparative intelligence tests on a total of 154 children raised in institutions and at home up to the third year of life, and found a constant relative retardation of the institutionalized children after the third month.

GRINDL's results are especially impressive. Using CH. BÜHLER and HETZER's tests (1932), he examined the developmental age of three groups of children: children in well-run institutions, children in good foster homes, and children who had been removed from their homes by the welfare department and transferred to institutions or foster homes. In these investigations, it was found that the institution affects the development even more unfavorably than living at home with parents whom the welfare department had deemed unsuitable (Table XVI). The children in institutions were retarded by one to two months compared to the children living in unsuitable parental homes, and these again were one to two months behind the children

Table XVI: Test Achievements in Percentages (according to GRINDL*)*

Children from	Age	
	1.3–1.5	1.9–1.11
Institutions	47	46
Unsuitable homes	75	80
Good foster homes	94	88

in good foster homes. The difference in development of a child in an institution and a child in a good foster home was from three to four months.

BOWLBY (1952) had similar experiences: A bad parental home is basically still better than a good institution; there is no perfect substitute for the parental home.

SPITZ reported on the development of children in an infant's home for imprisoned mothers; every mother cared for her own child and had few other duties. He compared the findings there with those in a foundling home where the children were cared for by pediatric nurses. While there were no infant deaths in the infant's home, the mortality rate was high in the foundling home, although sanitary conditions were approximately equal in both institutions. The children in the foundling home seemed to be depressed, indifferent to their environment, cried when anyone came near them, ate poorly, lost weight, did not sleep well, and were easily taken ill. These phenomena were very pronounced after they had been separated from their mothers for three to four months. If they remained in the institution, they displayed a frozen rigidity, wide-open and expressionless eyes, and autoerotic activity. If a motherly person entered their life, they improved again. Children in the second half of their first year of life were very susceptible to such damages.

If admission to an institution occurs too early or the children remain too long, their minds may be impaired.

VON HARNACK and OBERSCHELP (1957) found impairments from hospitalization in 20% of infants six to twelve months old, in 55% of young children after infancy, and in 6.5% of children of school age. Young children after infancy were particularly susceptible; they experience the hospital stay as an intrusion into their world. The disturbances usually decrease in intensity after several days, at the most after two to three weeks, and leave no permanent damage.

Lack of stimulation causes a decrease in the attention which is a prerequisite for the gain of new faculties. The child in an institution is quieter, more indifferent and has fewer friends and less capacity for suffering than the child in his family environment. Foolish pranks, which – let us be honest – every child should play once in a while, are hardly feasible in an institution and also are most undesirable there. Unfortunately, space is often also so restricted that the motility drive cannot be adequately satisfied. Thus, the institutionalized child shows a paucity of movements. In addition, there are special motor disturbances which appear particularly often in institutionalized children; however, they may occasionally also occur at home if intellectual stimulation is completely absent so that the child's management is similar to that of an institution. Children suffering from these disturbances are usually characterized by a certain dullness; whether this is congenital or acquired remains an unanswered question. Mental retardation is by no means a necessary prerequisite, although mental defectives are frequently afflicted in this fashion. In many cases, these children later develop in a completely satisfactory manner.

Nocturnal head shaking (jactatio capitis nocturna, ZAPPERT, 1904) is one of these disturbances. This consists of frequently repeated rhythmic movements in the lying or sitting position. In the lying position the head, and frequently the upper half of the body also, is alternately rolled to one side and then to the other; in the sitting position the upper half of the body rocks forward and backward. The movements appear before falling asleep and disappear during deep sleep, but recur when the child is half awake. They disappear temporarily if the child is disturbed. Young children after infancy are particularly afflicted. BAYER (1956) observed this phenomenon even in children six months old. It becomes more infrequent at about the tenth year of life. A pediatric nurse whom I knew persisted in this childhood habit until adulthood. BAYER observed rocking, playing with the curls on his forehead, and thumb-sucking in a boy; the boy's father had had the same habits until his tenth year of life.

Head shaking is an expression of the insufficiently satisfied motility drive and is therefore frequently found in institutions. One evening in 1920, I visited the small children's ward of an orphanage after the children were in bed. At least half of the children were shaking their heads to and fro. This was one of the institutions where nothing was done to stimulate the children intellectually; toys were completely absent. Feeble-minded children are particularly subject to head shaking.

Similar stereotype movements may appear in adults if they are restricted to a narrow space. Thus, the sailor walks to and fro aboard his ship; so does the prisoner in his cell. According to KRETSCHMER (1926), "relaxed daydreaming" at times causes beautiful motor stereotypes such as rocking on a chair, finger drumming, thumb wiggling, and monotonously repeated doodling.

Corresponding phenomena may be found in many animals which are kept in captivity and are thus very much inhibited in their movements. HOLZAPFEL (1939) described compulsively repeated movement sequences (running back and forth along the fence, running in circles or in figure 8's, jumping up and down in the same spot) in bears, elephants, wolves, jackals, foxes, and other mammals. These phenomena also occur frequently among our domestic animals. One such pattern, for example, is the "weaving" which frequently occurs in horses: The animal suddenly begins to make swinging to and fro movements with its head in which the neck and upper half of the body also participate; the front legs are held in a somewhat straddled position (HOLZAPFEL). The canary, which hops untiringly back and forth from one bar to the other in its narrow cage, should also be counted among these examples.

Nocturnal head shaking is a disturbance which occurs at the time of falling asleep, but spasmus nutans (see p. 66) is a light-deprivation illness caused by prolonged confinement to dark rooms. It is especially bad if the light source is in back of the child's head. Spasmus nutans is characterized by a shaking or turning of the head in limited excursion, usually accompanied by nystagmus. This restlessness is continuously present in the fully awake child in the erect posture, but subsides during sleep or when both eyes are closed. It develops in the course of several months and gradually disappears again when the light conditions have been drastically improved. FINKELSTEIN (1938, p. 567) noted this disorder frequently in institutions.

In basement apartments, which were formerly customary, spasmus nutans must have been a frequent occurrence (RAUDNITZ, 1897, 1911, 1918). I even observed it frequently 40 years ago in young Berlin children, but since then I have seen it only rarely. If it occurs only once, it may easily not be recognized. It should not be placed in the same category with nocturnal head shaking (WEISSE, 1957), for in every case it can be clearly distinguished from this disorder.

Another motor disturbance frequently seen in young children in institutions is tonic immobility. This phenomenon has been described previously (see p. 505). Like nocturnal head shaking, this disorder is also found predominantly in institutionalized children; however, it may also occur in dull, sluggish, and inert children who live with their own families.

The younger the child, the more evident it is that defective physical development retards the child's intellectual development. Conversely, however, a disturbance of the emotional balance may also inhibit the physical development. Even homesickness may disrupt food intake. Since it usually disappears rapidly, further consequences fail to appear. But this is different if there is a lasting lack of harmony between the child and his environment. Every small child demands gratification of his various needs. He not only wants to receive the food he likes when he is hungry, but in addition he wants a time to be merry, to be entertained, to practice his abilities, and a time to sleep; he wants to be comforted and reassured, taken up or placed in his bed, he wants to play, be taken outside, or be left by himself, everything at its proper time. Certainly the environment cannot and should not satisfy all his wishes. He has to be trained to adjust to a household schedule. An institution, however, can adapt itself only to the average needs of the children. Hence, some will not be satisfied, since their special wishes cannot be recognized and therefore cannot be fulfilled in the large institution.

A lack of harmony between mother and child may arise even at home. If this happens, the child – whether suffering from a nutritional disorder or not – is brought to an institution; here he does well without any difficulties, but after his return home he again no longer makes any progress, although no damaging influence can be determined. Rather, the mother does her best to give exactly the food ordered, and may even have brought a supply along from the institution so that no change would occur in this regard. Despite all this, the mother has to return the child to the institution, where he again does as well as before. These changes to and fro may have to be repeated several times until the child is older, and has no problems in doing well at home too. Usually in such instances we are dealing with nervous mothers and children who do not get along with each other at home.

A similar observation may also be made in an institution. An infant who has not made any progress in one department for some time does well without any difficulty after having been transferred to another ward – perhaps after a short period of adjustment. The nurse and the child are not always able to find the right relation to each other. Occasionally, the mother may prove superior in her approach to institutional care: A child who seems to make no progress in a hospital may finally be discharged prematurely and do better than expected at home, although the mother does not seem to be intelligent and knows nothing of baby care. The ability to care for infants is not solely a question of intelligence or education, but depends considerably on an innate ability for empathy. Naturally, I do not wish to advocate ignorance in infant care with this statement.

The physical separation of mother and child practiced in maternity hospitals today is also absolutely unnatural. SIMSARION and MCLENDON (1945) stated correctly that infants today are born twice, the first time in the delivery room and the second time at home after they are discharged from the hospital. Until that time the mother has no opportunity to become really acquainted with her child. As a rule he is brought to her 24 hours or even later after birth, and he is so bundled up that she can only see parts of him. His visits always last only a short time: 20 minutes at 4-hour intervals. She is assured on discharge that her child is healthy and strong, but she has never seen her child's body. She does not know how to recognize the early symptoms of pain, satisfaction, displeasure, hunger, or satiation. Every mammalian mother responds to its newborn's cry for aid – only the human mother in an institution does not. Visits so limited in time are of no help in this situation. Thus, today's maternity hospitals do not satisfy the needs of mother and child, who long to be together.

Since the damages from institutionalization depend on the building, the equipment, the administration, the available economic resources, and last but not least, on the sympathetic attitude of the nursing personnel, they vary in different institutions. The more the institution resembles a home, the better the ratio of nursing personnel to children – an economic question – and the better the nurses are able to adapt themselves to the individual children, the more limited such damages will be. But as yet it is impossible to avoid them completely.

Grossly Neglected Children

(136)

Anyone who is acquainted with the child's natural environment must be surprised by the recurring reports that a young child has survived in a completely different setting, perhaps even without human assistance and by its own power. It is unimportant that old fairy tales and new fiction tell about such events. However, many serious reports from centuries ago as well as from contemporary scientific reports affirm such occurrences. From these reports two groups of children can be distinguished:

1. Children who do not obtain the needed care at home because of gross negligence. I call these children "Kaspar Hauser children" after their best-known representative.
2. Children who have had no human home and have been raised by animals. Among these feral children, "wolf children" are especially prominent; thus, I name the whole group after them.

Kaspar Hauser Children

Gross negligence inhibits the physical and intellectual development in their mutual interdependence. Since ancient times, the most important of these problems has been the question of intellectual development. It was hoped that the observation of such children would reveal which components of behavior depend on innate endowment and which are acquired from the influence of the environment. The fairy tales belonging to this category reported by HERODOTUS and SALIMBENE VON PARMA have been mentioned previously (see pp. 6 and 627).

Science encountered the question of innate behavior in all its importance for the first time in the mysterious person of Kaspar Hauser. He was born about 1812 and appeared in Nürnberg in 1828 (VON FEUERBACH, 1832, 1852; WAGNER, 1928, 1931; E. ENGEL, 1931; PIES). According to his later statements, as far back as he could remember he had sat on the floor of a dark room with a low ceiling, without hearing a sound and without seeing a human being. He said food—bread and water—was placed next to him while he was asleep.

Kaspar Hauser attracted great attention by his report. He died in Ansbach in 1833 of a deep stab wound which caused injuries to his heart, stomach, and liver. Despite unending efforts, the riddle of his origin could not be solved. Sometimes he was taken for an impostor and sometimes as the heir apparant of Baden, who had been cheated of his heritage. That he had big vaccination scars at a time when vaccination was not compulsory is in itself not exactly a sign of neglect, and gives food for thought. It is also uncertain whether his life ended by murder or suicide. In any case, the great volume of work by the "Kaspar Hauser investigators" is out of proportion to the meager results. Science did not gain any new knowledge from Kaspar Hauser's appearance.

JESPERSEN (1925) reported on twins who in 1903 were grossly neglected by their mother and who were completely left to themselves. While their mother was sick they lived with an almost deaf woman, who did little for them. Not until they were four years old were they admitted to a children's home, where they received good care. There they learned to speak Danish, but between themselves they talked without constraint in a completely incomprehensible gibberish. Most expressions were mutilated Danish words. Thus, it was not a new language. "If they were actually able at the beginning of the twentieth century and in a civilized country to create something which deviated so greatly from the language all around them... then children left even more to themselves, in an uninhabited area where they do not immediately starve or freeze to death, should be able to develop a language for their mutual understanding."

According to MIERKE (1955), a mother in a village in Holstein had kept her illegitimate child hidden for nearly nine years; she had neglected him in the most inhuman way. When the authorities were notified and took care of the neglected child, he recovered but remained feeble-minded. Whether he was feeble-minded from birth, as the mother maintained and the medical expert assumed, or whether the neglect had retarded his mental development could not be ascertained beyond doubt.

According to MASON (1942), a deaf-mute mother who had fled with her illegitimate child in her arms was discovered. After the birth the two had lived locked in a room together for six and a half years. Lack of sunshine, fresh air, and suitable food had made Isabelle, as the child was called, severely rachitic.

Isabelle was at first completely negativistic and could not speak a word. In the beginning it was hard to decide whether she was deaf, feeble-minded, or both. Factors contributing to this condition were probably not only the isolation but also the severe rachitis, which causes prolonged and severe retardation of cerebral function which, however, can be compensated for.

According to the test results, Isabelle had — disregarding her inability to speak — the mental age of a three-year-old at a chronological age of six and one half years. One week after MASON's first visit, she spoke her first word, "buh" for "ball." The beginnings of language developed exclusively by experiences with objects available in her present surroundings. Isabelle passed through the succeeding stages of language development far more rapidly than a mentally normal child does.

After one and one half years her vocabulary consisted of 1500 to 2000 words. At the age of eight she was well adapted to the living, listening world around her. She asked intelligent questions, could recite children's rhymes, tell stories, sing songs, count up to 100, solve arithmetic problems within the numbers 1 to 10, and tell the value of coins. After having lived in a world of silence for six and one half years, she had become a girl of average intelligence.

The children described, though grossly neglected, remained in a human environment. The question of whether children can create a language of their own must remain unanswered. After all, mankind needed several hundred thousand years for this task.

The observation of these children stimulated science to pose a new question. Today behavioral science uses the "Kaspar Hauser test," i.e., the training of an animal from its birth in a surrounding alien to its species, to differentiate between the inherited components of its behavior which are characteristic for its species and those which are acquired from the environment.

W. DENNIS (1941) conducted corresponding not too commendable experiments on human children: A pair of twins was raised from their second to fifteenth month of life under conditions which restricted the relations with their two nurses as much as possible. Though physically well cared for, the children were never spoken to, smiled at or caressed, never encouraged or scolded. According to the experience gained in this experiment, all responses characteristic for infants during their first year of life also appear without external stimuli; they develop without instruction, reward, or example, and they do not consist exclusively of unlearned components. Maturation is a prerequisite for learning. Those responses which are learned only by imitation of others are very limited in number before the second year of life.

Feral Children (Wolf Children)

Children who lack not only human care but also the usual home environment are even worse off than Kaspar Hauser children. At first sight, it appears improbable that they could have endured such an unnatural situation for any length of time. But might it perhaps be possible under special conditions? Exciting adventure sagas and epics would have us believe that the child is not necessarily doomed. However, do examples of such children exist in real life?

In his Systema Naturae (1735), LINNAEUS contrasts *Homo sapiens (nosce te ipsum)* with *Homo ferus (tetrapus, mutus, hirsutus)* and cites a number of examples of the "savage" from ancient times. BLUMENBACH (1811), RAUBER (1885), and ZINGG (see SINGH, 1942) compiled these reports. BLUMENBACH, who thinks quite critically, speaks just as LINNAEUS did of "wild" (savage) man, RAUBER of "feral" man (who has become savage).

According to ZINGG, a story of more recent date tells of a boy ("Tarzanito") who grew up as a savage in the tropics and who was found in 1933 in El Salvador (Central America). From a lonely hacienda a child had been repeatedly observed roaming alone through the surrounding woods. He had been caught several times, but always managed to free himself again. One night when he approached the hacienda he was trapped by dogs, against which he defended himself very skillfully with a stick; this time, however, he could not free himself. The boy, then five years old, was naked and had long hair and nails; he could not speak, but uttered only some yelling sounds. As he later recounted, he had slept in caves or in trees because he was afraid of wild animals, and had lived on wild berries and raw fish which he caught himself.

After overcoming initial difficulties, the boy became friendly and pleasant, learned to speak, and attended school successfully. He continued to climb trees skilfully and swung from branch to branch like a wild animal in an astonishing way.

He was said to have been in the wilderness several years. It is hard to attribute such a performance to a five-year-old boy, even more to one still younger. How he came to be in the woods remained unknown.

In 1798, a great sensation was caused by the "wild boy of Aveyron." He was picked up by hunters in the forest at the age of eleven or twelve years; he may have lived there for years, separated from human beings. J. M. G. ITARD (1775–1838), a physician and teacher of the deaf, tried with unending effort to educate him and attempted to stimulate his attention in ever new ways. The results were modest; the "wild boy" never learned to speak. He may have been an escaped or abandoned feeble-minded child.

Old sagas and more recent reports as well tell of female beasts of prey which stole human babies, dragged them into their caves to eat them, then did not eat them but instead nursed them and raised them for years. Wolves are usually said to be the child abductors; occasionally lions, leopards, eagles, and dragons take their place (SCHELUDKO, 1930).

Zarathustra as well as the ancestors of the Turks and Mongols are all said to have been raised by wolves, according to the legends (E. HOFFMANN and KRAYER).

An ancient bronze figure in the Capitoline Museum shows the abandoned twins Romulus and Remus, later the founders of Rome, sucking on a female wolf who is showing her teeth. This saga is depicted on the shield of Aeneas forged by Vulcan (Virgil, The Aeneid, 8, 630).

The old German thirteenth century saga Wolfdietrich tells of such a wolf child:

Ein neugeborenes Kind war ausgesetzt worden.
"Ein wolf nach siner spise in dem hage gie,
dar inne er hüner und kappen é vil dicke gevie.
(in dem er oft Hühner und Kapaune gefangen hatte).
der nam daz kindel kleine und vazte ez in den munt:
er truoc ez hin ze walde an der selben stunt.
gèn einem hôhen berge, der was innen hol.
der alten wáren zwéne, daz sage ich iu wol.
sie heten vier jungen, diu wárn einer wochen alt:
der witze und ouch des kindes waren geliche gestalt.
(deren Verstand und der des Kindes waren gleich geartet).
Für sie leite der alte daz kleine kindelin.
ez solte der jungen wolve spise gewesen sin.
dô schuof ez diu jungent daz si dannoch waren blint:
daz half die küniginne daz ir genas daz kint."

In SHAKESPEARE's Winter's Tale (II, 3) Antigonus says to a child whom he is supposed to abandon:

"Come on poor babe,
Some powerful spirit instruct the kites and ravens
To be thy nurses! Wolves and bears, they say,
Casting their savageness aside have done
Like offices of pity."

In 1891, ULRICH JAHN recorded a Pomeranian fairy tale of a wolf child:

Little John ran away from his father in the forest. When he became tired, he fell asleep under a tree. "There an old female wolf spotted him, and since she took a liking to the child she did not eat him up, but grasped his jacket with her sharp teeth and then ran with mighty leaps to her cave. There she neatly put the child down, put him to rest on a dry bed of moss and fallen leaves, and gave raw sheep meat to him to satisfy his hunger. When night came and the boy became tired, the female wolf lay down at his side and warmed him with her thick fur." After twelve years John returned to the world of man.

Several reports on German wolf children which are told in a similar vein and were intended to be taken seriously are given in RAUBER's book.

In Germany, wolf children disappeared at the same time as wolves. In India, however, wolves exist even today – and wolf children. DICKENS (1853) wrote:

In 1843 in the kingdom of Oudh, a boy approximately eighteen months old was missed by his parents. It was assumed that wolves had dragged him away and had torn him to pieces.

Seven years later a hunter saw a wolf in the jungle; the wolf was accompanied by several cubs and by a creature the likes of which he had never seen. It looked like a boy, but walked on all fours. The man pursued the creature to a cave, but could not keep up with it. Some days later he caught it alive. It barked, or rather growled like a wolf, and tried to bite the man who had caught it. The female wolf and her cubs followed the man from a distance as if they wished to regain the creature. But finally they returned to the jungle.

The creature was displayed in Lucknow and, locked in a cage, stirred up a great sensation. There was no doubt that it was a human being, although it never stood erect or uttered any sounds other than growling or a hoarse bark. It rejected any cooked food and would eat only raw meat, which it avidly tore to pieces. It also tore clothes with its teeth. A disagreeable smell emanated from it; its skin was covered with short, thick hair. It smelled like the wolf that had raised it, and it liked to take hard bones, which it chewed like a dog and digested. In a word, the creature had adopted all the habits of its stepmother, the female wolf. Every day, hosts of natives came to look at the strange creature, and finally the woman who had lost her child appeared, too. She recognized her missing child by certain marks. However she did not intend to take him back, but looked at him with utter terror and disgust.

All attempts to tame the child were unsuccessful. Locked in its cage, the creature seemed to suffer terribly; it ate only when it was forced by hunger. It would have been dangerous to release it from the cage, for it was as unruly as a wild animal in the desert. Many attempts were made to teach it to speak, but it uttered only the sounds mentioned before. It lived for approximately a year, and during this time became emaciated and dwindled away to a living skeleton.

Based on such reports, the English author RUDYARD KIPLING (1894) wrote the exciting story of Mowgli's adventures among Indian wolves. Since then the reports on Indian wolf children have increased in number. But now they are supposed to be taken seriously: The wolves are supposed to have actually kidnapped the children and to have given them shelter in their dens for years. When the children were found they walked on all fours, were unable to talk but could only howl or even bark, they lapped up water like a wolf, etc.

The most famous wolf children of our time were found by the Indian missionary SINGH in 1920 and were described in 1942. According to his report, he saw two human children playing with wolves in front of a wolves' den. When the den was dug out, two girls were found. The missionary raised the two children in an orphanage.

When they were found the age of the younger girl, who died in 1921, was estimated to be one and one half years and that of the older girl, whose name was Kamala and who died in 1929 from uremia, to be eight years. At first Kamala moved on all fours so rapidly that she could not be overtaken in an open field. Her knees, elbows, and palms were covered with callusses. She grasped all food with her mouth, and lapped water up with her mouth as dogs do. Her ability to hear, smell, and see in the dark was developed to an extraordinary degree. During great effort or in hot weather, Kamala let her tongue hang out of her mouth, panting. At certain times she howled like a wolf; she avoided adults and children, but liked to play with animals. When frightened she reacted with temper tantrums and frequently also with the death-feigning reflex, and at the end of these emotional reactions showed a sullen shyness and refusal to eat. Only slowly and with much effort did she become adjusted to human habits under the devoted guidance of Mrs. Singh, for whom she soon showed great affection. Only after three years of care did she start to walk erect, and she did not begin to talk before the fourth year. After the fifth year she knew something over 30 words. From the sixth year on she showed distinct social traits: She played with other children and ate in a civilized way; she cried when left alone and was ashamed when she was undressed. She helped young children, followed simple orders, and enjoyed praise. Her vocabulary increased, but remained within moderate limits.

According to SINGH, Kamala had spent eight years among the wolves; afterward she was observed by him for nine years until her death. Her behavior at the age of seventeen corresponded approximately to that of a child two or three years old; thus, she was considerably retarded. The other child, who had lived with the wolves for a year and a half according to SINGH, was also greatly retarded in her development.

Doubts as to the authenticity of this report were repeatedly raised. The zoologist KOEHLER (1950) raised the following objections: At best, a nursing female wolf would take an infant and carry him to her den, where he might perhaps be protected by the truce existing within its confines. Even if the child was abandoned at a time when the female wolf's nursing period – perhaps four months – had been taken advantage of, the child would still have to adjust very quickly after this time to a pure carrion or meat diet. What, asked KOEHLER, would become of the infants when the wolves roamed around in packs and did not remain in the cave?

Furthermore, the following question arises: Is a wolf really able to carry a human infant in its mouth for any length of time without seriously injuring him? Customarily, Indian children are practically naked. Will not the flowing blood stir up instincts other than motherly love?

SINGH's report was published by the American professor of anthropology, ZINGG. According to the professor of zoology PORTMANN in Basel (1949), this publication is one of the "great documents of mankind." "The scientific value of this extraordinary document has not yet been fully recognized." The child psychologist GESELL, Yale University (1941), devoted a special book to these two Indian wolf children; the French biologist ROSTAND (1956) characterizes the report on these children as very informative. But what information did we actually obtain?

At the end of April 1957, Indian newspapers reported on a new wolf child in the village of Jar-Ka-Nagla near Agra. The child's name was Parshuram, and his father's Babulal Jatava. According to the newspapers the child had been stolen by a wolf six years earlier, at the age of 18 months, when he was asleep next to his mother in the open air. When the mother called for help, people came with lanterns and tried to chase the wolf, but all was in vain. The next day they searched for the child's corpse, but found no trace. Thus, they were convinced that the boy had been eaten by wolves.

"The circumstances of his rescue from the wolves six months ago are still shrouded in secrecy." According to his father's statement, the child was picked up by some soldiers when he left the cave to drink from a river in the jungle. He was taken successively to a hospital, to the police, and finally to a Brahman, but nobody could handle him. He ate only raw meat, and howled like a wolf day and night. He was placed in chains as a precautionary measure and handed over to a maid to be cared for. Since he did not calm down, the maid took him to her house. A neighbor recognized the boy as his grandson and informed his daughter. She confirmed that the boy was her son. "A mother can never forget her child, no matter what happens."

This report stirred up a great sensation in India. It was postulated that any female wolf who had lost her cub remained in a special state of excitement for a week. If she found a child to pet and to nurse during this time, she accepted the child as a substitute.

OGBURN (1959) of Washington visited this "wolf child" and shed some light on the story. The first impression he had was that of a spoiled and exhibitionistic child who showed his tricks to visitors. He had a strong motility drive, ran aimlessly around without any inhibitions, and paid no attention to people or objects. He could speak only a few words. From this short description it can easily be seen that the child was feeble-minded.

The story told by Babulal Jatava conformed only in part to the newspaper reports. One day a child had been stolen from him by a wolf, which not infrequently happens in India. Now he had recognized the child by two scars on the scalp in the same places where the stolen child had had abscesses. A blood test to prove the kinship was not done because Babulal Jatava had asked for an exorbitant sum of money for permitting this test. He was a poor peasant who expected to get money from the visitors.

In Ferozabad, OGBURN succeeded in finding and in interviewing the Brahman with whom the child had stayed previously. This man, who seemed quite reliable, had, he assured OGBURN, found the deserted child himself without seeing any wolves, had taken him into his house, and had finally handed him over to a maid. The child had lived in Ferozabad for four months without having been in any way connected with wolves. Finally, the Brahman announced the news of the child over the radio. Thereupon, Babulal Jatava came, said he was the father, and took the child with him. The story of the child's association with a wolf originated with the father. The place where the child was found was at least 24 miles from the home of his "parents" – "rather far for a wolf to carry a child or for a child to walk." Thus, no proof could be found "that this wolf child had ever seen a wolf."

According to BETTELHEIM (1959), wolf children behave very similarly to severely autistic children he has observed, who show animal-like behavior; the latter children, of course, have never in their lives had anything to do with animals. Their behavior can be explained better as a result of the inhumanity of certain persons in their environment, usually their parents, than as a result of the humanity of animals.

The belief in wolf children is based on the fact that human babies can be nursed by animals; however, for obvious reasons beasts of prey have never been chosen for this task. On the other hand, animal babies have occasionally been nursed by human wet nurses, for example, when it was necessary to maintain their milk secretion if the child was unable to suck or in order to raise the animal baby after it had lost its mother.

The fairy tales of wolf children will not die rapidly. The tendency of newspapers to create sensations does not favor clarification of the real facts. Thus, the wolf child has replaced the once famous sea serpent for the newspapers.

Physically and mentally, the young child depends entirely on his mother. Daily experience shows how difficult it is to replace her. Nobody who knows an infant's needs would believe that a female wolf, bear, leopard, even less eagles and crows, could substitute for a mother beginning in infancy and for many years thereafter.

As early as 1874, DR. WISHAW, who was associated with the insane asylum in Lucknow, India, wrote (cited according to SINGH and ZINGG, p. 170):

"I will never, never believe that in even a single case a child could have been raised by wolves, and I cannot understand how anyone can believe such a thing. Most of the wolf children were idiots who were abandoned by their parents near a police station."

SUMMARY

Even the young infant has a constant and close mutual relationship to his natural environment, i.e., the environment which corresponds to his vital needs. This relationship has developed in the course of evolution.

The baby monkey, which constantly clings to its mother's body, has special reflexes which assure its hold on the fur and its finding of the mother's breast. The human infant, who involuntarily has become a bed baby, still has the same reflexes without having adapted himself to the bed. The longing for his mother is expressed in his restlessness and his crying and can be satisfied by taking him out of the bed and carrying him around.

The environment of the infant and young child in the sense of VON UEXKÜLL's hypothesis differs fundamentally from that of adults: During the first months of his life it consists exclusively of his mother; vice versa, the infant is a part of the environment of his mother, whose essential attribute is her love for her child. The infant is by no means – in contrast to statements made by child psychologists – poor in instincts; neither is he biologically poorly equipped or helpless, but instead he is merely attuned to his mother in all his needs. Only in her arms, the only place nature has assigned to him, does he find all he needs: warmth, food, and protection. The mother and her child complement each other physically and in their functions to form a higher unity of life; they fit to each other as a key into a keyhole. When separated from each other, they strive to be united again.

In the phylogenetic past, when separation of the child from his mother would have been fatal, the child acquired the instinct of continuously monitoring his mother's presence. This instinct, which has now become superfluous and is misunderstood, but which is still effective, is often quite annoying. The child cries again and again and remains restless until he is calmed by his mother. The cradle and the nipple serve as make-believe devices by simulating his mother's presence; the rocking of the cradle substitutes for the movements of his mother's body and the nipple for the mother's breast.

At first the father is much more distant from his children. The mutual links are not a gift of nature to the same extent, but develop only gradually. They are, however, among the deepest human relations.

Homesickness in a young child means only the longing for his mother, who is the living embodiment of his home; only the older child longs for "home". Almost every child who is admitted to an institution suffers from homesickness during the first days. This has great educational value.

Children do not do so well in institutions as at home since their natural environment, i.e., their mother, is absent. Infants could be successfully brought up in institutions only after the proper methods of feeding them and preventing the spread of contagious diseases had been learned. If these aspects are neglected, institutionalized children are very much endangered even today. In the regulated but monotonous life of institutions, the many stimulations which children receive at home are missing. Nocturnal head shaking and tonic rigidity are often found in institutionalized children.

Old fairy tales and recent reports claim that female beasts of prey, particularly she-wolves, have kidnapped babies and instead of eating them have raised them for many years. There are no valid proofs for these statements.

We may conclude as follows: The mother, and she alone, has been determined by nature to be the one to care for her child. It is true that efforts to replace her have had some success, but even today the child does best physically and mentally with his mother.

REFERENCES

ALDRICH, C. A.: J. Pediatr. *15*, 578 (1939).

–, C. SUNG, and C. KNOP: J. Pediatr. *26*, 313 (1945).

VON ALLESCH, G. J.: Preuß. Akad. Wiss., Sitzb. der Phil.-hist. Abt., July 14, 1921; Naturwiss. *1921*, 774.

AUENBRUGGER, L.: Neue Erfindung (New Invention) (1761). Vom Wirken berühmter Ärzte aus vier Jahrhunderten (On the Works of Famous Physicians from Four Centuries). Publisher: A. G. Knoll Chemical Factories. p. 153. Ludwigshafen am Rhein, 1936.

BAYER, W.: Heilkunst *69*, No. 9 (1956).

BERSOT, H.: Le rôle de la musculature dans le développement du petit enfant (The Role of the Muscles in the Development of the Young Infant). p. 120. Vevey, Montreux, Berne, 1926.

BETTELHEIM, B. R.: Amer. J. Sociol. *64*, 455 (1959).

BIRK, W.: Mschr. Kinderhk. *12*, 1 (1919).

BLUMENBACH, J. F.: Beitr. z. Naturgesch. (Contributions to Natural History). Pt. 2. Göttingen, 1811.

BOLK, L.: Das Problem der Menschwerdung (The Problem of Humanization). Jena, 1926.

BOWLBY, J.: Maternal Care and Mental Health. World Health Org. Monograph, Series No. 2. Geneva, 1952.

–, and ROBERTSON: Lancet *1952*, II, 1122.

BRANDES, G.: Buschi. Leipzig, 1939.

BROCK, F.: Verh. dtsch. zoolog. Ges. *1939*, 16; Typenlehre und Umweltforschung (Theory of Types and Ecology). Bios. Vol. 9. Leipzig, 1939.

BÜHLER, C., and H. HETZER: Kleinkindertests vom 1.–6. Lebensjahr (Tests for Small Children from the First to Sixth Year of Life). Leipzig, 1932.

BÜHLER, K.: Die Seele des Kindes (The Mind of the Child). 3rd ed. Jena, 1929.

BUSEMANN, A.: Handbuch der pädagogischen Milieukunde (Handbook of the Pedagogic Theory of Environment). Halle, 1932.

BUYTENDIJK, F. J. J.: Naturwissensch. u. Theologie (Natural Science and Theology). No. 1, p. 42. Munich, 1957.

CZERNY, A.: Festschrift zur Eröffnung des Kaiserin-Auguste-Viktoria-Hauses (Commemorative Volume for the Opening of the Empress Auguste Victoria House). p. 57 Berlin, 1909; Pädiatrie meiner Zeit (Pediatrics of My Time). p. 8. Berlin, 1939; Der Arzt als Erzieher des Kindes (The Physician as Educator of the Child). 9th ed. Vienna, 1942.

Deutsches Heldenbuch (Book of German Heroes): Edited by A. Amelung and O. Jänicke. p. 152. "Wolfdietrich." Berlin, 1871.

DICKENS, CHARLES: Household Words. Vol. 18, p. 297. Leipzig, 1853. (Tauchnitz Collection of British Authors, Vol. 215.)

DIX, W. C.: Seelenleben des Kindes im ersten Lebensjahr (The Infant's Intellectual Life During His First Year). Jena, 1939.

DOFLEIN. F.: In R. Hesse u. F. Doflein, Tierbau und Tierleben (In R. Hesse and F. Doflein's Animal Structure and Animal Life). Vol. 2, p. 659. Leipzig and Berlin, 1914.

ELSENHANS, T.: Lehrbuch der Psychologie (Textbook of Psychology). 3rd ed. Tübingen, 1939.

ENGEL, E.: Kaspar Hauser. Braunschweig-Hamburg. 4th ed. Leipzig, 1931.

ERIKSSON, Z.: Acta paediatr., Stockholm *4*, Suppl. 1925; Zschr. Kinderhk. *39*, 349 (1925).

FEER, E.: Beih. der Med. Klin. *1909*, No. 1, p. 25.

VON FEUERBACH, A.: Beispiel eines Verbrechens am Seelenleben des Menschen (Example of a Crime on the Intellectual Life of Man). Ansbach, 1832; Leben. Von seinem Sohn (Life. By his Son). Vol. 2, p. 319. 1852.

FINKELSTEIN, H.: Zschr. Hyg. *28*, 125 (1898).

–, Lehrbuch der Säuglingskrankheiten (Textbook of Diseases of Infancy). 4th ed., p. 283. Amsterdam, 1938.

FRÖBEL, F.: H. Volkelt, Friedrich Fröbel, Ein Führer aus den Nöten der Gegenwart (H. Volkelt's Friedrich Fröbel, A Leader for the Difficulties of Our Time). p. 173. Weimar, 1932.

GEHLEN, A.: Der Mensch (Man). 6th ed. Bonn, 1958.

GERLACH, R.: Deutsche Allg. Zeitung, July 1 (1942).

GESELL, A.: Wolf Child and Human Child – The Life History of Kamala, the Wolf Girl. London, 1941 (According to Koehler and Mierke).

GLASER, K., and L. EISENBERG: Pediatrics *18*, 626 (1956).

GRIMM, J.: Kleinere Schriften (Minor Writings). Vol. 5, p. 480. Berlin, 1871.

GRINDL, H.: According to HETZER.

GUENTHER, K.: Mutterliebe im Tierreich (Mother Love in the Animal Kingdom). Ulm, 1936.

HAMBURGER, F.: Die Neurosen des Kindesalters (The Neuroses of Childhood). p. 21. Stuttgart, 1939.

VON HARNACK, G. A., and M. OBERSCHELP: Dtsch. med. Wschr. *1957*, p. 1916.

HEDIGER, H.: Ciba-Zschr., No. 59, 1965 (1953).

HERDER, J. G.: Sämtliche Werke, herausgegeben von B. Suphon. Abhandlung über den Ursprung der Sprache (1772) [Complete Works, edited by B. Suphon. Treatise on the Origin of Language (1772)]. Vol. 5, p. 113. Berlin, 1891.

HETZER, H.: Die Entwicklung des Kindes in der Anstalt (The Development of the Child in the Institution). In BUSEMANN (1932).

–, Zschr. angew. Psychol. *50*, 209 (1936).

–, and B. H. TUDOR-HART: Quellen und Studien zur Jugendkunde (Sources and Studies on Adolescence). No. 5. Jena, 1927.

HODEK, L.: Dtsch. Schwesternzeitung *10*, 264 (1957).

HOFFMANN, E., and KRAYER: Handwörterbuch des deutschen Aberglaubens (Brief Dictionary of German Superstitions). Published by H. Bächtholdt-Stäuble. Vol. 9, 738. Berlin, 1883–1941.

HOLZAPFEL, M.: Zschr. Tierpsych. *2*, 46 (1939).

HOMBURGER, A.: Psychopathologie des Kindesalters (Psychopathology of Childhood). pp. 586 and 700. Berlin, 1926.

IBRAHIM, J.: Über die Mütter. Antrittsvorlesung (On Mothers. Introductory Lecture). Jena, 1917.

ILLINGWORTH, R. S.: Brit. Med. J., No. 4905, 75 (1955).

JACOBSEN, C. F., M. M. JACOBSEN, and J. G. YOSHIOKA: Comp. Psychol. Monograph, Baltimore *9*, 1 (1932).

JASPERS, K.: Heimweh und Verbrechen (Homesickness and Crime). Thesis, Heidelberg, 1909.

JAHN, ULRICH: Volksmärchen aus Pommern und Rügen. Forschungen, hrsg. vom Verein für niederdeutsche Sprachforschung (Popular Fairy Tales from Pomerania and Rügen.

Investigations, published by the Society for Study of the Low German Language). Vol. 2, p. 107. Norden and Leipzig, 1891.

JESPERSEN, O.: Die Sprache. Indogermanische Bibliothek (Language. Indogermanic Library). Sect. 4, Vol. 9, p. 167. Heidelberg, 1925.

KELLOGG, W. N., and L. A. KELLOGG: The Ape and the Child. New York and London, 1933.

KIENE, M.: Das Kind im Kindergarten (The Child in Kindergarten). p. 10. Freiburg i. B. 1953.

KIPLING, RUDYARD: The Jungle Books. London, 1894.

KOEHLER, O.: Zschr. Tierpsychol. *7*, 148 (1950); 1. Arbeitstagung über zentrale Regulation (First Worshop on Central Regulation). p. 12. 1955.

—, and A. ZAGARUS: Beitr. Fortpfl. biol. Vögel *13*, 1 (1937).

KOFFKA, K.: Die Grundlagen der psychischen Entwicklung (The Foundations of Psychological Development). 2nd ed. Osterwieck, 1925.

KRETSCHMER, E.: Medizinische Psychologie (Medical Psychology). 3rd ed., p. 117. Leipzig, 1926.

KUSSMAUL, A.: Jugenderinnerungen eines alten Arztes (Youthful Memoirs of an Old Physician). 11th–13th printing, p. 283. Stuttgart, 1922.

LINDNER, J.: Heimweh, eine pathopsychologische Untersuchung (Homesickness, A Psychopathological Study). Amsterdam, 1940; Ref. Zbl. Neurol. *101*, 212.

LINNAEUS: Systema Naturae. German translation. Halle, 1740.

LORENZ, K.: Naturw. *25*, 298 (1937); J. Ornith. *83*, 137 (1935); Tiergeschichten (Animal Stories). 5th ed., p. 137. Vienna, 1951; Zschr. Tierpsychol. *5*, 235 (1942–1943); Dtsch. med. Wschr. *1953*, 1601.

—, In G. Heberer, Die Evolution der Organismen (In G. Heberer's The Evolution of Organisms). pp. 105 and 122. Jena, 1943.

MARBE, K.: Arch. ges. Psychologie *50*, 513 (1925).

MASON, M. K.: J. of Speech Disorders *7*, 295 (1942).

MCCANN, W. H.: Psychol. Bull. *38*, 165 (1941).

MIERKE, K.: Wille und Leistung (Will and Action). p. 191. Göttingen, 1955.

MITCHELL, P. C.: The Childhood of Animals. p. 168. London, New York, 1912.

MOLL, L.: Beiträge zur sozialen Hygiene des Säuglings- und Kleinkindesalters (Contributions to the Social Hygiene of Infancy and Early Childhood). Edited by Rott. p. 248. Berlin, 1920; Kinderärztl. Prax. *1*, 7 (1930).

MÜLLER, H. W.: Jb. Kinderhk. *109*, 345 (1925).

NEUMANN, H.: Dtsch. med. Wschr. *1903*, 542.

NITSCHKE, A.: Dtsch. med. Wschr. *1955*, 1901.

OGBURN, W. F.: Amer. J. Sociol. *63*, 549 (1959).

PARACELSUS: Nach Lejeune, Leitfaden der Geschichte der Medizin (According to Lejeune's Guide to the History of Medicine). p. 96. Leipzig, 1943.

PIES, H.: Kaspar Hauser. Stuttgart.

PESTALOZZI, J. H.: Über die Idee der Elementarbildung. § 249. Ausgewählte Werke (On the Idea of Elementary Education. Sect. 249. Selected Works). 5th ed., Vol. 3, p. 524. Langensalza, 1900; Mutter und Kind. Eine Abhandlung über die Erziehung kleiner Kinder (1827) [Mother and Child. A Treatise on the Education of Small Children (1827)]. p. 61. Zürich and Leipzig, 1925.

VON PFAUNDLER, M.: 16. Vers. Ges. Kinderhk. p. 38. Munich (1899); Südd. Monatshefte, September (1909); Döderleins Handbuch der Geburtshilfe (Döderlein's Handbook of Obstetrics). Vol. 1, p. 725. Wiesbaden, 1915; Mschr. Kinderhk. *29*, 611 (1925); Gesundheitsfürsorge im Kindesalter (Health Guidance in Childhood). Vol. 1, p. 3, 1925—26.

PLOSS, H.: Das kleine Kind vom Tragbett bis zum ersten Schritt (The Small Child from the Carried Bed to His First Step). p. 24. Leipzig, 1881.

— Das Kind in Brauch und Sitte der Völker (The Child in Folk Ways and Mores). Vol. 2, 3rd ed. Leipzig, 1912.

PORTMANN, A.: Kinder, die bei Wölfen aufwachsen (Children Raised by Wolves). Die Weltwoche. Zürich, August 19 (1949); Biologische Fragmente zu einer Lehre vom Menschen (Biological Fragments for A Theory of Man). Basel, 1944; Vom Ursprung des Menschen (On the Origin of Man). 4th ed. Basel, 1944; Verh. Schweiz. Naturforsch. Ges. *125*, 44 (1945); Das Tier als soziales Wesen (The Animal as a Social Creature). p. 100. Zürich, 1953; Ciba-Zschr., Vol. 5, No. 59, 1974 (1953); Zoologie und das neue Bild vom Menschen (Zoology and the New Image of Man). p. 49. Hamburg, 1956; in Goma, das Basler Gorillakind (in Goma, the Basel Gorilla Child). Bulletins No. 2 and 3. Basel, 1960.

PRECHTL, H., and W. M. SCHLEIDT: Zschr. vgl. Physiol. *33*, 53 (1951).

RAUBER, A., Homo sapiens oder die Zustände der Verwilderten (*Homo sapiens*, or the Conditions of Feral Man). Leipzig, 1885.

RAUDNITZ, W.: Jb. Kinderhk. *45*, 145 (1897); *73*, 259 (1911); *87*, 15 (1918).

REICHENOW, E.: Naturw. *1921*, 73.

REIMARUS, H. S.: Allgemeine Betrachtungen über die Triebe der Tiere, hauptsächlich über ihre Kunsttriebe (General Observations on the Instincts of Animals, Especially on Their Art Instincts). 3rd ed., p. 343. Hamburg, 1773.

REMPLEIN, H.: Die seelische Entwicklung in der Kindheit (Intellectual Development in Childhood). 3rd ed., p. 79. Munich and Basel, 1952.

RIESEN, A. H., and E. F. KINDER: Postural Development of Infant Chimpanzees. pp. 13, 14, and 168. Yale University Press. New York, 1952.

ROBERTSON, J., and J. BOWLBY: Courier *2*, 131 (1952).

ROSEN VON ROSENSTEIN, N.: Anweisung zur Kenntnis und Cur der Kinderkrankheiten (1764) [Instructions in the Knowledge and Cure of Children's Diseases (1764)]. German translation. Vienna, 1793.

ROSTAND, J.: Das Abenteuer des Lebens (The Adventure of Life). p. 47. Frankfurt am Main, 1956.

ROUSSEAU, J.: Julie oder die neue Heloise (1762) [Julie, or the New Heloise (1762)]. German translation. Leipzig; Emil oder über die Erziehung (Emil, or On Education). German translation. 2nd ed. Langensalza, 1882.

RÜMKE, H. C.: Ref. Zbl. ges. Neurol. *99*, 392 (1941).

SALIMBENE VON PARMA: Chronik. Bearbeitet von A. Doren (Chronicle. Adapted by A. Doren). Vol. 1, p. 359. Leipzig, 1914.

SCHELBLE, H.: Mschr. Kinderhk. *8*, 611 (1909).

SCHELUDKO: Zschr. deutsche Philologie *55*, 2 (1930).

SCHMIDT, T.: Prax. Kinderpsychol. *4*, 250 (1955).

SCHLOSSMANN, A.: Mschr. Kinderhk. *11*, 545 (1913); Münch. med. Wschr. *1920*, 1318.

SCHNEIDER, K. M.: Mutterliebe bei Tieren (Mother Love among Animals). Wittenberg, 1956.

SCHREIBER, G.: Mutter und Kind in der Kultur der Kirche (Mother and Child in the Culture of the Church). Freiburg i. Br., 1918.

SCHWAB, G.: Jb. Kinderhk. *108*, 15 (1925).

SCHWIDETZKY, I.: Das Menschenbild der Biologie (The Image of Man in Biology). p. 55. Stuttgart, 1959.

SIMSARION, F. P., and P. A. MCLENDON: J. Pediatr. *27*, 109 (1945).

SINGH, J. A., and R. M. ZINGG: Wolf Children and Feral Man. IV. Contrib. of the University of Denver. New York and London, 1945.

SOMBART, W.: Vom Menschen (On Man). p. 395. Berlin, 1938.

SPEMANN, H.: Experimentelle Beiträge zu einer Theorie der Entwicklung (Experimental Contributions to a Theory of Development). Berlin, 1936.

SPITZ, R. A.: Die Entstehung der ersten Objektbeziehungen (The Origin of the First Object Relations). Stuttgart, 1957.

–, According to GLASER and EISENBERG.

STEINBACHER, G.: Zschr. Tierpsychol. *4*, 188 (1940–1941).

STEINER, H.: Arch. d. Klaus-Stiftung *24*, 217 (1949).

STEINERT, E.: Zschr. Kinderhk. *28*, 255 (1921).
STERN, W.: Psychologie der frühen Kindheit (Psychology of Early Childhood). 3rd ed. Leipzig, 1923.
STIRNIMANN, F.: Das Kind und seine früheste Umwelt (The Child and His Earliest Environment). p. 30. Basel, 1947.
STURM, M.: According to HETZER (1932).
TEMBROCK, G.: Grundzüge der Schimpansenpsychologie (Principles of the Psychology of Chimpanzees). Berlin, 1949.
TINBERGEN, N.: The Study of Instinct. p. 45. Oxford, 1951.
—, Tiere untereinander (Animals among Themselves). Berlin—Hamburg, 1955.
VON UEXKÜLL, J.: Theoretische Biologie (Theoretical Biology). Berlin, 1920; Umwelt und Innenwelt der Tiere (The Environment and Inner World of Animals). Berlin, 1921; Zschr. Tierpsychol. *2*, 101 (1939); Bedeutungslehre (Theory of Significance). Bios. Vol. 10. Hamburg, 1940.
—, and G. KRISZAT: Streifzüge durch die Umwelt von Menschen und Tieren. Verständliche Wissenschaft (Excursions through the Environment of Men and Animals. The Science of Understanding). Vol. 21. Berlin, 1934.
VILLINGER, W.: Regensburger Jahrbücher ärztl. Fortbildg. *4*, 548 (1956).
WAGNER, L.: Die Bilanz einer hundertjährigen Hauser-Forschung (The Balance of A Century of Hauser Investigation). Nürnberg, 1928; Ein Danaergeschenk E. Engels an die Hauser-Literatur (E. Engel's Greek Gift to the Hauser Literature). 1931.
WEISSE, K., H. Opitz and B. de Rudder, Pädiatrie (In H. Opitz and B. de Rudder's Pediatrics). p. 928. Berlin-Göttingen-Heidelberg, 1957.
YERKES, R. M.: Chimpanzees. New Haven, 1945.
—, and M. L. TOMILIN: J. Comp. Psychol. *20*, 231 (1935).
ZAPPERT, J.: Wien. klin. Wschr. *1904*, 1305.

15. RETROSPECT AND OUTLOOK

The anatomist cannot determine by looking at maturing nervous tissue when it will be able to function.

The awakening of cerebral function proceeds according to definite laws which may be observed again and again at the beginning of each individual function.

The different portions of the brain become active in the same sequence in which they were formed during phylogenesis and ontogenesis. Starting out from the brain stem, which in the mature newborn infant contains active centers up to and including the pallidum, functional ability ascends to the higher brain portions. As soon as they begin to function, they direct or inhibit the lower centers. Thus, in the course of the development, which mostly occurs during the first year of life, layers of overlapping brain functions of various ages are formed. In this development the youngest layer is always the highest and the guiding stratum; it functions better than the lower center. The entire cerebral function remains homogeneous and harmonious at each stage of development; the considerable inhibition of the older layers by the activity of the higher centers contributes to this.

The evolution of the brain function of different overlapping layers can still be demonstrated in the adult's cerebral function. If pathology leads to the dissolution of the higher brain functions (injuries, inflammations, especially encephalitis or meningitis, tumors, poisoning, etc.), the elimination of higher inhibiting layers may release the lower centers so that brain function regresses to the stage of infancy. On the other hand, we can observe how in the course of cerebral development the different strata, which more or less correspond to phylogenetic stages of development down to the reptiles, amphibia, and fish, rest one on top of the other. The lower levels are still free of the inhibitory activity of the higher centers, and thus the unchecked function of the lower centers can be observed.

According to Wieser (1957, pp. 213 and 261), the amount of movement in the normal adult can be reduced by slowly and steadily progressive cerebral diseases. This dissolution increasingly curtails the voluntary movements and more and more allows the predominance of uniform motion patterns which have the characteristics of stereotyped movement forms. In this group belong grasp phenomena, flexion and extension synergies with or without external stimulation, turning and rolling movements around the longitudinal axis of the body, quadruped rhythms, motor phenomena in the oral region, muscular synergies which correspond to acquired accomplishments, defense reflexes, etc. When a motor disorder arising from such a dissolution is interpreted on the basis of phylogenesis, a species-preserving behavior which has been established to reach a certain goal is compared to a stereotyped pattern which is contained in the behavior as a nucleus.

As long as we are dealing with species-specific and functioning innate sensorimotor accomplishments not based on experience, there is no basic difference between the oral functions of the healthy premature or newborn infant and the disinhibitions due to cerebral damage. Whereas the innate qualities of the stimulators remain unaffected, the conditioned, acquired afferents are regularly dissolved (WIESER).

The neurological process corresponds to the theory of levels, which was introduced by KRAUS (1919, 1926) together with the theory of the deep personality rooted in the brain stem and the cortical personality.

Similarly, STUTTE (1954) calls psychic development an ascending development in the here archaically arranged order of psychic life. Parallel with the progressive corticalization of the nervous system, a stepwise transfer of the leading role to higher personality layers occurs — however, without elimination of the lower layers. Newborns and infants are still completely contained at the level of the vital instincts.

The physiologists try to recognize the functions of certain brain parts in experiments on animals by observing the change in behavior after these parts are eliminated or inactivated by removal, freezing, or poisoning. These experiments, which were made by GOLTZ (1869), TRENDELENBURG (1910), MAGNUS (1924), RADEMAKER (1931), and many others, have no doubt enlarged our knowledge but are subject to criticism in certain details. The sequelae of the operation per se and the shock effect cannot be clearly differentiated from the consequences of the elimination of the brain portion. Moreover, it must appear questionable to what extent results in animals can be applied to man. However, in the infant, nature has presented us with intact developmental stages in which the higher centers are not yet functioning because of their immaturity. Infants correspond in many ways in their movements, locomotor development, reflex excitability, and respiration to the stages which are produced by the elimination of higher brain portions in the physiologists' experiments on animals.

Stimulus receptors, nerves, and the lower centers are already functioning in the newborn infant. The reflexes which can be elicited pass through the brain stem at this stage. The progressively better utilization of sensory impressions cannot generally be attributed to the improved function of the sensory organs, but rather to the progressively better evaluation of sensory impressions in the brain because of the gradual ascent of the relay station.

Special conditions exist in the eye: Although PURKINJE's phenomenon can be found even in the premature infant, the function of the cones matures phylogenetically and ontogenetically before that of the rods.

The nervous behavior of the newborn infant is determined by the lower senses, i.e., the tactile and gustatory senses. Their influence will be superseded, without disappearing, in the course of the first year of life by the higher senses of vision and hearing. This developmental sequence corresponds to phylogenesis.

The sensory reactions from which the facial expressions originate at first either improve or worsen the receptive ability of the sensory organs of the face. The reaction frequently spreads from a stimulated to a nonstimulated sensory organ, which in turn assumes a receptive or defensive position. This spreading reaction may even occur when a sensory organ is voluntarily opened or closed, so that the hands and also the feet to a certain degree act like sensory organs. Spreading reactions are also frequently observed in facial expressions. Most important for the facial expression are the eyes and the mouth; the nose participates in it only to a minor degree. The most important expressive movements originate by different combinations of the reactions of the facial muscles and their greatly varied sequences, as well as their associations with respiration. The facial expressions, as well as the motility, emotions, and affects, are more vivid and more uninhibited in childhood than in adulthood.

As even the accompanying expressive movements show, man's entire higher brain function has developed from the senses. This knowledge, which was already expressed by the philosophers and psychologists of antiquity, can only be confirmed by natural science.

During the static and locomotor development of the infant there appear reflex groups which MAGNUS, RADEMAKER, and others investigated in animals by the removal of higher brain portions. From the basic agreement of these experiments on animals with the findings in the human infant, it can be concluded that the human brain is ontogenetically and phylogenetically based on the same blueprint as the animal brain.

The stepping, ascending, crawling, and swimming movements which can be elicited regularly in the first months of life have no connection at all with independent locomotion. The latter develops to the erect gait in the course of the first year of life by passing through the stages of creeping, crawling, and sliding. The head and limbs join the coordinated motion mechanism successively in the descending direction. Phylogenetically, locomotion developed by passing through the same stages as it does in the infant.

In general, the development of human fetuses progresses anatomically and physiologically in the cervicocaudal direction from the brain stem to the spinal cord, and in the cervicorostral direction in the brain from the brain stem to the cerebral hemispheres. The ontogenetically and phylogenetically younger centers break down prior to the older centers if a lack of oxygen occurs. Correspondingly, the local reflexes elicited by stimulation of the trigeminus region are the last to be extinguished in human fetuses during progressive oxygen deprivation. They are the first to be evoked in the maturing fetus.

Extrapyramidal mass movements which are directed by the pallidum and not yet inhibited by the striatum are characteristic for the young infant. With the progressing development of the brain, these mass movements are inhibited and are gradually replaced by goal-directed voluntary individual movements; these movements are regulated by the motor centers of the cerebral cortex by way of the pyramidal tracts. After their physiological disappearance, mass movements recur in diseases affecting the striatum. The motility drive is considerably greater in childhood than it is in the adult; the younger the child, the greater this urge. With progressive growth, the movements are usually more and more restricted. Stimuli which affect the young child tend to cause responses in larger areas than they do later in life. An example of this general rule is the patellar reflex, which during the first six months of life is accompanied by an adductor contraction of the opposite side.

The vital centers of the brain stem, the respiratory and sucking centers, are fully functional in the mature newborn infant, i.e., at the stage of the lower cerebral function. In very immature infants, the respiratory center is only insufficiently equipped to fulfill its task and the sucking center, which developed after the respiratory center phylogenetically, fails altogether.

Many phylogenetic memories express themselves in respiration. In the course of individual development, the respiratory center of man is assembled successively from partial centers: the swallowing center, which makes "swallowing respiration" possible; the gasping center, which causes single breaths at irregular intervals accompanied by wide opening of the mouth, jerking back of the head, and retraction of the tongue; the periodic respiratory center, which regulates periodically intermittent breaths; and the highest component, which triggers rhythmic respiration. The higher center inhibits the lower center by its function, but in cerebral immaturity it is so sensitive that it easily fails under the influence of the greatest variety of general afflictions. I call this process "breakdown of the respiratory center." This breakdown releases a lower partial center which maintains an emergency respiration until the higher center has recovered or until death has occurred. The death of premature infants almost always occurs with the symptomatology of suffocation attacks, i.e., attacks of gasping respiration (erroneously termed "asphyxia," i.e., pulselessness).

Gasping respiration appears after an extended block of oxygen supply, for example, following birth, after the "breath holding" of the young child, after an attack of whooping cough in a young child, or after oxygen deprivation in experiments on animals.

Amphibians, for example, frogs, cannot yet inhale air into their lungs, but swallow air from the pharynx into the lungs after it has been mixed with outside fresh air by floor-of-the-mouth movements. Yawning reminds us of this "swallowing respiration" because it is terminated by a swallowing movement. The yawning center is released by fatigue. The swallowing movements during suffocation and during dying also belong in this category; they may still be present when even gasping respiration can no longer be maintained. During the infant's food intake, respiration and swallowing are still closely coupled. Floor-of-the-mouth respiration in the severely ill older child reminds us of the laryngeal respiratory movements of the lower lung breathers. The same is true of alae nasi respiration. The singultus center is a lower part of the gasping center and directs the diaphragmatic movements characteristic of gasping respiration; the accompanying movements of the tongue and the mandible are still suggested in infancy, but they disappear later. A slow withdrawal of oxygen may evoke singultus in experiments on animals.

The mature sucking center effects rhythmic sucking movements; periodic sucking movements occur at a lower stage of excitability.

Immature centers which have just been formed are not fully functional because their nervous energy is not yet sufficient for uninterrupted activity. They are very sensitive, need a large amount of oxygen, and therefore are the first to be damaged by the progressive lack of oxygen which occurs in many diseases; they also break down if excessive demands are made on them otherwise, so that the functions they regulate (respiration, sucking) regress to lower stages. In premature infants, the higher components of the respiratory center may even fail after extended activity — for example, coughing, crying, food intake, vomiting, and fright reactions — because of the expenditure of the limited amount of nervous energy available, so that periodic or gasping respiration occurs. In more pronounced immaturity, the nervous energy is not sufficient for the simultaneous activity of the respiratory and the sucking centers; in this case life may still be maintained by pendular induction — alternate breathing and sucking.

During the food intake of healthy infants, an interplay between sucking and respiratory centers occurs; in this process the phylogenetically and ontogenetically younger sucking center inhibits the older respiratory center or forces its own rhythm upon this center (relative coordination as described by von Holst, 1939). The swallowing center also comes into close relation with the respiratory center, since the food and respiratory pathways cross in the pharynx. The interplay between sucking and respiratory centers is much more easily disturbed in the sick infant than in the healthy baby. Frequently a sucking disorder develops in which the sucking center loses its influence on the respiratory center (block between sucking and respiratory centers).

Aspiration is a severe disorder which frequently results in death. It occurs in the newborn infant as a nonphysiological aspiration of amniotic fluid; in infancy, aspiration occurs mostly in severe alimentary disorders, but it may also be seen as an aspiration of vomited stomach contents in other serious illnesses. The signs of aspiration may frequently be recognized even before death.

The infant becomes immobile during feeding because the sucking center draws nervous energy away from the motility center. The reflexes of food intake become most distinct during hunger. When the infant is satiated, these reflexes disappear and are replaced by reflexes of rejection.

Neurological examination shows us whether certain reflexes are present or absent. However, the process of the disappearance of a reflex cannot be observed. On the other hand, the movement rhythms caused by the function of the respiratory and sucking centers may be exactly observed and recorded. We can determine the moment when a higher center fails and a lower center is reactivated, how inhibitions occur, and how the various centers influence each other. A good opportunity for these observations exists in young infants, whose cerebral function is still being formed, since the higher centers which have just been established are not yet at the height of their faculties. When these centers fail or are again reactivated, a to and fro play between the higher and lower brain functions is expressed in the rhythms which they regulate, a phenomenon which afterward can still be followed in the tracings. Thus, observation of the motion rhythms supplements examination of the reflexes, which in themselves also give a picture of a higher and lower brain function.

Central respiratory disturbances may frequently be decisive for the life or the death of the premature infant. Since these disturbances may be improved by giving oxygen, today many wards for premature infants are equipped with an oxygen center from which each room can be supplied. Sucking and swallowing weaknesses have been overcome by tube feeding for a long time. The emptying of the newborn infant's air passages and stomach, when indicated, counteracts the aspiration of amniotic fluid, mucus, and blood. Frequently, constant observation of an endangered premature infant by special nurses is necessary. With the aid of the procedures mentioned, supplemented by many others, we are successful today in keeping alive many premature infants who would formerly have died. In infants suffering from severe nutritional disturbances, death from aspiration of stomach contents formerly occurred frequently; today the incidence can be reduced to a great extent by continuous gastric drip infusion.

Central disturbances of the rhythm occur in the adult — with little attention given to them — as attacks of singultus, CHEYNE-STOKES respiration, and the last gasp. Aspiration is more important also for the adult.

The younger the child, the greater his need for sleep. During infancy, the periods of being awake and of sleeping alternate more frequently than later in life. The preferred position of the infant disappears at the end of the first year of life, but it can be released again later by diseases, especially by meningitis.

Many of the procedures of "animal hypnosis" can be applied to the infant. Some months after birth, conditioned reflexes can be established from all sensory organs. It is uncertain whether or not they can be demonstrated before that time. They do not attain their full development before the second year of life. With their aid, the discriminatory ability of the different senses can be tested in the infant. Conditioned associations form the basis for the first training. The first words are acquired as conditioned reflexes. Some disease entities are due to a disturbance of existing conditioned reflexes or the formation of pathological conditioned reflexes.

The phenomenon of dominance can be demonstrated in the sucking center of the young infant. It consists of the phenomenon of an excited focus diverting and attracting the excitations which should actually pass through another tract, thus reinforcing its own state of excitation.

The mature newborn infant is a pallidum creature without functional cerebral hemispheres. Only this fact, and not the long overemphasized cerebral hemorrhages caused by birth trauma, is the basis of the characteristics of his lower brain function. The most important neurological differences between newborn infants and adults have already disappeared at the end of the first year of life.

The physiological testing of the infant's senses usually done in the laboratory must absolutely be supplemented by the investigation of the mutual relations between the child and his natural environment.

The monkey baby lives as a pure "breast baby" on its mother's body. The human baby, however, who has been involuntarily made a "bed baby," has still preserved the adaptations of the monkey baby to its mother's body and has not yet developed new adaptations to the bed. Thus, a discrepancy between his needs and their satisfaction easily arises.

The infant possesses his own subjective environment, as defined by VON UEXKÜLL (1920), which is fundamentally different from the adult's and changes in the course of development. During the first months of life, this environment consists exclusively of his mother. The child and his mother are exactly attuned to each other and mutually supplement each other to a higher unity of life; this unity was formed phylogenetically when the child still lived on his mother's body as the baby monkey does even today.

In the arms of his mother, the only place which nature has destined for him, the infant finds everything he needs for life, i.e., warmth, food, and protection. The mother sees the happiness of her life in the care of her child. Whoever calls the infant helpless or even biologically inferior does not see him in his natural environment. The infant — in contrast to the adult — is adjusted to an extremely narrow environment which is of vital importance to him. The younger he is, the more easily he is exposed to vital dangers if he loses this environment.

The relations between the father and his children are not given in the same way by nature; they develop only later in life.

Homesickness in the young child is almost exclusively the longing for his mother; he longs for home only when he is older.

Infants do not do as well in institutions as they do at home because their mothers, their natural environment, are missing. Therefore, institutionalized children are physically and mentally retarded compared to children raised at home.

The neurological testing procedures available to us today do not yet enable us to comprehend even the young infant's entire behavior. As much as he depends on the stimuli of his environment, he nevertheless can act independently from birth. At no time is he a mere reflex creature whose behavior is completely dependent on outside stimuli, as Besser (1878), Rothacker (1941), Tumlirz (1955, p. 106), Villinger (1956), and Ewald (1959) have asserted. The mere observation of every infant proves that he does not behave in a purely passive way toward his environment at all; on the contrary, he can become independently active or inactive with no immediately preceding change in his environment. He can move, fall asleep or wake up, change his mood, utter sounds, and — depending on his state of satiation — turn to certain stimuli and neglect others. According to Jennings's (1910) classic investigations, even the Infusoria are able to move without external stimuli and solely because of internal processes. How much more should the human infant be able to do this! A mere reflex creature has not yet been observed in living nature; it exists only in the phantasy of the theorist where, together with the idea of the "completely helpless" newborn infant, it will certainly still live for a long time.

The characteristics of the cerebral function of the infant and child are by no means exhausted with the characterization of their reflexes. Since these are comparatively easy to demonstrate, there is the danger that the essentials of the child's inner self will be overlooked. Actually, however, all environmental stimuli are met by an independently acting being with his own impulses, his own motility, his own moods and affects, which are by no means always responses to external stimuli, but also may absolutely originate internally. The mother, who watches her child continuously, who is aroused at night by his crying, would never conceive the idea that everything is related to stimuli from the environment. However, the investigator, who observes certain reflexes or any other behavioral responses for a certain period and under exactly determined and maintained conditions, is apt to overlook all vital processes outside the problem under investigation.

During the fight over the theory of evolution the statement was made that it was impossible and even ridiculous to believe that the human intellect could have developed from the animal stage in the course of phylogenesis. However, every infant develops from the animal to the human stage right before our eyes. The transition from animal to man was reached in phylogenesis at the end of the Pliocene Period. The infant reaches this transition period at the end of his first year of life.

REFERENCES

BESSER, L.: Arch. Psychiatr. *8*, 460 (1878).

EWALD, G.: Neurologie und Psychiatrie (Neurology and Psychiatry). 4th ed., p. 218. Munich, Berlin, 1959.

GOLTZ, F.: Beiträge zur Lehre von den Funktionen der Nervenzentren des Frosches (Contributions to the Theory of the Functions of the Nerve Centers of the Frog). Berlin, 1869.

VON HOLST, E.: Erg. Physiol. *42*, 228 (1939).

JENNINGS, H. H.: Das Verhalten der niederen Organismen (The Behavior of Lower Organisms). German translation. Leipzig and Berlin, 1910.

KRAUS, F.: Die allgemeine und spezielle Pathologie der Person (The General and Special Pathology of the Person). Leipzig, 1919 and 1926.

MAGNUS, R.: Körperstellung (Body Posture). Berlin, 1924.

RADEMAKER, G. G. J.: Das Stehen (Standing). Berlin, 1931.

ROTHACKER, E.: Die Schichten der Persönlichkeit (The Layers of Personality). 2nd ed. Leipzig, 1941.

STUTTE, H.: Brocks Biologische Daten f. d. Kinderarzt (In Brock's Biological Data for the Pediatrician). 2nd ed., Vol. 2. Berlin-Göttingen-Heidelberg, 1954.

TRENDELENBURG, W.: Erg. Physiol. *10*, 454 (1910).

TUMLIRZ, O.: Anthropologische Psychologie (Anthropological Psychology). 2nd ed., pp. 76, 102, 108, and 141. Munich, 1955.

VON UEXKÜLL, J.: Theoretische Biologie (Theoretical Biology). Berlin, 1920.

VILLINGER, W.: Regensburger Jahrb. f. ärztl. Fortbildung *4*, 546 (1956).

WIESER, S.: Ann. Univers. Saraviensis *5*, 213 and 261 (1957).

PUBLICATIONS BY ALBRECHT PEIPER AND HIS CO-WORKERS ON CEREBRAL FUNCTION IN INFANCY AND CHILDHOOD

I. Summaries

1. Die Hirntätigkeit des Säuglings (Cerebral Function in the Infant). Berlin, 1928.
2. Die Hirntätigkeit des Säuglings (Cerebral Function in the Infant). Russian translation. Moscow and Leningrad, 1929.
3. Das Nervensystem, in Brocks Biologischen Daten für den Kinderarzt (The Nervous System, in Brock's Biological Data for the Pediatrician). Vol. 2. Berlin, 1934. 2nd ed., p. 679. Berlin-Göttingen-Heidelberg, 1954.
4. Unreife und Lebensschwäche (Immaturity and Vital Debility). Leipzig, 1937.
5. Die neurologischen Grundlagen der psychischen Entwicklung (The Neurological Foundations of Mental Development). Mschr. Kinderhk. *87*, 179 (1941).

II. Individual Publications

6. Beiträge zur Sinnesphysiologie der Frühgeburt (Contributions to the Sensory Physiology of the Premature Infant). Jb. Kinderhk. *104*, 195 (1924).
7. Untersuchungen über den galvanischen Hautreflex (psychogalvanischen Reflex) im Kindesalter (Investigations of the Psychogalvanic Skin Reflex in Childhood). Jb. Kinderhk. *107*, 139 (1924).
8. Über die Erregbarkeit des autonomen Nervensystems im Schlafe (On the Excitability of the Autonomic Nervous System during Sleep). Jb. Kinderhk. *107*, 191 (1924).
9. Über die Reizbarkeit im Schlafe (On excitability during Sleep). Med. Klin. *1924*, No. 45.
10. Sinnesempfindungen des Kindes vor seiner Geburt (Sensory Perceptions of the Child Before Birth). Mschr. Kinderhk. *29*, 236 (1924).
11. Die Hirntätigkeit des Neugeborenen (Cerebral Function in the Newborn Infant). Jb. Kinderhk. *111*, 290 (1926).
12. Untersuchungen über die Reaktionszeit auf Schallreiz (Investigations of the Reaction Time to Sound Stimuli). Mschr. Kinderhk. *31*, 491 (1926).
13. Untersuchungen über die Reaktionszeit auf Schmerzreiz (Investigations of the Reaction Time to Pain Stimuli). Mschr. Kinderhk. *32*, 136 (1926).
14. Über das Pupillenspiel des Säuglings (On Pupillary Movements in the Infant). Jb. Kinderhk. *112*, 179 (1926).
15. Über einen Augenreflex auf den Hals im frühen Säuglingsalter (On the Eye-Neck Reflex in Early Infancy). Jb. Kinderhk. *113*, 87 (1926).
16. Über die Helligkeits- und Farbenempfindungen der Frühgeburten (On Light and Color Perception in Premature Infants). Arch. Kinderhk. *80*, 1 (1927).
17. Über die Körperstellung des Säuglings (On Body Posture in the Infant). Jb. Kinderhk. *115*, 142 (1927) (with ISBERT).
18. Über die klinische Bedeutung des Geburtstraumas für den Neugeborenen (On the Clinical Significance of Birth Trauma in Newborn Infants). Dtsch. med. Wschr. *1927*, No. 11.

19. Über das Unterscheidungsvermögen des Kleinkindes (On the Differentiating Ability of the Young Child). Jb. Kinderhk. *117,* 350 (1927).

Movements of the Gastrointestinal Tract in Infancy (with ISBERT)

20. Der Magen (The Stomach). Jb. Kinderhk. *118,* 291 (1928).
21. Die Speiseröhre (The Esophagus). Jb. Kinderhk. *120,* 48 (1928).
22. Der Dünndarm (The Small Intestine). Jb. Kinderhk. *120,* 306 (1928).
23. Der Dickdarm (The Colon). Jb. Kinderhk. *120,* 312 (1928).
24. Enthemmte paläenzephale Reaktionen im Kindesalter (Disinhibited Paleoencephalic Reactions in Childhood). Mschr. Kinderhk. *39,* 286 (1928) (with ISBERT).
25. Tetanie der Neugeborenen (Tetany of Newborn Infants). Mschr. Kinderhk. *42,* 415 (1929).
26. Erfahrungen bei der Aufzucht von Frühgeburten (Experiences in the Raising of Premature Infants). Dtsch. med. Wschr. *1929,* No. 8.
27. Die apnoischen Anfälle (Apnoic Attacks). Jb. Kinderhk. *123,* 129 (1929).
28. Die angeborene Lebensschwäche und ihre Bekämpfung (Congenital Vital Debility and How to Combat It). Dtsch. med. Wschr. *1929,* No. 17.
29. Die Krampfbereitschaft des Säuglings (The Tendency to Convulsions in the Infant). Jb. Kinderhk. *125,* 194 (1929).
30. Die Schreitbewegungen der Neugeborenen (The Striding Movements of the Newborn Infant). Mschr. Kinderhk. *45,* 444 (1929).
31. Sinnesreaktionen der Neugeborenen (Sensory Reactions in the Newborn Infant). Zschr. Psychol., Leipzig *114,* 363 (1930).
32. Untersuchungen über die neuromuskuläre Erregbarkeit im Kindesalter (Investigations of Neuromuscular Excitability in Childhood). Mschr. Kinderhk. *46,* 237 (1930).
33. Die Mechanik des Sterbens (The Mechanics of Dying). Jb. Kinderhk. *127,* 157 (1930).
34. Der Zerfall des Atemzentrums (The Dissolution of the Respiratory Center). Mschr. Kinderhk. *47,* 189 (1930).
35. Der Singultus (Singultus). Mschr. Kinderhk. *48,* 445 (1930).
36. Beiträge zur Neurologie des jungen Säuglings (Essays on the Neurology of the Young Infant). Mschr. Kinderhk. *49,* 265 (1931).
37. Bewegungs- und Atemstörungen schwachsinniger Säuglinge (Motor and Respiratory Disorders in Feeble-Minded Infants). Jb. Kinderhk. *131,* 129 (1931).
38. Die Nahrungsaufnahme des Säuglings (Food Intake in the Infant). Mschr. Kinderhk. *50,* 20 (1931).
39. Die Atemstorungen der Fruhgeburten (Respiratory Disorders in Premature Infants). Erg. inn. Med. *40,* 1 (1931).
40. Aufstoßen und Erbrechen (Belching and Vomiting). Jb. Kinderhk. *133,* 301 (1931).
41. Das Stehen im Säuglingsalter (Standing in Infancy). Jb. Kinderhk. *134,* 149 (1932).
42. Untersuchungen über die Todesursache der Frühgeburten (Investigations of the Causes of Death in Premature Infants). Mschr. Kinderhk. *52,* 24 (1932) (with CREUTZFELDT).
43. Das Gähnen (Yawning). Dtsch. med. Wschr. *1932,* No. 18.
44. Das Erwachen der Hirntätigkeit in der Säuglingszeit (The Awakening of Cerebral Function in Infancy). Zschr. Neurol. *139,* 781 (1932).
45. Kohlensäure bei den Atemstörungen der Neugeborenen und Frühgeburten (Carbon Dioxide in the Respiratory Disorders of Newborn and Premature Infants). Mschr. Kinderhk. *53,* 323 (1932) (with CAMMANN).
46. Der Zerfall des Atemzentrums im Tierversuch (The Dissolution of the Respiratory Center in Animal Experimentation). Jb. Kinderhk. *136,* 165 (1942) (with CAMMANN).
47. Agonie und Tod im frühen Säuglingsalter (Agony and Death in Early Infancy). Dtsch. med. Wschr. *1933,* No. 4.
48. Diagnostisch wichtige Reflexe des Kindesalters (Diagnostically Important Reflexes in Childhood). Kinderärztl. Praxis *4,* 189 (1933).

49. Die Atembewegungen des Unterkiefers (Respiratory Movements of the Lower Jaw). Jb. Kinderhk. *139*, 117 (1933).
50. Die Herztätigkeit während des Zerfalls des Atemzentrums (Cardiac Function during the Dissolution of the Respiratory Center). Jb. Kinderhk. *143*, 1 (1934) (with GOOD).
51. Erregung und Hemmung (Excitation and Inhibition). Jb. Kinderhk. *143*, 143 (1934).
52. Stammesgeschichtliche Erinnerungen in der Hirntätigkeit des Säuglings (Phylogenetic Memories in the Cerebral Function of the Infant). Med. Welt *1934*, No. 47.
53. Die bedingten Reflexe im Kindesalter (Conditioned Reflexes in Childhood). Kinderärztl. Praxis *5*, 551 (1934).
54. Die Entwicklung des Mienenspieles (Development of the Facial Expression). Mschr. Kinderhk. *63*, 39 (1935).
55. Rückfälle in die Hirntätigkeit der Säuglingszeit (Relapses into the Cerebral Function of Infancy). Med. Welt *1935*, No. 46.
56. Röntgenkymographie des Saugvorganges (Radiokymography of the Sucking Act). Klin. Wschr. *1935*, 1723 (with E. HOFMANN).
57. Die "Instinkte" des Neugeborenen (The "Instincts" of the Newborn Infant). Zschr. Psychol., Leipzig *136*, 299 (1935).
58. Sinnesphysiologie und Hirntätigkeit (Sensory Physiology and Cerebral Function). Mschr. Kinderhk. *64*, 385 (1936).
59. Hautschutzreflexe (Protective Skin Reflexes). Jb. Kinderhk. *146*, 233 (1936).
60. Reflektorische Mitbewegungen bei dem Mienenspiel (Reflex Movements Accompanying Facial Expression). Mschr. Kinderhk. *65*, 294 (1936).
61. Über Poliomyelitis (On Poliomyelitis). Jahreskurse ärztl. Fortbild. *1936*, June issue, p. 17.
62. Der Saugvorgang (The Sucking Act). Erg. Inn. Med. *50*, 527 (1936).
63. Säuglingspsychologie (Psychology of Infants). Zschr. Psychol., Leipzig *138*, 97 (1936).
64. Sinnesphysiologie und Hirntätigkeit (Sensory Physiology and Cerebral Function). Mschr. Kinderhk. *69*, 327 (1937).
65. Der Schluckvorgang (The Swallowing Act). Mschr. Kinderhk. *70*, 54 (1937) (with E. HOFMANN).
66. Die bedingten Reflexe in der Kinderpsychiatrie (Conditioned Reflexes in Child Psychiatry). Prem. Congrès internat. de Psychiatr. infant. *1*, 87. Lille, 1937.
67. Zentrale Atemstörungen jenseits der Säuglingszeit (Central Respiratory Disorders After Infancy). Kinderärztl. Praxis *8*, 321 (1937).
68. Die Erscheinung der Dominanz und die Erregungsstufen des Saugzentrums (The Phenomenon of Dominance and the Excitatory Stages of the Sucking Center). Jb. Kinderhk. *149*, 201 (1937).
69. Sinnesphysiologie und Hirntätigkeit (Sensory Physiology and Cerebral Function). Mschr. Kinderhk. *71*, 448 (1937).
70. Comments on J. M. Smith's Work: "The Relative Brightness Values of Three Hues for Newborn Infants." Child Development *8*, 299 (1937).
71. Unreife und Lebensschwäche (Immaturity and Vital Debility). Leipzig, 1937.
72. Die Erscheinung der Dominanz bei Reizlöschung (The Phenomenon of Dominance in Stimulus Extinction). Jb. Kinderhk. *151*, 1 (1938).
73. Der Bewegungsdrang des Kindes (The Motility Drive, in the Child). Münch. med. Wschr. *1938*, 634.
74. Das Zusammenspiel des Saugzentrums mit dem Atemzentrum beim menschlichen Säugling (The Interplay of the Sucking and Respiratory Centers in the Human Infant). Pflügers Arch. Physiol. *240*, 312 (1938).
75. Pendelinduktion zwischen Atemzentrum und Saugzentrum (Pendular Induction between Respiratory and Sucking Centers). Mschr. Kinderhk. *75*, 78 (1938).
76. Der Schlaf des Kindes (The Sleep of the Child). Münch. med. Wschr. *1938*, 1585.
77. Sinnesphysiologie und Hirntätigkeit (Sensory Physiology and Cerebral Function). Mschr. Kinderhk. *77*, 146 (1939).

78. "Tierische" Hypnose am menschlichen Kinde ("Animal Hypnosis" in Human Children). Dtsch. med. Wschr. *1939*, 559.
79. Das "Wegbleiben" ("Breath Holding"). Mschr. Kinderhk. *79*, 236 (1939).
80. Die Saugstörung (Sucking Disorders). Mschr. Kinderhk. *79*, 241 (1939).
81. Naschende Kinder (Nibbling Children). Med. Welt *1939*, 1234.
82. Der Saugakt, in Brocks Biologischen Daten für den Kinderarzt (The Sucking Act, in Brock's Biological Data for the Pediatrician). Vol. 3, p. 359. Berlin, 1939.
83. Die Führung des Saugzentrums durch das Schluckzentrum (Direction of the Sucking Center by the Swallowing Center). Pflügers Arch. Physiol. *242*, 751 (1939).
84. Unreife des Frühgeborenen (Immaturity of the Premature Infant). Mschr. Kinderhk. *81*, 321 (1940).
85. Sinnesphysiologie und Hirntätigkeit (Sensory and Cerebral Function). Mschr. Kinderhk. *82*, 374 (1940).
86. Die Erstickung (die sog. Asphyxie) des Neugeborenen [Suffocation (So-Called Asphyxia) in the Newborn Infant]. Dtsch. med. Wschr. *1940*, 495.
87. Die Mutter und ihr Kind (The Mother and Her Child). Kinderärztl. Praxis *12*, 79 (1941).
88. Der Keuchhustenanfall (The Whooping Cough Attack). Mschr. Kinderhk. *86*, 17 (1941).
89. Betäubungsfreie Schmerzstillung im Kindesalter (Non-Narcotic Pain Relief in Childhood). Kinderärztl. Praxis *12*, 148 (1941).
90. Die neurologischen Erscheinungen bei Meningismus und Meningitis (The Neurological Phenomena in Meningism and Meningitis). Dtsch. med. Wschr. *1941*, 541.
91. Sinnesphysiologie und Hirntätigkeit (Sensory Physiology and Cerebral Function). Mschr. Kinderhk. *88*, 356 (1941).
92. Das Heimweh (Homesickness). Dtsch. med. Wschr. *1942*, 121.
93. Die Schluckstörung (Swallowing Disorders). Mschr. Kinderhk. *90*, 37 (1942).
94. Das Stillen (Breast Feeding). Kinderärztl. Praxis *13*, 236 (1942).
95. Krabbeln, Kriechen, Rutschen und Gehen (Creeping, Crawling, Sliding, and Walking). Mschr. Kinderhk. *91*, 187 (1942).
96. Die Umwelt des Säuglings (The Infant's Environment). Med. Klin. *1942*, No. 43.
97. Die Umwelt des Säuglings (The Infant's Environment). Arch. Psychol. *111*, 1 (1942).
98. Sinnesphysiologie und Hirntätigkeit (Sensory Physiology and Cerebral Function). Mschr. Kinderhk. *91*, 390 (1942).
99. Unreife, Lebensschwäche und Geburtstrauma (Immaturity, Vital Debility, and Birth Trauma). Mschr. Kinderhk. *92*, 188 (1943).
100. Der Vater und seine Kinder (The Father and His Children). Kinderärztl. Praxis *14*, 35 (1943).
101. Die Umwelt des jungen Säuglings (The Young Infant's Environment). Forsch. u. Fschr. *19*, 278 (1943).
102. The Environment of the Unweaned Child. Research and Progress *10*, 73 (1944).
103. Das Sauerstoffbett für atemgestörte Säuglinge (The Oxygen Tent for Infants with Respiratory Disorders). Mschr. Kinderhk. *94*, 221 (1944).
104. Hunger und Sättigung (Hunger and Satiation). Mschr. Kinderhk. *94*, 385 (1945).
105. Der Leistenbeugereflex, ein meningitisches Krankheitszeichen (The Inguinal Reflex, A Sign of Meningitis). Dtsch. Gesd.wes. *1*, 145 (1946).
106. Atavismus im Säuglingsalter (Atavism in Infancy). Ärztl. Forsch. *1*, 330 (1947).
107. Das Wegbleiben des Kleinkindes (Breath Holding in the Small Child). Kinderärztl. Praxis *16*, 12 (1948).
108. Die meningitischen Krankheitszeichen (The Signs of Meningitis). Mschr. Kinderhk. *96*, 115 (1948).
109. Das Vorkauen (Premastication). Arch. Kinderhk. *135*, 67 (1948).
110. Sauger, Milchflaschen und Milchpumpen (Nipples, Baby Bottles, and Breast Pumps). Kinderärztl. Praxis *1950*. Special Issue, p. 223.

111. Mutter und Kind (Mother and Child). Kinderärztl. Praxis *18*, 424 (1950).
112. Die Atem-, Saug- und Schluckstörungen der Frühgeburten. 6. Internationaler Kongreß für Kinderheilkunde (Respiratory, Sucking, and Swallowing Disorders in Premature Infants. Sixth International Congress for Pediatrics). Zürich, 1950.
113. Mutterliebe (Mother Love). Arch. Kinderhk. *141*, 174 (1951).
114. Die Schnappatmung in Filmausschnitten (Gasping Respiration in Film Clips). Kinderärztl. Praxis *19*, 272 (1951) (with HEMPEL and THOMAS).
115. Instinkt und angeborenes Schema beim Säugling (Instinct and Innate Blueprint in the Infant). Zschr. Tierpsychol. *8*, 449 (1951).
116. Die Aspiration (Aspiration). Mschr. Kinderhk. *99*, 377 (1951) (with H. THOMAS).
117. Chronik der Kinderheilkunde (Chronicle of Pediatrics). Leipzig, 1951; 3rd ed. 1958.
118. Anwendung der Reflexologie auf das Kindesalter. Tagung der med.-wiss. Ges. an der Univ. Leipzig, Tochtergesellschaft Dresden (The Application of Reflexology to Childhood. Session of the Medical-Scientific Society at the University of Leipzig, Dresden Affiliate). p. 55. February 9, 1952.
119. Die gesetzmäßige Entwicklung des aufrechten Ganges (Development of the Erect Gait according to Innate Laws) (A Film). Kinderärztl. Praxis Special Issue (1953), p. 135 (with H. THOMAS).
120. Die Lehre Pawlows und die Entwicklung des Kindes. Tagungsbericht der Pawlow-Tagung 15. und 16. 1. 1953 in Leipzig (Pavlov's Theory and the Development of the Child. Report of the Pavlov Session in Leipzig, January 15 and 16, 1953). p. 108. Berlin, 1953.
121. Die akute Ernährungsstörung des Säuglings (Acute Alimentary Disorders in the Infant). Mschr. Kinderhk. *100*, 369 (1952).
122. Störungen der Reizbildung und Reizleitung im Herzen frühgeborener Kinder. Bemerkungen zu der gleichnamigen Arbeit von H. U. Köttgen und H. Feyerabend (Disorders of Cardiac Stimulus Formation and Stimulus Conduction in Premature Infants. Comments on the paper of the same title by H. U. Köttgen and H. Feyerabend). Mschr. Kinderhk. *101*, 15 (1953).
123. Woran sterben die Frühgeburten? (What Causes the Death of Premature Infants?). Zschr. Kinderhk. *73*, 188 (1953) (with H. EßBACH).
124. Leerlaufendes Brustsuchen (Rooting Reflexes in Vacuum Activity). Mschr. Kinderhk. *101*, 303 (1953) (with H. THOMAS).
125. Die Schreit- und die Steigbewegungen der Neugeborenen (Striding and Climbing Movements in Newborn Infants). Arch. Kinderhk. *147*, 135 (1953).
126. Die Beratung der Eltern schwachsinniger Kinder (Counseling Parents of Feeble-Minded Children). Dtsch. med. Wschr. *1954*, 1585.
127. Klinik und Pathologie der akuten Ernährungsstörung (Treatment and Pathology of Acute Alimentary Disorders). Mschr. Kinderhk. *103*, 330 (1955) (with F. BOLCK).
128. Der menschliche Säugling als Nesthocker und Nestflüchter (The Human Infant as an Insessorial and an Autophagous Animal). Kinderärztl. Praxis *23*, 507 (1955).
129. Dyspepsie, Subtoxikose und Toxikose (Dyspepsia, Subtoxicosis, and Toxicosis) Ärztl. Wschr. *1956*, 145.
130. Vorstufen der selbständigen Fortbewegung im Säuglingsalter (Preliminary Stages of Independent Locomotion in Infancy). Wiss. Ann. *5*, 556 (1956).
131. Frühgeborene, in Opitz u. de Rudder: Pediatrie (Premature Infants, in Opitz and de Rudder: Pediatrics). p. 199. Berlin-Göttingen-Heidelberg, 1957.
132. Die Grenze der Lebensfähigkeit Frühgeborener (The Limits of Viability in Premature Infants). Ärztl. Wschr. *1957*, p. 103 (with H. THOMAS).
133. Die Sorge für die Kinder werktätiger Mütter (The Care of Children of Working Mothers). Dtsch. Gesd.wes. *1957*, 420.
134. Die Nahrungsaufnahme des Säuglings (Food Intake in the Infant). Münch. med. Wschr. *1958*, 249.

135. Intracranielle Blutungen, in A. PEIPER: Krankheiten der Neugeborenen (Intracranial Hemorrhages, in A. PEIPER: Diseases of Newborn Infants). p. 23. Leipzig, 1958 (with H. THOMAS).
136. Kaspar-Hauser-Kinder, verwilderte Kinder (Wolfskinder) [Kaspar Hauser Children, Feral Children (Wolf Children)]. Die Medizinische *1958*, p. 1411.
137. Unbedingte und bedingte Reflexe der Nahrungsaufnahme (Unconditioned and Conditioned Reflexes of Food Intake). Kinderärztl. Praxis *26*, 507 (1958).
138. Der unbekannte Säugling (The Unknown Infant). Ärztl. Wschr. *1959*, 297.
139. Die Schwerkraftreflexe des Säuglings (Gravity Reflexes in the Infant). Abhandl. Sächs. Akad. Wiss. Leipzig, Math.nat.wiss. Klasse, Vol. 46, No. 4. Berlin, 1959.
140. Der Gampersche Verbeugungsreflex (Gamper's Bowing Reflex). Mschr. Kinderhk. *107*, 393 (1959) (with HEMPEL and WÜNSCHER).
141. Der Aufstieg zum Menschen (The Ascent to Man). Wien. med. Wschr. *1960*, p. 1039.
142. The Basic Ideas in Leipzig: The Relations between the Sucking Center and the Respiratory Center; The Development of the Upright Gait. The Little Club, Summer Meeting. Groningen, 1960 (unpublished).
143. Der Mutterglücksauger (The Nipple of Maternal Happiness). Dtsch. Gesd.wes. *1960*, p. 2128.

Co-Workers

1a. BOLL, FRITZ: Der gekreuzte Patellarsehnenreflex (The Crossed Patellar Reflex). Thesis, Greifswald (1945), Mschr. Kinderhk. *95*, 30 (1944).
2a. BRUNN, EINHART: Die Wirkung des Coramins auf das Saugvermögen trinkschwacher Frühgeburten (The Effect of Nikethamide on the Sucking Ability of Premature Infants with Drinking Weakness). Kinderärztl. Praxis *15*, 131 (1944) and Thesis, Greifswald, 1945.
3a. DIETEL, KLAUS: Die Atmung vor und nach der Geburt (Respiration Before and After Birth). Thesis, Leipzig (1954); Die Atemzahlen des Kindes vor und nach der Geburt (The Child's Respiratory Rate Before and After Birth). Mschr. Kinderheilk. *103*, 449 (1951).
4a. DIETEL, VOLKER: Das Auftreten der höchsten Atemform als Ausdruck der Reifung des Atemzentrums (The Appearance of the Highest Respiratory Pattern as an Expression of the Maturation of the Respiratory Center). Thesis, Leipzig (1952); Die Reifung des Atemzentrums bei Frühgeborenen (The Maturation of the Respiratory Center in Premature Infants). Zschr. Kinderhk. *73*, 463 (1953).
5a. DITTRICH, J. K.: Beiträge zur Pathogenese der akuten Ernährungsstörungen: Die Darmlähmung (Essays on the Pathogenesis of Acute Alimentary Disorders: Paralytic Ileus). Zschr. Kinderhk. *72*, 29 (1952).
6a. DOERFEL, GÜNTER: Die Wirkung des Rauchens in der Schwangerschaft auf die Herzfrequenz des Feten (Effect of Smoking during Pregnancy on the Fetal Heart Rate). Thesis, Leipzig (1951); Zschr. inn. Med., Leipzig *7*, 227 (1952).
7a. FINK, JOHANNES: Die Pupillenreaktion, die Pupillenweite, der Patellarsehnenreflex, das Babinski-Phänomen, das Mundsuchen beim schlafenden Kinde (Pupillary Reaction, Pupillary Width, Patellar Tendon Reflex, Babinski's Sign, and Rooting Reflex in the Sleeping Child). Thesis, Greifswald (1945).
8a. HEMPEL, H. C., and H. THOMAS: Die Bedeutung der inspiratorischen Einziehungen für die Diagnose kindlicher Lungenerkrankungen (The Importance of Inspiratory Pulling-In for the Diagnosis of Pulmonary Diseases in Children). Kinderärztl. Praxis *19*, 401 (1951).
9a. KÖDDING, I.: Der Lidschlag im Kindesalter (Blinking in Childhood). Mschr. Kinderhk. *84*, 212 (1940).
10a. KRUMPEL, J.: Schnappatmung durch Rachenstenose (Gasping Respiration from Pharyngeal Stenosis). Kinderärztl. Praxis *20*, 391 (1952).

11a. LOTH, G.: Neurologische Untersuchungen an Frühgeborenen und jungen Säuglingen (Neurological Examinations of Premature and Young Infants). Zschr. Kinderhk. *72*, 42 (1952).

12a. MAYERHOFER, ALFRED: Schwimmbewegungen bei Säuglingen (Swimming Movements in Infants). Thesis, Leipzig (1952); Arch. Kinderhk. *146*, 137 (1953).

13a. ROSENBAUM, JUDITH: Frühgeburt und soziale Lage (Premature Birth and Social Conditions). Thesis, Berlin (1931).

14a. SCHAAF, REINHARD: Übererregbarkeitserscheinungen am Neugeborenen (Phenomena of Hyperirritability in Newborn Infants). Thesis, Greifswald (1945).

15a. SCHENDEL, FRITZ: Wärme- und Kältereaktionen von neugeborenen Säuglingen (Reactions to Hot and Cold Stimuli in Newborn Infants). Thesis, Greifswald (1944).

16a. SCHMIDT, L.: Der "erste" Atemzug (The "First" Breath). Mschr. Kinderhk. *98*, 213 (1950).

17a. SCHMÖGER, R.: Die Pupillarmembran als Zeichen der Unreife (The Pupillary Membrane as a Sign of Immaturity). Kinderärztl. Praxis *23*, 433 (1955).

18a. STENGEL, ROBERT: Der gleichsinnige gekreuzte Beugereflex der Beine (The Homonymous Crossed Flexor Reflex of the Legs). Thesis, Greifswald (1945).

19a. THUROW, ILSE: Die Säuglingsschlafhaltung (The Infant's Sleeping Posture). Thesis, Leipzig (1956).

20a. THOMAS, H.: Absaugen zur Wiederbelebung und Behandlung von Frühgeborenen (Suction in the Resuscitation and Treatment of Premature Infants). Mschr. Kinderhk. *104*, 236 (1956).

21a. THOMAS, H.: Das klinische Bild der Ösophagusmißbildungen beim Neugeborenen (The Clinical Picture of Esophageal Malformations in the Newborn Infant). Z. Kinderhk. *75*, 465 (1954).

22a. LIPPMANN, C.: Über den Babkinschen Reflex (Babkin's Reflex). Doctoral thesis, Leipzig (1958); Arch. Kinderheilk. *157*, 234 (1958).

23a. NAUMANN, U.: Die Säuglingsschlafhaltung (The Infant's Sleeping Posture). Doctoral thesis, Leipzig (1958).

24a. DITTRICH, J. K.: Die Magen-Darmtätigkeit bei gesunden und bei ernährungsgestörten Säuglingen im Röntgenbild (Radiography of Gastrointestinal Function in Healthy Infants and Infants with Alimentary Disorders). Habil.-Schrift. Leipzig, 1957 (unpublished).

25a. FRIEDRICH, S.: Das Krankheitsbild der hyalinen Membranen bei Frühgeborenen (Pathology of the Hyaline Membranes in Premature Infants). Doctoral thesis, Leipzig (1958); Zschr. Kinderhk. *82*, 367 (1959).

AUTHOR INDEX

The italic numbers refer to literature citations

SUBJECT INDEX

MIX
Papier aus verantwortungsvollen Quellen
Paper from responsible sources
FSC® C105338
FSC
www.fsc.org

If you have any concerns about our products,
you can contact us on
ProductSafety@springernature.com

In case Publisher is established outside the EU,
the EU authorized representative is:
Springer Nature Customer Service Center GmbH
Europaplatz 3, 69115 Heidelberg, Germany

Printed by Libri Plureos GmbH
in Hamburg, Germany